A Brief Table of Integrals

(*An arbitrary constant may be added to each integral.*)

1. $\int x^n\,dx = \dfrac{1}{n+1}x^{n+1}\ (n \neq -1)$
2. $\int \dfrac{1}{x}\,dx = \ln|x|$
3. $\int e^x\,dx = e^x$
4. $\int a^x\,dx = \dfrac{a^x}{\ln a}$
5. $\int \sin x\,dx = -\cos x$
6. $\int \cos x\,dx = \sin x$
7. $\int \tan x\,dx = -\ln|\cos x|$
8. $\int \cot x\,dx = \ln|\sin x|$
9. $\int \sec x\,dx = \ln|\sec x + \tan x|$
   $= \ln\left|\tan\left(\dfrac{1}{2}x + \dfrac{1}{4}\pi\right)\right|$
10. $\int \csc x\,dx = \ln|\csc x - \cot x|$
    $= \ln\left|\tan\dfrac{1}{2}x\right|$
11. $\int \sin^{-1}\dfrac{x}{a}\,dx = x\sin^{-1}\dfrac{x}{a} + \sqrt{a^2 - x^2} \qquad (a > 0)$
12. $\int \cos^{-1}\dfrac{x}{a}\,dx = x\cos^{-1}\dfrac{x}{a} - \sqrt{a^2 - x^2} \qquad (a > 0)$
13. $\int \tan^{-1}\dfrac{x}{a}\,dx = x\tan^{-1}\dfrac{x}{a} - \dfrac{a}{2}\ln(a^2 + x^2) \qquad (a > 0)$
14. $\int \sin^2 mx\,dx = \dfrac{1}{2m}(mx - \sin mx\cos mx)$
15. $\int \cos^2 mx\,dx = \dfrac{1}{2m}(mx + \sin mx\cos mx)$
16. $\int \sec^2 x\,dx = \tan x$
17. $\int \csc^2 x\,dx = -\cot x$
18. $\int \sin^n x\,dx = -\dfrac{\sin^{n-1}x\cos x}{n} + \dfrac{n-1}{n}\int \sin^{n-2}x\,dx$
19. $\int \cos^n x\,dx = \dfrac{\cos^{n-1}x\sin x}{n} + \dfrac{n-1}{n}\int \cos^{n-2}x\,dx$
20. $\int \tan^n x\,dx = \dfrac{\tan^{n-1}x}{n-1} - \int \tan^{n-2}x\,dx \qquad (n \neq 1)$
21. $\int \cot^n x\,dx = -\dfrac{\cot^{n-1}x}{n-1} - \int \cot^{n-2}x\,dx \qquad (n \neq 1)$
22. $\int \sec^n x\,dx = \dfrac{\tan x\sec^{n-2}x}{n-1} + \dfrac{n-2}{n-1}\int \sec^{n-2}x\,dx \qquad (n \neq 1)$
23. $\int \csc^n x\,dx = -\dfrac{\cot x\csc^{n-2}x}{n-1} + \dfrac{n-2}{n-1}\int \csc^{n-2}x\,dx \qquad (n \neq 1)$
24. $\int \sinh x\,dx = \cosh x$
25. $\int \cosh x\,dx = \sinh x$
26. $\int \tanh x\,dx = \ln|\cosh x|$
27. $\int \coth x\,dx = \ln|\sinh x|$
28. $\int \operatorname{sech} x\,dx = \tan^{-1}(\sinh x)$

This table is continued on the endpapers at the back.

Derivatives

1. $\dfrac{d(au)}{dx} = a\dfrac{du}{dx}$

2. $\dfrac{d(u+v-w)}{dx} = \dfrac{du}{dx} + \dfrac{dv}{dx} - \dfrac{dw}{dx}$

3. $\dfrac{d(uv)}{dx} = u\dfrac{dv}{dx} + v\dfrac{du}{dx}$

4. $\dfrac{d(u/v)}{dx} = \dfrac{v(du/dx) - u(dv/dx)}{v^2}$

5. $\dfrac{d(u^n)}{dx} = nu^{n-1}\dfrac{du}{dx}$

6. $\dfrac{d(u^v)}{dx} = vu^{v-1}\dfrac{du}{dx} + u^v(\ln u)\dfrac{dv}{dx}$

7. $\dfrac{d(e^u)}{dx} = e^u\dfrac{du}{dx}$

8. $\dfrac{d(e^{au})}{dx} = ae^{au}\dfrac{du}{dx}$

9. $\dfrac{da^u}{dx} = a^u(\ln a)\dfrac{du}{dx}$

10. $\dfrac{d(\ln u)}{dx} = \dfrac{1}{u}\dfrac{du}{dx}$

11. $\dfrac{d(\log_a u)}{dx} = \dfrac{1}{u(\ln a)}\dfrac{du}{dx}$

12. $\dfrac{d\sin u}{dx} = \cos u\dfrac{du}{dx}$

13. $\dfrac{d\cos u}{dx} = -\sin u\dfrac{du}{dx}$

14. $\dfrac{d\tan u}{dx} = \sec^2 u\dfrac{du}{dx}$

15. $\dfrac{d\cot u}{dx} = -\csc^2 u\dfrac{du}{dx}$

16. $\dfrac{d\sec u}{dx} = \tan u\sec u\dfrac{du}{dx}$

17. $\dfrac{d\csc u}{dx} = -(\cot u)(\csc u)\dfrac{du}{dx}$

18. $\dfrac{d\sin^{-1}u}{dx} = \dfrac{1}{\sqrt{1-u^2}}\dfrac{du}{dx}$

19. $\dfrac{d\cos^{-1}u}{dx} = \dfrac{-1}{\sqrt{1-u^2}}\dfrac{du}{dx}$

20. $\dfrac{d\tan^{-1}u}{dx} = \dfrac{1}{1+u^2}\dfrac{du}{dx}$

21. $\dfrac{d\cot^{-1}u}{dx} = \dfrac{-1}{1+u^2}\dfrac{du}{dx}$

22. $\dfrac{d\sec^{-1}u}{dx} = \dfrac{1}{u\sqrt{u^2-1}}\dfrac{du}{dx}$

23. $\dfrac{d\csc^{-1}u}{dx} = \dfrac{-1}{u\sqrt{u^2-1}}\dfrac{du}{dx}$

24. $\dfrac{d\sinh u}{dx} = \cosh u\dfrac{du}{dx}$

25. $\dfrac{d\cosh u}{dx} = \sinh u\dfrac{du}{dx}$

26. $\dfrac{d\tanh u}{dx} = \operatorname{sech}^2 u\dfrac{du}{dx}$

27. $\dfrac{d\coth u}{dx} = -(\operatorname{csch}^2 u)\dfrac{du}{dx}$

28. $\dfrac{d\operatorname{sech} u}{dx} = -(\operatorname{sech} u)(\tanh u)\dfrac{du}{dx}$

29. $\dfrac{d\operatorname{csch} u}{dx} = -(\operatorname{csch} u)(\coth u)\dfrac{du}{dx}$

30. $\dfrac{d\sinh^{-1}u}{dx} = \dfrac{1}{\sqrt{1+u^2}}\dfrac{du}{dx}$

31. $\dfrac{d\cosh^{-1}u}{dx} = \dfrac{1}{\sqrt{u^2-1}}\dfrac{du}{dx}$

32. $\dfrac{d\tanh^{-1}u}{dx} = \dfrac{1}{1-u^2}\dfrac{du}{dx}$

33. $\dfrac{d\coth^{-1}u}{dx} = \dfrac{1}{1-u^2}\dfrac{du}{dx}$

34. $\dfrac{d\operatorname{sech}^{-1}u}{dx} = \dfrac{-1}{u\sqrt{1-u^2}}\dfrac{du}{dx}$

35. $\dfrac{d\operatorname{csch}^{-1}u}{dx} = \dfrac{-1}{|u|\sqrt{1+u^2}}\dfrac{du}{dx}$

Continued on overle

Undergraduate Texts in Mathematics

Springer
New York
Berlin
Heidelberg
Barcelona
Hong Kong
London
Milan
Paris
Singapore
Tokyo

(continued after index)

To Nancy and Margo

Preface

The goal of this text is to help students learn to use calculus intelligently for solving a wide variety of mathematical and physical problems.

This book is an outgrowth of our teaching of calculus at Berkeley, and the present edition incorporates many improvements based on our use of the first edition. We list below some of the key features of the book.

Examples and Exercises

The exercise sets have been carefully constructed to be of maximum use to the students. With few exceptions we adhere to the following policies.

- The section *exercises are graded* into three consecutive groups:

(a) The first exercises are routine, modelled almost exactly on the examples; these are intended to give students confidence.

(b) Next come exercises that are still based directly on the examples and text but which may have variations of wording or which combine different ideas; these are intended to train students to think for themselves.

(c) The last exercises in each set are difficult. These are marked with a star (★) and some will challenge even the best students. Difficult does not necessarily mean theoretical; often a starred problem is an interesting application that requires insight into what calculus is really about.

- The *exercises come in groups* of two and often four similar ones.
- *Answers* to odd-numbered exercises are available in the back of the book, and every other odd exercise (that is, Exercise 1, 5, 9, 13, . . .) has a complete solution in the student guide. Answers to even-numbered exercises are not available to the student.

Placement of Topics

Teachers of calculus have their own pet arrangement of topics and teaching devices. After trying various permutations, we have arrived at the present arrangement. Some highlights are the following.

- *Integration* occurs early in Chapter 4; *antidifferentiation* and the $\int$ notation with motivation already appear in Chapter 2.

- *Trigonometric functions* appear in the first semester in Chapter 5.
- The *chain rule* occurs early in Chapter 2. We have chosen to use rate-of-change problems, square roots, and algebraic functions in conjunction with the chain rule. Some instructors prefer to introduce $\sin x$ and $\cos x$ early to use with the chain rule, but this has the penalty of fragmenting the study of the trigonometric functions. We find the present arrangement to be smoother and easier for the students.
- *Limits* are presented in Chapter 1 along with the derivative. However, while we do not try to hide the difficulties, technicalities involving epsilonics are deferred until Chapter 11. (Better or curious students can read this concurrently with Chapter 2.) Our view is that it is very important to teach students to differentiate, integrate, and solve calculus problems as quickly as possible, without getting delayed by the intricacies of limits. After some calculus is learned, the details about limits are best appreciated in the context of l'Hôpital's rule and infinite series.
- *Differential equations* are presented in Chapter 8 and again in Sections 12.7, 12.8, and 18.3. Blending differential equations with calculus allows for more interesting applications early and meets the needs of physics and engineering.

Prerequisites and Preliminaries

A historical introduction to calculus is designed to orient students before the technical material begins.

Prerequisite material from algebra, trigonometry, and analytic geometry appears in Chapters R, 5, and 14. These topics are treated completely: in fact, analytic geometry and trigonometry are treated in enough detail to serve as a first introduction to the subjects. However, high school algebra is only lightly reviewed, and knowledge of some plane geometry, such as the study of similar triangles, is assumed.

Several *orientation quizzes* with answers and a *review section* (Chapter R) contribute to bridging the gap between previous training and this book. Students are advised to assess themselves and to take a pre-calculus course if they lack the necessary background.

Chapter and Section Structure

The book is intended for a three-semester sequence with six chapters covered per semester. (Four semesters are required if pre-calculus material is included.)

The length of chapter sections is guided by the following typical course plan: If six chapters are covered per semester (this typically means four or five student contact hours per week) then approximately two sections must be covered each week. Of course this schedule must be adjusted to students' background and individual course requirements, but it gives an idea of the pace of the text.

Proofs and Rigor

Proofs are given for the most important theorems, with the customary omission of proofs of the intermediate value theorem and other consequences of the completeness axiom. Our treatment of integration enables us to give particularly simple proofs of some of the main results in that area, such as the fundamental theorem of calculus. We de-emphasize the theory of limits, leaving a detailed study to Chapter 11, after students have mastered the fundamentals of calculus—differentiation and integration. Our book *Calculus Unlimited* (Benjamin/Cummings) contains all the proofs omitted in this text and additional ideas suitable for supplementary topics for good students. Other references for the theory are Spivak's *Calculus* (Benjamin/Cummings & Publish or Perish), Ross' *Elementary Analysis: The Theory of Calculus* (Springer) and Marsden's *Elementary Classical Analysis* (Freeman).

Calculators

Calculator applications are used for motivation (such as for functions and composition on pages 40 and 112) and to illustrate the numerical content of calculus (see, for instance, p. 405 and Section 11.5). Special calculator discussions tell how to use a calculator and recognize its advantages and shortcomings.

Applications

Calculus students should not be treated as if they are already the engineers, physicists, biologists, mathematicians, physicians, or business executives they may be preparing to become. Nevertheless calculus is a subject intimately tied to the physical world, and we feel that it is misleading to teach it any other way. Simple examples related to distance and velocity are used throughout the text. Somewhat more special applications occur in examples and exercises, some of which may be skipped at the instructor's discretion. Additional connections between calculus and applications occur in various section supplements throughout the text. For example, the use of calculus in the determination of the length of a day occurs at the end of Chapters 5, 9, and 14.

Visualization

The ability to visualize basic graphs and to interpret them mentally is very important in calculus and in subsequent mathematics courses. We have tried to help students gain facility in forming and using visual images by including plenty of carefully chosen artwork. This facility should also be encouraged in the solving of exercises.

Computer-Generated Graphics

Computer-generated graphics are becoming increasingly important as a tool for the study of calculus. High-resolution plotters were used to plot the graphs of curves and surfaces which arose in the study of Taylor polynomial approximation, maxima and minima for several variables, and three-dimensional surface geometry. Many of the computer drawn figures were kindly supplied by Jerry Kazdan.

Supplements

Student Guide Contains

- Goals and guides for the student
- Solutions to every other odd-numbered exercise
- Sample exams

Instructor's Guide Contains

- Suggestions for the instructor, section by section
- Sample exams
- Supplementary answers

Misprints

Misprints are a plague to authors (and readers) of mathematical textbooks. We have made a special effort to weed them out, and we will be grateful to the readers who help us eliminate any that remain.

Acknowledgments

We thank our students, readers, numerous reviewers and assistants for their help with the first and current edition. For this edition we are especially grateful to Ray Sachs for his aid in matching the text to student needs, to Fred Soon and Fred Daniels for their unfailing support, and to Connie Calica for her accurate typing. Several people who helped us with the first edition deserve our continued thanks. These include Roger Apodaca, Grant Gustafson, Mike Hoffman, Dana Kwong, Teresa Ling, Tudor Ratiu, and Tony Tromba.

Jerry Marsden
Alan Weinstein

Berkeley, California

How to Use this Book: A Note to the Student

Begin by orienting yourself. Get a rough feel for what we are trying to accomplish in calculus by rapidly reading the Introduction and the Preface and by looking at some of the chapter headings.

Next, make a preliminary assessment of your own preparation for calculus by taking the quizzes on pages 13 and 14. If you need to, study Chapter R in detail and begin reviewing trigonometry (Section 5.1) as soon as possible.

You can learn a little bit about calculus by reading this book, but you can learn to use calculus only by practicing it yourself. You should do many more exercises than are assigned to you as homework. The answers at the back of the book and solutions in the student guide will help you monitor your own progress. There are a lot of examples with complete solutions to help you with the exercises. The end of each example is marked with the symbol ▲.

Remember that even an experienced mathematician often cannot "see" the entire solution to a problem at once; in many cases it helps to begin systematically, and then the solution will fall into place.

Instructors vary in their expectations of students as far as the degree to which answers should be simplified and the extent to which the theory should be mastered. In the book we have arranged the theory so that only the proofs of the most important theorems are given in the text; the ends of proofs are marked with the symbol ■. Often, technical points are treated in the starred exercises.

In order to prepare for examinations, try reworking the examples in the text and the sample examinations in the Student Guide without looking at the solutions. Be sure that you can do all of the assigned homework problems.

When writing solutions to homework or exam problems, you should use the English language liberally and correctly. A page of disconnected formulas with no explanatory words is incomprehensible.

We have written the book with your needs in mind. Please inform us of shortcomings you have found so we can correct them for future students. We wish you luck in the course and hope that you find the study of calculus stimulating, enjoyable, and useful.

Jerry Marsden
Alan Weinstein

Contents

Chapter 16
Gradients, Maxima, and Minima

Chapter 17
Multiple Integration

Chapter 18
Vector Analysis

Contents of Volume I

Contents of Volume II

Chapter 13

Vectors

Vectors are arrows which have a definite magnitude and direction.

Many interesting mathematical and physical quantities depend upon more than one independent variable. The length z of the hypotenuse of a right triangle, for instance, depends upon the lengths x and y of the two other sides; the dependence is given by the Pythagorean formula $z = \sqrt{x^2 + y^2}$. Similarly, the growth rate of a plant may depend upon the amounts of sunlight, water, and fertilizer it receives; such a dependence relation may be determined experimentally or predicted by a theory.

The calculus of functions of a single variable, which we have been studying since the beginning of this book, is not enough for the study of functions which depend upon several variables—what we require is the *calculus of functions of several variables*. In the final six chapters, we present this general calculus.

In this chapter and the next, we set out the algebraic and geometric preliminaries for the calculus of several variables. This material is thus analogous to Chapter R, but not so elementary. Chapters 15 and 16 are devoted to the differential calculus, and Chapters 17 and 18 to the integral calculus, of functions of several variables.

13.1 Vectors in the Plane

The components of vectors in the plane are ordered pairs.

An ordered pair (x, y) of real numbers has been considered up to now as a point in the plane—that is, as a *geometric* object. We begin this section by giving the number pairs an *algebraic* structure.[1] Next we introduce the notion of a vector. In Section 13.2, we discuss the representation of points in space by triples (x, y, z) of real numbers, and we extend the vector concept to three dimensions. In Section 13.3, we apply the algebra of vectors to the solution of geometric problems.

The authors of this book, and probably many of its readers, were brought up mathematically on the precept that "you cannot add apples to oranges." If we have x apples and y oranges, the number $x + y$ represents the number of

[1] The reader who has studied Section 12.6 on complex numbers will have seen some of this algebraic structure already.

pieces of fruit but confounds the apples with the oranges. By using the ordered pair (x, y) instead of the sum $x + y$, we can keep track of both our apples and our oranges without losing any information in the process. Furthermore, if someone adds to our fruit basket X apples and Y oranges, which we might denote by (X, Y), our total accumulation is now $(x + X, y + Y)$—that is, $x + X$ apples and $y + Y$ oranges. This addition of items kept in separate categories is useful in many contexts.

Addition of Ordered Pairs

If (x_1, y_1) and (x_2, y_2) are ordered pairs of real numbers, the ordered pair $(x_1 + x_2, y_1 + y_2)$ is called their *sum* and is denoted by $(x_1, y_1) + (x_2, y_2)$. Thus, $(x_1, y_1) + (x_2, y_2) = (x_1 + x_2, y_1 + y_2)$.

Example 1 (a) Calculate $(-3, 2) + (4, 6)$.
(b) Calculate $(1, 4) + (1, 4) + (1, 4)$.
(c) Given pairs (a, b) and (c, d), find (x, y) such that $(a, b) + (x, y) = (c, d)$.

Solution (a) $(-3, 2) + (4, 6) = (-3 + 4, 2 + 6) = (1, 8)$.
(b) We have not yet defined the sum of three ordered pairs, so we take the problem to mean $[(1,4) + (1,4)] + (1,4)$, which is $(2, 8) + (1, 4) = (3, 12)$. Notice that this is $(1 + 1 + 1, 4 + 4 + 4)$, or $(3 \cdot 1, 3 \cdot 4)$.
(c) The equation $(a, b) + (x, y) = (c, d)$ means $(a + x, b + y) = (c, d)$. Since two ordered pairs are equal only when their corresponding components are equal, the last equation is equivalent to the two numerical equations

$$a + x = c \quad \text{and} \quad b + y = d.$$

Solving these equations for x and y gives $x = c - a$ and $y = d - b$, or $(x, y) = (c - a, d - b)$. ▲

Following Example 1(b), we may observe that the sum $(x, y) + (x, y) + \cdots + (x, y)$, with n terms, is equal to (nx, ny). Thinking of the sum as "n times (x, y)," we denote it by $n(x, y)$, so we have the equation

$$n(x, y) = (nx, ny).$$

Noting that the right-hand side of this equation makes sense when n is any real number, not just a positive integer, we take this as a definition.

Multiplication of Ordered Pairs by Numbers

If (x, y) is an ordered pair and r is a real number, the ordered pair (rx, ry) is called the *product* of r and (x, y) and is denoted by $r(x, y)$. Thus, $r(x, y) = (rx, ry)$.

To distinguish ordinary numbers from ordered pairs, we sometimes call numbers *scalars*. The operation just defined is called *scalar multiplication*. Notice that we have not defined the product of two ordered pairs—we will do so in Section 13.4.

Example 2 (a) Calculate $4(2,-3)+4(3,5)$. (b) Calculate $4[(2,-3)+(3,5)]$.

Solution (a) $4(2,-3)+4(3,5)=(8,-12)+(12,20)=(20,8)$.
(b) $4[(2,-3)+(3,5)]=4(5,2)=(20,8)$. ▲

It is sometimes a useful shorthand to denote an ordered pair by a single letter such as $A=(x,y)$. This makes the algebra of ordered pairs look more like the algebra of ordinary numbers. The results in Example 2 illustrate the following general rule.

Example 3 Show that if a is a number and A_1 and A_2 are ordered pairs, then $a(A_1+A_2) = aA_1+aA_2$.

Solution We write $A_1=(x_1,y_1)$ and $A_2=(x_2,y_2)$. Then $A_1+A_2=(x_1+x_2,y_1+y_2)$, and so

$$\begin{aligned} a(A_1+A_2) &= (a(x_1+x_2),a(y_1+y_2))=(ax_1+ax_2,ay_1+ay_2)\\ &= (ax_1,ay_1)+(ax_2,ay_2)=a(x_1,y_1)+a(x_2,y_2)\\ &= aA_1+aA_2, \end{aligned}$$

as required. ▲

All the usual algebraic identities which make sense for numbers and ordered pairs are true, and they can be used in computations with ordered pairs (see Exercises 17–22).

Example 4 (a) Find real numbers a_1,a_2,a_3 such that $a_1(3,1)+a_2(6,2)+a_3(-1,1)=(5,6)$.
(b) Is the solution in part (a) unique?
(c) Can you find a solution in which a_1, a_2, and a_3 are integers?

Solution (a) $a_1(3,1)+a_2(6,2)+a_3(-1,1)=(3a_1+6a_2-a_3,a_1+2a_2+a_3)$; for this to equal $(5,6)$, we must solve the equations

$$\begin{aligned} 3a_1+6a_2-a_3&=5,\\ a_1+2a_2+a_3&=6. \end{aligned}$$

A solution of these equations is $a_1=0$, $a_2=\frac{11}{8}$, $a_3=\frac{13}{4}$. (Part (b) explains where this solution came from.)
(b) We can rewrite the equations as

$$\begin{aligned} 6a_2-a_3&=5-3a_1\\ 2a_2+a_3&=6-a_1. \end{aligned}$$

We may choose a_1 at will, obtaining a pair of linear equations in a_2 and a_3 which always have a solution, since the lines in the (a_2,a_3) plane which they represent have different slopes. The choice $a_1=0$ led us to the equations $6a_2-a_3=5$ and $2a_2+a_3=6$, with the solution as given in part (a). The choice $a_1=6$, for instance, leads to the new solution $a_1=6$, $a_2=-\frac{13}{8}$, $a_3=\frac{13}{4}$, so the solution is not unique.
(c) We notice that the sums $3+1=4$, $6+2=8$, and $-1+1=0$ of the components in each of the ordered pairs are even. Thus the sum $4a_1+8a_2$ of the components of $a_1(3,1)+a_2(6,2)+a_3(-1,1)$ is even if a_1, a_2, and a_3 are integers; but $5+6=11$ is odd, so there is no solution with a_1, a_2, and a_3 being integers. ▲

Example 5 Interpret the chemical equation $2NH_2 + H_2 = 2NH_3$ as a relation in the algebra of ordered pairs.

Solution We think of the molecule N_xH_y (x atoms of nitrogen, y atoms of hydrogen) as represented by the ordered pair (x, y). Then the chemical equation given is equivalent to $2(1,2) + (0,2) = 2(1,3)$. Indeed, both sides are equal to $(2,6)$. ▲

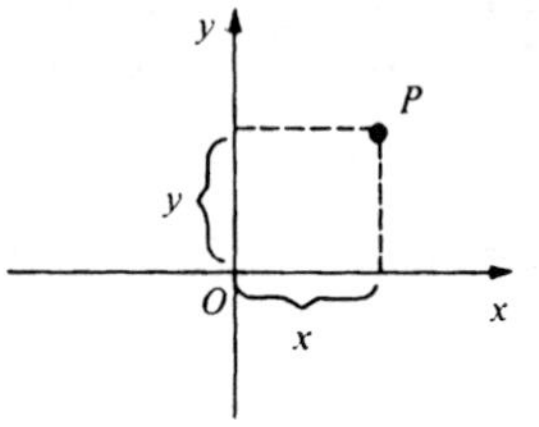

Figure 13.1.1. The point P has coordinates (x, y) relative to the given axes.

We know from Section R.4 how to represent points P in the plane by ordered pairs by selecting an origin O and two perpendicular lines throught it. Relative to these axes, P is assigned the coordinates (x, y), as in Fig. 13.1.1. If the axes are changed, the coordinates of P change as well. (In Section 14.2, we will study how coordinates change when the axes are rotated.) When a definite coordinate system is understood, we refer to "the point (x, y)" when x and y are the coordinates in that system.

We turn now from the algebra of ordered pairs to the related geometric concept of a vector.

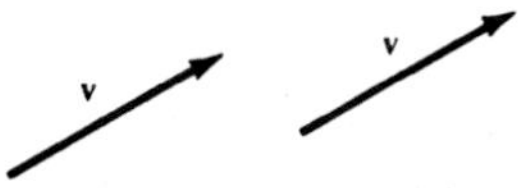

Figure 13.1.2. A vector **v** is an arrow with definite length and direction. The same vector is represented by the two arrows in this figure.

Vectors in the Plane

A *vector* in the plane is a directed line segment in the plane and is drawn as an arrow.

Vectors are denoted by boldface symbols such as **v**. Two directed line segments will be said to be *equal* when they have the same length and direction (as in Fig. 13.1.2).[2]

The vector represented by the arrow from a point P to a point Q is denoted $\overrightarrow{PQ}$. (Figure 13.1.3). If the arrows from P_1 to Q_1 and P_2 to Q_2 represent the same vector, we write $\overrightarrow{P_1Q_1} = \overrightarrow{P_2Q_2}$.

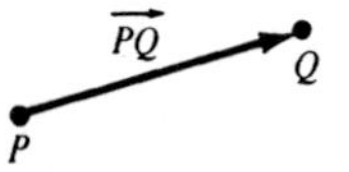

Figure 13.1.3. The vector from P to Q is denoted $\overrightarrow{PQ}$.

Ordered pairs are related to vectors in the following way. We first choose a set of x and y axes. Given a vector **v**, we drop perpendiculars from its head and tail to the x and y axes, as shown in Fig. 13.1.4, producing two signed numbers x and y equaling the directed lengths of the vector in the x and y directions. These numbers are called the *components* of the vector. Notice that once the x and y axes are chosen, the components do not depend on where the arrow representing the vector **v** is placed; they depend only on the magnitude and direction of **v**. Thus, for any vector **v**, we get an ordered pair (x, y). Conversely, given an ordered pair (x, y), we can construct a vector with these components; for example, the vector from the origin to the point with coordinates (x, y). The arrow representing this vector can be relocated as long as its magnitude and direction are preserved.

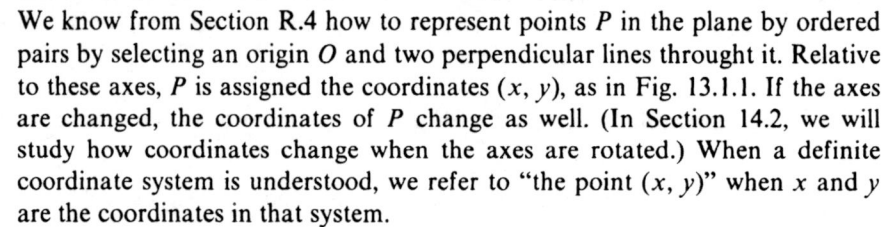

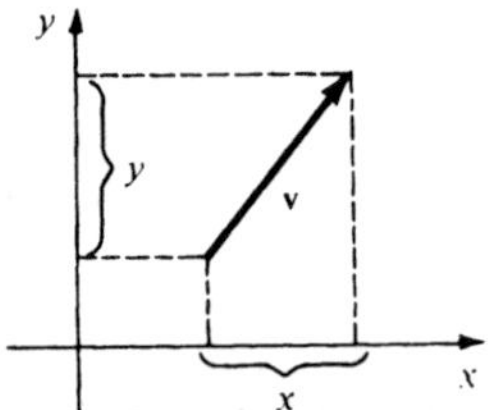

Figure 13.1.4. The components of **v** are x and y.

Operations of vector addition and scalar multiplication are defined in the following box. These geometric definitions will be seen later to be related to the algebraic ones we studied earlier.

[2] Strictly speaking, this definition does not make sense. The two directed line segments in Fig. 13.1.2 are clearly *not* equal—that is, they are not identical. However, it is very convenient to have the *set* of all directed line segments with the same magnitude and direction represent a single geometric entity—a *vector*. A convenient way to do this is to *regard* two such segments as equal.

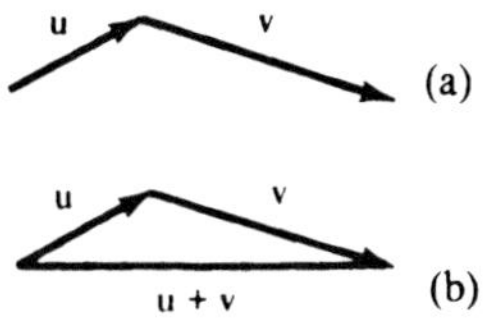

Figure 13.1.5. The geometric construction of **u** + **v**.

Vector Addition and Scalar Multiplication

Addition. Let **u** and **v** be vectors. Their sum is the vector represented by the arrow from the tail of **u** to the tip of **v** when the tail of **v** is placed at the tip of **u** (Fig. 13.1.5).
Scalar Multiplication. Let **u** be a vector and r a number. The vector r**u** is an arrow with length $|r|$ times the length of **u**. It has the same direction as **u** if $r > 0$ and the opposite direction if $r < 0$ (Fig. 13.1.6).

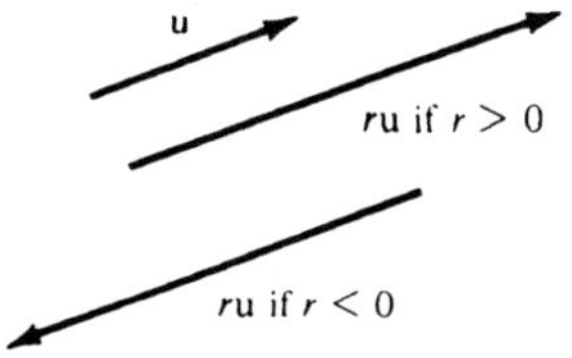

Figure 13.1.6. The product r**u**.

Example 6 In Fig. 13.1.7, which vector is (a) **u** + **v**?, (b) 3**u**?, (c) − **v**?

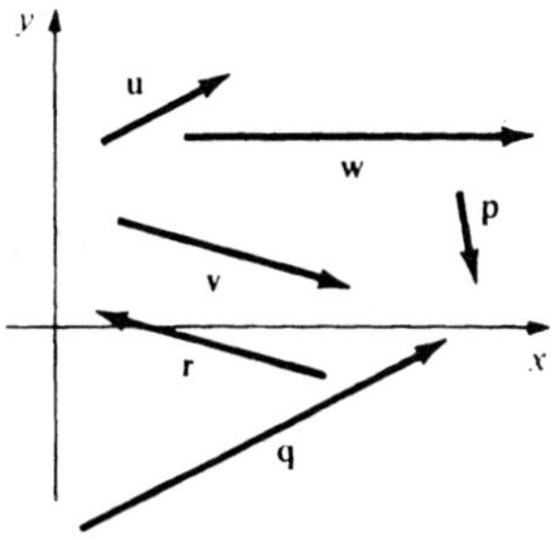

Figure 13.1.7. Find **u** + **v**, 3**u**, and −**v**.

Solution (a) To construct **u** + **v**, we represent **u** and **v** by directed line segments so that the head of the first coincides with the tail of the second. We fill in the third side of the triangle to obtain **u** + **v** (see Fig. 13.1.5(b)). Comparing Fig. 13.1.5(b) with Fig. 13.1.7, we find that **u** + **v** = **w**.
(b) 3**u** = **q** (see Fig. 13.1.8).
(c) −**v** = (−1)**v** = **r** (see Fig. 13.1.8). ▲

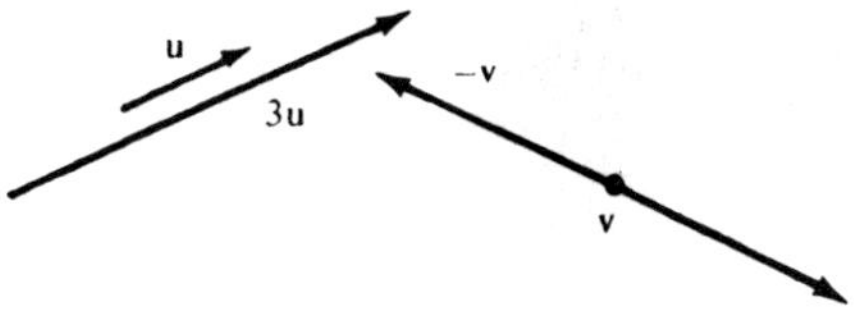

Figure 13.1.8. To find 3**u**, draw a vector in the same direction as **u**, three times as long; −**v** is a vector having the same length as **v**, pointing in the opposite direction.

Example 7 Let **u** and **v** be the vectors shown in Fig. 13.1.9. Draw **u** + **v** and −2**u**. What are their components?

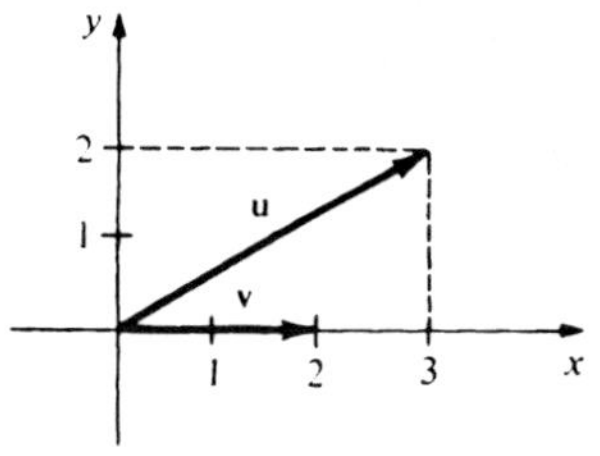

Figure 13.1.9. Find **u** + **v** and −2**u**.

Solution We place the tail of $\mathbf{v}$ at the tip of $\mathbf{u}$ to obtain the vector shown in Fig. 13.1.10.

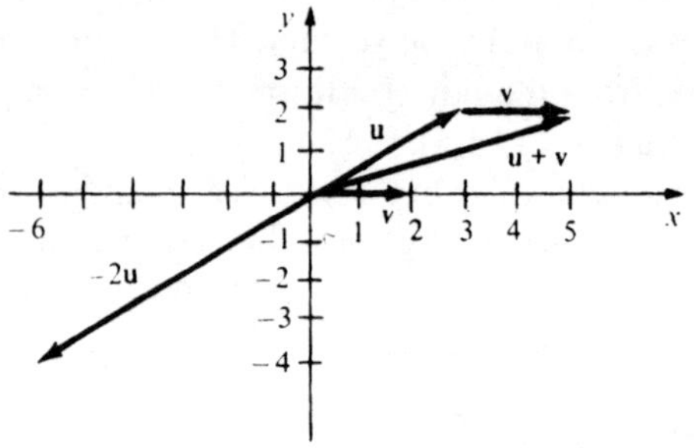

Figure 13.1.10. Computing $\mathbf{u}+\mathbf{v}$ and $-2\mathbf{u}$.

The vector $-2\mathbf{u}$, also shown, has length twice that of $\mathbf{u}$ and points in the opposite direction. From the figure, we see that $\mathbf{u}+\mathbf{v}$ has components $(5,2)$ and $-2\mathbf{u}$ has components $(-6,-4)$. ▲

The results in Example 7 are illustrations of the following general rules which relate the geometric operations on vectors to the algebra of ordered pairs.

Vectors and Ordered Pairs

Addition. If $\mathbf{u}$ has components (x_1, y_1) and $\mathbf{v}$ has components (x_2, y_2), then $\mathbf{u}+\mathbf{v}$ has components $(x_1 + x_2, y_1 + y_2)$.
Scalar Multiplication. If $\mathbf{u}$ has components (x, y), then $r\mathbf{u}$ has components (rx, ry).

The statements in this box may be proved by plane geometry. For example, the addition rule follows by an examination of Fig. 13.1.11(a), and the one for scalar multiplication follows from the similarity of the triangles in Fig. 13.1.11(b).

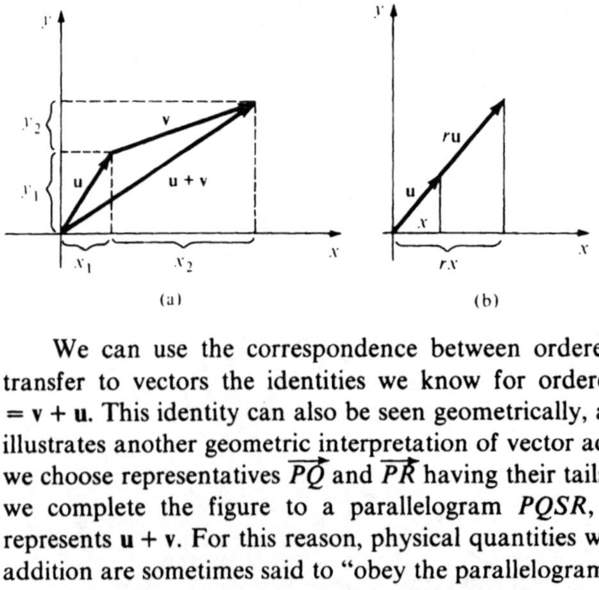

Figure 13.1.11. The geometry which relates vector algebra to the algebra of ordered pairs.

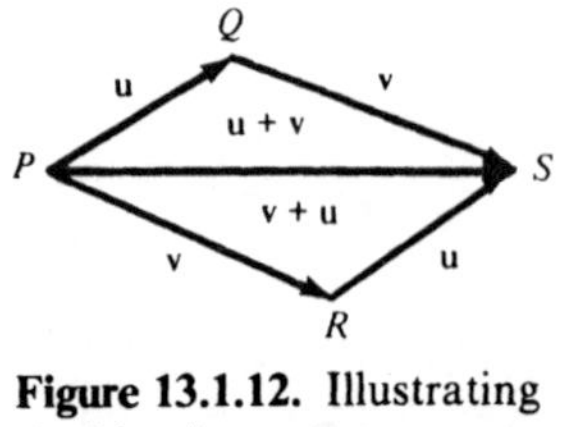

Figure 13.1.12. Illustrating the identity $\mathbf{u}+\mathbf{v}=\mathbf{v}+\mathbf{u}$ and the parallelogram law of addition.

We can use the correspondence between ordered pairs and vectors to transfer to vectors the identities we know for ordered pairs, such as $\mathbf{u}+\mathbf{v} = \mathbf{v}+\mathbf{u}$. This identity can also be seen geometrically, as in Fig. 13.1.12, which illustrates another geometric interpretation of vector addition. To add $\mathbf{u}$ and $\mathbf{v}$, we choose representatives $\overrightarrow{PQ}$ and $\overrightarrow{PR}$ having their tails at the same point P. If we complete the figure to a parallelogram $PQSR$, then the diagonal $\overrightarrow{PS}$ represents $\mathbf{u}+\mathbf{v}$. For this reason, physical quantities which combine by vector addition are sometimes said to "obey the parallelogram law."

If $\mathbf{v}$ and $\mathbf{w}$ are vectors, their difference $\mathbf{v}-\mathbf{w}$ is the vector such that $(\mathbf{v}-\mathbf{w})+\mathbf{w}=\mathbf{v}$. It follows from the "triangle" construction of vector sums that if we draw $\mathbf{v}$ and $\mathbf{w}$ with a common tail, $\mathbf{v}-\mathbf{w}$ is represented by the

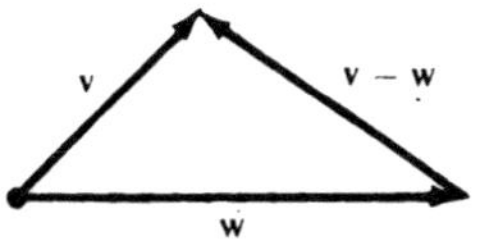

Figure 13.1.13. Geometric interpretation of vector subtraction.

directed line segment from the head of **w** to the head of **v**. (See Fig. 13.1.13.) The components of $\mathbf{v} - \mathbf{w}$ are obtained by subtracting the corresponding ordered pairs. Since $\overrightarrow{PQ} = \overrightarrow{OQ} - \overrightarrow{OP}$ (Fig. 13.1.14), we obtain the following.

Vectors and Directed Line Segments

If the point P has coordinates (x_1, y_1) and Q has coordinates (x_2, y_2), then the vector $\overrightarrow{PQ}$ has components $(x_2 - x_1, y_2 - y_1)$.

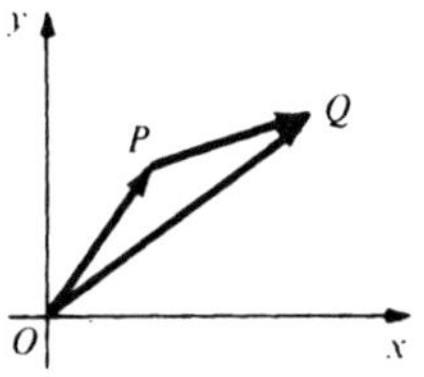

Figure 13.1.14. $\overrightarrow{PQ} = \overrightarrow{OQ} - \overrightarrow{OP}$.

Example 8 (a) Find the components of the vector from (3,5) to (4,7).
(b) Add the vector **v** from $(-1,0)$ to $(2,-3)$ and the vector **w** from $(2,0)$ to $(1,1)$.
(c) Multiply the vector **v** in (b) by 8. If this vector is represented by the directed line segment from (5,6) to Q, what is Q?

Solution (a) By the preceding box, we subtract the ordered pairs $(4,7) - (3,5) = (1,2)$. Thus the required components are $(1,2)$.
(b) The vector **v** has components $(2,-3) - (-1,0) = (3,-3)$ and **w** has components $(1,1) - (2,0) = (-1,1)$. Therefore, the vector $\mathbf{v} + \mathbf{w}$ has components $(3,-3) + (-1,1) = (2,-2)$.
(c) The vector $8\mathbf{v}$ has components $8(3,-3) = (24,-24)$. If this vector is represented by the directed line segment from (5,6) to Q, and Q has coordinates (x, y), then $(x, y) - (5,6) = (24,-24)$, so $(x, y) = (5,6) + (24,-24) = (29,-18)$. ▲

Exercises for Section 13.1

Complete the computations in Exercises 1–4.

1. $(1,2) + (3,7) =$
2. $(-2,6) - 6(2,-10) =$
3. $3[(1,1) - 2(3,0)] =$
4. $2[(8,6) - 4(2,-1)] =$

Solve for the unknown quantities, if possible, in Exercises 5–16.

5. $(1,2) + (0, y) = (1,3)$
6. $r(7,3) = (14,6)$
7. $a(2,-1) = (6,-\pi)$
8. $(7,2) + (x, y) = (3,10)$
9. $2(1,b) + (b,4) = (3,4)$
10. $(x,2) + (-3)(x, y) = (-2x, 1)$
11. $0(3,a) = (3,a)$
12. $6(1,0) + b(0,1) = (6,2)$
13. $a(1,1) + b(1,-1) = (3,5)$
14. $(a,1) - (2,b) = (0,0)$
15. $(3a,b) + (b,a) = (1,1)$
16. $a(3a,1) + a(1,-1) = (1,0)$

In Exercises 17–22, A, B, and C denote ordered pairs; O is the pair $(0,0)$; if $A = (x_1, y_1)$, then $-A = (-x_1, -y_1)$; and a and b are numbers. Show the following.

17. $A + O = A$
18. $A + (-A) = O$
19. $(A + B) + C = A + (B + C)$
20. $A + B = B + A$
21. $a(bA) = (ab)A$
22. $(a + b)A = aA + bA$

23. Describe geometrically the set of all points with coordinates of the form $m(0,1) + n(1,1)$, where m and n are integers. (A sketch will do.)
24. Describe geometrically the set of all points with coordinates of the form $m(0,1) + r(1,1)$, where m is an integer and r is a real number.
25. (a) Write the chemical equation $kSO_3 + lS_2 = mSO_2$ as an equation in ordered pairs.

(b) Write the equation in part (a) as a pair of simultaneous equations in k, l, and m.
(c) Solve the equation in part (b) for the smallest positive integer values of k, l, and m.

26. Illustrate the solution of Exercise 25 by a vector diagram in the plane, with SO_3, S_2, and SO_2 represented as vectors.
27. In Fig. 13.1.15, which vector is (a) $\mathbf{a} - \mathbf{b}$?, (b) $\frac{1}{2}\mathbf{a}$?
28. In Fig. 13.1.15, find the number r such that $\mathbf{c} - \mathbf{a} = r\mathbf{b}$.
29. Trace Fig. 13.1.15 and draw the vectors (a) $\mathbf{c} + \mathbf{d}$, (b) $-2\mathbf{e} + \mathbf{a}$. What are their components?
30. Trace Fig. 13.1.15 and draw the vectors (a) $3(\mathbf{e} - \mathbf{d})$, (b) $-\frac{2}{3}\mathbf{c}$. What are their components?

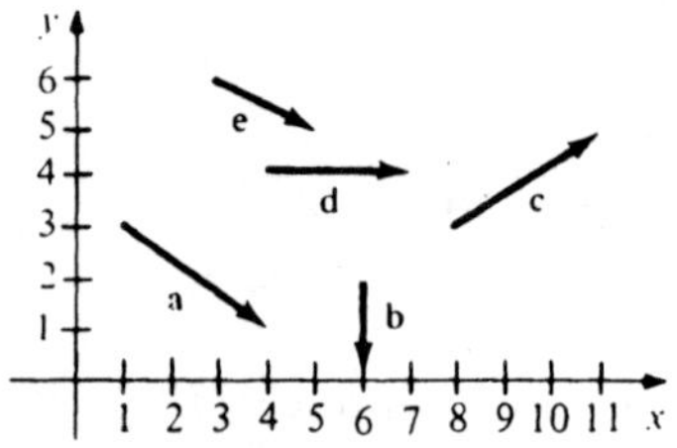

Figure 13.1.15. Compute with these vectors in Exercises 27–30.

In Exercises 31–36, let $\mathbf{u}$ have components $(2, 1)$ and $\mathbf{v}$ have components $(1, 2)$. Draw each of the indicated vectors.

31. $\mathbf{u} + \mathbf{v}$
32. $\mathbf{u} - \mathbf{v}$
33. $2\mathbf{u}$
34. $-4\mathbf{v}$
35. $2\mathbf{u} - 4\mathbf{v}$
36. $-\mathbf{u} + 2\mathbf{v}$
37. Let $P = (2, 1)$, $Q = (3, 3)$, and $R = (4, 1)$ be points in the xy plane.
(a) Draw (on the same diagram) these vectors: $\mathbf{v}$ joining P to Q; $\mathbf{w}$ joining Q to R; $\mathbf{u}$ joining R to P.
(b) What are the components of $\mathbf{v}$, $\mathbf{w}$, and $\mathbf{u}$?
(c) What is $\mathbf{v} + \mathbf{w} + \mathbf{u}$?
38. Answer the questions in Exercise 37 for $P = (-2, -1)$, $Q = (-3, -3)$, and $R = (-1, -4)$.
39. (a) Draw the vector $\mathbf{v}_1$ joining $(1, 0)$ to $(1, 1)$.
(b) What are the components of $\mathbf{v}_1$?
(c) Draw $\mathbf{v}_2$ joining $(1, 0)$ to $(1, \frac{5}{2})$ and find the components of $\mathbf{v}_2$.
(d) Draw the vector $\mathbf{v}_3$ joining $(1, 0)$ to $(1, -2)$.
(e) What are the coordinates of an arbitrary point on the vertical (that is, parallel to the y axis) line through $(1, 0)$?
(f) What are the components of the vector $\mathbf{v}$ joining $(1, 0)$ to such a point?
40. (a) Draw a vector $\mathbf{v}$ joining $(-1, 1)$ to $(1, 1)$.
(b) What are the components of $\mathbf{v}$?
(c) Sketch the vectors $\mathbf{v}_t = (-1, 1) + t\mathbf{v}$ when $t = 0, \frac{1}{4}, \frac{1}{2}, \frac{3}{4}$, and 1.
(d) Describe, geometrically, the set of vectors $\mathbf{v}_t = (-1, 1) + t\mathbf{v}$, where t takes on all values between 0 and 1. (Assume that all the vectors have their tails at the origin.)

★41. We say that $\mathbf{v}$ and $\mathbf{w}$ are *linearly dependent* if there are numbers r and s, not both zero, such that $r\mathbf{v} + s\mathbf{w} = \mathbf{0}$. Otherwise $\mathbf{v}$ and $\mathbf{w}$ are called *linearly independent*.
(a) Are $(0, 0)$ and $(1, 1)$ linearly dependent?
(b) Show that two non-zero vectors are linearly dependent if and only if they are parallel.
(c) Let $\mathbf{v}$ and $\mathbf{w}$ be vectors in the plane given by $\mathbf{v} = (a, b)$ and $\mathbf{w} = (c, d)$. Show that $\mathbf{v}$ and $\mathbf{w}$ are linearly dependent if and only if $ad = bc$. [*Hint*: For one implication, you might use three cases: $b \neq 0$, $d \neq 0$, and $b = d = 0$.]
(d) Suppose that $\mathbf{v}$ and $\mathbf{w}$ are vectors in the plane which are linearly independent. Show that for any vector $\mathbf{u}$ in the plane there are numbers x and y such that $x\mathbf{v} + y\mathbf{w} = \mathbf{u}$.

★42. Let $P = (a, b)$ and $Q = (c, d)$ be points in the plane. (You may assume that $0 < a < c$ and $b > d > 0$ to make the picture unambiguous.) Compute the area of the parallelogram with vertices at O, P, Q, and $P + Q$. Comment on the relationship between this and Exercise 41(d).

13.2 Vectors in Space

A vector in space has three components.

The plane is two-dimensional, but space is three-dimensional—that is, it requires *three* numbers to specify the position of the point in space. For instance, the location of a bird is specified not only by the two coordinates of the point on the ground directly below it, but also by its height. Accordingly, our algebraic model for space will be the set of *triples* (x, y, z) rather than pairs of real numbers.

If one starts with abstract "space" as studied in elementary solid geome-

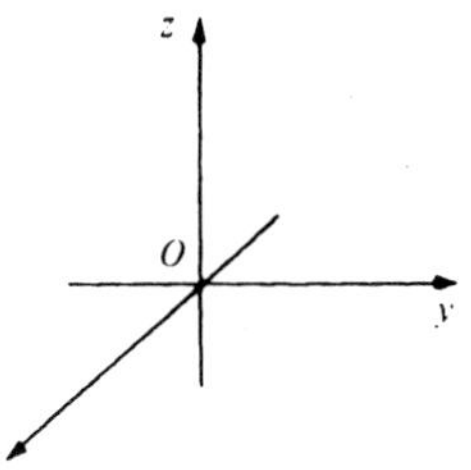

Figure 13.2.1. Coordinate axes in space.

try, the first step in the introduction of coordinates is the choice of an origin O and three directed lines, each perpendicular to the other two, called the x, y, and z axes. We will usually draw figures in space with the axes oriented as in Fig. 13.2.1.

Think of the x axis as pointing toward you, out of the paper. Notice that if you wrap the fingers of your right hand around the z axis, with your fingers curling in the usual (counterclockwise) direction of rotation in the xy plane, then your thumb points toward the positive z axis. For this reason, we say that the choice of axes obeys the *right-hand rule*. For example, the coordinate axes (a) and (d) in Fig. 13.2.2 obey the right-hand rule, but (b) and (c) do not. (Think of all horizontal and vertical arrows as being in the plane of the paper, while slanted arrows point out toward you.)

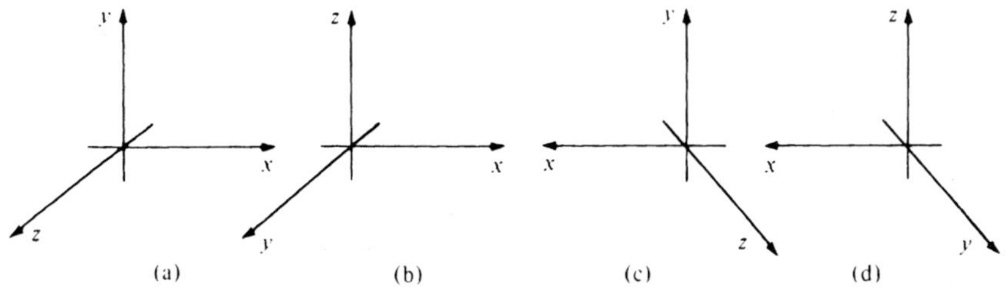

Figure 13.2.2. Which axes obey the right-hand rule?

Given a point P in space, drop a perpendicular from P to each of the axes. By measuring the (directed) distance from the origin to the foot of each of these perpendiculars, we obtain numbers (x, y, z) which we call the *coordinates* of P (see Fig. 13.2.3).

If you cannot see the lines through P in Fig. 13.2.3 as being perpendicular to the axis it may help to draw in some additional lines *parallel* to the axes, using the convention that lines which are parallel to one another in space are drawn parallel. This simple convention, sometimes called the *rule of parallel projection*, does not conform to ordinary rules of perspective (think of railroad tracks "converging at infinity"), but it is reasonably accurate if your distance from an object is great compared to the size of the object.

Now look at Fig. 13.2.4. Observe that the point Q, obtained by dropping a perpendicular from P to the xy plane, has coordinates $(x, y, 0)$. Similarly, the points R and S, obtained by dropping perpendiculars from P to the yz and xz planes, have coordinates $(0, y, z)$ and $(x, 0, z)$, respectively. The coordinates of T, U, and V are $(x, 0, 0)$, $(0, y, 0)$, and $(0, 0, z)$.

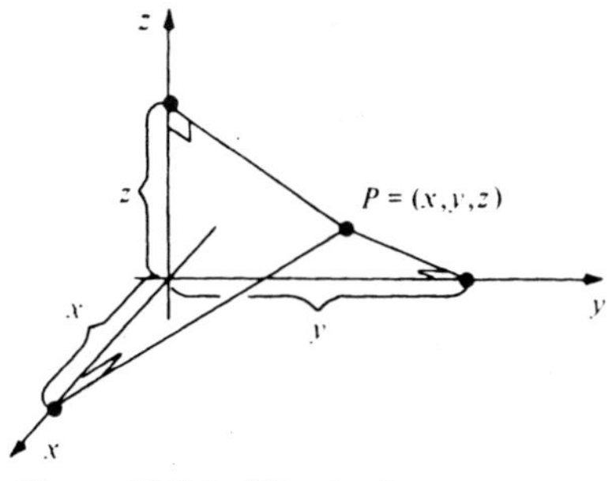

Figure 13.2.3. We obtain the coordinates of the point P by dropping perpendiculars to the x, y, and z axes.

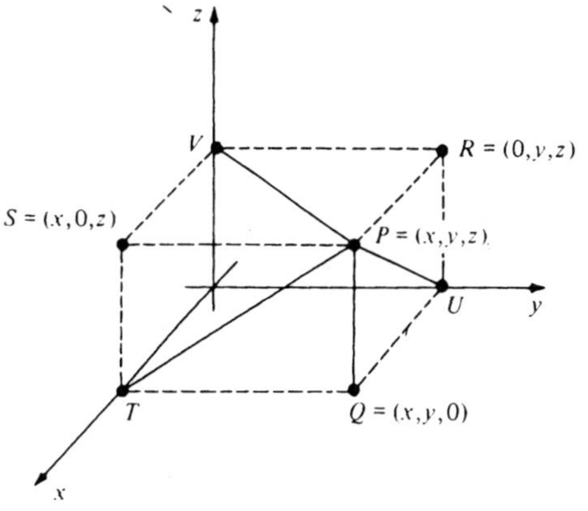

Figure 13.2.4. Lines which are parallel in space are drawn parallel.

As in the plane, we can find the unique point which has a given ordered triple (x, y, z) as its coordinates. To do so, we begin by finding the point $Q = (x, y, 0)$ in the xy plane. By drawing a line through Q and parallel to the z axis, we can locate the point P at a (directed) distance of z units from Q along this line. The process just described can be carried out graphically, as in the following example.

Example 1 Plot the point $(1, 2, -3)$.

Solution We begin by plotting $(1, 2, 0)$ in the xy plane (Fig. 13.2.5(a)). Then we draw the line through this point parallel to the z axis and measure 3 units downward (Fig. 13.2.5(b)). ▲

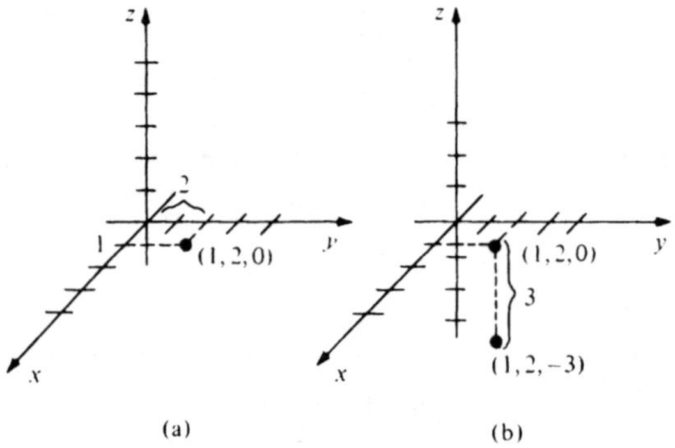

Figure 13.2.5. Plotting the point $(1, 2, -3)$.

Warning If you are given a picture consisting simply of three axes (with units of measure) and a point, it is not possible to determine the coordinates of the point from these data alone, since some information must be lost in making a two-dimensional picture of the three-dimensional space. (See Review Exercise 86.)

Addition and scalar multiplication are defined for ordered triples just as for pairs.

The Algebra of Ordered Triples

1. If (x_1, y_1, z_1) and (x_2, y_2, z_2) are ordered triples of real numbers, the ordered triple $(x_1 + x_2, y_1 + y_2, z_1 + z_2)$ is called their *sum* and is denoted by $(x_1, y_1, z_1) + (x_2, y_2, z_2)$.
2. If (x, y, z) is an ordered triple and r is a real number, the triple (rx, ry, rz) is called the *product* of r and (x, y, z) and is denoted by $r(x, y, z)$.

Example 2 Find $(3, 2, -2) + (-1, -2, -1)$ and $(-6)(2, -1, 1)$.

Solution We have $(3, 2, -2) + (-1, -2, -1) = (3 - 1, 2 - 2, -2 - 1) = (2, 0, -3)$ and $(-6)(2, -1, 1) = (-12, 6, -6)$. ▲

Now we look at vectors in space, the geometric objects which correspond to ordered triples.

Vectors in Space

A *vector* in space is a directed line segment in space and is drawn as an arrow. Two directed line segments will be regarded as equal when they have the same length and direction.

Vectors are denoted by boldface symbols. The vector represented by the arrow from a point P to a point Q is denoted $\overrightarrow{PQ}$. If the arrows from P_1 to Q_1 and P_2 to Q_2 represent the same vectors, we write $\overrightarrow{P_1Q_1} = \overrightarrow{P_2Q_2}$.

Vectors in space are related to ordered triples as follows. We choose x, y, and z axes and drop perpendiculars to the three axes. The directed distances obtained are called the *components* of the vector (see Fig. 13.2.6).

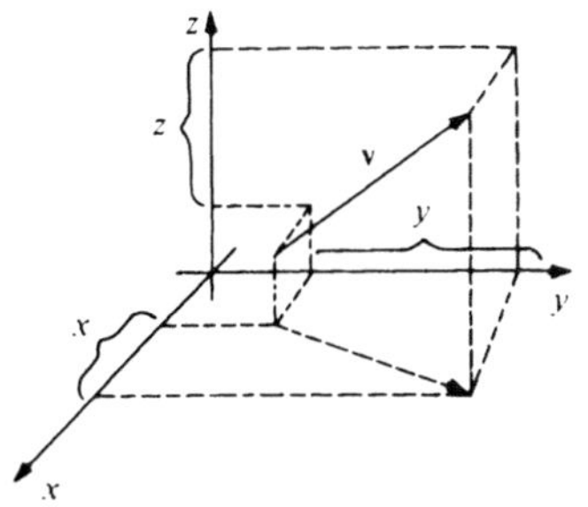

Figure 13.2.6. The vector $\mathbf{v}$ has components (x, y, z).

Vector addition and scalar multiplication for vectors in space are defined just as in the plane. The student should reread the corresponding development in Section 13.1, replacing the plane by space.

Vectors and Ordered Triples

1. The algebra of vectors corresponds to the algebra of ordered triples.
2. If P has coordinates (x_1, y_1, z_1) and Q has coordinates (x_2, y_2, z_2), then the vector $\overrightarrow{PQ}$ has components $(x_2 - x_1, y_2 - y_1, z_2 - z_1)$.

Example 3 (a) Sketch $-2\mathbf{v}$, where $\mathbf{v}$ has components $(-1, 1, 2)$. (b) If $\mathbf{v}$ and $\mathbf{w}$ are any two vectors, show that $\mathbf{v} - \frac{1}{3}\mathbf{w}$ and $3\mathbf{v} - \mathbf{w}$ are parallel.

Solution (a) The vector $-2\mathbf{v}$ is twice as long as $\mathbf{v}$ but points in the opposite direction (see Fig. 13.2.7). (b) $\mathbf{v} - \frac{1}{3}\mathbf{w} = \frac{1}{3}(3\mathbf{v} - \mathbf{w})$; vectors which are multiples of one another are parallel. ▲

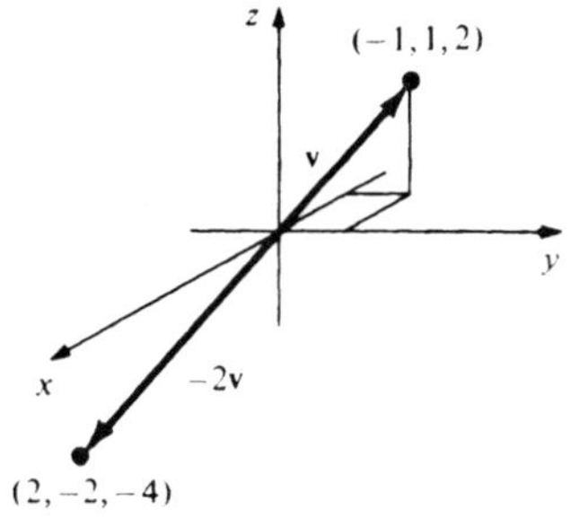

Figure 13.2.7. Multiplying $(-1, 1, 2)$ by -2.

Example 4 Let $\mathbf{v}$ be the vector with components $(3, 2, -2)$ and let $\mathbf{w}$ be the vector from the point $(2, 1, 3)$ to the point $(-1, 0, -1)$. Find $\mathbf{v} + \mathbf{w}$. Illustrate with a sketch.

Solution Since $\mathbf{w}$ has components $(-1, 0, -1) - (2, 1, 3) = (-3, -1, -4)$, we find that $\mathbf{v} + \mathbf{w}$ has components $(3, 2, -2) + (-3, -1, -4) = (0, 1, -6)$, as illustrated in Fig. 13.2.8. ▲

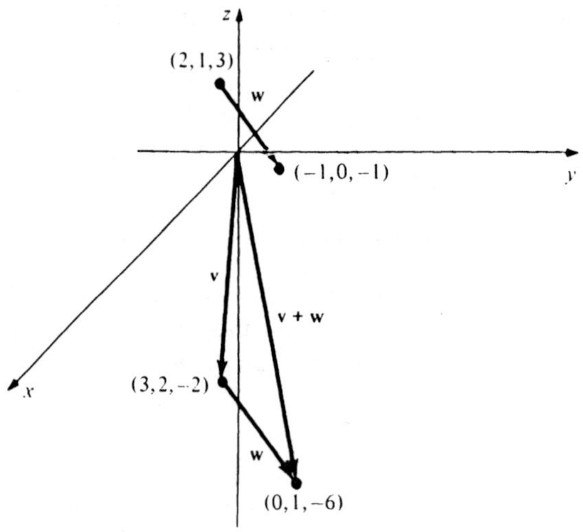

Figure 13.2.8. Adding $\mathbf{v} = (3, 2, -2)$ to $\mathbf{w}$, the vector from $(2, 1, 3)$ to $(-1, 0, -1)$.

To describe vectors in space, it is convenient to introduce three special vectors along the x, y, and z axes.

$\mathbf{i}$: the vector with components $(1, 0, 0)$;
$\mathbf{j}$: the vector with components $(0, 1, 0)$;
$\mathbf{k}$: the vector with components $(0, 0, 1)$.

These *standard basis vectors* are illustrated in Fig. 13.2.9. In the plane one has, analogously, $\mathbf{i}$ and $\mathbf{j}$ with components $(1, 0)$ and $(0, 1)$.

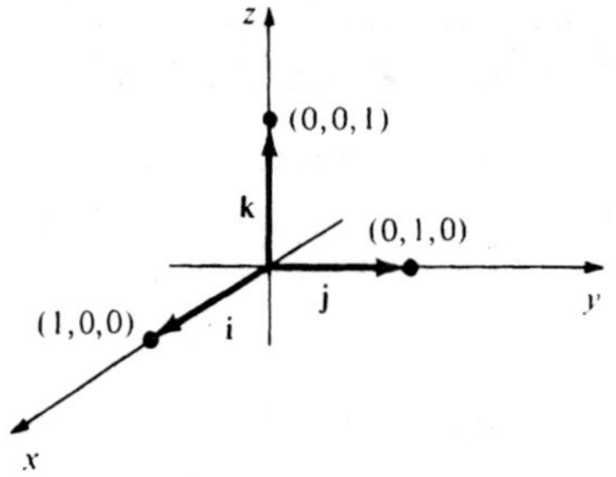

Figure 13.2.9. The standard basis vectors.

Now, let $\mathbf{v}$ be any vector, and let (a, b, c) be its components. Then

$$\mathbf{v} = a\mathbf{i} + b\mathbf{j} + c\mathbf{k},$$

since the right-hand side is given in components by

$$a(1, 0, 0) + b(0, 1, 0) + c(0, 0, 1) = (a, 0, 0) + (0, b, 0) + (0, 0, c) = (a, b, c).$$

Thus we can express every vector as a sum of scalar multiples of $\mathbf{i}$, $\mathbf{j}$, and $\mathbf{k}$.

The Standard Basis Vectors

1. The vectors $\mathbf{i}$, $\mathbf{j}$, and $\mathbf{k}$ are unit vectors along the three coordinate axes, as shown in Fig. 13.2.9.
2. If $\mathbf{v}$ has components (a, b, c), then

$$\mathbf{v} = a\mathbf{i} + b\mathbf{j} + c\mathbf{k}.$$

Example 5 (a) Express the vector whose components are $(e, \pi, -\sqrt{3})$ in the standard basis. (b) Express the vector $\mathbf{v}$ joining $(2,0,1)$ to $(\frac{3}{2}, \pi, -1)$ by using the standard basis.

Solution (a) $\mathbf{v} = e\mathbf{i} + \pi\mathbf{j} - \sqrt{3}\,\mathbf{k}$. (b) The vector $\mathbf{v}$ has components $(\frac{3}{2}, \pi, -1) - (2,0,1) = (-\frac{1}{2}, \pi, -2)$, so $\mathbf{v} = -\frac{1}{2}\mathbf{i} + \pi\mathbf{j} - 2\mathbf{k}$. ▲

Now we turn to some physical applications of vectors.[3] A simple example of a physical quantity represented by a vector is a displacement. Suppose that, on a part of the earth's surface small enough to be considered flat, we introduce coordinates so that the x axis points east, the y axis points north, and the unit of length is the kilometer. If we are at a point P and wish to get to a point Q, the *displacement vector* $\mathbf{u}$ joining P to Q tells us the direction and distance we have to travel. If x and y are the components of this vector, the displacement of Q from P is "x kilometers east, y kilometers north."

Example 6 Suppose that two navigators, who cannot see one another but can communicate by radio, wish to determine the relative position of their ships. Explain how they can do this if they can each determine their displacement vector to the same lighthouse.

Solution Let P_1 and P_2 be the positions of the ships and Q be the position of the lighthouse. The displacement of the lighthouse from the ith ship is the vector $\mathbf{u}_i$ joining P_i to Q. The displacement of the second ship from the first is the vector $\mathbf{v}$ joining P_1 to P_2. We have $\mathbf{v} + \mathbf{u}_2 = \mathbf{u}_1$ (Fig. 13.2.10), so $\mathbf{v} = \mathbf{u}_1 - \mathbf{u}_2$.

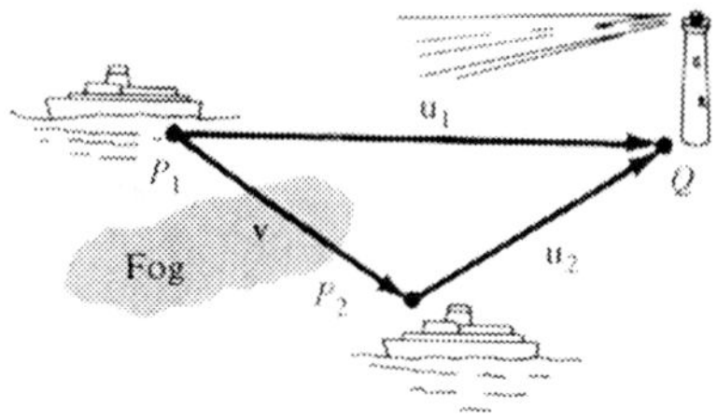

Figure 13.2.10. Vector methods can be used to locate objects.

That is, the displacement from one ship to the other is the difference of the displacements from the ships to the lighthouse. ▲

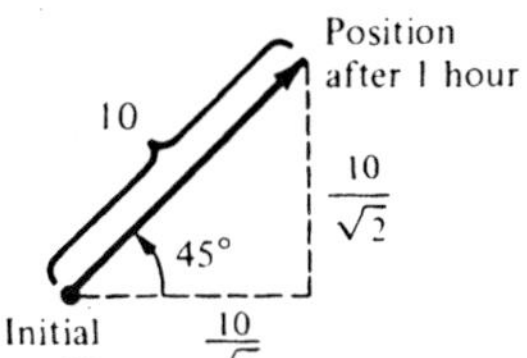

Figure 13.2.11. If an object moves northeast at 10 kilometers per hour, its velocity vector has components $(10/\sqrt{2}, 10/\sqrt{2})$.

We can also represent the velocity of a moving object as a vector. For the moment, we will consider only objects moving at uniform speed along straight lines—the general case is discussed in Section 14.6. Suppose, for example, that a boat is steaming across a lake at 10 kilometers per hour in the northeast direction. After 1 hour of travel, the displacement is $(10/\sqrt{2}, 10/\sqrt{2}) \approx$ $\approx (7.07, 7.07)$ (see Fig. 13.2.11). The vector whose components are $(10/\sqrt{2}, 10/\sqrt{2})$ is called the *velocity vector* of the boat. In general, if an object is moving uniformly along a straight line, *its velocity vector is the displacement vector from the position at any moment to the position 1 unit of time later.* If a

[3] Historical note: Many scientists resisted the use of vectors in favor of the more complicated theory of quaternions until around 1900. The book which popularized vector methods was *Vector Analysis*, by E. B. Wilson (reprinted by Dover in 1960), based on lectures of J. W. Gibbs at Yale in 1899–1900. Wilson was reluctant to take Gibbs' course since he had just completed a full-year course in quaternions at Harvard under J. M. Pierce, a champion of quaternionic methods, but was forced by a dean to add the course to his program. (For more details, see M. J. Crowe, *A History of Vector Analysis*, University of Notre Dame Press, 1967.)

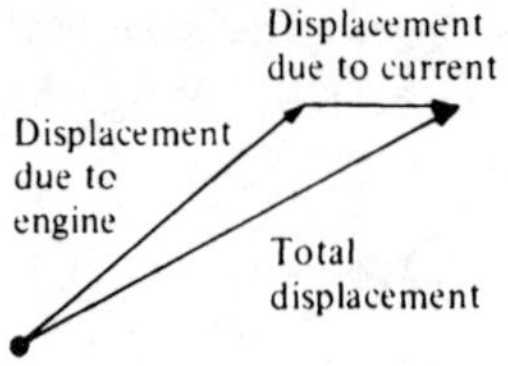

Figure 13.2.12. The total displacement is the sum of the displacements due to the engine and the current.

current appears on the lake, moving due eastward at 2 kilometers per hour, and the boat continues to point in the same direction with its engine running at the same rate, its displacement after 1 hour will have components given by $(10/\sqrt{2}+2, 10/\sqrt{2})$. (See Fig. 13.2.12.) The new velocity vector, therefore, has components $(10/\sqrt{2}+2, 10/\sqrt{2})$. We note that this is the sum of the original velocity vector $(10/\sqrt{2}, 10/\sqrt{2})$ of the boat and the velocity vector $(2,0)$ of the current.

Similarly, consider a seagull which flies in calm air with velocity vector $\mathbf{v}$. If a wind comes up with velocity $\mathbf{w}$ and the seagull continues flying "the same way," its actual velocity will be $\mathbf{v}+\mathbf{w}$. One can "see" the direction of the vector $\mathbf{v}$ because it points along the "axis" of the seagull; by comparing the direction of actual motion with the direction of $\mathbf{v}$, you can get an idea of the wind direction (see Fig. 13.2.13).

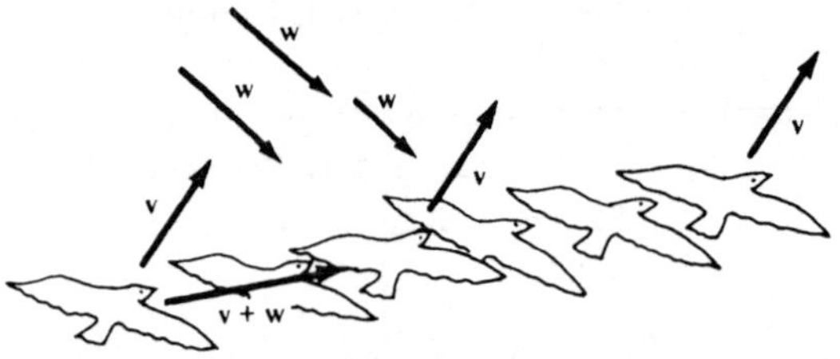

Figure 13.2.13. The velocity $\mathbf{w}$ of the wind can be estimated by comparing the "wingflap" velocity $\mathbf{v}$ with the actual velocity $\mathbf{v}+\mathbf{w}$.

Another example comes from medicine. An electrocardiograph detects the flow of electricity in the heart. Both the magnitude and the direction of the net flow are of importance. This information can be summarized at every instant by means of a vector called the *cardiac vector*. The motion of this vector (see Fig. 13.2.14) gives physicians useful information about the heart's function.[4]

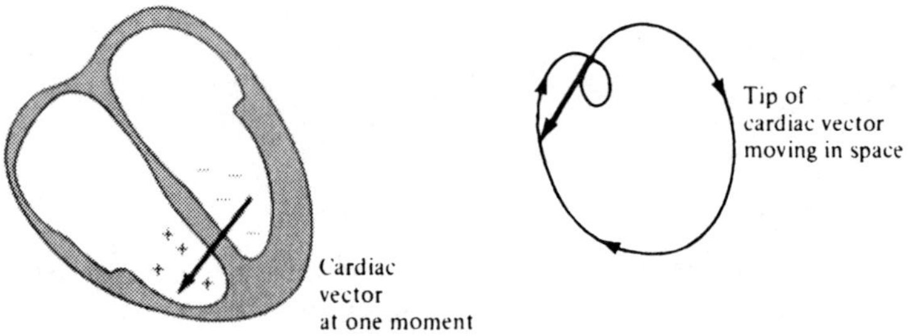

Figure 13.2.14. The magnitude and direction of electrical flow in the heart are indicated by the cardiac vector.

Example 7 A bird is flying in a straight line with velocity vector $10\mathbf{i}+6\mathbf{j}+\mathbf{k}$ (in kilometers per hour). Suppose that (x, y) are coordinates on the ground and z is the height above the ground.

(a) If the bird is at position $(1,2,3)$ at a certain moment, where is it 1 minute later?
(b) How many seconds does it take the bird to climb 10 meters?

Solution (a) The displacement vector from $(1,2,3)$ is $\frac{1}{60}(10\mathbf{i}, 6\mathbf{j}, \mathbf{k}) = \frac{1}{6}\mathbf{i}+\frac{1}{10}\mathbf{j}+\frac{1}{60}\mathbf{k}$, so the new position is $(1,2,3)+(\frac{1}{6},\frac{1}{10}\,\frac{1}{60}) = (1\frac{1}{6}, 2\frac{1}{10}, 3\frac{1}{60})$.
(b) After t seconds ($= t/3600$ hours), the displacement vector from $(1,2,3)$ is $(t/3600)(10\mathbf{i}+6\mathbf{j}+\mathbf{k}) = (t/360)\mathbf{i}+(t/600)\mathbf{j}+(t/3600)\mathbf{k}$. The increase in altitude is the z component $t/3600$. This will equal 10 meters ($=\frac{1}{100}$ kilometer) when $t/3600 = 1/100$—that is, when $t = 36$ seconds. ▲

[4] See M. J. Goldman, *Principles of Clinical Electrocardiography*, 8th edition, Lange, 1973. Chapters 14 and 19.

Example 8 Physical forces have magnitude and direction and may thus be represented by vectors. If several forces act at once on an object, the resultant force is represented by the sum of the individual force vectors. Suppose that forces $\mathbf{i}+\mathbf{k}$ and $\mathbf{j}+\mathbf{k}$ are acting on a body. What third force must we impose to counteract these two—that is, to make the total force equal to zero?

Solution The force $\mathbf{v}$ should be chosen so that $(\mathbf{i}+\mathbf{k})+(\mathbf{j}+\mathbf{k})+\mathbf{v}=\mathbf{0}$; that is $\mathbf{v}=-(\mathbf{i}+\mathbf{k})-(\mathbf{j}+\mathbf{k})=-\mathbf{i}-\mathbf{j}-2\mathbf{k}$. (Here $\mathbf{0}$ is the *zero vector*, the vector whose components are all zero.) ▲

Exercises for Section 13.2

Plot the points in Exercises 1–4.

1. $(1,0,0)$
2. $(0,2,4)$
3. $(3,-1,5)$
4. $(2,-1,\frac{1}{2})$

Complete the computations in Exercises 5–8.

5. $(6,0,5)+(5,0,6)=$
6. $(0,0,0)+(0,0,0)=$
7. $(1,3,5)+4(-1,-3,-5)=$
8. $(2,0,1)-8(3,-\frac{1}{2},\frac{1}{4})=$

9. Sketch $\mathbf{v}$, $2\mathbf{v}$, and $-\mathbf{v}$, where $\mathbf{v}$ has components $(1,-1,-1)$.
10. Sketch $\mathbf{v}$, $3\mathbf{v}$, and $-\frac{1}{2}\mathbf{v}$, where $\mathbf{v}$ has components $(2,-1,1)$.
11. Let $\mathbf{v}$ have components $(0,1,1)$ and $\mathbf{w}$ have components $(1,1,0)$. Find $\mathbf{v}+\mathbf{w}$ and sketch.
12. Let $\mathbf{v}$ have components $(2,-1,1)$ and $\mathbf{w}$ have components $(1,-1,-1)$. Find $\mathbf{v}+\mathbf{w}$ and sketch.

In Exercises 13–20, express the given vector in terms of the standard basis.

13. The vector with components $(-1,2,3)$.
14. The vector with components $(0,2,2)$.
15. The vector with components $(7,2,3)$.
16. The vector with components $(-1,2,\pi)$.
17. The vector from $(0,1,2)$ to $(1,1,1)$.
18. The vector from $(3,0,5)$ to $(2,7,6)$.
19. The vector from $(1,0,0)$ to $(2,-1,1)$.
20. The vector from $(1,0,0)$ to $(3,-2,2)$.

21. A ship at position $(1,0)$ on a nautical chart (with north in the positive y direction) sights a rock at position $(2,4)$. What is the vector joining the ship to the rock? What angle does this vector make with due north? This is called the *bearing* of the rock from the ship.
22. Suppose that the ship in Exercise 21 is pointing due north and travelling at a speed of 4 knots relative to the water. There is a current flowing due east at 1 knot. (The units on the chart are nautical miles; 1 knot = 1 nautical mile per hour.)
    (a) If there were no current, what vector $\mathbf{u}$ would represent the velocity of the ship relative to the sea bottom?
    (b) If the ship were just drifting with the current, what vector $\mathbf{v}$ would represent its velocity relative to the sea bottom?
    (c) What vector $\mathbf{w}$ represents the total velocity of the ship?
    (d) Where would the ship be after 1 hour?
    (e) Should the captain change course?
    (f) What if the rock were an iceberg?
23. An airplane is located at position $(3,4,5)$ at noon and travelling with velocity $400\mathbf{i}+500\mathbf{j}-\mathbf{k}$ kilometers per hour. The pilot spots an airport at position $(23,29,0)$.
    (a) At what time will the plane pass directly over the airport? (Assume that the earth is flat and that the vector $\mathbf{k}$ points straight up.)
    (b) How high above the airport will the plane be when it passes?
24. The wind velocity $\mathbf{v}_1$ is 40 miles per hour from east to west while an airplane travels with air speed $\mathbf{v}_2$ of 100 miles per hour due north. The speed of the airplane relative to the earth is the vector sum $\mathbf{v}_1+\mathbf{v}_2$.
    (a) Find $\mathbf{v}_1+\mathbf{v}_2$.
    (b) Draw a figure to scale.
25. A force of 50 lbs is directed 50° above horizontal, pointing to the right. Determine its horizontal and vertical components. Display all results in a figure.
26. Two persons pull horizontally on ropes attached to a post, the angle between the ropes being 60°. A pulls with a force of 150 lbs, while B pulls with a force of 110 lbs.
    (a) The resultant force is the vector sum of the two forces in a conveniently chosen coordinate system. Draw a figure to scale which graphically represents the three forces.
    (b) Using trigonometry, determine formulas for the vector components of the two forces in a conveniently chosen coordinate system. Perform the algebraic addition, and find the angle the resultant force makes with A.
27. What restrictions must be placed on x, y, and z so that the triple (x,y,z) will represent a point

on the y axis? On the z axis? In the xy plane? In the xz plane?

28. Plot on one set of axes the eight points of the form (a,b,c), where a, b, and c are each equal to 1 or -1. Of what geometric figure are these the vertices?
29. Let $\mathbf{u} = 2\mathbf{i} + 3\mathbf{j} + \mathbf{k}$. Sketch the vectors $\mathbf{u}$, $2\mathbf{u}$, and $-3\mathbf{u}$ on the same set of axes.

In Exercises 30–34, consider the vectors $\mathbf{v} = 3\mathbf{i} + 4\mathbf{j} + 5\mathbf{k}$ and $\mathbf{w} = \mathbf{i} - \mathbf{j} + \mathbf{k}$. Express the given vector in terms of $\mathbf{i}$, $\mathbf{j}$ and $\mathbf{k}$.

30. $\mathbf{v} + \mathbf{w}$
31. $3\mathbf{v}$
32. $-2\mathbf{w}$
33. $6\mathbf{v} + 8\mathbf{w}$
34. the vector $\mathbf{u}$ from the tip of $\mathbf{w}$ to the tip of $\mathbf{v}$. (Assume that the tails of $\mathbf{w}$ and $\mathbf{v}$ are at the same point.)

In Exercises 35–37, let $\mathbf{v} = \mathbf{i} + \mathbf{j}$ and $\mathbf{w} = -\mathbf{i} + \mathbf{j}$. Find numbers a and b such that $a\mathbf{v} + b\mathbf{w}$ is the given vector.

35. $\mathbf{i}$
36. $\mathbf{j}$
37. $3\mathbf{i} + 7\mathbf{j}$
38. Let $\mathbf{u} = \mathbf{i} + \mathbf{j} + \mathbf{k}$, $\mathbf{v} = \mathbf{i} + \mathbf{j}$, and $\mathbf{w} = \mathbf{i}$. Given numbers r, s, and t, find a, b, and c such that $a\mathbf{u} + b\mathbf{v} + c\mathbf{w} = r\mathbf{i} + s\mathbf{j} + t\mathbf{k}$.
39. A 1-kilogram mass located at the origin is suspended by ropes attached to the points $(1,1,1)$ and $(-1,-1,1)$. If the force of gravity is pointing in the direction of the vector $-\mathbf{k}$, what is the vector describing the force along each rope? [*Hint*: Use the symmetry of the problem. A 1-kilogram mass weighs 9.8 newtons.]
40. Write the chemical equation $CO + H_2O = H_2 + CO_2$ as an equation in ordered triples, and illustrate it by a vector diagram in space.
41. (a) Write the chemical equation $pC_3H_4O_3 + qO_2 = rCO_2 + sH_2O$ as an equation in ordered triples with unknown coefficients p, q, r, and s.
    (b) Find the smallest integer solution for p, q, r, and s.
    (c) Illustrate the solution by a vector diagram in space.
42. Suppose that the cardiac vector is given by $\cos t\mathbf{i} + \sin t\mathbf{j} + \mathbf{k}$ at time t.
    (a) Draw the cardiac vector for $t = 0, \pi/4, \pi/2, 3\pi/4, \pi, 5\pi/4, 3\pi/2, 7\pi/4, 2\pi$.
    (b) Describe the motion of the tip of the cardiac vector in space if the tail is fixed at the origin.

★43. Let $P_t = (1,0,0) + t(2,1,1)$, where t is a real number.
    (a) Compute the coordinates of P_t for $t = -1$, 0, 1, and 2.
    (b) Sketch these four points on the same set of axes.
    (c) Try to describe geometrically the set of all the P_t.

★44. The z coordinate of the point P in Fig. 13.2.15 is 3. What are the x and y coordinates?

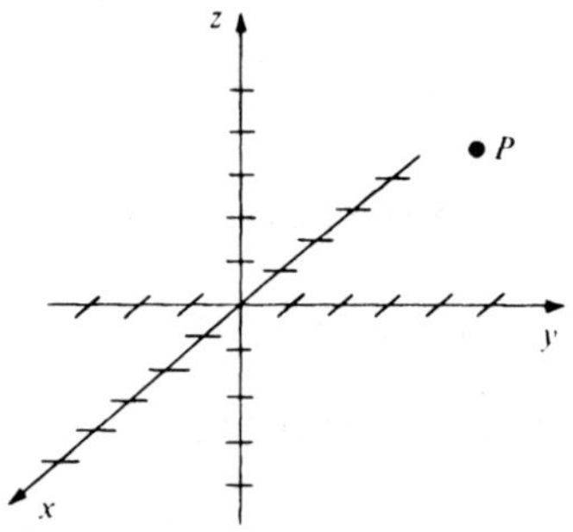

Figure 13.2.15. Let $P = (x, y, 3)$. What are x and y?

13.3 Lines and Distance

Algebraic operations on vectors can be used to solve geometric problems.

In this section we apply the algebra of vectors to the description of lines and planes in space and to the solution of other geometric problems.

The invention of analytic geometry made it possible to solve geometric problems in the plane or space by reducing them to algebraic problems involving number pairs or triples. Vector methods also convert geometric problems to algebraic ones; moreover, the vector calculations are often simpler than those from analytic geometry, since we do not need to write down all the components.

Example 1 Use vector methods to prove that the diagonals of a parallelogram bisect each other.

Solution Let $PQRS$ be the parallelogram, $\mathbf{w}$ the vector $\overrightarrow{PQ}$ and $\mathbf{v}$ the vector $\overrightarrow{PS}$ (see Fig.

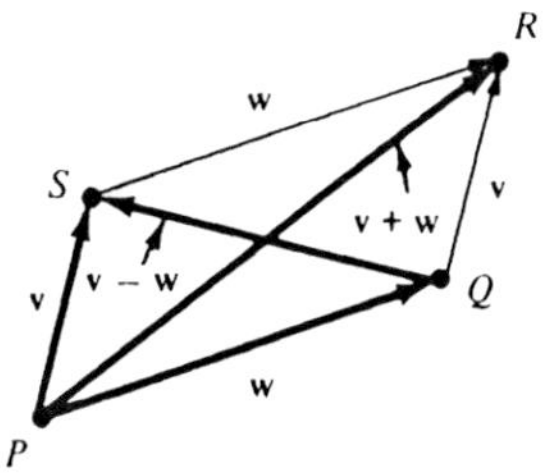

Figure 13.3.1. The diagonals of a parallelogram bisect each other.

13.3.1). Since $PQRS$ is a parallelogram, $\mathbf{v} + \mathbf{w}$ is the vector $\overrightarrow{PR}$.

The vector joining P to the midpoint M_1 of the diagonal PR is thus $\frac{1}{2}(\mathbf{v} + \mathbf{w})$. On the other hand, the vector $\overrightarrow{QS}$ is $\mathbf{v} - \mathbf{w}$, so the vector joining Q to the midpoint M_2 of the diagonal QS is $\frac{1}{2}(\mathbf{v} - \mathbf{w})$.

To show that the diagonals bisect each other, it is enough to show that the midpoints M_1 and M_2 are the same. The vector $\overrightarrow{PM_2}$ is the sum

$$\mathbf{w} + \tfrac{1}{2}(\mathbf{v} - \mathbf{w}) = \mathbf{w} + \tfrac{1}{2}\mathbf{v} - \tfrac{1}{2}\mathbf{w} = \mathbf{w} - \tfrac{1}{2}\mathbf{w} + \tfrac{1}{2}\mathbf{v}$$
$$= \tfrac{1}{2}\mathbf{w} + \tfrac{1}{2}\mathbf{v} = \tfrac{1}{2}(\mathbf{v} + \mathbf{w})$$

which is the same as the vector $\overrightarrow{PM_1}$. It follows that M_1 and M_2 are the same point. ▲

Example 2 Consider the cube in space with vertices at (0,0,0), (1,0,0), (1,1,0), (0,1,0), (0,0,1), (1,0,1), (1,1,1), and (0,1,1). Use vector methods to locate the point one-third of the way from the origin to the middle of the face whose vertices are (0,1,0), (0,1,1), (1,1,1), and (1,1,0).

Solution Refer to Fig. 13.3.2. The vector $\overrightarrow{OP}$ is $\mathbf{j}$, and vector $\overrightarrow{OQ}$ is $\mathbf{i} + \mathbf{j} + \mathbf{k}$. The vector

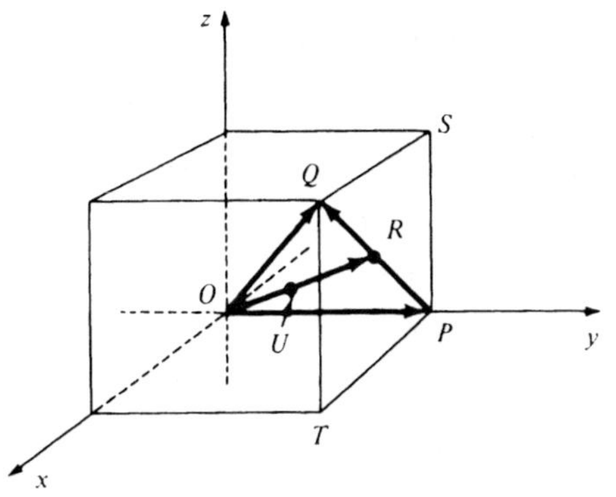

Figure 13.3.2. The point U is one-third of the way from O to the midpoint of the face $PSQT$.

$\overrightarrow{PQ}$ is the difference $\mathbf{i} + \mathbf{j} + \mathbf{k} - \mathbf{j} = \mathbf{i} + \mathbf{k}$; the vector joining P to the midpoint R of PQ (and hence of the face $PSQT$) is one-half of this, that is, $\frac{1}{2}(\mathbf{i} + \mathbf{k})$; the vector $\overrightarrow{OR}$ is then $\mathbf{j} + \frac{1}{2}(\mathbf{i} + \mathbf{k})$, and the vector joining O to the point U one-third of the way from O to R is $\frac{1}{3}[\mathbf{j} + \frac{1}{2}(\mathbf{i} + \mathbf{k})] = \frac{1}{3}\mathbf{j} + \frac{1}{6}\mathbf{i} + \frac{1}{6}\mathbf{k} = \frac{1}{6}\mathbf{i} + \frac{1}{3}\mathbf{j} + \frac{1}{6}\mathbf{k}$. It follows that the coordinates of U are $(\frac{1}{6}, \frac{1}{3}, \frac{1}{6})$. ▲

Example 3 Prove that the figure obtained by joining the midpoints of successive sides of any quadrilateral is a parallelogram.

Solution Refer to Fig. 13.3.3. Let $PQRS$ be the quadrilateral, $\mathbf{v} = \overrightarrow{PS}$, $\mathbf{w} = \overrightarrow{SR}$, $\mathbf{t} = \overrightarrow{PQ}$,

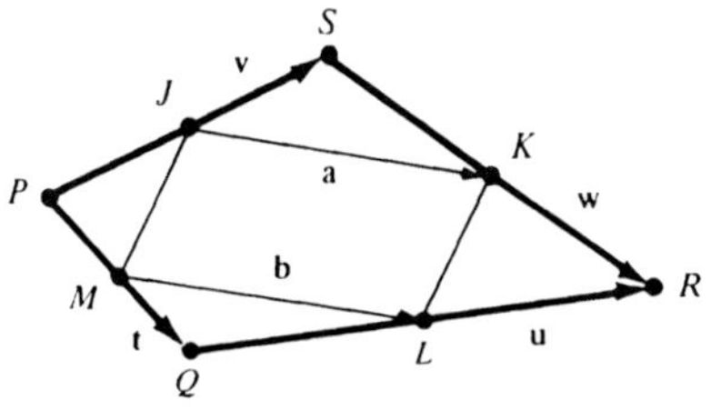

Figure 13.3.3. The figure obtained by joining the midpoints of successive sides of $PQRS$ is the parallelogram $JKLM$.

and $\mathbf{u} = \overrightarrow{QR}$. The vector $\mathbf{a}$ from the midpoint J of PS to the midpoint K of SR satisfies $\frac{1}{2}\mathbf{v} + \mathbf{a} = \mathbf{v} + \frac{1}{2}\mathbf{w}$; solving for $\mathbf{a}$ gives $\mathbf{a} = \frac{1}{2}\mathbf{v} + \frac{1}{2}\mathbf{w} = \frac{1}{2}(\mathbf{v} + \mathbf{w})$. Similarly, the vector $\mathbf{b}$ from the midpoint M of PQ to the midpoint L of QR satisfies $\frac{1}{2}\mathbf{t} + \mathbf{b} = \mathbf{t} + \frac{1}{2}\mathbf{u}$, so $\mathbf{b} = \frac{1}{2}\mathbf{t} + \frac{1}{2}\mathbf{u} = \frac{1}{2}(\mathbf{t} + \mathbf{u})$.

To show that $JKLM$ is a parallelogram, it suffices to show that the vectors $\mathbf{a}$ and $\mathbf{b}$ are equal, but $\mathbf{v}+\mathbf{w}=\mathbf{t}+\mathbf{u}$, since both sides are equal to the vector from P to R, so $\mathbf{a}=\frac{1}{2}(\mathbf{v}+\mathbf{w})=\frac{1}{2}(\mathbf{t}+\mathbf{u})=\mathbf{b}$. ▲

In Section R.4 we discussed the equations of lines in the plane. These equations can be conveniently described in terms of vectors, and this description is equally applicable whether the line is the plane or in space. We will now find such equations in parametric form (see Section 10.4 for a discussion of parametric curves).

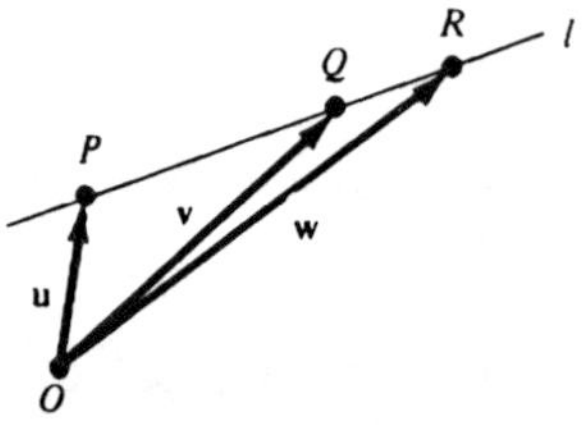

Figure 13.3.4. Since P, Q, and R lie on a line, the vector $\mathbf{w}-\mathbf{u}$ is a multiple of $\mathbf{v}-\mathbf{u}$.

Suppose that we wish to find the equation of the line l passing through the two points P and Q. Let O be the origin and let $\mathbf{u}$ and $\mathbf{v}$ be the vectors $\overrightarrow{OP}$ and $\overrightarrow{OQ}$ as in Fig. 13.3.4. Let R be an arbitrary point on l and let $\mathbf{w}$ be the vector $\overrightarrow{OR}$. Since R is on l, the vector $\mathbf{w}-\mathbf{u}=\overrightarrow{PR}$ is a multiple of the vector $\mathbf{v}-\mathbf{u}=\overrightarrow{PQ}$—that is, $\mathbf{w}-\mathbf{u}=t(\mathbf{v}-\mathbf{u})$ for some number t. This gives $\mathbf{w}=\mathbf{u}+t(\mathbf{v}-\mathbf{u})=(1-t)\mathbf{u}+t\mathbf{v}$.

The coordinates of the points P, Q, and R are the same as the components of the vectors $\mathbf{u}$, $\mathbf{v}$, and $\mathbf{w}$, so we obtain the parametric equation $R=(1-t)P+tQ$ for the line l.

Parametric Equation of a Line: Point-Point Form

The equation of the line l through the points $P=(x_1,y_1,z_1)$ and $Q=(x_2,y_2,z_2)$ is

$$R=(1-t)P+tQ.$$

In coordinate form, one has the three equations

$$x=(1-t)x_1+tx_2,$$

$$y=(1-t)y_1+ty_2,$$

$$z=(1-t)z_1+tz_2,$$

where $R=(x,y,z)$ is the typical point of l, and the parameter t takes on all real values.

Example 4 Find the equation of the line through $(2,1,-3)$ and $(6,-1,-5)$.

Solution We have $P=(2,1,-3)$ and $Q=(6,-1,-5)$, so

$$\begin{aligned}(x,y,z)=R&=(1-t)P+tQ=(1-t)(2,1,-3)+t(6,-1,-5)\\&=(2-2t,1-t,-3+3t)+(6t,-t,-5t)\\&=(2+4t,1-2t,-3-2t)\end{aligned}$$

or, since corresponding entries of equal ordered triples are equal,

$$x=2+4t,\qquad y=1-2t,\qquad z=-3-2t. ▲$$

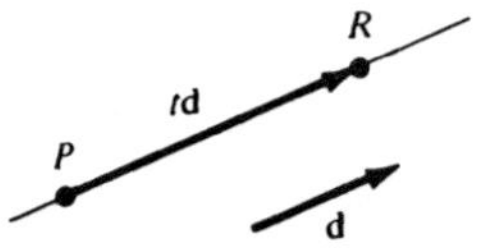

Figure 13.3.5. If the line through P and R has the direction of the vector $\mathbf{d}$, then the vector from P to R is a multiple of $\mathbf{d}$.

We can also ask for the equation of the line which passes through a given point P in the direction of a given vector $\mathbf{d}$. A point R lies on the line (see Fig. 13.3.5) if and only if the vector $\overrightarrow{PR}$ is a multiple of $\mathbf{d}$. Thus we can describe all points R on the line by $\overrightarrow{PR}=t\mathbf{d}$ for some number t. As t varies, R moves on the line; when $t=0$, R coincides with P. Since $\overrightarrow{PR}=\overrightarrow{OR}-\overrightarrow{OP}$, we can rewrite the equation as $\overrightarrow{OR}=\overrightarrow{OP}+t\mathbf{d}$. This reasoning leads to the following conclusion.

Parametric Equation of a Line: Point-Direction Form

The equation of the line through the point $P = (x_0, y_0, z_0)$ and pointing in the direction of the vector $\mathbf{d} = a\mathbf{i} + b\mathbf{j} + c\mathbf{k}$ is $\overrightarrow{PR} = t\mathbf{d}$ or equivalently $\overrightarrow{OR} = \overrightarrow{OP} + t\mathbf{d}$.

In coordinate form, the equations are

$$x = x_0 + at,$$
$$y = y_0 + bt,$$
$$z = z_0 + ct,$$

where $R = (x, y, z)$ is the typical point on l and the parameter t takes on all real values.

For lines in the xy plane, the z component is not present; otherwise, the results are the same.

Example 5 (a) Find the equations of the line in space through the point $(3, -1, 2)$ in the direction $2\mathbf{i} - 3\mathbf{j} + 4\mathbf{k}$.

(b) Find the equation of the line in the plane through the point $(1, -6)$ in the direction of $5\mathbf{i} - \pi\mathbf{j}$.

(c) In what direction does the line $x = -3t + 2$, $y = -2(t - 1)$, $z = 8t + 2$ point?

Solution (a) Here $P = (3, -1, 2) = (x_0, y_0, z_0)$ and $\mathbf{d} = 2\mathbf{i} - 3\mathbf{j} + 4\mathbf{k}$, so $a = 2$, $b = -3$, and $c = 4$. Thus the equations are

$$x = 3 + 2t, \qquad y = -1 - 3t, \qquad z = 2 + 4t.$$

(b) Here $P = (1, -6)$ and $\mathbf{d} = 5\mathbf{i} - \pi\mathbf{j}$, so the line is

$$R = (1, -6) + (5t, -\pi t) = (1 + 5t, -6 - \pi t)$$

or

$$x = 1 + 5t, \qquad y = -6 - \pi t.$$

(c) Using the preceding box, we construct the direction $\mathbf{d} = a\mathbf{i} + b\mathbf{j} + c\mathbf{k}$ from the coefficients of t: $a = -3$, $b = -2$, $c = 8$. Thus the line points in the direction of $\mathbf{d} = -3\mathbf{i} - 2\mathbf{j} + 8\mathbf{k}$. ▲

Example 6 (a) Do the lines $R_1 = (t, -6t + 1, 2t - 8)$ and $R_2 = (3t + 1, 2t, 0)$ intersect?

(b) Find the "equation" of the line segment between $(1, 1, 1)$ and $(2, 1, 2)$.

Solution (a) If the lines intersect, there must be numbers t_1 and t_2 such that the corresponding points are equal: $(t_1, -6t_1 + 1, 2t_1 - 8) = (3t_2 + 1, 2t_2, 0)$; that is,

$$t_1 = 3t_2 + 1,$$
$$-6t_1 + 1 = 2t_2,$$
$$2t_1 - 8 = 0.$$

From the third equation we have $t_1 = 4$. The first equation then becomes $4 = 3t_2 + 1$ or $t_2 = 1$. We must check whether these values satisfy the middle equation:

$$-6t_1 + 1 \stackrel{?}{=} 2t_2, \quad \text{i.e.,}$$

$$-6 \cdot 4 + 1 \stackrel{?}{=} 2 \cdot 1, \quad \text{i.e.,}$$

$$-24 + 1 \stackrel{?}{=} 2.$$

The answer is no; the lines do not intersect.

(b) The *line* through (1, 1, 1) and (2, 1, 2) is described in parametric form by $R = (1 - t)(1,1,1) + t(2,1,2) = (1 + t, 1, 1 + t)$, as t takes on all real values. The point R lies between (1, 1, 1) and (2, 1, 2) only when $0 \leqslant t \leqslant 1$, so the line *segment* is described by $R = (1 + t, 1, 1 + t)$, $0 \leqslant t \leqslant 1$. ▲

Since all the line segments representing a given vector $\mathbf{v} = a\mathbf{i} + b\mathbf{j} + c\mathbf{k}$ have the same length, we may define the *length* of $\mathbf{v}$ to be the length of any of these segments. To calculate the length of $\mathbf{v}$, it is convenient to use the segment $\overrightarrow{OP}$, where $P = (a, b, c)$, so that the length of $\mathbf{v}$ is just the distance from (0, 0, 0) to (a, b, c). We apply the Pythagorean theorem twice to calculate this distance. (See Fig. 13.3.6.) Let $Q = (a, b, 0)$ and $R = (a, 0, 0)$. Then $|OR| = |a|$ and $|RQ| = |b|$, so $|OQ| = \sqrt{a^2 + b^2}$. Now $|QP| = |c|$, so applying Pythagoras' theorem again, this time to the right triangle OQP, we obtain $|OP| = \sqrt{a^2 + b^2 + c^2}$. We denote the length of a vector $\mathbf{v}$ by $\|\mathbf{v}\|$; it is sometimes called the *magnitude* of $\mathbf{v}$ as well.

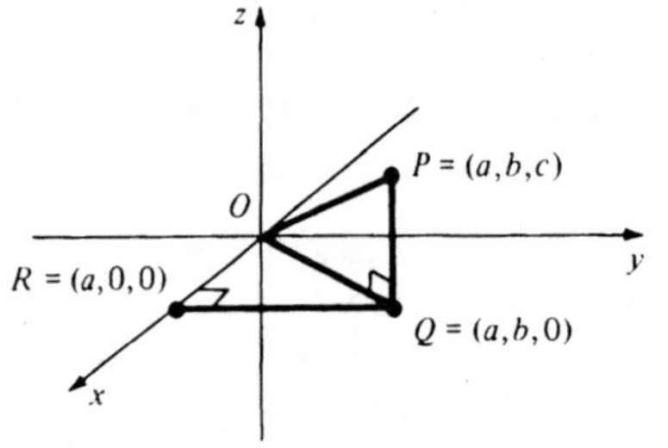

Figure 13.3.6. $|OP| = \sqrt{a^2 + b^2 + c^2}$.

Length of a Vector

The length $\|\mathbf{v}\|$ of a vector $\mathbf{v}$ is the square root of the sum of the squares of the components of $\mathbf{v}$:

$$\|a\mathbf{i} + b\mathbf{j} + c\mathbf{k}\| = \sqrt{a^2 + b^2 + c^2}.$$

Example 7 (a) Find the length of $\mathbf{v} = 2\mathbf{i} - 6\mathbf{j} + 7\mathbf{k}$. (b) Find the values of c for which $\|\mathbf{i} + \mathbf{j} + c\mathbf{k}\| = 4$.

Solution (a) $\|\mathbf{v}\| = \sqrt{2^2 + (-6)^2 + 7^2} = \sqrt{4 + 36 + 49} = \sqrt{89} \approx 9.434$.

(b) We have $\|\mathbf{i} + \mathbf{j} + c\mathbf{k}\| = \sqrt{1 + 1 + c^2} = \sqrt{2 + c^2}$. This equals 4 when $2 + c^2 = 16$, or $c = \pm\sqrt{14}$. ▲

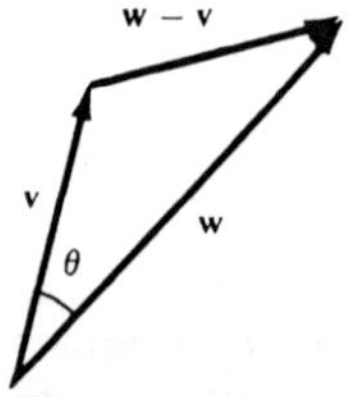

Figure 13.3.7. The law of cosines applied to vectors.

Some basic properties of length may be deduced from the law of cosines (see Section 5.1 for its proof). In terms of vectors, the law of cosines states that

$$\|\mathbf{w} - \mathbf{v}\|^2 = \|\mathbf{v}\|^2 + \|\mathbf{w}\|^2 - 2\|\mathbf{v}\|\|\mathbf{w}\|\cos\theta,$$

where θ is the angle between the vectors $\mathbf{v}$ and $\mathbf{w}$, $0 \leqslant \theta \leqslant \pi$. (See Fig. 13.3.7.)

In particular, since $\cos\theta \leqslant 1$, we get

$$\begin{aligned}\|\mathbf{w}-\mathbf{v}\|^2 &= \|\mathbf{w}\|^2+\|\mathbf{v}\|^2-2\|\mathbf{v}\|\|\mathbf{w}\|\cos\theta\\ &\geqslant \|\mathbf{w}\|^2+\|\mathbf{v}\|^2-2\|\mathbf{v}\|\|\mathbf{w}\|\\ &= (\|\mathbf{w}\|-\|\mathbf{v}\|)^2.\end{aligned}$$

Taking square roots and remembering that $\sqrt{x^2} = |x|$, we get

$$\|\mathbf{w}-\mathbf{v}\| \geqslant |\|\mathbf{w}\|-\|\mathbf{v}\||.$$

Hence

$$-(\|\mathbf{w}-\mathbf{v}\|) \leqslant \|\mathbf{w}\|-\|\mathbf{v}\| \leqslant \|\mathbf{w}-\mathbf{v}\|.$$

In particular, from the right-hand inequality, we get

$$\|\mathbf{w}\| \leqslant \|\mathbf{w}-\mathbf{v}\| + \|\mathbf{v}\|.$$

That is, the length of one side of a triangle is less than or equal to the sum of the lengths of the other sides. If we write $\mathbf{u} = \mathbf{w} - \mathbf{v}$, then $\mathbf{w} = \mathbf{u} + \mathbf{v}$, and the inequality above takes the useful form

$$\|\mathbf{u}+\mathbf{v}\| \leqslant \|\mathbf{u}\| + \|\mathbf{v}\|,$$

which is called the *triangle inequality*.

The relation between length and scalar multiplication is given by

$$\|r\mathbf{v}\| = |r|\,\|\mathbf{v}\|$$

since, if $\mathbf{v} = a\mathbf{i} + b\mathbf{j} + c\mathbf{k}$, then

$$\|r\mathbf{v}\| = \sqrt{(ra)^2+(rb)^2+(rc)^2} = \sqrt{r^2}\,\sqrt{a^2+b^2+c^2} = |r|\|\mathbf{v}\|.$$

Properties of Length

If $\mathbf{u}$, $\mathbf{v}$, and $\mathbf{w}$ are any vectors and r is any number:

(1) $\|\mathbf{u}\| \geqslant 0$;
(2) $\|\mathbf{u}\| = 0$ if and only if $\mathbf{u} = \mathbf{0}$;
(3) $\|r\mathbf{u}\| = |r|\,\|\mathbf{u}\|$;
(4) $\|\mathbf{u}+\mathbf{v}\| \leqslant \|\mathbf{u}\| + \|\mathbf{v}\|$; } Triangle inequality
(4′) $\|\mathbf{w}-\mathbf{v}\| \geqslant |\|\mathbf{w}\| - \|\mathbf{v}\||$. } Triangle inequality

Example 8 (a) Verify the triangle inequality (4) for $\mathbf{u} = \mathbf{i} + \mathbf{j}$ and $\mathbf{v} = 2\mathbf{i} + \mathbf{j} + \mathbf{k}$.
(b) Prove that $\|\mathbf{u}-\mathbf{v}\| \leqslant \|\mathbf{u}-\mathbf{w}\| + \|\mathbf{w}-\mathbf{v}\|$ for any vectors $\mathbf{u}, \mathbf{v}, \mathbf{w}$. Illustrate with a figure in which $\mathbf{u}$, $\mathbf{v}$, and $\mathbf{w}$ are drawn with the same base point, that is, the same "tail."

Solution (a) We have $\mathbf{u} + \mathbf{v} = 3\mathbf{i} + 2\mathbf{j} + \mathbf{k}$, so $\|\mathbf{u}+\mathbf{v}\| = \sqrt{9+4+1} = \sqrt{14}$. On the other hand, $\|\mathbf{u}\| = \sqrt{2}$ and $\|\mathbf{v}\| = \sqrt{6}$, so the triangle inequality asserts that $\sqrt{14} \leqslant \sqrt{2} + \sqrt{6}$. This is indeed true, since $\sqrt{14} \approx 3.74$, while $\sqrt{2} + \sqrt{6} \approx 1.41 + 2.45 = 3.86$.
(b) We find that $\mathbf{u} - \mathbf{v} = (\mathbf{u} - \mathbf{w}) + (\mathbf{w} - \mathbf{v})$, so the result follows from the triangle inequality with $\mathbf{u}$ replaced by $\mathbf{u} - \mathbf{w}$ and $\mathbf{v}$ replaced by $\mathbf{w} - \mathbf{v}$. Geometrically, we are considering the shaded triangle in Fig. 13.3.8. ▲

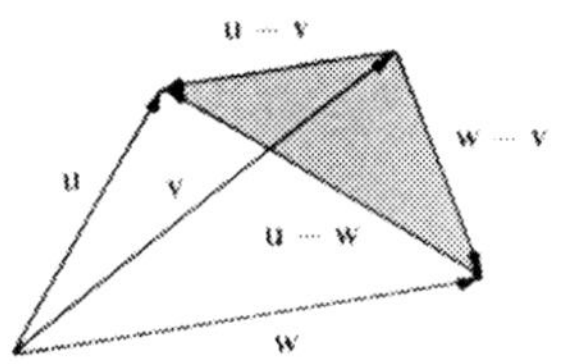

Figure 13.3.8. Illustrating the inequality $\|\mathbf{u}-\mathbf{v}\| \leqslant \|\mathbf{u}-\mathbf{w}\| + \|\mathbf{w}-\mathbf{v}\|$.

The length of a vector can have interpretations other than the geometric one given above. For example, suppose that an object is moving uniformly along a straight line. What physical quantity is represented by the length of its velocity vector? To answer this, let $\mathbf{v}$ be the velocity vector. The displacement vector from its position P at any time to its position Q, t units of time later, is $t\mathbf{v}$. The distance between P and Q is then $|t|\|\mathbf{v}\|$, so the length $\|\mathbf{v}\|$ of the velocity vector represents the ratio of distance travelled to elapsed time—it is called the *speed*.

A vector $\mathbf{u}$ is called a *unit vector* if its length is equal to 1. If $\mathbf{v}$ is any nonzero vector, $\|\mathbf{v}\| \neq 0$ then we can obtain a unit vector pointing in the direction of $\mathbf{v}$ by taking $\mathbf{u} = (1/\|\mathbf{v}\|)\mathbf{v}$. In fact,

$$\|\mathbf{u}\| = \left\| \frac{1}{\|\mathbf{v}\|}\mathbf{v} \right\| = \frac{1}{\|\mathbf{v}\|}\|\mathbf{v}\| = 1.$$

We call $\mathbf{u}$ the *normalization* of $\mathbf{v}$.

Example 9 (a) Normalize $\mathbf{v} = 2\mathbf{i} + 3\mathbf{j} - \frac{1}{2}\mathbf{k}$. (b) Find unit vectors $\mathbf{u}$, $\mathbf{v}$, and $\mathbf{w}$ in the plane such that $\mathbf{u} + \mathbf{v} = \mathbf{w}$.

Solution (a) We have $\|\mathbf{v}\| = \sqrt{2^2 + 3^2 + 1/2^2} = (1/2)\sqrt{53}$, so the normalization of $\mathbf{v}$ is

$$\mathbf{u} = \frac{1}{\|\mathbf{v}\|}\mathbf{v} = \frac{4}{\sqrt{53}}\mathbf{i} + \frac{6}{\sqrt{53}}\mathbf{j} - \frac{1}{\sqrt{53}}\mathbf{k}.$$

Figure 13.3.9. The vectors $\mathbf{u}$, $\mathbf{v}$, and $\mathbf{w}$ are represented by the sides of an equilateral traingle.

(b) A triangle whose sides represent $\mathbf{u}$, $\mathbf{v}$, and $\mathbf{w}$ must be equilateral as in Fig. 13.3.9. Knowing this, we may take $\mathbf{w} = \mathbf{i}$, $\mathbf{u} = \frac{1}{2}\mathbf{i} + (\sqrt{3}/2)\mathbf{j}$, $\mathbf{v} = \frac{1}{2}\mathbf{i} - (\sqrt{3}/2)\mathbf{j}$. Check that $\|\mathbf{u}\| = \|\mathbf{v}\| = \|\mathbf{w}\| = 1$ and that $\mathbf{u} + \mathbf{v} = \mathbf{w}$. ▲

Finally, we can use the formula for the length of vectors to obtain a formula for the distance between any two points in space. If $P_1 = (x_1, y_1, z_1)$ and $P_2 = (x_2, y_2, z_2)$, then the distance between P_1 and P_2 is the length of the vector from P_2 to P_1; that is,

$$|P_1P_2| = \|(x_1 - x_2)\mathbf{i} + (y_1 - y_2)\mathbf{j} + (z_1 - z_2)\mathbf{k}\|$$

$$= \sqrt{(x_1 - x_2)^2 + (y_1 - y_2)^2 + (z_1 - z_2)^2}\,.$$

Distance Formula

If P_1 has coordinates (x_1, y_1, z_1) and P_2 has coordinates (x_2, y_2, z_2), then the distance between P_1 and P_2 is

$$\sqrt{(x_1 - x_2)^2 + (y_1 - y_2)^2 + (z_1 - z_2)^2}\,.$$

Example 10 (a) Find the distance between $(2, 1, 0)$ and $(3, -2, 6)$.
(b) Let $P_t = t(1, 1, 1)$.
(i) What is the distance from P_t to $(3, 0, 0)$?
(ii) For what value of t is the distance shortest?
(iii) What is the shortest distance?

Solution (a) The distance is

$$\sqrt{(2-3)^2 + [1-(-2)]^2 + (0-6)^2} = \sqrt{(-1)^2 + 3^2 + (-6)^2}$$

$$= \sqrt{1 + 9 + 36} = \sqrt{46}\,.$$

(b) (i) By the distance formula, the distance is

$$\sqrt{(t-3)^2+(t-0)^2+(t-0)^2}$$
$$=\sqrt{t^2-6t+9+t^2+t^2}=\sqrt{3t^2-6t+9}\,.$$

(ii) The distance is shortest when $3t^2-6t+9$ is least—that is, when $(d/dt)(3t^2-6t+9)=6t-6=0$, or $t=1$.

(iii) For $t=1$, the distance in (i) is $\sqrt{6}$. ▲

Exercises for Section 13.3

Use vector methods in Exercises 1–6.

1. Show that the line segment joining the midpoints of two sides of a triangle is parallel to and has half of the length of the third side.
2. Prove that the medians of the triangle intersect in a point two-thirds of the way along any median from a vertex to the midpoint of the opposite side.
3. Prove that if PQR is a triangle in space and $b>0$ is a number, then there is a triangle with sides parallel to those of PQR and side lengths b times those of PQR.
4. Prove that if the corresponding sides of two triangles are parallel, then the lengths of corresponding sides have a common ratio. (Assume that the triangles are not degenerated into lines.)
5. Find the point in the plane two-thirds of the way from the origin to the midpoint of the line segment between $(1,1)$ and $(2,-2)$.
6. Let $P=(3,5,2)$ and $Q=(2,5,3)$. Find the point R such that Q is the midpoint of the line segment PR.

Write equations for the lines in Exercises 7–10.

7. The line through $(1,1,0)$ and $(0,0,1)$.
8. The line through $(2,0,0)$ and $(0,1,0)$.
9. The line through $(0,0,0)$ and $(1,1,1)$.
10. The line through $(-1,-1,0)$ and $(1,8,-4)$.

Write parametric equations for the lines in Exercises 11–14.

11. The line through the point $(1,1,0)$ in the direction of vector $-\mathbf{i}-\mathbf{j}+\mathbf{k}$.
12. The line through $(0,1,0)$ in the direction $\mathbf{j}$.
13. The line in the plane through $(-1,-2)$ and in direction $3\mathbf{i}-2\mathbf{j}$.
14. The line in the plane through $(2,-1)$ and in direction $-\mathbf{i}-\mathbf{j}$.
15. At what point does the line through $(0,1,2)$ with direction $\mathbf{i}+\mathbf{j}+\mathbf{k}$ cross the xy plane?
16. Where does the line through $(3,4,5)$ and $(6,7,8)$ meet the yz plane?
17. Do the lines given by $R_1=(t,3t-1,4t)$ and $R_2=(3t,5,1-t)$ intersect?
18. Find the unique value of c for which the lines $R_1=(t,-6t+c,2t-8)$ and $R_2=(3t+1,2t,0)$ intersect.

Compute the length of the vectors in Exercises 19–24.

19. $\mathbf{i}+\mathbf{j}+\mathbf{k}$
20. $2\mathbf{i}+\mathbf{j}$
21. $\mathbf{i}+\mathbf{k}$
22. $3\mathbf{i}+4\mathbf{j}$
23. $2\mathbf{i}+2\mathbf{k}$
24. $\mathbf{i}-\mathbf{j}-3\mathbf{k}$
25. For what a is $\|a\mathbf{i}-3\mathbf{j}+\mathbf{k}\|=16$?
26. For what b is $\|\mathbf{i}-b\mathbf{j}+2\mathbf{k}\|=3$?
27. Verify the triangle inequality for the vectors in Exercises 19 and 21.
28. Verify the triangle inequality for the vectors in Exercises 20 and 22.
29. Find unit vectors $\mathbf{u}$, $\mathbf{v}$, and $\mathbf{w}$ in the plane such that $\mathbf{u}+2\mathbf{v}+\mathbf{w}=\mathbf{0}$.
30. Find unit vectors $\mathbf{u}$, $\mathbf{v}$, $\mathbf{w}$, and $\mathbf{z}$ in the plane such that $\mathbf{u}+\mathbf{v}+\mathbf{w}+\mathbf{z}=\mathbf{0}$.
31. Show that $(0,0,1)$ $(0,1,0)$, and $(1,0,0)$ are the vertices of an equilateral triangle. How long is each side?
32. Find an equilateral triangle in space which shares just one side with the one in Exercise 31.
33. Normalize the vectors in Exercises 19 and 21.
34. Normalize the vectors in Exercises 20 and 22.

Find the distance between the pairs of points in Exercises 35–38.

35. $(1,1,3)$ and $(2,2,2)$
36. $(2,0,0)$ and $(2,1,2)$
37. $(1,1,2)$ and $(1,2,3)$
38. $(1,2,3)$ and $(3,2,1)$
39. Let $P_t=t(3,2,1)$.
    (i) What is the distance from P_t to $(2,0,0)$?
    (ii) For what value of t is the distance shortest?
    (iii) What is that shortest distance?
40. Draw a figure, similar to Fig. 13.3.6, to illustrate the distance formula on p. 666.
41. A boat whose top speed in still water is 12 knots points north and steams at full power. If there is an eastward current of 5 knots, what is the speed of the boat?
42. A ship starting at $(0,0)$ proceeds at a speed of 10 knots directly toward a buoy located at $(3,4)$. (The chart is measured in nautical miles; a knot equals 1 nautical mile per hour.)
    (a) What is the ship's point of closest approach to a rock located at $(2,2)$?
    (b) After how long does it reach this point?
    (c) How far is this point from the rock?

★43. Derive the point-point form of the equation of a line obtained in Section R.4 from the parametric form obtained in this section. Comment on the case in which $x_1 = x_2$.

★44. Derive the point-direction form for the parametric equation of a line from the point-point form. [*Hint*: If a line through P is to have direction $\mathbf{d}$, what other point must lie on the line?]

★45. When does equality hold in the triangle inequality? (You might try using the law of cosines as was done in the text.) Test your conclusion on the vectors in Exercises 21 and 23.

★46. The potential V produced at (x, y, z) by charges q_1 and q_2 of opposite sign placed at distances x_1 and x_2 from the origin along the x axis is given by

$$V = \frac{q_1}{4\pi\epsilon_0\|\mathbf{r} - \mathbf{r}_1\|} + \frac{q_2}{4\pi\epsilon_0\|\mathbf{r} - \mathbf{r}_2\|}.$$

In this formula, $\mathbf{r}$ is the vector from the origin to the point (x, y, z). The vectors $\mathbf{r}_1$ and $\mathbf{r}_2$ are vectors from the origin to the respective charges q_1 and q_2.

(a) Express the formula for V entirely in terms of the scalar quantities $x, y, z, x_1, x_2, q_1, q_2, \epsilon_0$.

(b) Show that the locus of points (x, y, z) for which $V = 0$ is a plane or a sphere whose radius is $|q_1q_2(x_1 - x_2)/(q_1^2 - q_2^2)|$ and whose center is on the x axis or is a plane.

13.4 The Dot Product

The dot product of two unit vectors is the cosine of the angle between them.

To introduce the dot product, we will calculate the angle θ between two vectors in terms of the components of the vectors.

If $\mathbf{v}_1$ and $\mathbf{v}_2$ are two vectors, we have seen (Fig. 13.3.7) that

$$\|\mathbf{v}_2 - \mathbf{v}_1\|^2 = \|\mathbf{v}_1\|^2 + \|\mathbf{v}_2\|^2 - 2\|\mathbf{v}_1\|\,\|\mathbf{v}_2\|\cos\theta,$$

where θ is the angle between $\mathbf{v}_1$ and $\mathbf{v}_2$; $0 \leqslant \theta \leqslant \pi$. Therefore,

$$2\|\mathbf{v}_1\|\,\|\mathbf{v}_2\|\cos\theta = \|\mathbf{v}_1\|^2 + \|\mathbf{v}_2\|^2 - \|\mathbf{v}_2 - \mathbf{v}_1\|^2. \tag{1}$$

If $\mathbf{v}_1 = a_1\mathbf{i} + b_1\mathbf{j} + c_1\mathbf{k}$ and $\mathbf{v}_2 = a_2\mathbf{i} + b_2\mathbf{j} + c_2\mathbf{k}$, then the right-hand side of equation (1) is

$$\begin{aligned}&(a_1^2 + b_1^2 + c_1^2) + (a_2^2 + b_2^2 + c_2^2) - \left[(a_1 - a_2)^2 + (b_1 - b_2)^2 + (c_1 - c_2)^2\right]\\ &\quad = 2(a_1a_2 + b_1b_2 + c_1c_2).\end{aligned}$$

Thus we have proved that

$$\|\mathbf{v}_1\|\,\|\mathbf{v}_2\|\cos\theta = a_1a_2 + b_1b_2 + c_1c_2. \tag{2}$$

This very convenient formula enables us to compute $\cos\theta$ and hence θ; thus the quantity on the right-hand side deserves a special name. If $\mathbf{v}_1 = a_1\mathbf{i} + b_1\mathbf{j} + c_1\mathbf{k}$ and $\mathbf{v}_2 = a_2\mathbf{i} + b_2\mathbf{j} + c_2\mathbf{k}$ are two vectors, the number $a_1a_2 + b_1b_2 + c_1c_2$ is called their *dot product* and is denoted by $\mathbf{v}_1 \cdot \mathbf{v}_2$. The dot product in the plane is defined analogously; just think of c_1 and c_2 as being zero.

Notice that the dot product of two vectors is a number, not a vector. It is sometimes called the *scalar product* (do not confuse this with scalar *multiplication*) or the *inner product*.

Example 1 (a) If $\mathbf{v}_1 = 3\mathbf{i} + \mathbf{j} - 2\mathbf{k}$ and $\mathbf{v}_2 = \mathbf{i} - \mathbf{j} + \mathbf{k}$, calculate $\mathbf{v}_1 \cdot \mathbf{v}_2$.
(b) Calculate $(2\mathbf{i} + \mathbf{j} - \mathbf{k}) \cdot (3\mathbf{k} - 2\mathbf{j})$.

Solution (a) $\mathbf{v}_1 \cdot \mathbf{v}_2 = 3 \cdot 1 + 1 \cdot (-1) + (-2) \cdot 1 = 3 - 1 - 2 = 0$.
(b) $(2\mathbf{i} + \mathbf{j} - \mathbf{k}) \cdot (3\mathbf{k} - 2\mathbf{j}) = (2\mathbf{i} + \mathbf{j} - \mathbf{k}) \cdot (0\mathbf{i} - 2\mathbf{j} + 3\mathbf{k}) = 2 \cdot 0 - 1 \cdot 2 - 1 \cdot 3 = -5$. ▲

Combining formula (2) with the definition of the dot product gives

$$\|\mathbf{v}_1\|\,\|\mathbf{v}_2\|\cos\theta = \mathbf{v}_1 \cdot \mathbf{v}_2, \tag{3}$$

where θ is the angle between $\mathbf{v}_1$ and $\mathbf{v}_2$.

We may solve (3) for $\cos\theta$ to obtain the formula

$$\cos\theta = \frac{\mathbf{v}_1 \cdot \mathbf{v}_2}{\|\mathbf{v}_1\|\,\|\mathbf{v}_2\|};$$

$$\text{i.e.,}\quad \theta = \cos^{-1}\left(\frac{\mathbf{v}_1 \cdot \mathbf{v}_2}{\|\mathbf{v}_1\|\,\|\mathbf{v}_2\|}\right)\text{if}\quad \mathbf{v}_1 \neq \mathbf{0} \quad\text{and}\quad \mathbf{v}_2 \neq \mathbf{0}.$$

Example 2 (a) Find the angle between the vectors $\mathbf{i}+\mathbf{j}+\mathbf{k}$ and $\mathbf{i}+\mathbf{j}-\mathbf{k}$. (b) Find the angle between $3\mathbf{i}+\mathbf{j}-2\mathbf{k}$ and $\mathbf{i}-\mathbf{j}+\mathbf{k}$.

Solution (a) Let $\mathbf{v}_1 = \mathbf{i}+\mathbf{j}+\mathbf{k}$ and $\mathbf{v}_2 = \mathbf{i}+\mathbf{j}-\mathbf{k}$. Then $\|\mathbf{v}_1\| = \sqrt{3}$, $\|\mathbf{v}_2\| = \sqrt{3}$, and $\mathbf{v}_1 \cdot \mathbf{v}_2 = 1\cdot 1 + 1 \cdot 1 - 1 \cdot 1 = 1$. Hence $\cos\theta = \frac{1}{3}$, so $\theta = \cos^{-1}(\frac{1}{3}) \approx 1.23$ radians (70°32′).
(b) From Example 1(a), $(3\mathbf{i}+\mathbf{j}-2\mathbf{k})\cdot(\mathbf{i}-\mathbf{j}+\mathbf{k}) = 0$, so $\cos\theta = 0$ and hence $\theta = \pi/2$. ▲

From (3) we get

$$|\mathbf{v}_1 \cdot \mathbf{v}_2| = \|\mathbf{v}_1\|\,\|\mathbf{v}_2\|\,|\cos\theta|.$$

However, $|\cos\theta| \leqslant 1$, so we have

$$|\mathbf{v}_1 \cdot \mathbf{v}_2| \leqslant \|\mathbf{v}_1\|\,\|\mathbf{v}_2\|.$$

This is a useful inequality called the Schwarz inequality (and sometimes the Cauchy–Schwarz–Buniakowski inequality).

From either (2) or (3), we notice that if $\mathbf{v} = a\mathbf{i} + b\mathbf{j} + c\mathbf{k}$, then

$$\mathbf{v}\cdot\mathbf{v} = a^2 + b^2 + c^2 = \|\mathbf{v}\|^2.$$

Since two nonzero vectors are perpendicular when $\theta = \pi/2$—that is, when $\cos\theta = 0$—we have an algebraic test for perpendicularity: the nonzero vectors $\mathbf{v}_1$ and $\mathbf{v}_2$ are perpendicular when $\mathbf{v}_1 \cdot \mathbf{v}_2 = 0$. (We adopt the convention that the zero vector is perpendicular to every vector.) The synonyms "orthogonal" or "normal" are also used for "perpendicular."

The Dot Product

Algebraic definition:

$$(a_1\mathbf{i} + b_1\mathbf{j} + c_1\mathbf{k})\cdot(a_2\mathbf{i} + b_2\mathbf{j} + c_2\mathbf{k}) = a_1a_2 + b_1b_2 + c_1c_2.$$

Geometric interpretation:

$$\mathbf{v}_1 \cdot \mathbf{v}_2 = \|\mathbf{v}_1\|\,\|\mathbf{v}_2\|\cos\theta,$$

where θ is the angle between $\mathbf{v}_1$ and $\mathbf{v}_2$, $0 \leqslant \theta \leqslant \pi$. In particular,

$$\|\mathbf{v}\|^2 = \mathbf{v}\cdot\mathbf{v}.$$

Properties:

1. $\mathbf{u}\cdot\mathbf{u} \geqslant 0$ for any vector $\mathbf{u}$.
2. $\mathbf{u}\cdot\mathbf{u} = 0$ only if $\mathbf{u} = \mathbf{0}$.
3. $\mathbf{u}\cdot\mathbf{v} = \mathbf{v}\cdot\mathbf{u}$.
4. $(\mathbf{u}+\mathbf{v})\cdot\mathbf{w} = \mathbf{u}\cdot\mathbf{w} + \mathbf{v}\cdot\mathbf{w}$.
5. $(a\mathbf{u})\cdot\mathbf{v} = a(\mathbf{u}\cdot\mathbf{v})$.
6. $|\mathbf{u}\cdot\mathbf{v}| \leqslant \|\mathbf{u}\|\,\|\mathbf{v}\|$ (Schwarz inequality).
7. $\mathbf{u}$ and $\mathbf{v}$ are perpendicular when $\mathbf{u}\cdot\mathbf{v} = 0$.

Example 3 Find a unit vector in the plane which is orthogonal to $\mathbf{v} = \mathbf{i} - 3\mathbf{j}$.

Solution If $\mathbf{w} = a\mathbf{i} + b\mathbf{j}$ is perpendicular to $\mathbf{i} - 3\mathbf{j}$, we must have $0 = \mathbf{v} \cdot \mathbf{w} = a - 3b$; that is, $a = 3b$. If $\mathbf{w}$ is to be a unit vector, we must also have $1 = a^2 + b^2 = (3b)^2 + b^2 = 10b^2$, so $b = \pm 1/\sqrt{10}$ and $a = \pm 3/\sqrt{10}$. Thus there are two possible solutions: $(\pm 1/\sqrt{10})(3\mathbf{i} + \mathbf{j})$. (See Fig. 13.4.1.) ▲

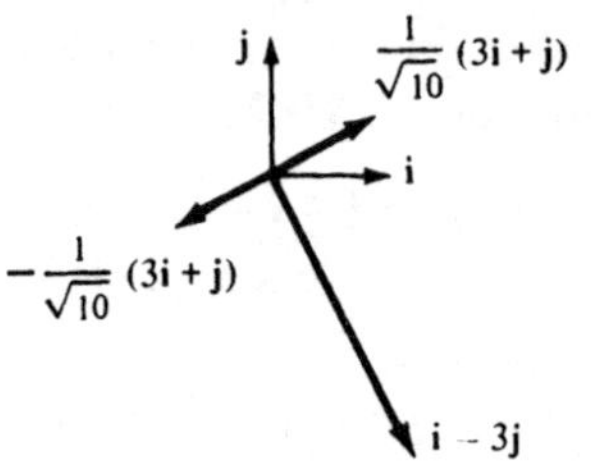

Figure 13.4.1. There are two unit vectors orthogonal to $\mathbf{i} - 3\mathbf{j}$.

Example 4 Let $\mathbf{u}$ and $\mathbf{v}$ be vectors in the plane; assume that $\mathbf{u}$ is nonzero.

(a) Show that $\mathbf{w} = \mathbf{v} - (\mathbf{v} \cdot \mathbf{u}/\mathbf{u} \cdot \mathbf{u})\mathbf{u}$ is orthogonal to $\mathbf{u}$.
(b) Sketch the vectors $\mathbf{u}$, $\mathbf{v}$, $\mathbf{w}$, and $\mathbf{v} - \mathbf{w}$. The vector $\mathbf{v} - \mathbf{w} = (\mathbf{v} \cdot \mathbf{u}/\mathbf{u} \cdot \mathbf{u})\mathbf{u}$ is called the *orthogonal projection* of $\mathbf{v}$ on $\mathbf{u}$. Why?
(c) Find the orthogonal projection of $\mathbf{i} + \mathbf{j}$ on $\mathbf{i} - 2\mathbf{j}$.

Solution (a) We compute

$$\mathbf{u} \cdot \mathbf{w} = \mathbf{u} \cdot \left(\mathbf{v} - \frac{\mathbf{v} \cdot \mathbf{u}}{\mathbf{u} \cdot \mathbf{u}}\mathbf{u}\right).$$

By the algebraic properties of the dot product, this is equal to

$$\mathbf{u} \cdot \mathbf{v} - \frac{\mathbf{v} \cdot \mathbf{u}}{\mathbf{u} \cdot \mathbf{u}}\mathbf{u} \cdot \mathbf{u} = \mathbf{u} \cdot \mathbf{v} - \mathbf{v} \cdot \mathbf{u} = 0,$$

so $\mathbf{u} \cdot \mathbf{w} = 0$ and $\mathbf{w}$ is orthogonal to $\mathbf{u}$.

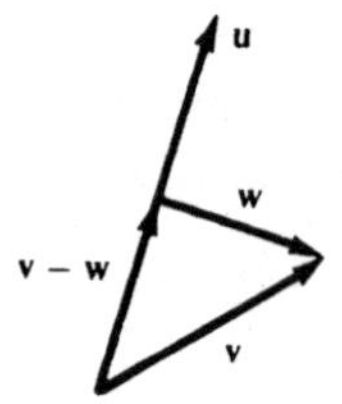

Figure 13.4.2. The vector $\mathbf{v} - \mathbf{w} = (\mathbf{v} \cdot \mathbf{u}/\mathbf{u} \cdot \mathbf{u})\mathbf{u}$ is the orthogonal projection of $\mathbf{v}$ on $\mathbf{u}$.

(b) We note that $\mathbf{v} - \mathbf{w} = (\mathbf{v} \cdot \mathbf{u}/\mathbf{u} \cdot \mathbf{u})\mathbf{u}$ is a multiple of $\mathbf{u}$. Thus the configuration of vectors must be as in Fig. 13.4.2. The vector $\mathbf{v} - \mathbf{w}$ is called the *orthogonal projection* of $\mathbf{v}$ on $\mathbf{u}$ because it is obtained by dropping a perpendicular from the "tip" of $\mathbf{v}$ to the line determined by $\mathbf{u}$. (The base points of $\mathbf{u}$ and $\mathbf{v}$ must be the same for this construction.)
(c) With $\mathbf{u} = \mathbf{i} - 2\mathbf{j}$ and $\mathbf{v} = \mathbf{i} + \mathbf{j}$, the orthogonal projection of $\mathbf{v}$ on $\mathbf{u}$ is

$$\frac{\mathbf{u} \cdot \mathbf{v}}{\mathbf{u} \cdot \mathbf{u}}\mathbf{u} = \frac{1-2}{1+4}(\mathbf{i} - 2\mathbf{j}) = -\frac{1}{5}(\mathbf{i} - 2\mathbf{j}).$$

(See Fig. 13.4.3.) ▲

We can use the dot product to find the distance from a point $Q = (x_1, y_1, z_1)$ to the line l which passes through a point $P = (x_0, y_0, z_0)$ and has the direction $\mathbf{d} = a\mathbf{i} + b\mathbf{j} + c\mathbf{k}$. Indeed, in Fig. 13.4.4 the distance from Q to the line is the

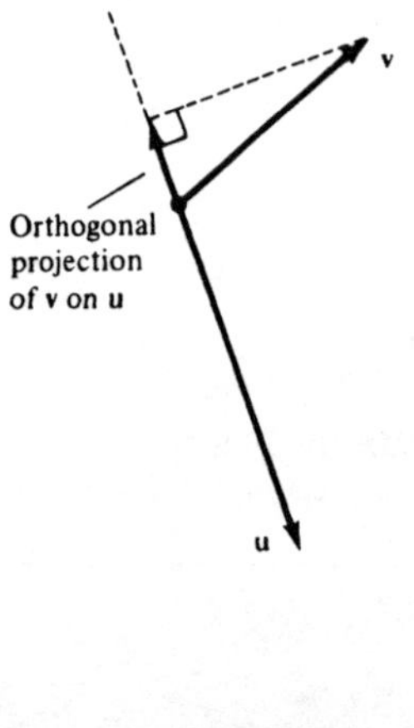

Figure 13.4.3. The orthogonal projection of $\mathbf{v}$ on $\mathbf{u}$ equals $-\frac{1}{5}\mathbf{u}$.

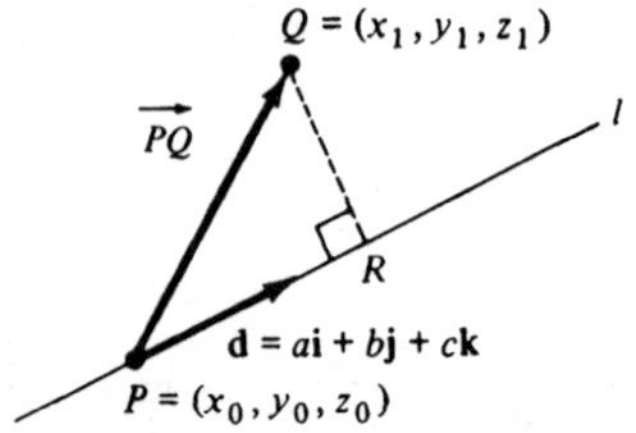

Figure 13.4.4. $\|\overrightarrow{QR}\|$ is the distance from Q to l.

distance between Q and R, where R is chosen on l in such a way that $\overrightarrow{PR}$ and $\overrightarrow{QR}$ are orthogonal. Then $\overrightarrow{PR}$ is the orthogonal projection of $\overrightarrow{PQ}$ on the line l. Thus, by Example 4,

$$\overrightarrow{PR} = \frac{\overrightarrow{PQ}\cdot\mathbf{d}}{\mathbf{d}\cdot\mathbf{d}}\mathbf{d} = \frac{a(x_1 - x_0) + b(y_1 - y_0) + c(z_1 - z_0)}{a^2 + b^2 + c^2}\mathbf{d}.$$

By Pythagoras' theorem, $\|\overrightarrow{RQ}\| = \sqrt{\|\overrightarrow{PQ}\|^2 - \|\overrightarrow{PR}\|^2}$ which gives

$$\operatorname{dist}(Q,l) = \left\{(x_1 - x_0)^2 + (y_1 - y_0)^2 + (z_1 - z_0)^2 - \frac{\left[a(x_1 - x_0) + b(y_1 - y_0) + c(z_1 - z_0)\right]^2}{a^2 + b^2 + c^2}\right\}^{1/2} \quad (4)$$

as the distance from Q to the line l.

Example 5 Find the distance from $(1,1,2)$ to the line through $(2,0,0)$ in the direction $(1/\sqrt{2})\mathbf{i} - (1/\sqrt{2})\mathbf{j}$.

Solution In formula (4), we set $(x_1, y_1, z_1) = (1,1,2)$, $(x_0, y_0, z_0) = (2,0,0)$ and obtain a, b, c from $a\mathbf{i} + b\mathbf{j} + c\mathbf{k} = (1/\sqrt{2})\mathbf{i} - (1/\sqrt{2})\mathbf{j}$ to be $a = 1/\sqrt{2}$, $b = -1\sqrt{2}$, $c = 0$. Thus,

$$\operatorname{dist}(Q,l) = \left\{(1-2)^2 + 1^2 + 2^2 - \frac{\left[\frac{1}{\sqrt{2}}(1-2) - \frac{1}{\sqrt{2}}1\right]^2}{\frac{1}{2} + \frac{1}{2}}\right\}^{1/2}$$

$$= (6-2)^{1/2} = 2. \blacktriangle$$

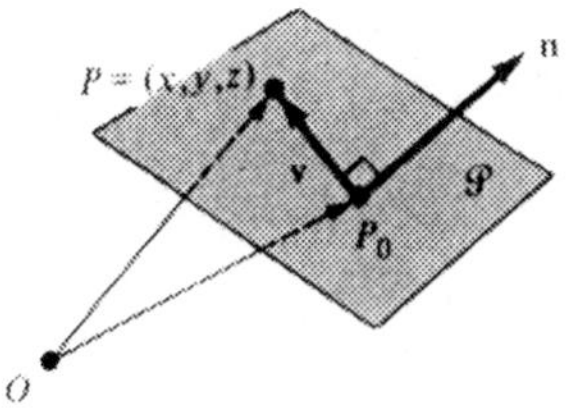

Figure 13.4.5. The plane $\mathscr{P}$ is perpendicular to the vector **n**.

The dot product makes it simple to determine the equation of a plane. Suppose that a plane $\mathscr{P}$ passes through a point $P_0 = (x_0, y_0, z_0)$ and is perpendicular to a vector $\mathbf{n} = A\mathbf{i} + B\mathbf{j} + C\mathbf{k}$. (See Fig. 13.4.5.)

Let $P = (x, y, z)$ be a point on $\mathscr{P}$. Then **n** must be perpendicular to the vector **v** from P_0 to P; that is, $\mathbf{n}\cdot\mathbf{v} = 0$, or

$$(A\mathbf{i} + B\mathbf{j} + C\mathbf{k})\cdot\left[(x - x_0)\mathbf{i} + (y - y_0)\mathbf{j} + (z - z_0)\mathbf{k}\right] = 0.$$

Hence

$$A(x - x_0) + B(y - y_0) + C(z - z_0) = 0.$$

We call **n** the *normal vector* of the plane. If we let $D = -(Ax_0 + By_0 + Cz_0)$, the equation of the plane becomes

$$Ax + By + Cz + D = 0.$$

Example 6 Find the equation of the plane through $(1,1,1)$ with normal vector $2\mathbf{i} + \mathbf{j} - 2\mathbf{k}$.

Solution Here $A\mathbf{i} + B\mathbf{j} + C\mathbf{k} = 2\mathbf{i} + \mathbf{j} - 2\mathbf{k}$, and $(x_0, y_0, z_0) = (1,1,1)$, so we get

$$A(x-1) + B(y-1) + C(z-1) = 0.$$

Hence

$$2(x-1) + (y-1) - 2(z-1) = 0,$$

$$2x + y - 2z = 1. \blacktriangle$$

Example 7 Find a unit normal vector to the plane $3x + y - z = 10$. Sketch the plane.

Solution A normal vector is obtained by making a vector out of the coefficients of x, y, and z; that is, $(3, 1, -1)$. Normalizing, we get

$$(3,1,-1)/\sqrt{9+1+1} = \left(3/\sqrt{11}\,, 1/\sqrt{11}\,, -1/\sqrt{11}\right);\ \text{i.e.}\ \frac{1}{\sqrt{11}}(3\mathbf{i} + \mathbf{j} - \mathbf{k}).$$

We may sketch the plane by noting where it meets the coordinate axes. For example, setting $y = z = 0$, we see that $(\frac{10}{3}, 0, 0)$ lies on the plane (see Fig. 13.4.6). ▲

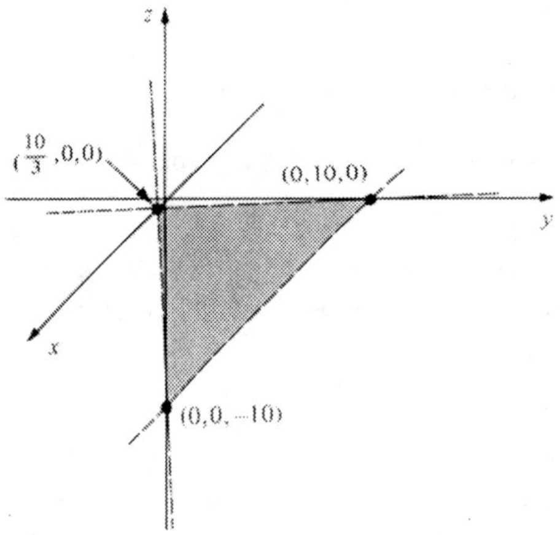

Figure 13.4.6. The plane $3x + y - z = 10$.

Equation of a Plane in Space

The equation of the plane through (x_0, y_0, z_0) with normal vector $\mathbf{n} = A\mathbf{i} + B\mathbf{j} + C\mathbf{k}$ is

$$A(x - x_0) + B(y - y_0) + C(z - z_0) = 0 \tag{5}$$

or

$$Ax + By + Cz + D = 0. \tag{6}$$

Example 8 (a) Find the equation of the plane passing through the point $(3, -1, -1)$ and perpendicular to the vector $\mathbf{i} - 2\mathbf{j} + \mathbf{k}$. (b) Find the equation of the plane containing the points $(1, 1, 1)$, $(2, 0, 0)$, and $(1, 1, 0)$.

Solution (a) We use the first displayed equation (5) in the preceding box, choosing the point $(x_0, y_0, z_0) = (3, -1, -1)$ and components of the normal vector to be $A = 1$, $B = -2$, $C = 1$ to give

$$1(x - 3) - 2(y + 1) + 1(z + 1) = 0$$

which simplifies to $x - 2y + z = 4$.

(b) The general equation of a plane has the form (6) $Ax + By + Cz + D = 0$. Since the points $(1, 1, 1)$, $(2, 0, 0)$, and $(1, 1, 0)$ lie on the plane, the coefficients A, B, C, D satisfy the three equations:

$$\begin{aligned} A + B + C + D &= 0, \\ 2A \qquad\quad + D &= 0, \\ A + B \qquad + D &= 0. \end{aligned}$$

Proceeding by elimination, we reduce this system to the form

$$2A + D = 0 \qquad \text{(second equation)},$$
$$2B + D = 0 \qquad \text{(twice the third equation minus the second)},$$
$$C = 0 \qquad \text{(first equation minus the third)}.$$

Since the numbers A, B, C, and D are determined only up to a common factor, we can fix the value of one of them and then the others will be determined uniquely. If we let $D = -2$, then $A = 1$, $B = 1$, $C = 0$. Thus $x + y - 2 = 0$ is an equation of the plane that contains the given points. (You may go back and verify that the given points actually satisfy this equation.) ▲

Example 9 Where does the line through the origin in the direction of $\mathbf{i} + \mathbf{j} + 2\mathbf{k}$ meet the plane $x + y + 2z = 5$? Use your answer to find the distance from the origin to this plane. Sketch.

Solution The line has parametric equations $x = t, y = t, z = 2t$. It meets the plane when $x + y + 2z = t + t + 4t = 5$; that is, when $t = \frac{5}{6}$. The point of intersection is $P_1 = (\frac{5}{6}, \frac{5}{6}, \frac{5}{3})$.

Since a normal to the plane is $\mathbf{n} = \mathbf{i} + \mathbf{j} + 2\mathbf{k}$, which is the same as the direction vector of this line, we see that the line is perpendicular to the plane at P_1. If P is also in the plane, consideration of the right triangle OP_1P shows that $\overrightarrow{OP_1}$ must be shorter than $\overrightarrow{OP}$ (see Fig. 13.4.7). Thus the distance from the origin to the plane is the length of $\overrightarrow{OP_1}$:

$$\sqrt{\frac{25}{36} + \frac{25}{36} + \frac{25}{9}} = \frac{\sqrt{150}}{6} = \frac{5\sqrt{6}}{6}. \; ▲$$

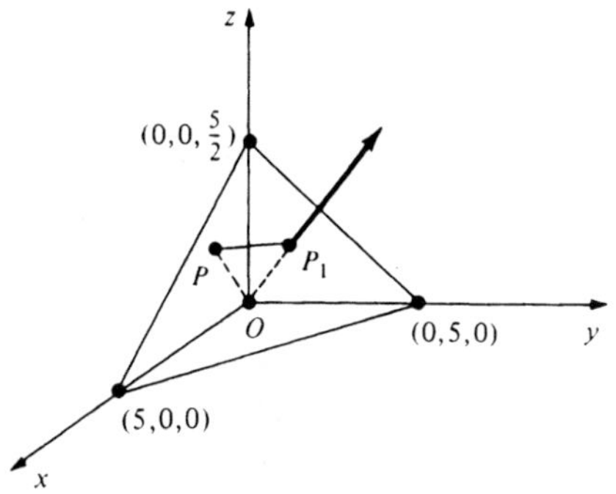

Figure 13.4.7. The vector from O to the closest point P_1 on a plane is perpendicular to the plane.

Let $A(x - x_0) + B(y - y_0) + C(z - z_0) = 0$ be the equation of a plane $\mathcal{P}$ through the point $P = (x_0, y_0, z_0)$ in space. Let us use the basic ideas of the preceding example to determine the distance from a point $Q = (x_1, y_1, z_1)$ to the plane (see Figure 13.4.8). Consider the vector

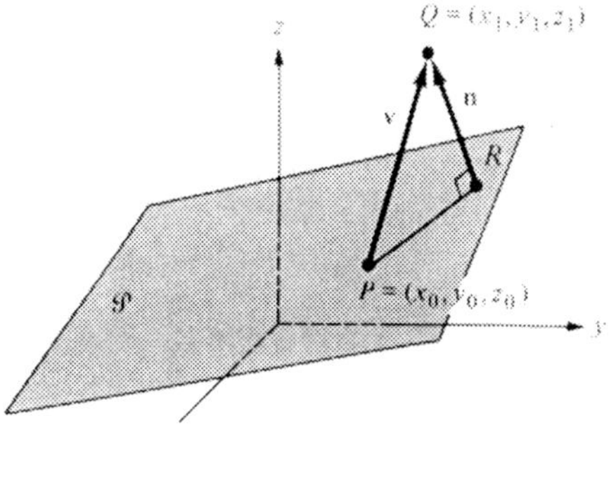

Figure 13.4.8. The geometry for determining the distance from a point to a plane.

$$\mathbf{n} = \frac{A\mathbf{i} + B\mathbf{j} + C\mathbf{k}}{\sqrt{A^2 + B^2 + C^2}},$$

which is a unit vector normal to the plane. Next drop a perpendicular from Q to the plane and construct the triangle PQR shown in Figure 13.4.8. The distance $d = \overrightarrow{RQ}$ is the length of the projection of $\mathbf{v} = \overrightarrow{PQ}$ (the vector from P to Q) onto $\mathbf{n}$; thus

$$\begin{aligned} \text{distance} &= |\mathbf{v} \cdot \mathbf{n}| = \left|\left[(x_1 - x_0)\mathbf{i} + (y_1 - y_0)\mathbf{j} + (z_1 - z_0)\mathbf{k}\right] \cdot \mathbf{n}\right| \\ &= \frac{|A(x_1 - x_0) + B(y_1 - y_0) + C(z_1 - z_0)|}{\sqrt{A^2 + B^2 + C^2}}. \end{aligned}$$

If the plane is given in the form $Ax + By + Cz + D = 0$, choose a point (x_0, y_0, z_0) on it and note that $D = -(Ax_0 + By_0 + Cz_0)$. Substituting in the previous formula gives

$$\operatorname{dist}(Q,\mathscr{P}) = \frac{|Ax_1 + By_1 + Cz_1 + D|}{\sqrt{A^2 + B^2 + C^2}} \tag{7}$$

for the distance from Q to $\mathscr{P}$.

Example 10 Find the distance from $Q = (2, 0, -1)$ to the plane $\mathscr{P}$: $3x - 2y + 8z + 1 = 0$.

Solution We substitute into (7) the values $x_1 = 2$, $y_1 = 0$, $z_1 = -1$ (from the point) and $A = 3$, $B = -2$, $C = 8$, $D = 1$ (from the plane) to give

$$\operatorname{dist}(Q,\mathscr{P}) = \frac{|3 \cdot 2 + (-2) \cdot 0 + 8(-1) + 1|}{\sqrt{3^2 + (-2)^2 + 8^2}} = \frac{|-1|}{\sqrt{77}} = \frac{1}{\sqrt{77}}. \ \blacktriangle$$

Exercises for Section 13.4

Compute the dot products in Exercises 1–4.

1. $(\mathbf{i} + \mathbf{j} + \mathbf{k}) \cdot (\mathbf{i} + \mathbf{j} + 2\mathbf{k})$
2. $(\mathbf{i} + \mathbf{j} + \mathbf{k}) \cdot (\mathbf{i} + \mathbf{k})$
3. $\mathbf{i} \cdot \mathbf{j}$
4. $(3\mathbf{i} + 4\mathbf{j}) \cdot (3\mathbf{j} + 4\mathbf{k})$
5. Find the angle between the pair of vectors in Exercise 1.
6. Find the angle between the pair of vectors in Exercise 2.
7. Find the angle between the pair of vectors in Exercise 3.
8. Find the angle between the pair of vectors in Exercise 4.
9. Find a unit vector in the xy plane which is orthogonal to $2\mathbf{i} - \mathbf{j}$.
10. Find a unit vector in the xy plane which is orthogonal to $3\mathbf{j} - 5\mathbf{i}$.
11. Use the formula $(\mathbf{i} + \mathbf{j} + \mathbf{k}) \cdot \mathbf{i} = 1$ to find the angle between the diagonal of a cube and one of its edges. Sketch.
12. (a) Show that if $\|\mathbf{u}\| = \|\mathbf{v}\|$, and $\mathbf{u}$ and $\mathbf{v}$ are not parallel, then $\mathbf{u} + \mathbf{v}$ and $\mathbf{u} - \mathbf{v}$ are perpendicular. (b) Use the result of part (a) to prove that any triangle inscribed in a circle, with one side of the triangle as a diameter, is a right triangle.
13. Show that the length of the orthogonal projection of $\mathbf{v}$ on $\mathbf{u}$ is equal to $\|\mathbf{v}\|\,|\cos\theta|$, where θ is the angle between $\mathbf{v}$ and $\mathbf{u}$.
14. Use vector methods to prove that a triangle is isosceles if and only if its base angles are equal.
15. Find the distance from $(2, 8, -1)$ to the line through $(1, 1, 1)$ in the direction of the vector $(1/\sqrt{3})\mathbf{i} + (1/\sqrt{3})\mathbf{j} + (1/\sqrt{3})\mathbf{k}$.
16. Find the distance from $(1, 1, -1)$ to the line through $(2, -1, 2)$ in the direction of $\mathbf{k}$.
17. Find the distance from $(1, 1, 2)$ to the line $x = 3t + 2$, $y = -t - 1$, $z = t + 1$.
18. Find the distance from $(1, 1, 0)$ to the line through $(1, 0, -1)$ and $(2, 3, 1)$.

Give the equation for each of the planes in Exercises 19–24.

19. The plane through the origin orthogonal to the vector $\mathbf{i} + \mathbf{j} + \mathbf{k}$.
20. The plane through $(1, 0, 0)$ orthogonal to the vector $\mathbf{i} + \mathbf{j} + \mathbf{k}$.

21. The plane through the origin orthogonal to **i**.
22. The plane containing (a, b, c) with normal vector $a\mathbf{i} + b\mathbf{j} + c\mathbf{k}$.
23. The plane containing the points $(0,0,1)$, $(1,1,1)$ and $(0,1,0)$.
24. The plane containing the points $(1,0,0)$, $(0,2,0)$, and $(0,0,3)$.

Find a unit vector orthogonal to each of the planes in Exercises 25–28.

25. The plane given by $2x + 3y + z = 0$.
26. The plane given by $8x - y - 2z + 10 = 0$.
27. The plane through the origin containing the points $(1,1,1)$ and $(1,1,-1)$.
28. The plane containing the line $(1+t, 1-t, t)$ and the point $(1,1,1)$.

Find the equation of the objects in Exercises 29–32.

29. The plane containing $(0,1,0)$, $(1,0,0)$, $(0,0,1)$.
30. The line through $(1,2,1)$ and $(-1,1,0)$.
31. The line through $(1,1,1)$ and orthogonal to the plane in Exercise 29.
32. The line through $(0,0,0)$ which passes through and is orthogonal to the line in Exercise 30.

33. Where does the line through the origin in the direction of $2\mathbf{i} - \mathbf{j} + 3\mathbf{k}$ meet the plane $2x - y + 3z = 7$? Find the distance from the origin to this plane.
34. Where does the line through $P = (1,1,1)$ in the direction of $2\mathbf{i} - \mathbf{j} + 3\mathbf{k}$ meet the plane $2x - y + 3z = 7$? Find the distance from P to this plane.
35. The planes $3x + 4y + 5z = 6$ and $x - y + z = 4$ meet in a line. Find the parametric equations of this line.
36. Find the parametric equations of the line where the plane $x + y = z$ meets the plane $y + z = x$.
37. Find the distance from the point $(1,1,1)$ to the plane $x - y - z + 10 = 0$.
38. Find the distance from the point $(2,-1,2)$ to the plane $2x - y + z = 5$.
39. Find the distance from the origin to the plane through $(1,2,3)$, $(-1,2,3)$, and $(0,0,1)$.
40. Find the distance from the point $(4,2,0)$ to the plane through $(0,0,0)$, $(1,1,1)$, and $(1,1,2)$.
41. Show that the locus of points in the plane equidistant from two given points is a line, and give an equation for that line in terms of the coordinates of the two points.
42. Use vector methods to show that if three parallel lines in the plane cut off equal segments on one transversal, then they do so on any transversal.
43. Compute the following:
    (a) $(3\mathbf{i} + 2\mathbf{j} - \mathbf{k}) \cdot (\mathbf{j} - \mathbf{k})$.
    (b) $[(3\mathbf{i} - \mathbf{j} - \mathbf{k}) - (\mathbf{i} + \mathbf{j})] \cdot \mathbf{j}$.
    (c) The distance between $(1,0,2)$ and $(3,2,4)$.
    (d) The length of $(\mathbf{i} - \mathbf{j} - \mathbf{k}) + (2\mathbf{j} - \mathbf{k} + \mathbf{i})$.
44. Find the following:
    (a) A unit normal to the plane $x - 2y + z = 0$.
    (b) A vector orthogonal to the vectors $\mathbf{i} - \mathbf{j} + \mathbf{k}$ and $\mathbf{i} + \mathbf{j} + \mathbf{k}$.
    (c) The angle between $2\mathbf{i} + \mathbf{j} + \mathbf{k}$ and $\mathbf{k} - \mathbf{i}$.
    (d) A vector in space making an angle of 45° with **i** and 60° with **j**.
45. Let P_1 and P_2 be points in the plane. Give an equation of the form $ax + by = c$ for the perpendicular bisector of the line segment between P_1 and P_2.
46. Given nonzero vectors **a** and **b**, show that the vector $\mathbf{v} = \|\mathbf{a}\|\mathbf{b} + \|\mathbf{b}\|\mathbf{a}$ bisects the angle between **a** and **b**.

Exercises 47–50 form a unit.

47. Suppose that $\mathbf{e}_1$ and $\mathbf{e}_2$ are perpendicular unit vectors in the plane, and let **v** be an arbitrary vector. Show that $\mathbf{v} = (\mathbf{v} \cdot \mathbf{e}_1)\mathbf{e}_1 + (\mathbf{v} \cdot \mathbf{e}_2)\mathbf{e}_2$. The numbers $\mathbf{v} \cdot \mathbf{e}_1$ and $\mathbf{v} \cdot \mathbf{e}_2$ are called the *components* of **v** in the directions of $\mathbf{e}_1$ and $\mathbf{e}_2$. This expression of **v** as a sum of vectors pointing in the directions of $\mathbf{e}_1$ and $\mathbf{e}_2$ is called the *orthogonal decomposition* of **v** relative to $\mathbf{e}_1$ and $\mathbf{e}_2$.
48. Consider the vectors $\mathbf{e}_1 = (1/\sqrt{2})(\mathbf{i} + \mathbf{j})$ and $\mathbf{e}_2 = (1/\sqrt{2})(\mathbf{i} - \mathbf{j})$ in the plane. Check that $\mathbf{e}_1$ and $\mathbf{e}_2$ are unit vectors perpendicular to each other and express each of the following vectors in the form $\mathbf{v} = a_1\mathbf{e}_1 + a_2\mathbf{e}_2$ (that is, as a *linear combination* of $\mathbf{e}_1$ and $\mathbf{e}_2$):
    (a) $\mathbf{v} = \mathbf{i}$, (b) $\mathbf{v} = \mathbf{j}$,
    (c) $\mathbf{v} = 2\mathbf{i} + \mathbf{j}$, (d) $\mathbf{v} = -2\mathbf{i} - \mathbf{j}$.
49. Suppose that a force **F** (for example, gravity) is acting vertically downward on an object sitting on a plane which is inclined at an angle of 45° to the horizontal. Express this force as a sum of a force acting parallel to the plane and one acting perpendicular to it.
50. Suppose that an object moving in direction $\mathbf{i} + \mathbf{j}$ is acted on by a force given by the vector $2\mathbf{i} + \mathbf{j}$. Express this force as a sum of a force in the direction of motion and a force perpendicular to the direction of motion.

51. A force of 6 newtons makes an angle of $\pi/4$ radians with the y axis, pointing to the right. The force acts against the movement of an object along the straight line connecting $(1,2)$ to $(5,4)$.
    (a) Find a formula for the force vector **F**.
    (b) Find the angle θ between the displacement direction $\mathbf{D} = (5-1)\mathbf{i} + (4-2)\mathbf{j}$ and the force direction **F**.
    (c) The *work* done is $\mathbf{F} \cdot \mathbf{D}$, or equivalently, $\|\mathbf{F}\|\,\|\mathbf{D}\|\cos\theta$. Compute the work from both formulas and compare.
52. A fluid flows across a plane surface with uniform vector velocity **v**. Let **n** be a unit normal to the plane surface. Show that $\mathbf{v} \cdot \mathbf{n}$ is the volume of fluid that passes through a unit area of the plane in unit time.

53. Establish the following properties of the dot product:
 (a) $\mathbf{u}\cdot\mathbf{u} \geqslant 0$ for any vector $\mathbf{u}$.
 (b) If $\mathbf{u}\cdot\mathbf{u} = 0$, then $\mathbf{u} = \mathbf{0}$.
 (c) $\mathbf{u}\cdot\mathbf{v} = \mathbf{v}\cdot\mathbf{u}$ for any vector $\mathbf{u}$ and $\mathbf{v}$.
 (d) $(a\mathbf{u} + b\mathbf{v})\cdot\mathbf{w} = a(\mathbf{u}\cdot\mathbf{w}) + b(\mathbf{v}\cdot\mathbf{w})$ for any numbers a and b and any vectors $\mathbf{u}, \mathbf{v}, \mathbf{w}$.

54. Let:
 L_1 = the line $(2,1,1) + t(1,1,1)$;
 L_2 = the line $(1 + 7t, 7t - 2, 2 + 7t)$;
 L_3 = the line $(1,0,8) + t(1,1,9)$;
 L_4 = the line through the points $(-1,0,1)$ and $(1,2,19)$.
 (a) Determine whether each of the following pairs of lines is parallel or intersects. If the lines intersect, find the point of intersection.
  (i) L_1 and L_2
  (ii) L_2 and L_3
  (iii) L_1 and L_3
  (iv) L_2 and L_4
  (v) L_1 and L_4
  (vi) L_3 and L_4
 (b) For each pair of lines in part (a) which lie in a plane (that is, are not skew), find an equation for that plane.
 (c) For each of the lines L_1 to L_4, find the point of the closest approach to the origin and an equation for the plane perpendicular to that line through that point.

55. A construction worker is checking the architect's plans for some sheet metal construction. One diagram contains a triangle with sides 12.5, 16.7, 20.9, but no angles have been included. The worker gets out a calculator to check that $(12.5)^2 + (16.7)^2 \approx (20.9)^2$, then marks the angle opposite the long side as 90°.
 (a) Explain from the law of cosines the reason why the worker's actions are essentially correct.
 (b) The angle is not exactly 90°, from the data given. What percentage error is present?

★56. Let P_1 and P_2 be points in the plane with *polar* coordinates (r_1, θ_1) and (r_2, θ_2), respectively, and let $\mathbf{u}_1$ be the vector from O to P_1 and $\mathbf{u}_2$ the vector from O to P_2. Show that $\mathbf{u}_1\cdot\mathbf{u}_2 = r_1 r_2\cos(\theta_1 - \theta_2)$. [*Hint*: Use a trigonometric identity.]

★57. Suppose that $R = P_0 + t(a,b,c)$ is the line through P_0 in the direction $\mathbf{d} = a\mathbf{i} + b\mathbf{j} + c\mathbf{k}$. Let $\mathbf{u} = \mathbf{d}/\|\mathbf{d}\| = (\mu, \lambda, \nu)$. Let:

 α = angle from $\mathbf{i}$ to $\mathbf{d}$;
 β = angle from $\mathbf{j}$ to $\mathbf{d}$;
 γ = angle from $\mathbf{k}$ to $\mathbf{d}$.

 These are called the *direction angles* of the line. The numbers $\cos\alpha$, $\cos\beta$, and $\cos\gamma$ are called its *direction cosines* (see Fig. 13.4.9).

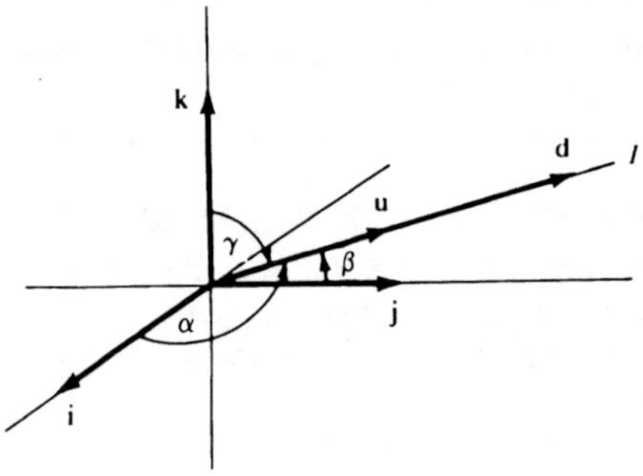

Figure 13.4.9. The direction angles of the line l are α, β, and γ.

 (a) Show that $P_0 + s(\mu, \lambda, \nu)$ gives the same line. (What values of s and t correspond to the same points on the line?)
 (b) Show that $\cos\alpha = \mathbf{i}\cdot\mathbf{u} = \mu$; $\cos\beta = \mathbf{j}\cdot\mathbf{u} = \lambda$; $\cos\gamma = \mathbf{k}\cdot\mathbf{u} = \nu$.
 (c) Show that $\cos^2\alpha + \cos^2\beta + \cos^2\gamma = 1$.
 (d) Determine the direction angles and cosines of each of the lines in Exercise 54.
 (e) Which lines through the origin have direction angles $\alpha = \beta = \gamma$?

★58. Imagine that you look to the side as you walk in the rain. Now you stop walking.
 (a) How does the (apparent) direction of the falling rain change? (The rain may be falling at an angle because of wind.)
 (b) Explain the observation in (a) in terms of vectors.
 (c) Suppose that there is no wind and that you know your walking speed. How could you measure the speed at which the rain is falling?
 (d) Do part (c) if the rain is falling at an angle.

13.5 The Cross Product

The cross product of two vectors in space is a new vector that is perpendicular to the first two.

What is the velocity of a point on a rotating object? Let $\mathbf{v}_1$ be a vector which points in the direction of the axis of rotation and whose length equals the rotation rate (in radians per unit time). Let $\mathbf{v}_2$ be a vector from a point O on the axis of rotation to a point P on the object (see Fig. 13.5.1). A little thought shows that the velocity $\mathbf{v}$ of the point P has the following properties:

1. $\|\mathbf{v}\| = \|\mathbf{v}_1\|\,\|\mathbf{v}_2\|\sin\theta$, where θ is the angle between $\mathbf{v}_1$ and $\mathbf{v}_2$.
2. If $\theta \neq 0$ (so that $\mathbf{v} \neq \mathbf{0}$), $\mathbf{v}$ is perpendicular to both $\mathbf{v}_1$ and $\mathbf{v}_2$, and the triple $(\mathbf{v}_1, \mathbf{v}_2, \mathbf{v})$ of vectors obeys the right-hand rule (see Fig. 13.5.2).

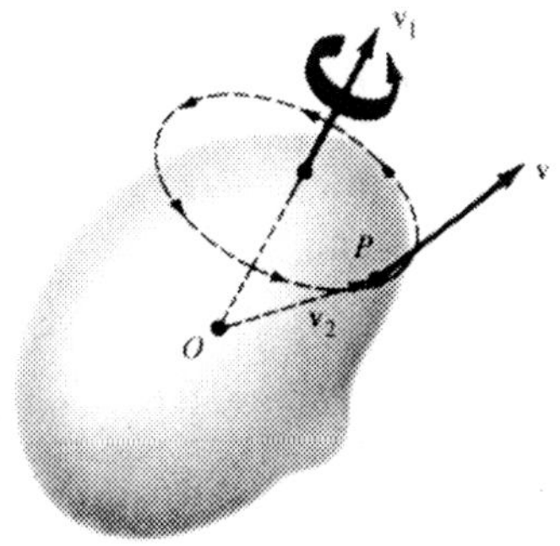

Figure 13.5.1. The point P has velocity vector $\mathbf{v}$.

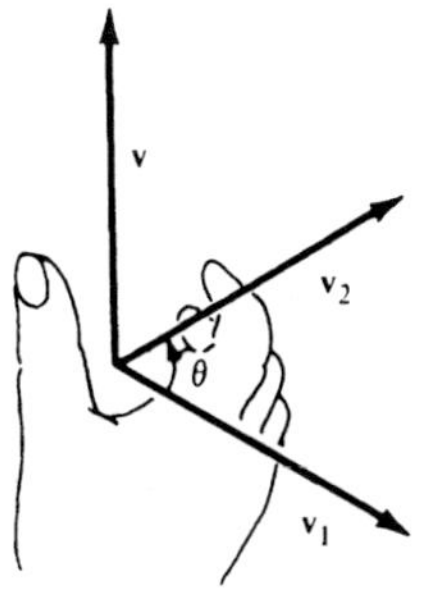

Figure 13.5.2. Right-hand rule: Place the palm of your hand so that your fingers curl from $\mathbf{v}_1$ in the direction of $\mathbf{v}_2$ through the angle θ. Then your thumb points in the direction of $\mathbf{v}$.

Note that condition 1 says the magnitude of the velocity of P is proportional to the product of the magnitude of the rotation rate and the distance of P from the axis of rotation; furthermore, for fixed $\|\mathbf{v}_2\|$, the velocity is greatest when $\mathbf{v}_2$ is perpendicular to the axis.

Conditions 1 and 2 determine $\mathbf{v}$ uniquely in terms of $\mathbf{v}_1$ and $\mathbf{v}_2$; $\mathbf{v}$ is called the *cross product* (or *vector product*) of $\mathbf{v}_1$ and $\mathbf{v}_2$ and is denoted by $\mathbf{v}_1 \times \mathbf{v}_2$.

We will now determine some properties of the cross product operation. Our ultimate goal is to find a formula for the components of $\mathbf{v}_1 \times \mathbf{v}_2$ in terms of the components of $\mathbf{v}_1$ and $\mathbf{v}_2$. Let us first show that $\|\mathbf{v}_1 \times \mathbf{v}_2\|$ is equal to the area of the parallelogram spanned by $\mathbf{v}_1$ and $\mathbf{v}_2$. Drop the perpendicular PS as shown in Fig. 13.5.3. Then $A = |OR|\,|PS| = |OR|\,|OP|\sin\theta = \|\mathbf{v}_1\|\,\|\mathbf{v}_2\|\sin\theta = \|\mathbf{v}_1 \times \mathbf{v}_2\|$ by condition 1, proving our claim.

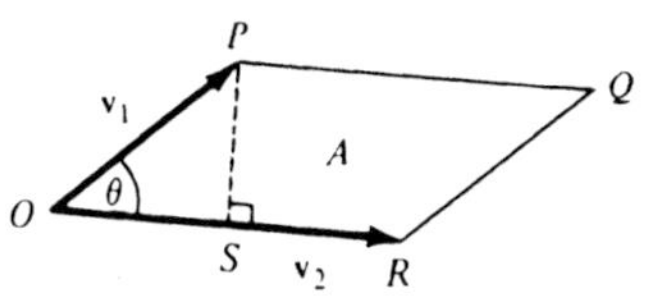

Figure 13.5.3. Calculating the area of $ORQP$.

Example 1 Find all the cross products between the standard basis vectors **i**, **j**, and **k**.

Solution We observe first that $\mathbf{i} \times \mathbf{i} = \mathbf{j} \times \mathbf{j} = \mathbf{k} \times \mathbf{k} = \mathbf{0}$, because the angle between any vector and itself is zero, and $\sin 0 = 0$. Next we observe that $\mathbf{i} \times \mathbf{j}$ must be a multiple of $\mathbf{k}$, since it is perpendicular to $\mathbf{i}$ and $\mathbf{j}$. On the other hand, $\|\mathbf{i} \times \mathbf{j}\| = \|\mathbf{i}\|\,\|\mathbf{j}\|\sin 90° = 1$, so $\mathbf{i} \times \mathbf{j}$ must be $\mathbf{k}$ or $-\mathbf{k}$. The right-hand rule then shows that $\mathbf{i} \times \mathbf{j} = \mathbf{k}$ (see Fig. 13.5.4). Next, $\mathbf{j} \times \mathbf{i}$ must be $\mathbf{k}$ or $-\mathbf{k}$; this time the right-hand rule gives $-\mathbf{k}$ as the answer. Similarly, $\mathbf{j} \times \mathbf{k} = \mathbf{i}$, $\mathbf{k} \times \mathbf{i} = \mathbf{j}$, $\mathbf{k} \times \mathbf{j} = -\mathbf{i}$, and $\mathbf{i} \times \mathbf{k} = -\mathbf{j}$. ▲

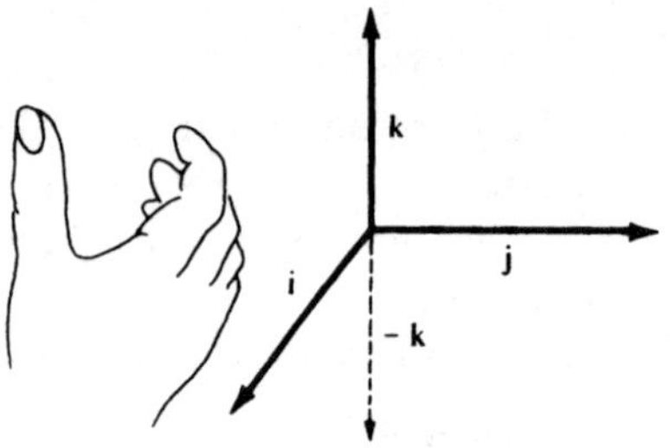

Figure 13.5.4. The right-hand rule requires that $\mathbf{i} \times \mathbf{j}$ equal $\mathbf{k}$, not $-\mathbf{k}$.

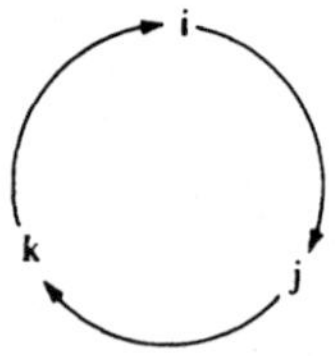

Figure 13.5.5. As we go around the circle, the cross product of any two consecutive vectors equals the third vector. Going backwards produces the negative of the preceding vector.

A good way to remember these products is to write $\mathbf{i}$, $\mathbf{j}$, and $\mathbf{k}$ in a circle as in Fig. 13.5.5.

We will now obtain a general formula for the cross product

$$(a_1\mathbf{i} + b_1\mathbf{j} + c_1\mathbf{k}) \times (a_2\mathbf{i} + b_2\mathbf{j} + c_2\mathbf{k}).$$

If we assumed that the usual rules of algebra apply to the cross product, we could use the result of Example 1 to make the following calculation:

$$\begin{aligned}(a_1\mathbf{i} + b_1\mathbf{j} + c_1\mathbf{k}) &\times (a_2\mathbf{i} + b_2\mathbf{j} + c_2\mathbf{k}) \\ &= a_1\mathbf{i} \times (a_2\mathbf{i} + b_2\mathbf{j} + c_2\mathbf{k}) + b_1\mathbf{j} \times (a_2\mathbf{i} + b_2\mathbf{j} + c_2\mathbf{k}) + c_1\mathbf{k} \times (a_2\mathbf{i} + b_2\mathbf{j} + \\ &= a_1b_2\mathbf{k} + a_1c_2(-\mathbf{j}) + b_1a_2(-\mathbf{k}) + b_1c_2\mathbf{i} + c_1a_2\mathbf{j} + c_1b_2(-\mathbf{i}).\end{aligned}$$

Collecting terms, we have

$$\begin{aligned}(a_1\mathbf{i} + b_1\mathbf{j} + c_1\mathbf{k}) &\times (a_2\mathbf{i} + b_2\mathbf{j} + c_2\mathbf{k}) \\ &= (b_1c_2 - c_1b_2)\mathbf{i} + (c_1a_2 - a_1c_2)\mathbf{j} + (a_1b_2 - b_1a_2)\mathbf{k}. \qquad (1)\end{aligned}$$

Although to derive (1) we made the unjustified assumption that some laws of algebra hold for the cross product, it turns out that the result is correct. To see this, we shall show that $\mathbf{u} = (b_1c_2 - c_1b_2)\mathbf{i} + (c_1a_2 - a_1c_2)\mathbf{j} + (a_1b_2 - b_1a_2)\mathbf{k}$ is indeed the cross product of $\mathbf{v}_1 = a_1\mathbf{i} + b_1\mathbf{j} + c_1\mathbf{k}$ and $\mathbf{v}_2 = a_2\mathbf{i} + b_2\mathbf{j} + c_2\mathbf{k}$. We accomplish this by verifying that $\mathbf{u}$ satisfies conditions 1 and 2 in the definition of the cross product.

First, we consider the squared length of $\mathbf{u}$:

$$\begin{aligned}&(b_1c_2 - c_1b_2)^2 + (c_1a_2 - a_1c_2)^2 + (a_1b_2 - b_1a_2)^2 \\ &= b_1^2c_2^2 - 2b_1c_1b_2c_2 + c_1^2b_2^2 + c_1^2a_2^2 - 2a_1c_1a_2c_2 + a_1^2c_2^2 + a_1^2b_2^2 - 2a_1b_1a_2b_2 + b_1^2a_2^2\end{aligned}$$

Now we compute the square of $\|\mathbf{v}_1\|\,\|\mathbf{v}_2\|\sin\theta$:

$$\begin{aligned}\|\mathbf{v}_1\|^2\|\mathbf{v}_2\|^2\sin^2\theta &= \|\mathbf{v}_1\|^2\|\mathbf{v}_2\|^2(1 - \cos^2\theta) \\ &= \|\mathbf{v}_1\|^2\|\mathbf{v}_2\|^2 - (\|\mathbf{v}_1\|\,\|\mathbf{v}_2\|\cos\theta)^2 \\ &= \|\mathbf{v}_1\|^2\|\mathbf{v}_2\|^2 - (\mathbf{v}_1 \cdot \mathbf{v}_2)^2 \\ &= (a_1^2 + b_1^2 + c_1^2)(a_2^2 + b_2^2 + c_2^2) - (a_1a_2 + b_1b_2 + c_1c_2)^2\end{aligned}$$

which, when it is multiplied out and terms are collected, is the same as $\|\mathbf{u}\|^2$, so $\|\mathbf{u}\| = \|\mathbf{v}_1\|\,\|\mathbf{v}_2\|\sin\theta$.

Next we check that $\mathbf{u}$ is perpendicular to $\mathbf{v}_1$ and $\mathbf{v}_2$. We have

$$\begin{aligned}\mathbf{u} \cdot \mathbf{v}_1 &= (b_1c_2 - c_1b_2)a_1 + (c_1a_2 - a_1c_2)b_1 + (a_1b_2 - b_1a_2)c_1 \\ &= b_1c_2a_1 - c_1b_2a_1 + c_1a_2b_1 - a_1c_2b_1 + a_1b_2c_1 - b_1a_2c_1 \\ &= 0,\end{aligned}$$

since the terms cancel in pairs. Similarly, $\mathbf{u} \cdot \mathbf{v}_2 = 0$, so $\mathbf{u}$ is perpendicular to $\mathbf{v}_1$ and $\mathbf{v}_2$.

To check the right-hand rule would require a precise mathematical definition of that rule, which we will not attempt to give. Instead, we will merely remark that the rule is satisfied for all products of standard basis vectors; see Example 1.

We have now shown that the vector $\mathbf{u}$ on the right-hand side of (1) satisfies the conditions in the definition of $\mathbf{v}_1 \times \mathbf{v}_2$, so it must be $\mathbf{v}_1 \times \mathbf{v}_2$. The algebraic rules in the following display may then be verified as a consequence of formula (1) (see Example 7).

The Cross Product

Geometric definition: $\mathbf{v}_1 \times \mathbf{v}_2$ is the vector such that:

1. $\|\mathbf{v}_1 \times \mathbf{v}_2\| = \|\mathbf{v}_1\| \, \|\mathbf{v}_2\| \sin\theta$, the area of the parallelogram spanned by $\mathbf{v}_1$ and $\mathbf{v}_2$ (θ is the angle between $\mathbf{v}_1$ and $\mathbf{v}_2$; $0 \leqslant \theta \leqslant \pi$).
2. $\mathbf{v}_1 \times \mathbf{v}_2$ is perpendicular to $\mathbf{v}_1$ and $\mathbf{v}_2$, and the triple $(\mathbf{v}_1, \mathbf{v}_2, \mathbf{v}_1 \times \mathbf{v}_2)$ obeys the right-hand rule.

Component formula:

$$(a_1\mathbf{i} + b_1\mathbf{j} + c_1\mathbf{k}) \times (a_2\mathbf{i} + b_2\mathbf{j} + c_2\mathbf{k})$$
$$= (b_1c_2 - c_1b_2)\mathbf{i} + (c_1a_2 - a_1c_2)\mathbf{j} + (a_1b_2 - b_1a_2)\mathbf{k}$$

Algebraic rules:

1. $\mathbf{v}_1 \times \mathbf{v}_2 = \mathbf{0}$ if and only if $\mathbf{v}_1$ and $\mathbf{v}_2$ are parallel or $\mathbf{v}_1$ or $\mathbf{v}_2$ is zero.
2. $\mathbf{v}_1 \times \mathbf{v}_2 = -\mathbf{v}_2 \times \mathbf{v}_1$.
3. $\mathbf{v}_1 \times (\mathbf{v}_2 + \mathbf{v}_3) = \mathbf{v}_1 \times \mathbf{v}_2 + \mathbf{v}_1 \times \mathbf{v}_3$.
4. $(\mathbf{v}_1 + \mathbf{v}_2) \times \mathbf{v}_3 = \mathbf{v}_1 \times \mathbf{v}_3 + \mathbf{v}_2 \times \mathbf{v}_3$.
5. $(a\mathbf{v}_1) \times \mathbf{v}_2 = a(\mathbf{v}_1 \times \mathbf{v}_2)$.

Multiplication table (see Fig. 13.5.5):

		Second Factor		
	$\times$	**i**	**j**	**k**
First Factor	**i**	**0**	**k**	−**j**
	j	−**k**	**0**	**i**
	k	**j**	−**i**	**0**

Example 2 (a) Compute $(3\mathbf{i} + 2\mathbf{j} - \mathbf{k}) \times (\mathbf{j} - \mathbf{k})$. (b) Find $\mathbf{i} \times (\mathbf{i} \times \mathbf{j})$ and $(\mathbf{i} \times \mathbf{i}) \times \mathbf{j}$. Are they equal?

Solution (a) We use the products $\mathbf{i} \times \mathbf{j} = \mathbf{k}$, etc. and the algebraic rules as follows:

$$\begin{aligned}(3\mathbf{i} + 2\mathbf{j} - \mathbf{k}) \times (\mathbf{j} - \mathbf{k}) &= (3\mathbf{i} + 2\mathbf{j} - \mathbf{k}) \times \mathbf{j} - (3\mathbf{i} + 2\mathbf{j} - \mathbf{k}) \times \mathbf{k} \\ &= 3\mathbf{i} \times \mathbf{j} + 2\mathbf{j} \times \mathbf{j} - \mathbf{k} \times \mathbf{j} - 3\mathbf{i} \times \mathbf{k} - 2\mathbf{j} \times \mathbf{k} + \mathbf{k} \times \mathbf{k} \\ &= 3\mathbf{k} + \mathbf{0} + \mathbf{i} + 3\mathbf{j} - 2\mathbf{i} + \mathbf{0} \\ &= -\mathbf{i} + 3\mathbf{j} + 3\mathbf{k}.\end{aligned}$$

This can be checked using the component formula.

(b) We find that $\mathbf{i} \times (\mathbf{i} \times \mathbf{j}) = \mathbf{i} \times \mathbf{k} = -\mathbf{j}$, while $(\mathbf{i} \times \mathbf{i}) \times \mathbf{j} = \mathbf{0} \times \mathbf{j} = \mathbf{0}$, so the two expressions are *not* equal. This example means that the cross product is *not associative*—one cannot move parentheses as in ordinary multiplication. ▲

Example 3 Find the area of the parallelogram spanned by the vectors $\mathbf{v}_1 = \mathbf{i} + 2\mathbf{j} + 3\mathbf{k}$ and $\mathbf{v}_2 = -\mathbf{i} - \mathbf{k}$.

Solution We calculate the cross product of $\mathbf{v}_1$ and $\mathbf{v}_2$ by applying the component formula, with $a_1 = 1$, $b_1 = 2$, $c_1 = 3$, $a_2 = -1$, $b_2 = 0$, $c_2 = -1$:

$$\begin{aligned}\mathbf{v}_1 \times \mathbf{v}_2 &= [(2)(-1) - (3)(0)]\mathbf{i} + [(3)(-1) - (1)(-1)]\mathbf{j} + [(1)(0) - (2)(-1)]\mathbf{k} \\ &= -2\mathbf{i} - 2\mathbf{j} + 2\mathbf{k}.\end{aligned}$$

Thus the area is

$$\begin{aligned}\|\mathbf{v}_1 \times \mathbf{v}_2\| &= \sqrt{(-2)^2 + (-2)^2 + (2)^2} \\ &= 2\sqrt{3}\,. \blacktriangle\end{aligned}$$

Comparing the methods of Examples 2 and 3 shows that it is often easier to use the algebraic rules and the multiplication table directly, rather than using the component formula.

Example 4 Find a unit vector which is orthogonal to the vectors $\mathbf{i} + \mathbf{j}$ and $\mathbf{j} + \mathbf{k}$.

Solution A vector perpendicular to both $\mathbf{i} + \mathbf{j}$ and $\mathbf{j} + \mathbf{k}$ is the vector

$$\begin{aligned}(\mathbf{i} + \mathbf{j}) \times (\mathbf{j} + \mathbf{k}) &= \mathbf{i} \times \mathbf{j} + \mathbf{i} \times \mathbf{k} + \mathbf{j} \times \mathbf{j} + \mathbf{j} \times \mathbf{k} \\ &= \mathbf{k} - \mathbf{j} + \mathbf{0} + \mathbf{i} \\ &= \mathbf{i} - \mathbf{j} + \mathbf{k}.\end{aligned}$$

Since $\|\mathbf{i} - \mathbf{j} + \mathbf{k}\| = \sqrt{3}$, the vector $(1/\sqrt{3})(\mathbf{i} - \mathbf{j} + \mathbf{k})$ is a unit vector perpendicular to $\mathbf{i} + \mathbf{j}$ and $\mathbf{j} + \mathbf{k}$. ▲

Example 5 Use the cross product to find the equation of the plane containing the points $(1,1,1)$, $(2,0,0)$, and $(1,1,0)$. (Compare Example 8 of Section 13.4.)

Solution The normal to the plane is perpendicular to any vector which joins two points in the plane, so it is perpendicular to $\mathbf{v}_1 = (1,1,1) - (2,0,0) = (-1,1,1) = -\mathbf{i} + \mathbf{j} + \mathbf{k}$ and $\mathbf{v}_2 = (1,1,1) - (1,1,0) = (0,0,1) = \mathbf{k}$. A vector perpendicular to $\mathbf{v}_1$ and $\mathbf{v}_2$ is $\mathbf{v}_1 \times \mathbf{v}_2 = (-\mathbf{i} + \mathbf{j} + \mathbf{k}) \times \mathbf{k} = -\mathbf{i} \times \mathbf{k} + \mathbf{j} \times \mathbf{k} + \mathbf{k} \times \mathbf{k} = \mathbf{j} + \mathbf{i} + \mathbf{0} = \mathbf{i} + \mathbf{j}$, so the equation of the plane has the form $x + y + D = 0$. Since $(1,1,1)$ lies in the plane, $2 + D = 0$, and the equation is $x + y - 2 = 0$. (In Example 8 of Section 13.4, we obtained this result by solving a system of simultaneous equations. Here the cross product does the solving for us.) ▲

Example 6 Find the area of the triangle with vertices $P_1 = (1,1,2)$, $P_2 = (2,-1,0)$, and $P_3 = (1,-1,3)$.

Solution The area of a triangle is half that of the parallelogram spanned by two of its sides. As sides we take the vectors $\mathbf{v}_1 = \overrightarrow{P_1P_2} = \mathbf{i} - 2\mathbf{j} - 2\mathbf{k}$ and $\mathbf{v}_2 = \overrightarrow{P_1P_3} = -2\mathbf{j} + \mathbf{k}$. Then

$$\mathbf{v}_1 \times \mathbf{v}_2 = (\mathbf{i} - 2\mathbf{j} - 2\mathbf{k}) \times (-2\mathbf{j} + \mathbf{k}) = -6\mathbf{i} - \mathbf{j} - 2\mathbf{k}.$$

The area of the triangle is thus

$$\begin{aligned}\tfrac{1}{2}\|\mathbf{v}_1 \times \mathbf{v}_2\| &= \tfrac{1}{2}(6^2 + 1^2 + 2^2)^{1/2} \\ &= \tfrac{1}{2}\sqrt{41} \approx 3.20. \blacktriangle\end{aligned}$$

Example 7 Prove the algebraic rule 3 by using the component formula.

Solution Let $\mathbf{v}_i = a_i\mathbf{i} + b_i\mathbf{j} + c_i\mathbf{k}$ $(i = 1, 2, 3)$. Then

$$\mathbf{v}_2 + \mathbf{v}_3 = (a_2 + a_3)\mathbf{i} + (b_2 + b_3)\mathbf{j} + (c_2 + c_3)\mathbf{k}$$

and

$$\begin{aligned}\mathbf{v}_1 \times (\mathbf{v}_2 + \mathbf{v}_3) &= \left[b_1(c_2 + c_3) - c_1(b_2 + b_3) \right]\mathbf{i} \\ &\quad + \left[c_1(a_2 + a_3) - a_1(c_2 + c_3) \right]\mathbf{j} + \left[a_1(b_2 + b_3) - b_1(a_2 + a_3) \right]\mathbf{k} \\ &= (b_1c_2 - c_1b_2)\mathbf{i} + (c_1a_2 - a_1c_2)\mathbf{j} + (a_1b_2 - b_1a_2)\mathbf{k} \\ &\quad + (b_1c_3 - c_1b_3)\mathbf{i} + (c_1a_3 - a_1c_3)\mathbf{j} + (a_1b_3 - a_3b_1)\mathbf{k} \\ &= \mathbf{v}_1 \times \mathbf{v}_2 + \mathbf{v}_1 \times \mathbf{v}_3 . \ \blacktriangle\end{aligned}$$

Example 8 Show that $\|\mathbf{u} \times \mathbf{v}\|^2 = \|\mathbf{u}\|^2\|\mathbf{v}\|^2 - (\mathbf{u} \cdot \mathbf{v})^2$.

Solution Let θ be the angle between $\mathbf{u}$ and $\mathbf{v}$. Then

$$\|\mathbf{u} \times \mathbf{v}\| = \|\mathbf{u}\|\,\|\mathbf{v}\|\sin\theta \quad \text{and} \quad \mathbf{u} \cdot \mathbf{v} = \|\mathbf{u}\|\,\|\mathbf{v}\|\cos\theta.$$

Squaring both equations and summing gives

$$\begin{aligned}\|\mathbf{u} \times \mathbf{v}\|^2 + (\mathbf{u} \cdot \mathbf{v})^2 &= \|\mathbf{u}\|^2\|\mathbf{v}\|^2(\sin^2\theta + \cos^2\theta) \\ &= \|\mathbf{u}\|^2\|\mathbf{v}\|^2,\end{aligned}$$

so $\|\mathbf{u} \times \mathbf{v}\|^2 = \|\mathbf{u}\|^2\|\mathbf{v}\|^2 - (\mathbf{u} \cdot \mathbf{v})^2$. ▲

Exercises for Section 13.5

Calculate the cross products in Exercises 1–10.

1. $(\mathbf{i} - \mathbf{j} + \mathbf{k}) \times (\mathbf{j} - \mathbf{k})$.
2. $\mathbf{i} \times (\mathbf{j} - \mathbf{k})$.
3. $(\mathbf{i} + \mathbf{j}) \times [(\mathbf{k} - \mathbf{j}) + (3\mathbf{j} - 2\mathbf{i} + \mathbf{k})]$.
4. $(a\mathbf{i} + \mathbf{j} - \mathbf{k}) \times \mathbf{i}$.
5. $[(3\mathbf{i} + 2\mathbf{j}) \times 3\mathbf{j}] \times (2\mathbf{i} - \mathbf{j} + \mathbf{k})$.
6. $(\mathbf{i} \times \mathbf{j}) \times (\mathbf{i} + \mathbf{j} + \mathbf{k})$.
7. $(\mathbf{i} + 2\mathbf{j} + 3\mathbf{k}) \times (\mathbf{i} + 3\mathbf{k})$.
8. $(\mathbf{i} + \mathbf{j} + \mathbf{k}) \times (\mathbf{i} + \mathbf{k})$.
9. $(\mathbf{i} + \mathbf{k}) \times (\mathbf{i} + \mathbf{j} + \mathbf{k})$.
10. $(3\mathbf{i} - 2\mathbf{k}) \times (3\mathbf{i} - \mathbf{j} - \mathbf{k})$.

Find the area of the parallelogram spanned by the vectors in Exercises 11–14.

11. $\mathbf{i} - 2\mathbf{j} + \mathbf{k}$ and $\mathbf{i} + \mathbf{j} + \mathbf{k}$.
12. $\mathbf{i} - \mathbf{j}$ and $\mathbf{i} + \mathbf{j}$.
13. $\mathbf{i}$ and $\mathbf{i} - 2\mathbf{j}$.
14. $\mathbf{i} - \mathbf{j} - \mathbf{k}$ and $\mathbf{i} + \mathbf{j} + \mathbf{k}$.

Find a unit vector orthogonal to the pairs of vectors in Exercises 15–18.

15. $\mathbf{i}$ and $\mathbf{i} + \mathbf{j} + \mathbf{k}$.
16. $\mathbf{i} - \mathbf{j}$ and $\mathbf{i} + \mathbf{j}$.
17. $\mathbf{i} - \mathbf{j} - \mathbf{k}$ and $2\mathbf{i} - 2\mathbf{j} + \mathbf{k}$.
18. $\mathbf{i} + 2\mathbf{j} + \mathbf{k}$ and $3\mathbf{i} - \mathbf{j}$.
19. Find a unit vector perpendicular to $\mathbf{i} + \mathbf{j}$ and to $\mathbf{i} - \mathbf{j} - \frac{1}{2}\mathbf{k}$ and with a positive $\mathbf{k}$ component.
20. Find a unit vector perpendicular to $\mathbf{i} - \mathbf{j}$ and to $\mathbf{i} + \mathbf{k}$ and with a positive $\mathbf{k}$ component.
21. Find the equation of the plane passing through the points $(0, 0, 0)$, $(2, 0, -1)$, and $(0, 4, -3)$.
22. Find the equation of the plane through the points $(1, 2, 0)$, $(0, 1, -2)$, and $(4, 0, 1)$.
23. Find the equation of the plane through $(1, 1, 1)$ and containing the line which is the intersection of the planes $x - y = 2$ and $y - z = 1$.
24. Find the equation of the plane through the point $(2, 1, 1)$ and containing the line $x = t - 1$, $y = 2t + 1$, $z = -t - 1$.
25. Find the area of the triangle whose vertices are $(0, 1, 2)$, $(3, 4, 5)$, and $(-1, -1, 0)$.
26. Find the area of the triangle whose vertices are $(0, 1, 2)$, $(1, 1, 1)$, and $(2, 1, 0)$.
27. Find the area of the triangle whose vertices are $(0, 0, 0)$, $(0, -1, 1)$, and $(0, 1, -1)$.
28. Find the area of the triangle whose vertices are $(-1, -1, -1)$, $(-1, 0, 1)$, and $(1, 0, -1)$.
29. Prove algebraic rule 5 by using the component formula.
30. Prove the formula in Example 8 by using the component formula.
31. By using the cross product of the vectors $\cos\theta\mathbf{i} + \sin\theta\mathbf{j}$ and $\cos\psi\mathbf{i} + \sin\psi\mathbf{j}$, verify that $\sin(\theta - \psi) = \sin\theta\cos\psi - \cos\theta\sin\psi$.

32. Let ℓ be a line through a point P_0 in direction $\mathbf{d}$. Show that the distance from a point P to ℓ is given by $\|(\overrightarrow{P_0P}) \times \mathbf{d}\| / \|\mathbf{d}\|$.
33. Let a line in the plane be given by the equation $ax + by = c$. Use the cross product to show that the distance from a point $P = (x, y)$ to this line is given by

$$\frac{|ax + by - c|}{\sqrt{a^2 + b^2}}.$$

34. Use the cross product to find a solution of the following simultaneous equations: $x + y = 0$ and $x - y - 2z = 0$.
35. In mechanics, the *moment M of a force* $\mathbf{F}$ *about a point O* is defined to be the magnitude of $\mathbf{F}$ times the perpendicular distance $\mathbf{d}$ from O to the line of action of $\mathbf{F}$. The *vector moment* $\mathbf{M}$ is the vector of magnitude M whose direction is perpendicular to the plane of O and $\mathbf{F}$, determined by the right-hand rule. Show that $\mathbf{M} = \mathbf{R} \times \mathbf{F}$, where $\mathbf{R}$ is any vector from O to the line of action of $\mathbf{F}$. (See Fig. 13.5.6.)

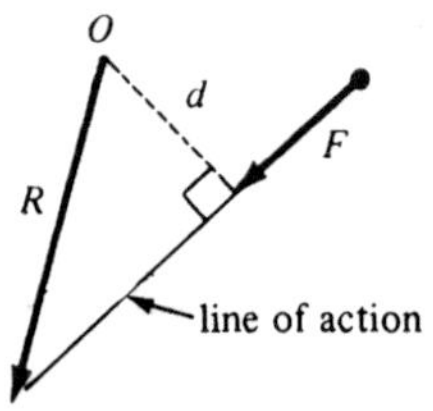

Figure 13.5.6. Moment of a force.

36. The angular velocity $\mathbf{\Omega}$ of rotation of a rigid body has direction equal to the axis of rotation and magnitude equal to the angular velocity in radians per second. The sense of $\mathbf{\Omega}$ is determined by the right-hand rule.
    (a) Let $\mathbf{r}$ be a vector from the axis to a point P on the rigid body. Show that the quantity $\mathbf{v} = \mathbf{\Omega} \times \mathbf{r}$ is the velocity of $\mathbf{P}$, as in Fig. 13.5.1, with $\mathbf{\Omega} = \mathbf{v}_1$ and $\mathbf{r} = \mathbf{v}_2$.
    (b) Interpret the result for the rotation of a carousel about its axis, with P a point on the circumference.
37. Two media with indices of refraction n_1 and n_2 are separated by a plane surface perpendicular to the unit vector $\mathbf{N}$. Let $\mathbf{a}$ and $\mathbf{b}$ be unit vectors along the incident and refracted rays, respectively, their directions being those of the light rays. Show that $n_1(\mathbf{N} \times \mathbf{a}) = n_2(\mathbf{N} \times \mathbf{b})$ by using *Snell's law* $\sin\theta_1/\sin\theta_2 = n_2/n_1$, where θ_1 and θ_2 are the angles of incidence and refraction, respectively. (See Fig. 13.5.7.)

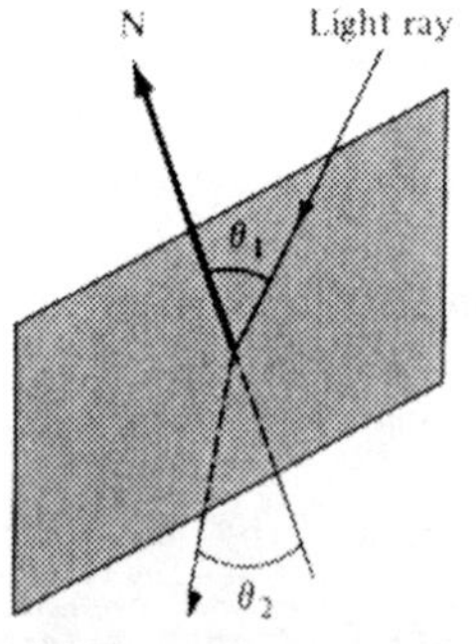

Figure 13.5.7. Snell's law.

★38. Prove the following:
    (a) $(\mathbf{u} \times \mathbf{v}) \cdot (\mathbf{a} \times \mathbf{b}) = (\mathbf{u} \cdot \mathbf{a})(\mathbf{v} \cdot \mathbf{b}) - (\mathbf{v} \cdot \mathbf{a})(\mathbf{u} \cdot \mathbf{b})$.
    (b) *The Jacobi identity*:

$$(\mathbf{u} \times \mathbf{v}) \times \mathbf{w} + (\mathbf{v} \times \mathbf{w}) \times \mathbf{u} + (\mathbf{w} \times \mathbf{u}) \times \mathbf{v} = \mathbf{0}.$$

★39. (a) Using vector methods, show that the distance between two nonparallel lines l_1 and l_2 is given by

$$d = \frac{|(\mathbf{v}_2 - \mathbf{v}_1) \cdot (\mathbf{a}_1 \times \mathbf{a}_2)|}{\|\mathbf{a}_1 \times \mathbf{a}_2\|},$$

where $\mathbf{v}_1$, $\mathbf{v}_2$ are vectors from the origin to points on l_1 and l_2, respectively, and $\mathbf{a}_1$ and $\mathbf{a}_2$ are the directions of l_1 and l_2. [*Hint:* Consider the plane through l_2 which is parallel to l_1. Show that $(\mathbf{a}_1 \times \mathbf{a}_2)/\|\mathbf{a}_1 \times \mathbf{a}_2\|$ is a unit normal for this plane; now project $\mathbf{v}_2 - \mathbf{v}_1$ onto this normal direction.]
    (b) Find the distance between the line l_1 determined by the two points $(-1, -1, 1)$ and $(0, 0, 0)$ and the line l_2 determined by the points $(0, -2, 0)$ and $(2, 0, 5)$.

★40. Use properties of the cross product to explain why, in the discussion of rotation at the beginning of this section, the resulting vector $\mathbf{v}$ does not depend on the choice of the origin O; i.e., what happens if O is replaced by another point O' on the axis of rotation?

★41. When a gyroscope rotating about an axis $\mathbf{\Omega}$, as in Fig. 13.5.8, is subject to a force $\mathbf{F}$, the gyroscope responds by moving in the direction $\mathbf{\Omega} \times \mathbf{F}$.[5] Show that this fact is consistent with the gyroscopic precession you actually observe in toy gyroscopes.

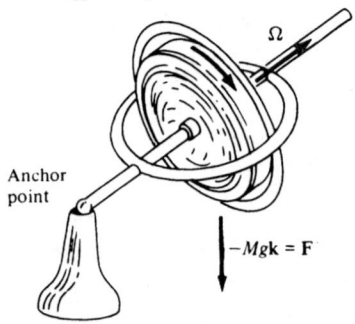

Figure 13.5.8. Gyroscope and cross products.

[5] This relationship is easiest to see in an orbiting earth satellite, where the effects of gravity do not complicate the issue. Indeed, the Skylab astronauts had great fun carrying out such experiments. (See Henry S. Cooper, Jr., *A House in Space*, Holt, Rinehart, and Winston, 1976).

13.6 Matrices and Determinants

The cross product can be expressed as a 3×3 *determinant.*

From the point of view of geometry and vectors, we will consider, in turn, 2×2 determinants and matrices, and then the 3×3 case.

If $\mathbf{v}_1 = a_1\mathbf{i} + b_1\mathbf{j}$ and $\mathbf{v}_2 = a_2\mathbf{i} + b_2\mathbf{j}$ are vectors in the plane, to compute the area of the parallelogram spanned by $\mathbf{v}_1$ and $\mathbf{v}_2$, we may consider them as vectors in space (with $c_1 = c_2 = 0$) and take the cross-product $\mathbf{v}_1 \times \mathbf{v}_2 = (a_1b_2 - b_1a_2)\mathbf{k}$. The area is then $\|\mathbf{v}_1 \times \mathbf{v}_2\| = |a_1b_2 - b_1a_2|$, but the sign of $a_1b_2 - b_1a_2$ also gives us some information: it is positive if the sense of (shortest) rotation from $\mathbf{v}_1$ to $\mathbf{v}_2$ is counterclockwise and negative if the sense of rotation is clockwise (see Fig. 13.6.1). We may say that the sign of $a_1b_2 - b_1a_2$ determines the *orientation* of the ordered pair of vectors $(\mathbf{v}_1, \mathbf{v}_2)$.

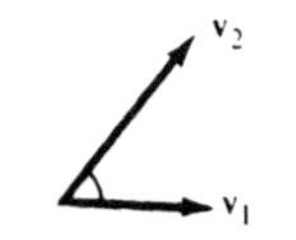

(a) $a_1b_2 - b_1a_2 > 0$

(b) $a_1b_2 - b_1a_2 < 0$

Figure 13.6.1. The sign of $a_1b_2 - b_1a_2$ determines the orientation of the ordered pair $(a_1\mathbf{i} + b_1\mathbf{j}, a_2\mathbf{i} + b_2\mathbf{j})$.

The combination $a_1b_2 - b_1a_2$ of the four numbers a_1, b_1, a_2, and b_2 is denoted by

$$\begin{vmatrix} a_1 & b_1 \\ a_2 & b_2 \end{vmatrix}$$

and is called the *determinant* of these four numbers.

2×2 Determinants

If a, b, c, d are any four numbers, we write

$$\begin{vmatrix} a & b \\ c & d \end{vmatrix} = ad - bc.$$

The absolute value of

$$\begin{vmatrix} a & b \\ c & d \end{vmatrix}$$

equals the area of the parallelogram spanned by the vectors $\mathbf{v} = a\mathbf{i} + b\mathbf{j}$ and $\mathbf{w} = c\mathbf{i} + d\mathbf{j}$. The sign of

$$\begin{vmatrix} a & b \\ c & d \end{vmatrix}$$

gives the orientation of the pair $(\mathbf{v}, \mathbf{w})$.

Example 1 (a) Find the determinant $\begin{vmatrix} -\pi & 3 \\ \pi/2 & 6 \end{vmatrix}$.

(b) Show that $\begin{vmatrix} 1 & 2 \\ 3 & 4 \end{vmatrix} = \begin{vmatrix} 1 & 2 \\ 4 & 6 \end{vmatrix} = \begin{vmatrix} 1 & 4 \\ 2 & 6 \end{vmatrix}$. What is their common value?

(c) Prove that $\begin{vmatrix} a & b \\ c & d \end{vmatrix} = -\begin{vmatrix} b & a \\ d & c \end{vmatrix}$ (the two columns are interchanged).

Solution (a)

$$\begin{vmatrix} -\pi & 3 \\ \pi/2 & 6 \end{vmatrix} = -6\pi - \frac{3\pi}{2} = -\frac{15\pi}{2}.$$

(b) $\begin{vmatrix} 1 & 2 \\ 3 & 4 \end{vmatrix} = (1)(4) - (2)(3) = -2;$

$$\begin{vmatrix} 1 & 2 \\ 4 & 6 \end{vmatrix} = (1)(6) - (2)(4) = -2;$$

$$\begin{vmatrix} 1 & 4 \\ 2 & 6 \end{vmatrix} = (1)(6) - (4)(2) = -2.$$

Their common value is -2.

(c) $\begin{vmatrix} a & b \\ c & d \end{vmatrix} = ad - bc$, and

$$-\begin{vmatrix} b & a \\ d & c \end{vmatrix} = -(bc - ad) = ad - bc. \blacktriangle$$

From the previous section we recall that

$$(a_1\mathbf{i} + b_1\mathbf{j} + c_1\mathbf{k}) \times (a_2\mathbf{i} + b_2\mathbf{j} + c_2\mathbf{k})$$
$$= (b_1c_2 - c_1b_2)\mathbf{i} + (c_1a_2 - a_1c_2)\mathbf{j} + (a_1b_2 - b_1a_2)\mathbf{k}.$$

All the components on the right-hand side are determinants. We have:

$$(a_1\mathbf{i} + b_1\mathbf{j} + c_1\mathbf{k}) \times (a_2\mathbf{i} + b_2\mathbf{j} + c_2\mathbf{k})$$
$$= \begin{vmatrix} b_1 & c_1 \\ b_2 & c_2 \end{vmatrix}\mathbf{i} + \begin{vmatrix} c_1 & a_1 \\ c_2 & a_2 \end{vmatrix}\mathbf{j} + \begin{vmatrix} a_1 & b_1 \\ a_2 & b_2 \end{vmatrix}\mathbf{k}. \qquad (1)$$

The middle term can also be written

$$-\begin{vmatrix} a_1 & c_1 \\ a_2 & c_2 \end{vmatrix}\mathbf{j}.$$

Shortly we shall see how to write the cross product in terms of a single 3×3 determinant.

Sometimes we wish to refer to the array

$$\begin{matrix} a & b \\ c & d \end{matrix}$$

of numbers without taking the combination $ad - bc$. In this case, we use the notation

$$\begin{bmatrix} a & b \\ c & d \end{bmatrix}$$

and refer to this object as a 2×2 *matrix*. Two matrices are considered equal only when all their corresponding entries are equal; thus, in contrast to the equalities in Example 1(b), the matrices

$$\begin{bmatrix} 1 & 2 \\ 3 & 4 \end{bmatrix}, \quad \begin{bmatrix} 1 & 2 \\ 4 & 6 \end{bmatrix}, \quad \begin{bmatrix} 1 & 4 \\ 2 & 6 \end{bmatrix}$$

are all different. The determinant

$$\begin{vmatrix} a & b \\ c & d \end{vmatrix}$$

is a *single* number obtained by combining the *four* numbers in the matrix

$$\begin{bmatrix} a & b \\ c & d \end{bmatrix}.$$

Geometrically, we may think of a matrix as representing a parallelogram (that is, a geometric figure), while the determinant represents only the area of the parallelogram (that is, a number).

Example 2 The transpose of a matrix $\begin{bmatrix} a & b \\ c & d \end{bmatrix}$ is defined to be the matrix $\begin{bmatrix} a & c \\ b & d \end{bmatrix}$ obtained by reflection across the main (upper left to lower right) diagonal.

(a) Find the transpose of $\begin{bmatrix} 1 & 5 \\ -3 & 2 \end{bmatrix}$.

(b) Show that the determinant of a matrix is equal to the determinant of its transpose: $\begin{vmatrix} a & b \\ c & d \end{vmatrix} = \begin{vmatrix} a & c \\ b & d \end{vmatrix}$.

(c) Check (b) for the matrix in (a).

Solution (a) The transpose of $\begin{bmatrix} 1 & 5 \\ -3 & 2 \end{bmatrix}$ is $\begin{bmatrix} 1 & -3 \\ 5 & 2 \end{bmatrix}$.

(b) $$\begin{vmatrix} a & b \\ c & d \end{vmatrix} = ad - bc = \begin{vmatrix} a & c \\ b & d \end{vmatrix}.$$

(c) $$\begin{vmatrix} 1 & 5 \\ -3 & 2 \end{vmatrix} = 2 + 15 = 17;$$

$$\begin{vmatrix} 1 & -3 \\ 5 & 2 \end{vmatrix} = 2 + 15 = 17. \blacktriangle$$

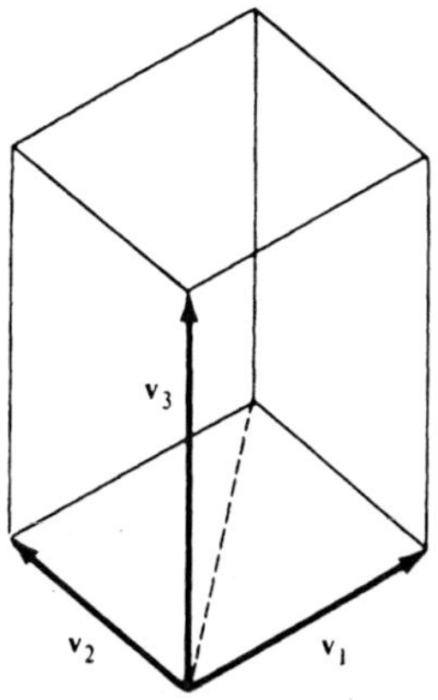

Figure 13.6.2. The absolute value of a determinant equals the volume of the parallelepiped spanned by the rows.

A 3×3 *matrix* consists of nine numbers in a square array, such as

$$\begin{bmatrix} 1 & 4 & 6 \\ 0 & 2 & 9 \\ 3 & 1 & -5 \end{bmatrix}.$$

To define the determinant $D = \begin{vmatrix} a_1 & b_1 & c_1 \\ a_2 & b_2 & c_2 \\ a_3 & b_3 & c_3 \end{vmatrix}$ of a 3×3 matrix $\begin{bmatrix} a_1 & b_1 & c_1 \\ a_2 & b_2 & c_2 \\ a_3 & b_3 & c_3 \end{bmatrix}$, we proceed by analogy with the 2×2 case. The rows of the matrix give us three vectors in space:

$$\mathbf{v}_1 = a_1\mathbf{i} + b_1\mathbf{j} + c_1\mathbf{k},$$
$$\mathbf{v}_2 = a_2\mathbf{i} + b_2\mathbf{j} + c_2\mathbf{k},$$
$$\mathbf{v}_3 = a_3\mathbf{i} + b_3\mathbf{j} + c_3\mathbf{k}.$$

Since the determinant of a 2×2 matrix represents an area, we should define the determinant in such a way that its absolute value is the *volume* of the parallelepiped spanned by $\mathbf{v}_1$, $\mathbf{v}_2$, and $\mathbf{v}_3$. (See Fig. 13.6.2.)

To compute this volume in terms of the nine entries in the matrix, we drop a perpendicular PQ from the tip of $\mathbf{v}_3$ to the plane spanned by $\mathbf{v}_1$ and $\mathbf{v}_2$. It is a theorem of elementary solid geometry that the volume of the parallelepiped is equal to the length of PQ times the area of the parallelogram spanned by $\mathbf{v}_1$ and $\mathbf{v}_2$ (see Fig. 13.6.3).

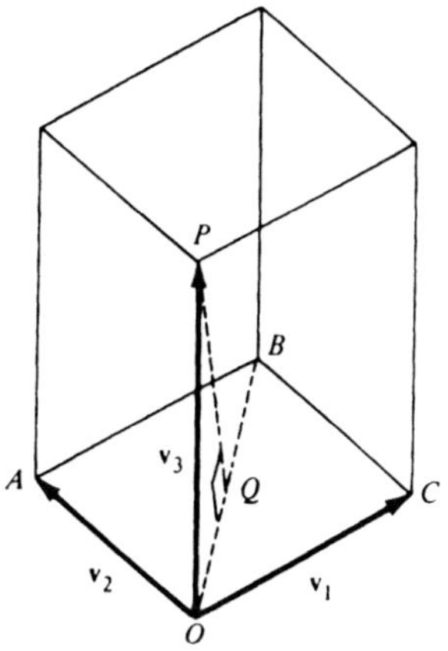

Figure 13.6.3. The volume of the parallelepiped is its height $|PQ|$ times the area of its base.

Looking at the right angle OQP, we find that the length of $\overrightarrow{PQ}$ is $\|\mathbf{v}_3\|\,|\cos\theta|$, where θ is the angle between $\mathbf{v}_3$ and $\overrightarrow{PQ}$. On the other hand, PQ is perpendicular to $\mathbf{v}_1$ and $\mathbf{v}_2$, so it is parallel to $\mathbf{v}_1 \times \mathbf{v}_2$, so θ is also the angle between $\mathbf{v}_3$ and $\mathbf{v}_1 \times \mathbf{v}_2$. Now we have:

$$\begin{aligned}\text{Volume} &= (\text{area of base})(\text{height})\\ &= \|\mathbf{v}_1 \times \mathbf{v}_2\|\,\|\mathbf{v}_3\|\,|\cos\theta|\\ &= \big|\,\|\mathbf{v}_1 \times \mathbf{v}_2\|\,\|\mathbf{v}_3\|\cos\theta\,\big|\\ &= |(\mathbf{v}_1 \times \mathbf{v}_2)\cdot\mathbf{v}_3|.\end{aligned}$$

Since the volume is to be the absolute value of the determinant, we define the determinant to be the expression inside the bars:

$$\begin{vmatrix} a_1 & b_1 & c_1\\ a_2 & b_2 & c_2\\ a_3 & b_3 & c_3\end{vmatrix} = \big[(a_1\mathbf{i} + b_1\mathbf{j} + c_1\mathbf{k}) \times (a_2\mathbf{i} + b_2\mathbf{j} + c_2\mathbf{k})\big]\cdot(a_3\mathbf{i} + b_3\mathbf{j} + c_3\mathbf{k}).$$

By using the component formula for the cross product in equation (1), we find

$$\begin{vmatrix} a_1 & b_1 & c_1\\ a_2 & b_2 & c_2\\ a_3 & b_3 & c_3\end{vmatrix} = \begin{vmatrix} b_1 & c_1\\ b_2 & c_2\end{vmatrix}a_3 + \begin{vmatrix} c_1 & a_1\\ c_2 & a_2\end{vmatrix}b_3 + \begin{vmatrix} a_1 & b_1\\ a_2 & b_2\end{vmatrix}c_3$$

or

$$\begin{vmatrix} a_1 & b_1 & c_1\\ a_2 & b_2 & c_2\\ a_3 & b_3 & c_3\end{vmatrix} = \begin{vmatrix} b_1 & c_1\\ b_2 & c_2\end{vmatrix}a_3 - \begin{vmatrix} a_1 & c_1\\ a_2 & c_2\end{vmatrix}b_3 + \begin{vmatrix} a_1 & b_1\\ a_2 & b_2\end{vmatrix}c_3. \tag{2}$$

We take equation (2) as the definition of the determinant of a 3×3 matrix; the volume of the parallelepiped spanned by three vectors is thus equal to the absolute value of the determinant of the matrix whose rows are the components of the vectors. The sign of the determinant is interpreted geometrically in Example 9.

Example 3 Evaluate the determinant $\begin{vmatrix} 0 & 0 & 4\\ 2 & -1 & 6\\ 3 & 1 & 2\end{vmatrix}$.

Solution By equation (2),

$$\begin{aligned}\begin{vmatrix} 0 & 0 & 4\\ 2 & -1 & 6\\ 3 & 1 & 2\end{vmatrix} &= \begin{vmatrix} 0 & 4\\ -1 & 6\end{vmatrix}(3) - \begin{vmatrix} 0 & 4\\ 2 & 6\end{vmatrix}(1) + \begin{vmatrix} 0 & 0\\ 2 & -1\end{vmatrix}(2)\\ &= (4)(3) - (-8)(1) + (0)(2)\\ &= 12 + 8 = 20. \blacktriangle\end{aligned}$$

We can express the cross product of two vectors as a single 3×3 determinant. In fact, comparing equation (2), with (1),

$$(a_1\mathbf{i} + b_1\mathbf{j} + c_1\mathbf{k}) \times (a_2\mathbf{i} + b_2\mathbf{j} + c_2\mathbf{k}) = \begin{vmatrix} a_1 & b_1 & c_1\\ a_2 & b_2 & c_2\\ \mathbf{i} & \mathbf{j} & \mathbf{k}\end{vmatrix}. \tag{3}$$

Example 4 Write $(\mathbf{i} + \mathbf{k}) \times (\mathbf{j} - 2\mathbf{k})$ as a determinant.

Solution

$$(\mathbf{i} + \mathbf{k}) \times (\mathbf{j} - 2\mathbf{k}) = \begin{vmatrix} 1 & 0 & 1\\ 0 & 1 & -2\\ \mathbf{i} & \mathbf{j} & \mathbf{k}\end{vmatrix}. \blacktriangle$$

Formula (2) is worth memorizing. To do so, notice that the ith entry in the third row is multiplied by the determinant obtained by crossing out the ith column and third row of the original matrix. Such a 2×2 determinant is called a *minor*, and formula (2) is called the *expansion by minors of the third row*.

It turns out that a determinant can be evaluated by expanding in minors of any row or column. (We shall verify this for the second column in Example 7.) To do the expansion, multiply each entry in a given row or column by the 2×2 determinant obtained by crossing out the row and column of the given entry. Signs are assigned to the products according to the checkerboard pattern:

$$\begin{vmatrix} + & - & + \\ - & + & - \\ + & - & + \end{vmatrix}.$$

(Remember the plus sign in the upper left-hand corner, and you can always reconstruct this pattern.) Thus, the cross product (3) can also be written

$$\begin{vmatrix} \mathbf{i} & \mathbf{j} & \mathbf{k} \\ a_1 & b_1 & c_1 \\ a_2 & b_2 & c_2 \end{vmatrix}$$

Example 5 (a) Evaluate the determinant

$$\begin{vmatrix} 0 & 0 & 4 \\ 2 & -1 & 6 \\ 3 & 1 & 2 \end{vmatrix}$$

of Example 3 by expanding in minors of the first row.
(b) Find $(2\mathbf{i} - \mathbf{j} + \mathbf{k}) \times (\mathbf{i} - \mathbf{j} - 3\mathbf{k})$ using a 3×3 determinant.

Solution (a)

$$\begin{vmatrix} 0 & 0 & 4 \\ 2 & -1 & 6 \\ 3 & 1 & 2 \end{vmatrix} = \begin{vmatrix} -1 & 6 \\ 1 & 2 \end{vmatrix}(0) - \begin{vmatrix} 2 & 6 \\ 3 & 2 \end{vmatrix}(0) + \begin{vmatrix} 2 & -1 \\ 3 & 1 \end{vmatrix}4$$
$$= 0 - 0 + (5)(4) = 20.$$

Since we do not need to evaluate the minors of the zero entries, the expansion by minors of the first row results in a simpler calculation (for this particular matrix) than the expansion by minors of the third row.

(b) $$(2\mathbf{i} - \mathbf{j} + \mathbf{k}) \times (\mathbf{i} - \mathbf{j} - 3\mathbf{k}) = \begin{vmatrix} \mathbf{i} & \mathbf{j} & \mathbf{k} \\ 2 & -1 & 1 \\ 1 & -1 & -3 \end{vmatrix}$$
$$= \begin{vmatrix} -1 & 1 \\ -1 & -3 \end{vmatrix}\mathbf{i} - \begin{vmatrix} 2 & 1 \\ 1 & -3 \end{vmatrix}\mathbf{j} + \begin{vmatrix} 2 & -1 \\ 1 & -1 \end{vmatrix}\mathbf{k}$$
$$= 4\mathbf{i} + 7\mathbf{j} - \mathbf{k}. \blacktriangle$$

Example 6 Find the volume of the parallelepiped spanned by the following vectors: $\mathbf{i} + 3\mathbf{k}$, $2\mathbf{i} + \mathbf{j} - 2\mathbf{k}$, and $5\mathbf{i} + 4\mathbf{k}$.

Solution The volume is the absolute value of

$$\begin{vmatrix} 1 & 0 & 3 \\ 2 & 1 & -2 \\ 5 & 0 & 4 \end{vmatrix}.$$

If we expand this by minors of the second column, the only nonzero term is

$$\begin{vmatrix} 1 & 3 \\ 5 & 4 \end{vmatrix}(1) = -11,$$

so the volume equals 11. ▲

To prove that the expansions by minors of rows or columns all give the same result, it is sufficient to compare these expansions for the "general" 3×3 determinant.

Example 7 Show that expanding the determinant

$$\begin{vmatrix} a_1 & b_1 & c_1 \\ a_2 & b_2 & c_2 \\ a_3 & b_3 & c_3 \end{vmatrix}$$

by minors of the second column gives the correct result.

Solution The expansion by minors of the second column is

$$-\begin{vmatrix} a_2 & c_2 \\ a_3 & c_3 \end{vmatrix} b_1 + \begin{vmatrix} a_1 & c_1 \\ a_3 & c_3 \end{vmatrix} b_2 - \begin{vmatrix} a_1 & c_1 \\ a_2 & c_2 \end{vmatrix} b_3 .$$

Expanding the 2×2 determinants gives

$$-(a_2c_3 - c_2a_3)b_1 + (a_1c_3 - c_1a_3)b_2 - (a_1c_2 - c_1a_2)b_3$$
$$= -b_1a_2c_3 + b_1c_2a_3 + a_1b_2c_3 - c_1b_2a_3 - a_1c_2b_3 + c_1a_2b_3 .$$

Collecting the terms in a_3, b_3, and c_3, we get

$$(b_1c_2 - c_1b_2)a_3 - (a_1c_2 - c_1a_2)b_3 + (a_1b_2 - b_1a_2)c_3$$
$$= \begin{vmatrix} b_1 & c_1 \\ b_2 & c_2 \end{vmatrix} a_3 - \begin{vmatrix} a_1 & c_1 \\ a_2 & c_2 \end{vmatrix} b_3 + \begin{vmatrix} a_1 & b_1 \\ a_2 & b_2 \end{vmatrix} c_3$$
$$= \begin{vmatrix} a_1 & b_1 & c_1 \\ a_2 & b_2 & c_2 \\ a_3 & b_3 & c_3 \end{vmatrix} . \ ▲$$

The proof that we get the same result by expanding in any row or column is similar.

If $\mathbf{v}_1 = a_1\mathbf{i} + b_1\mathbf{j} + c_1\mathbf{k}$, $\mathbf{v}_2 = a_2\mathbf{i} + b_2\mathbf{j} + c_2\mathbf{k}$, and $\mathbf{v}_3 = a_3\mathbf{i} + b_3\mathbf{j} + c_3\mathbf{k}$, equation (2) can be written

$$\begin{vmatrix} a_1 & b_1 & c_1 \\ a_2 & b_2 & c_2 \\ a_3 & b_3 & c_3 \end{vmatrix} = (\mathbf{v}_1 \times \mathbf{v}_2) \cdot \mathbf{v}_3 .$$

This is called the *triple product* of $\mathbf{v}_1$, $\mathbf{v}_2$, and $\mathbf{v}_3$.

Example 8 Show that $(\mathbf{v}_1 \times \mathbf{v}_2) \cdot \mathbf{v}_3 = (\mathbf{v}_3 \times \mathbf{v}_1) \cdot \mathbf{v}_2$. In fact, the numbers 1, 2, 3 can be moved cyclically without changing the value of the triple product:

$$\begin{matrix} & \curvearrowright 1 \searrow & \\ 3 & & 2 \\ & \curvearrowleft & \end{matrix}$$

Solution The triple product $(\mathbf{v}_3 \times \mathbf{v}_1) \cdot \mathbf{v}_2$ is equal to the determinant

$$\begin{vmatrix} a_3 & b_3 & c_3 \\ a_1 & b_1 & c_1 \\ a_2 & b_2 & c_2 \end{vmatrix} .$$

Expanding this by minors of the first row gives

$$\begin{vmatrix} b_1 & c_1 \\ b_2 & c_2 \end{vmatrix} a_3 - \begin{vmatrix} a_1 & c_1 \\ a_2 & c_2 \end{vmatrix} b_3 + \begin{vmatrix} a_1 & b_1 \\ a_2 & b_2 \end{vmatrix} c_3 .$$

Expanding the determinant for $(\mathbf{v}_1 \times \mathbf{v}_2) \cdot \mathbf{v}_3$ by the *third* row gives the same result. ▲

Example 9 Let $\mathbf{v}_1 = a_1\mathbf{i} + b_1\mathbf{j} + c_1\mathbf{k}$, $\mathbf{v}_2 = a_2\mathbf{i} + b_2\mathbf{j} + c_2\mathbf{k}$, and $\mathbf{v}_3 = a_3\mathbf{i} + b_3\mathbf{j} + c_3\mathbf{k}$. Show that the triple $(\mathbf{v}_1, \mathbf{v}_2, \mathbf{v}_3)$ is right-handed (left-handed) if the determinant

$$\begin{vmatrix} a_1 & b_1 & c_1 \\ a_2 & b_2 & c_2 \\ a_3 & b_3 & c_3 \end{vmatrix}$$

is positive (negative). What does it mean if the determinant is zero?

Solution Let $\mathscr{P}$ be the plane spanned by $\mathbf{v}_1$ and $\mathbf{v}_2$. (We assume that $\mathbf{v}_1$ and $\mathbf{v}_2$ are not parallel; otherwise $\mathbf{v}_1 \times \mathbf{v}_2 = \mathbf{0}$, and the determinant is zero.) Then $(\mathbf{v}_1, \mathbf{v}_2, \mathbf{v}_3)$ is right handed (left handed) if and only if $\mathbf{v}_3$ lies on the same (opposite) side of $\mathscr{P}$ as $\mathbf{v}_1 \times \mathbf{v}_2$; that is, if and only if $(\mathbf{v}_1 \times \mathbf{v}_2) \cdot \mathbf{v}_3$ is positive (negative); but $(\mathbf{v}_1 \times \mathbf{v}_2) \cdot \mathbf{v}_3$ is the determinant

$$\begin{vmatrix} a_1 & b_1 & c_1 \\ a_2 & b_2 & c_2 \\ a_3 & b_3 & c_3 \end{vmatrix} .$$

If the determinant is zero (but $\mathbf{v}_1 \times \mathbf{v}_2$ is not zero), then $\mathbf{v}_3$ must lie *in* the plane $\mathscr{P}$. In general, we may say that the determinant is zero when the vectors $\mathbf{v}_1$, $\mathbf{v}_2$, and $\mathbf{v}_3$ fail to span a solid parallelepiped, but instead lie in a plane, lie on a line, or are all zero. Such triples of vectors are said to be *linearly dependent* and are neither right-handed nor left-handed. ▲

3 × 3 Determinants

$$\begin{vmatrix} a_1 & b_1 & c_1 \\ a_2 & b_2 & c_2 \\ a_3 & b_3 & c_3 \end{vmatrix} = \begin{vmatrix} b_1 & c_1 \\ b_2 & c_2 \end{vmatrix} a_3 - \begin{vmatrix} a_1 & c_1 \\ a_2 & c_2 \end{vmatrix} b_3 + \begin{vmatrix} a_1 & b_1 \\ a_2 & b_2 \end{vmatrix} c_3$$

The determinant can be expanded by minors of any row or column. If

$$\mathbf{v}_1 = a_1\mathbf{i} + b_1\mathbf{j} + c_1\mathbf{k},$$
$$\mathbf{v}_2 = a_2\mathbf{i} + b_2\mathbf{j} + c_2\mathbf{k},$$
$$\mathbf{v}_3 = a_3\mathbf{i} + b_3\mathbf{j} + c_3\mathbf{k},$$

then the determinant equals $(\mathbf{v}_1 \times \mathbf{v}_2) \cdot \mathbf{v}_3$ which is also called the *triple product* of $\mathbf{v}_1$, $\mathbf{v}_2$, and $\mathbf{v}_3$.

The absolute value of the determinant equals the volume of the parallelepiped spanned by $\mathbf{v}_1$, $\mathbf{v}_2$, and $\mathbf{v}_3$.

The sign of the determinant tells whether the triple $(\mathbf{v}_1, \mathbf{v}_2, \mathbf{v}_3)$ is right- or left-handed.

Exercises for Section 13.6

Evaluate the determinants in Exercises 1–10.

1. $\begin{vmatrix} 1 & 1 \\ -1 & 1 \end{vmatrix}$
2. $\begin{vmatrix} 1 & 2 \\ -1 & 0 \end{vmatrix}$
3. $\begin{vmatrix} 6 & 5 \\ 12 & 10 \end{vmatrix}$
4. $\begin{vmatrix} 0 & 0 \\ 3 & 17 \end{vmatrix}$
5. $\begin{vmatrix} 1 & 2 \\ 3 & 4 \end{vmatrix}$
6. $\begin{vmatrix} 1 & 2 \\ 2 & 4 \end{vmatrix}$
7. $\begin{vmatrix} 4 & 3 \\ 1 & 7 \end{vmatrix}$
8. $\begin{vmatrix} 1-x & -1 \\ 1 & -1-x \end{vmatrix}$
9. $\begin{vmatrix} a & b \\ 0 & c \end{vmatrix}$
10. $\begin{vmatrix} a & b \\ -b & a \end{vmatrix}$

Prove the identities in Exercises 11–14.

11. $\begin{vmatrix} a & b \\ c & d \end{vmatrix} = -\begin{vmatrix} c & d \\ a & b \end{vmatrix}$ (the two rows are interchanged).
12. $\begin{vmatrix} a & b \\ c & d \end{vmatrix} = \begin{vmatrix} a+c & b+d \\ c & d \end{vmatrix}$ (the last row is added to the first).
13. $\begin{vmatrix} ra & rb \\ c & d \end{vmatrix} = r\begin{vmatrix} a & b \\ c & d \end{vmatrix}$ (the first row is multiplied by a constant).
14. $\begin{vmatrix} a-sb & b \\ c-sd & d \end{vmatrix} = \begin{vmatrix} a & b \\ c & d \end{vmatrix}$ (a constant times the second column is subtracted from the first).

Evaluate the determinants in Exercises 15–24.

15. $\begin{vmatrix} 1 & 0 & 1 \\ 0 & 1 & 0 \\ 1 & 0 & 1 \end{vmatrix}$
16. $\begin{vmatrix} 1 & 1 & 0 \\ 0 & 1 & 1 \\ 0 & 1 & 1 \end{vmatrix}$
17. $\begin{vmatrix} 2 & -1 & 0 \\ 4 & 3 & 2 \\ 3 & 0 & 1 \end{vmatrix}$
18. $\begin{vmatrix} 1 & 1 & 2 \\ 2 & -1 & 1 \\ 1 & 0 & 0 \end{vmatrix}$
19. $\begin{vmatrix} 1 & 2 & 3 \\ -1 & -1 & 2 \\ 0 & 1 & -1 \end{vmatrix}$
20. $\begin{vmatrix} -1 & 0 & 1 \\ 2 & 1 & 3 \\ 0 & 1 & 2 \end{vmatrix}$
21. $\begin{vmatrix} 2 & 1 & 0 \\ 0 & 2 & 1 \\ 1 & 0 & 2 \end{vmatrix}$
22. $\begin{vmatrix} -1 & 0 & 1 \\ 2 & 1 & 3 \\ 3 & 1 & 2 \end{vmatrix}$
23. $\begin{vmatrix} a & d & e \\ 0 & b & f \\ 0 & 0 & c \end{vmatrix}$
24. $\begin{vmatrix} a & 0 & 0 \\ e & b & d \\ 0 & 0 & c \end{vmatrix}$

Write the cross products in Exercises 25–30 as determinants and evaluate.

25. $(3\mathbf{i}-\mathbf{j})\times(\mathbf{j}+\mathbf{k})$.
26. $(\mathbf{i}+3\mathbf{j}-\mathbf{k})\times(\mathbf{i}-3\mathbf{j}+\mathbf{k})$.
27. $(\mathbf{i}+\mathbf{j})\times(\mathbf{j}+\mathbf{k})$.
28. $(2\mathbf{i}+4\mathbf{j}+6\mathbf{k})\times\mathbf{k}$.
29. $(\mathbf{i}-\mathbf{k})\times(\mathbf{i}+\mathbf{k})$.
30. $(\mathbf{j}+2\mathbf{k})\times(32\mathbf{j}+64\mathbf{k})$.
31. Find the volume of the parallelepiped spanned by the vectors $\mathbf{i}+\mathbf{j}+\mathbf{k}$, $\mathbf{i}-\mathbf{j}+\mathbf{k}$, and $3\mathbf{k}$.
32. Find the volume of the parallelepiped spanned by $\mathbf{i}-\mathbf{j}$, $\mathbf{j}-\mathbf{k}$, and $\mathbf{k}+\mathbf{i}$.
33. Find the volume of the parallelepiped with one vertex at $(1,1,2)$ and three adjacent vertices at $(2,0,2)$, $(3,1,3)$, and $(2,2,-3)$.
34. Find the other four vertices of the parallelepiped in Exercise 33 and use them to recompute the volume.
35. Check that expanding the determinant

$$\begin{vmatrix} a_1 & b_1 & c_1 \\ a_2 & b_2 & c_2 \\ a_3 & b_3 & c_3 \end{vmatrix}$$

by (a) minors of the second row and (b) minors of the third column gives the correct result.

36. Show that if the first two rows of a 3×3 matrix are interchanged, the determinant changes sign.
37. Show that if the first two columns of a matrix are interchanged, the determinant changes sign.
38. Use Exercise 36 to verify that $(\mathbf{v}_1\times\mathbf{v}_2)\cdot\mathbf{v}_3 = -(\mathbf{v}_2\times\mathbf{v}_1)\cdot\mathbf{v}_3$.
39. Verify that $(\mathbf{v}_1\times\mathbf{v}_2)\cdot\mathbf{v}_3 = (\mathbf{v}_2\times\mathbf{v}_3)\cdot\mathbf{v}_1$.
40. Verify that $(\mathbf{v}_1-\mathbf{v}_2)\times(\mathbf{v}_1+\mathbf{v}_2) = 2\mathbf{v}_1\times\mathbf{v}_2$.
41. Show by drawing appropriate diagrams that if $(\mathbf{v}_1,\mathbf{v}_2,\mathbf{v}_3)$ is a right-handed triple, then so is $(\mathbf{v}_2,\mathbf{v}_3,\mathbf{v}_1)$.
42. Show that if two rows or columns of a 3×3 matrix are equal, then the determinant of the matrix is zero.
43. Show that if

$$\begin{vmatrix} a & b \\ c & d \end{vmatrix} \neq 0,$$

then the solutions to the equations

$$ax+by=e,$$
$$cx+dy=f$$

are given by the formulas

$$x=\frac{\begin{vmatrix} e & b \\ f & d \end{vmatrix}}{\begin{vmatrix} a & b \\ c & d \end{vmatrix}},\qquad y=\frac{\begin{vmatrix} a & e \\ c & f \end{vmatrix}}{\begin{vmatrix} a & b \\ c & d \end{vmatrix}}$$

This result is called *Cramer's rule*[6] (We have already used it without the language of determinants in our discussion of Wronskians in the Supplement to Section 12.7.)

44. Use Cramer's rule (Exercise 43) to solve the equations $4x+3y=2$; $2x-6y=1$.

★45. Suppose that the determinant

$$D=\begin{vmatrix} a_1 & b_1 & c_1 \\ a_2 & b_2 & c_2 \\ a_3 & b_3 & c_3 \end{vmatrix}$$

[6] Gabriel Cramer (1704–1752) published this rule in his book, *Introduction à l'analyse des lignes courbes algébriques* (1750). However, it was probably known to Maclaurin in 1729. For systems of n equations in n unknowns, there is a generalization of this rule, but it can be inefficient to use on a computer when n is large. (See Exercises 45 for the case $n = 3$.)

is unequal to zero. Show that the solution of the equations

$$a_1x + b_1y + c_1z = d_1,$$
$$a_2x + b_2y + c_2z = d_2,$$
$$a_3x + b_3y + c_3z = d_3$$

is given by the formulas

$$x = \frac{\begin{vmatrix} d_1 & b_1 & c_1 \\ d_2 & b_2 & c_2 \\ d_3 & b_3 & c_3 \end{vmatrix}}{D}, \quad y = \frac{\begin{vmatrix} a_1 & d_1 & c_1 \\ a_2 & d_2 & c_2 \\ a_3 & d_3 & c_3 \end{vmatrix}}{D},$$

$$z = \frac{\begin{vmatrix} a_1 & b_1 & d_1 \\ a_2 & b_2 & d_2 \\ a_3 & b_3 & d_3 \end{vmatrix}}{D}.$$

This result is called Cramer's rule for 3×3 systems.

46. Use Cramer's rule (Exercise 45) to solve

$$\begin{aligned} -x+y \quad\; &= 14, \\ 2x+y+z &= 8, \\ x+y+5z &= -1. \end{aligned}$$

Use Cramer's rule to solve the systems in Exercises 47 and 48.

47. $2x + 3y = 5$; $3x - 2y = 9$.
48. $x + y + z = 3$; $x - y + z = 4$; $x + y - z = 5$.
49. Check that

$$\begin{vmatrix} a & d & g \\ b & e & h \\ c & f & i \end{vmatrix} = \begin{vmatrix} a & b & c \\ d & e & f \\ g & h & i \end{vmatrix}.$$

That is, check that the determinant of the transpose of a 3×3 matrix is equal to the determinant of the original matrix.

★50. Show that adding a multiple of the first row of a matrix to the second row leaves the determinant unchanged; that is,

$$\begin{vmatrix} a_1 & b_1 & c_1 \\ a_2 + \lambda a_1 & b_2 + \lambda b_1 & c_2 + \lambda c_1 \\ a_3 & b_3 & c_3 \end{vmatrix} = \begin{vmatrix} a_1 & b_1 & c_1 \\ a_2 & b_2 & c_2 \\ a_3 & b_3 & c_3 \end{vmatrix}.$$

[In fact, adding a multiple of any row (column) of a matrix to another row (column) leaves the determinant unchanged.]

★51. Justify the steps in the following computation:

$$\begin{vmatrix} 1 & 2 & 3 \\ 4 & 5 & 6 \\ 7 & 8 & 10 \end{vmatrix} = \begin{vmatrix} 1 & 2 & 3 \\ 0 & -3 & -6 \\ 7 & 8 & 10 \end{vmatrix}$$
$$= \begin{vmatrix} 1 & 2 & 3 \\ 0 & -3 & -6 \\ 0 & -6 & -11 \end{vmatrix}$$
$$= \begin{vmatrix} -3 & -6 \\ -6 & -11 \end{vmatrix} = 33 - 36 = -3.$$

★52. Follow the technique of Exercise 51 to evaluate the determinant

$$\begin{vmatrix} 1 & 2 & 5 \\ 2 & 3 & 6 \\ -1 & 2 & 1 \end{vmatrix}.$$

(Add -2 times the first row to the second row, then add the first row to the third row.)

★53. Use the technique in Exercise 51 to evaluate the 3×3 determinants in Exercises 18 and 19.

★54. Show that the plane which passes through the three points $A = (a_1, a_2, a_3)$, $B = (b_1, b_2, b_3)$, and $C = (c_1, c_2, c_3)$ consists of the points $P = (x, y, z)$ given by

$$\begin{vmatrix} a_1 - x & a_2 - y & a_3 - z \\ b_1 - x & b_2 - y & b_3 - z \\ c_1 - x & c_2 - y & c_3 - z \end{vmatrix} = 0.$$

[*Hint:* Write the determinant as a triple product.]

Review Exercises for Chapter 13

Complete the calculations in Exercises 1–16.

1. $(3,2) + (-1,6) =$
2. $2(1,-4) + (1,5) =$
3. $(1,2,3) + 2(-1,-2,7) =$
4. $2[(-1,0,1) + (6,0,2)] - (0,0,1) =$
5. $(3\mathbf{i} + 2\mathbf{j}) + (8\mathbf{i} - \mathbf{j} - \mathbf{k}) =$
6. $(8\mathbf{i} + 3\mathbf{j} - \mathbf{k}) - 6(\mathbf{i} - \mathbf{j} - \mathbf{k}) =$
7. $(8\mathbf{i} + 3\mathbf{j} - \mathbf{k}) \times (\mathbf{i} - \mathbf{j} - \mathbf{k}) =$
8. $(\mathbf{i} - \mathbf{j} - \mathbf{k}) \cdot (\mathbf{i} + \mathbf{j} + 2\mathbf{k}) =$
9. $(8\mathbf{i} + 3\mathbf{j} - \mathbf{k}) \cdot (\mathbf{i} - \mathbf{j} - \mathbf{k}) =$
10. $(\mathbf{i} + \mathbf{j}) \cdot (\mathbf{i} - \mathbf{j}) =$
11. $(\mathbf{i} + \mathbf{j}) \times (\mathbf{i} - \mathbf{j}) =$
12. $[(2\mathbf{i} - \mathbf{j}) \times (3\mathbf{i} + \mathbf{j})] \cdot (2\mathbf{j} + \mathbf{k}) =$
13. $\mathbf{u} \times \mathbf{v} = ?$, where $\mathbf{u} = 2\mathbf{i} + \mathbf{j}$ and $\mathbf{v} = \mathbf{k}$.
14. $\mathbf{u} \cdot \mathbf{v} = ?$, where $\mathbf{u} = 3\mathbf{j} - \mathbf{k}$ and $\mathbf{v} = 2\mathbf{j} + \mathbf{i}$.
15. $\mathbf{u} - 3\mathbf{v} = ?$, where $\mathbf{u}$ and $\mathbf{v}$ are as in Exercise 13.
16. $\mathbf{u} + 6\mathbf{v} = ?$, where $\mathbf{u}$ and $\mathbf{v}$ are as in Exercise 14.
17. Find a unit vector orthogonal to $3\mathbf{i} + 2\mathbf{k}$ and $\mathbf{j} - \mathbf{k}$.
18. Find a unit vector orthogonal to $\mathbf{j} + \mathbf{k}$ and $2\mathbf{i} + \mathbf{k}$.
19. Find the volume of the parallelepiped spanned by $2\mathbf{i} - \mathbf{j} + \mathbf{k}$, $\mathbf{i}$, and $\mathbf{j} - \mathbf{k}$.
20. Find the volume of the parallelepiped spanned by $\mathbf{i} + \mathbf{j}$, $\mathbf{i} - \mathbf{j}$, and $\mathbf{i} + \mathbf{k}$.
21. (a) Draw the vector $\mathbf{v}$ joining $(-2,0)$ to $(4,6)$ and find the components of $\mathbf{v}$. (b) Add $\mathbf{v}$ to the vector joining $(-2,0)$ to $(1,1)$.
22. Find the intersection of the medians of the triangle with vertices at $(0,0)$, $(1,\frac{1}{2})$, and $(2,0)$.
23. Let PQR be a triangle in the plane. For each side of the triangle, construct the vector perpendicular to that side, pointing into the triangle, and having the same length as the side. Prove that the sum of the three vectors is zero.

24. Show that the diagonals of a rhombus are perpendicular to each other.
25. A bird is headed northeast with speed 40 kilometers per hour. A wind from the north at 15 kilometers per hour begins to blow, but the bird continues to head northeast and flies at the same rate relative to the air. Find the speed of the bird relative to the earth's surface.
26. An airplane flying in a straight line at 500 miles per hour for 12 minutes moves 35 miles north and 93.65 miles east. How much does its altitude change? Can you determine whether the airplane is climbing or descending? (Ignore the curvature of the earth.)
27. The work W done in moving an object from $(0,0)$ to $(7,2)$ subject to a force $\mathbf{F}$ is $W = \mathbf{F}\cdot\mathbf{r}$ where $\mathbf{r}$ is the vector with head at $(7,2)$ and tail at $(0,0)$. The units are feet and pounds.
    (a) Suppose the force $\mathbf{F} = 10\cos\theta\mathbf{i} + 10\sin\theta\mathbf{j}$. Find W in terms of θ.
    (b) Suppose the force $\mathbf{F}$ has magnitude 6 lbs and makes an angle of $\pi/6$ radian with the horizontal, pointing right. Find W in feet-lbs.
28. If a particle with mass m moves with velocity $\mathbf{v}$, its *momentum* is $\mathbf{p} = m\mathbf{v}$. In a game of marbles, a marble with mass 2 grams is shot with velocity 2 meters per second, hits two marbles with mass 1 gram each, and comes to a dead halt. One of the marbles flies off with a velocity of 3 meters per second at an angle of 45° to the incident direction of the larger marble as in Fig. 13.R.1. Assuming that the total momentum before and after the collision is the same (law of conservation of momentum), at what angle and speed does the second marble move?

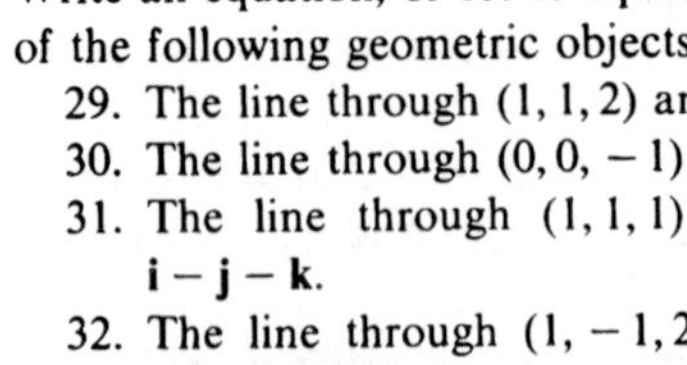

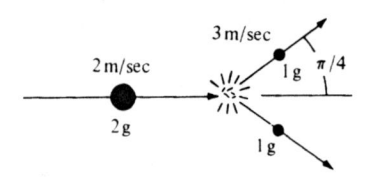

Figure 13.R.1. Momentum and marbles.

Write an equation, or set of equations, to describe each of the following geometric objects in Exercises 29–40.

29. The line through $(1,1,2)$ and $(2,2,3)$.
30. The line through $(0,0,-1)$ and $(1,1,3)$.
31. The line through $(1,1,1)$ in the direction of $\mathbf{i}-\mathbf{j}-\mathbf{k}$.
32. The line through $(1,-1,2)$ in the direction of $\mathbf{i}+\mathbf{j}+3\mathbf{k}$.
33. The plane through $(1,1,2)$, $(2,2,3)$, and $(0,0,0)$.
34. The plane through the points $(1,2,3)$, $(1,-1,1)$, and $(-1,1,1)$.
35. The plane through $(1,1,-1)$ and orthogonal to $\mathbf{i}-\mathbf{j}-\mathbf{k}$.
36. The plane through $(1,-1,6)$ and orthogonal to $\mathbf{i}+\mathbf{j}+\mathbf{k}$.
37. The line perpendicular to the plane in Exercise 33 and passing through $(0,0,3)$.
38. The line perpendicular to the plane in Exercise 34 and passing through $(1,1,1)$.
39. The line perpendicular to the plane in Exercise 35 and passing through $(2,3,1)$.
40. The line perpendicular to the plane in Exercise 36 and passing through the origin.

In Exercises 41–46, find a unit vector which has the given property.

41. Orthogonal to the plane $x-6y+z=12$.
42. Parallel to the line $x=3t+1$, $y=16t-2$, $z=-(t+2)$.
43. Orthogonal to $\mathbf{i}+2\mathbf{j}-\mathbf{k}$ and to $\mathbf{k}$.
44. Parallel to both the planes $8x+y+z=1$ and $x-y-z=0$.
45. At an angle of 30° to $\mathbf{i}$ and makes equal angles with $\mathbf{j}$ and $\mathbf{k}$.
46. Orthogonal to the line $x=2t-1$, $y=-t-1$, $z=t+2$, and the vector $\mathbf{i}-\mathbf{j}$.

47. Suppose that $\mathbf{v}$ and $\mathbf{w}$ each are parallel to the (x,y) plane. What can you say about $\mathbf{v}\times\mathbf{w}$?
48. Suppose that $\mathbf{u}$, $\mathbf{v}$ and $\mathbf{w}$ are three vectors. Explain how to find the angle between $\mathbf{w}$ and the plane determined by $\mathbf{u}$ and $\mathbf{v}$.
49. Describe the set of all lines through the origin in space which make an angle of $\pi/3$ with the x axis.
50. Consider the set of all points P in space such that the vector from O to P has length 2 and makes an angle of 45° with $\mathbf{i}+\mathbf{j}$.
    (a) What kind of geometric object is this set?
    (b) Describe this set using equation(s) in x, y, and z.
51. Let a triangle have adjacent sides $\mathbf{a}$ and $\mathbf{b}$.
    (a) Show that $\mathbf{c}=\mathbf{b}-\mathbf{a}$ is the third side.
    (b) Show that $\mathbf{c}\times\mathbf{a}=\mathbf{c}\times\mathbf{b}$.
    (c) Derive the law of sines (see p. 263).
52. Find the equation of the plane through $(1,2,-1)$ which is parallel to both $\mathbf{i}-\mathbf{j}+2\mathbf{k}$ and $\mathbf{i}-3\mathbf{k}$.
53. Thales' theorem states that the angle θ in Fig. 13.R.2(a) is $\pi/2$. Prove this using the vectors $\mathbf{a}$ and $\mathbf{b}$ shown in Fig. 13.R.2(b).

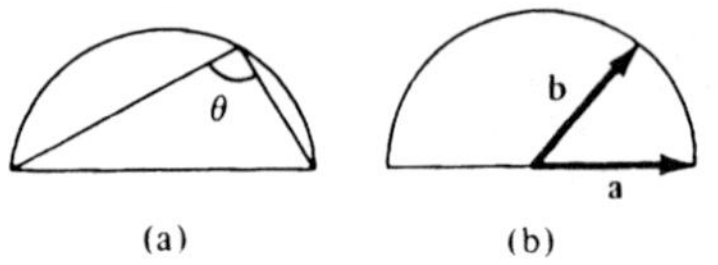

Figure 13.R.2. Triangle inscribed in a semicircle.

54. Show that the midpoint of the hypotenuse of a right triangle is equidistant from all three vertices.

Evaluate the determinants in Exercises 55–62.

55. $\begin{vmatrix} 1 & 2 \\ -1 & 1 \end{vmatrix}$

56. $\begin{vmatrix} 2 & -1 \\ 0 & 1 \end{vmatrix}$

57. $\begin{vmatrix} -1 & -1 \\ 2 & 1 \end{vmatrix}$

58. $\begin{vmatrix} 0 & 1 \\ -1 & 0 \end{vmatrix}$

59. $\begin{vmatrix} 1 & -1 & 1 \\ 2 & 1 & 1 \\ 1 & 3 & -1 \end{vmatrix}$

60. $\begin{vmatrix} 1 & 0 & 0 \\ 0 & -1 & 1 \\ 0 & 1 & 1 \end{vmatrix}$

61. $\begin{vmatrix} 1 & 1 & 1 \\ 2 & 2 & 2 \\ 3 & 3 & 3 \end{vmatrix}$

62. $\begin{vmatrix} -1 & 0 & 1 \\ 1 & 0 & 0 \\ -1 & 0 & -1 \end{vmatrix}$

63. Find the area of the parallelogram spanned by $3\mathbf{i} - 2\mathbf{j} + \mathbf{k}$ and $8\mathbf{i} - \mathbf{k}$.

64. Find the area of the parallelogram spanned by $2\mathbf{i} - \mathbf{j}$ and $3\mathbf{i} - 2\mathbf{j}$.

65. Find the volume of the parallelepiped spanned by $\mathbf{i} - \mathbf{j} - \mathbf{k}$, $2\mathbf{i} + \mathbf{j} - 5\mathbf{k}$, and $\frac{8}{3}\mathbf{i} - \mathbf{j} + \frac{1}{2}\mathbf{k}$.

66. Find the volume of the parallelepiped spanned by $2\mathbf{j} + \mathbf{i}$, $\mathbf{i} - \mathbf{j}$, and $\mathbf{k}$.

67. The volume of a *tetrahedron* with concurrent edges $\mathbf{a}$, $\mathbf{b}$, $\mathbf{c}$ is given by $V = (1/6)\mathbf{a} \cdot (\mathbf{b} \times \mathbf{c})$.
   (a) Express the volume as a determinant.
   (b) Evaluate V when $\mathbf{a} = \mathbf{i} + \mathbf{j} + \mathbf{k}$, $\mathbf{b} = \mathbf{i} - \mathbf{j} + \mathbf{k}$, $\mathbf{c} = \mathbf{i} + \mathbf{j}$.

68. A tetrahedron sits in xyz coordinates with one vertex at $(0,0,0)$, and the three edges concurrent at $(0,0,0)$ are coincident with the vectors $\mathbf{a}$, $\mathbf{b}$, $\mathbf{c}$.
   (a) Draw a figure and label the heads of the vectors as $\mathbf{a}$, $\mathbf{b}$, $\mathbf{c}$.
   (b) Find the center of mass of each of the four triangular faces of the tetrahedron.

69. Let $\mathbf{r}_1, \ldots, \mathbf{r}_n$ be vectors from 0 to the masses $m_1, \ldots, m_n$. The *center of mass* is the vector

$$\mathbf{c} = \left(\sum_{i=1}^{n} m_i\mathbf{r}_i\right) \Big/ \left(\sum_{i=1}^{n} m_i\right)$$

Show that for any vector $\mathbf{r}$,

$$\sum_{i=1}^{n} m_i\|\mathbf{r} - \mathbf{r}_i\|^2 = \sum_{i=1}^{n} m_i\|\mathbf{r}_i - \mathbf{c}\|^2 + m\|\mathbf{r} - \mathbf{c}\|^2,$$

where $m = \sum_{i=1}^{n} m_i$ is the total mass of the system.

70. Solve the following equations using Cramer's rule (Exercise 43, Section 13.6): $x + y = 2$; $3x - y = 4$.

71. Solve by using determinants (Exercise 45, Section 13.6): $x - y + 2z = 4$; $3x + y + z = 1$; $4x - y - z = 2$.

72. Use Exercise 50, Section 13.6 to show that

$$\begin{vmatrix} 66 & 628 & 246 \\ 88 & 435 & 24 \\ 2 & -1 & 1 \end{vmatrix} = \begin{vmatrix} 68 & 627 & 247 \\ 86 & 436 & 23 \\ 2 & -1 & 1 \end{vmatrix}.$$

73. Evaluate

$$\begin{vmatrix} 6 & 2 & -3 \\ 2 & 2 & 3 \\ 4 & 8 & -1 \end{vmatrix}$$

using Exercise 50, Section 13.6.

74. Use Exercise 50, Section 13.6 to show that

$$\begin{vmatrix} n & n+1 & n+2 \\ n+3 & n+4 & n+5 \\ n+6 & n+7 & n+8 \end{vmatrix}$$

has the same value no matter what n is. What is this value?

75. Show that for all x, y, z,

$$\begin{vmatrix} x+2 & y & z \\ z & y+1 & 10 \\ 5 & 5 & 2 \end{vmatrix} = -\begin{vmatrix} y & x+2 & z \\ 1 & z-x-2 & 10-z \\ 5 & 5 & 2 \end{vmatrix}$$

76. Show that

$$\begin{vmatrix} 1 & x & x^2 \\ 1 & y & y^2 \\ 1 & z & z^2 \end{vmatrix} \neq 0$$

if x, y, and z are all different.

77. If the triple product $(\mathbf{v} \times \mathbf{j}) \cdot \mathbf{k}$ is zero, what can you say about the vector $\mathbf{v}$?

78. Suppose that the three vectors $a_i\mathbf{i} + b_i\mathbf{j} + c_i\mathbf{k}$ for $i = 1$, 2, and 3 are unit vectors, each orthogonal to the other two. Find the value of

$$\begin{vmatrix} a_1 & b_1 & c_1 \\ a_2 & b_2 & c_2 \\ a_3 & b_3 & c_3 \end{vmatrix}.$$

79. A sphere of radius 10 centimeters with center at $(0,0,0)$ rotates about the z axis with angular velocity 4 radians per second such that the rotation looks counterclockwise from the positive z axis.
   (a) Find the rotation vector $\boldsymbol{\Omega}$ (see Section 13.5, Exercise 36).
   (b) Find the velocity $\mathbf{v} = \boldsymbol{\Omega} \times \mathbf{r}$ when $\mathbf{r} = 5\sqrt{2}(\mathbf{i} - \mathbf{j})$ is on the "equator."

80. A pair of dipoles are located at a distance r from each other. The magnetic potential energy P is given by $P = -\mathbf{m}_1 \cdot \mathbf{B}_2$ (*dipole-dipole interaction potential*), where the first dipole has moment $\mathbf{m}_1$ in the external field $\mathbf{B}_2$ of the second dipole. In MKS units,

$$\mathbf{B}_2 = \mu_0 \frac{-\mathbf{m}_2 + 3(\mathbf{m}_2 \cdot \mathbf{1}_r)\mathbf{1}_r}{4\pi r^3},$$

where $\mathbf{1}_r$ is a unit vector, and μ_0 is a scalar constant.
   (a) Show that

$$P = \mu_0 \frac{\mathbf{m}_1 \cdot \mathbf{m}_2 - 3(\mathbf{m}_2 \cdot \mathbf{1}_r)(\mathbf{m}_1 \cdot \mathbf{1}_r)}{4\pi r^3}.$$

   (b) Find P when $\mathbf{m}_1$ and $\mathbf{m}_2$ are perpendicular.

81. (a) Suppose that $\mathbf{v} \cdot \mathbf{w} = 0$ for all vectors $\mathbf{w}$. Show that $\mathbf{v} = \mathbf{0}$. [*Note:* This is not the same thing as showing that $\mathbf{0} \cdot \mathbf{w} = 0$.]
(b) Suppose that $\mathbf{u} \cdot \mathbf{w} = \mathbf{v} \cdot \mathbf{w}$ for all vectors $\mathbf{w}$. Show that $\mathbf{u} = \mathbf{v}$.
(c) Suppose that $\mathbf{v} \cdot \mathbf{i} = \mathbf{v} \cdot \mathbf{j} = \mathbf{v} \cdot \mathbf{k} = 0$. Show that $\mathbf{v} = \mathbf{0}$.
(d) Suppose that $\mathbf{u} \cdot \mathbf{i} = \mathbf{v} \cdot \mathbf{i}$, $\mathbf{u} \cdot \mathbf{j} = \mathbf{v} \cdot \mathbf{j}$, and $\mathbf{u} \cdot \mathbf{k} = \mathbf{v} \cdot \mathbf{k}$. Show that $\mathbf{u} = \mathbf{v}$.

82. Let A, B, C, D be four points in space. Consider the tetrahedron bounded by the four triangles $\triangle_1 = BCD$, $\triangle_2 = ACD$, $\triangle_3 = ABD$, and $\triangle_4 = ABC$. The triangle $\triangle_i$ is called the *ith face of the tetrahedron*. For each i, there is a unique vector $\mathbf{v}_i$ defined as follows: $\mathbf{v}_i$ is perpendicular to the face $\triangle_i$ and points *into* the tetrahedron; the length of $\mathbf{v}_i$ is equal to the area of $\triangle_i$.
(a) Prove that for any tetrahedron $ABCD$, the sum $\mathbf{v}_1 + \mathbf{v}_2 + \mathbf{v}_2 + \mathbf{v}_4$ is zero. [*Hint:* Use algebraic properties of the cross-product.]
★(b) Try to generalize the result of part (a) to more complicated polyhedra. (A polyhedron is a solid which is bounded by planar figures.) In other words, show that the sum of the inward normals, with lengths equal to areas of the sides, is zero. You may want to do some numerical calculations if you cannot prove anything, or you may want to restrict yourself to a special class of figures (distorted cubes, decapitated tetrahedra, figures with five vertices, and so forth).
★(c) There is a physical interpretation to the results in parts (a) and (b). If the polyhedron is immersed in a fluid under *constant* pressure p, then the force acting on the ith face is $p\mathbf{v}_i$. Interpret the result $\sum \mathbf{v}_i = 0$ in this context. Does this contradict the fact that water pressure tends to buoy up an immersed object? [*Note:* There is a version of all this material for smooth surfaces. It is related to a result called the *divergence theorem* and involves partial differentiation and surface integrals. See Chapter 18.]

83. Let $P = (1, 2)$ and $Q = (2, 1)$. Sketch the set of points in the plane of the form $rP + sQ$, where r and s are:
(a) positive integers;
(b) integers;
(c) positive real numbers;
(d) real numbers.

84. Repeat Exercise 83 with $P = (1, 2)$ and $Q = (2, 4)$.

★85. Repeat Exercise 83 using the points $P = (1, 2)$ and $Q = (\pi, 2\pi)$. (You may have to guess parts (a) and (b).)

★86. There are two unit vectors such that if they were drawn on the axes of Fig. 13.R.3, their heads and tails would appear to be at the same point (that is, they would be viewed head on). Approximately what are these vectors? [*Hint:* Suppose that when you tried to plot a point P, the resulting dot on the paper fell right where the axes cross.]

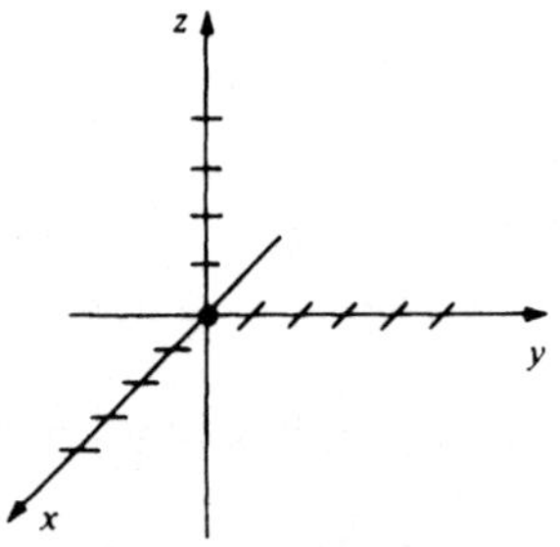

Figure 13.R.3. Which unit vectors would be drawn as the dot at the orign?

★87. A *regular tetrahedron* is a solid bounded by four equilateral triangles. Use vector methods to find the angle between the planes containing two of the faces.

★88. In integrating by partial fractions, we are led to the problem of expressing a rational function

$$\frac{c_2x^2 + c_1x + c_0}{(x - r_1)(x - r_2)(x - r_3)}$$

as a sum

$$\frac{a_2}{x - r_1} + \frac{a_2}{x - r_2} + \frac{a_3}{x - r_3}.$$

(a) Show that the undetermined coefficients a_1, a_2, a_3 satisfy a system of three simultaneous linear equations.
(b) Applying Cramer's rule (Exercise 45, Section 13.6) to this system, show that the determinant D is equal to

$$\begin{vmatrix} 1 & 1 & 1 \\ r_2 + r_3 & r_3 + r_1 & r_1 + r_2 \\ r_2r_3 & r_3r_1 & r_1r_2 \end{vmatrix}.$$

(c) Evaluate the determinant in (b) and show that it is nonzero whenever r_1, r_2, and r_3 are all different.
(d) Conclude that the decomposition into partial fractions is always possible if r_1, r_2, and r_3 are all different.
(e) What happens if $r_1 = r_2$? Give an example.

★89. Read Chapter 5 of Friedrichs' book, *From Pythagoras to Einstein* (Mathematical Association of America, New Mathematical Library, **16** (1965)) on the application of vectors to the study of elastic impacts, and prepare a two-page written report on your findings.

Chapter 14

Curves and Surfaces

Some three-dimensional geometry is needed for understanding functions of two variables.

The main subject of this chapter is surfaces in three-dimensional space. In preparation for this, we begin with a study of some special curves in the plane —the conic sections. In the last two sections, we will do some calculus with curves in space. Applications of calculus to surfaces are given in Chapters 15 and 16.

14.1 The Conic Sections

All the curves described by quadratic equations in two variables can be obtained by cutting a cone with planes.

The ellipse, hyperbola, parabola, and circle are called *conic sections* because they can all be obtained by slicing a cone with a plane (see Fig. 14.1.1). The

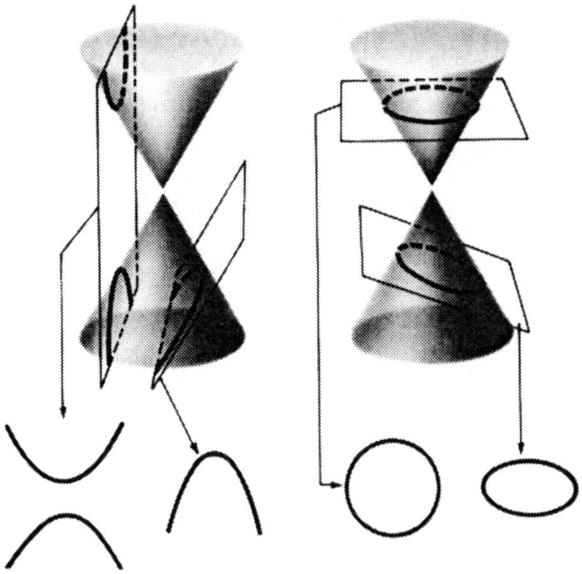

Figure 14.1.1. Conic sections are obtained by slicing a cone with a plane; which conic section is obtained depends on the direction of the slicing plane.

theory of these curves, developed by Apollonius of Perga (262–200 B.C.), is a masterwork of Greek geometry. We will return to the three-dimensional origin of the conics in Section 14.4, after we have studied some analytic geometry in space. For now, we will treat these curves, beginning with the ellipse, purely as objects in the plane.

Definition of Ellipse

An *ellipse* is the set of points in the plane for which the sum of the distances from two fixed points is constant. These two points are called the *foci* (plural of *focus*).

An ellipse can be drawn with the aid of a string tacked at the foci, as shown in Fig. 14.1.2.

To find an equation for the ellipse, we locate the foci on the x axis at the points $F' = (-c, 0)$ and $F = (c, 0)$. Let $2a > 0$ be the sum of the distances from a point on the ellipse to the foci. Since the distance between the foci is $2c$, and the length of a side of a triangle is less than the sum of the lengths of the other sides, we must have $2c < 2a$; i.e., $c < a$. Referring to Fig. 14.1.3, we

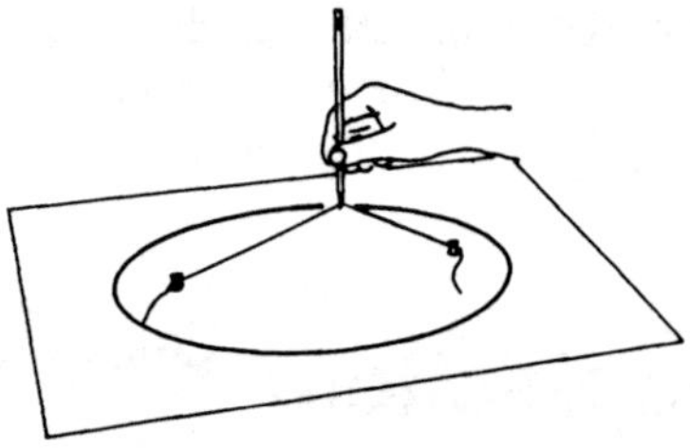

Figure 14.1.2. Mechanical construction of an ellipse.

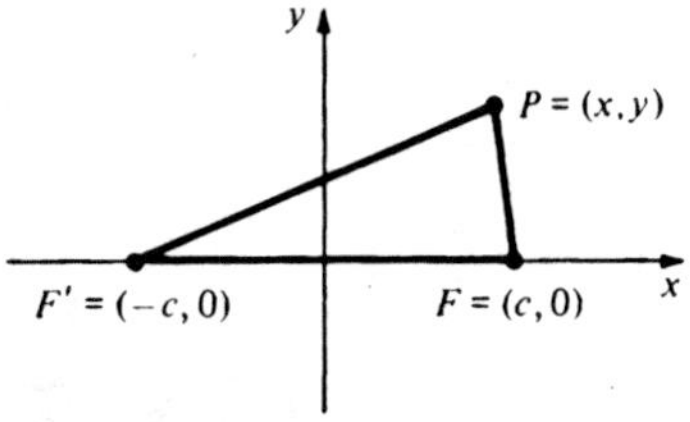

Figure 14.1.3. P is on the ellipse when $|FP| + |F'P| = 2a$.

see that a point $P = (x, y)$ is on the ellipse precisely when

$$|FP| + |F'P| = 2a.$$

That is,

$$\sqrt{(x+c)^2 + y^2} + \sqrt{(x-c)^2 + y^2} = 2a.$$

Transposing $\sqrt{(x-c)^2 + y^2}$, squaring, simplifying, and squaring again yields

$$(a^2 - c^2)x^2 + a^2y^2 = a^2(a^2 - c^2).$$

Let $a^2 - c^2 = b^2$ (remember that $a > c > 0$ and so $a^2 - c^2 > 0$). Then, after division by a^2b^2, the equation becomes

$$\frac{x^2}{a^2} + \frac{y^2}{b^2} = 1.$$

This is the *equation of an ellipse in standard form.*

Since $b^2 = a^2 - c^2 < a^2$, we have $b < a$. If we had put the foci on the y axis, we would have obtained an equation of the same form with $b > a$; the length of the "string" would now be $2b$ rather than $2a$. (See Fig. 14.1.4.) In either case, the length of the long axis of the ellipse is called the *major axis*, and the length of the short axis is the *minor axis*.

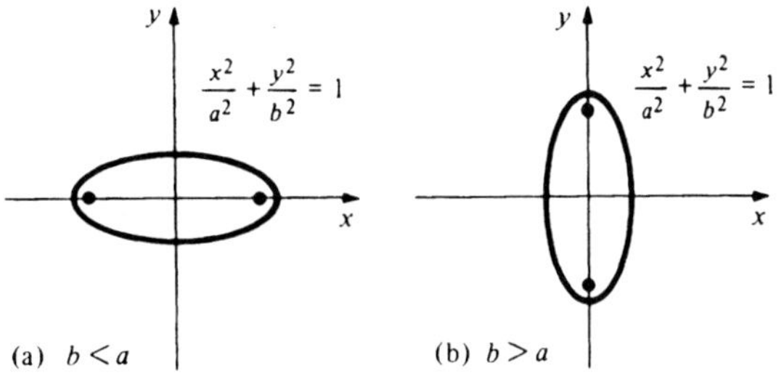

Figure 14.1.4. The appearance of an ellipse in the two cases $b < a$ and $b > a$.

Example 1 Sketch the graph of $4x^2 + 9y^2 = 36$. Where are the foci? What are the major and minor axes?

Solution Dividing both sides of the equation by 36, we obtain the standard form

$$\frac{x^2}{9} + \frac{y^2}{4} = 1.$$

Hence $a = 3$, $b = 2$, and $c = \sqrt{a^2 - b^2} = \sqrt{5}$. The foci are $(\pm\sqrt{5}, 0)$, the y intercepts are $(0, \pm 2)$, and the x intercepts are $(\pm 3, 0)$. The major axis is 6 and the minor axis is 4. The graph is shown in Fig. 14.1.5.

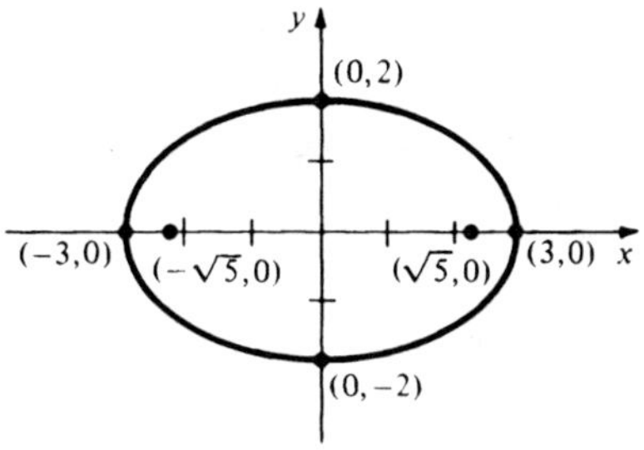

Figure 14.1.5. The graph of $4x^2 + 9y^2 = 36$.

Example 2 Sketch the graph of $9x^2 + y^2 = 81$. Where are the foci?

Solution Dividing by 81, we obtain the standard form $x^2/3^2 + y^2/9^2 = 1$. The graph is sketched in Fig. 14.1.6. The foci are at $(0, \pm 6\sqrt{2})$. ▲

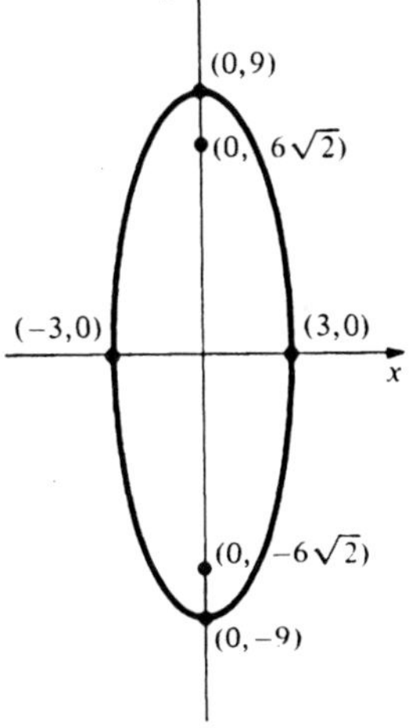

Figure 14.1.6. The ellipse $9x^2 + y^2 = 81$.

Ellipse

Equation: $\frac{x^2}{a^2} + \frac{y^2}{b^2} = 1$ (standard form).

Foci: $(\pm c, 0)$ where $c = \sqrt{a^2 - b^2}$ if $a > b$.

$(0, \pm c)$ where $c = \sqrt{b^2 - a^2}$ if $a < b$.

If $a = b$, the ellipse is a circle.

x intercepts: $(a, 0)$ and $(-a, 0)$.

y intercepts: $(0, b)$ and $(0, -b)$.

If P is any point on the ellipse, the sum of its distances from the foci is $2a$ if $b < a$ or $2b$ if $b > a$.

The second type of conic section, to which we now turn, is the hyperbola.

Definition of Hyperbola

A hyperbola is the set of points in the plane for which the *difference* of the distances from two fixed points is constant. These two points are called the *foci*.

To draw a hyperbola requires a mechanical device more elaborate than the one for the ellipse (see Fig. 14.1.7); however, we can obtain the equation in the same way as we did for the ellipse. Again let the foci be placed at $F' = (-c, 0)$

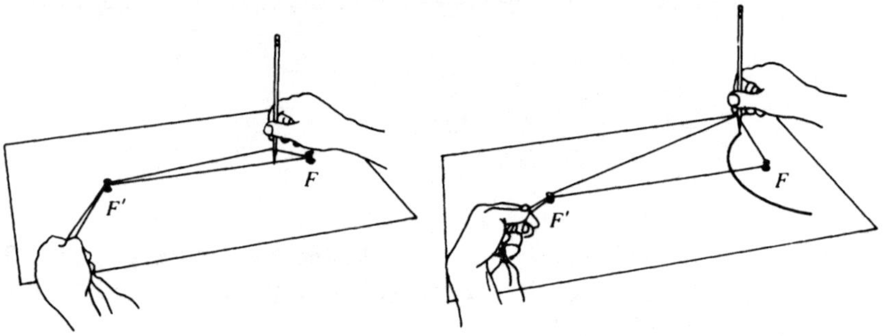

Figure 14.1.7. Mechanical construction of a hyperbola.

and $F = (c, 0)$, and let the difference in question be $2a$, $a > 0$. Since the difference of the distances from the two foci is $2a$ and we must have $|F'P| < |FP| + |F'F|$, it follows that $|F'P| - |FP| < |F'F|$, and so $2a < 2c$. Thus we must have $a < c$ (see Fig. 14.1.8). The point $P = (x, y)$ lies on the

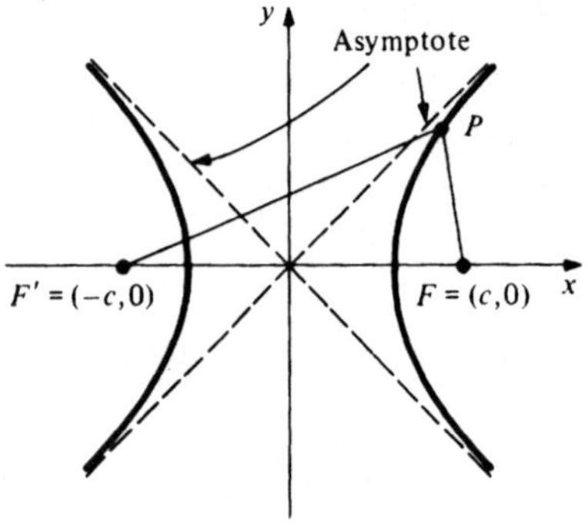

Figure 14.1.8. P is on the hyperbola when $|F'P| - |FP| = \pm 2a$.

hyperbola exactly when

$$\sqrt{(x+c)^2 + y^2} - \sqrt{(x-c)^2 + y^2} = \pm 2a.$$

After some calculations (squaring, simplifying and squaring again), we get

$$(a^2 - c^2)x^2 + a^2y^2 = a^2(a^2 - c^2).$$

If we let $c^2 - a^2 = b^2$ (since $a < c$), we get

$$\frac{x^2}{a^2} - \frac{y^2}{b^2} = 1$$

which is the equation of a hyperbola in standard form.

For x large in magnitude, the hyperbola approaches the two lines $y = \pm(b/a)x$, which are called the *asymptotes* of the hyperbola. To see this, for x and y positive, we first solve for y in the equation of the hyperbola, obtaining $y = (b/a)\sqrt{x^2 - a^2}$. Subtracting this from the linear function $(b/a)x$, we find that the vertical distance from the hyperbola to the line $y = (b/a)x$ is given by

$$d = \frac{b}{a}\left(x - \sqrt{x^2 - a^2}\right).$$

To study the behavior of this expression as x becomes large, we multiply by $(x + \sqrt{x^2 - a^2})/(x + \sqrt{x^2 - a^2})$ and simplify to obtain $ab/(x + \sqrt{x^2 - a^2})$. As x becomes larger and larger, the denominator increases as well, so the quantity d approaches zero. Thus the hyperbola comes closer and closer to the line. The other quadrants are treated similarly. (See Fig. 14.1.9.)

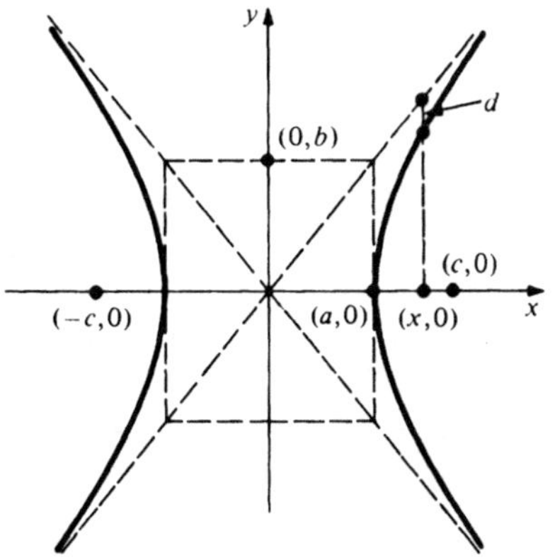

Figure 14.1.9. The vertical distance d from the hyperbola to its asymptote $y = (b/a)x$ is

$$\frac{b}{a}\left(x - \sqrt{x^2 - a^2}\right) = \frac{ab}{x + \sqrt{x^2 - a^2}}.$$

Example 3 Sketch the curve $25x^2 - 16y^2 = 400$.

Solution Dividing by 400, we get the standard form $x^2/16 - y^2/25 = 1$, so $a = 4$ and $b = 5$. The asymptotes are $y = \pm\frac{5}{4}x$, and the curve intersects the x axis at $(\pm 4, 0)$ (see Fig. 14.1.10). ▲

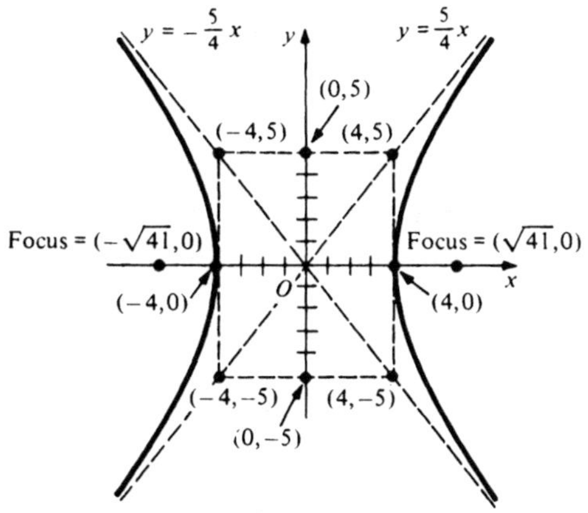

Figure 14.1.10. The hyperbola $25x^2 - 16y^2 = 400$.

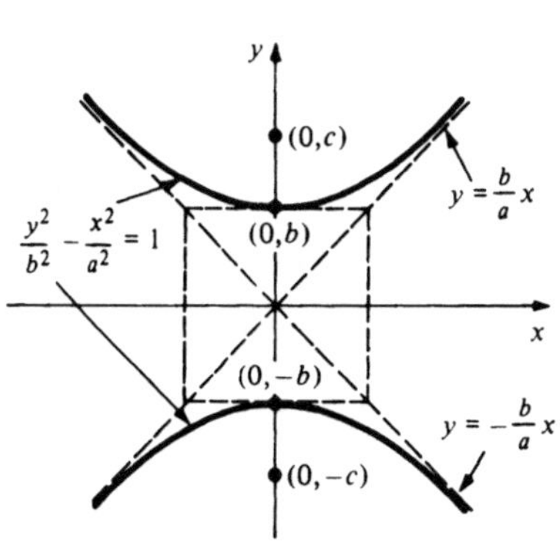

Figure 14.1.11. A hyperbola with foci on the y axis.

If the foci are located on the y axis, the equation of the hyperbola takes the second standard form $y^2/b^2 - x^2/a^2 = 1$ (see Fig. 14.1.11).

Notice that if we draw the rectangle with $(\pm a, 0)$ and $(0, \pm b)$ at the midpoints of its sides, then the asymptotes are the lines through opposite corners, as shown in Figs. 14.1.10 and 14.1.11.

Example 4 Sketch the graph of $4y^2 - x^2 = 4$.

Solution Dividing by 4, we get $y^2 - x^2/2^2 = 1$, which is in the second standard form with $a = 2$ and $b = 1$. The hyperbola and its asymptotes are sketched in Fig. 14.1.12. ▲

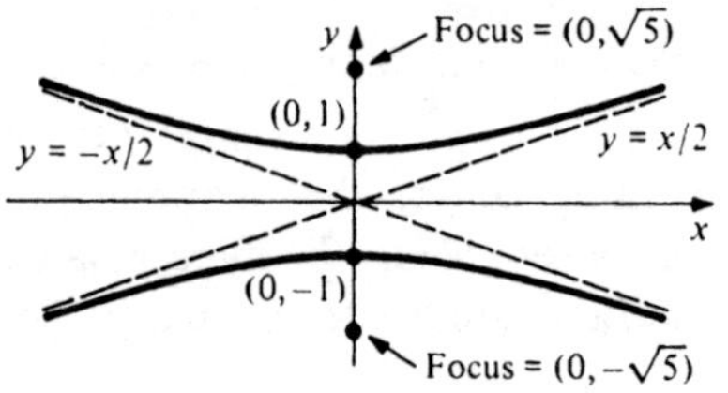

Figure 14.1.12. The hyperbola $4y^2 - x^2 = 4$.

Hyperbola

Case 1: Foci on x axis	**Case 2:** Foci on y axis
Equation: $\dfrac{x^2}{a^2} - \dfrac{y^2}{b^2} = 1$	$\dfrac{y^2}{b^2} - \dfrac{x^2}{a^2} = 1$
Foci: $(\pm c, 0)$, $c = \sqrt{a^2 + b^2}$	$(0, \pm c)$, $c = \sqrt{a^2 + b^2}$
x intercepts: $(\pm a, 0)$	none
y intercepts: none	$(0, \pm b)$
Asymptotes: $y = \pm \dfrac{b}{a} x$	$y = \pm \dfrac{b}{a} x$

If P is any point on the hyperbola, the difference between its distances from the two foci is $2a$ in case 1 and $2b$ in case 2.

We are already familiar with the circle and parabola from Section R.5. The circle is a special case of an ellipse in which $a = b$; that is, the foci coincide. The parabola can be thought of as a limiting case of the ellipse or hyperbola, in which one of the foci has moved to infinity. It can also be described as follows:

Definition of Parabola

A parabola is the set of points in the plane for which the distances from a fixed point, the *focus*, and a fixed line, the *directrix*, are equal.

Placing the focus at $(0, c)$ and the directrix at the line $y = -c$ leads, as above, to an equation relating x and y. Here we have (see Fig. 14.1.13) $|PF| = |PG|$.

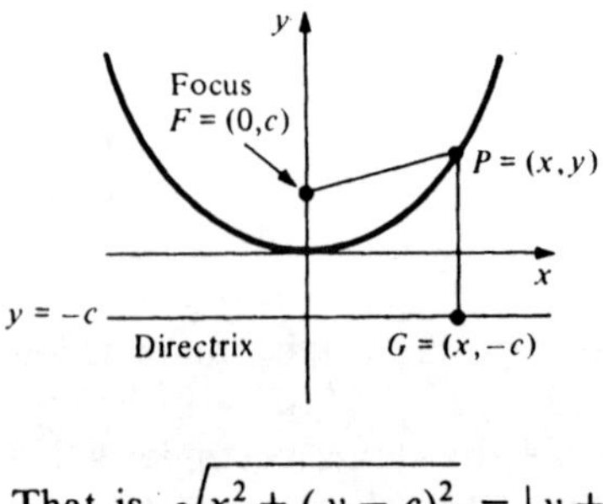

Figure 14.1.13. P is on the parabola when $|PF| = |PG|$.

That is, $\sqrt{x^2 + (y - c)^2} = |y + c|$, so $x^2 + (y - c)^2 = (y + c)^2$, which gives

$$x^2 - 4cy = 0,$$

$$y = \frac{x^2}{4c}$$

which is the form of a parabola as given in Section R.5.

If we place the focus at $(c, 0)$ on the axis and use $x = -c$ as the directrix, we get the "horizontal" parabola $x = y^2/4c$.

Parabola

Case 1: Focus on y axis	**Case 2:** Focus on x axis
Equation: $y = ax^2 \quad \left(a = \frac{1}{4c}\right)$	$x = by^2 \quad \left(b = \frac{1}{4c}\right)$
Focus: $(0, c)$	$(c, 0)$
Directrix: $y = -c$	$x = -c$

If P is any point on the parabola, its distances from the focus and directrix are equal.

Example 5 (a) Find the equation of the parabola with focus $(0, 2)$ and directrix $y = -2$.
(b) Find the focus and directrix of the parabola $x = 10y^2$.

Solution (a) Here $c = 2$, so $a = 1/4c = 1/8$, and so the parabola is $y = x^2/8$.
(b) Here $b = 10 = 1/4c$, so $c = 1/40$. Thus the focus is $(1/40, 0)$ and the directrix is the line $x = -1/40$. ▲

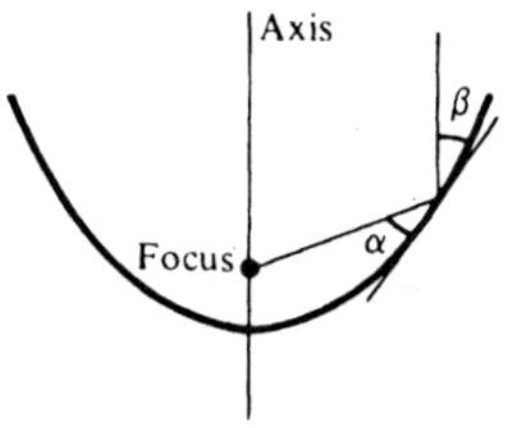

Figure 14.1.14. The angles α and β are equal.

The conic sections appear in a number of physical problems, two of which will be mentioned here; we will see additional ones in later sections. The first application we discuss is to parabolic mirrors. The parabola has the property that the angles α and β shown in Fig. 14.1.14 are equal. This fact, called the *reflecting property* of the parabola, was demonstrated in Review Exercise 86 of Chapter 1. Since the angles of incidence and reflection are equal for a beam of light, this implies that a parallel beam of light impinging on a parabolic mirror will converge at the focus. This is the basis of parabolic telescopes (visual and radio) as well as solar-energy collectors. Similarly, a searchlight will produce a parallel beam of light if a light source is placed at the focus of a parabolic mirror.

Example 6 A parabolic mirror for a searchlight is to be constructed with width 1 meter and depth 0.2 meter. Where should the light source be placed?

Solution We set up the parabola on the coordinate axes as shown in Fig. 14.1.15. The equation of the parabola is $y = ax^2$. Since $y = 0.2$ when $x = 0.5$, we get

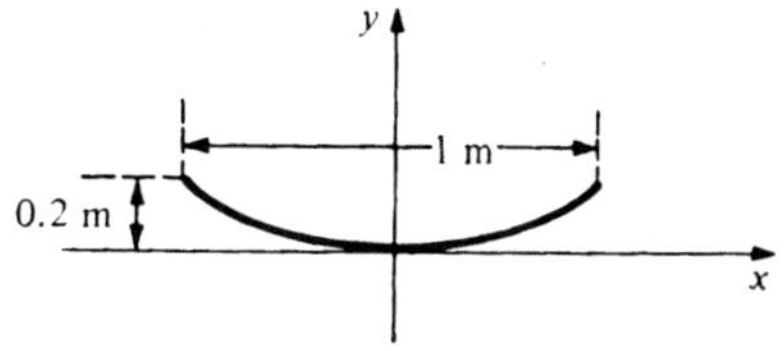

Figure 14.1.15. Find the focus of the searchlight.

$a = 0.2/0.25 = 0.8$. The focus is at $(0, c)$, where $a = 1/4c$, so $c = 1/4a = 0.3125$. Thus the light source should be placed on the axis, 0.3125 meters from the mirror. ▲

In courses in mechanics, it is shown that bodies revolving about the sun (planets, asteroids, and comets) do so in elliptic, parabolic, or hyperbolic orbits with the sun at one focus. We shall see part of the derivation of this fact in Section 14.7. Most planetary orbits are nearly circular. To measure the departure from circularity, the *eccentricity* is introduced. It is defined by

$$e = \frac{c}{a},$$

where a, b, and c are defined as on p. 697, with $a > b$. Thus a and b are the semi-major and semi-minor axes and c is the distance of a focus from the center; $c = \sqrt{a^2 - b^2}$. An ellipse is circular when $e = 0$, and as e approaches 1, the ellipse grows longer and narrower.

Example 7 The eccentricity of Mercury's orbit is 0.21. How wide is its orbit compared to its length?

Solution Since $e = 0.21$, $c = 0.21a$, so $c^2 = a^2 - b^2$, and therefore $b^2 = a^2 - c^2 = a^2(1 - (0.21)^2) = 0.9559a^2$. Hence $b \approx 0.9777a$, so the orbit is 0.9777 times as wide as it is long. ▲

Exercises for Section 14.1

1. Sketch the graph of $x^2 + 9y^2 = 36$. Where are the foci?
2. Sketch the graph of $x^2 + \frac{1}{9}y^2 = 1$. Where are the foci?
3. Sketch the graphs of $x^2 + 4y^2 = 4$, $x^2 + y^2 = 4$, and $4x^2 + 4y^2 = 4$ on the same set of axes.
4. Sketch the graphs of $x^2 + 9y^2 = 9$, $9x^2 + y^2 = 9$, and $9x^2 + 9y^2 = 9$ on the same set of axes.
5. Sketch the graph of $y^2 - x^2 = 2$, showing its asymptotes and foci.
6. Sketch the graph of $3x^2 = 2 + y^2$, showing asymptotes and foci.
7. Sketch the graphs $x^2 + 4y^2 = 4$ and $x^2 - 4y^2 = 4$ on the same set of axes.
8. Sketch the graphs of $x^2 - y^2 = 4$ and $x^2 + y^2 = 4$ on the same set of axes.

Find the equation of the parabolas in Exercises 9 and 10 with the given focus and directrix.

9. Focus $(0, 4)$, directrix $y = -4$
10. Focus $(0, 3)$, directrix $y = -3$.

Find the focus and directrix of the parabolas in Exercises 11–14.

11. $y = x^2$
12. $y = 5x^2$
13. $x = y^2$
14. $x = 4y^2$

Find the equations of the curves described in Exercises 15–20.

15. The circle with center $(0, 0)$ and radius 5.
16. The ellipse consisting of those points whose distances from $(-2, 0)$ and $(2, 0)$ sum to 8.
17. A parabola with vertex at $(0, 0)$ and passing through $(2, 1)$.
18. The circle centered at $(0, 0)$ and passing through $(1, 1)$.
19. The hyperbola with foci at $(0, 2)$ and $(0, -2)$ and passing though $(0, 1)$.
20. The ellipse with x intercept $(1, 0)$ and foci $(0, -2)$ and $(0, 2)$.

21. A parabolic mirror to be used in a searchlight has width 0.8 meters and depth 0.3 meters. Where should the light source be placed?
22. A parabolic disk 10 meters in diameter and 5 meters deep is to be used as a radio telescope. Where should the receiver be placed?
23. The eccentricity of Pluto's orbit is 0.25. What is the ratio of the length to width of this orbit?
24. A comet has an orbit 20 times as long as it is wide. What is the eccentricity of the orbit?

★25. Prove the reflecting property of the ellipse: light originating at one focus converges at the other (*Hint*: Use implicit differentiation.)

★26. A planet travels around its sun on the polar path $r = 1/(2 + \cos\theta)$, the sun at the origin.
   (a) Verify that the path is an ellipse by changing to (x, y) coordinates.
   (b) Compute the *perihelion* distance (minimum distance from the sun to the planet).

14.2 Translation and Rotation of Axes

Whatever their position or orientation, conics are still described by quadratic equations.

In Section R.5 we studied the shifted parabola: if we move the origin to (p,q), $y = ax^2$ becomes $(y-q) = a(x-p)^2$. We can do the same for the other conic sections:

Shifted Conic Sections

Shifted ellipse: $$\frac{(x-p)^2}{a^2} + \frac{(y-q)^2}{b^2} = 1 \quad \text{(shifted circle if } a = b\text{).}$$

Shifted hyperbola: $$\frac{(x-p)^2}{a^2} - \frac{(y-q)^2}{b^2} = 1 \quad \text{(horizontal);}$$

$$\frac{(y-q)^2}{b^2} - \frac{(x-p)^2}{a^2} = 1 \quad \text{(vertical).}$$

Shifted parabola: $$y - q = a(x-p)^2 \quad \text{(vertical);}$$

$$x - p = b(y-q)^2 \quad \text{(horizontal).}$$

Example 1 Graph the ellipse $\frac{x^2}{9} + \frac{y^2}{4} = 1$ and shifted ellipse $\frac{(x-5)^2}{9} + \frac{(y-4)^2}{4} = 1$ on the same xy axes.

Solution The graph of $x^2/9 + y^2/4 = 1$ may be found in Fig. 14.1.5. If (x, y) is any point on this graph, then the point $(x+5, y+4)$ satisfies the equation $(x-5)^2/9 + (y-4)^2/4 = 1$; thus the graph of $(x-5)^2/9 + (y-4)^2/4 = 1$ is obtained by shifting the original ellipse 5 units to the right and 4 units upward. (See Fig. 14.2.1.) ▲

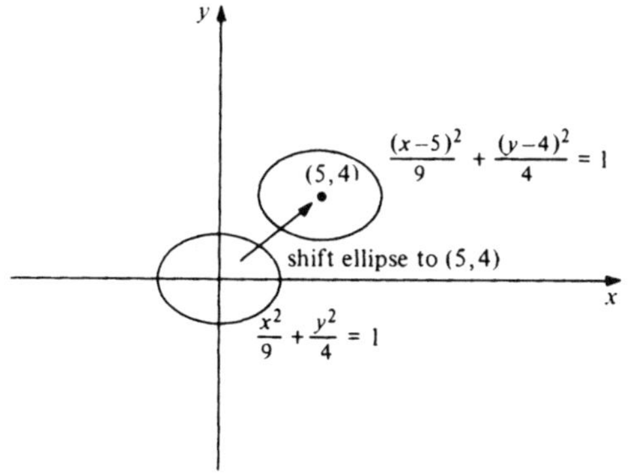

Figure 14.2.1. The graph $(x-5)^2/9 + (y-4)^2/4 = 1$ is an ellipse centered at $(5, 4)$.

Although we referred to the second graph in Example 1 as a "shifted ellipse," it is really just an ellipse, since it satisfies the geometric definition given in Section 14.1. (Can you locate the foci?) To emphasize this, we may introduce new "shifted variables," $X = x - 5$ and $Y = y - 4$, for which the equation becomes $X^2/9 + Y^2/4 = 1$. If we superimpose X and Y axes on our graph as in Fig. 14.2.2, the "shifted" ellipse is now centered at the origin of our new coordinate system. We refer to this process as *translation of axes*.

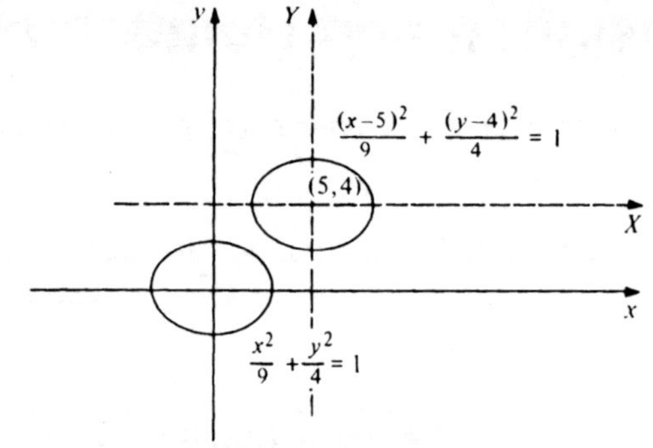

Figure 14.2.2. The ellipse $(x-5)^2/9 + (y-4)^2/4 = 1$ is centered at the origin in a shifted coordinate system.

The importance of translation of axes is that it is possible to bring any equation of the form

$$Ax^2 + Cy^2 + Dx + Ey + F = 0 \tag{1}$$

into the simpler form

$$AX^2 + CY^2 + G = 0 \tag{2}$$

of a conic by letting $X = x - a$ and $Y = y - b$ for suitable choices of constants a and b. Thus, (1) always describes a shifted conic. The way to find the quantities a and b by which the axes are to be shifted is by completing the square, as was done for circles and parabolas in Chapter R. (Notice that in equation (1) there is no xy term. We shall deal with such terms by means of rotation of axes in the second half of this section.)

Example 2 Sketch the graph of $x^2 - 4y^2 - 2x + 16y = 19$.

Solution We complete the square twice:

$$x^2 - 2x = (x-1)^2 - 1,$$

$$-4y^2 + 16y = -4(y^2 - 4y) = -4\left[(y-2)^2 - 4\right].$$

Thus

$$0 = x^2 - 4y^2 - 2x + 16y - 19 = (x-1)^2 - 1 - 4\left[(y-2)^2 - 4\right] - 19$$
$$= (x-1)^2 - 4(y-2)^2 - 4.$$

Hence our equation is

$$\frac{(x-1)^2}{4} - (y-2)^2 = 1$$

which is the hyperbola $x^2/4 - y^2 = 1$ shifted over to $(1,2)$. (See Fig. 14.2.3.)

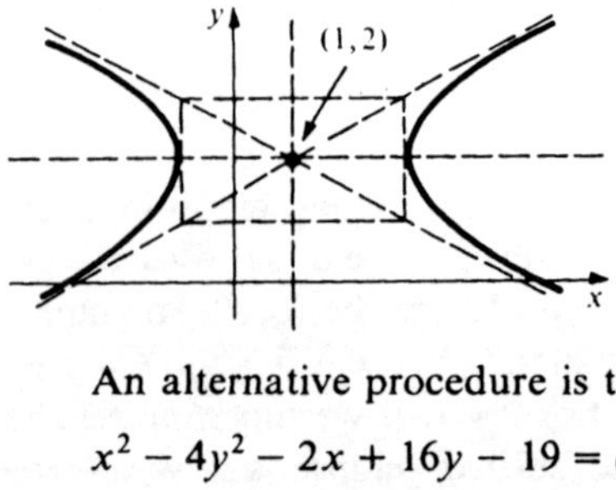

Figure 14.2.3. The hyperbola $x^2 - 4y^2 - 2x + 16y = 19$.

An alternative procedure is to write

$$x^2 - 4y^2 - 2x + 16y - 19 = (x-a)^2 - 4(y-b)^2 + G.$$

Expanding and simplifying, we get

$$-2x + 16y - 19 = -2ax + a^2 + 8by - 4b^2 + G.$$

We find a, b, and G by comparing both sides, which gives $a = 1$, $b = 2$, and $-19 = a^2 - 4b^2 + G$, or $G = -19 - 1 + 16 = -4$. This gives the same answer as above. ▲

Example 3 Sketch the curve $y^2 + x + 3y - 8 = 0$.

Solution Completing the square, we get $y^2 + 3y = (y + \frac{3}{2})^2 - \frac{9}{4}$, so that $y^2 + x + 3y - 8 = 0$ becomes $(y + \frac{3}{2})^2 + x - \frac{41}{4} = 0$; that is, $x - \frac{41}{4} = -(y + \frac{3}{2})^2$. This is a shifted parabola opening to the left, as in Fig. 14.2.4. ▲

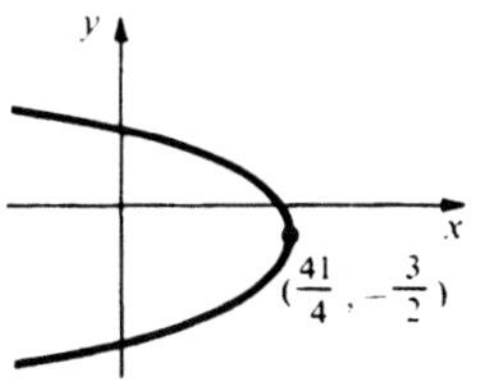

Figure 14.2.4. The parabola $y^2 + x + 3y - 8 = 0$.

We next turn our attention to rotation of axes. The geometric definitions of the ellipse, hyperbola, and parabola given in Section 14.1 do not depend on how these figures are shifted or oriented with respect to the coordinate axes. In the preceding examples we saw how the equations are changed when the coordinate axes are shifted; now we examine how they are changed when the axes are rotated.

In Figure 14.2.5 we have drawn a new set of XY axes which have been rotated by an angle α relative to the old xy axes. The corresponding unit vectors along the axes are denoted $\mathbf{i}, \mathbf{j}$ and $\mathbf{I}, \mathbf{J}$, as shown in the figure.

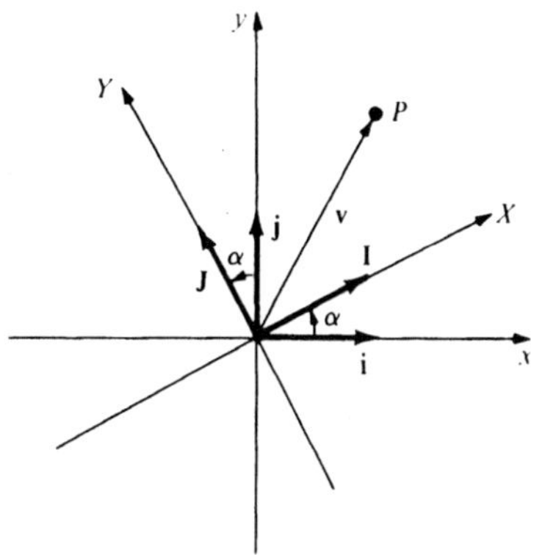

Figure 14.2.5. The XY coordinate system is obtained by rotating the xy system through an angle α.

To understand how to change coordinates from the xy to XY systems, we will use vector methods. Note that as vectors in the plane,

$$\begin{aligned}\mathbf{I} &= \mathbf{i}\cos\alpha + \mathbf{j}\sin\alpha\\ \mathbf{J} &= -\mathbf{i}\sin\alpha + \mathbf{j}\cos\alpha.\end{aligned} \tag{3}$$

Observe that either a direct examination of Fig. 14.2.5 or the fact that $\mathbf{J} = \mathbf{k} \times \mathbf{I}$ can be used to derive the formula for $\mathbf{J}$.

Now consider a point P in the plane and the vector $\mathbf{v}$ from O to P. The coordinates of P relative to the two systems are denoted (x, y) and (X, Y), respectively, and satisfy

$$\mathbf{v} = x\mathbf{i} + y\mathbf{j} = X\mathbf{I} + Y\mathbf{J}. \tag{4}$$

Substituting (3) into (4), we get

$$x\mathbf{i} + y\mathbf{j} = X(\mathbf{i}\cos\alpha + \mathbf{j}\sin\alpha) + Y(-\mathbf{i}\sin\alpha + \mathbf{j}\cos\alpha).$$

Comparing coefficients of $\mathbf{i}$ and $\mathbf{j}$ on both sides gives

$$\begin{aligned}x &= X\cos\alpha - Y\sin\alpha\\ y &= X\sin\alpha + Y\cos\alpha.\end{aligned} \tag{5}$$

To solve these equations for X, Y in terms of x, y, we notice that the roles of (X, Y) and (x, y) are reversed if we change α to $-\alpha$. In other words, the xy

axes are obtained from the XY axes by a rotation through an angle $-\alpha$. Thus we can interchange (x, y) and (X, Y) in (5) if we switch the sign of α:

$$\begin{aligned} X &= x\cos\alpha + y\sin\alpha \\ Y &= -x\sin\alpha + y\cos\alpha. \end{aligned} \tag{6}$$

This conclusion can be verified by substituting (6) into (5) or (5) into (6).

Example 4 Write down the change of coordinates corresponding to a rotation of 30°.

Solution We have $\cos 30° = \sqrt{3}/2$ and $\sin 30° = 1/2$, so (5) and (6) become

$$x = \frac{\sqrt{3}}{2}X - \frac{1}{2}Y,$$

$$y = \frac{1}{2}X + \frac{\sqrt{3}}{2}Y$$

and $X = \dfrac{\sqrt{3}}{2}x + \dfrac{1}{2}y,$

$$Y = -\frac{1}{2}x + \frac{\sqrt{3}}{2}y. \blacktriangle$$

Now suppose we have a rotated conic, such as the ellipse shown in Fig. 14.2.6. In the XY coordinate system, such a conic has the form given by (1):

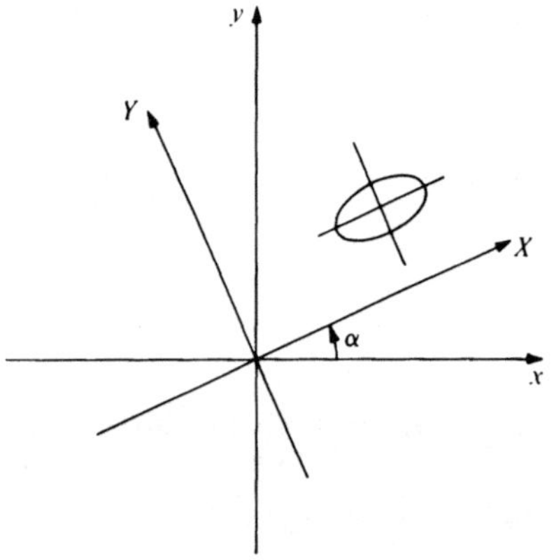

Figure 14.2.6. The conic is aligned with the rotated coordinate system (X, Y) but is rotated relative to the (x, y) coordinate system.

$$\bar{A}X^2 + \bar{C}Y^2 + \bar{D}X + \bar{E}Y + \bar{F} = 0. \tag{7}$$

Substituting (6) into (7) gives

$$\bar{A}(x\cos\alpha + y\sin\alpha)^2 + \bar{C}(-x\sin\alpha + y\cos\alpha)^2$$
$$+ \bar{D}(x\cos\alpha + y\sin\alpha) + \bar{E}(-x\sin\alpha + y\cos\alpha) + \bar{F} = 0.$$

Expanding, we find

$$Ax^2 + Bxy + Cy^2 + Dx + Ey + F = 0, \tag{8}$$

where

$$\left.\begin{aligned} A &= \bar{A}\cos^2\alpha + \bar{C}\sin^2\alpha, \\ B &= (\bar{A} - \bar{C})\cdot 2\cos\alpha\sin\alpha = (\bar{A} - \bar{C})\sin 2\alpha, \\ C &= \bar{A}\sin^2\alpha + \bar{C}\cos^2\alpha, \\ D &= \bar{D}\cos\alpha - \bar{E}\sin\alpha, \\ E &= \bar{D}\sin\alpha + \bar{E}\cos\alpha, \\ F &= \bar{F}. \end{aligned}\right\} \tag{9}$$

Notice the introduction of the xy term in (8). If we are *given* an equation of the form (8), we may determine the type of conic it is and the rotation angle α by finding the rotated form (7). To accomplish this, we notice from (9) that

$$A - C = \bar{A}(\cos^2\alpha - \sin^2\alpha) - \bar{C}(\cos^2\alpha - \sin^2\alpha)$$
$$= (\bar{A} - \bar{C})\cos 2\alpha.$$

Therefore,

$$B = (\bar{A} - \bar{C})\sin 2\alpha = (A - C)\tan 2\alpha.$$

Thus,

$$\tan 2\alpha = \frac{B}{A - C} \tag{10}$$

($\alpha = 45°$ if $A = C$). Equation (10) enables one to solve for α given equation (8).

Equation (8) will describe an ellipse only when (7) does, i.e., when $\bar{A}$ and $\bar{C}$ have the same sign, or $\bar{A}\bar{C} > 0$. To recognize this condition directly from (8), we use (9) to obtain

$$AC = (\bar{A}\cos^2\alpha + \bar{C}\sin^2\alpha)(\bar{A}\sin^2\alpha + \bar{C}\cos^2\alpha)$$
$$= (\bar{A}^2 + \bar{C}^2)\cos^2\alpha\sin^2\alpha + \bar{A}\bar{C}(\cos^4\alpha + \sin^4\alpha).$$

However, $B = (\bar{A} - \bar{C})2\cos\alpha\sin\alpha$, so $\frac{1}{4}B^2 = (\bar{A}^2 + \bar{C}^2 - 2\bar{A}\bar{C})(\cos^2\alpha\sin^2\alpha)$, and thus

$$AC - \tfrac{1}{4}B^2 = \bar{A}\bar{C}(\cos^4\alpha + \sin^4\alpha + 2\cos^2\alpha\sin^2\alpha)$$
$$= \bar{A}\bar{C}(\cos^2\alpha + \sin^2\alpha)^2 = \bar{A}\bar{C}.$$

Thus (8) is an ellipse if $AC - \frac{1}{4}B^2 > 0$; i.e., $B^2 - 4AC < 0$. The other conics are identified in a similar way, as described in the following box.

Rotation of Axes

The equation

$$Ax^2 + Bxy + Cy^2 + Dx + Ey + F = 0$$

(with A, B, and C not all zero) is a conic; it is

an ellipse if $B^2 - 4AC < 0$;
a hyperbola if $B^2 - 4AC > 0$;
a parabola if $B^2 - 4AC = 0$.

To graph this conic, proceed as follows:

1. Find α from $\alpha = \frac{1}{2}\tan^{-1}\left[\dfrac{B}{A - C}\right]$.
2. Let $x = X\cos\alpha - Y\sin\alpha$, $y = X\sin\alpha + Y\cos\alpha$, and substitute into the given equation. You will get an equation of the form

   $$\bar{A}X^2 + \bar{C}Y^2 + \bar{D}X + \bar{E}Y + \bar{F} = 0.$$

3. This is a shifted conic in XY coordinates which may be plotted by completing the square (as in Examples 2 and 3).
4. Place your conic in the XY coordinates in the xy plane by rotating the axes through an angle α, as in Figure 14.2.6.

Example 5 What type of conic is given by $x^2 + 3y^2 - 2\sqrt{3}\,xy + 2\sqrt{3}\,x + 2y = 0$?

Solution This is a rotated conic because it has an xy term. Here $A = 1$, $B = -2\sqrt{3}$, $C = 3$, $D = 2\sqrt{3}$, and $E = 2$. To find the type, we compute the quantity $B^2 - 4AC = 4\cdot 3 - 4\cdot 3 = 0$, so this is a parabola. ▲

Example 6 Sketch the graph of the conic in Example 5.

Solution We follow the four steps in the preceding box:

$$1.\ \alpha = \tfrac{1}{2}\tan^{-1}\big[B/(A-C)\big] = \tfrac{1}{2}\tan^{-1}\big[-2\sqrt{3}/(1-3)\big] = \tfrac{1}{2}\tan^{-1}\sqrt{3} = \pi/6.$$

Thus $\alpha = \pi/6$ or $30°$.

2. As in Example 4, we have

$$x = \frac{\sqrt{3}}{2}X - \frac{1}{2}Y \quad\text{and}\quad y = \frac{1}{2}X + \frac{\sqrt{3}}{2}Y.$$

Substituting into $x^2 + 3y^2 - 2\sqrt{3}\,xy + 2\sqrt{3}\,x + 2y = 0$, we get

$$\left(\frac{\sqrt{3}}{2}X - \frac{1}{2}Y\right)^2 + 3\left(\frac{1}{2}X + \frac{\sqrt{3}}{2}Y\right)^2 - 2\sqrt{3}\left(\frac{\sqrt{3}}{2}X - \frac{1}{2}Y\right)\left(\frac{1}{2}X + \frac{\sqrt{3}}{2}Y\right) + 2\sqrt{3}\left(\frac{\sqrt{3}}{2}X - \frac{1}{2}Y\right) + 2\left(\frac{1}{2}X + \frac{\sqrt{3}}{2}Y\right) = 0.$$

Expanding, we get

$$\left(\frac{3}{4}X^2 + \frac{1}{4}Y^2 - \frac{\sqrt{3}}{2}XY\right) + \left(\frac{3}{4}X^2 + \frac{9}{4}Y^2 + \frac{3\sqrt{3}}{2}XY\right) - 2\sqrt{3}\left(\frac{\sqrt{3}}{4}X^2 + \frac{1}{2}XY - \frac{\sqrt{3}}{4}Y^2\right) + 3X - \sqrt{3}\,Y + X + \sqrt{3}\,Y = 0$$

which simplifies to $4Y^2 + 4X = 0$ or $X = -Y^2$.

3. The conic $X = -Y^2$ is a parabola opening to the left in XY coordinates.

4. We plot the graph in Fig. 14.2.7. ▲

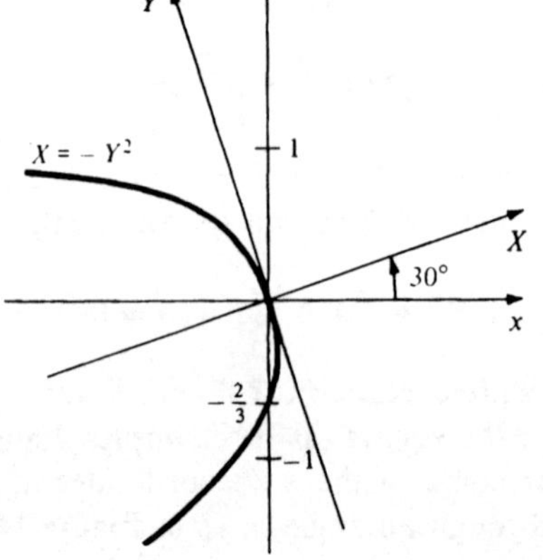

Figure 14.2.7. The graph of $x^2 + 3y^2 - 2\sqrt{3}\,xy + 2\sqrt{3}\,x + 2y = 0$.

Example 7 Sketch the graph of $3x^2 + 3y^2 - 10xy + 18\sqrt{2}\,x - 14\sqrt{2}\,y + 38 = 0$.

Solution Let us first determine the type of conic. Here $B^2 - 4AC = 100 - 4 \cdot 3 \cdot 3 = 100 - 36 = 64 > 0$, so it is a hyperbola.

1. $\alpha = \frac{1}{2}\tan^{-1}[B/(A-C)] = \frac{1}{2}\tan^{-1}\infty = \pi/4$ or $45°$.
2. $x = (1/\sqrt{2})(X - Y)$, $y = (1/\sqrt{2})(X + Y)$; substituting and simplifying, one arrives at

$$X^2 - 4Y^2 - 2X + 16Y - 19 = 0.$$

3. This is the hyperbola in Fig. 14.2.3.
4. See Fig. 14.2.8. ▲

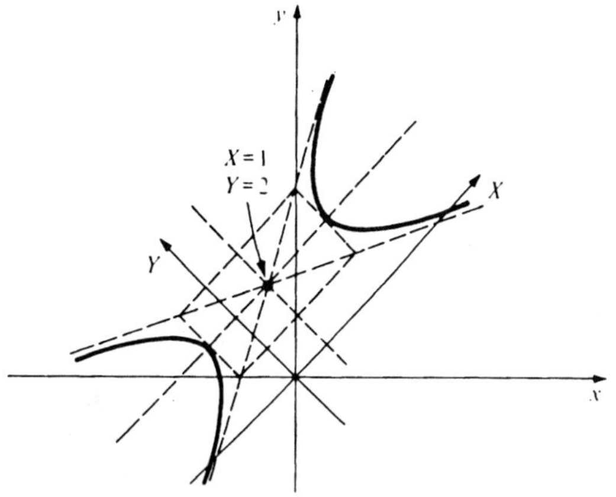

Figure 14.2.8. The graph of $3x^2 + 3y^2 - 10xy + 18\sqrt{2}\,x - 14\sqrt{2}\,y + 38 = 0$.

Exercises for Section 14.2

In Exercises 1–4, graph the conics and shifted conics on the same xy axes.

1. $y = -x^2$, $y - 2 = -(x+1)^2$.
2. $x^2 - y^2 = 1$, $(x-2)^2 - (y+3)^2 = 1$.
3. $x^2 + y^2 = 4$, $(x+3)^2 + (y-8)^2 = 4$.
4. $x^2/9 + y^2/16 = 1$, $(x-1)^2/9 + (y-2)^2/16 = 1$.

Identify the equations in Exercises 5–10 as shifted conic sections and sketch their graphs.

5. $x^2 + y^2 - 2x = 0$
6. $x^2 + 4y^2 - 8y = 0$
7. $2x^2 + 4y^2 - 6y = 8$
8. $x^2 + 2x + y^2 - 2y = 2$
9. $x^2 + 2x - y^2 - 2y = 1$
10. $3x^2 - 6x + y - 7 = 0$

In Exercises 11–14, write down the transformation of coordinates corresponding to a rotation through the given angle.

11. $60°$
12. $\pi/4$
13. $15°$
14. $2\pi/3$

In Exercises 15–18, determine the type of conic.

15. $xy = 2$.
16. $x^2 + xy + y^2 = 4$.
17. $\frac{19}{4}x^2 + \frac{43}{12}y^2 + \frac{7\sqrt{3}}{6}xy = 48$.
18. $3x^2 + 3y^2 - 2xy - \frac{6}{\sqrt{2}}x - \frac{6}{\sqrt{2}}y = 8$.

Sketch the graphs of the conics in Exercises 19–22.

19. The conic in Exercise 15.
20. The conic in Exercise 16.
21. The conic in Exercise 17.
22. The conic in Exercise 18.

Find the equations of the curves described in Exercises 23–28.

23. The circle with center $(2,3)$ and radius 5.
24. The ellipse consisting of those points whose distances from $(0,0)$ and $(2,0)$ sum to 8.
25. The parabola with vertex at $(1,0)$ and passing through $(0,1)$ and $(2,1)$.
26. The circle passing through $(0,0)$, $(1,\frac{1}{2})$, and $(2,0)$.
27. The hyperbola with foci at $(0,-1)$ and $(0,3)$ and passing through $(0,2)$.
28. The ellipse with x intercept $(1,0)$ and foci $(0,0)$ and $(0,2)$.

29. Find the equation of the conic in Exercise 8 rotated through $\pi/3$ radians.
30. Find the equation of the conic in Exercise 9 rotated through $45°$.
31. Show that $A + C$ is unchanged under a rotation or translation of axes.
32. Show that $D^2 + E^2$ is unchanged under a rotation of axis, but not under translation.

★33. Show that if $B^2 - 4AC < 0$, the area of the ellipse $Ax^2 + Bxy + Cy^2 = 1$ is $2\pi/\sqrt{4AC - B^2}$.

14.3 Functions, Graphs, and Level Surfaces

The graph of a function of two variables is a surface in space.

The daily weather map of North America shows the temperatures of various locations at a fixed time. If we let x be the longitude and y the latitude of a point, the temperature T at that point may be thought of as a function of the *pair* (x, y). Weather maps often contain curves through points with the same temperature. These curves, called *isotherms*, help us to visualize the temperature function; for instance, in Fig. 14.3.1 they help us to locate a hot spot in the southwestern U.S. and a cold spot in Canada.

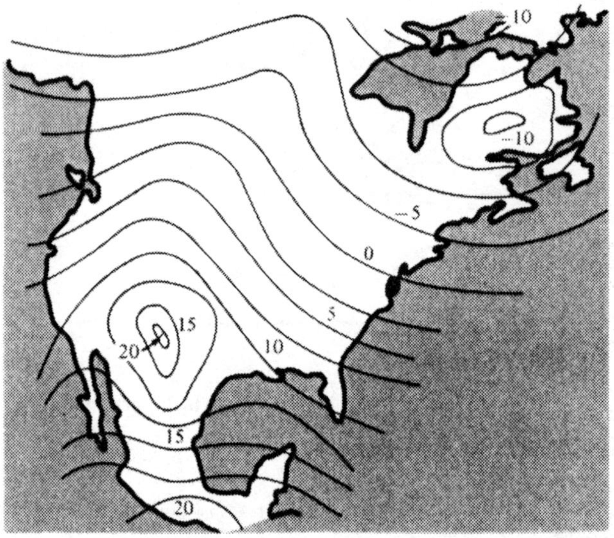

Figure 14.3.1. Isotherms are lines of constant temperature (in degrees Celsius).

Functions of two variables arise in many other contexts as well. For instance, in topography the height h of the land depends on the two coordinates that give the location. The reaction rate σ of two chemicals A and B depends on their concentrations a and b. The altitude α of the sun in the sky on June 21 depends on the latitude l and the number of hours t after midnight.

Many quantities depend on more than two variables. For instance, the temperature can be regarded as a function of the time t as well as of x and y to give a function of three variables. (Try to imagine visualizing this function by watching the isotherms move and wiggle as the day progresses.) The rate of a reaction involving 10 chemicals is a function of 10 variables.

In this book we limit our attention to functions of two and three variables. Readers who have mastered this material can construct for themselves, or find in a more advanced work,[1] the generalizations of the concepts presented here to functions of four and more variables.

[1] See, for example, J. Marsden and A. Tromba, *Vector Calculus*, Second Edition, W. H. Freeman and Co., 1980.

The mathematical development of functions of several variables begins with some definitions.

Functions of Two Variables

A *function of two variables* is a rule which assigns a number $f(x, y)$ to each point (x, y) of a domain in the xy plane.

Example 1 Describe the domain of $f(x, y) = x/(x^2 + y^2)$. Evaluate $f(1,0)$ and $f(1,1)$.

Solution As given, this function is defined as long as $x^2 + y^2 \neq 0$, that is, as long as $(x, y) \neq (0,0)$. We have

$$f(1,0) = \frac{1}{1^2 + 0^2} = 1 \quad \text{and} \quad f(1,1) = \frac{1}{1^2 + 1^2} = \frac{1}{2}. \blacktriangle$$

The Graph of a Function

The *graph* of a function $f(x, y)$ of two variables consists of all points (x, y, z) in space such that (x, y) is in the domain of the function and $z = f(x, y)$.

Some particularly simple graphs can be drawn on the basis of our work in earlier chapters.

Example 2 Sketch the graph of (a) $f(x, y) = x - y + 2$ and (b) $f(x, y) = 3x$.

Solution (a) We recognize $z = x - y + 2$ (that is, $x - y - z + 2 = 0$) as the equation of a plane. Its normal is $(1, -1, -1)$ and it meets the axes at $(-2,0,0), (0,2,0), (0,0,2)$. From this information we sketch its graph in Fig. 14.3.2.
(b) The graph of $f(x, y) = 3x$ is the plane $z = 3x$. It contains the y axis and is shown in Fig. 14.3.3. ▲

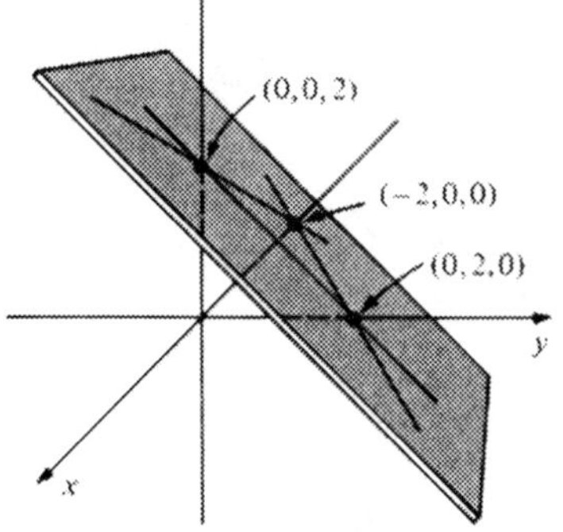

Figure 14.3.2. The graph of $z = x - y + 2$ is a plane.

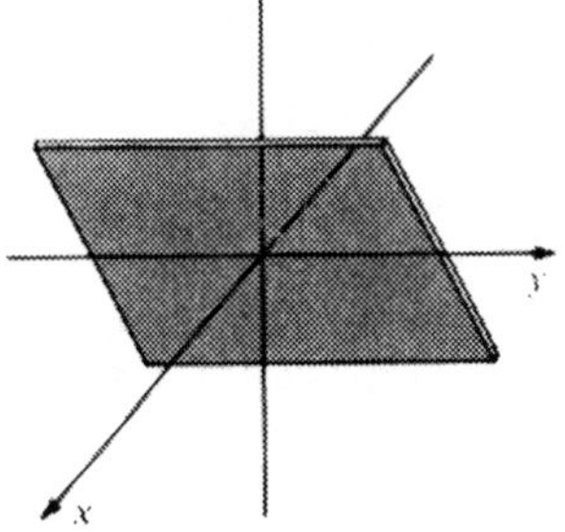

Figure 14.3.3. The graph of $z = 3x$.

Using level curves instead of graphs makes it possible to visualize a function of two variables by a two-dimensional rather than a three-dimensional picture.

Level Curves

Let f be a function of two variables and let c be a constant. The set of all (x, y) in the plane such that $f(x, y) = c$ is called a *level curve* of f (with value c).

Isotherms are just the level curves of a temperature function, and a contour plot of a mountain consists of representative level curves of the height function.

Example 3 Sketch the level curves with values $-1, 0, 1$ for $f(x, y) = x - y + 2$.

Solution The level curve with value -1 is obtained by setting $f(x, y) = -1$; that is,

$$x - y + 2 = -1, \qquad \text{that is,} \quad x - y + 3 = 0,$$

which is a straight line in the plane (see Fig. 14.3.4). The level curve with value zero is the line

$$x - y + 2 = 0,$$

and the curve with value 1 is the line

$$x - y + 2 = 1, \qquad \text{that is,} \quad x - y + 1 = 0. \blacktriangle$$

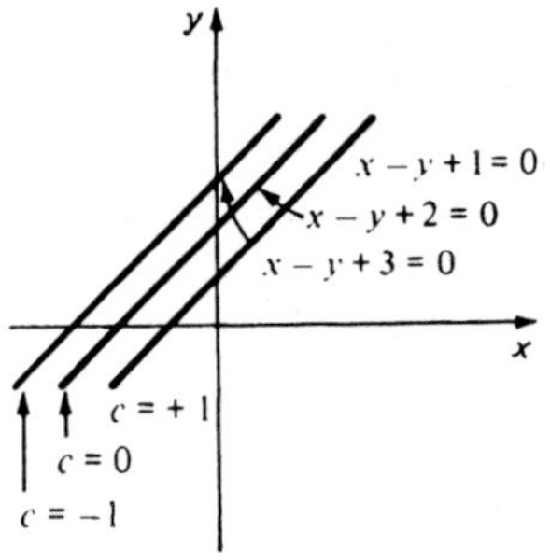

Figure 14.3.4. Three level curves of the function $f(x, y) = x - y + 2$.

Example 4 How is the intersection of the plane $z = c$ with the graph of $f(x, y)$ related to the level curves of f? Sketch.

Solution The intersection of the plane $z = c$ and the graph of f consists of the points (x, y, c) in space such that $f(x, y) = c$. This set has the same shape as the level curve with value c, but it is moved from the xy plane up to the plane $z = c$. (See Fig. 14.3.5.) $\blacktriangle$

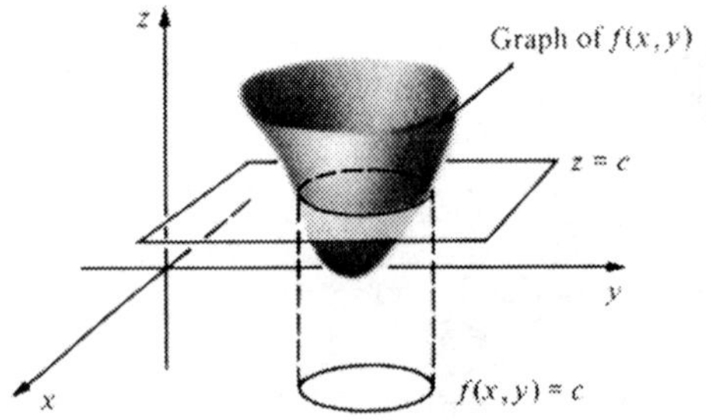

Figure 14.3.5. The level curve of $f(x, y)$ with value c is obtained by finding the intersection of the graph of f with the plane $z = c$ and moving it down to the (x, y) plane.

We turn now to functions of three variables.

Functions of Three Variables

A *function of three variables* is a rule which assigns a number $f(x, y, z)$ to each point (x, y, z) of a domain in (x, y, z) space.

The graph of a function $w = f(x, y, z)$ of three variables would have to lie in four-dimensional space, so we cannot visualize it; but the concept of level curve has a natural extension.

Level Surfaces

Let f be a function of three variables and let c be a constant. The set of all points (x, y, z) in space such that $f(x, y, z) = c$ is called a *level surface* of f (with value c).

Example 5 (a) Let $f(x, y, z) = x - y + z + 2$. Sketch the level surfaces with values 1, 2, 3.
(b) Sketch the level surface of $f(x, y, z) = x^2 + y^2 + z^2 - 8$ with value 1.

Solution (a) In each case we set $f(x, y, z) = c$:

$$c = 1: \quad x - y + z + 2 = 1 \qquad (\text{that is, } x - y + z + 1 = 0),$$
$$c = 2: \quad x - y + z + 2 = 2 \qquad (\text{that is, } x - y + z = 0),$$
$$c = 3: \quad x - y + z + 2 = 3 \qquad (\text{that is, } x - y + z - 1 = 0).$$

These surfaces are parallel planes and are sketched in Fig. 14.3.6.

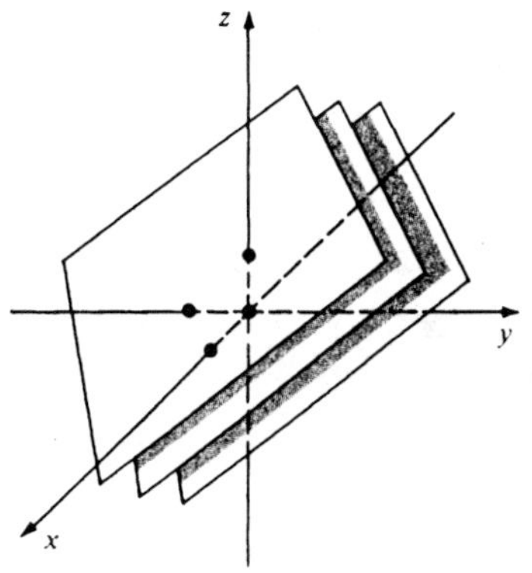

Figure 14.3.6. Three level surfaces of the function $f(x, y, z) = x - y + z + 2$.

(b) The surface $x^2 + y^2 + z^2 - 8 = 1$ (that is, $x^2 + y^2 + z^2 = 9$) is the set of points (x, y, z) whose distance from the origin is $\sqrt{9} = 3$; it is a sphere with radius 3 and center at the origin. (See Fig. 14.3.7.) ▲

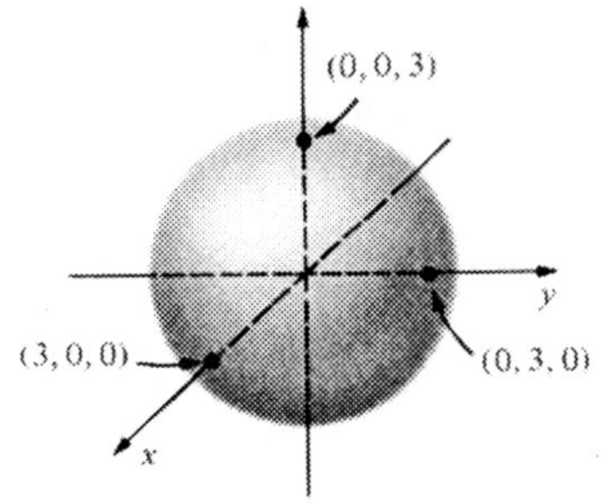

Figure 14.3.7. The level surface of $x^2 + y^2 + z^2 - 8$ with value 1 is a sphere of radius 3.

Plotting surfaces in space is usually more difficult than plotting curves in the plane. It is rare that plotting a few points on a surface will give us enough information to sketch the surface. Instead we often plot several *curves* on the surface and then interpolate between the curves. This technique, called the *method of sections*, is useful for plotting surfaces in space, whether they be graphs of functions of two variables or level surfaces of functions of three variables. The idea behind the method of sections is to obtain a picture of the surface in space by looking at its slices by planes parallel to one of the coordinate planes. For instance, for a graph $z = f(x, y)$ the section $z = c$ is illustrated in Fig. 14.3.8.

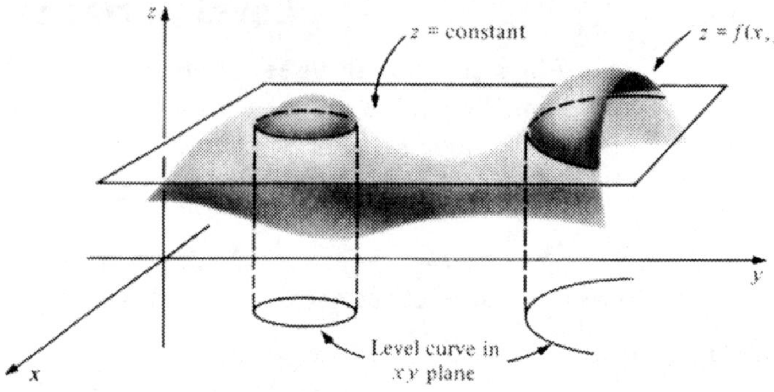

Figure 14.3.8. The section $z = c$ of the graph $z = f(x, y)$.

Example 6 Sketch the surfaces in xyz space given by (a) $z = -y^2$ and (b) $x^2 + y^2 = 25$.

Solution (a) Since x is missing, all sections x = constant look the same; they are copies of the parabola $z = -y^2$. Thus we draw the parabola $z = -y^2$ in the yz plane and extend it parallel to the x axis as shown in Fig. 14.3.9. The surface is called a *parabolic cylinder*.

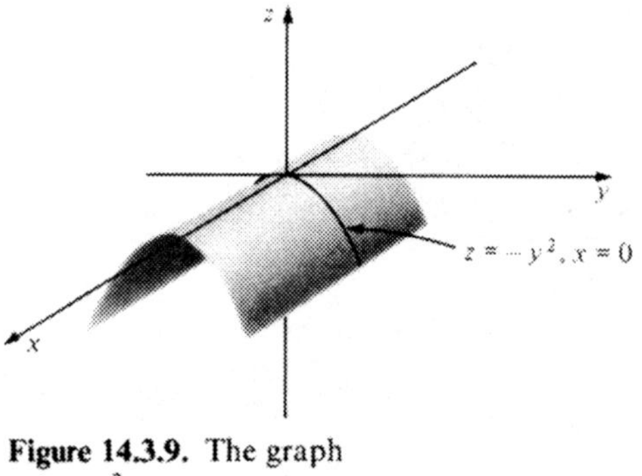

Figure 14.3.9. The graph $z = -y^2$ is a parabolic cylinder.

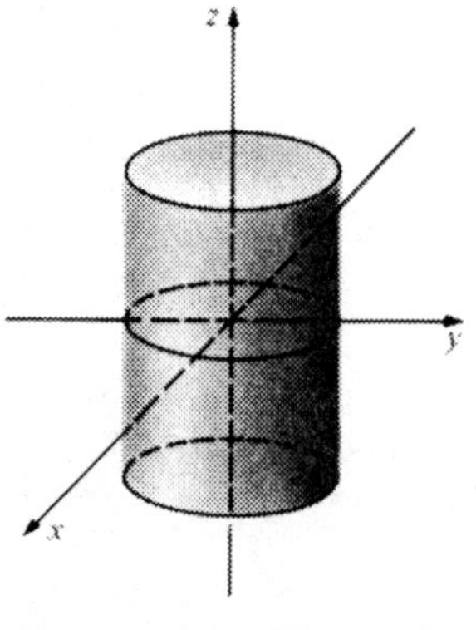

Figure 14.3.10. The graph $x^2 + y^2 = 25$ is a right circular cylinder.

(b) The variable z does not occur in the equation, so the surface is a cylinder parallel to the z axis. Its cross section is the plane curve $x^2 + y^2 = 25$, which is a circle of radius 5, so the surface is a right circular cylinder, as shown in Fig. 14.3.10. ▲

Example 7 (a) Sketch the graph of $f(x, y) = x^2 + y^2$ (this graph is called a *paraboloid of revolution*). (b) Sketch the surface $z = x^2 + y^2 - 4x - 6y + 13$. [*Hint:* Complete the square.]

Solution (a) If we set z = constant, we get $x^2 + y^2 = c$, a circle. Taking $c = 1^2, 2^2, 3^2, 4^2$, we get circles of radius 1, 2, 3, and 4. These are placed on the planes $z = 1^2 = 1$, $z = 2^2 = 4$, $z = 3^2 = 9$, and $z = 4^2 = 16$ to give the graph shown in Fig. 14.3.11.

If we set $x = 0$, we obtain the parabola $z = y^2$; if we set $y = 0$, we obtain the parabola $z = x^2$. The graph is symmetric about the z axis since z depends only on $r = \sqrt{x^2 + y^2}$.

(b) Completing the square, we write $z = x^2 + y^2 - 4x - 6y + 13$ as

$$\begin{aligned} z &= x^2 - 4x + \quad y^2 - 6y \quad + 13 \\ &= x^2 - 4x + 4 + y^2 - 6y + 9 + 13 - 13 \\ &= (x-2)^2 + (y-3)^2. \end{aligned}$$

The level surface for value c is thus the circle $(x-2)^2 + (y-3)^2 = c$ with center $(2,3)$ and radius $\sqrt{c}$. Comparing this result with (a) we find that the surface is again a paraboloid of revolution, with its axis shifted to the line $(x, y) = (2, 3)$. (See Fig. 14.3.12.) ▲

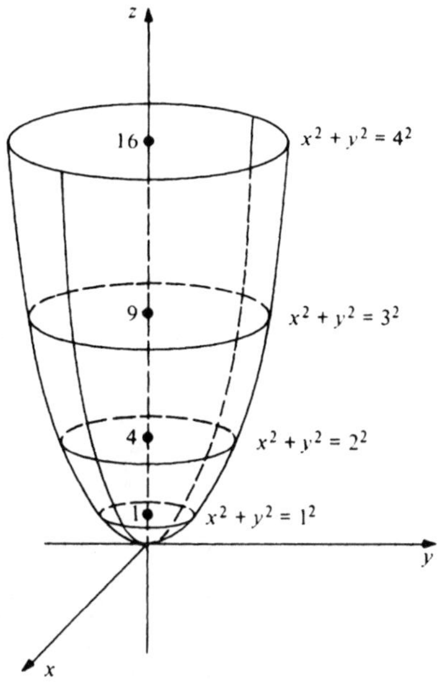

Figure 14.3.11. The sections of the graph $z = x^2 + y^2$ by planes $z = c$ are circles.

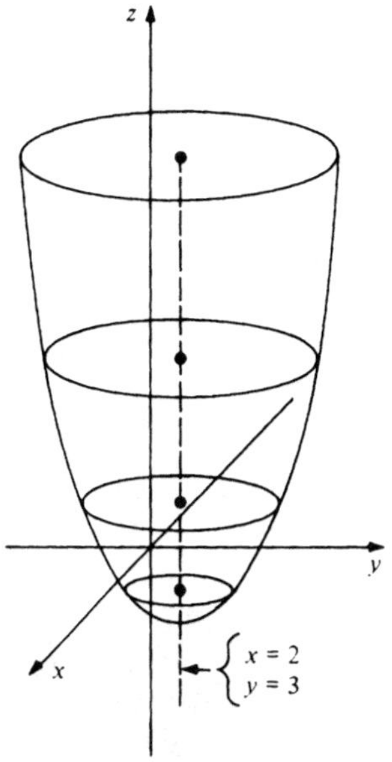

Figure 14.3.12. The graph $z = x^2 + y^2 - 4x - 6y + 13$ is a shifted paraboloid of revolution.

Plotting Surfaces: Methods of Sections

1. Note any symmetries of the graph.
2. See if any variables x, y, or z are missing from the equation. If so, the surface is a "cylinder" parallel to the axis of the missing variable, and its cross section is the curve in the other variables (see Example 6).
3. If the surface is a graph $z = f(x, y)$, find the level curves $f(x, y) = c$ for various convenient values of c and draw these curves on the planes $z = c$. Smoothly join these curves with a surface in space. Draw the curves obtained by setting $x = 0$ and $y = 0$ or other convenient values to help clarify the picture.
4. If the surface has the form $F(x, y, z) = c$, then either:
   (a) Solve for one of the variables in terms of the other two and use step 2 if it is convenient to do so.
   (b) Set x equal to various constant values to obtain curves in y and z; draw these curves on the corresponding $x =$ constant planes. Repeat with $y =$ constant or $z =$ constant or both. Fill in the curves obtained with a surface.

In the next section, we will use our knowledge of conic sections to plot the graphs of more complicated quadratic functions.

The computer can help us graph surfaces that may be tedious or impossible to plot by hand. The computer draws the graph either by drawing sections perpendicular to the x and y axes or by sections perpendicular to the z axis—that is, level curves lifted to the graph. When this is done for the function $z = x^2 + y^2$ (Example 7), Figs. 14.3.13(a) and 14.3.13(b) result. (The pointed tips appear because a rectangular domain has been chosen for the function.)[2]

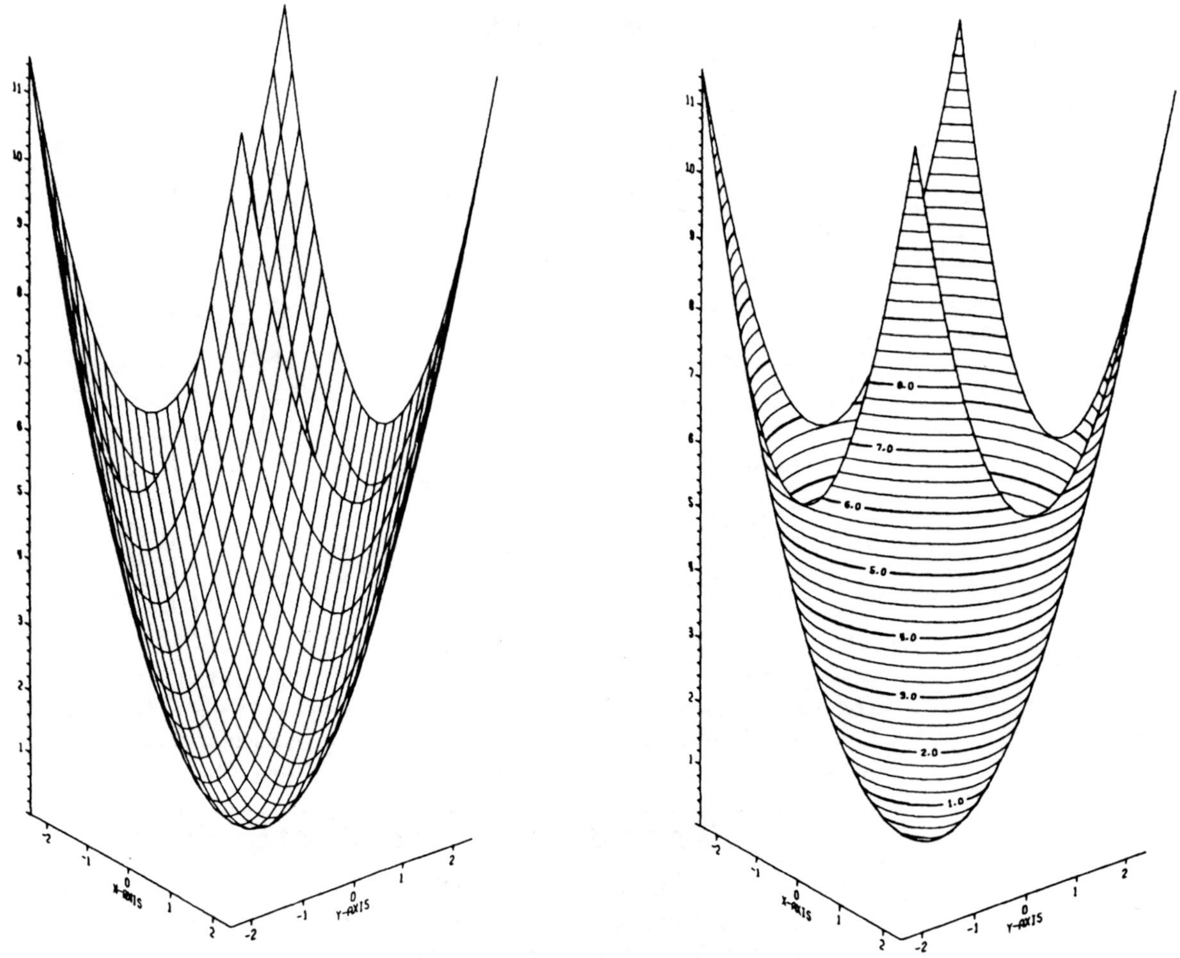

Figure 14.3.13. The graph of $z = x^2 + y^2$ drawn by computer in two ways.

The computer-generated graph in Fig. 14.3.14 shows the function

$$z = (x^2 + 3y^2)e^{1-(x^2+y^2)}.$$

Fig. 14.3.15 shows the level curves of this function in the xy plane, viewed first from an angle and then from above. Study these pictures to help develop your powers of three-dimensional visualization; attempt to reconstruct the graph in your mind by looking at the level curves.

[2] The authors are indebted to Jerry Kazdan for preparing most of the computer-generated graphs in this book.

(a)

(b)

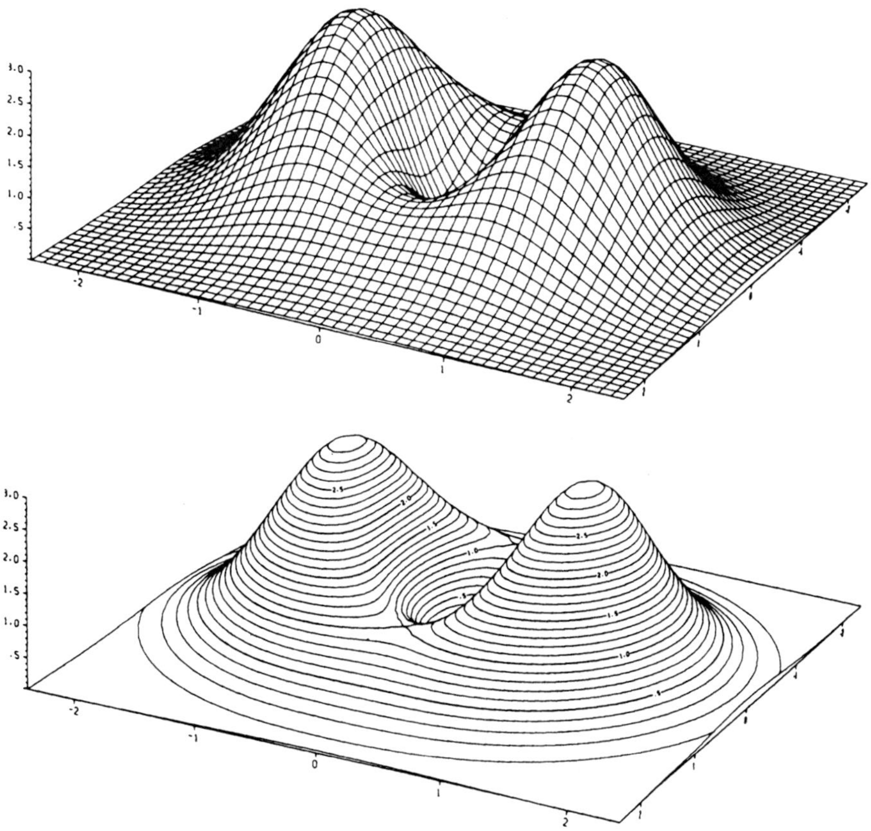

Figure 14.3.14. Computer-generated graphs of $z = (x^2 + 3y^2)e^{1-(x^2+y^2)}$.

(a)

(b)

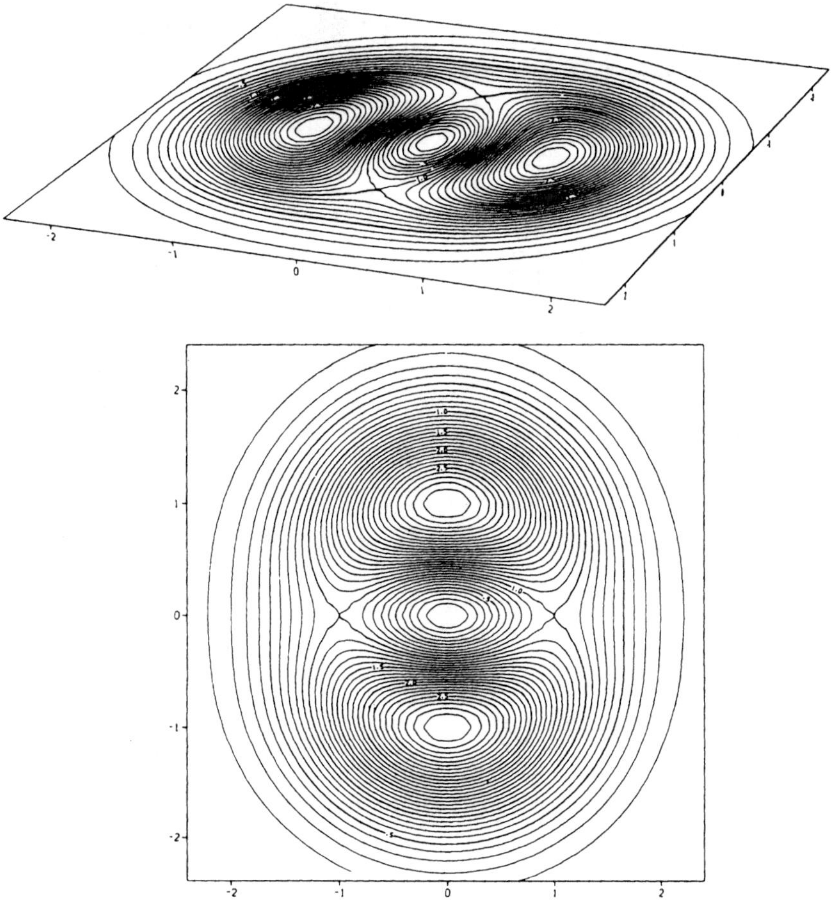

Figure 14.3.15. Level curves for the function $z = (x^2 + 3y^2)e^{1-(x^2+y^2)}$.

Exercises for Section 14.3

In Exercises 1–8, describe the domain of each of the given functions and evaluate the function at the indicated points.

1. $f(x, y) = \frac{y}{x}$; $(1,0)$, $(1,1)$.
2. $f(x, y) = \frac{x+y}{x-y}$; $(1,-1)$, $(1,0.9)$.
3. $f(x, y) = \frac{x+y}{x^2+y^2-1}$; $(1,1)$, $(-1,1)$.
4. $f(x, y) = \frac{2xy}{x^2+y^2}$; $(1,1)$, $(-1,1)$.
5. $f(x, y, z) = \frac{2x+y-z}{x^2+y^2+z^2-1}$; $(1,1,1)$, $(0,0,2)$.
6. $f(x, y, z) = \frac{z}{x^2-4y^2-1}$; $(1,0.5,1)$, $(0.1,0.5,1)$.
7. $f(x, y) = \frac{2x - \sin y}{1+\cos x}$; $(0, \pi/4)$, $(\pi/4, \pi)$.
8. $f(x, y) = \frac{e^x - e^y}{1+\sin x}$; $(0,1)$; $(\pi/4, -1)$.

Sketch the graphs of the functions in Exercises 9–12.

9. $f(x, y) = 1 - x - y$
10. $f(x, y) = -1 - x - y$
11. $z = x - y$
12. $z = x + 2$

Sketch the level curves for the indicated functions and values in Exercises 13–18.

13. f in Exercise 9, values $1, -1$.
14. f in Exercise 10, values $1, -1$.
15. f in Exercise 3, values $-2, -1, 1, 2$ and describe the level curve for the general value.
16. f in Exercise 4, values $-2, -1, 1, 2$ and describe the level curve for the general value.
17. f in Exercise 2, values $-2, -1, 1, 2$ and describe the level curve for the general value.
18. $f(x, y) = 3^{-1/(x^2+y^2)}$, value $1/e$.

Sketch the level surfaces in Exercises 19–22.

19. $x + y - 2z = 8$
20. $3x - 2y - z = 4$
21. $x^2 + y^2 + z^2 = 4$
22. $x^2 + y^2 - z = 4$

Draw the level curves $f(x, y) = c$—first in the xy plane and then lifted to the graph in space—for the functions and values in Exercises 23–26.

23. $f(x, y) = x^2 + 2y^2$; $c = 0, 1, 2$.
24. $f(x, y) = x^2 - y^2$; $c = -1, 0, 1$.
25. $f(x, y) = x - y^2$; $c = -2, 0, 2$.
26. $f(x, y) = y - x^2$; $c = -1, 0, 1$.

Sketch the surface in space defined by each of the equations in Exercises 27–40.

27. $z = x^2 + 2$
28. $z = |y|$
29. $z^2 + x^2 = 4$
30. $x^2 + y = 2$
31. $z = (x-1)^2 + y^2$
32. $x = -8z^2 + z$
33. $z = x^2 + y^2 - 2x + 8$.
34. $z = 3x^2 + 3y^2 - 6x + 12y + 15$.
35. $z = \sqrt{x^2 + y^2}$
36. $z = \max(|x|, |y|)$. [*Note:* $\max(|x|, |y|)$ is the maximum of $|x|$ and $|y|$.]
37. $z = \sin x$ (the "washboard").
38. $z = 1/(1 + y^2)$.
39. $4x^2 + y^2 + 9z^2 = 1$.
40. $x^2 + 4y^2 + 16z^2 = 1$.
41. Let $f(x, y) = e^{-1/(x^2+y^2)}$; $f(0,0) = 0$.
    (a) Sketch the level curve $f(x, y) = c$ for $c = 0.001$, $c = 0.01$, $c = 0.5$, and $c = 0.9$.
    (b) What happens if c is less than zero or greater than 1?
    (c) Sketch the cross section of the graph in the vertical plane $y = 0$ (that is, the intersection of the graph with the xz plane).
    (d) Argue that this cross section looks the same in any vertical plane through the origin.
    (e) Describe the graph in words and sketch it.
42. The formula

$$z = \frac{2x}{(y - y^{-1})^2 + (2x)^2}$$

appears in the study of steady state motions of a mechanical system with viscous damping subjected to a harmonic external force. The *average power input* by the external force is proportional to the variable z (with proportionality constant $k > 0$). The variable y is the ratio of input frequency to natural frequency. The variable x measures the viscous damping constant.
    (a) Plot z versus y for $x = 0.2, 0.5, 2.0$ on the same axes. Use the range of values $0 < y \leqslant 2.0$.
    (b) The average power input is a maximum when $y = 1$, that is, when the input and natural frequencies are the same. Verify this both graphically and algebraically.
43. The potential difference E between electrolyte solutions separated by a membrane is given by

$$E = \frac{RT}{F}\,\frac{x-y}{x+y}\ln z.$$

(The symbols R, T, F are the universal gas constant, absolute temperature, and Faraday unit, respectively—these are constants. The symbols x and y are the mobilities of Na^+ and Cl^- respectively. The symbol z is c_1/c_2, where c_1 and c_2 are the mean salt (NaCl) concentrations on each side of the membrane.) Assume hereafter that $RT/F = 25$.
    (a) Write the level surface $E = -12$ in the form $z = f(x, y)$.
    (b) In practice, $y = 3x/2$. Plot E versus z in this case.
★44. Describe the behavior, as c varies, of the level curve $f(x, y) = c$ for each of these functions:
    (a) $f(x, y) = x^2 + y^2 + 1$;
    (b) $f(x, y) = 1 - x^2 - y^2$;
    (c) $f(x, y) = x^2 + xy$;
    (d) $f(x, y) = x^3 - x$.

14.4 Quadric Surfaces

Quadric surfaces are defined by quadratic equations in x, y, and z.

The methods of Section 14.3, together with our knowledge of conics, enable us to graph a number of interesting surfaces defined by quadratic equations.

Example 1 Sketch the graph of

$$z = f(x, y) = x^2 - y^2 \qquad \text{(a } \textit{hyperbolic paraboloid}\text{).}$$

Solution To visualize this surface, we first draw the level curves $x^2 - y^2 = c$ for $c = 0, \pm 1, \pm 4$. For $c = 0$ we have $y^2 = x^2$ (that is, $y = \pm x$), so this level set consists of two straight lines through the origin. For $c = 1$ the level curve is $x^2 - y^2 = 1$, which is a hyperbola that passes vertically through the x axis at the points $(\pm 1, 0)$ (see Fig. 14.4.1). Similarly, for $c = 4$ the level curve is $x^2/4 - y^2/4 = 1$, the hyperbola passing vertically through the x axis at $(\pm 2, 0)$. For $c = -1$ we obtain the hyperbola $x^2 - y^2 = -1$ passing horizontally through the y axis at $(0, \pm 1)$, and for $c = -4$ the hyperbola through $(0, \pm 2)$ is obtained. These level curves are shown in Fig. 14.4.1. To aid us in visualizing the graph of f, we will also compute two sections. First, set $x = 0$ to obtain $z = -y^2$, a parabola opening downward. Second, setting $y = 0$ gives the parabola $z = x^2$ opening upward.

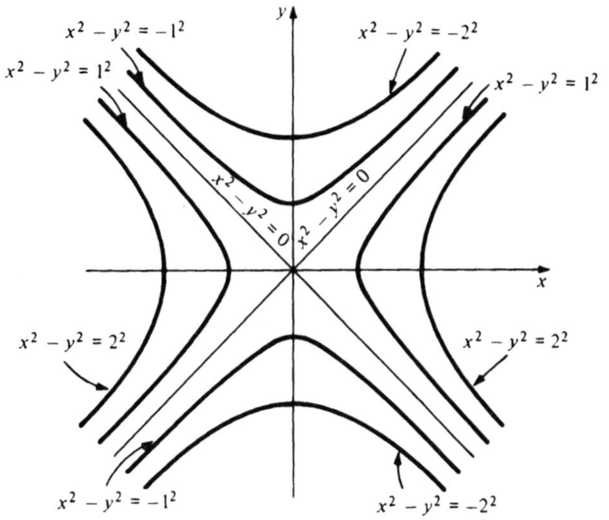

Figure 14.4.1. Some level curves of $f(x) = x^2 - y^2$.

The graph may now be visualized if we lift the level curves to the appropriate heights and smooth out the resulting surface. The placement of the lifted curves is aided by the use of the parabolic sections. This procedure generates the saddle-shaped surface indicated in Fig. 14.4.2. The graph is unchanged under reflection in the yz plane and in the xz plane. When accurately plotted by a computer, this graph has the appearance of Fig. 14.4.3; the level curves are shown in Fig. 14.4.4. (The graph has been rotated by 90° about the z axis.) ▲

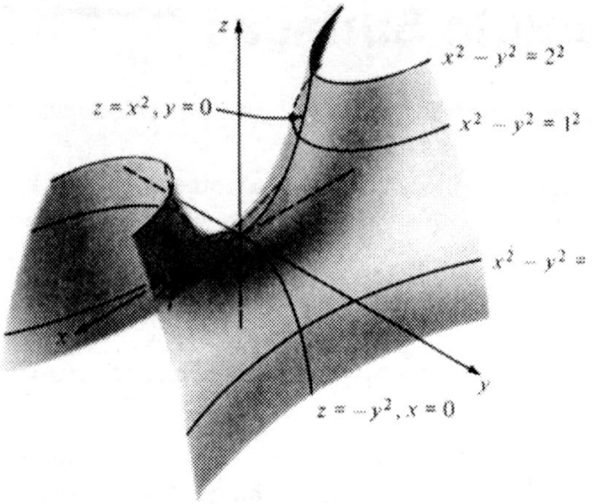

Figure 14.4.2. The graph $z = x^2 - y^2$ is a hyperbolic paraboloid, or "saddle."

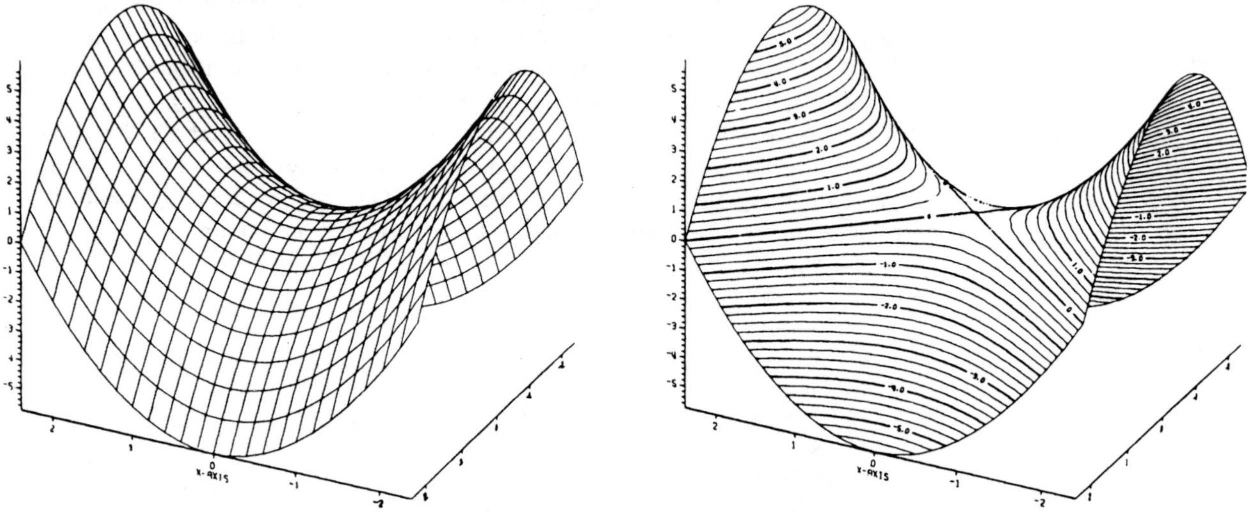

(a)

(b)

Figure 14.4.3. Computer-generated graph of $z = x^2 - y^2$.

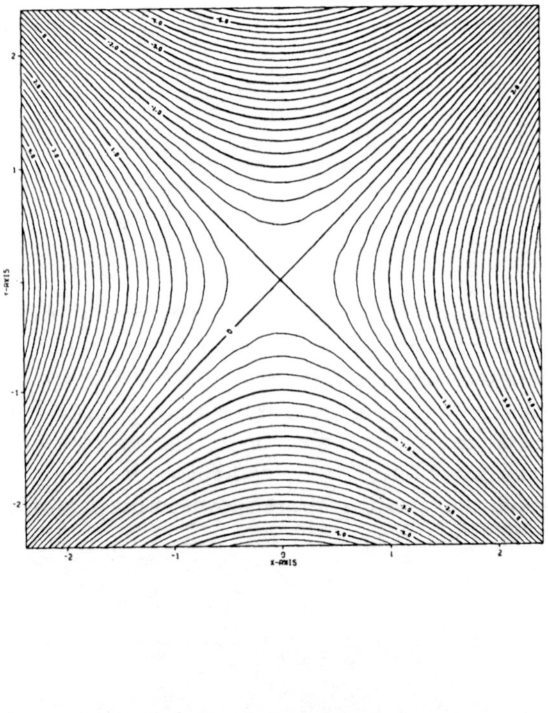

Figure 14.4.4. Level curves of $z = x^2 - y^2$ drawn by computer.

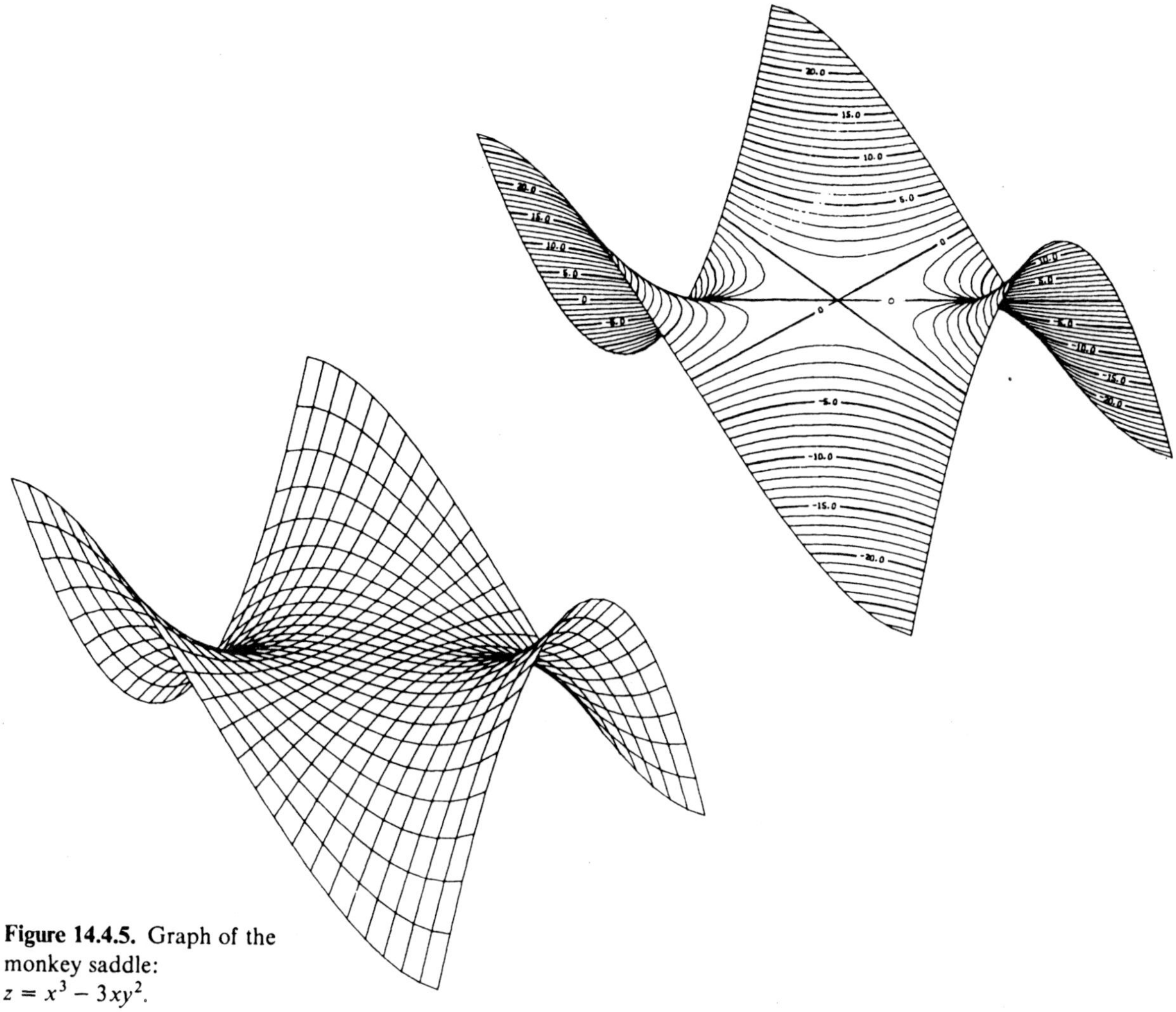

Figure 14.4.5. Graph of the monkey saddle: $z = x^3 - 3xy^2$.

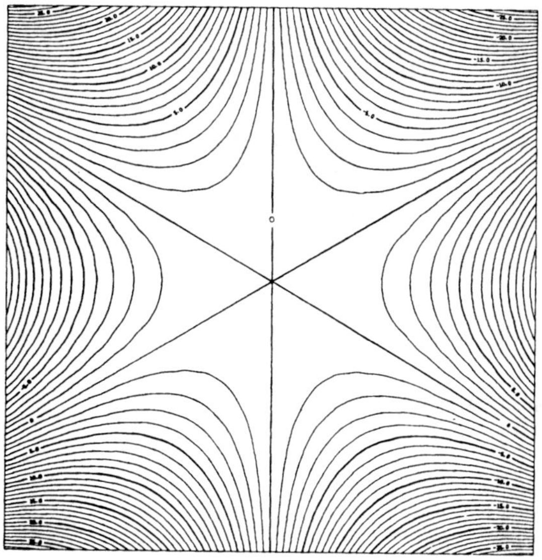

Figure 14.4.6. Level curves for the monkey saddle.

The origin is called a *saddle point* for the function $z = x^2 - y^2$ because of the appearance of the graph. We will return to the study of saddle points in Chapter 16, but it is worth noting another kind of saddle here. Figure 14.4.5 on the preceding page shows the graph of $z = x^3 - 3xy^2$, again plotted by a computer using sections and level curves. The origin now is called a *monkey saddle*, since there are two places for the legs and one for the tail. Figure 14.4.6 shows the contour lines in the plane. Figure 14.4.7 shows the four-legged or *dog saddle*: $z = 4x^3y - 4xy^3$.

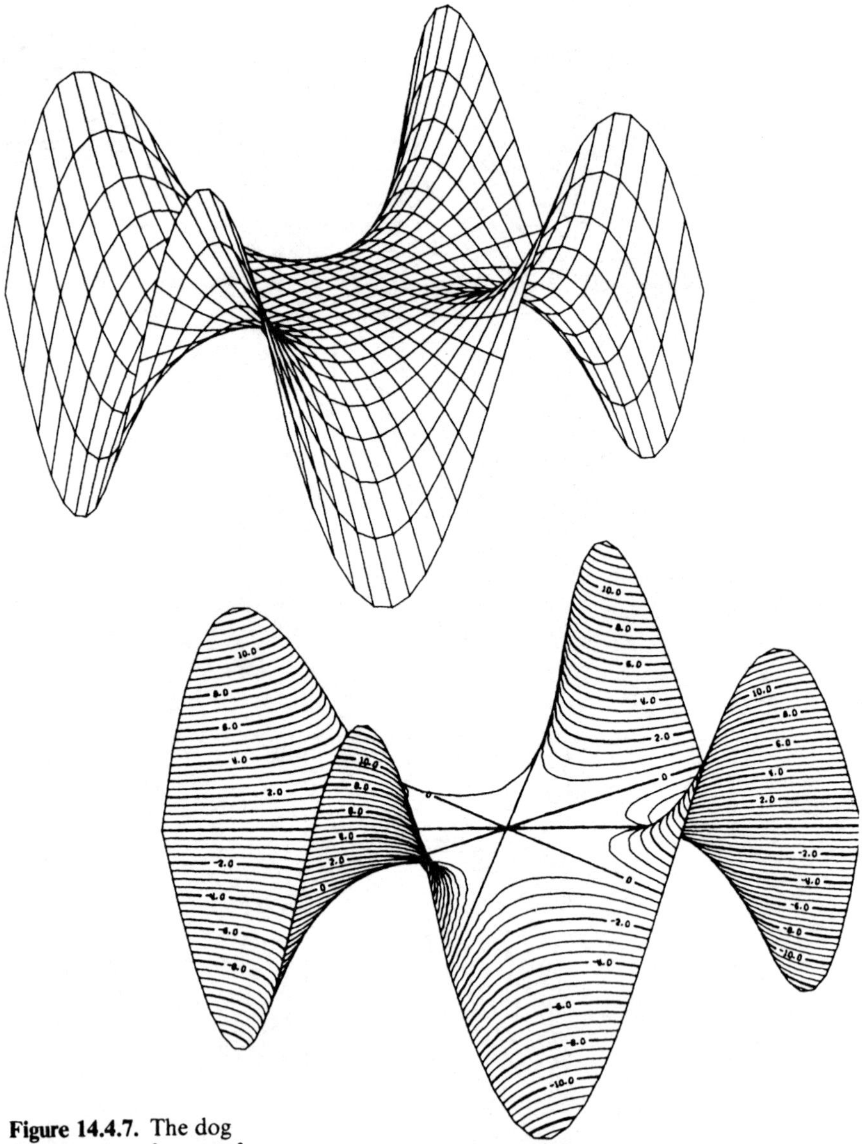

Figure 14.4.7. The dog saddle: $z = 4x^3y - 4xy^3$.

A *quadric surface* is a three-dimensional figure defined by a quadratic equation in three variables:

$$ax^2 + by^2 + cz^2 + dxy + exz + fyz + gx + hy + kz + m = 0.$$

The quadric surfaces are the three-dimensional versions of the conic sections, studied in Section 14.1, which were defined by quadratic equations in two variables.

Example 2 Particular conic sections can degenerate to points or lines. Similarly, some quadric surfaces can degenerate to points, lines, or planes. Match the sample equations to the appropriate descriptions.

(a) $x^2 + 3y^2 + z^2 = 0$	(1) No points at all
(b) $z^2 = 0$	(2) A single point
(c) $x^2 + y^2 = 0$	(3) A line
(d) $x^2 + y^2 + z^2 + 1 = 0$	(4) One plane
(e) $x^2 - y^2 = 0$	(5) Two planes

Solution Equation (a) matches (2) since only $(0,0,0)$ satisfies the equation; (b) matches (4) since this is the plane $z = 0$; (c) matches (3) since this is the z axis, where $x = 0$ and $y = 0$; (d) matches (1) since a non-negative number added to 1 can never be zero; (e) matches (5) since the equation $x^2 - y^2 = 0$ is equivalent to the two equations $x + y = 0$ or $x - y = 0$, which define two planes. ▲

If one variable is missing from an equation, we only have to find a curve in one plane and then extend it parallel to the axis of the missing variable. This procedure produces a generalized *cylinder*, either *elliptic*, *parabolic*, or *hyperbolic*.

Example 3 Sketch the surface $z = y^2 + 1$.

Solution The intersection of this surface with a plane $x =$ constant is a parabola of the form $z = y^2 + 1$. The surface, a *parabolic cylinder*, is sketched in Fig. 14.4.8. (See also Example 6, Section 14.3). ▲

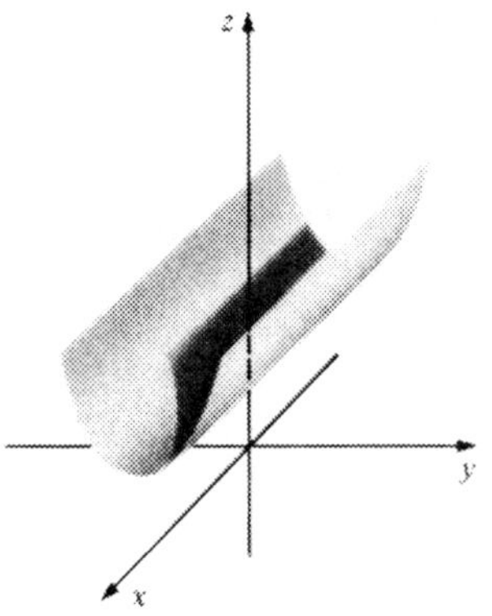

Figure 14.4.8. The surface $z = y^2 + 1$ is a parabolic cylinder.

Example 4 The surface defined by an equation of the form $x^2/a^2 + y^2/b^2 - z^2/c^2 = -1$ is called a *hyperboloid of two sheets*. Sketch the surface $x^2 + 4y^2 - z^2 = -4$.

Solution The section by the plane $z = c$ has the equation $x^2 + 4y^2 = c^2 - 4$. This is an ellipse when $|c| > 2$, a point when $c = \pm 2$, and is empty when $|c| < 2$. The section with the xz plane is the hyperbola $x^2 - z^2 = -4$, and the section with the yz plane is the hyperbola $4y^2 - z^2 = -4$. The surface is symmetric with respect to each of the coordinate planes. A sketch is given in Fig. 14.4.9. ▲

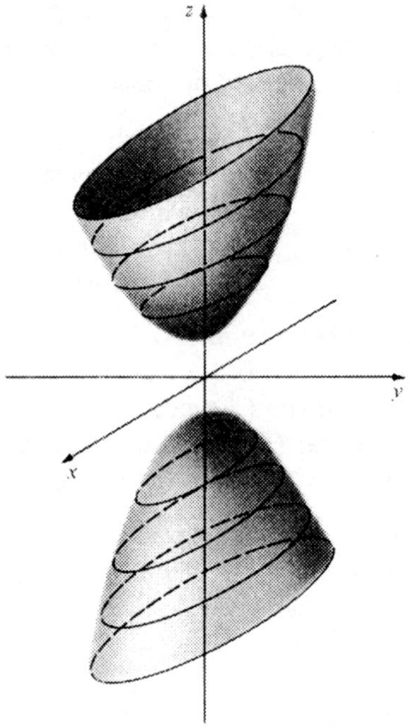

Figure 14.4.9. The surface $x^2 + 4y^2 - z^2 = -4$ is a hyperboloid of two sheets (shown with some of its sections by planes of the form z = constant).

Example 5 The surface defined by an equation of the form $x^2/a^2 + y^2/b^2 + z^2/c^2 = 1$ is called an *ellipsoid*. Sketch the surface $x^2/9 + y^2/16 + z^2 = 1$.

Solution First, let z be constant. Then we get $x^2/9 + y^2/16 = 1 - z^2$. This is an ellipse centered at the origin if $-1 < z < 1$. If $z = 1$, we just get a point $x = 0, y = 0$. Likewise, $(0, 0, -1)$ is on the surface. If $|z| > 1$ there are no (x, y) satisfying the equation.

Setting x = constant or y = constant, we also get ellipses. We must have $|x| \leqslant 3$ and, likewise, $|y| \leqslant 4$. The surface, shaped like a stepped-on football, is easiest to draw if the intersections with the three coordinate planes are drawn first. (See Fig. 14.4.10.) ▲

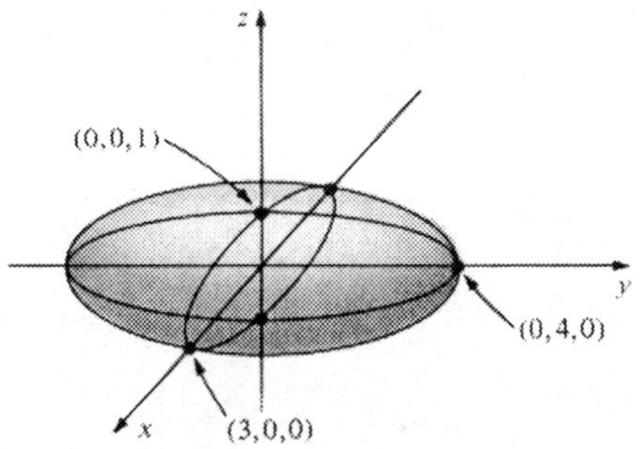

Figure 14.4.10. The surface $(x^2/9) + (y^2/16) + z^2 = 1$ is an ellipsoid.

Example 6 The surface defined by an equation of the form $x^2/a^2 + y^2/b^2 - z^2/c^2 = 1$ is called a *hyperboloid of one sheet*. Sketch the surface $x^2 + y^2 - z^2 = 4$.

Solution If z is a constant, then $x^2 + y^2 = 4 + z^2$ is a circle. Thus, in any plane parallel to the xy plane, we get a circle. Our job of drawing the surface is simplified if we note right away that the surface is rotationally invariant about the z axis (since z depends only on $r^2 = x^2 + y^2$). Thus we can draw the curve traced by the surface in the yz plane (or xz plane) and revolve it about the z axis. Setting $x = 0$, we get $y^2 - z^2 = 4$, a hyperbola. Hence we get the surface shown in Fig. 14.4.11, a one-sheeted hyperboloid. Since this surface is symmetric about the z axis, it is also called a hyperboloid of revolution. ▲

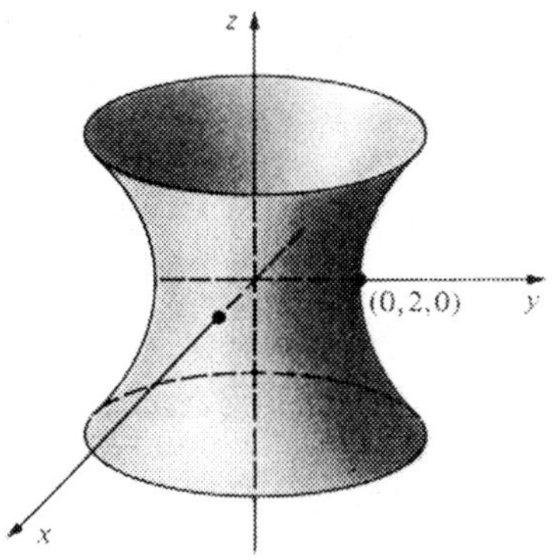

Figure 14.4.11. The surface $x^2 + y^2 - z^2 = 4$ is a one-sheeted hyperboloid of revolution.

The hyperboloid of one sheet has the property that it is *ruled*: that is, the surface is composed of straight lines (see Review Exercise 76). It is therefore easy to make with string models and is useful in architecture. (See Fig. 14.4.12.)

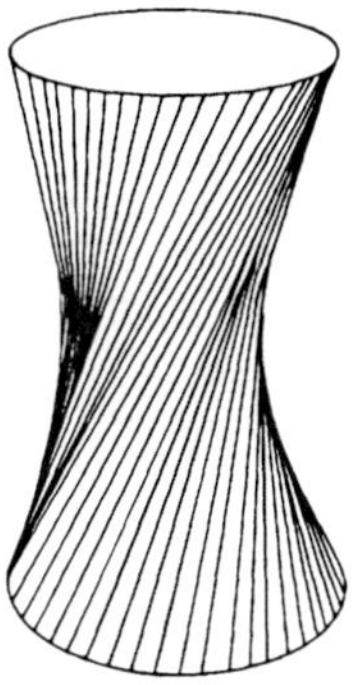

Figure 14.4.12. One can make a hyperboloid with a wire frame and string.

Example 7 Consider the equation $x^2 + y^2 - z^2 = 0$.

(a) What are the horizontal cross sections for $z = \pm 1, \pm 2, \pm 3$?

(b) What are the vertical cross sections for $x = 0$ or $y = 0$? (Sketch and describe.)

(c) Show that this surface is a cone by showing that any straight line through the origin making a 45° angle with the z axis lies in the surface.

(d) Sketch this surface.

Solution (a) Rewriting the equation as $x^2 + y^2 = z^2$ shows that the horizontal cross sections are circles centered around the z axis with radius $|z|$. Therefore, for $z = \pm 1$, ± 2, and ± 3, the cross sections are circles of radius 1, 2, and 3.

(b) When $x = 0$, the equation is $y^2 - z^2 = 0$ or $y^2 = z^2$ or $y = \pm z$, whose graph is two straight lines. When $y = 0$, the equation is $x^2 - z^2 = 0$ or $x = \pm z$, again giving two straight lines.

(c) Any point on a straight line through the origin making a 45° angle with the z axis satisfies $|z|/\sqrt{x^2 + y^2 + z^2} = \cos 45° = 1/\sqrt{2}$. Squaring gives $1/2 = z^2/(x^2 + y^2 + z^2)$, or $x^2 + y^2 + z^2 = 2z^2$, or $x^2 + y^2 - z^2 = 0$, which is the original equation.

(d) Draw a line as described in part (c) and rotate it around the z axis (see Figure 14.4.13). ▲

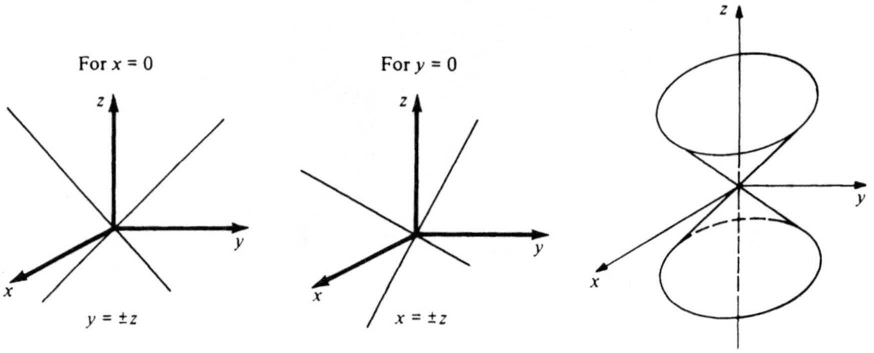

Figure 14.4.13. The cone $x^2 + y^2 - z^2 = 0$.

We now discuss how the conic sections, as introduced in the first section of this chapter, can actually be obtained by slicing a cone.

Example 8 Show that the intersection of the cone $x^2 + y^2 = z^2$ and the plane $y = 1$ is a hyperbola (see Figure 14.4.14).

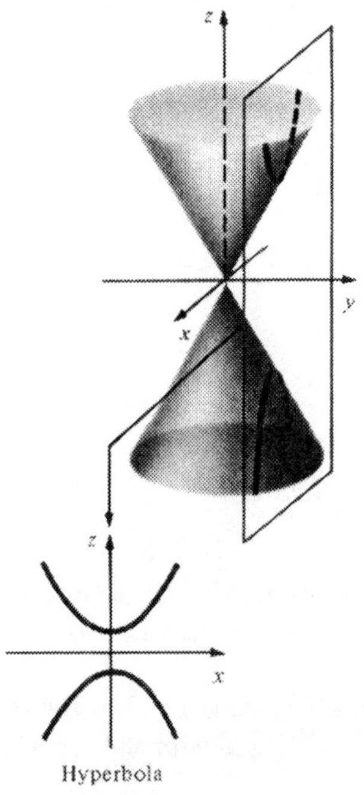

Figure 14.4.14. The intersection of this vertical plane and the cone is a hyperbola.

Solution The intersection of these two surfaces consists of all points (x, y, z) such that $x^2 + y^2 = z^2$ and $y = 1$. We can use x and z as coordinates to describe points in the plane. Thus, eliminating y, we get $x^2 + 1 = z^2$ or $z^2 - x^2 = 1$. From Section 14.1, we recognize this as a hyperbola with foci at $x = 0$, $z = \pm\sqrt{2}$ in the plane $y = 1$, with the branches opening vertically as in the figure. ▲

Example 9 Show that the intersection of the cone $x^2 + y^2 = z^2$ and the plane $z = y - 1$ is a parabola (see Figure 14.4.15).

Solution We introduce rectangular coordinates on the plane as follows. A normal vector to the plane is $\mathbf{n} = (0, 1, -1)$, and so a vector $\mathbf{w} = (a, b, c)$ is parallel to the plane if $0 = \mathbf{n} \cdot \mathbf{w} = b - c$. Two such vectors that are orthogonal and of unit length are

$$\mathbf{u} = \mathbf{i} \quad \text{and} \quad \mathbf{v} = \frac{1}{\sqrt{2}}(\mathbf{j} + \mathbf{k}).$$

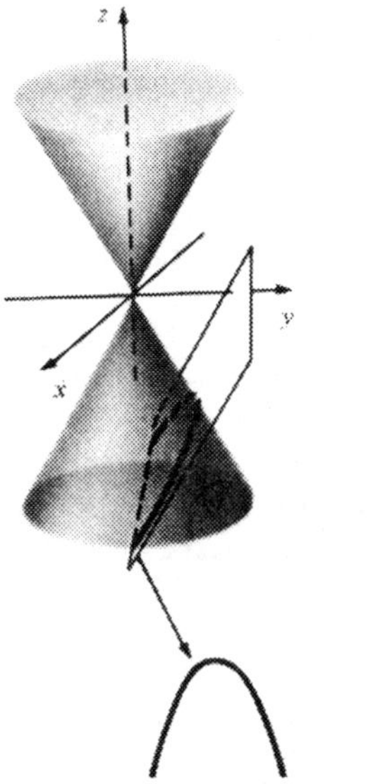

Figure 14.4.15. The intersection of the plane tilted at 45° and the cone is a parabola.

Pick a point on the plane, say $P_0 = (0, 0, -1)$, and write points $P = (x, y, z)$ in the plane in terms of coordinates (ξ, η) by writing

$$\overrightarrow{P_0P} = \xi\mathbf{u} + \eta\mathbf{v}$$

(see Fig. 14.4.16). In terms of (x, y, z), this reads

$$x = \xi, \quad y = \frac{\eta}{\sqrt{2}}, \quad \text{and} \quad z = \frac{\eta}{\sqrt{2}} - 1.$$

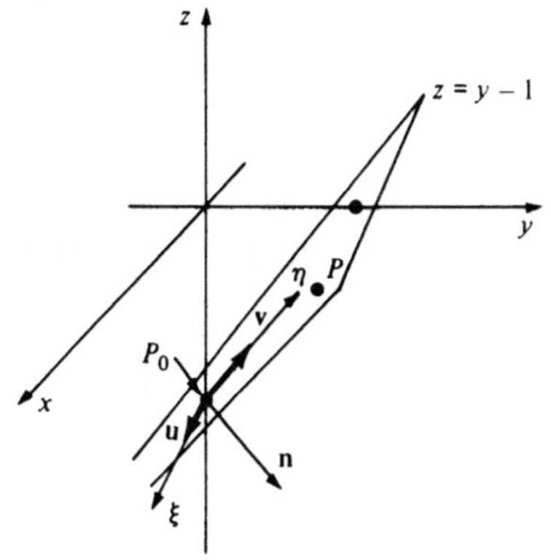

Figure 14.4.16. Coordinates (ξ, η) in the plane $z = y - 1$.

Substitution into $x^2 + y^2 = z^2$ gives

$$\xi^2 + \frac{\eta^2}{2} = \left(\frac{\eta}{\sqrt{2}} - 1\right)^2 = \frac{\eta^2}{2} - \sqrt{2}\eta + 1,$$

or

$$\xi^2 = -\sqrt{2}\,\eta + 1,$$

or

$$\eta = -\frac{1}{\sqrt{2}}\xi^2 + \frac{1}{\sqrt{2}}.$$

This, indeed, is a parabola opening downwards in the $\xi\eta$ plane. ▲

Other sections of the cone can be analyzed in a similar way, and one can prove that a conic will always result (see Exercise 27).

Exercises for Section 14.4

Sketch the surfaces in three-dimensional space defined by each of the equations in Exercises 1–16.

1. $y^2 + z^2 = 1$
2. $x^2 + y^2 = 0$
3. $9x^2 + 4z^2 = 36$
4. $4x^2 + y^2 = 2$
5. $z^2 - 8y^2 = 0$
6. $x^2 - z^2 = 1$
7. $8x^2 + 3z^2 = 0$
8. $x^2 = 4z^2 + 9$
9. $x^2 + y^2 + \frac{z^2}{4} = 1$
10. $\frac{x^2}{4} + y^2 + \frac{z^2}{4} = 1$
11. $x = 2z^2 - y^2$
12. $y = 4x^2 - z^2$
13. $x^2 + 9y^2 - z^2 = 1$
14. $x^2 + y^2 + 4z^2 = 1$
15. $16z^2 = 4x^2 + y^2 + 16$
16. $z^2 + 4y^2 = x^2 + 4$
17. This problem concerns the *hyperbolic paraboloid*. (A surface of this kind was studied in Example 1.) A standard form for the equation is $z = ax^2 - by^2$, with a and b both positive or both negative.
    (a) Sketch the graph of $z = x^2 - 2y^2$.
    (b) Show that $z = xy$ also determines a hyperbolic paraboloid. Sketch some of its level curves.
18. This exercise concerns the *elliptic paraboloid*:
    (a) Sketch the graph of $z = 2x^2 + y^2$.
    (b) Sketch the surface given by
    $$x = -3y^2 - 2z^2.$$
    (c) Consider the equation $z = ax^2 + by^2$, where a and b are both positive or both negative. Describe the horizontal cross sections where z = constant. Describe the sections obtained in the planes $x = 0$ and $y = 0$. What is the section obtained in the vertical plane $x = c$? (The special case in which $a = b$ is a *paraboloid of revolution* as in Example 7(a), Section 14.3.)
19. Sketch the cone $z^2 = 3x^2 + 3y^2$.
20. Sketch the cone $(z-1)^2 = x^2 + y^2$.
21. Sketch the cone $z^2 = x^2 + 2y^2$.
22. Sketch the cone $z^2 = x^2/4 + y^2/9$.
23. Show that the intersection of the cone $x^2 + y^2 = z^2$ and the plane $z = 1$ is a circle.
24. Show that the intersection of the cone $x^2 + y^2 = z^2$ and the plane $2z = y + 1$ is an ellipse.
25. This problem concerns the *elliptic cone*. Consider the equation $x^2/a^2 + y^2/b^2 - z^2/c^2 = 0$.
    (a) Describe the horizontal cross sections z = constant.
    (b) Describe the vertical cross sections $x = 0$ and $y = 0$.
    (c) Show that this surface has the property that if it contains the point (x_0, y_0, z_0), then it contains the whole line through $(0,0,0)$ and (x_0, y_0, z_0).

★26. The quadric surfaces may be shifted and rotated in space just as the conic sections may be shifted in the plane. These transformations will produce more complicated cases of the general quadratic equation in three variables. Complete squares to bring the following to one of the standard forms (shifted) and sketch the resulting surfaces:
    (a) $4x^2 + y^2 + 4z^2 + 8x - 4y - 8z + 8 = 0$;
    (b) $2x^2 + 3y^2 - 4z^2 + 4x + 9y - 8z + 10 = 0$.

★27. Show that the intersection of the cone $x^2 + y^2 = z^2$ and any plane is a conic section as follows. Let $\mathbf{u}$ and $\mathbf{v}$ be two orthonormal vectors and P_0 a point. Consider the plane described by points P such that $\overrightarrow{P_0P} = \xi\mathbf{u} + \eta\mathbf{v}$, which introduces rectangular coordinates (ξ, η) in the plane. Substitute an expression for (x, y, z) in terms of (ξ, η) into $x^2 + y^2 = z^2$ and show that the result is a conic section in the $\xi\eta$ plane.

14.5 Cylindrical and Spherical Coordinates

There are two ways to generalize polar coordinates to space.

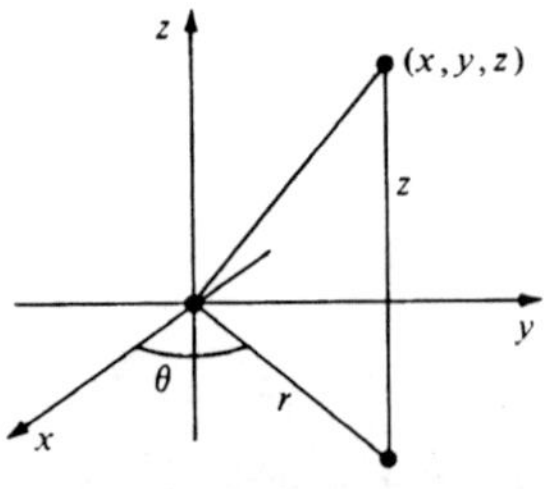

Figure 14.5.1. The cylindrical coordinates of the point (x, y, z).

In Sections 5.1, 5.6, and 10.5, we saw the usefulness of polar coordinates in the plane. In space there are two different coordinate systems analogous to polar coordinates, called cylindrical and spherical coordinates.

The *cylindrical coordinates* of a point (x, y, z) in space are the numbers (r, θ, z), where r and θ are the polar coordinates of (x, y); that is,

$$x = r\cos\theta, \quad y = r\sin\theta, \quad \text{and} \quad z = z.$$

See Fig. 14.5.1. As with polar coordinates, we can solve for r and θ in terms of x and y: squaring and adding gives

$$x^2 + y^2 = r^2(\cos^2\theta + \sin^2\theta) = r^2, \qquad \text{so} \quad r = \pm\sqrt{x^2 + y^2}\,.$$

Dividing gives

$$\frac{y}{x} = \frac{\sin\theta}{\cos\theta} = \tan\theta.$$

As in polar coordinates, it is sometimes convenient to allow negative r; thus (r,θ) and $(-r,\theta+\pi)$ represent the same point. Also, we recall that (r,θ) and $(r,\theta+2\pi)$ represent the same point. Sometimes we specify $r \geqslant 0$ (with $r=0$ corresponding to the z-axis) and a definite range for θ. If we choose θ between π and $-\pi$ and choose $\tan^{-1}u$ between $-\pi/2$ and $\pi/2$, then the solution of $y/x = \tan\theta$ is $\theta = \tan^{-1}(y/x)$ if $x>0$, and $\theta = \tan^{-1}(y/x)+\pi$ if $x<0$ ($\theta = \pi/2$ if $x=0$ and $y>0$, and $\theta = -\pi/2$ if $x=0$ and $y<0$).

Cylindrical Coordinates

If the cartesian coordinates of a point in space are (x, y, z), then the cylindrical coordinates of the point are (r,θ,z), where

$$x = r\cos\theta, \qquad y = r\sin\theta, \qquad z = z;$$

or, if we choose $r \geqslant 0$ and $-\pi < \theta \leqslant \pi$,

$$r = \sqrt{x^2+y^2},$$

$$\theta = \begin{cases} \tan^{-1}(y/x) & \text{if } x \geqslant 0, \\ \tan^{-1}(y/x)+\pi & \text{if } x < 0. \end{cases}$$

Example 1 (a) Find the cylindrical coordinates of $(6,6,8)$. Plot.

(b) If a point has cylindrical coordinates $(3, -\pi/6, -4)$, what are its cartesian coordinates? Plot.

(c) Let a point have cartesian coordinates $(2,-3,6)$. Find its cylindrical coordinates and plot.

(d) Let a point have cylindrical coordinates $(2, 3\pi/4, 1)$. Find its cartesian coordinates and plot.

Solution (a) Here $r = \sqrt{6^2+6^2} = 6\sqrt{2}$ and $\theta = \tan^{-1}(\frac{6}{6}) = \tan^{-1}(1) = \pi/4$. Thus the cylindrical coordinates are $(6\sqrt{2}, \pi/4, 8)$. See Fig. 14.5.2(a).

(b) $x = r\cos\theta = 3\cos(-\pi/6) = 3\sqrt{3}/2$, and $y = r\sin\theta = 3\sin(-\pi/6) = -3/2$. Thus the cartesian coordinates are $(3\sqrt{3}/2, -3/2, -4)$. See Fig. 14.5.2(b).

(c) $r = \sqrt{x^2+y^2} = \sqrt{2^2+(-3)^2} = \sqrt{13}$; $\theta = \tan^{-1}(-\frac{3}{2}) \approx -0.983 \approx -56.31°$; $z=6$. See Fig. 14.5.3(a).

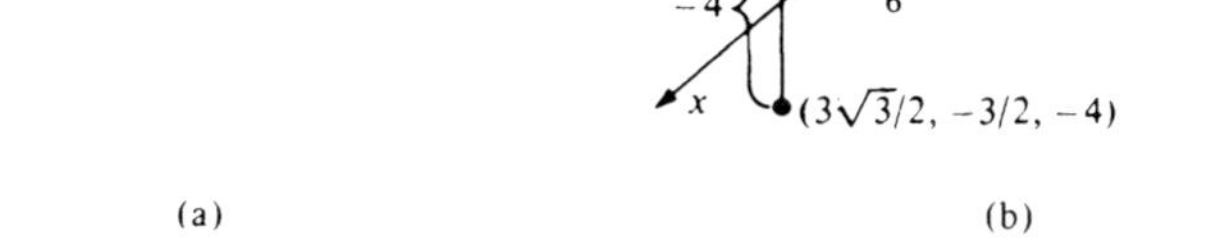

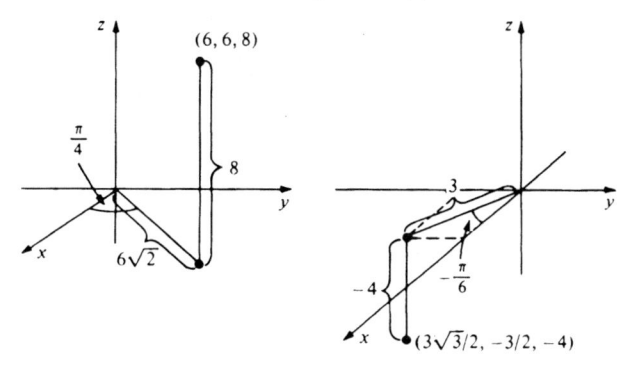

Figure 14.5.2. Comparing the cylindrical and cartesian coordinates of two points.

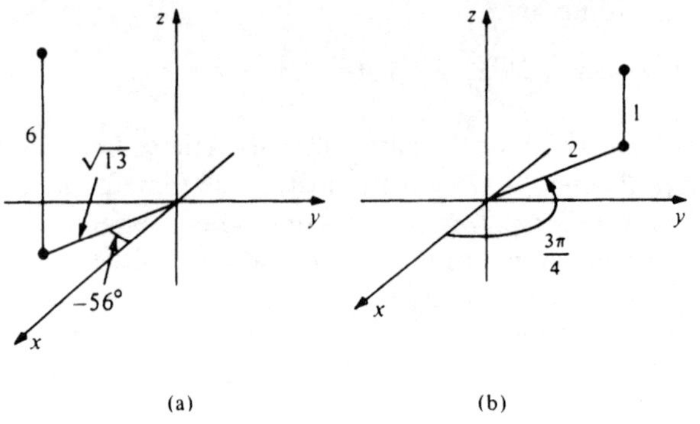

Figure 14.5.3. Two points in cylindrical coordinates.

(d) $x = r\cos\theta = 2\cos(3\pi/4) = 2\cdot\left(-\sqrt{2}/2\right) = -\sqrt{2}$;

$y = r\sin\theta = 2\sin(3\pi/4) = 2\cdot\left(\sqrt{2}/2\right) = \sqrt{2}$; $z = 1$.

See Fig. 14.5.3(b). ▲

Many surfaces are easier to describe in cylindrical than in cartesian coordinates, just as many curves are easier to work with using polar rather than cartesian coordinates.

Example 2 Plot the two surfaces described in cylindrical coordinates by (a) $r = 3$ and (b) $r = \cos 2\theta$.

Solution (a) Note that r is the distance from the given point to the z axis. Therefore the points with $r = 3$ lie on a cylinder of radius 3 centered on the z axis. See Fig. 14.5.4.

(b) The curve $r = \cos 2\theta$ in the xy plane is a four-petaled rose (see Example 1, Section 5.6). Thus in cylindrical coordinates we obtain a vertical cylinder with the four-leafed rose as a base, as shown in Fig. 14.5.5. ▲

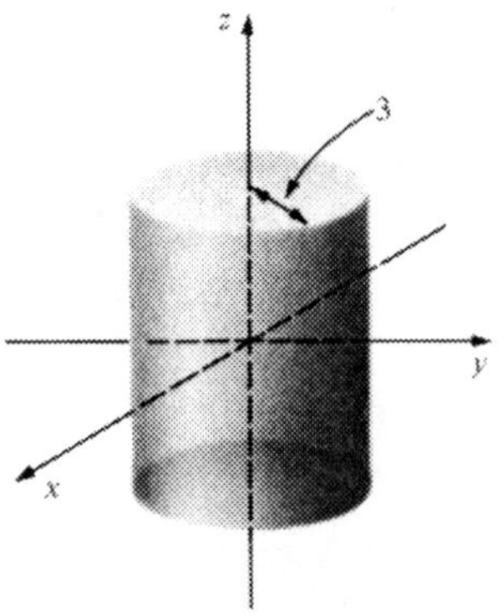

Figure 14.5.4. The cylinder has a very simple equation in cylindrical coordinates.

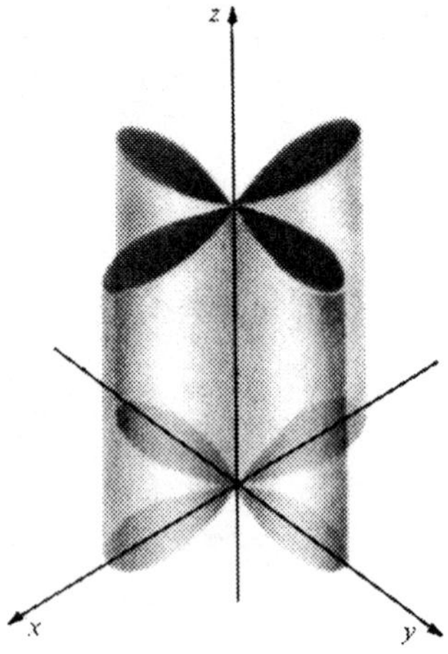

Figure 14.5.5. The surface $r = \cos 2\theta$ is a cylinder with a four-petaled rose as its base.

Example 3 Describe the geometric meaning of replacing (r, θ, z) by $(r, \theta + \pi, -z)$.

Solution Increasing θ by π is a rotation through 180° about the z axis. Switching z to $-z$ reflects in the xy plane (see Fig. 14.5.6). Combining the two operations results in reflection through the origin. ▲

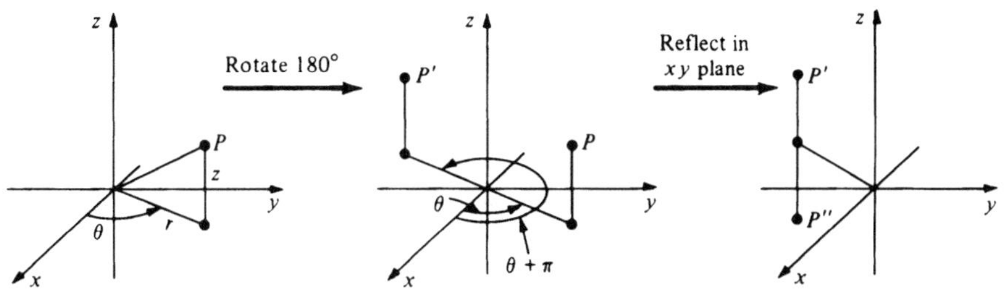

Figure 14.5.6. The effect of replacing (r, θ, z) by $(r, \theta + \pi, -z)$ is to replace P by $-P$.

Example 4 Show that the surface $r = f(z)$ is a surface of revolution.

Solution If we set $y = 0$ and take $x \geqslant 0$, then $r = x$ and $r = f(z)$ becomes $x = f(z)$; the remaining points satisfying $r = f(z)$ are then obtained by revolving the graph $x = f(z)$ about the z axis; note that $r = c$, $z = d$ is a circle centered on the z axis. Thus we get a surface of revolution with symmetry about the z axis. ▲

Cylindrical coordinates are best adapted to problems which have cylindrical symmetry—that is, a symmetry about the z axis. Similarly, for problems with spherical symmetry—that is, symmetry with respect to all rotations about the origin in space—the *spherical coordinate system* is useful.

The *spherical coordinates* of a point (x, y, z) in space are the numbers (ρ, θ, ϕ) defined as follows (see Fig. 14.5.7).

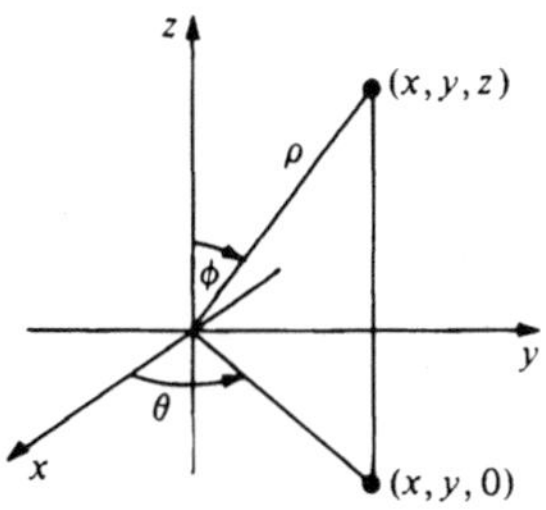

Figure 14.5.7. Spherical coordinates.

ρ = distance from (x, y, z) to the origin;
θ = cylindrical coordinate θ (angle from the positive x axis to the point (x, y));
ϕ = the angle (in $[0, \pi]$) from the positive z axis to the line from origin to (x, y, z).

To express the cartesian coordinates in terms of spherical coordinates, we first observe that the cylindrical coordinate $r = \sqrt{x^2 + y^2}$ is equal to $\rho \sin\phi$ and that $z = \rho\cos\phi$ (see Fig. 14.5.7). Therefore

$$x = r\cos\theta = \rho\sin\phi\cos\theta, \quad y = r\sin\theta = \rho\sin\phi\sin\theta, \quad z = \rho\cos\phi.$$

We may solve these equations for ρ, θ, and ϕ. The results are given in the following box.

Spherical Coordinates

If the cartesian coordinates of a point in space are (x, y, z), then the spherical coordinates of the point are (ρ, θ, ϕ), where

$$x = \rho\sin\phi\cos\theta,$$
$$y = \rho\sin\phi\sin\theta,$$
$$z = \rho\cos\phi,$$

or, if we choose $\rho \geqslant 0$, $-\pi < \theta \leqslant \pi$ and $0 \leqslant \phi \leqslant \pi$,

$$\rho = \sqrt{x^2 + y^2 + z^2},$$

$$\theta = \begin{cases} \tan^{-1}(y/x) & \text{if } x \geqslant 0, \\ \tan^{-1}(y/x) + \pi & \text{if } x < 0, \end{cases}$$

$$\phi = \cos^{-1}\frac{z}{\sqrt{x^2 + y^2 + z^2}}.$$

Notice that the spherical coordinates θ and ϕ are similar to the geographic coordinates of longitude and latitude if we take the earth's axis to be the z axis. There are differences, though: the geographical longitude is $|\theta|$ and is called east or west longitude according to whether θ is positive or negative; the geographical latitude is $|\pi/2-\phi|$ and is called north or south latitude according to whether $\pi/2-\phi$ is positive or negative.

Example 5 (a) Find the spherical coordinates of $(1,-1,1)$ and plot.
(b) Find the cartesian coordinates of $(3,\pi/6,\pi/4)$ and plot.
(c) Let a point have cartesian coordinates $(2,-3,6)$. Find its spherical coordinates and plot.
(d) Let a point have spherical coordinates $(1,-\pi/2,\pi/4)$. Find its cartesian coordinates and plot.

Solution (a) $\rho=\sqrt{x^2+y^2+z^2}=\sqrt{1^2+(-1)^2+1^2}=\sqrt{3}$,

$$\theta=\tan^{-1}\left(\frac{y}{x}\right)=\tan^{-1}\left(\frac{-1}{1}\right)=-\frac{\pi}{4},$$

$$\phi=\cos^{-1}\left(\frac{z}{\rho}\right)=\cos^{-1}\left(\frac{1}{\sqrt{3}}\right)\approx 0.955\approx 54.74°.$$

See Fig. 14.5.8(a).

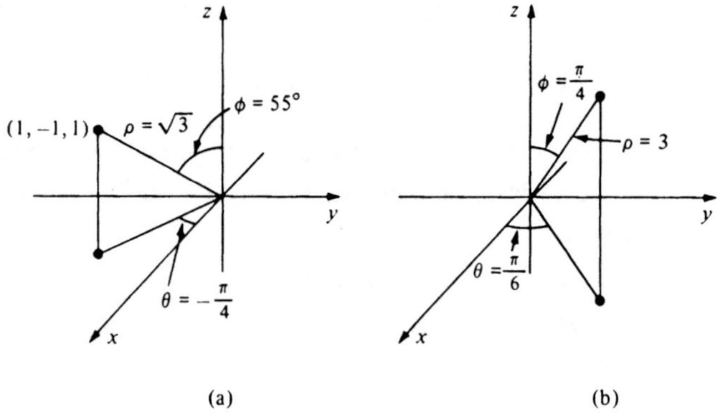

Figure 14.5.8. Finding the spherical coordinates of the point $(1,-1,1)$ and the cartesian coordinates of $(3,\pi/6,\pi/4)$.

(b) $x=\rho\sin\phi\cos\theta=3\sin\left(\frac{\pi}{4}\right)\cos\left(\frac{\pi}{6}\right)=3\left(\frac{1}{\sqrt{2}}\right)\frac{\sqrt{3}}{2}=\frac{3\sqrt{3}}{2\sqrt{2}}$,

$$y=\rho\sin\phi\sin\theta=3\sin\left(\frac{\pi}{4}\right)\sin\left(\frac{\pi}{6}\right)=3\left(\frac{1}{\sqrt{2}}\right)\left(\frac{1}{2}\right)=\frac{3}{2\sqrt{2}},$$

$$z=\rho\cos\phi=3\cos\left(\frac{\pi}{4}\right)=\frac{3\sqrt{2}}{2}.$$

See Fig. 14.5.8(b).

(c) $\rho=\sqrt{x^2+y^2+z^2}=\sqrt{2^2+(-3)^2+6^2}=\sqrt{49}=7$,

$$\theta=\tan^{-1}\left(\frac{y}{x}\right)=\tan^{-1}\left(\frac{-3}{2}\right)\approx -0.983\approx -56.31°$$

$$\phi=\cos^{-1}\left(\frac{z}{\rho}\right)=\cos^{-1}\left(\frac{6}{7}\right)\approx 0.541\approx 31.0°.$$

See Fig. 14.5.9(a) (the point is the same as in Example 1(d)).

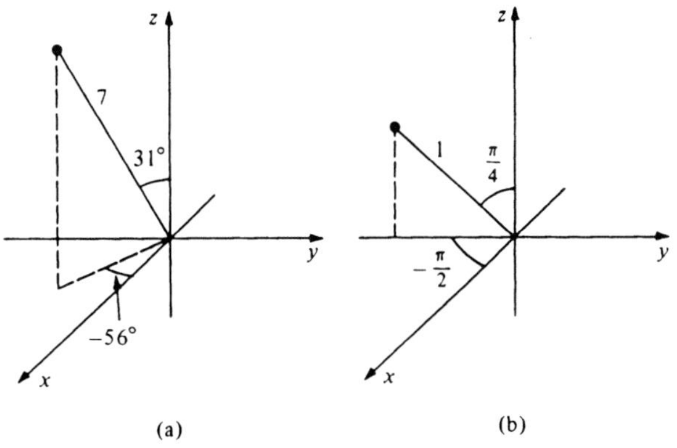

Figure 14.5.9. Two points in spherical coordinates.

(d) $x = \rho \sin\phi \cos\theta = 1 \sin\left(\frac{\pi}{4}\right)\cos\left(\frac{-\pi}{2}\right) = \left(\frac{\sqrt{2}}{2}\right)\cdot 0 = 0,$

$$y = \rho \sin\phi \sin\theta = 1 \sin\left(\frac{\pi}{4}\right)\sin\left(\frac{-\pi}{2}\right) = \left(\frac{\sqrt{2}}{2}\right)(-1) = -\frac{\sqrt{2}}{2},$$

$$z = \rho \cos\phi = 1 \cos\left(\frac{\pi}{4}\right) = \frac{\sqrt{2}}{2}.$$

See Fig. 14.5.9(b). ▲

Example 6 Find the equation in spherical coordinates of $x^2 + y^2 - z^2 = 4$ (a hyperboloid of revolution).

Solution To take advantage of the relationship $x^2 + y^2 + z^2 = \rho^2$, write

$$x^2 + y^2 - z^2 = (x^2 + y^2 + z^2) - 2z^2 = \rho^2 - 2\rho^2\cos^2\phi,$$

since $z = \rho\cos\phi$. Also, we can note that

$$\rho^2 - 2\rho^2\cos^2\phi = \rho^2(1 - 2\cos^2\phi) = -\rho^2\cos 2\phi.$$

Thus the surface is

$$\rho^2\cos 2\phi + 4 = 0. \ ▲$$

Example 7 (a) Describe the surface given in spherical coordinates by $\rho = 3$. (b) Describe the geometric meaning of replacing (ρ, θ, ϕ) by $(\rho, \theta + \pi, \phi)$.

Solution (a) In spherical coordinates, ρ is the distance from the point (x, y, z) to the origin. Thus $\rho = 3$ consists of all points a distance 3 from the origin—that is, a sphere of radius 3 centered at the origin. (b) Increasing θ by π has the effect of rotating about the z axis through an angle of 180°. ▲

Example 8 Show that the surface $\rho = f(\phi)$ is a surface of revolution.

Solution The equation $\rho = f(\phi)$ does not involve θ and hence is independent of rotations about the z axis; thus it is a surface of revolution. If we set $y = 0$, then $\rho = \sqrt{x^2 + z^2}$ and $\phi = \cos^{-1}(z/\sqrt{x^2 + z^2})$. Thus the surface $\rho = f(\phi)$ is obtained by revolving the curve in the xz plane given by

$$\sqrt{x^2 + z^2} = f\left(\cos^{-1}\left(\frac{z}{\sqrt{x^2 + z^2}}\right)\right),$$

about the z axis. ▲

Exercises for Section 14.5

In Exercises 1–6, convert from cartesian to cylindrical coordinates and plot:

1. $(1,-1,0)$
2. $(\sqrt{2},1,1)$
3. $(3,-2,1)$
4. $(0,6,-2)$
5. $(6,0,-2)$
6. $(-1,1,1)$

In Exercises 7–12, convert from cylindrical to cartesian coordinates and plot.

7. $(1,\pi/2,0)$
8. $(3,45°,8)$
9. $(-1,\pi/6,4)$
10. $(2,0,1)$
11. $(0,\pi/18,6)$
12. $(2,-\pi/4,3)$
13. Sketch the surface described in cylindrical coordinates by $r = 1 + 2\cos\theta$.
14. Sketch the surface given in cylindrical coordinates by $r = 1 + \cos\theta$.

In Exercises 15–18, describe the geometric meaning of the stated replacement.

15. (r,θ,z) by $(r,\theta,-z)$
16. (r,θ,z) by $(2r,\theta,z)$
17. (r,θ,z) by $(2r,\theta,-z)$
18. (r,θ,z) by $(2r,\theta+\pi,z)$
19. Describe the surfaces $r=$ constant, $\theta=$ constant, and $z=$ constant in cylindrical coordinates.
20. Describe the surface given in cylindrical coordinates by $z=\theta$.

In Exercises 21–26, convert from cartesian to spherical coordinates and plot.

21. $(0,1,1)$
22. $(1,0,1)$
23. $(-2,1,-3)$
24. $(1,2,3)$
25. $(-3,-2,-4)$
26. $(1,1,1)$

In Exercises 27–32, convert from spherical to cartesian coordinates and plot.

27. $(3,\pi/3,\pi)$
28. $(2,-\pi/6,\pi/3)$
29. $(3,2\pi,0)$
30. $(1,\pi/6,\pi/3)$
31. $(8,-\pi/3,\pi)$
32. $(1,\pi/2,\pi/2)$
33. Express the surface $xz=1$ in spherical coordinates.
34. Express the surface $z = x^2 + y^2$ in spherical coordinates.
35. Describe the surface given in spherical coordinates by $\theta = \pi/4$.
36. Describe the surface given in spherical coordinates by $\rho = \phi$.
37. Describe the geometric meaning of replacing (ρ,θ,ϕ) by $(2\rho,\theta,\phi)$.
38. Describe the geometric meaning of replacing (ρ,θ,ϕ) by $(\rho,\theta,\phi+\pi/2)$ in spherical coordinates.
39. Describe the curve given in spherical coordinates by $\rho = 1$, $\phi = \pi/2$.
40. Describe the curve given in spherical coordinates by $\rho = 1$, $\theta = 0$.

In Exercises 41–46, convert each of the points from cartesian to cylindrical and spherical coordinates and plot.

41. $(0,3,4)$
42. $(-\sqrt{2},1,0)$
43. $(0,0,0)$
44. $(-1,0,1)$
45. $(-2\sqrt{3},-2,3)$
46. $(-1,1,0)$

In Exercises 47–52, the points are given in cylindrical coordinates. Convert to cartesian and spherical coordinates:

47. $(1,\pi/4,1)$
48. $(3,\pi/6,-4)$
49. $(0,\pi/4,1)$
50. $(2,-\pi/2,1)$
51. $(-2,-\pi/2,1)$
52. $(1,-\pi/6,2)$

In Exercises 53–58, the points are given in spherical coordinates. Convert to cartesian and cylindrical coordinates and plot.

53. $(1,\pi/2,\pi)$
54. $(2,-\pi/2,\pi/6)$
55. $(0,\pi/8,\pi/35)$
56. $(2,-\pi/2,-\pi)$
57. $(-1,\pi,\pi/6)$
58. $(-1,-\pi/4,\pi/2)$
59. Express the surface $z = x^2 - y^2$ (a hyperbolic paraboloid) in (a) cylindrical and (b) spherical coordinates.
60. Express the plane $z = x$ in (a) cylindrical and (b) spherical coordinates.
61. Show that in spherical coordinates:
    (a) ρ is the length of $x\mathbf{i} + y\mathbf{j} + z\mathbf{k}$;
    (b) $\phi = \cos^{-1}(\mathbf{v}\cdot\mathbf{k}/\|\mathbf{v}\|)$, where $\mathbf{v} = x\mathbf{i} + y\mathbf{j} + z\mathbf{k}$;
    (c) $\theta = \cos^{-1}(\mathbf{u}\cdot\mathbf{i}/\|\mathbf{u}\|)$, where $\mathbf{u} = x\mathbf{i} + y\mathbf{j}$.
62. Two surfaces are described in spherical coordinates by the equations $\rho = f(\theta,\phi)$ and $\rho = -2f(\theta,\phi)$, where $f(\theta,\phi)$ is a function of two variables. How is the second surface obtained geometrically from the first?
63. A circular membrane in space lies over the region $x^2 + y^2 \leqslant a^2$. The maximum deflection z of the membrane is b. Assume that (x,y,z) is a point on the deflected membrane. Show that the corresponding point (r,θ,z) in cylindrical coordinates satisfies the conditions
    $0 \leqslant r \leqslant a,\ 0 \leqslant \theta \leqslant 2\pi,\ |z| \leqslant b.$
64. A tank in the shape of a right circular cylinder of radius 10 feet and height 16 feet is half filled and lying on its side. Describe the air space inside the tank by suitably chosen cylindrical coordinates.
65. A vibrometer is to be designed which withstands the heating effects of its spherical enclosure of diameter d, which is buried to a depth $d/3$ in the earth, the upper portion being heated by the sun. Heat conduction analysis requires a description of the buried portion of the enclosure, in spherical coordinates. Find it.
66. An oil filter cartridge is a porous right circular cylinder inside which oil diffuses from the axis to the outer curved surface. Describe the cartridge in cylindrical coordinates, if the diameter of the filter is 4.5″, the height is 5.6″ and the center of the cartridge is drilled (all the way through) from the top to admit a $\frac{3}{8}$″ diameter bolt.

★67. Describe the surface given in spherical coordinates by $\rho = \cos 2\theta$.

14.6 Curves in Space

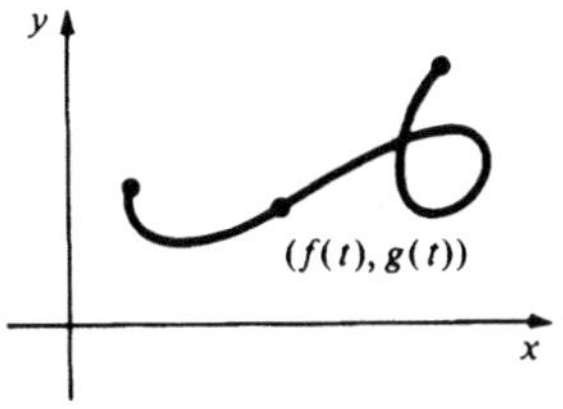

Figure 14.6.1. A parametric curve in the plane.

Tangents and velocities of curves in space can be computed by vector methods.

We continue our study of three-dimensional geometry by considering curves in space. We can consider tangents to these curves by using calculus, since only the calculus of functions of one variable and a knowledge of vectors are required. (To determine tangent planes to surfaces, we will need the calculus of functions of several variables.)

Recall from Section 2.4 that a parametric curve in the plane consists of a pair of functions $(x, y) = (f(t), g(t))$. As t ranges through some interval (on which f and g are defined), the point (x, y) traces out a curve in the plane; see Fig. 14.6.1.

Example 1 What curve is traced out by $(\sin t, 2\cos t)$, $0 \leqslant t \leqslant 2\pi$?

Solution Since $x = \sin t$ and $y/2 = \cos t$, (x, y) satisfies $x^2 + y^2/4 = 1$, so the curve traced out is an ellipse. As t goes from zero to 2π, the moving point goes once around the ellipse, starting and ending at P (Fig. 14.6.2). ▲

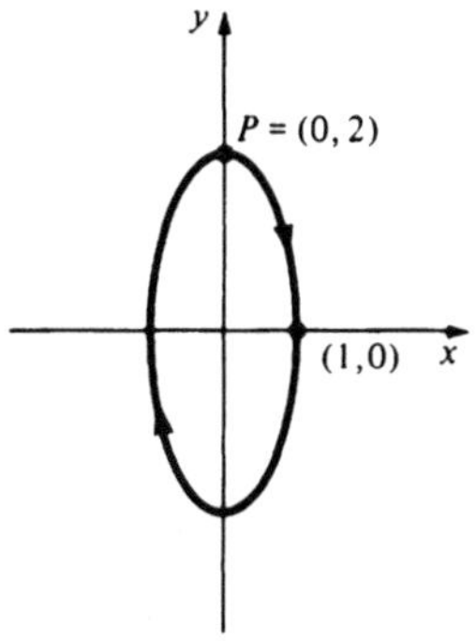

Figure 14.6.2. The ellipse traced out by $(\sin t, 2\cos t)$.

The step from two to three dimensions is accomplished by adding one more function; i.e., we state the following definition: A *parametric curve in space* consists of three functions $(x, y, z) = (f(t), g(t), h(t))$ defined for t in some interval on which f, g, and h are defined.

The curve we "see" is the path traced out by the point (x, y, z) as t varies, just as for curves in the plane.

Example 2 (a) Sketch the parametric curve $(x, y, z) = (3t + 2, 8t - 1, t)$. (b) Describe the curve $x = 3t^3 + 2$, $y = t^3 - 8$, $z = 4t^3 + 3$.

Solution (a) If we write $P = (x, y, z)$, then

$$P = (2, -1, 0) + t(3, 8, 1)$$

which is a straight line through $(2, -1, 0)$ in the direction $(3, 8, 1)$ (see Section 13.3). To sketch it, we pick the points obtained by setting $t = 0$ and $t = 1$, that is, $(2, -1, 0)$ and $(5, 7, 1)$; see Fig. 14.6.3 on the next page.

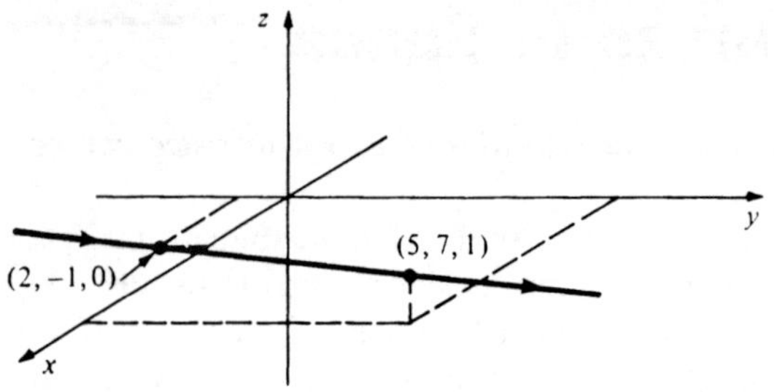

Figure 14.6.3. The parametric curve $(3t+2, 8t-1, t)$ is a straight line.

(b) We find

$$\begin{aligned}(x, y, z) &= (3t^3+2, t^3-8, 4t^3+3)\\ &= (2, -8, 3) + t^3(3, 1, 4);\end{aligned}$$

so the curve is a straight line through $(2, -8, 3)$ in the direction $(3, 1, 4)$. ▲

Example 3 (a) Sketch the curve given by $x = \cos t$, $y = \sin t$, $z = t$, where $-\infty < t < \infty$.
(b) Sketch the curve $(\cos t, 2\sin t, 2t)$.

Solution (a) As t varies, the point (x, y) traces out a circle in the plane. Thus (x, y, z) is a path which circles around the z axis, but at value t, its height above the xy plane is $z = t$. Thus we get the helix shown in Fig. 14.6.4. (It is called a *right circular helix*, since it lies on the right circular cylinder $x^2 + y^2 = 1$.)

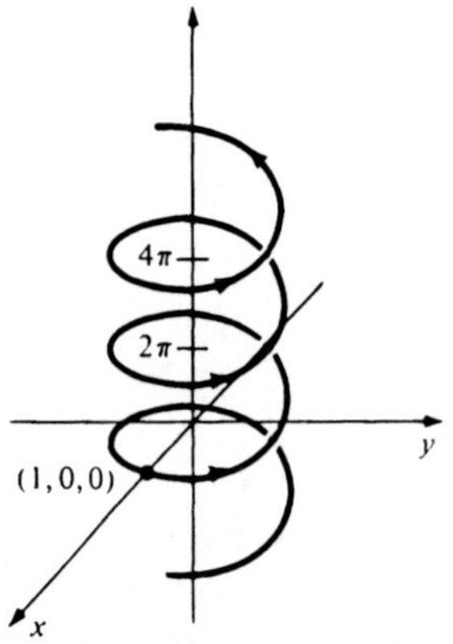

Figure 14.6.4. The curve $(\cos t, \sin t, t)$ is a helix.

In Fig. 14.6.4, the z axis has been drawn with a different scale than the x and y axes so that more coils of the helix can be shown. It is often useful to do something like this when displaying sketches of curves or graphs. You should be careful, however, not to give a false impression—label the axes to show the scale when necessary.

(b) Since $x = \cos t$ and $y/2 = \sin t$, the point $(x, y, 0)$ satisfies $x^2 + y^2/4 = 1$, so the curve lies over this ellipse in the xy plane. As t increases from zero to 2π, the projection in the xy plane goes once around the same ellipse as in Example 1 (Fig. 14.6.2), only now it starts at $(1, 0, 0)$ at $t = 0$ and proceeds counterclockwise since x behaves like $\cos t$ and y like $2\sin t$. Meanwhile, z increases steadily with t according to the formula $z = 2t$. The net result is a helix winding around the z axis, much like that of part (a), but no longer circular. It now lies on a cylinder of elliptical cross section (see Fig. 14.6.5). ▲

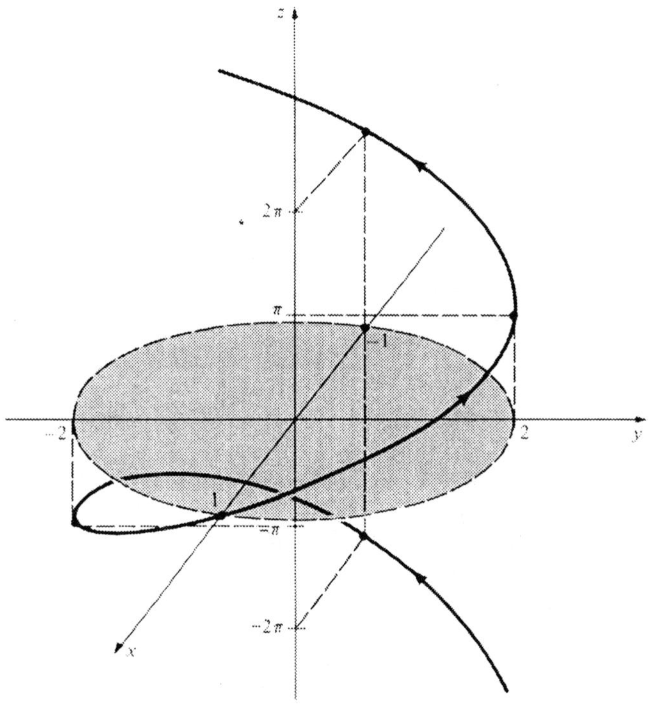

Figure 14.6.5. The curve $(\cos t, 2\sin t, 2t)$ is an elliptical helix.

Example 4 Sketch the curve $(t, 2t, \cos t)$.

Solution If we ignore z temporarily, we note that $(t, 2t)$ describes the line $y = 2x$ in the xy plane. As t varies, $(t, 2t)$ moves along this line. Thus $(t, 2t, \cos t)$ moves along a curve over this line with the z component oscillating as $\cos t$. Thus we get the curve shown in Fig. 14.6.6. ▲

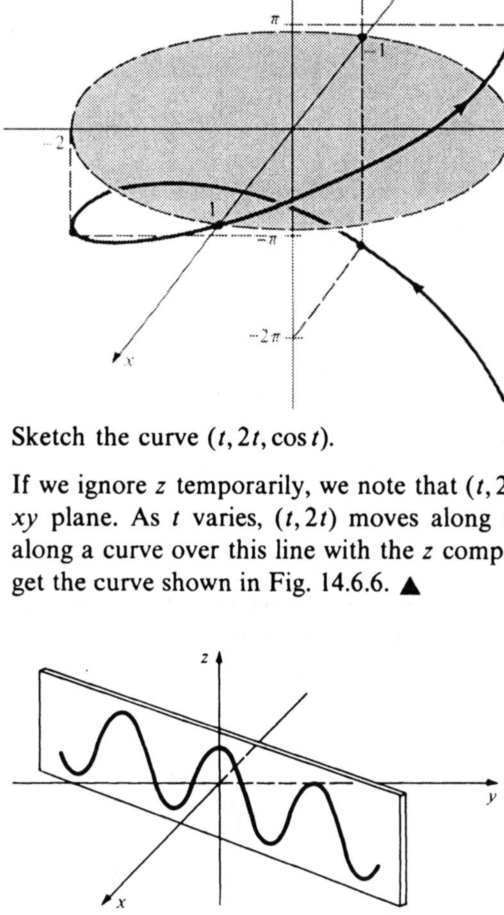

Figure 14.6.6. The curve $(t, 2t, \cos t)$ lies in the plane $y = 2x$.

In doing calculus with parametric curves, it is useful to identify the point $P = (x, y, z) = (f(t), g(t), h(t))$ with the vector

$$\mathbf{r} = x\mathbf{i} + y\mathbf{j} + z\mathbf{k} = f(t)\mathbf{i} + g(t)\mathbf{j} + h(t)\mathbf{k}.$$

This vector is a *function* of t, according to the following definition.

Vector Functions

A *vector function* of one variable is a rule $\boldsymbol{\sigma}$ which associates a vector $\mathbf{r} = \boldsymbol{\sigma}(t)$ in space (or the plane) to each real number t in some domain.

If $\boldsymbol{\sigma}$ is a vector function and t is in its domain, we can express $\boldsymbol{\sigma}(t)$ in terms of the standard basis vectors, $\mathbf{i}$, $\mathbf{j}$, and $\mathbf{k}$. The coefficients will themselves depend upon t, so we may write

$$\boldsymbol{\sigma}(t) = f(t)\mathbf{i} + g(t)\mathbf{j} + h(t)\mathbf{k},$$

where f, g, and h are scalar (real-valued) functions with the same domain as $\boldsymbol{\sigma}$. Notice that the functions $f(t)$, $g(t)$, $h(t)$ define a parametric curve such that the displacement vector from the origin to $(f(t), g(t), h(t))$ is just $\boldsymbol{\sigma}(t)$. The functions f, g, and h are called the *component functions* of the vector function $\boldsymbol{\sigma}(t)$. To summarize, we may say that parametric curves, vector functions, and triples of scalar functions are mathematically equivalent objects; we simply visualize them differently. For instance, the wind velocity at a fixed place on earth, or the cardiac vector (see Fig. 13.2.14), may be visualized as a vector depending on time.

Example 5 Let $\mathbf{u}$, $\mathbf{v}$, and $\mathbf{w}$ be three vectors such that $\mathbf{v}$ and $\mathbf{w}$ are perpendicular and have the same length r, and let[3] $\boldsymbol{\sigma}(t) = \mathbf{u} + \mathbf{v}\cos t + \mathbf{w}\sin t$.

(a) Describe the motion of the tip of $\boldsymbol{\sigma}(t)$ if the tail of $\boldsymbol{\sigma}(t)$ is fixed at the origin. (That is, describe the parametric curve corresponding to $\boldsymbol{\sigma}(t)$.)

(b) Find the component functions of $\boldsymbol{\sigma}(t)$ if $\mathbf{u} = 2\mathbf{i} + \mathbf{j}$, $\mathbf{v} = \mathbf{j} - \mathbf{k}$, and $\mathbf{w} = \mathbf{j} + \mathbf{k}$.

Solution (a) We observe first that the vector $\mathbf{v}\cos t + \mathbf{w}\sin t$ always lies in the plane spanned by $\mathbf{v}$ and $\mathbf{w}$ and that the square of its length is

$$\begin{aligned}(\mathbf{v}\cos t + \mathbf{w}\sin t)\cdot(\mathbf{v}\cos t + \mathbf{w}\sin t)\\ &= \mathbf{v}\cdot\mathbf{v}\cos^2 t + 2\mathbf{v}\cdot\mathbf{w}\sin t\cos t + \mathbf{w}\cdot\mathbf{w}\sin^2 t\\ &= r^2\cos^2 t + r^2\sin^2 t = r^2(\cos^2 t + \sin^2 t) = r^2,\end{aligned}$$

so the tip of the vector $\mathbf{v}\cos t + \mathbf{w}\sin t$ moves in a circle of radius r if its tail is fixed. Adding $\mathbf{u}$ to $\mathbf{v}\cos t + \mathbf{w}\sin t$ to get $\boldsymbol{\sigma}(t)$, we find that the tip of $\boldsymbol{\sigma}(t)$ moves in a circle of radius r whose center is at the tip of $\mathbf{u}$. (See Fig. 14.6.7.)

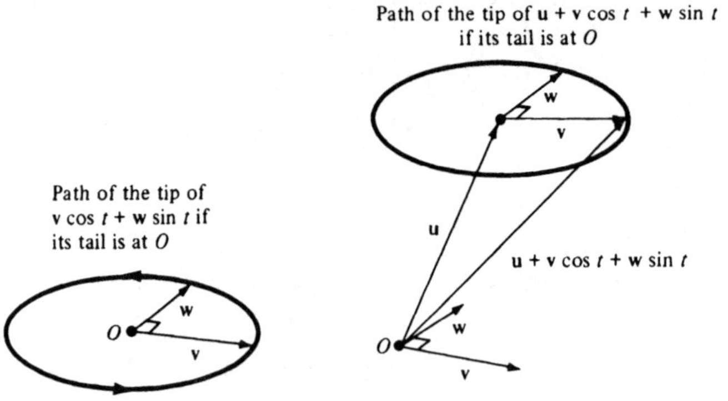

Figure 14.6.7. The tip of $\mathbf{u} + \mathbf{v}\cos t + \mathbf{w}\sin t$ moves in a circle of radius r with center at the tip of $\mathbf{u}$ and in a plane parallel to that spanned by $\mathbf{v}$ and $\mathbf{w}$.

(b) We have

$$\begin{aligned}\boldsymbol{\sigma}(t) &= \mathbf{u} + \mathbf{v}\cos t + \mathbf{w}\sin t = 2\mathbf{i} + \mathbf{j} + (\mathbf{j} - \mathbf{k})\cos t + (\mathbf{j} + \mathbf{k})\sin t\\ &= 2\mathbf{i} + (1 + \cos t + \sin t)\mathbf{j} + (-\cos t + \sin t)\mathbf{k},\end{aligned}$$

so the component functions are 2, $1 + \cos t + \sin t$, and $-\cos t + \sin t$. ▲

[3] Formulas involving vector functions are sometimes clearer to write and read if scalars are placed to the right of vectors. Any expression of the form $\mathbf{v}f(t)$ is to be interpreted as $f(t)\mathbf{v}$.

We now wish to define the rate of change, or *derivative*, of a vector function $\boldsymbol{\sigma}(t)$ with respect to t. If $\boldsymbol{\sigma}(t)$ is the displacement from a fixed origin to a moving point, this derivative will represent the velocity of the point. To see how the derivative should be defined, we examine the case of uniform rectilinear motion.

Example 6 Let $\boldsymbol{\sigma}(t) = \mathbf{u} + t\mathbf{v}$, so that $\boldsymbol{\sigma}(t)$ is the displacement from the origin to a point moving uniformly with velocity vector $\mathbf{v}$. Let $\mathbf{u} = a\mathbf{i} + b\mathbf{j} + c\mathbf{k}$ and $\mathbf{v} = l\mathbf{i} + m\mathbf{j} + n\mathbf{k}$.

(a) Find the component functions of $\boldsymbol{\sigma}(t)$.
(b) Show that the components of the velocity vector are obtained by differentiating the component functions of $\boldsymbol{\sigma}(t)$.

Solution (a) We have

$$\begin{aligned}\boldsymbol{\sigma}(t) &= \mathbf{u} + t\mathbf{v} = a\mathbf{i} + b\mathbf{j} + c\mathbf{k} + t(l\mathbf{i} + m\mathbf{j} + n\mathbf{k})\\ &= (a + lt)\mathbf{i} + (b + mt)\mathbf{j} + (c + nt)\mathbf{k},\end{aligned}$$

so the component functions are $a + lt$, $b + mt$, and $c + nt$.
(b) The derivatives of the component functions of $\boldsymbol{\sigma}(t)$ are the constants l, m, and n; these are precisely the components of the velocity vector $\mathbf{v}$. ▲

Example 7 Let $\boldsymbol{\sigma}(t) = f(t)\mathbf{i} + g(t)\mathbf{j}$ be a vector function in the plane. Show that the tangent line at time t_0 to the parametric curve corresponding to $\boldsymbol{\sigma}(t)$ (with the tail of $\boldsymbol{\sigma}(t)$ fixed at zero) has the direction of the vector $f'(t_0)\mathbf{i} + g'(t_0)\mathbf{j}$.

Solution Recall from Section 2.4 that if $(f(t), g(t))$ is a parametrized curve in the plane, then the slope of its tangent line at $(f(t_0), g(t_0))$ is $g'(t_0)/f'(t_0)$. A line in the direction of $f'(t_0)\mathbf{i} + g'(t_0)\mathbf{j}$ has slope $g'(t_0)/f'(t_0)$, so it is in the same direction as the tangent line. ▲

Guided by Examples 6 and 7, we make the following definition.

Derivative of a Vector Function

Let $\boldsymbol{\sigma}(t) = f(t)\mathbf{i} + g(t)\mathbf{j} + h(t)\mathbf{k}$ be a vector function. If the coordinate functions f, g, and h are all differentiable at t_0, then we say that $\boldsymbol{\sigma}$ is differentiable at t_0, and we define the *derivative* $\boldsymbol{\sigma}'(t_0)$ to be the vector $f'(t_0)\mathbf{i} + g'(t_0)\mathbf{j} + h'(t_0)\mathbf{k}$:

$$\boldsymbol{\sigma}'(t_0) = f'(t_0)\mathbf{i} + g'(t_0)\mathbf{j} + h'(t_0)\mathbf{k}.$$

The derivative of $\boldsymbol{\sigma}$ is a function of the value of t at which the derivative is evaluated. Thus $\boldsymbol{\sigma}'(t)$ is a new vector function, and we may consider the second derivative $\boldsymbol{\sigma}''(t)$, as well as higher derivatives.

We will sometimes use Leibniz notation for derivatives of vector functions: if $\mathbf{r} = \boldsymbol{\sigma}(t)$, we will write $d\mathbf{r}/dt$ for $\boldsymbol{\sigma}'(t)$ and $d^2\mathbf{r}/dt^2$ for $\boldsymbol{\sigma}''(t)$.

The derivative of a vector function can also be expressed as a limit of difference quotients. If $\mathbf{r} = \boldsymbol{\sigma}(t)$, we write $\Delta\mathbf{r} = \boldsymbol{\sigma}(t + \Delta t) - \boldsymbol{\sigma}(t)$. Then $\Delta\mathbf{r}/\Delta t$ (i.e., the scalar $1/\Delta t$ times the vector $\Delta\mathbf{r}$) is a vector which approaches $\boldsymbol{\sigma}'(t)$ as $\Delta t \to 0$. (See Fig. 14.6.8 on the next page and Exercise 52.)

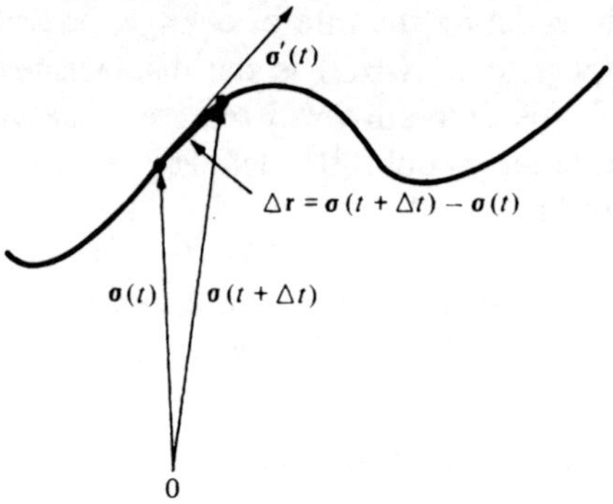

Figure 14.6.8. As $\Delta t \to 0$, the quotient $[\boldsymbol{\sigma}(t + \Delta t) - \boldsymbol{\sigma}(t)]/\Delta t$ approaches $\boldsymbol{\sigma}'(t)$; i.e., $\Delta \mathbf{r}/\Delta t \to d\mathbf{r}/dt$.

Example 8 Let $\boldsymbol{\sigma}(t)$ be the vector function of Example 5, with **u**, **v**, and **w** as in part (b) of that example. Find $\boldsymbol{\sigma}'(t)$ and $\boldsymbol{\sigma}''(t)$.

Solution In terms of components,

$$\boldsymbol{\sigma}(t) = 2\mathbf{i} + (1 + \cos t + \sin t)\mathbf{j} + (-\cos t + \sin t)\mathbf{k}.$$

Differentiating the components, we have

$$\boldsymbol{\sigma}'(t) = (-\sin t + \cos t)\mathbf{j} + (\sin t + \cos t)\mathbf{k}$$

and

$$\boldsymbol{\sigma}''(t) = (-\cos t - \sin t)\mathbf{j} + (\cos t - \sin t)\mathbf{k}. \ \blacktriangle$$

The differentiation of vector functions is facilitated by algebraic rules which follow from the corresponding rules for scalar functions. We list the rules in the following box.

Differentiation Rules for Vector Functions

To differentiate a vector function $\boldsymbol{\sigma}(t) = f(t)\mathbf{i} + g(t)\mathbf{j} + h(t)\mathbf{k}$, differentiate it component by component: $\boldsymbol{\sigma}'(t) = f'(t)\mathbf{i} + g'(t)\mathbf{j} + h'(t)\mathbf{k}$. Let $\boldsymbol{\sigma}(t)$, $\boldsymbol{\sigma}_1(t)$, and $\boldsymbol{\sigma}_2(t)$ be vector functions and let $p(t)$ and $q(t)$ be scalar functions.

Sum Rule: $\dfrac{d}{dt}\left[\boldsymbol{\sigma}_1(t) + \boldsymbol{\sigma}_2(t)\right] = \boldsymbol{\sigma}_1'(t) + \boldsymbol{\sigma}_2'(t).$

Scalar Multiplication Rule: $\dfrac{d}{dt}\left[p(t)\boldsymbol{\sigma}(t)\right] = p'(t)\boldsymbol{\sigma}(t) + p(t)\boldsymbol{\sigma}'(t).$

Dot Product Rule: $\dfrac{d}{dt}\left[\boldsymbol{\sigma}_1(t)\cdot\boldsymbol{\sigma}_2(t)\right] = \boldsymbol{\sigma}_1'(t)\cdot\boldsymbol{\sigma}_2(t) + \boldsymbol{\sigma}_1(t)\cdot\boldsymbol{\sigma}_2'(t).$

Cross Product Rule: $\dfrac{d}{dt}\left[\boldsymbol{\sigma}_1(t)\times\boldsymbol{\sigma}_2(t)\right] = \boldsymbol{\sigma}_1'(t)\times\boldsymbol{\sigma}_2(t) + \boldsymbol{\sigma}_1(t)\times\boldsymbol{\sigma}_2'(t).$

Chain Rule: $\dfrac{d}{dt}\left[\boldsymbol{\sigma}(q(t))\right] = q'(t)\boldsymbol{\sigma}'(q(t)).$

For example, to prove the dot product rule, let $\boldsymbol{\sigma}_1(t) = f_1(t)\mathbf{i} + g_1(t)\mathbf{j} + h_1(t)\mathbf{k}$ and $\boldsymbol{\sigma}_2(t) = f_2(t)\mathbf{i} + g_2(t)\mathbf{j} + h_2(t)\mathbf{k}$. Hence,

$$\boldsymbol{\sigma}_1(t)\cdot\boldsymbol{\sigma}_2(t) = f_1(t)f_2(t) + g_1(t)g_2(t) + h_1(t)h_2(t),$$

so by the sum and product rules for real-valued functions, we have

$$\frac{d}{dt}\left[\boldsymbol{\sigma}_1(t)\cdot\boldsymbol{\sigma}_2(t)\right] = \left[f_1'(t)f_2(t) + f_1(t)f_2'(t)\right] + \left[g_1'(t)g_2(t) + g_1(t)g_2'(t)\right] + \left[h_1'(t)h_2(t) + h_1(t)h_2'(t)\right].$$

Regrouping terms, we can rewrite this as

$$\begin{aligned}&[f_1'(t)f_2(t)+g_1'(t)g_2(t)+h_1'(t)h_2(t)]\\&\quad+[f_1(t)f_2'(t)+g_1(t)g_2'(t)+h_1(t)h_2'(t)]\\&=[f_1'(t)\mathbf{i}+g_1'(t)\mathbf{j}+h_1'(t)\mathbf{k}]\cdot[f_2(t)\mathbf{i}+g_2(t)\mathbf{j}+h_2(t)\mathbf{k}]\\&\quad+[f_1(t)\mathbf{i}+g_1(t)\mathbf{j}+h_1(t)\mathbf{k}]\cdot[f_2'(t)\mathbf{i}+g_2'(t)\mathbf{j}+h_2'(t)\mathbf{k}]\\&=\boldsymbol{\sigma}_1'(t)\cdot\boldsymbol{\sigma}_2(t)+\boldsymbol{\sigma}_1(t)\cdot\boldsymbol{\sigma}_2'(t),\end{aligned}$$

so the dot product rule is proved. The other rules are proved in a similar way (Exercises 53–56).

Example 9 Show that if $\boldsymbol{\sigma}(t)$ is a vector function such that $\|\boldsymbol{\sigma}(t)\|$ is constant, then $\boldsymbol{\sigma}'(t)$ is perpendicular to $\boldsymbol{\sigma}(t)$ for all t.

Solution Since $\|\boldsymbol{\sigma}(t)\|$ is constant, so is its square $\|\boldsymbol{\sigma}(t)\|^2=\boldsymbol{\sigma}(t)\cdot\boldsymbol{\sigma}(t)$. The derivative of this constant is zero, so by the dot product rule we have

$$0=\frac{d}{dt}[\boldsymbol{\sigma}(t)\cdot\boldsymbol{\sigma}(t)]=\boldsymbol{\sigma}'(t)\cdot\boldsymbol{\sigma}(t)+\boldsymbol{\sigma}(t)\cdot\boldsymbol{\sigma}'(t)=2\boldsymbol{\sigma}(t)\cdot\boldsymbol{\sigma}'(t);$$

so $\boldsymbol{\sigma}(t)\cdot\boldsymbol{\sigma}'(t)=0$; that is, $\boldsymbol{\sigma}'(t)$ is perpendicular to $\boldsymbol{\sigma}(t)$. ▲

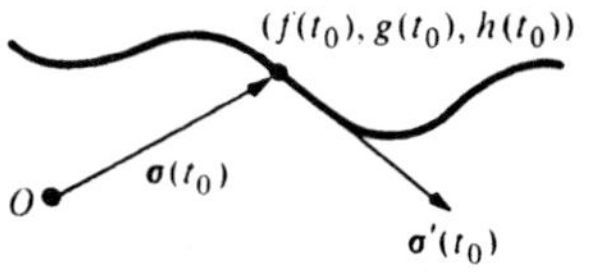

Figure 14.6.9. The velocity vector of a parametric curve is the derivative of the vector $\boldsymbol{\sigma}(t)$ from the origin to the curve.

Let $(f(t), g(t), h(t))$ be a parametric curve. If f, g, and h are differentiable at t_0, the vector $f'(t_0)\mathbf{i}+g'(t_0)\mathbf{j}+h'(t_0)\mathbf{k}$ is called the *velocity vector* of the curve at t_0. Notice that if $\boldsymbol{\sigma}(t)$ is the vector function corresponding to the curve $(f(t), g(t), h(t))$, then the velocity vector at t_0 is just $\boldsymbol{\sigma}'(t_0)$ (see Fig. 14.6.9). We often write $\mathbf{v}$ for the velocity vector—that is, $\mathbf{v}=\boldsymbol{\sigma}'(t)$. In Leibniz notation, if $\mathbf{r}=\boldsymbol{\sigma}(t)$, we have $\mathbf{v}=d\mathbf{r}/dt$.

Several other quantities of interest may be defined in terms of the velocity vector. If $\mathbf{v}=\boldsymbol{\sigma}'(t_0)$ is the velocity of a curve at t_0, then the length $v=\|\mathbf{v}\|=\|\boldsymbol{\sigma}'(t_0)\|$ is called the *speed* along the curve at t_0, and the line through $\boldsymbol{\sigma}(t_0)$ in the direction of $\boldsymbol{\sigma}'(t_0)$ (assuming $\boldsymbol{\sigma}'(t_0)\neq 0$) is called the *tangent line* to the curve (see Example 7). Thus the tangent line is given by $\mathbf{r}=\boldsymbol{\sigma}(t_0)+t\boldsymbol{\sigma}'(t_0)$.

For a curve describing uniform rectilinear motion, the velocity vector is constant (see Example 6). In general, the velocity vector is a vector function $\mathbf{v}=\boldsymbol{\sigma}'(t)$ which depends on t. The derivative $\mathbf{a}=d\mathbf{v}/dt=\boldsymbol{\sigma}''(t)$ is called the *acceleration vector* of the curve. Notice that if the curve is $(f(t), g(t), h(t))$, then the acceleration vector is

$$\mathbf{a}=f''(t)\mathbf{i}+g''(t)\mathbf{j}+h''(t)\mathbf{k}.$$

The terms *velocity*, *speed*, and *acceleration* come from physics, where parametric curves represent the motion of particles. These topics will be discussed in the next section.

Example 10 A particle moves in a helical path along the curve $(\cos t, \sin t, t)$. (a) Find its velocity and acceleration vectors. (b) Find its speed. (c) Find the tangent line at $t_0=\pi/4$.

Solution (a) Differentiating the components, we have $\mathbf{v}=-(\sin t)\mathbf{i}+(\cos t)\mathbf{j}+\mathbf{k}$, and $\mathbf{a}=d\mathbf{v}/dt=-(\cos t)\mathbf{i}-(\sin t)\mathbf{j}$. Notice that the acceleration vector points directly from $(\cos t, \sin t, t)$ to the z axis and is perpendicular to the axis as well as to the velocity vector (see Fig. 14.6.10).

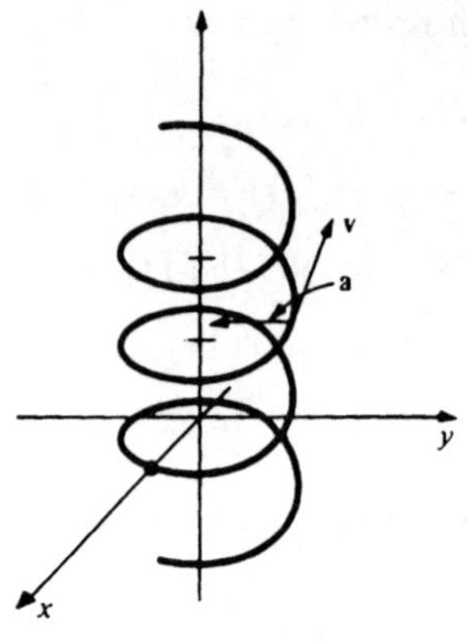

Figure 14.6.10. The velocity and acceleration of a particle moving on a helix.

(b) The velocity vector is $\mathbf{v} = -(\sin t)\mathbf{i} + (\cos t)\mathbf{j} + \mathbf{k}$, so the speed is

$$v = \|\mathbf{v}\| = \sqrt{(-\sin t)^2 + (\cos t)^2 + 1}$$
$$= \sqrt{\sin^2 t + \cos^2 t + 1} = \sqrt{2}\,.$$

(c) The tangent line is

$$\mathbf{r} = \boldsymbol{\sigma}(t_0) + t\boldsymbol{\sigma}'(t_0) = (\cos t_0)\mathbf{i} + (\sin t_0)\mathbf{j} + t_0\mathbf{k} + t\big[(-\sin t_0)\mathbf{i} + (\cos t_0)\mathbf{j} + \mathbf{k}\big].$$

At $t_0 = \pi/4$, we get

$$\mathbf{r} = \frac{1}{\sqrt{2}}(\mathbf{i} + \mathbf{j}) + \frac{\pi}{4}\mathbf{k} + t\left[-\frac{1}{\sqrt{2}}\mathbf{i} + \frac{1}{\sqrt{2}}\mathbf{j} + \mathbf{k}\right]$$
$$= \frac{1-t}{\sqrt{2}}\mathbf{i} + \frac{1+t}{\sqrt{2}}\mathbf{j} + \left(\frac{\pi}{4} + t\right)\mathbf{k}. \ \blacktriangle$$

Example 11 A particle moves in such a way that its acceleration is constantly equal to $-\mathbf{k}$. If the position when $t = 0$ is $(0, 0, 1)$ and the velocity at $t = 0$ is $\mathbf{i} + \mathbf{j}$, when and where does the particle fall below the plane $z = 0$? Describe the path travelled by the particle.

Solution Let $(f(t), g(t), h(t))$ be the parametric curve traced out by the particle, so that the velocity vector is $\boldsymbol{\sigma}'(t) = f'(t)\mathbf{i} + g'(t)\mathbf{j} + h'(t)\mathbf{k}$. The acceleration $\boldsymbol{\sigma}''(t)$ is equal to $-\mathbf{k}$, so we must have $f''(t) = 0$, $g''(t) = 0$, and $h''(t) = -1$. It follows that $f'(t)$ and $g'(t)$ are constant functions, and $h'(t)$ is a linear function with slope -1. Since $\boldsymbol{\sigma}'(0) = \mathbf{i} + \mathbf{j}$, we must have $\boldsymbol{\sigma}'(t) = \mathbf{i} + \mathbf{j} - t\mathbf{k}$. Integrating again and using the initial position $(0, 0, 1)$, we find that $(f(t), g(t), h(t)) = (t, t, 1 - \frac{1}{2}t^2)$. The particle drops below the plane $z = 0$ when $1 - \frac{1}{2}t^2 = 0$; that is, $t = \sqrt{2}$. At that time, the position is $(\sqrt{2}, \sqrt{2}, 0)$. The path travelled by the particle is a parabola in the plane $y = x$. (See Fig. 14.6.11.) $\blacktriangle$

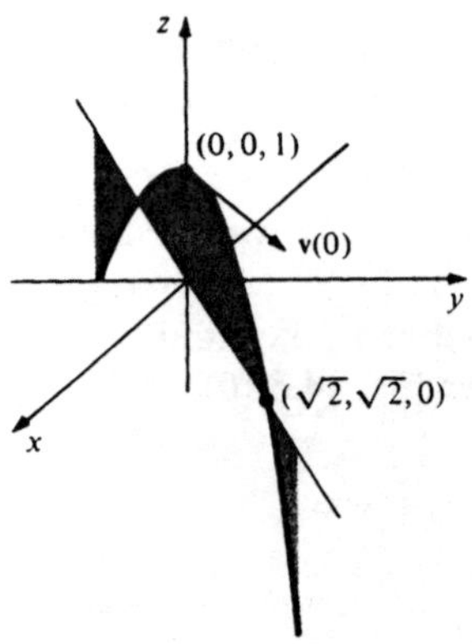

Figure 14.6.11. The path of the parabola with initial position $(0, 0, 1)$, initial velocity $\mathbf{i} + \mathbf{j}$, and constant acceleration $-\mathbf{k}$ is a parabola in the plane $y = x$.

Exercises for Section 14.6

Sketch the curves in Exercises 1–10.

1. $x = \sin t$, $y = 4\cos t$, $0 \leq t \leq 2\pi$.
2. $x = 2\sin t$, $y = 4\cos t$, $0 \leq t \leq 2\pi$.
3. $x = 2t - 1$; $y = t + 2$; $z = t$.
4. $x = -t$; $y = 2t$; $z = 1/t$; $1 \leq t \leq 3$.
5. $x = -t$; $y = t$; $z = t^2$; $0 \leq t \leq 3$.
6. $(t, -t, t^2)$; $0 \leq t \leq 2$.
7. $(4\cos t, 2\sin t, \mathrm{t})$; $0 \leq t \leq 2\pi$.
8. $x = \cos t$; $y = \sin t$; $z = t/2\pi$; $-2\pi \leq t \leq 2\pi$.
9. $(t, 1/t, t)$; $1 \leq t \leq 3$.
10. $(\cosh t, \sinh t, t)$; $-1 \leq t \leq 1$.

Let $\mathbf{u}$, $\mathbf{v}$, and $\mathbf{w}$ be three vectors such that $\mathbf{v}$ and $\mathbf{w}$ are perpendicular and have the same length r. In Exercises 11 and 12, (a) describe the motion of the tip of the vector $\boldsymbol{\sigma}(t)$ and (b) find the components of $\boldsymbol{\sigma}(t)$ if $\mathbf{u} = \mathbf{i} - \mathbf{j}$, $\mathbf{v} = 2(\mathbf{j} + \mathbf{k})$, $\mathbf{w} = 2(\mathbf{j} - \mathbf{k})$.

11. $\boldsymbol{\sigma}(t) = \mathbf{u} + 2\mathbf{v}\cos t + 4\mathbf{w}\sin t$.
12. $\boldsymbol{\sigma}(t) = \mathbf{u} + 3\mathbf{v}\cos t - 5\mathbf{w}\sin t$.

13. Let $\boldsymbol{\sigma}(t) = 3\cos t\mathbf{i} - 8\sin t\mathbf{j} + e^t\mathbf{k}$. Find $\boldsymbol{\sigma}'(t)$ and $\boldsymbol{\sigma}''(t)$.
14. (a) Give the "natural" domain for this vector function:
$$\boldsymbol{\sigma}(t) = \frac{1}{t}\mathbf{i} + \frac{1}{t-1}\mathbf{j} + \frac{1}{t-2}\mathbf{k}.$$
(b) Find $\boldsymbol{\sigma}'$ and $\boldsymbol{\sigma}''$.

In Exercises 15–20, let $\boldsymbol{\sigma}_1(t) = e^t\mathbf{i} + (\sin t)\mathbf{j} + t^3\mathbf{k}$ and $\boldsymbol{\sigma}_2(t) = e^{-t}\mathbf{i} + (\csc t)\mathbf{j} - 2t^3\mathbf{k}$. Find each of the stated derivatives in two different ways:

15. $\frac{d}{dt}[\boldsymbol{\sigma}_1(t) + \boldsymbol{\sigma}_2(t)]$
16. $\frac{d}{dt}[\boldsymbol{\sigma}_1(t) \cdot \boldsymbol{\sigma}_2(t)]$
17. $\frac{d}{dt}[\boldsymbol{\sigma}_1(t) \times \boldsymbol{\sigma}_2(t)]$
18. $\frac{d}{dt}\{\boldsymbol{\sigma}_1(t) \cdot [2\boldsymbol{\sigma}_2(t) + \boldsymbol{\sigma}_1(t)]\}$
19. $\frac{d}{dt}e^t\boldsymbol{\sigma}_1(t)$
20. $\frac{d}{dt}[\boldsymbol{\sigma}_1(t^2)]$

21. Show that if the acceleration of an object is always perpendicular to the velocity, then the speed of the object is constant. [*Hint:* See Example 9.]
22. Show that, at a local maximum or minimum of $\|\boldsymbol{\sigma}(t)\|$, $\boldsymbol{\sigma}'(t)$ is perpendicular to $\boldsymbol{\sigma}(t)$.

Compute (a) the velocity vector, (b) the acceleration vector, and (c) the speed for each of the curves in Exercises 23–32.

23. The curve in Exercise 1.
24. The curve in Exercise 2.
25. The curve in Exercise 3.
26. The curve in Exercise 4.
27. The curve in Exercise 5.
28. The curve in Exercise 6.
29. The curve in Exercise 7.
30. The curve in Exercise 8.
31. The curve in Exercise 9.
32. The curve in Exercise 10.

For each of the curves in Exercises 33–38, determine the velocity and acceleration vectors for all t and the equation for the tangent line at the specified value of t.

33. $(6t, 3t^2, t^3)$; $t = 0$.
34. $(\sin 3t, \cos 3t, 2t^{3/2})$; $t = 1$.
35. $(\cos^2 t, 3t - t^3, t)$; $t = 0$.
36. $(t\sin t, t\cos t, \sqrt{3}\,t)$; $t = 0$.
37. $(\sqrt{2}\,t, e^t, e^{-t})$; $t = 0$.
38. $(2\cos t, 3\sin t, t)$; $t = \pi$.
39. Suppose that a particle follows the path $(e^t, e^{-t}, \cos t)$ until it flies off on a tangent at $t = 1$. Where is it at $t = 2$?
40. If the particle in Exercise 39 flies off the path at $t = 0$ instead of $t = 1$, where is it at $t = 2$?
41. Describe and sketch the curves specified by the following data:
(a) $\boldsymbol{\sigma}'(t) = (1, 0, 1)$; $\boldsymbol{\sigma}(0) = (0, 0, 0)$,
(b) $\boldsymbol{\sigma}'(t) = (-1, 1, 1)$; $\boldsymbol{\sigma}(0) = (1, 2, 3)$,
(c) $\boldsymbol{\sigma}'(t) = (-1, 1, 1)$; $\boldsymbol{\sigma}(0) = (0, 0, 0)$.
42. Suppose that a curve $\boldsymbol{\sigma}(t)$ has the velocity vector $\boldsymbol{\sigma}'(t) = (a, b, \sin t)$, where a and b are constants. Sketch the curve if $a = -1$, $b = 2$, and assuming $\boldsymbol{\sigma}(0) = (0, 0, 1)$.
43. Suppose that a curve has the velocity vector $\mathbf{v} = \boldsymbol{\sigma}'(t) = (\sin t, -\cos t, d)$, where d is a constant. (a) Describe the curve. (b) Sketch the curve if you know that $\boldsymbol{\sigma}(0) = \mathbf{i}$. (c) What if in addition, $d = 0$?
44. Suppose that $\boldsymbol{\sigma}(t)$ is a vector function such that $\boldsymbol{\sigma}'(t) = -\boldsymbol{\sigma}(t)$. Show that $\boldsymbol{\sigma}(t) = e^{-t}\boldsymbol{\sigma}(0)$. (*Hint:* See Chapter 8.) What is the behavior of $\boldsymbol{\sigma}(t)$ as $t \to \infty$?
45. (a) Let $\boldsymbol{\sigma}_1(t)$ and $\boldsymbol{\sigma}_2(t)$ satisfy the differential equation $\boldsymbol{\sigma}''(t) = -\boldsymbol{\sigma}(t)$. Show that for any constants A_1 and A_2, $A_1\boldsymbol{\sigma}_1(t) + A_2\boldsymbol{\sigma}_2(t)$ satisfies the equation as well.
(b) Find as many solutions of $\boldsymbol{\sigma}''(t) = -\boldsymbol{\sigma}(t)$ as you can.
46. Suppose that $\boldsymbol{\sigma}(t)$ satisfies the differential equation $\boldsymbol{\sigma}''(t) + \omega^2\boldsymbol{\sigma}(t) = \mathbf{0}$. Describe and sketch the curve if $\boldsymbol{\sigma}(0) = (0, 0, 1)$ and $\boldsymbol{\sigma}'(0) = (0, \omega, \omega)$.
47. (a) Sketch the following curves. On each curve, indicate the points obtained when $t = 0, \frac{1}{4}, \frac{1}{2}, \frac{3}{4}, 1$.
(i) $x = -t$; $y = 2t$; $z = 3t$; $0 \leq t \leq 1$.
(ii) $x = -t^2$; $y = 2t^2$; $z = 3t^2$; $-1 \leq t \leq 1$.
(iii) $x = -\sin(\pi t/2)$; $y = 2\sin(\pi t/2)$; $z = 3\sin(\pi t/2)$; $0 \leq t \leq 1$.
(b) Show that the set of points in space covered by each of these curves is the same. Discuss differences between the curves thought of as functions of t. (How fast do you move along the curve as t changes? How many times is each point covered?)

48. For each curve in Exercise 47, find the velocity vector $\mathbf{v}$ and the speed v as functions of t. Compute $\int_a^b v\,dt$, where $[a, b]$ is the defining interval for t in each case. What should this number represent? Explain the difference between the result for part (ii) and that for (i) and (iii).

49. Let θ and ϕ be fixed angles, and consider the following two curves:

(a) $x = \sin\phi\cos t,$
$y = \sin\phi\sin t, \quad 0 \leqslant t \leqslant 2\pi$
$z = \cos\phi;$

(b) $x = \sin t\cos\theta,$
$y = \sin t\sin\theta, \quad 0 \leqslant t \leqslant 2\pi$
$z = \cos t.$

Show that each curve is a circle lying on the sphere of radius 1 centered at the origin. Find the center and radius of each circle. Sketch the curves for $\phi = 45°$ and for $\theta = 45°$.

50. Suppose that $0 \leqslant \theta \leqslant 2\pi$ and $0 \leqslant \phi \leqslant 2\pi$, and let

$$\boldsymbol{\sigma}(\theta, \phi) = ((2 + \cos\phi)\cos\theta, (2 + \cos\phi)\sin\theta, \sin\phi).$$

(Note that this is a vector function of two variables.)

(a) Describe each of the following curves:
(i) $\boldsymbol{\sigma}(\theta, 0);\quad 0 \leqslant \theta \leqslant 2\pi;$
(ii) $\boldsymbol{\sigma}(\theta, \pi);\quad 0 \leqslant \theta \leqslant 2\pi;$
(iii) $\boldsymbol{\sigma}(\theta, \pi/2);\ 0 \leqslant \theta \leqslant 2\pi;$
(iv) $\boldsymbol{\sigma}(0, \phi);\quad 0 \leqslant \phi \leqslant 2\pi;$
(v) $\boldsymbol{\sigma}(\pi/2, \phi);\ 0 \leqslant \phi \leqslant 2\pi;$
(vi) $\boldsymbol{\sigma}(\pi/4, \phi);\ 0 \leqslant \phi \leqslant 2\pi.$

(b) Show that the point $\boldsymbol{\sigma}(\theta, \phi)$ lies on the circle of radius $2 + \cos\phi$ parallel to the xy plane and centered at $(0, 0, \sin\phi)$.

(c) Show that $\boldsymbol{\sigma}(\theta, \phi)$ lies on the doughnut-shaped surface (a *torus*) shown in Fig. 14.6.12.

(d) Describe and sketch the curve
$((2 + \cos t)\cos t, (2 + \cos t)\sin t, \sin t)$.

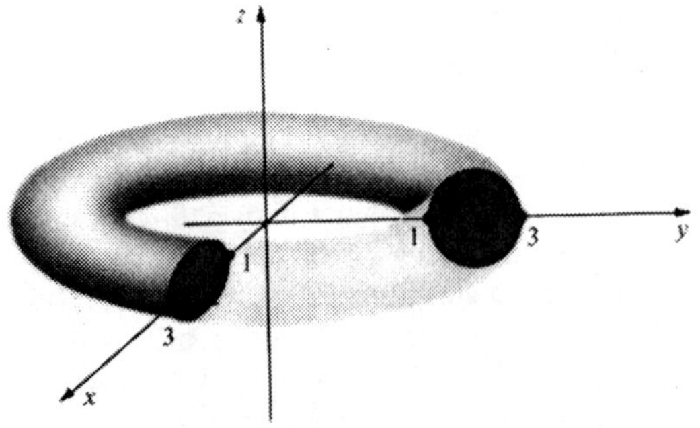

Figure 14.6.12. The points $\boldsymbol{\sigma}(\theta, \phi)$ (Exercise 50) lie on this surface.

51. Suppose that $P_0 = (x_0, y_0, 0)$ is a point on the unit circle in the xy plane. Describe the set of points lying directly above or below P_0 on the right circular helix of Example 3. What is the vertical distance between coils of the helix?

★52. If $\boldsymbol{\sigma}(t) = f(t)\mathbf{i} + g(t)\mathbf{j} + h(t)\mathbf{k}$ is a vector function, we may define $\lim_{t\to t_0}\boldsymbol{\sigma}(t)$ componentwise; that is,

$$\lim_{t\to t_0}\boldsymbol{\sigma}(t) = \left[\lim_{t\to t_0} f(t)\right]\mathbf{i} + \left[\lim_{t\to t_0} g(t)\right]\mathbf{j} + \left[\lim_{t\to t_0} h(t)\right]\mathbf{k}$$

if the three limits on the right-hand side all exist. Using this definition, show that

$$\lim_{\Delta t\to 0}\frac{1}{\Delta t}[\boldsymbol{\sigma}(t_0 + \Delta t) - \boldsymbol{\sigma}(t_0)] = \boldsymbol{\sigma}'(t_0).$$

Prove the rules in Exercises 53–56 for vector functions.

★53. The sum rule.

★54. The scalar multiplication rule.

★55. The cross product rule.

★56. The chain rule.

★57. Let $\mathbf{r} = \boldsymbol{\sigma}(t)$ be a parametric curve.

(a) Suppose there is a unit vector $\mathbf{u}$ (constant) such that $\boldsymbol{\sigma}(t)\cdot\mathbf{u} = 0$ for all values of t. What can you say about the curve $\boldsymbol{\sigma}(t)$?

(b) What can you say if $\boldsymbol{\sigma}(t)\cdot\mathbf{u} = c$ for some constant c?

(c) What can you say if $\boldsymbol{\sigma}(t)\cdot\mathbf{u} = b\|\boldsymbol{\sigma}(t)\|$ for some constant b with $0 < b < 1$?

★58. Consider the curve given by

$$x = r\cos\omega t, \quad y = r\sin\omega t, \quad \text{and} \quad z = ct,$$

where r, ω, and c are positive constants and $-\infty < t < \infty$.

(a) What path is traced out by (x, y) in the plane?

(b) The curve in space lies on what cylinder?

(c) For what t_0 does the curve trace out one coil of the helix as t goes through the interval $0 \leqslant t \leqslant t_0$?

(d) What is the vertical distance between coils?

(e) The curve is a right-circular helix. Sketch it.

14.7 The Geometry and Physics of Space Curves

Particles moving in space according to physical laws can trace out geometrically interesting curves.

This section is concerned with applications of calculus: arc length, Newton's second law, and some geometry of space curves.

In Section 10.4, we found the arc length formula

$$L = \int_a^b \sqrt{f'(t)^2 + g'(t)^2}\, dt = \int_a^b \sqrt{\left(\frac{dx}{dt}\right)^2 + \left(\frac{dy}{dt}\right)^2}\, dt$$

for a parametric curve in the plane. A similar formula, with one term added, applies to curves in space.

Arc Length

Let $(x, y, z) = (f(t), g(t), h(t))$ be a parametric curve in space. The *length* of the curve, for t in the interval $[a, b]$, is defined to be

$$L = \int_a^b \sqrt{(f'(t))^2 + (g'(t))^2 + (h'(t))^2}\, dt$$

or

$$L = \int_a^b \sqrt{\left(\frac{dx}{dt}\right)^2 + \left(\frac{dy}{dt}\right)^2 + \left(\frac{dz}{dt}\right)^2}\, dt. \qquad (1)$$

Example 1 Find the length of the helix $(\cos t, \sin t, t)$ for $0 \leqslant t \leqslant \pi$.

Example Here $f'(t) = -\sin t$, $g'(t) = \cos t$, and $h'(t) = 1$, so the integrand in the arc length formula (1) is $\sqrt{\sin^2 t + \cos^2 t + 1} = \sqrt{2}$, a constant. Thus the length is simply

$$L = \int_0^\pi \sqrt{2}\, dt = \pi\sqrt{2}\,. \blacktriangle$$

Notice that the integrand in the arc length formula is precisely the speed $\|\boldsymbol{\sigma}'(t)\|$ of a particle moving along the parametric curve. Thus the arc length, which can be written as $L = \int_a^b \|\boldsymbol{\sigma}'(t)\|\, dt$, is the integral of speed with respect to time and represents the total distance travelled by the particle between time a and time b.

Example 2 Find the arc length of $(\cos t, \sin t, t^2)$, $0 \leqslant t \leqslant \pi$.

Solution The curve $\boldsymbol{\sigma}(t) = (\cos t, \sin t, t^2)$ has velocity vector $\mathbf{v} = (-\sin t, \cos t, 2t)$. Since $\|\mathbf{v}\| = \sqrt{1 + 4t^2} = 2\sqrt{t^2 + (\frac{1}{2})^2}$, the arc length is

$$L = \int_0^\pi 2\sqrt{t^2 + (\tfrac{1}{2})^2}\, dt.$$

This integral may be evaluated using the formula (43) from the table of integrals:

$$\int \sqrt{x^2+a^2}\,dx = \tfrac{1}{2}\left[x\sqrt{x^2+a^2} + a^2\ln\left(x+\sqrt{x^2+a^2}\right)\right] + C.$$

Thus

$$\begin{aligned} L &= 2\cdot\tfrac{1}{2}\left[t\sqrt{t^2+(\tfrac{1}{2})^2} + (\tfrac{1}{2})^2\ln\left(t+\sqrt{t^2+(\tfrac{1}{2})^2}\right)\right]\Bigg|_{t=0}^{\pi} \\ &= \pi\sqrt{\pi^2+\tfrac{1}{4}} + \tfrac{1}{4}\ln\left(\pi+\sqrt{\pi^2+\tfrac{1}{4}}\right) - \tfrac{1}{4}\ln\left(\sqrt{\tfrac{1}{4}}\right) \\ &= \frac{\pi}{2}\sqrt{1+4\pi^2} + \tfrac{1}{4}\ln\left(2\pi+\sqrt{1+4\pi^2}\right) \\ &\approx 10.63. \ \blacktriangle \end{aligned}$$

Example 3 Find the arc length of (e^t, t, e^t), $0 \leqslant t \leqslant 1$. [*Hint:* Use $u = \sqrt{1+2e^{2t}}$ to evaluate the integral.]

Solution $\sigma(t) = (e^t, t, e^t)$, so $\mathbf{v} = (e^t, 1, e^t)$, and $\|\mathbf{v}\| = \sqrt{1+2e^{2t}}$; so $L = \int_0^1 \sqrt{1+2e^{2t}}\,dt$. To evaluate this integral, set $u = \sqrt{1+2e^{2t}}$, which leads to

$$\begin{aligned} \int \sqrt{1+2e^{2t}}\,dt &= \int u\frac{u\,du}{u^2-1} \\ &= \int\left[1+\frac{1}{2}\left(\frac{1}{u-1}\right) - \frac{1}{2}\left(\frac{1}{u+1}\right)\right]du \quad \text{(partial fractions)} \\ &= u + \frac{1}{2}\ln(u-1) - \frac{1}{2}\ln(u+1) + C \\ &= \sqrt{1+2e^{2t}} + \frac{1}{2}\ln\frac{\sqrt{1+2e^{2t}}-1}{\sqrt{1+2e^{2t}}+1} + C. \end{aligned}$$

(This result may be checked by differentiation.) To find L, we evaluate the last expression at $t = 0$ and $t = 1$ and subtract, to obtain

$$L = \sqrt{1+2e^2} + \frac{1}{2}\ln\frac{\sqrt{1+2e^2}-1}{\sqrt{1+2e^2}+1} - \sqrt{3} - \frac{1}{2}\ln\frac{\sqrt{3}-1}{\sqrt{3}+1} \approx 2.64. \ \blacktriangle$$

We turn next to the study of curves followed by physical particles subject to forces.

If a particle of mass m moves in space, the total force $\mathbf{F}$ acting on it at any time is a vector which is related to the acceleration by *Newton's second law* (see Section 8.1): $\mathbf{F} = m\mathbf{a}$.

In many situations, the force is a given function of position $\mathbf{r}$ (the "force law"), and the problem of interest is to find the vector function $\mathbf{r} = \sigma(t)$ describing a particle's motion, given the initial position and velocity. Thus, Newton's second law becomes a differential equation for $\sigma(t)$, and techniques of differential equations can be used to solve it (as we solved the spring equation in Section 8.1). For example, a planet moving around the sun (considered to be located at the origin) obeys to a high degree of accuracy *Newton's law of gravitation*:

$$\mathbf{F} = -\frac{GmM}{\|\mathbf{r}\|^3}\mathbf{r} = -\frac{GmM}{r^3}\mathbf{r},$$

where $\mathbf{r}$ is the vector pointing from the sun to the planet at time t, M is the mass of the sun, m that of the planet, $r = \|\mathbf{r}\|$, and G is the gravitational constant ($G = 6.67 \times 10^{-11}$ newton meter^2 per kilogram^2). The differential equation arising from this force law is

$$\frac{d^2\mathbf{r}}{dt^2} = -\frac{GM}{r^3}\mathbf{r}.$$

Rather than solving this equation here, we shall content ourselves with understanding its consequences for the case of circular motion.

Example 4 A particle of mass m is moving in the xy plane at constant speed v in a circular path of radius r_0. Find the acceleration of the particle and the force acting on it.

Solution Let $\mathbf{r}$ be the vector from the center of the circle to the particle at time t. Motion of the type described is given by

$$\mathbf{r} = r_0\cos\left(\frac{tv}{r_0}\right)\mathbf{i} + r_0\sin\left(\frac{tv}{r_0}\right)\mathbf{j}.$$

Differentiating twice, we see that

$$\mathbf{a} = \frac{d^2\mathbf{r}}{dt^2} = -\frac{v^2}{r_0}\cos\left(\frac{tv}{r_0}\right)\mathbf{i} - \frac{v^2}{r_0}\sin\left(\frac{tv}{r_0}\right)\mathbf{j} = -\frac{v^2}{r_0^2}\mathbf{r}.$$

The force acting on the particle is $\mathbf{F} = m\mathbf{a} = -(mv^2/r_0^2)\mathbf{r}$. ▲

Figure 14.7.1. The acceleration vector of a particle in uniform circular motion points to the center of the circle.

Example 4 shows that in uniform circular motion, the acceleration vector points in a direction opposite to $\mathbf{r}$—that is, it is directed toward the center of the circle (see Fig. 14.7.1). This acceleration, multiplied by the mass of the particle, is called the *centripetal force*. Note that even though the speed is constant, the *direction* of the velocity vector is continually changing, which is why there is an acceleration. By Newton's law, some force must cause the acceleration which keeps the particle moving in its circular path. In whirling a rock at the end of a string, you must constantly be pulling on the string. If you stop that force by releasing the string, the rock will fly off in a straight line tangent to the circle. The force needed to keep a planet or satellite bound into an elliptical or circular orbit is supplied by gravity. The force needed to keep a car going around a curve may be supplied by the friction of the tires against the road or by direct pressure if the road is banked (see Exercise 12).

Suppose that a satellite is moving with a speed v around a planet with mass M in a circular orbit of radius r_0. Then the force computed in Example 4 must equal that in Newton's law:

$$-\frac{v^2}{r_0^2}\mathbf{r} = -\frac{GM}{r_0^3}\mathbf{r}.$$

The lengths of the vectors on both sides of this equation must be equal. Hence

$$v^2 = \frac{GM}{r_0}. \tag{2}$$

If T is the period of one revolution, then $2\pi r_0/T = v$ (distance/time = speed); substituting this value for v in equation (2) and solving for T^2, we obtain the rule:

$$T^2 = r_0^3\frac{(2\pi)^2}{GM}. \tag{3}$$

The square of the period is proportional to the cube of the radius. This law is one

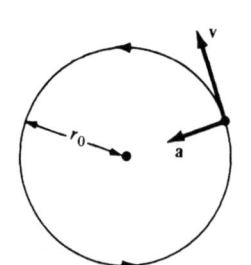

of the famous three which were discovered empirically by Kepler before Newton's laws were formulated; it enables one to compute the period of a satellite given the radius of its orbit or to determine the radius of the orbit if the period is prescribed. If both the radius and period are known, (3) can be used to determine GM, and hence if G is known, M can be computed.

Example 5 Suppose we want to have a satellite in circular orbit about the earth in such a way that it stays fixed in the sky over one point on the equator. What should be the radius of such an orbit? (The mass of the earth is 5.98×10^{24} kilograms.)

Solution The period of the satellite should be 1 day, so $T = 60 \times 60 \times 24 = 86{,}400$ seconds. By formula (3), the radius of the orbit should satisfy

$$r_0^3 = \frac{T^2 GM}{(2\pi)^2} = \frac{(86{,}400)^2 \times (6.67 \times 10^{-11}) \times (5.98 \times 10^{24})}{(2\pi)^2}$$

$$= 7.54 \times 10^{22} \text{ meters}^3,$$

so $r_0 = 4.23 \times 10^7$ meters $= 42.300$ kilometers
$= 26{,}200$ miles. ▲

Example 6 Let $\mathbf{r} = \boldsymbol{\sigma}(t)$ be the vector from a fixed point to the position of an object, $\mathbf{v}$ the velocity, and $\mathbf{a}$ the acceleration. Suppose that $\mathbf{F}$ is the force acting at time t.

(a) Prove that $(d/dt)(m\mathbf{r} \times \mathbf{v}) = \boldsymbol{\sigma} \times \mathbf{F}$, (that is, "*rate of change of angular momentum = torque*"). What can you conclude if $\mathbf{F}$ is parallel to $\mathbf{r}$? Is this the case in planetary motion?

(b) Prove that a planet moving about the sun does so in a fixed plane. (This is another of Kepler's laws.)

Solution (a) We use the rules of differentiation for vector functions:

$$\frac{d}{dt}(m\mathbf{r} \times \mathbf{v}) = m\frac{d\mathbf{r}}{dt} \times \mathbf{v} + m\mathbf{r} \times \frac{d\mathbf{v}}{dt} = m\mathbf{v} \times \mathbf{v} + m\mathbf{r} \times \mathbf{a}$$

$$= \mathbf{0} + \mathbf{r} \times m\mathbf{a} = \mathbf{r} \times \mathbf{F}.$$

If $\mathbf{F}$ is parallel to $\mathbf{r}$, then this last cross product is $\mathbf{0}$. Thus $m\mathbf{r} \times \mathbf{v}$ must be a constant vector. It represents the angular momentum, a quantity which measures the tendency of a spinning body to keep spinning. The magnitude of $m\mathbf{r} \times \mathbf{v}$ measures the amount of angular momentum, and the direction is along the axis of spin. If the derivative above is zero, it means that angular momentum is conserved; both its magnitude and its direction are preserved. This is the case for our model of planetary motion in which the sun is regarded as fixed and the gravitational force

$$\mathbf{F} = -\frac{GmM}{\|r\|^3}\mathbf{r}$$

is parallel to the vector $\mathbf{r}$ from the sun to the planet. (The actual situation is a bit more complicated than this: in fact, both the sun and the planet move around their common center of mass. However, the mass M of the sun is so much greater than the mass m of the planet that this center of mass is very close to the center of the sun, and our approximation is quite good. Things would be more complicated, for example, in a double-star system where the masses were more nearly the same and the center of gravity somewhere between. What is conserved is the total angular momentum of the whole system, taking both stars into account.)

(b) Let $\mathbf{l} = m\mathbf{r} \times \mathbf{v}$ be the angular momentum vector. Then certainly $\mathbf{r} \cdot \mathbf{l} = 0$. We argued above that $d\mathbf{l}/dt = \mathbf{0}$, so $\mathbf{l}$ is a constant vector. Since $\mathbf{r}$ satisfies $\mathbf{r} \cdot \mathbf{l} = 0$, the planet stays in the plane through the sun with normal vector $\mathbf{l}$. ▲

Our third and final application in this section is to the geometry of space curves.

Differential geometry is the branch of mathematics in which calculus is used to study the geometry of curves, surfaces, and higher dimensional objects. When we studied the arc length of curves, we were already doing differential geometry—now we will go further and introduce the important idea of *curvature.*

The curvature of a curve in the plane or in space is a measure of the rate at which the direction of motion along the curve is changing. A curve with curvature zero is just a straight line. We can define the curvature as the rate of change of the velocity vector, if the length of this vector happens to be 1; otherwise the change in length of the velocity vector confuses the issue. We therefore make the following definitions.

Parametrization by Arc Length

Let $\mathbf{r} = \boldsymbol{\sigma}(t)$ be a parametric curve.

1. The curve is called *regular* if $\mathbf{v} = \boldsymbol{\sigma}'(t)$ is not equal to $\mathbf{0}$ for any t.
2. If the curve is regular, the vector $\mathbf{T} = \mathbf{v}/\|\mathbf{v}\| = \boldsymbol{\sigma}'(t)/\|\boldsymbol{\sigma}'(t)\|$ is called the *unit tangent vector* to the curve.
3. If the length of $\boldsymbol{\sigma}'(t)$ is constant and equal to 1 (in which case $\mathbf{T} = \mathbf{v}$), the curve is said to be *parametrized by arc length.*

Example 7 Suppose that the curve $\mathbf{r} = \boldsymbol{\sigma}(t)$ is parametrized by arc length. Show that the length of the curve between $t = a$ and $t = b$ is simply $b - a$.

Solution The integrand in the arc length formula (1) is constant and equal to 1 if the curve is parametrized by arc length. Thus

$$L = \int_a^b 1\, dt = b - a. \ ▲$$

If a curve $\mathbf{r} = \boldsymbol{\sigma}(t)$, as it is presented to us, is regular but not parametrized by arc length, we can introduce a new independent variable so that the new curve *is* parametrized by arc length. In fact, we can choose a value a in the domain of the curve and define $s = p(t)$ to be the arc length $\int_a^t \|\boldsymbol{\sigma}'(u)\|\, du$ of the curve between a and t. We have $ds/dt = \|\boldsymbol{\sigma}'(t)\| > 0$ since the curve is regular, so the inverse function $t = q(s)$ exists (see Section 5.3). Now look at the new curve $\mathbf{r} = \boldsymbol{\sigma}_1(s) = \boldsymbol{\sigma}(q(s))$, which goes through the same points in space as the original curve but at a different speed. In fact, the new speed is

$$\begin{aligned} \|\boldsymbol{\sigma}_1'(s)\| &= \|q'(s)\boldsymbol{\sigma}'(q(s))\| \quad \text{(chain rule)} \\ &= q'(s)\|\boldsymbol{\sigma}'(q(s))\| \quad (q'(s) \text{ is positive}) \\ &= \frac{1}{p'(q(s))}\|\boldsymbol{\sigma}'(q(s))\|. \end{aligned}$$

However, by the fundamental theorem of calculus and the definition of p, $p'(t) = \|\boldsymbol{\sigma}'(t)\|$, so $\|\boldsymbol{\sigma}_1'(s)\| = 1$, and so the new curve is parametrized by arc length.

Example 8 Find the arc length parametrization for the helix $(\cos t, \sin t, t)$.

Solution We have $\|\mathbf{v}\| = \sqrt{\sin^2 t + \cos^2 t + 1} = \sqrt{2}$. Taking $a = 0$, we have $s = p(t) = \int_0^t \sqrt{2}\, du = \sqrt{2}\, t$, so $s = \sqrt{2}\, t$, and $t = s/\sqrt{2}$; the curve in arc length parametrization is therefore $(\cos(s/\sqrt{2}), \sin(s/\sqrt{2}), s/\sqrt{2})$. ▲

The arc length parametrization is mostly useful for theoretical purposes, since the integral in its definition is often impossible to evaluate. Still, the *existence* of this parametrization makes the definitions which follow much simpler.

Whenever a curve is parametrized by arc length, we will denote this parameter by s. Notice that in this case $\mathbf{T} = \mathbf{v} = d\mathbf{r}/ds$. Now we can define the curvature of a curve.

Curvature

Let $\mathbf{T}$ be the unit tangent vector of a curve parametrized by arc length. The scalar $k = \|d\mathbf{T}/ds\|$ is called the *curvature* of the curve. If $k \neq 0$, the unit vector $\mathbf{N} = (d\mathbf{T}/ds)/\|d\mathbf{T}/ds\|$ is called the *principal normal vector* to the curve.

Let us show that the principal normal vector is perpendicular to the unit tangent vector. Since $\mathbf{T}$ has constant length, we know, by Example 9 of the previous section, that $d\mathbf{T}/ds$ is perpendicular to $\mathbf{T}$. Since $\mathbf{N}$ has the same direction as $d\mathbf{T}/ds$, it is perpendicular to $\mathbf{T}$ as well.

Example 9 Compute the curvature and principal normal vector of the helix in Example 8.

Solution We have $\mathbf{T} = -(1/\sqrt{2})\sin(s/\sqrt{2})\mathbf{i} + (1/\sqrt{2})\cos(s/\sqrt{2})\mathbf{j} + (1/\sqrt{2})\mathbf{k}$, so $d\mathbf{T}/ds = -(1/2)\cos(s/\sqrt{2})\mathbf{i} - (1/2)\sin(s/\sqrt{2})\mathbf{j}$; the curvature is

$$\sqrt{\frac{1}{4}\cos^2\left(\frac{s}{\sqrt{2}}\right) + \frac{1}{4}\sin^2\left(\frac{s}{\sqrt{2}}\right)} = \sqrt{\frac{1}{4}} = \frac{1}{2},$$

and the principal normal vector is $-\cos(s/\sqrt{2})\mathbf{i} - \sin(s/\sqrt{2})\mathbf{j}$. ▲

If a curve is not parametrized by arc length, it is possible to compute th curvature and principal normal vector directly by the following formulas:

$$k = \frac{\|\mathbf{v} \times \mathbf{v}'\|}{\|\mathbf{v}\|^3}, \qquad (4$$

$$\mathbf{N}(t) = \frac{(\mathbf{v}\cdot\mathbf{v})\mathbf{v}' - (\mathbf{v}'\cdot\mathbf{v})\mathbf{v}}{\|(\mathbf{v}\cdot\mathbf{v})\mathbf{v}' - (\mathbf{v}'\cdot\mathbf{v})\mathbf{v}\|}. \qquad (5$$

We now prove formula (4). The curvature is defined as $\|d\mathbf{T}/ds\|$. W must use the chain rule to express this in terms of the original parametrizatic t. First of all, we have $\mathbf{T} = \mathbf{v}/\|\mathbf{v}\| = \mathbf{v}/(\mathbf{v}\cdot\mathbf{v})^{1/2}$, so

$$\frac{d\mathbf{T}}{ds} = \frac{d\mathbf{T}}{dt}\frac{dt}{ds} = \frac{1}{(ds/dt)}\frac{d\mathbf{T}}{dt}.$$

Now $ds/dt = \|\mathbf{v}\|$ and so

$$\begin{aligned}\frac{d\mathbf{T}}{dt} &= \frac{d}{dt}\left(\frac{\mathbf{v}}{(\mathbf{v}\cdot\mathbf{v})^{1/2}}\right)\\ &= \left(\frac{d}{dt}(\mathbf{v}\cdot\mathbf{v})^{-1/2}\right)\mathbf{v} + (\mathbf{v}\cdot\mathbf{v})^{-1/2}\frac{d\mathbf{v}}{dt}\\ &= -\frac{1}{2}(\mathbf{v}\cdot\mathbf{v})^{-3/2}2\left(\mathbf{v}\cdot\frac{d\mathbf{v}}{dt}\right)\mathbf{v} + (\mathbf{v}\cdot\mathbf{v})^{-1/2}\frac{d\mathbf{v}}{dt}\\ &= (\mathbf{v}\cdot\mathbf{v})^{-3/2}\left[-\left(\mathbf{v}\cdot\frac{d\mathbf{v}}{dt}\right)\mathbf{v} + (\mathbf{v}\cdot\mathbf{v})\frac{d\mathbf{v}}{dt}\right],\end{aligned}$$

so

$$\begin{aligned}\left\|\frac{d\mathbf{T}}{dt}\right\|^2 &= (\mathbf{v}\cdot\mathbf{v})^{-3}\left[-\left(\mathbf{v}\cdot\frac{d\mathbf{v}}{dt}\right)\mathbf{v} + (\mathbf{v}\cdot\mathbf{v})\frac{d\mathbf{v}}{dt}\right]\cdot\left[-\left(\mathbf{v}\cdot\frac{d\mathbf{v}}{dt}\right)\mathbf{v} + (\mathbf{v}\cdot\mathbf{v})\frac{d\mathbf{v}}{dt}\right]\\ &= (\mathbf{v}\cdot\mathbf{v})^{-3}\left[\left(\mathbf{v}\cdot\frac{d\mathbf{v}}{dt}\right)^2(\mathbf{v}\cdot\mathbf{v}) - 2\left(\mathbf{v}\cdot\frac{d\mathbf{v}}{dt}\right)^2(\mathbf{v}\cdot\mathbf{v}) + (\mathbf{v}\cdot\mathbf{v})^2\left(\frac{d\mathbf{v}}{dt}\cdot\frac{d\mathbf{v}}{dt}\right)\right]\\ &= (\mathbf{v}\cdot\mathbf{v})^{-2}\left[(\mathbf{v}\cdot\mathbf{v})\left(\frac{d\mathbf{v}}{dt}\cdot\frac{d\mathbf{v}}{dt}\right) - \left(\mathbf{v}\cdot\frac{d\mathbf{v}}{dt}\right)^2\right] = \|\mathbf{v}\|^{-4}\left\|\mathbf{v}\times\frac{d\mathbf{v}}{dt}\right\|^2.\end{aligned}$$

(See Exercise 38a, Section 13.5.) Thus

$$k = \left\|\frac{d\mathbf{T}}{ds}\right\| = \frac{1}{\|\mathbf{v}\|}\|\mathbf{v}\|^{-2}\left\|\mathbf{v}\times\frac{d\mathbf{v}}{dt}\right\| = \frac{\|\mathbf{v}\times d\mathbf{v}/dt\|}{\|\mathbf{v}\|^3},$$

which is formula (4).

In Exercise 20, the reader is asked to derive (5) using similar methods.

Example 10 Find the curvature of the exponential spiral $(e^{-t}\cos t, e^{-t}\sin t, 0)$ (Fig. 14.7.2). What happens as $t \to \infty$?

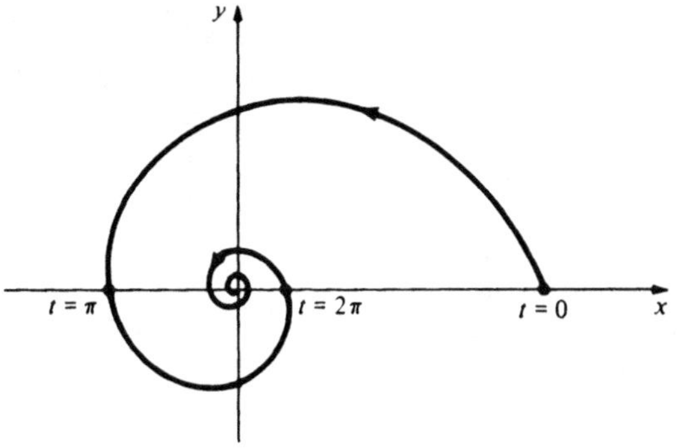

Figure 14.7.2. Graph of the exponential spiral in the (x, y) plane.

Solution We have

$$\mathbf{v} = (-e^{-t}\cos t - e^{-t}\sin t)\mathbf{i} + (-e^{-t}\sin t + e^{-t}\cos t)\mathbf{j}$$

and

$$\begin{aligned}\mathbf{v}' &= (e^{-t}\cos t + e^{-t}\sin t + e^{-t}\sin t - e^{-t}\cos t)\mathbf{i}\\ &\quad + (e^{-t}\sin t - e^{-t}\cos t - e^{-t}\cos t - e^{-t}\sin t)\mathbf{j}\\ &= 2e^{-t}(\sin t\,\mathbf{i} - \cos t\,\mathbf{j}).\end{aligned}$$

Then

$$\begin{aligned}\mathbf{v}\times\mathbf{v}' &= 2e^{-2t}\begin{vmatrix} -\cos t - \sin t & -\sin t + \cos t \\ \sin t & -\cos t \end{vmatrix}\mathbf{k}\\ &= 2e^{-2t}(\cos^2 t + \cos t\sin t + \sin^2 t - \cos t\sin t)\mathbf{k} = 2e^{-2t}\mathbf{k},\end{aligned}$$

and so $\|\mathbf{v} \times \mathbf{v}'\| = 2e^{-2t}$. Since

$$\|\mathbf{v}\| = \{[e^{-t}(\cos t + \sin t)]^2 + [e^{-t}(\cos t - \sin t)]^2\}^{1/2},$$

$$\|\mathbf{v}\|^3 = e^{-3t}\left[(\cos t + \sin t)^2 + (\cos t - \sin t)^2\right]^{3/2} = e^{-3t}2^{3/2}.$$

By formula (4),

$$k = \frac{2e^{-2t}}{e^{-3t}2^{3/2}} = \frac{e^t}{\sqrt{2}}.$$

As $t \to \infty$, the curvature approaches infinity as the spiral wraps more and more tightly about the origin. ▲

Exercises for Section 14.7

Find the arc length of the given curve on the specified interval in Exercises 1–6.

1. $(2\cos t, 2\sin t, t)$; $0 \leqslant t \leqslant 2\pi$.
2. $(1, 3t^2, t^3)$; $0 \leqslant t \leqslant 1$.
3. $(\sin 3t, \cos 3t, 2t^{3/2})$; $0 \leqslant t \leqslant 1$.
4. $\left(t + 1, \frac{2\sqrt{2}}{3}t^{3/2} + 7, \frac{1}{2}t^2\right)$ for $1 \leqslant t \leqslant 2$.
5. (t, t, t^2) for $1 \leqslant t \leqslant 2$.
6. $(t, t\sin t, t\cos t)$; $0 \leqslant t \leqslant \pi$.
7. A body of mass 2 kilograms moves in a circular path on a circle of radius 3 meters, making one revolution every 5 seconds. Find the centripetal force acting on the body.
8. Find the centripetal force acting on a body of mass 4 kilograms, moving on a circle of radius 10 meters with a frequency of 2 revolutions per second.
9. A satellite is in a circular orbit 500 miles above the surface of the earth. What is the period of the orbit? (See Example 5; 1 mile = 1.609 kilometer; the radius of the earth is 6370 km).
10. What is the gravitational acceleration on the satellite in Exercise 9? The centripetal acceleration?
11. For a falling body near the surface of the earth, the force of gravity can be approximated very well as a constant downward force with magnitude $F = GmM/R^2$, where G is the gravitational constant, M the mass of the earth, m the mass of the body, and R the radius of the earth (6.37×10^6 meters).
    (a) Show that this approximation means that any body falling freely (neglecting air resistance) near the surface of the earth experiences a constant acceleration of $g = 9.8$ meters per second per second. Note that this acceleration is independent of m: any two bodies fall at the same rate.
    (b) Show that the flight path of a projectile or a baseball is a parabola (see Example 11 in Section 14.6).
12. (a) Suppose that a car is going around a circular curve of radius r at speed v. It will then exert an outward horizontal force on the roadway due to the centripetal acceleration and a vertical force due to gravity. At what angle θ should the roadway be banked so that the total force tends to press the car directly into (perpendicular to) the roadway? (See Fig. 14.7.3.) How does the bank angle depend on r and on the speed v?

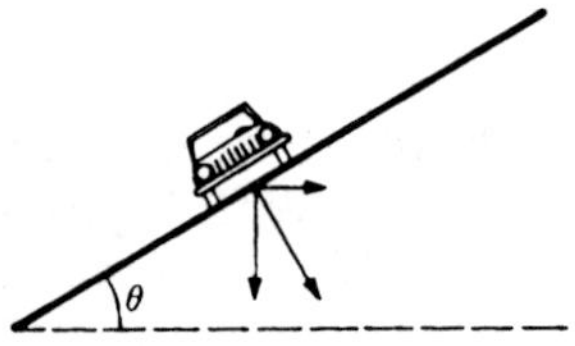

Figure 14.7.3. For what value of θ does the total force press directly into the roadway?

    (b) Discuss how you might treat the design problem in part (a) for a curve that is not part of a circle. Design an elliptical racetrack with major axis 800 meters, minor axis 500 meters, and speed 160 kilometers per hour.
13. A particle with charge q moving with velocity vector $\mathbf{v}$ through a magnetic field is acted on by the force $\mathbf{F} = (q/c)\mathbf{v} \times \mathbf{B}$, where c is the speed of light and $\mathbf{B}$ is a vector describing the magnitude and direction of the magnetic field. Suppose that:
    (1) The particle has mass m and is following a path $\mathbf{r} = \boldsymbol{\sigma}(t) = x\mathbf{i} + y\mathbf{j} + z\mathbf{k}$.
    (2) $\boldsymbol{\sigma}(0) = \mathbf{i}$; $\boldsymbol{\sigma}'(0) = a\mathbf{j} + c\mathbf{k}$.
    (3) The magnetic field is constant and uniform given by a vector $\mathbf{B} = b\mathbf{k}$.
    (a) Use the equation $\mathbf{F} = (q/c)\mathbf{v} \times \mathbf{B}$ to write differential equations relating the components of $\mathbf{a} = \boldsymbol{\sigma}''(t)$ and $\mathbf{v} = \boldsymbol{\sigma}'(t)$.

(b) Solve these equations to obtain the components of $\sigma(t)$. [*Hint:* Integrate the equations for d^2x/dt^2 once, use item (2) in the list above to determine the constant of integration, and substitute the resulting expression for dx/dt into the equation for d^2y/dt^2 to get an equation similar to the spring equation solved in Section 8.1.]

(c) Show that the path is a right circular helix. What are the radius and axis of the cylinder on which it lies? (The dimensions of the helix followed by a particle in a magnetic field in a bubble chamber are used to measure the charge to mass ratio of the particle.)

14. In Exercise 13, how does the geometry of the helix change if (a) m is doubled, (b) q is doubled, (c) $\|\sigma'(0)\|$ is doubled?

15. Show that a circle of radius r has constant curvature $1/r$.

16. Compute the curvature vector and principal normal vector for the helix $(r\cos wt, r\sin wt, ct)$ in terms of r, w, and c.

17. Find the curvature of the ellipse $x^2 + 2y^2 = 1$, $z = 0$. (Choose a suitable parametrization.)

18. Compute the curvature and principal normal vector of the elliptical helix $(\cos t, 2\sin t, t)$.

19. Show that if the curvature of a curve is identically zero, then the curve is a straight line.

★20. Derive formula (5) by using the methods used to derive (4).

★21. A particle is moving along a curve at constant speed. Express the magnitude of the force on the particle in terms of the mass of the particle, the speed of the particle, and the curvature of the curve.

★22. Let $\mathbf{T}$ and $\mathbf{N}$ be the unit tangent and principal normal vectors to a space curve $\mathbf{r} = \sigma(t)$. Define a third unit vector perpendicular to them by $\mathbf{B} = \mathbf{T} \times \mathbf{N}$. This is called the *binormal* vector. Together, $\mathbf{T}$, $\mathbf{N}$, and $\mathbf{B}$ form a right-handed system of mutually orthogonal unit vectors which may be thought of as moving along the curve (see Fig. 14.7.4.)

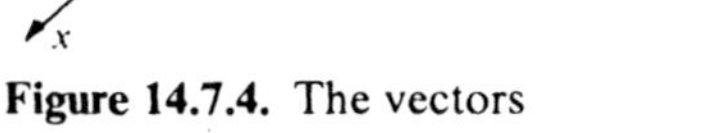

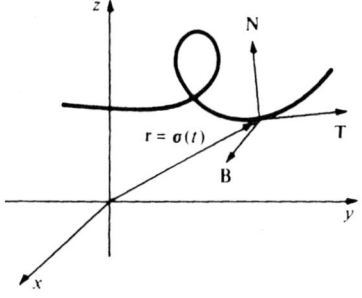

Figure 14.7.4. The vectors $\mathbf{T}$, $\mathbf{N}$, and $\mathbf{B}$ form a "moving basis" along the curve.

(a) Show that

$$\mathbf{B} = (\mathbf{v} \times \mathbf{a})/\|\mathbf{v} \times \mathbf{a}\| = (\mathbf{v} \times \mathbf{a})/(k\|\mathbf{v}\|^3),$$

where $\mathbf{a}$ is the acceleration vector.

(b) Show that $(d\mathbf{B}/dt) \cdot \mathbf{B} = 0$. [*Hint:* $\|\mathbf{B}\|^2 = 1$ is constant.]

(c) Show that $(d\mathbf{B}/dt) \cdot \mathbf{T} = 0$. [*Hint:* Take derivatives in $\mathbf{B} \cdot \mathbf{T} = 0$.]

(d) Show that $d\mathbf{B}/dt$ is a scalar multiple of $\mathbf{N}$.

(e) Using part (d) we can define a scalar-valued function τ called the *torsion* by $d\mathbf{B}/dt = -\tau\|\mathbf{v}\|\mathbf{N}$. Show that

$$\tau = \frac{[\sigma'(t) \times \sigma''(t)] \cdot \sigma'''(t)}{\|\sigma'(t) \times \sigma''(t)\|^2}.$$

★23. (a) Show that if a curve lies in a plane, then the torsion τ is identically zero. [*Hint*: The vector function $\sigma(t)$ must satisfy an equation of the form $\sigma(t) \cdot \mathbf{n} = 0$. By taking successive derivatives show that σ', σ'', and σ''' all lie in the same plane through the origin. What does this do to the triple product in Exercise 22(e)?]

(b) Show that $\mathbf{B}$ is then constant and is a normal vector to the plane in which the curve lies.

★24. If the torsion is not zero, it gives a measure of how fast the curve is tending to twist out of the plane. Compute the binormal vector and the torsion for the helix of Example 8.

★25. Using the results of Exercises 22 and 23, prove the following *Frenet formulas* for a curve parametrized by arc length:

$$\frac{d\mathbf{T}}{ds} = k\mathbf{N},$$

$$\frac{d\mathbf{N}}{ds} = -k\mathbf{T} + \tau\mathbf{B},$$

$$\frac{d\mathbf{B}}{ds} = -\tau\mathbf{N}.$$

[*Hint:* To get the second formula from the others, note that $\mathbf{N} \cdot \mathbf{N}$, $\mathbf{N} \cdot \mathbf{B}$, and $\mathbf{N} \cdot \mathbf{T}$ are constant. Take derivatives and use earlier formulas to get $(d\mathbf{N}/ds) \cdot \mathbf{B}$ and $(d\mathbf{N}/ds) \cdot \mathbf{T}$.]

★26. Kepler's first law of planetary motion states that *the orbit of each planet is an ellipse with the sun as one focus.* The origin $(0,0)$ is placed at the sun, and polar coordinates (r, θ) are introduced. The planet's motion is $r = r(t)$, $\theta = \theta(t)$, and these are related by $r(t) = l/[1 + e\cos\theta(t)]$, where $l = k^2/GM$ and $e^2 = 1 - (2k^2E/G^2M^2m)$; k is a constant, G is the universal gravitation constant, E is the energy of the system, M and m are the masses of the sun and planet, respectively

(a) Assume $e < 1$. Change to rectangular coordinates to verify that the planet's orbit is an ellipse.

(b) Let $\mu = 1/r$. Verify the *energy equation*

$$(d\mu/d\theta)^2 + \mu^2 = (2/k^2m)(GMm\mu - E).$$

Supplement to Chapter 14: Rotations and the Sunshine Formula

Rotations described in terms of the cross product are used to derive the sunshine formula.

The purpose of this section is to derive the sunshine formula, which has been stated and used in the supplements to Chapters 5 and 10. Before we begin the actual derivation, we will study some properties of rotations in preparation for the description of the earth rotating on its axis. The cross product, introduced in Section 13.5, will be used extensively here.

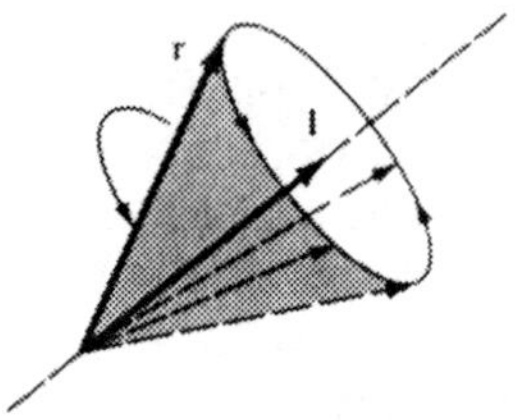

Figure 14.S.1. If $\mathbf{r}$ rotates about $\mathbf{l}$, its tip describes a circle.

Consider two unit vectors $\mathbf{l}$ and $\mathbf{r}$ in space with the same base point. If we rotate $\mathbf{r}$ about the axis through $\mathbf{l}$, then the tip of $\mathbf{r}$ describes a circle (Fig. 14.S.1). (Imagine $\mathbf{l}$ and $\mathbf{r}$ glued rigidly at their base points and then spun about the axis through $\mathbf{l}$.) Assume that the rotation is at a uniform rate counterclockwise (when viewed from the tip of $\mathbf{l}$), making a complete revolution in T units of time. The vector $\mathbf{r}$ now is a vector function of time, so we may write $\mathbf{r} = \boldsymbol{\sigma}(t)$. Our first aim is to find a convenient formula for $\boldsymbol{\sigma}(t)$ in terms of its starting position $\mathbf{r}_0 = \boldsymbol{\sigma}(0)$.

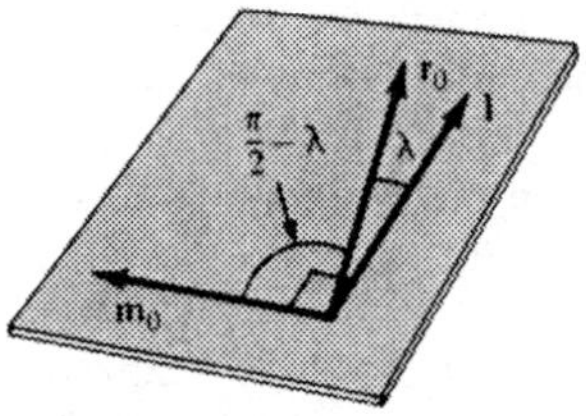

Figure 14.S.2. The vector $\mathbf{m}_0$ is in the plane of $\mathbf{r}_0$ and $\mathbf{l}$, is orthogonal to $\mathbf{l}$, and makes an angle of $(\pi/2) - \lambda$ with $\mathbf{r}_0$.

Let λ denote the angle between $\mathbf{l}$ and $\mathbf{r}_0$; we can assume that $\lambda \neq 0$ and $\lambda \neq \pi$, i.e., $\mathbf{l}$ and $\mathbf{r}_0$ are not parallel, for otherwise $\mathbf{r}$ would not rotate. Construct the unit vector $\mathbf{m}_0$ as shown in Fig. 14.S.2. From this figure we see that

$$\mathbf{r}_0 = \cos\lambda\mathbf{l} + \sin\lambda\mathbf{m}_0 . \tag{1}$$

(In fact, formula (1) can be taken as the algebraic definition of $\mathbf{m}_0$ by writing $\mathbf{m}_0 = (1/\sin\lambda)\mathbf{r}_0 - (\cos\lambda/\sin\lambda)\mathbf{l}$. We assumed that $\lambda \neq 0$, and $\lambda \neq \pi$, so $\sin\lambda \neq 0$.)

Now add to this figure the unit vector $\mathbf{n}_0 = \mathbf{l} \times \mathbf{m}_0$. (See Fig. 14.S.3.) The triple $(\mathbf{l}, \mathbf{m}_0, \mathbf{n}_0)$ consists of three mutually orthogonal unit vectors, just like $(\mathbf{i}, \mathbf{j}, \mathbf{k})$.

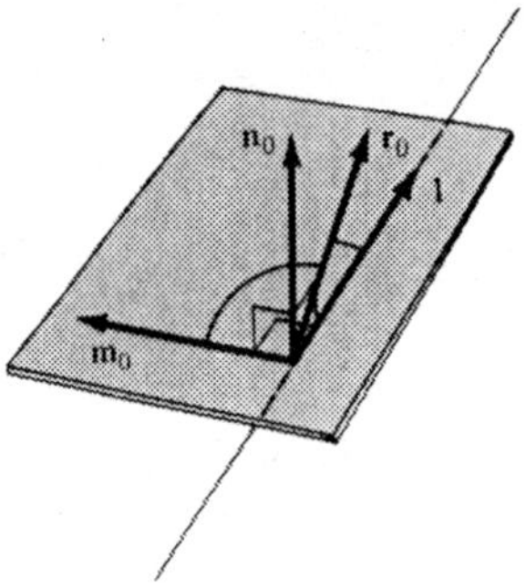

Figure 14.S.3. The triple $(\mathbf{l}, \mathbf{m}_0, \mathbf{n}_0)$ is a right-handed orthogonal set of unit vectors.

Example 1 Let $\mathbf{l} = (1/\sqrt{3})(\mathbf{i} + \mathbf{j} + \mathbf{k})$ and $\mathbf{r}_0 = \mathbf{k}$. Find $\mathbf{m}_0$ and $\mathbf{n}_0$.

Solution The angle between $\mathbf{l}$ and $\mathbf{r}_0$ is given by $\cos\lambda = \mathbf{l} \cdot \mathbf{r}_0 = 1/\sqrt{3}$. This wa determined by dotting both sides of formula (1) by $\mathbf{l}$ and using the fact that is a unit vector. Thus $\sin\lambda = \sqrt{1 - \cos^2\lambda} = \sqrt{2/3}$, and so from formula (1) w get

$$\begin{aligned}\mathbf{m}_0 &= \frac{1}{\sin\lambda}\boldsymbol{\sigma}(0) - \frac{\cos\lambda}{\sin\lambda}\mathbf{l}\\ &= \sqrt{\frac{3}{2}}\,\mathbf{k} - \frac{1}{\sqrt{3}}\sqrt{\frac{3}{2}} \cdot \frac{1}{\sqrt{3}}(\mathbf{i} + \mathbf{j} + \mathbf{k})\\ &= \frac{2}{\sqrt{6}}\mathbf{k} - \frac{1}{\sqrt{6}}(\mathbf{i} + \mathbf{j})\end{aligned}$$

and

$$\mathbf{n}_0 = \mathbf{l} \times \mathbf{m}_0 = \begin{vmatrix} \mathbf{i} & \mathbf{j} & \mathbf{k} \\ \frac{1}{\sqrt{3}} & \frac{1}{\sqrt{3}} & \frac{1}{\sqrt{3}} \\ -\frac{1}{\sqrt{6}} & -\frac{1}{\sqrt{2}} & -\frac{2}{\sqrt{6}} \end{vmatrix} = \frac{1}{\sqrt{2}}\mathbf{i} - \frac{1}{\sqrt{2}}\mathbf{j}. \blacktriangle$$

Return to Fig. 14.S.3 and rotate the whole picture about the axis **l**. Now **m** and **n** will vary with time as well. Since the angle λ remains constant, formula (1) applied after time t to **r** and **l** gives

$$\mathbf{m} = \frac{1}{\sin\lambda}\mathbf{r} - \frac{\cos\lambda}{\sin\lambda}\mathbf{l} \qquad (1')$$

(See Fig. 14.S.4.)

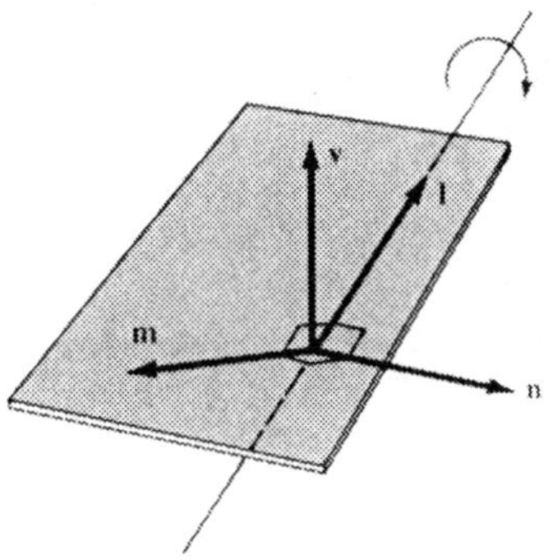

Figure 14.S.4. The three vectors **v**, **m**, and **n** all rotate about **l**.

On the other hand, since **m** is perpendicular to **l**, it rotates in a circle in the plane of $\mathbf{m}_0$ and $\mathbf{n}_0$. It goes through an angle 2π in time T, so it goes through an angle $2\pi t/T$ in t units of time, and so

$$\mathbf{m} = \cos\left(\frac{2\pi t}{T}\right)\mathbf{m}_0 + \sin\left(\frac{2\pi t}{T}\right)\mathbf{n}_0 .$$

Inserting this in formula (1′) and rearranging gives

$$\mathbf{r} = \boldsymbol{\sigma}(t) = (\cos\lambda)\mathbf{l} + \sin\lambda\cos\left(\frac{2\pi t}{T}\right)\mathbf{m}_0 + \sin\lambda\sin\left(\frac{2\pi t}{T}\right)\mathbf{n}_0 . \qquad (2)$$

This formula expresses explicitly how **r** changes in time as it is rotated about **l**, in terms of the basic trihedral $(\mathbf{l}, \mathbf{m}_0, \mathbf{n}_0)$.

Example 2 Express the function $\boldsymbol{\sigma}(t)$ explicitly in terms of **l**, $\mathbf{r}_0$, and T.

Solution We have $\cos\lambda = \mathbf{l} \cdot \mathbf{r}_0$ and $\sin\lambda = \|\mathbf{l} \times \mathbf{r}_0\|$. Furthermore $\mathbf{n}_0$ is a unit vector perpendicular to both **l** and $\mathbf{r}_0$, so we must have

$$\mathbf{n}_0 = \frac{\mathbf{l} \times \mathbf{r}_0}{\|\mathbf{l} \times \mathbf{r}_0\|} .$$

Thus $(\sin\lambda)\mathbf{n}_0 = \mathbf{l} \times \mathbf{r}_0$. Finally, from formula (1), we obtain $(\sin\lambda)\mathbf{m}_0 = \mathbf{r}_0 - (\cos\lambda)\mathbf{l} = \mathbf{r}_0 - (\mathbf{r}_0 \cdot \mathbf{l})\mathbf{l}$. Substituting all this into formula (2),

$$\mathbf{r} = (\mathbf{r}_0 \cdot \mathbf{l})\mathbf{l} + \cos\left(\frac{2\pi t}{T}\right)\left[\mathbf{r}_0 - (\mathbf{r}_0 \cdot \mathbf{l})\mathbf{l}\right] + \sin\left(\frac{2\pi t}{T}\right)(\mathbf{l} \times \mathbf{r}_0). \blacktriangle$$

Example 3 Show by a direct geometric argument that the speed of the tip of $\mathbf{r}$ is $(2\pi/T)\sin\lambda$. Verify that equation (2) gives the same formula.

Solution The tip of $\mathbf{r}$ sweeps out a circle of radius $\sin\lambda$, so it covers a distance $2\pi\sin\lambda$ in time T. Its speed is therefore $(2\pi\sin\lambda)/T$ (Fig. 14.S.5). From formula (2), we find the velocity vector to be

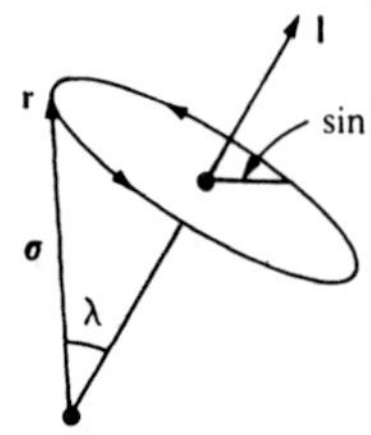

Figure 14.S.5. The tip of $\mathbf{r}$ sweeps out a circle of radius $\sin\lambda$.

$$\frac{d\mathbf{r}}{dt} = -\sin\lambda\cdot\frac{2\pi}{T}\sin\left(\frac{2\pi t}{T}\right)\mathbf{m}_0 + \sin\lambda\cdot\frac{2\pi}{T}\cos\left(\frac{2\pi t}{T}\right)\mathbf{n}_0,$$

and its length is (since $\mathbf{m}_0$ and $\mathbf{n}_0$ are unit orthogonal vectors)

$$\left\|\frac{d\mathbf{r}}{dt}\right\| = \sqrt{\sin^2\lambda\cdot\left(\frac{2\pi}{T}\right)^2\sin^2\left(\frac{2\pi t}{T}\right) + \sin^2\lambda\cdot\left(\frac{2\pi}{T}\right)^2\cos^2\left(\frac{2\pi t}{T}\right)}$$
$$= \sin\lambda\cdot\left(\frac{2\pi}{T}\right), \text{ as above. } \blacktriangle$$

Now we apply our study of rotations to the motion of the earth about the sun, incorporating the rotation of the earth about its own axis as well. We will use a simplified model of the earth–sun system, in which the sun is fixed at the origin of our coordinate system and the earth moves at uniform speed around a circle centered at the sun. Let $\mathbf{u}$ be a unit vector pointing *from* the sun *to* the earth; we have $\mathbf{u} = \cos(2\pi t/T_y)\mathbf{i} + \sin(2\pi t/T_y)\mathbf{j}$, where T_y is the length of a year (t and T_y measured in the same units). See Fig. 14.S.6. Notice that the unit vector pointing from the earth to the sun is $-\mathbf{u}$ and that we have oriented our axes so that $\mathbf{u} = \mathbf{i}$ when $T = 0$.

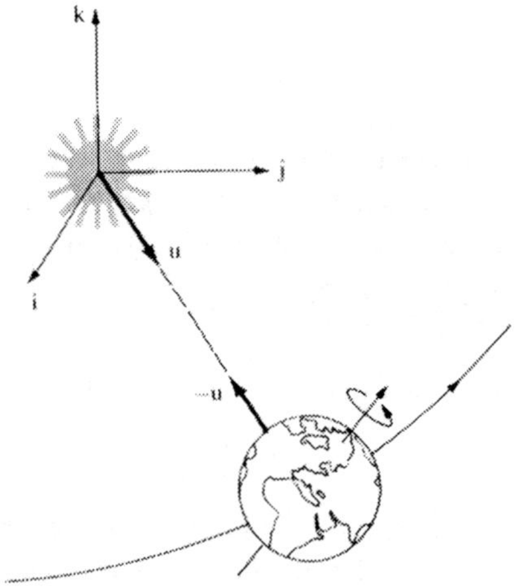

Figure 14.S.6. The unit vector $\mathbf{u}$ points from the sun to the earth at time t.

Next we wish to take into account the rotation of the earth. The earth rotates about an axis which we represent by a unit vector $\mathbf{l}$ pointing from the center of the earth to the North Pole. We will assume that $\mathbf{l}$ is fixed[4] with respect to $\mathbf{i}$, $\mathbf{j}$, and $\mathbf{k}$; astronomical measurements show that the inclination of $\mathbf{l}$ (the angle between $\mathbf{l}$ and $\mathbf{k}$) is presently about 23.5°. We will denote this angle by α. If we measure time so that the first day of summer in the northern hemisphere occurs when $t = 0$, then the axis $\mathbf{l}$ must tilt in the direction $-\mathbf{i}$, and so we must have $\mathbf{l} = \cos\alpha\mathbf{k} - \sin\alpha\mathbf{i}$. (See Fig. 14.S.7.)

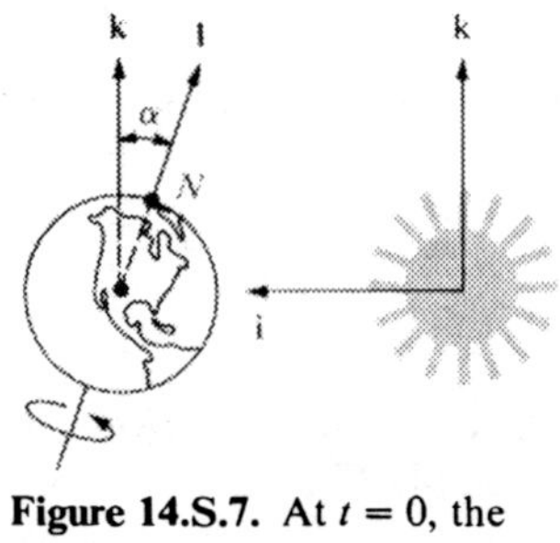

Figure 14.S.7. At $t = 0$, the earth's axis is tilted toward the sun.

Now let $\mathbf{r}$ be the unit vector at time t from the center of the earth to a fixed point P on the earth's surface. Notice that if $\mathbf{r}$ is located with its base

[4] Actually, the axis $\mathbf{l}$ is known to rotate about $\mathbf{k}$ once every 21,000 years. This phenomenon, called precession or wobble, is due to the irregular shape of the earth and may play a role in long-term climatic changes, such as ice ages. See pp. 130–134 of *The Weather Machine* by Nigel Calder, Viking (1974).

point at P, then it represents the local *vertical* direction. We will assume that P is chosen so that at $t = 0$, it is noon at the point P; then $\mathbf{r}$ lies in the plane of $\mathbf{l}$ and $\mathbf{i}$ and makes an angle of less than 90° with $-\mathbf{i}$. Referring to Fig. 14.S.8, we introduce the unit vector $\mathbf{m}_0 = -(\sin\alpha)\mathbf{k} - (\cos\alpha)\mathbf{i}$ orthogonal to $\mathbf{l}$. We then have $\mathbf{r}_0 = (\cos\lambda)\mathbf{l} + (\sin\lambda)\mathbf{m}_0$, where λ is the angle between $\mathbf{l}$ and $\mathbf{r}_0$. Since $\lambda = 90° - l$, where l is the latitude of the point P, we obtain the expression $\mathbf{r}_0 = (\sin l)\mathbf{l} + (\cos l)\mathbf{m}_0$. As in Fig. 14.S.3, let $\mathbf{n}_0 = \mathbf{l} \times \mathbf{m}_0$.

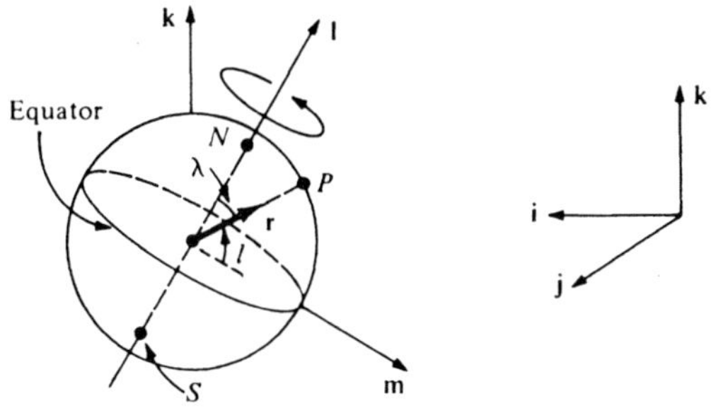

Figure 14.S.8. The vector $\mathbf{r}$ is the vector from the center of the earth to a fixed location P. The latitude of P is l and the colatitude is $\lambda = 90° - l$. The vector $\mathbf{m}_0$ is a unit vector in the plane of the equator (orthogonal to $\mathbf{l}$) and in the plane of $\mathbf{l}$ and $\mathbf{r}_0$.

Example 4 Prove that $\mathbf{n}_0 = \mathbf{l} \times \mathbf{m}_0 = -\mathbf{j}$.

Solution Geometrically, $\mathbf{l} \times \mathbf{m}_0$ is a unit vector orthogonal to $\mathbf{l}$ and $\mathbf{m}_0$ pointing in the sense given by the right-hand rule. But $\mathbf{l}$ and $\mathbf{m}_0$ are both in the $\mathbf{ik}$ plane, so $\mathbf{l} \times \mathbf{m}_0$ points orthogonal to it in the direction $-\mathbf{j}$ (see Fig. 14.S.8).

Algebraically, $\mathbf{l} = (\cos\alpha)\mathbf{k} - (\sin\alpha)\mathbf{i}$ and $\mathbf{m}_0 = -(\sin\alpha)\mathbf{k} - (\cos\alpha)\mathbf{i}$, so

$$\mathbf{l} \times \mathbf{m}_0 = \begin{vmatrix} \mathbf{i} & \mathbf{j} & \mathbf{k} \\ -\sin\alpha & 0 & \cos\alpha \\ -\cos\alpha & 0 & -\sin\alpha \end{vmatrix} = -\mathbf{j}(\sin^2\alpha + \cos^2\alpha) = -\mathbf{j}. \ \blacktriangle$$

Now we apply formula (2) to get

$$\mathbf{r} = (\cos\lambda)\mathbf{l} + \sin\lambda\cos\left(\frac{2\pi t}{T_d}\right)\mathbf{m}_0 + \sin\lambda\sin\left(\frac{2\pi t}{T_d}\right)\mathbf{n}_0,$$

where T_d is the length of time it takes for the earth to rotate once about its axis (with respect to the "fixed stars"—i.e., our $\mathbf{i}, \mathbf{j}, \mathbf{k}$ vectors).[5] Substituting the expressions derived above for λ, $\mathbf{l}$, $\mathbf{m}_0$, and $\mathbf{n}_0$, we get

$$\mathbf{r} = \sin l(\cos\alpha\mathbf{k} - \sin\alpha\mathbf{i}) + \cos l\cos\left(\frac{2\pi t}{T_d}\right)(-\sin\alpha\mathbf{k} - \cos\alpha\mathbf{i}) - \cos l\sin\left(\frac{2\pi t}{T_d}\right)\mathbf{j}.$$

Hence

$$\mathbf{r} = -\left[\sin l\sin\alpha + \cos l\cos\alpha\cos\left(\frac{2\pi t}{T_d}\right)\right]\mathbf{i} - \cos l\sin\left(\frac{2\pi t}{T_d}\right)\mathbf{j}$$
$$+\left[\sin l\cos\alpha - \cos l\sin\alpha\cos\left(\frac{2\pi t}{T_d}\right)\right]\mathbf{k}. \qquad (3)$$

Example 5 What is the speed (in kilometers per hour) of a point on the equator due to the rotation of the earth? A point at latitude 60°? (The radius of the earth is 6371 kilometers.)

Solution From Example 3, the speed is $s = (2\pi R/T_d)\sin\lambda = (2\pi R/T_d)\cos l$, where R is the radius of the earth and l is the latitude. (The factor R is inserted since $\mathbf{r}$ is a *unit* vector; the actual vector from the earth's center to a point P is $R\mathbf{r}$).

[5] T_d is called the length of the sidereal day. It differs from the ordinary, or solar, day by about 1 part in 365. (Can you explain why?) In fact, $T_d \approx 23.93$ hours.

Using $T_d = 23.93$ hours and $R = 6371$ kilometers, we get $s = 1673 \cos l$ kilometers per hour. At the equator $l = 0$, so the speed is 1673 kilometers per hour; at $l = 60°$, $s = 836.4$ kilometers per hour. ▲

With formula (3) at our disposal, we are now ready to derive the sunshine formula. The intensity of light on a portion of the earth's surface (or at the top of the atmosphere) is proportional to $\sin A$, where A is the angle of elevation of the sun above the horizon (see Fig. 14.S.9). (At night $\sin A$ is negative, and the intensity then is of course zero.)

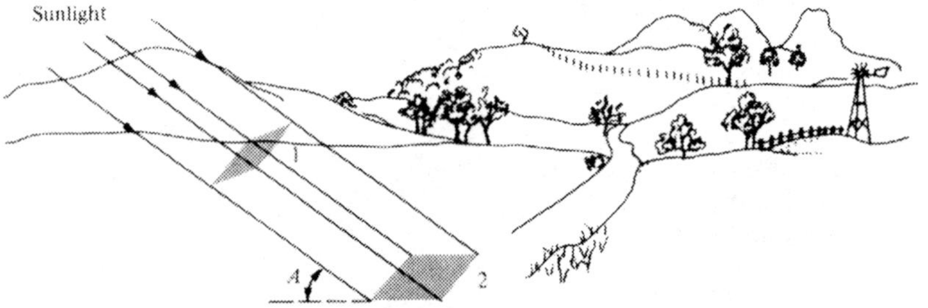

Figure 14.S.9. The intensity of sunlight is proportional to $\sin A$. The ratio of area 1 to area 2 is $\sin A$.

Thus we want to compute $\sin A$. From Fig. 14.S.10 we see that $\sin A = -\mathbf{u} \cdot \mathbf{r}$. Substituting $\mathbf{u} = \cos(2\pi t/T_y)\mathbf{i} + \sin(2\pi t/T_y)\mathbf{j}$ and formula (3) into this formula for $\sin A$ and taking the dot product gives

$$\begin{aligned}\sin A &= \cos\left(\frac{2\pi t}{T_y}\right)\left[\sin l \sin\alpha + \cos l \cos\alpha \cos\left(\frac{2\pi t}{T_d}\right)\right]\\ &\quad + \sin\left(\frac{2\pi t}{T_y}\right)\left[\cos l \sin\left(\frac{2\pi t}{T_d}\right)\right]\\ &= \cos\left(\frac{2\pi t}{T_y}\right)\sin l \sin\alpha\\ &\quad + \cos l\left[\cos\left(\frac{2\pi t}{T_y}\right)\cos\alpha\cos\left(\frac{2\pi t}{T_d}\right) + \sin\left(\frac{2\pi t}{T_y}\right)\sin\left(\frac{2\pi t}{T_d}\right)\right]. \qquad (4)\end{aligned}$$

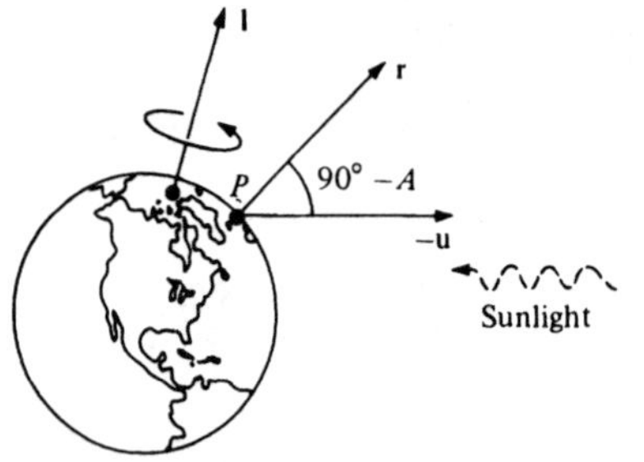

Figure 14.S.10. The geometry for the formula $\sin A = \cos(90° - A) = -\mathbf{u} \cdot \mathbf{r}$.

Example 6 Set $t = 0$ in formula (4). For what l is $\sin A = 0$? Interpret your result.

Solution With $t = 0$ we get

$$\sin A = \sin l \sin\alpha + \cos l \cos\alpha = \cos(l - \alpha).$$

This is zero when $l - \alpha = \pm\pi/2$. Now $\sin A = 0$ corresponds to the sun on the horizon (sunrise or sunset), when $A = 0$ or π. Thus, at $t = 0$, this occurs when $l = \alpha \pm (\pi/2)$. The case $\alpha + (\pi/2)$ is impossible, since l lies between $-\pi/2$ and $\pi/2$. The case $l = \alpha - (\pi/2)$ corresponds to a point on the Antarctic Circle; indeed at $t = 0$ (corresponding to noon on the first day of northern summer) the sun is just on the horizon at the Antarctic Circle. ▲

Our next goal is to describe the variation of $\sin A$ with time *on a particular day*. For this purpose, the time variable t is not very convenient; it will be better to measure time from noon on the day in question.

To simplify our calculations, we will assume that the expressions $\cos(2\pi t/T_y)$ and $\sin(2\pi t/T_y)$ are constant over the course of any particular day; since T_y is 365 times as large as the change in t, this is a reasonable approximation. On the nth day (measured from June 21), we may replace $2\pi t/T_y$ by $2\pi n/365$, and formula (4) gives

$$\sin A = (\sin l)P + (\cos l)\left[Q\cos\left(\frac{2\pi t}{T_d}\right) + R\sin\left(\frac{2\pi t}{T_d}\right)\right], \tag{5}$$

where $P = \cos(2\pi n/365)\sin\alpha$, $Q = \cos(2\pi n/365)\cos\alpha$, and $R = \sin(2\pi n/365)$.

We will write the expression $Q\cos(2\pi t/T_d) + R\sin(2\pi t/T_d)$ in the form $U\cos[2\pi(t - t_n)/T_d]$, where t_n is the time of noon of the nth day. To find U, we use the addition formula to expand the cosine:

$$U\cos\left(\frac{2\pi t}{T_d} - \frac{2\pi t_n}{T_d}\right) = U\left[\cos\left(\frac{2\pi t}{T_d}\right)\cos\left(\frac{2\pi t_n}{T_d}\right) + \sin\left(\frac{2\pi t}{T_d}\right)\sin\left(\frac{2\pi t_n}{T_d}\right)\right].$$

Setting this equal to $Q\cos(2\pi t/T_d) + R\sin(2\pi t/T_d)$ and comparing coefficients of $\cos 2\pi t/T_d$ and $\sin 2\pi t/T_d$ gives

$$U\cos\frac{2\pi t_n}{T_d} = Q \quad \text{and} \quad U\sin\frac{2\pi t_n}{T_d} = R.$$

Squaring the two equations and adding gives

$$U^2 = Q^2 + R^2 \quad \text{or} \quad U = \sqrt{Q^2 + R^2}\,,$$[6]

while dividing the second equation by the first gives $\tan(2\pi t_n/T_d) = R/Q$. We are interested mainly in the formula for U; substituting for Q and R gives

$$\begin{aligned} U &= \sqrt{\cos^2\left(\frac{2\pi n}{365}\right)\cos^2\alpha + \sin^2\left(\frac{2\pi n}{365}\right)} \\ &= \sqrt{\cos^2\left(\frac{2\pi n}{365}\right)(1 - \sin^2\alpha) + \sin^2\left(\frac{2\pi n}{365}\right)} \\ &= \sqrt{1 - \cos^2\left(\frac{2\pi n}{365}\right)\sin^2\alpha}\,. \end{aligned}$$

Letting τ be the time in hours from noon on the nth day so that $(t - t_n)/T_d = \tau/24$, we may substitute into formula (5) to obtain the final formula:

$$\sin A = \sin l\cos\left(\frac{2\pi n}{365}\right)\sin\alpha + \cos l\sqrt{1 - \cos^2\left(\frac{2\pi n}{365}\right)\sin^2\alpha}\,\cos\left(\frac{2\pi\tau}{24}\right), \tag{6}$$

which is identical (after some changes in notation) to formula (1) on page 301.

Example 7 How high is the sun in the sky in Edinburgh (latitutde 56°) at 2 P.M. on February 1?

Solution We plug into formula (6): $\alpha = 23.5°$, $l = 90 - 56 = 34°$, n = number of days after June 21 = 225, and $\tau = 2$ hours. We get

$$\sin A = 0.5196,$$

so $A = 31.3°$. ▲

[6] We take the positive square root because $\sin A$ should have a local maximum when $t = t_n$.

Exercises for The Supplement to Chapter 14

1. Let $\mathbf{l} = (\mathbf{j} + \mathbf{k})/\sqrt{2}$ and $\mathbf{r}_0 = (\mathbf{i} - \mathbf{j})/\sqrt{2}$. (a) Find $\mathbf{m}_0$ and $\mathbf{n}_0$. (b) Find $\mathbf{r} = \boldsymbol{\sigma}(t)$ if $T = 24$. (c) Find the equation of the line tangent to $\boldsymbol{\sigma}(t)$ at $t = 12$ and $T = 24$.
2. From formula (2), verify that $\boldsymbol{\sigma}(T/2) \cdot \mathbf{n} = 0$. Also, show this geometrically. For what values of t is $\boldsymbol{\sigma}(t) \cdot \mathbf{n} = 0$?
3. If the earth rotated in the opposite direction about the sun, would T_d be longer or shorter than 24 hours? (Assume the solar day is fixed at 24 hours.)
4. Show by a direct geometric construction that $\mathbf{r} = \boldsymbol{\sigma}(T_d/4) = -\sin l \sin \alpha \mathbf{i} - \cos l \mathbf{j} + \sin l \cos \alpha \mathbf{k}$. Does this formula agree with formula (3)?
5. Derive an "exact" formula for the time of sunset from formula (4).
6. Why does formula (6) for $\sin A$ not depend on the radius of the earth? The distance of the earth from the sun?
7. How high is the sun in the sky in Paris at 3 P.M. on January 15? (The latitude of Paris is 49°N).
8. How much solar energy (relative to a summer day at the equator) does Paris receive on January 15? (The latitude of Paris is 49° N).
9. How would your answer in Exercise 8 change if the earth were to roll to a tilt of 32° instead of 23.5°?

Review Exercises for Chapter 14

Sketch the graphs of the conics in Exercises 1–8.

1. $4x^2 + 9y^2 = 36$
2. $x = 2y^2$
3. $x^2 - 4y^2 = 16$
4. $4x^2 + 16y^2 = 81$
5. $100x^2 + 100y^2 = 1$
6. $y^2 = 16 + 4x^2$
7. $x^2 - y = 14$
8. $2x^2 + 2y^2 = 80$

Sketch the graphs of the conics in Exercises 9–12.

9. $9x^2 - 18x + y^2 - 4y + 4 = 0$.
10. $9x^2 + 18x - y^2 + 2y - 8 = 0$.
11. $x^2 + 2xy + 3y^2 = 14$.
12. $x^2 - 2xy - 3y^2 = 14$.

Sketch or describe the level curves for the functions and values in Exercises 13–16.

13. $f(x, y) = 3x - 2y$; $c = 2$
14. $f(x, y) = x^2 - y^2$; $c = -1$
15. $f(x, y) = x^2 + xy$; $c = 2$
16. $f(x, y) = x^2 + 4$; $c = 85$

Describe the level surfaces $f(x, y, z) = c$ for each of the functions in Exercises 17–20. Sketch for $c = 1$ and $c = 5$.

17. $f(x, y, z) = x - y - z$
18. $f(x, y, z) = x + y - 2z$
19. $f(x, y, z) = x^2 + y^2 + z^2 + 1$
20. $f(x, y, z) = x^2 + 2y^2 + 3z^2$

Sketch and describe the surfaces in Exercises 21–28.

21. $x^2 + 4y^2 + z^2 = 1$
22. $x^2 + 4y^2 - z^2 = 0$
23. $x^2 + 4y^2 - z^2 = 1$
24. $x^2 + 4y^2 + z^2 = 0$
25. $x^2 + 4y^2 - z = 1$
26. $x^2 + 4y^2 - z = 0$
27. $x^2 + 4y^2 + z = 1$
28. $x^2 + 4y^2 + z = 0$

29. This exercise concerns the *elliptic hyperboloid of one sheet*. An example of this type was studied in Example 6, Section 14.4. A standard form for the equation is
$$\frac{x^2}{a^2} + \frac{y^2}{b^2} - \frac{z^2}{c^2} = 1 \quad (a, b, \text{and } c \text{ positive}).$$
   (a) What are the horizontal cross sections obtained by holding z constant?
   (b) What are the vertical cross sections obtained by holding either x or y constant?
   (c) Sketch the surface defined by
$$\frac{x^2}{4} + \frac{y^2}{1} - \frac{z^2}{1} = 1.$$
   In a yz plane, sketch the cross-section curves obtained from this surface when x is held constantly equal to 0, 1, 2, and 3 ($x = 2$ is especially interesting).
30. (a) Describe the level surfaces of the function $f(x, y, z) = x^2 + y^2 - z^2$. In particular, discuss the surface $f(x, y, z) = c$ when c is positive, when c is negative, and when c is zero.
   (b) Several level surfaces of f are sketched in Fig. 14.R.1. Find the value of c associated with each.
   (c) Describe how the appearance of the level surfaces changes if we consider instead the function $g(x, y, z) = x^2 + 2y^2 - z^2$.
31. Let $f(x, y) = x^2 + 2y^2 + 1$
   (a) Sketch the level curves $f(x, y) = c$ for $c = -10, -1, 0, 1, 2$, and 10.
   (b) Describe the intersection of the graph of f with the vertical planes $x = 1$, $x = -1$, $x = 2$, $y = 1$, $y = 2$, $y = -1$.
   (c) Sketch the graph of f.

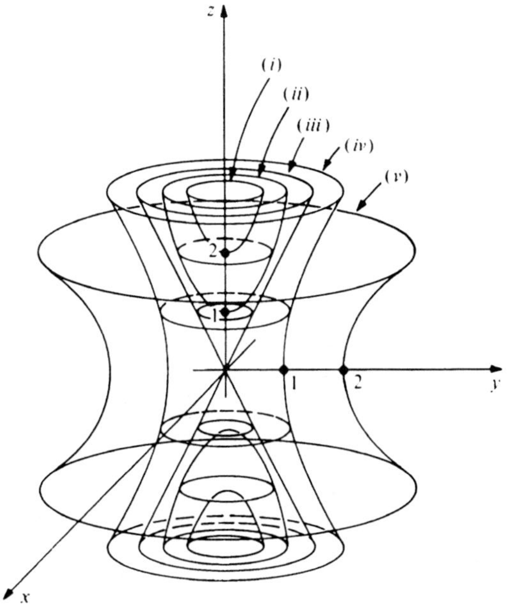

Figure 14.R.1. Level surfaces of $x^2 + y^2 - z^2$.

32. Do as in Exercise 31 for $f(x, y) = y/x$, and describe the intersection of the graph of f with the cylinder of radius R (that is, $r = R$ in cylindrical coordinates).

In Exercises 33–38, fill in the blanks and plot.

	Rectangular coordinates	*Cylindrical coordinates*	*Spherical coordinates*
33.	$(1, -1, 1)$		
34.	$(1, 0, 3)$		
35.		$(5, \pi/12, 4)$	
36.		$(8, 3\pi/2, 2)$	
37.			$(3, -\pi/6, \pi/4)$
38.			$(10, \pi/4, \pi/2)$

39. A surface is described in cylindrical coordinates by $3r^2 = z^2 + 1$. Convert to rectangular coordinates and plot.
40. Show that a surface described in spherical coordinates by $f(\rho, \phi) = 0$ is a surface of revolution.
41. Describe the geometric meaning of replacing (ρ, θ, ϕ) by $(\rho, \theta + \pi, \phi + \pi/2)$ in spherical coordinates.
42. Describe the geometric meaning of replacing (ρ, θ, ϕ) by $(4\rho, \theta, \phi)$ in spherical coordinates.
43. Describe by means of cylindrical coordinates a solenoid consisting of a copper rod of radius 5 centimeters and length 15 centimeters wound on the outside with copper wire to a thickness of 1.2 centimeters. Give separate descriptions of the rod and the winding.
44. A gasoline storage tank has two spherical cap ends of arc length 56.55 feet. The cylindrical part of the tank has length 16 feet and circumference 113.10 feet. Let $(0, 0, 0)$ be the geometric center of the tank. See Fig. 14.R.2.

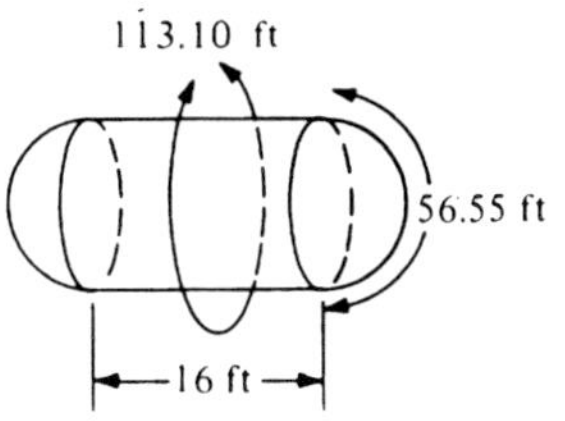

Figure 14.R.2. The gasoline storage tank for Exercise 44.

(a) Describe the cylindrical part of the tank via cylindrical coordinates.
(b) Describe the hemispherical end caps with spherical coordinates. (Set up spherical coordinates using the centers of the cap ends as the origin.)

Sketch the curves or surfaces given by the equations in Exercises 45–52.

45. $z = x + y$
46. $x^2 + 2xz + z^2 = 0$
47. $\boldsymbol{\sigma}(t) = 3\sin t\,\mathbf{i} + t\,\mathbf{j} + \cos t\,\mathbf{k}$
48. $\boldsymbol{\sigma}(t) = (\sin t, t + 1, 2t - 1)$
49. $z^2 = -(x^2 + y^2/4)$
50. $z = -(x^2 + y^2)$
51. $z^2 = -x^2 - 3y^2 + 2$
52. $z^2 = x^2 - 4y^2$

Find the equation of the line tangent to each of the curves at the indicated point in Exercises 53 and 54.

53. $(t^3 + 1, e^{-t}, \cos(\pi t/2))$; $t = 1$
54. $(t^2 - 1, \cos t^2, t^4)$; $t = \sqrt{\pi}$

Find the velocity and acceleration vectors for the curves in Exercises 55–58.

55. $\boldsymbol{\sigma}_1(t) = e^t\mathbf{i} + \sin t\,\mathbf{j} + \cos t\,\mathbf{k}$.
56. $\boldsymbol{\sigma}_2(t) = \dfrac{t^2}{1 + t^2}\mathbf{i} + t\mathbf{j} + \mathbf{k}$.
57. $\boldsymbol{\sigma}(t) = \boldsymbol{\sigma}_1(t) + \boldsymbol{\sigma}_2(t)$, where $\boldsymbol{\sigma}_1$ and $\boldsymbol{\sigma}_2$ are given in Exercises 55 and 56.
58. $\boldsymbol{\sigma}(t) = \boldsymbol{\sigma}_1(t) \times \boldsymbol{\sigma}_2(t)$, where $\boldsymbol{\sigma}_1$ and $\boldsymbol{\sigma}_2$ are given in Exercises 55 and 56.
59. Write in parametric form the curve described by the equations $x - 1 = 2y + 1 = 3z + 2$.
60. Write the curve $x = y^3 = z^2 + 1$ in parametric form.
61. Find the arc length of $\boldsymbol{\sigma}(t) = t\,\mathbf{i} + \ln t\,\mathbf{j} + 2\sqrt{2t}\,\mathbf{k}$; $1 \leqslant t \leqslant 2$.
62. Express as an integral the arc length of the curve $x^2 = y^3 = z^5$ between $x = 1$ and $x = 4$. (Find a parametrization.)
63. A particle moving on the curve $\boldsymbol{\sigma}(t) = 3t^2\mathbf{i} - \sin t\mathbf{j} - e^t\mathbf{k}$ is released at time $t = \frac{1}{2}$ and flies off on a tangent. What are its coordinates at time $t = 1$?
64. A particle is constrained to move around the unit circle in the xy plane according to the formula $(x, y, z) = (\cos(t^2), \sin(t^2), 0)$, $t \geqslant 0$.

(a) What are the velocity vector and speed of the particle as functions of t?

(b) At what point on the circle should the particle be released to hit a target at $(2,0,0)$? (Be careful about which direction the particle is moving around the circle.)

(c) At what time t should the release take place? (Use the smallest $t > 0$ which will work.)

(d) What are the velocity and speed at the time of release?

(e) At what time is the target hit?

65. A particle of mass m is subject to the force law $\mathbf{F} = -k\mathbf{r}$, where k is a constant.

(a) Write down differential equations for the components of $\mathbf{r}(t)$.

(b) Solve the equations in (a) subject to the initial conditions $\mathbf{r}(0) = \mathbf{0}$, $\mathbf{r}'(0) = 2\mathbf{j} + \mathbf{k}$.

66. Show that the quantity

$$\frac{m}{2}\left\|\frac{d\mathbf{r}}{dt}\right\|^2 + \frac{k}{2}\mathbf{r}\cdot\mathbf{r}$$

is independent of time when a particle moves under the force law in Exercise 65.

67. Find the curvature of the ellipse $4x^2 + 9y^2 = 16$.

★ 68. Let $\mathbf{r} = \boldsymbol{\sigma}(t)$ be a curve in space and $\mathbf{N}$ be its principal normal vector. Consider the "parallel curve" $\mathbf{r} = \boldsymbol{\mu}(t) = \boldsymbol{\sigma}(t) + \mathbf{N}(t)$, where $\boldsymbol{\sigma}(t)$ is the displacement vector to $P(t)$ from a fixed origin.

(a) Under what conditions does $\boldsymbol{\mu}(t)$ have zero velocity for some t_0? [Hint: Use the Frenet formulas, Exercise 25, Section 14.7]

(b) Find the parametric equation of the parallel curve to the ellipse $(\frac{1}{4}\cos t, 4\sin t, 0)$.

69. Find a formula for the curvature of the graph $y = f(x)$ in terms of f and its derivatives.

70. The contour lines on a topographical map are the level curves of the function giving height above sea level as a function of position. Figure 14.R.3 is a portion of the U.S. Geological Survey map of Yosemite Valley. There is a heavy contour line for every 200 feet of elevation and a lighter line at each 40-foot interval between these.

(a) What does it mean in terms of the terrain when these contour lines are far apart?

(b) What if they are close together?

(c) What does it mean when several contour lines seem to merge for a distance? Is eleva-

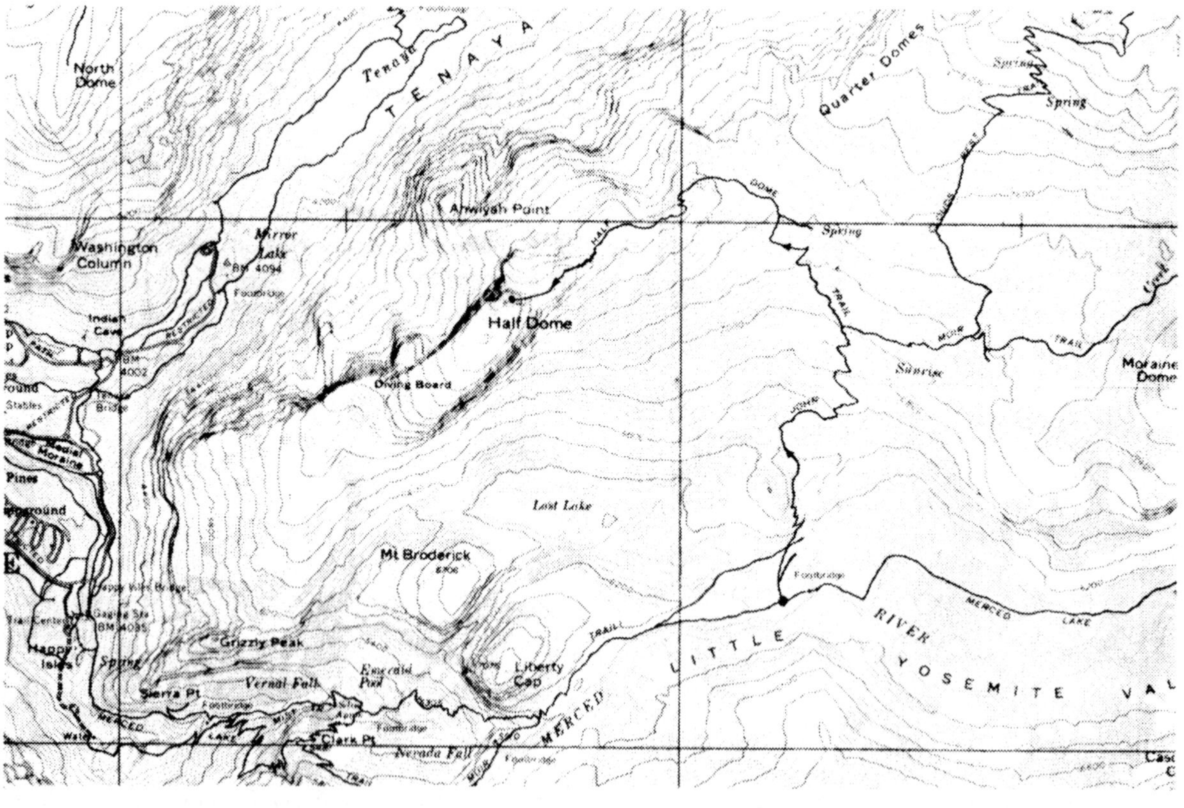

Figure 14.R.3. Yosemite Valley (portion). (U.S. Department of Interior Geological Survey.)

tion really a function of position at such points? (Look at the west face of Half Dome. Does this seem like a good direction from which to climb it?)

(d) Sketch a cross section of terrain along a north–south line through the top of Half Dome.

(e) A hiker is not likely to follow the straight north–south path in part (d) and is probably more interested in the behavior of the terrain along the trail he or she will follow. The 3-mile route from the Merced River (altitude approximately 6100 feet) to the top of Half Dome (approximately 8842 feet) along the John Muir and Half Dome Trails has been emphasized in Fig. 14.R.3. Show how a cross section of the terrain along this trail behaves by plotting altitude above sea level as a function of miles along the trail. (A piece of string or flexible wire may be of aid in measuring distances along the trail.)

★71. Find the curvature of the "helical spiral" $(t, t\cos t, t\sin t)$ for $t > 0$. Sketch.

★72. Describe the level curves $f(x, y) = c$ for each of the following functions. In particular, discuss any special values of c at which the behavior of the level curves changes suddenly. Sketch the curves for $c = -1, 0$, and 1.

(a) $f(x, y) = x + 2y$;
(b) $f(x, y) = x^2 - y^2$;
(c) $f(x, y) = y^2 - x^2$;
(d) $f(x, y) = x^2 + y^2$;
(e) $f(x, y) = xy$;
(f) $f(x, y) = y - 2x^2$.

★73. (a) Write in parametric form the curve which is the intersection of the surfaces $x^2 + y^2 + z^2 = 3$ and $y = 1$.

(b) Find the equation of the line tangent to this curve at $(1, 1, 1)$.

(c) Write an integral expression for the arc length of this curve. What is the value of this integral?

★74. Let n be a positive integer and consider the curve

$$\left.\begin{aligned} x &= \cos t\cos(4nt) \\ y &= \cos t\sin(4nt) \\ z &= \sin t \end{aligned}\right\} \quad -\frac{\pi}{2} \leqslant t \leqslant \frac{\pi}{2},$$

(a) Show that the path traced out lies on the surface of the sphere of radius 1 centered at the origin.

(b) How many times does the curve wind around the z axis?

(c) Where does the curve cross the xy plane?

(d) Sketch the curve when $n = 1$ and when $n = 2$.

★75. Let $\mathbf{u}_1$ and $\mathbf{u}_2$ be unit vectors, and define the curve $\boldsymbol{\sigma}(t)$ by

$$\boldsymbol{\sigma}(t) = \frac{\mathbf{u}_1\cos t + \mathbf{u}_2\sin t}{\|\mathbf{u}_1\cos t + \mathbf{u}_2\sin t\|}, \qquad 0 \leqslant t \leqslant \frac{\pi}{2}.$$

(a) Find $\boldsymbol{\sigma}(0)$ and $\boldsymbol{\sigma}(\pi/2)$.

(b) On what surfaces does $\boldsymbol{\sigma}(t)$ lie for all t?

(c) Find, by geometry, the arc length of the curve $\boldsymbol{\sigma}(t)$ for $0 \leqslant t \leqslant \pi/2$.

(d) Express the arc length of the curve $\boldsymbol{\sigma}(t)$ for $0 \leqslant t \leqslant \pi/2$ as an integral.

(e) Find a curve $\boldsymbol{\sigma}_1(t)$ which traverses the same path as $\boldsymbol{\sigma}(t)$ for $0 \leqslant t \leqslant \pi/2$ and such that the speed $\|\boldsymbol{\sigma}_1'(t)\|$ is constant.

★76 (a) Show that the hyperboloid $x^2 + y^2 - z^2 = 4$ is a ruled surface by finding two straight lines lying in the surface through each point. [*Hint:* Let (x_0, y_0, z_0) lie on the surface; write the equation of the line in the form $x = x_0 + at$, $y = y_0 + bt$, $z = z_0 + t$; write out $x^2 + y^2 - z^2 = 4$ using $x_0^2 + y_0^2 - z_0^2 = 4$ to obtain two equations for a and b representing a line and a circle in the (a, b) plane. Show that these equations have two solutions by showing that the distance from the origin to the line is less than the radius of the circle.]

(b) Is the hyperboloid $x^2 + y^2 - z^2 = -4$ a ruled surface? Explain.

(c) Generalize the results of parts (a) and (b).

Chapter 15

Partial Differentiation

A function of several variables can be differentiated with respect to one variable at a time.

The rate of change of a function of several variables is not just a single function, since the independent variables may vary in different ways. All the rates of change, for a function of n variables, are described by n functions called its *partial derivatives*. This chapter begins with the definition and basic properties of partial derivatives. Methods for computation, including the chain rule, are presented along with a geometric interpretation in terms of tangent planes. The next chapter continues the development with topics including implicit differentiation, gradients, and maxima and minima.

15.1 Introduction to Partial Derivatives

The partial derivatives of a function of several variables are its ordinary derivatives with respect to each variable separately.

In this section we define partial derivatives and practice computing them. The geometric significance of partial derivatives and their use in computing tangent planes are explained in the next section.

Consider a function $f(x, y)$ of two variables. If we treat y as a constant, f may be differentiated with respect to x. The result is called the *partial derivative of f with respect to x* and is denoted by f_x. If we let $z = f(x, y)$, we write

$$f_x = \frac{\partial z}{\partial x} .$$

These symbols[1] are analogous to those we used in one-variable calculus: $f'(x) = dy/dx$.

The partial derivative with respect to y is similarly defined by treating x as a constant and differentiating $f(x, y)$ with respect to y.

[1] The symbol ∂ seems to have first been used by Clairaut and Euler around 1740 to avoid confusion with d. The notation $D_x f$ or $D_1 f$ for f_x is also used.

Example 1 (a) If $f(x, y) = xy + e^x \cos y$, compute f_x and f_y.
(b) For f as in (a), calculate $f_x(1, \pi/2)$.
(c) If $z = x^2y^3 + x^3y^4 - e^{xy^2}$, calculate $\dfrac{\partial z}{\partial x}$ and $\dfrac{\partial z}{\partial y}$.

Solution (a) Treating y as a constant and differentiating with respect to x, we get

$$f_x(x, y) = y + e^x \cos y.$$

Differentiating with respect to y and considering x as a constant gives

$$f_y(x, y) = x - e^x \sin y.$$

(b) Substituting $x = 1$ and $y = \pi/2$, we get

$$f_x(1, \pi/2) = \pi/2 + e^1 \cos(\pi/2) = \pi/2.$$

(c) Here we again hold y constant and calculate the x derivative:

$$\frac{\partial z}{\partial x} = 2xy^3 + 3x^2y^4 - y^2e^{xy^2}.$$

Similarly,

$$\frac{\partial z}{\partial y} = 3x^2y^2 + 4x^3y^3 - 2yxe^{xy^2}. \ \blacktriangle$$

In terms of limits, partial derivatives are given by

$$f_x(x, y) = \lim_{\Delta x \to 0} \frac{f(x + \Delta x, y) - f(x, y)}{\Delta x}$$

and

$$f_y(x, y) = \lim_{\Delta y \to 0} \frac{f(x, y + \Delta y) - f(x, y)}{\Delta y}.$$

See Fig. 15.1.1.

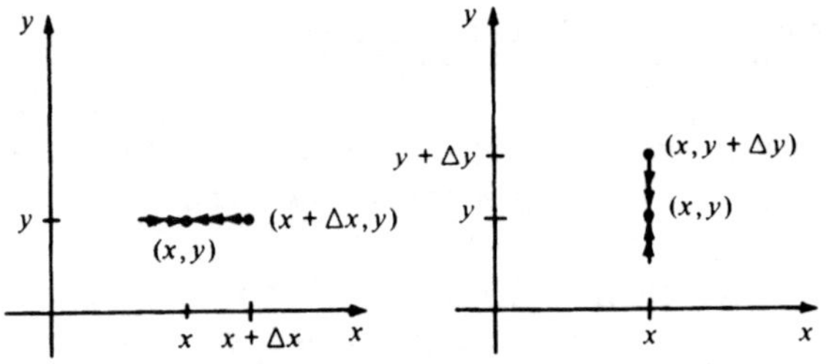

Figure 15.1.1. The partial derivatives f_x and f_y are limits of difference quotients along the horizontal and vertical paths shown here.

Partial derivatives of functions $f(x, y, z)$ of three variables are defined similarly. Two variables are treated as constant while we differentiate with respect to the third.

Example 2 (a) Let $f(x, y, z) = \sin(xy/z)$. Calculate $f_z(x, y, z)$ and $f_z(1, 2, 3)$.
(b) Evaluate

$$\frac{\partial}{\partial y} \frac{1}{\sqrt{x^2 + y^2 + z^2}}$$

at $(0, 1, 1)$.
(c) Write the result in (b) as a limit.

Solution (a) We differentiate $\sin(xy/z)$ with respect to z, thinking of x and y as constants; the result is

$$f_z(x, y, z) = \cos(xy/z)(-xy/z^2) = -(xy/z^2)\cos(xy/z).$$

Substituting (1, 2, 3) for (x, y, z) gives

$$f_z(1,2,3) = -\frac{1\cdot 2}{3^2}\cos\left(\frac{1\cdot 2}{3}\right) = -\frac{2}{9}\cos\left(\frac{2}{3}\right).$$

(b) Treating x and z as constants, and using the chain rule of one-variable calculus,

$$\begin{aligned}\frac{\partial}{\partial y}\frac{1}{\sqrt{x^2+y^2+z^2}} &= \frac{\partial}{\partial y}(x^2+y^2+z^2)^{-1/2}\\ &= \left(-\frac{1}{2}\right)(x^2+y^2+z^2)^{-3/2}\cdot 2y\\ &= \frac{-y}{(x^2+y^2+z^2)^{3/2}}.\end{aligned}$$

At (0, 1, 1) this becomes

$$\frac{-1}{(0^2+1^2+1^2)^{3/2}} = \frac{-1}{2\sqrt{2}}.$$

(c) In general,

$$\lim_{\Delta y\to 0}\frac{f(x, y+\Delta y, z)-f(x,y,z)}{\Delta y} = f_y(x,y,z).$$

In case (b), this becomes

$$\lim_{\Delta y\to 0}\frac{1}{\Delta y}\left[\frac{1}{\sqrt{(1+\Delta y)^2+1}}-\frac{1}{\sqrt{2}}\right] = -\frac{1}{2\sqrt{2}}. \blacktriangle$$

Partial Differentiation

If f is a function of several variables, to calculate the partial derivative with respect to a certain variable, treat the remaining variables as constants and differentiate as usual by using the rules of one-variable calculus.

If $z = f(x, y)$ is a function of two variables, the partial derivatives are denoted $f_x = \partial z/\partial x$ and $f_y = \partial z/\partial y$.

If $u = f(x, y, z)$ is a function of three variables, the partial derivatives are denoted $f_x = \partial u/\partial x$, $f_y = \partial u/\partial y$, and $f_z = \partial u/\partial z$.

As in one-variable calculus, the letters for the variables do not always have to be x, y, z.

Example 3 If $h = rs^2\sin(r^2+s^2)$, find $\partial h/\partial s$.

Solution Holding r constant, we get

$$\begin{aligned}\frac{\partial h}{\partial s} &= 2rs\sin(r^2+s^2) + rs^2\cdot 2s\cdot\cos(r^2+s^2)\\ &= 2rs\left[\sin(r^2+s^2)+s^2\cos(r^2+s^2)\right]. \blacktriangle\end{aligned}$$

Partial derivatives may be interpreted in terms of rates of change, just as derivatives of functions of one variable.

Example 4 The temperature (in degrees Celsius) near Dawson Creek at noon on April 15, 1901 is given by $T = -(0.0003)x^2y + (0.9307)y$, where x and y are the latitude and longitude (in degrees). At what rate is the temperature changing if we proceed directly north? (The latitude and longitude of Dawson Creek are $x = 55.7°$ and $y = 120.2°$.)

Solution Proceeding directly north means increasing the latitude x. Thus we calculate

$$\frac{\partial T}{\partial x} = -(0.0003) \cdot 2xy = -(0.0003) \cdot 2 \cdot (55.7) \cdot (120.2) \approx -4.017.$$

So the temperature drops as we proceed north from Dawson Creek, at the instantaneous rate of 4.017°C per degree of latitude. ▲

Since the partial derivatives are themselves functions, we can take their partial derivatives to obtain higher derivatives. For a function of two variables, there are four ways to take a second derivative. If $z = f(x, y)$, we may compute

$$f_{xx}(x, y) = \frac{\partial}{\partial x}\left(\frac{\partial z}{\partial x}\right) = \frac{\partial^2 z}{\partial x^2}, \qquad f_{yy}(x, y) = \frac{\partial}{\partial y}\left(\frac{\partial z}{\partial y}\right) = \frac{\partial^2 z}{\partial y^2},$$

$$f_{xy}(x, y) = \frac{\partial}{\partial y}\left(\frac{\partial z}{\partial x}\right) = \frac{\partial^2 z}{\partial y\,\partial x}, \qquad f_{yx}(x, y) = \frac{\partial}{\partial x}\left(\frac{\partial z}{\partial y}\right) = \frac{\partial^2 z}{\partial x\,\partial y}.$$

Example 5 Compute the second partial derivatives of $z = xy^2 + ye^{-x} + \sin(x - y)$.

Solution We compute the first partials:

$$\frac{\partial z}{\partial x} = y^2 - ye^{-x} + \cos(x - y)$$

and

$$\frac{\partial z}{\partial y} = 2xy + e^{-x} - \cos(x - y).$$

Now we differentiate again:

$$\frac{\partial^2 z}{\partial x^2} = ye^{-x} - \sin(x - y), \qquad \frac{\partial^2 z}{\partial y^2} = 2x - \sin(x - y),$$

$$\frac{\partial^2 z}{\partial y\,\partial x} = \frac{\partial}{\partial y}\left(\frac{\partial z}{\partial x}\right) = 2y - e^{-x} + \sin(x - y),$$

$$\frac{\partial^2 z}{\partial x\,\partial y} = \frac{\partial}{\partial x}\left(\frac{\partial z}{\partial y}\right) = 2y - e^{-x} + \sin(x - y). \ ▲$$

Example 6 (a) If $u = y\cos(xz) + x\sin(yz)$, calculate $\partial^2 u/\partial x\,\partial z$ and $\partial^2 u/\partial z\,\partial x$. (b) Let $f(x, y, z) = e^{xy} + z\cos x$. Find f_{zx} and f_{xz}.

Solution (a) We find $\partial u/\partial x = -yz\sin(xz) + \sin(yz)$ and $\partial u/\partial z = -xy\sin(xz) + xy\cos(yz)$. Thus

$$\frac{\partial^2 u}{\partial z\,\partial x} = -y\sin(xz) - xyz\cos(xz) + y\cos(yz).$$

Differentiation of $\partial u/\partial z$ with respect to x yields

$$\frac{\partial^2 u}{\partial x\,\partial z} = -y\sin(xz) - xyz\cos(xz) + y\cos(yz).$$

(b) $f_x(x, y, z) = ye^{xy} - z\sin x$; $f_z(x, y, z) = \cos x$; $f_{zx}(x, y, z) = (\partial/\partial x)(\cos x) = -\sin x$; $f_{xz}(x, y, z) = (\partial/\partial z)(ye^{xy} - z\sin x) = -\sin x$. ▲

In the preceding examples, note that the mixed partials taken in different orders, like $\partial^2 z/\partial x\,\partial y$ and $\partial^2 z/\partial y\,\partial x$, or f_{xz} and f_{zx}, are equal. This is no accident.

Theorem: Equality of Mixed Partial Derivatives

If $u = f(x, y)$ has continuous second partial derivatives, then the mixed partial derivatives are equal; that is,

$$\frac{\partial^2 u}{\partial x\,\partial y} = \frac{\partial^2 u}{\partial y\,\partial x} \quad \text{or} \quad f_{yx} = f_{xy}.$$

Similar equalities hold for mixed partial derivatives of functions of three variables.

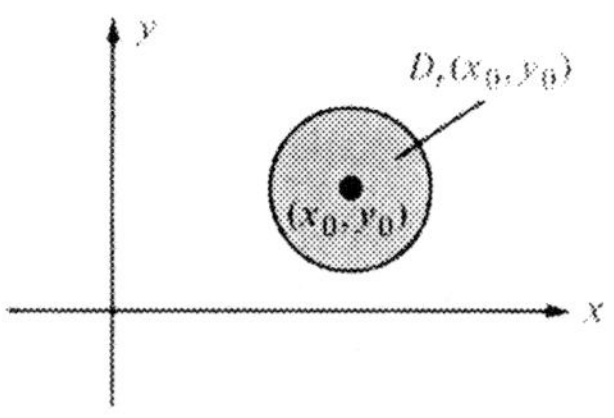

Figure 15.1.2. The disk $D_r(x_0, y_0)$ consists of the shaded region (excluding the solid circle).

L. Euler discovered this result around 1734 in connection with problems in hydrodynamics. To prove it requires the notions of continuity and limit for functions of two variables.[2]

Let us write $d((x, y),(x_0, y_0)) = \sqrt{(x - x_0)^2 + (y - y_0)^2}$ for the distance between (x, y) and (x_0, y_0), with a similar notation $d((x, y, z),(x_0, y_0, z_0))$ in space. The *disk* $D_r(x_0, y_0)$ of radius r centered at (x_0, y_0) is, by definition, the set of all (x, y) such that $d((x, y),(x_0, y_0)) < r$, as shown in Fig. 15.1.2. The limit concept now can be defined by the same ε, δ technique as in one variable calculus.

The ε, δ Definition of Limit

Suppose that f is defined on a region which includes a disk about (x_0, y_0), but need not include (x_0, y_0) itself. We write

$$\lim_{(x,y)\to(x_0,y_0)} f(x, y) = l$$

if, for every $\varepsilon > 0$, there is a $\delta > 0$ such that $|f(x, y) - l| < \varepsilon$ whenever $0 < d((x, y),(x_0, y_0)) < \delta$. A similar definition is made for functions of three variables.

The ε, δ definition of limit may be rephrased as follows: for every $\varepsilon > 0$, there is a $\delta > 0$ such that $|f(x, y) - l| < \varepsilon$ if (x, y) lies in $D_\delta(x_0, y_0)$.

The similarity between this definition and the one in Chapter 11 should be evident. The rules for limits, including rules for sums, products, and quotients, are analogous to those for functions of one variable.

Example 7 Prove the "obvious" limit, $\lim_{(x,y)\to(x_0,\, y_0)} x = x_0$, using ε's and δ's.

Solution Let $\varepsilon > 0$ be given and let $f(x, y) = x$ and $l = x_0$. We seek a number $\delta > 0$ such that $|f(x, y) - l| < \varepsilon$ whenever $d((x, y),(x_0, y_0)) < \delta$, that is, such that $|x - x_0| < \varepsilon$ whenever $\sqrt{(x - x_0)^2 + (y - y_0)^2} < \delta$. However, note that

$$|x - x_0| = \sqrt{(x - x_0)^2} \leqslant \sqrt{(x - x_0)^2 + (y - y_0)^2},$$

so if we choose $\delta = \varepsilon$, $d((x, y),(x_0, y_0)) < \delta$ will imply $|x - x_0| < \varepsilon$. ▲

[2] If you are not interested in the theory of calculus, you may skip to p. 772. Consult your instructor.

Example 8 Find $\lim_{(x,y)\to(0,0)} \dfrac{x^3+2x^2+xy^2+2y^2}{x^2+y^2}$.

Solution The numerator and denominator vanish when $(x, y) = (0,0)$. The numerator factors as $(x^2+y^2)(x+2)$, so we may use the replacement rule and algebraic rules to get

$$\lim_{(x,y)\to(0,0)} \frac{(x^2+y^2)(x+2)}{x^2+y^2} = \lim_{(x,y)\to(0,0)} (x+2) = 0+2 = 2. \blacktriangle$$

Example 9 Show that

$$\lim_{(x,y)\to(0,0)} \frac{\partial}{\partial x}\sqrt{x^2+y^2}$$

does not exist. [*Hint*: Look at the limits along the x and y axes.]

Solution $(\partial/\partial x)\sqrt{x^2+y^2} = x/\sqrt{x^2+y^2}$ if $(x, y) \neq (0,0)$. Thus

$$\lim_{(x,y)\to(0,0)} \frac{\partial}{\partial x}\sqrt{x^2+y^2} = \lim_{(x,y)\to(0,0)} \frac{x}{\sqrt{x^2+y^2}}.$$

If we approach $(0,0)$ on the y axis—that is, along points $(0, y)$—we get zero. Thus the limit, if it exists, is zero. On the other hand, if we approach $(0,0)$ along the positive x axis, we have $y = 0$ and $x > 0$; then $x/\sqrt{x^2+y^2} = 1$ because $x/\sqrt{x^2} = 1$, so the limit is 1. Since we obtain different answers in different directions, the actual limit cannot exist. ▲

We can base the concept of continuity on that of limits, just as we did in Section 1.2.

Definition of Continuity

Let f be defined in a disk about (x_0, y_0). Then we say f is *continuous* at (x_0, y_0) if

$$\lim_{(x,y)\to(x_0,y_0)} f(x, y) = f(x_0, y_0).$$

There is a similar definition for functions of three variables.

Most "reasonable" functions of several variables are continuous, although this may not be simple to prove from the definition. Here is an example of how to do this.

Example 10 (a) If $f(x)$ and $g(y)$ are continuous functions of x and y, respectively, show that $h(x, y) = f(x)g(y)$ is continuous. (b) Use (a) to show that $e^x \cos y$ is continuous.

Solution (a) We must show that for any (x_0, y_0), $\lim_{(x,y)\to(x_0,y_0)} f(x)g(y) = f(x_0)g(y_0)$. To this end, we manipulate the difference:

$$\begin{aligned}&|f(x)\,g(y) - f(x_0)\,g(y_0)|\\ &\quad = |f(x)\,g(y) - f(x_0)\,g(y) + f(x_0)\,g(y) - f(x_0)\,g(y_0)|\end{aligned}$$

$$\begin{aligned}
&\leqslant |(f(x)-f(x_0))g(y)| + |f(x_0)(g(y)-g(y_0))| \\
&= |f(x)-f(x_0)|\,|g(y)| + |f(x_0)|\,|g(y)-g(y_0)| \\
&\leqslant |f(x)-f(x_0)|(|g(y_0)| + |g(y)-g(y_0)|) + |f(x_0)|\,|g(y)-g(y_0)| \\
&= |f(x)-f(x_0)|\,|g(y_0)| + |f(x_0)|\,|g(y)-g(y_0)| \\
&\quad + |f(x)-f(x_0)|\,|g(y)-g(y_0)|.
\end{aligned}$$

Now let $\varepsilon > 0$ be given. We may choose $\varepsilon_1 > 0$ so small that we have the inequality $\varepsilon_1(|g(y_0)| + |f(y_0)|) + \varepsilon_1^2 < \varepsilon$, by letting ε_1 be the smaller of

$$\frac{\varepsilon}{|g(y_0)| + |f(y_0)| + 1} \quad \text{and} \quad 1.$$

Since f and g are continuous, there exists $\delta_1 > 0$ such that, when $|x - x_0| < \delta_1$, $|f(x) - f(x_0)| < \varepsilon_1$, and there exists δ_2 such that when $|y - y_0| < \delta_2$, we have the inequality $|g(x) - g(x_0)| < \varepsilon_1$. Let δ be the smaller of δ_1 and δ_2.

Now if $d((x, y), (x_0, y_0)) < \delta$, we have $|x - x_0| < \delta \leqslant \delta_1$ and $|y - y_0| < \delta \leqslant \ell_2$. so

$$\begin{aligned}
|f(x)g(y) - f(x_0)g(y_0)| &\leqslant |f(x)-f(x_0)|\,|g(y_0)| + |f(x_0)|\,|g(y)-g(y_0)| \\
&\quad + |f(x)-f(x_0)|\,|g(y)-g(y_0)| \\
&\leqslant \varepsilon_1|g(y_0)| + |f(x_0)|\varepsilon_1 + \varepsilon_1\cdot\varepsilon_1 < \varepsilon.
\end{aligned}$$

Thus we have proven that $\lim_{(x,y)\to(x_0,y_0)} = f(x_0)g(y_0)$, so $f(x)g(y)$ is continuous.

(b) We know from one-variable calculus that the functions $f(x) = e^x$ and $g(y) = \cos y$ are differentiable and hence continuous, so by part (a), $f(x)g(y)$ is continuous. ▲

Using the ideas of limit and continuity, we can now give the proof of the equality of mixed partials; it uses the mean value theorem for functions of one variable.

Proof of the equality of mixed partial derivatives

Consider the expression

$$f(x_0 + \Delta x, y_0 + \Delta y) - f(x_0 + \Delta x, y_0) - f(x_0, y_0 + \Delta y) + f(x_0, y_0). \tag{1}$$

We fix y_0 and Δy and introduce the function

$$g(x) = f(x, y_0 + \Delta y) - f(x, y_0),$$

so that the expression (1) equals $g(x_0 + \Delta x) - g(x_0)$. By the mean value theorem for functions of one variable, this equals $g'(\bar{x})\Delta x$ for some $\bar{x}$ between x_0 and $x_0 + \Delta x$. Hence (1) equals

$$\left[\frac{\partial z}{\partial x}(\bar{x}, y_0 + \Delta y) - \frac{\partial z}{\partial x}(\bar{x}, y_0)\right]\Delta x.$$

Applying the mean value theorem again, we get, for (1),

$$\frac{\partial^2 z}{\partial y\,\partial x}(\bar{x}, \bar{y})\Delta x\,\Delta y.$$

Since $\partial^2 z/\partial y\,\partial x$ is continuous and $(\bar{x}, \bar{y}) \to (x_0, y_0)$ as $(\Delta x, \Delta y) \to (0,0)$, it follows that

$$\frac{\partial^2 z}{\partial y\,\partial x}(x_0, y_0)$$

$$= \lim_{(\Delta x,\Delta y)\to(0,0)} \frac{[f(x_0 + \Delta x, y_0 + \Delta y) - f(x_0 + \Delta x, y_0) - f(x_0, y_0 + \Delta y) + f(x_0, y_0)]}{\Delta x\,\Delta y}. \tag{2}$$

The right-hand side of formula (2) is symmetric in x and y, so that in this derivation we can reverse the roles and x and y. In other words, in the same manner we prove that $\partial^2 z/\partial x\,\partial y$ is given by the same limit, and so we obtain the desired result: the mixed partials are equal. ■

Supplement to Section 15.1: Partial Derivatives and Wave Motion

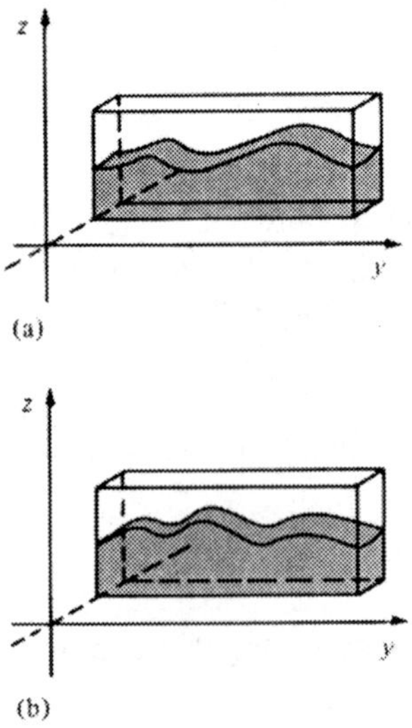

Figure 15.1.3. Moving water in a narrow tank shown at two different instants of time.

Two of the most important problems in the historical development of partial differentiation concerned *wave motion* and *heat conduction*. Here we concentrate on the first of these problems. (See also Exercises 71 and 72.)

Consider water in motion in a narrow tank, as illustrated in Fig. 15.1.3. We will assume that the motion of the water is gentle enough so that, at any instant of time, the height z of the water above the bottom of the tank is a function of the position y measured along the long direction of the tank; this means that there are no "breaking waves" and that the height of water is constant along the short direction of the tank. Since the water is in motion, the height z depends on the time as well as on y, so we may write $z = f(t, y)$; the domain of the function f consists of all pairs (t, y) such that t lies in the interval of time relevant for the experiment, and $a \leqslant y \leqslant b$, where a and b mark the ends of the tank.

We can graph the entire function f as a surface in (t, y, z) space lying over the strip $a \leqslant y \leqslant b$ (see Fig. 15.1.4); the section of this surface by a plane

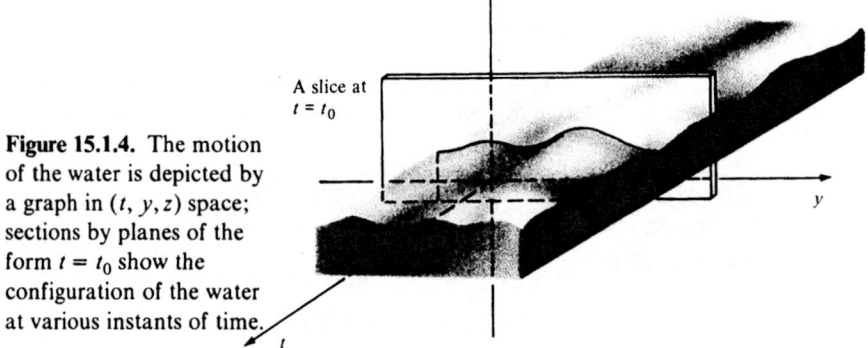

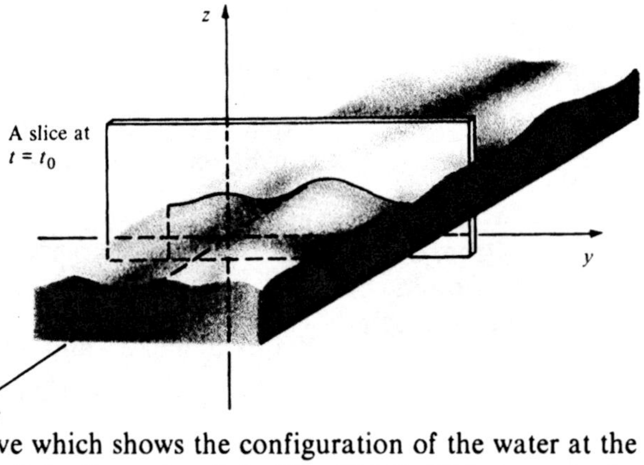

Figure 15.1.4. The motion of the water is depicted by a graph in (t, y, z) space; sections by planes of the form $t = t_0$ show the configuration of the water at various instants of time.

of the form $t = t_0$ is a curve which shows the configuration of the water at the moment t_0 (such as each of the "snapshots" in Fig. 15.1.3). This curve is the graph of a function of *one* variable, $z = g(y)$, where g is defined by $g(y) = f(t_0, y)$. If we take the derivative of the function g at a point y_0 in (a, b), we get a number $g'(y_0)$ which represents the slope of the water's surface at the time t_0 and at the location y_0. (See Fig. 15.1.5.) It could be observed as the slope of a small stick parallel to the sides of the tank floating on the water at that time and position.

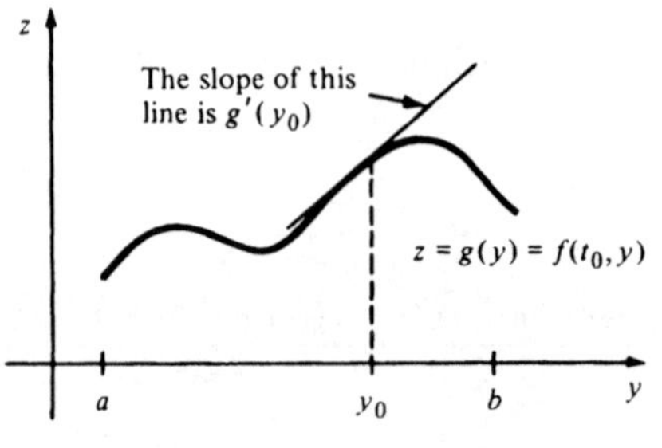

Figure 15.1.5. The derivative $g'(y_0)$ of the function $g(y) = f(t_0, y)$ represents the slope of the water's surface at time t_0 and position y_0.

This number, $g'(y_0)$, is obtained from the function f by:
1. Fixing t at the value t_0.
2. Differentiating the resulting function of y.
3. Setting y equal to y_0.

The number $g'(y_0)$ is just $f_y(t_0, y_0)$.

We can also define the partial derivative of f with respect to t at (t_0, y_0); it is obtained by:
1. Fixing y at the value y_0.
2. Differentiating the resulting function of t.
3. Setting t equal to t_0.

The result is $f_t(t_0, y_0)$. In the first step, we obtain the function $h(t) = f(t, y_0)$, which represents the vertical motion of the water's surface observed at the fixed position y_0. The derivative with respect to t is, therefore, the *vertical velocity* of the surface at the position y_0. It could be observed as the vertical velocity of a cork floating on the water at that position. Finally, setting t equal to t_0 merely involves observing the velocity at the specific time t_0.

Exercises for Section 15.1

Compute f_x and f_y for the functions in Exercises 1–8 and evaluate them at the indicated points.

1. $f(x, y) = xy$; $(1, 1)$
2. $f(x, y) = x/y$; $(1, 1)$
3. $f(x, y) = \tan^{-1}(x - 3y^2)$; $(1, 0)$
4. $f(x, y) = \sqrt{x^2 + y^2}$; $(1, -1)$
5. $f(x, y) = e^{xy}\sin(x + y)$; $(0, 0)$
6. $f(x, y) = \ln(x^2 + y^2 + 1)$; $(0, 0)$
7. $f(x, y) = 1/(x^3 + y^3)$; $(-1, 2)$
8. $f(x, y) = e^{-x^2-y^2}$; $(1, -1)$

Compute f_x, f_y, and f_z for the functions in Exercises 9–12, and evaluate them at the indicated points.

9. $f(x, y, z) = xyz$; $(1, 1, 1)$
10. $f(x, y, z) = \sqrt{x^2 + y^2 + z^2}$; $(3, 0, 4)$
11. $f(x, y, z) = \cos(xy^2) + e^{3xyz}$; $(\pi, 1, 1)$
12. $f(x, y, z) = x^{yz}$; $(1, 1, 0)$

Find the partial derivatives $\partial z/\partial x$ and $\partial z/\partial y$ for the functions in Exercises 13–16.

13. $z = 3x^2 + 2y^2$
14. $z = \sin(x^2 - 3xy)$
15. $z = (2x^2 + 7x^2y)/3xy$
16. $z = x^2y^2e^{2xy}$

Find the partial derivatives $\partial u/\partial x$, $\partial u/\partial y$, and $\partial u/\partial z$ in Exercises 17–20.

17. $u = e^{-xyz}(xy + xz + yz)$
18. $u = \sin(xy^2z^3)$
19. $u = e^x\cos(yz^2)$
20. $u = (xy^3 + e^z)/(x^3y - e^z)$

Compute the indicated partial derivatives in Exercises 21–24.

21. $\dfrac{\partial}{\partial y}\left(\dfrac{xe^y - 1}{ye^x + 1}\right)$
22. $\dfrac{\partial}{\partial u}(uvw - \sin(uvw))$
23. $\dfrac{\partial}{\partial b}(mx + b^2)^8$
24. $\dfrac{\partial}{\partial m}(mx + b^2)^8$

In Exercises 25–28, let

$$f(x, y) = 3x^2 + 2\sin(x/y^2) + y^3(1 - e^x)$$

and find the indicated quantities.

25. $f_x(2, 3)$
26. $f_x(0, 1)$
27. $f_y(1, 1)$
28. $f_y(-1, -1)$
29. Let $z = (\sin x)e^{-xy}$.
    (a) Find $\partial z/\partial y$.
    (b) Evaluate $\partial z/\partial y$ at the following four points: $(0, 0)$, $(0, \pi/2)$, $(\pi/2, 0)$, and $(\pi/2, \pi/2)$.
30. Let $u = (xy/z)\cos(yz)$.
    (a) Find $\partial u/\partial z$.
    (b) Evaluate $\partial u/\partial z$ at $(1, \pi, 1)$, $(0, \pi/2, 1)$, and $(1, \pi, 1/2)$.

Let $g(t, u, v) = \ln(t + u + v) - \tan(tuv)$ and find the indicated quantities in Exercises 31–36.

31. $g_t(0, 0, 1)$
32. $g_t(1, 0, 0)$
33. $g_u(1, 2, 3)$
34. $g_u(2, 3, 1)$
35. $g_v(-1, 3, 5)$
36. $g_v(-1, 5, 3)$

In Exercises 37–40, compute the indicated partial derivatives

37. $\dfrac{\partial}{\partial s}e^{stu^2}$
38. $\dfrac{\partial}{\partial r}\left(\dfrac{1}{3}\pi r^2h\right)$
39. $\dfrac{\partial}{\partial\lambda}\left(\dfrac{\cos\lambda\mu}{1 + \lambda^2 + \mu^2}\right)$
40. $\dfrac{\partial}{\partial a}(bcd)$
41. If $f(x, y, z)$ is a function of three variables, express f_z as a limit.
42. Find
$$\lim_{\Delta y \to 0}\frac{3 + (x + y + \Delta y)^2z - \left(3 + (x + y)^2z\right)}{\Delta y}$$
43. In the situation of Example 4, how fast is the temperature changing if we proceed directly west? (The longitude y is increasing as we go west.)

44. Chicago Skate Company produces three kinds of roller skates. The cost in dollars for producing x, y, and z units of each, respectively, is $c(x, y, z) = 3000 + 27x + 36y + 47z$.
    (a) The value of $\partial c/\partial x$ is the change in cost due to a one unit increase in production of the least expensive skate, the levels of production of the higher-priced units being held fixed. Find it.
    (b) Find $\partial c/\partial z$, and interpret.
45. If three resistors R_1, R_2, and R_3 are connected in parallel, the total electrical resistance is determined by the equation
$$\frac{1}{R} = \frac{1}{R_1} + \frac{1}{R_2} + \frac{1}{R_3}.$$
    (a) What is $\partial R/\partial R_1$?
    (b) Suppose that R_1, R_2, and R_3 are variable resistors set at 100, 200, and 300 ohms, respectively. How fast is R changing with respect to R_1?
46. Consider the topographical map of Yosemite Valley in Fig. 14.R.3. Let r represent the east–west coordinate on the map, increasing from west to east. Let s be the north–south coordinate, increasing as you go north. (East is the positive r direction, north the positive s direction.) Let h be the elevation above sea level.
    (a) Explain how $\partial h/\partial r$ and $\partial h/\partial s$ are related to the distances between contour lines and their directions.
    (b) At the center of the letter *o* in Half Dome, what is the sign of $\partial h/\partial r$? of $\partial h/\partial s$?

In Exercises 47–50, find the partial derivatives $\partial^2 z/\partial x^2$, $\partial^2 z/\partial x\,\partial y$, $\partial^2 z/\partial y\,\partial x$ and $\partial^2 z/\partial y^2$ for each of the functions in the indicated exercise.

47. Exercise 13
48. Exercise 14
49. Exercise 15
50. Exercise 16
51. Let $f(x, y, z) = x^2y + xy^2 + yz^2$. Find f_{xy}, f_{yz}, f_{zx}, and f_{xyz}.
52. Let $z = x^4y^3 - x^8 + y^4$.
    (a) Compute $\partial^3 z/\partial y\,\partial x\,\partial x$, $\partial^3 z/\partial x\,\partial y\,\partial x$, and $\partial^3 z/\partial x\,\partial x\,\partial y$.
    (b) Compute $\partial^3 z/\partial x\,\partial y\,\partial y$, $\partial^3 z/\partial y\,\partial x\,\partial y$, and $\partial^3 z/\partial y\,\partial y\,\partial x$.

Compute $\partial^2 u/\partial x^2$, $\partial^2 u/\partial y\,\partial x$, $\partial^2 u/\partial y^2$, and $\partial^2 u/\partial x\,\partial y$ for each of the functions in Exercises 53–56. Check directly the equality of mixed partials.

53. $u = 2xy/(x^2 + y^2)^2$
54. $u = \cos(xy^2)$
55. $u = e^{-xy^2} + y^3x^4$
56. $u = 1/(\cos^2 x + e^{-y})$
57. Prove, using ε's and δ's, that $\lim_{(x,y)\to(x_0,y_0)} y = y_0$.
58. Prove, using ε's and δ's, that $\lim_{(x,y)\to(x_0,y_0)}(x + y) = x_0 + y_0$.

In Exercises 59–66, evaluate the given limits if they exist (do not attempt a precise justification).

59. $\displaystyle\lim_{(x,y)\to(0,0)} \frac{x^2y + y^3}{x^2 + y^2}$
60. $\displaystyle\lim_{(x,y)\to(1,1)} \frac{x + y}{(x-1)^2 + 1}$
61. $\displaystyle\lim_{(x,y)\to(2,3)} \frac{x^2y + y^3 + x^2 + y^2}{\sqrt{x^2 + y^2}}$
62. $\displaystyle\lim_{(x,y)\to(0,0)} \frac{4x^2 + 3y^2 + x^3y^3}{x^2 + y^2 + x^4y^4}$
63. $\displaystyle\lim_{(x,y)\to(1,1)} e^x\cos(\pi y)$
64. $\displaystyle\lim_{(x,y)\to(0,1)} e^{xy}\cos(\pi xy)$
65. $\displaystyle\lim_{(x,y)\to(0,0)} \sin(xy)$
66. $\displaystyle\lim_{(x,y)\to(0,0)} \frac{1}{1 + \ln(1 + 1/(x^2 + y^2))}$
67. Let $f(x, y) = x^2 + y^2$ and suppose that (x, y) moves along the curve $(x(t), y(t)) = (\cos t, e^t)$.
    (a) Find $g(t) = f(x(t), y(t))$ and use your formula to compute $g'(t_0)$.
    (b) Show that this is the same as
$$f_x(x(t_0), y(t_0))\cdot x'(t_0) + f_y(x(t_0), y(t_0))\cdot y'(t_0).$$
68. Let $f(x, y, z) = x^2 + 2y - z$ and suppose the point (x, y, z) moves along the parametric curve $(1, t, t^2)$.
    (a) Let $g(t) = f(1, t, t^2)$ and compute $g'(t)$.
    (b) Show that your answer in (a) is equal to
$$f_x\frac{dx}{dt} + f_y\frac{dy}{dt} + f_z\frac{dz}{dt}.$$
69. A function $u = f(x, y)$ with continuous second partial derivatives satisfying *Laplace's equation*
$$\frac{\partial^2 u}{\partial x^2} + \frac{\partial^2 u}{\partial y^2} = 0$$
is called a *harmonic function*. Show that the function $u(x, y) = x^3 - 3xy^2$ is harmonic.
70. Which of the following functions satisfy Laplace's equation? (See Exercise 69).
    (a) $f(x, y) = x^2 - y^2$;
    (b) $f(x, y) = x^2 + y^2$;
    (c) $f(x, y) = xy$;
    (d) $f(x, y) = y^3 + 3x^2y$;
    (e) $f(x, y) = \sin x \cosh y$;
    (f) $f(x, y) = e^x \sin y$.
71. Let f and g be differentiable functions of one variable. Set $\varphi = f(x - t) + g(x + t)$.
    (a) Prove that φ satisfies the *wave equation*: $\partial^2\varphi/\partial t^2 = \partial^2\varphi/\partial x^2$.
    (b) Sketch the graph of φ against t and x if $f(x) = x^2$ and $g(x) = 0$.

72. (a) Show that function $g(x,t) = 2 + e^{-t}\sin x$ satisfies the *heat equation*: $g_t = g_{xx}$. (Here $g(x,t)$ represents the temperature in a rod at position x and time t.)
(b) Sketch the graph of g for $t \geqslant 0$. [*Hint*: Look at sections by the planes $t = 0$, $t = 1$, and $t = 2$.]
(c) What happens to $g(x,t)$ as $t \to \infty$? Interpret this limit in terms of the behavior of heat in a rod.

73. The productivity z per employee per week of a company depends on the size x of the labor force and the amount y of investment capital in millions of dollars. A typical formula is $z(x, y) = 60xy - x^2 - 4y^2$.
(a) The value of $\partial z/\partial x$ at $x = 5$, $y = 3$ is the marginal productivity of labor per worker at a labor force of 5 people and investment level of 3 million dollars. Find it.
(b) Find $\partial z/\partial y$ at $x = 5$, $y = 3$, and interpret.

74. The productivity z of a company is given by $z(x, y) = 100xy - 2x^2 - 6y^2$ where $x \times 10^3$ people work for the company, and the capital investment of the company is y million dollars.
(a) Find the marginal productivity of labor $\partial z/\partial x$. This number is the expected change in production for an increase of 1000 staff with fixed capital investment.
(b) Find $\partial z/\partial y$ when $x = 5$ and $y = 3$. Interpret.

★75. Show that

$$\lim_{(x,y,z)\to(0,0,0)} \frac{\partial}{\partial z}(x^2 + y^2 + z^2)^{1/3}$$

does not exist.

★76. Let

$$f(x, y) = \begin{cases} \dfrac{xy(x^2 - y^2)}{x^2 + y^2}, & (x, y) \neq (0,0), \\ 0, & (x, y) = (0,0). \end{cases}$$

(a) If $(x, y) \neq (0,0)$, compute f_x and f_y.
(b) What is the value of $f(x, 0)$ and $f(0, y)$?
(c) Show that $f_x(0,0) = 0 = f_y(0,0)$.

★77 Consider the function f in Exercise 76.
(a) Show that $f_x(0, y) = -y$ when $y \neq 0$.
(b) What is $f_y(x, 0)$ when $x \neq 0$?
(c) Show that $f_{yx}(0,0) = 1$ and $f_{xy}(0,0) = -1$. [*Hint*: Express them as limits.]
(d) What went wrong? Why are the mixed partials not equal?

★78. Suppose that f is continuous at (x_0, y_0) and $f(x_0, y_0) > 0$. Show that there is a disk about (x_0, y_0) on which $f(x, y) > 0$.

15.2 Linear Approximations and Tangent Planes

The plane tangent to the graph of a function of two variables has two slopes.

In the calculus of functions of one variable, the simplest functions are the linear functions $l(x) = mx + b$. The derivative of such a function is the constant m, which is the slope of the graph or the rate of change of y with respect to x. If $f(x)$ is any differentiable function, its tangent line at x_0 is the graph of the *linear approximation* $y = f(x_0) + f'(x_0)(x - x_0)$.

To extend these ideas to functions of two variables, we begin by looking at linear functions of the form $z = l(x, y) = ax + by + c$, whose graphs are planes. Such a plane has two "slopes," the numbers a and b, which determine the direction of its normal vector $-a\mathbf{i} - b\mathbf{j} + \mathbf{k}$ (see Section 13.4). These slopes can be recovered from the function l as the partial derivatives $l_x = a$ and $l_y = b$. By analogy with the situation for one variable, we define the *linear approximation* at (x_0, y_0) for a general function f of two variables to be the linear function

$$l(x, y) = f(x_0, y_0) + f_x(x_0, y_0)(x - x_0) + f_y(x_0, y_0)(y - y_0),$$

which is of the form $ax + by + c$ with

$$a = f_x(x_0, y_0), \quad b = f_y(x_0, y_0), \quad \text{and}$$
$$c = f(x_0, y_0) - x_0 f_x(x_0, y_0) - y_0 f_y(x_0, y_0).$$

The function l is the unique linear function which has at (x_0, y_0) the same value *and* the same partial derivatives as f. The graph

$$z = f(x_0, y_0) + f_x(x_0, y_0)(x - x_0) + f_y(x_0, y_0)(y - y_0) \qquad (1)$$

of the linear approximation is a plane through $(x_0, y_0, f(x_0, y_0))$, with normal vector $-f_x(x_0, y_0)\mathbf{i} - f_y(x_0, y_0)\mathbf{j} + \mathbf{k}$; it is called the *tangent plane* at (x_0, y_0) to the graph of f (see Fig. 15.2.1).

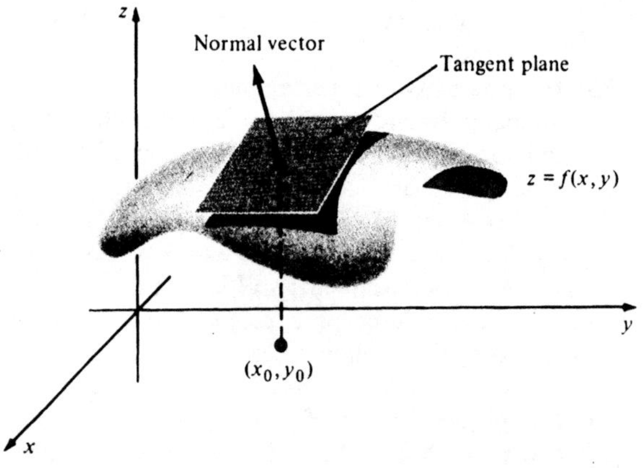

Figure 15.2.1. The tangent plane at (x_0, y_0) to the graph $z = f(x, y)$ has the equation $z = f(x_0, y_0) + f_x(x_0, y_0)(x - x_0) + f_y(x_0, y_0)(y - y_0)$. A vector normal to the plane is

$$\mathbf{n} = -f_x(x_0, y_0)\mathbf{i} - f_y(x_0, y_0)\mathbf{j} + \mathbf{k}.$$

Example 1 Find the equation of the plane tangent to the hemisphere $z = \sqrt{1 - x^2 - y^2}$ at a point (x_0, y_0). Interpret your result geometrically.

Solution Letting $f(x, y) = \sqrt{1 - x^2 - y^2}$, we have $f_x(x, y) = -x/\sqrt{1 - x^2 - y^2}$ and $f_y(x, y) = -y/\sqrt{1 - x^2 - y^2}$. The equation of the tangent plane at (x_0, y_0, z_0) is obtained from (1) to be

$$z = \sqrt{1 - x_0^2 - y_0^2} - \frac{x_0}{\sqrt{1 - x_0^2 - y_0^2}}(x - x_0) - \frac{y_0}{\sqrt{1 - x_0^2 - y_0^2}}(y - y_0),$$

$$\text{or} \quad z = z_0 - \frac{x_0}{z_0}(x - x_0) - \frac{y_0}{z_0}(y - y_0).$$

A normal vector is thus $\frac{x_0}{z_0}\mathbf{i} + \frac{y_0}{z_0}\mathbf{j} + \mathbf{k}$. Multiplying by z_0, we find that another normal vector is $x_0\mathbf{i} + y_0\mathbf{j} + z_0\mathbf{k}$. Thus we have recovered the geometric result that the tangent plane at a point P of a sphere is perpendicular to the vector from the center of the sphere to P. ▲

The linear approximation may be defined as well for a function of three variables. We include its definition in the following box.

Linear Approximation and Tangent Plane

The linear approximation at (x_0, y_0) of $f(x, y)$ is the linear function:

$$l(x, y) = f(x_0, y_0) + f_x(x_0, y_0)(x - x_0) + f_y(x_0, y_0)(y - y_0). \qquad (2)$$

The graph $z = l(x, y)$ is called the *tangent plane* to the graph of f at (x_0, y_0). It has normal vector $-f_x(x_0, y_0)\mathbf{i} - f_y(x_0, y_0)\mathbf{j} + \mathbf{k}$.

The linear approximation at (x_0, y_0, z_0) to $f(x, y, z)$ is the linear function:

$$l(x, y, z) = f(x_0, y_0, z_0) + f_x(x_0, y_0, z_0)(x - x_0) + f_y(x_0, y_0, z_0)(y - y_0) + f_z(x_0, y_0, z_0)(z - z_0). \qquad (3)$$

Example 2 Find the equation of the plane tangent to the graph of

$$f(x, y) = (x^2 + y^2)/xy$$

at $(x_0, y_0) = (1, 2)$.

Solution Here $x_0 = 1$, $y_0 = 2$, and $f(1, 2) = \frac{5}{2}$. The partial derivative with respect to x is

$$f_x(x, y) = \frac{2x \cdot xy - (x^2 + y^2)y}{(xy)^2} = \frac{x^2y - y^3}{(xy)^2} = \frac{x^2 - y^2}{x^2y},$$

which is $-\frac{3}{2}$ at $(1, 2)$. Similarly,

$$f_y(x, y) = \frac{y^2x - x^3}{(xy)^2} = \frac{y^2 - x^2}{xy^2},$$

which is $\frac{3}{4}$ at $(1, 2)$. Thus the tangent plane is given by the equation (1)

$$z = -\tfrac{3}{2}(x - 1) + \tfrac{3}{4}(y - 2) + \tfrac{5}{2},$$

i.e., $4z = -6x + 3y + 10$. ▲

Example 3 Find a formula for a unit normal vector to the graph of the function $f(x, y) = e^x y$ at the point $(-1, 1)$.

Solution Since a normal vector is $-f_x(x_0, y_0)\mathbf{i} - f_y(x_0, y_0)\mathbf{j} + \mathbf{k}$, a unit normal is obtained by normalizing:

$$\mathbf{n} = \frac{-f_x(x_0, y_0)\mathbf{i} - f_y(x_0, y_0)\mathbf{j} + \mathbf{k}}{\sqrt{[f_x(x_0, y_0)]^2 + [f_y(x_0, y_0)]^2 + 1}}.$$

In this case, $f_x(x, y) = e^x y$ and $f_y(x, y) = e^x$. Evaluating the partial derivatives at $(-1, 1)$, we find a normal to be $-e^{-1}\mathbf{i} - e^{-1}\mathbf{j} + \mathbf{k}$, and so a unit normal is

$$\frac{-e^{-1}\mathbf{i} - e^{-1}\mathbf{j} + \mathbf{k}}{\sqrt{e^{-2} + e^{-2} + 1}} = \frac{-e^{-1}}{\sqrt{2e^{-2} + 1}}\mathbf{i} - \frac{e^{-1}}{\sqrt{2e^{-2} + 1}}\mathbf{j} + \frac{1}{\sqrt{2e^{-2} + 1}}\mathbf{k}. \ ▲$$

Just as in one-variable calculus, we can use the linear approximation for approximate numerical computations. Suppose that the number $z = f(x, y)$ depends on both x and y and we want to know how much z changes as x and y are changed a little. The partial derivative $f_x(x_0, y_0)$ gives the rate of change of z with respect to x at (x_0, y_0). Thus the change in z which results from a change Δx in x should be about

$$f_x(x_0, y_0)\,\Delta x.$$

Similarly, the change in z caused by a shift in y by Δy should be about

$$f_y(x_0, y_0)\,\Delta y.$$

Thus the total change in z should be approximately

$$\Delta z \approx f_x(x_0, y_0)\,\Delta x + f_y(x_0, y_0)\,\Delta y. \tag{4}$$

Notice that the change in z is obtained by simply *adding* the changes due to Δx and Δy. If we write $\Delta x = x - x_0$ and $\Delta y = y - y_0$, then the expression for Δz is the linear approximation to $f(x, y) - f(x_0, y_0)$ at (x_0, y_0).

Example 4 Calculate an approximate value for $(0.99e^{0.02})^8$. Compare with the value from a calculator.

Solution Let $z = f(x, y) = (xe^y)^8$ and let $x_0 = 1$ and $y_0 = 0$, so $f(1,0) = 1$. We get

$$\frac{\partial z}{\partial x} = 8x^7e^{8y}, \qquad \text{which is 8 at } (1,0),$$

and

$$\frac{\partial z}{\partial y} = 8x^8e^{8y}, \qquad \text{which is 8 at } (1,0).$$

Thus if we let $x = 0.99$ and $y = 0.02$ so that $x - x_0 = -0.01$ and $y - y_0 = 0.02$, the linear approximation is (by (2) or (4))

$$1 + 8(-0.01) + 8(0.02) = 1.08.$$

The value for $(0.99e^{0.02})^8$ obtained on our calculator is 1.082850933. ▲

Example 5 Find an approximate value for $\sin(0.01)\cdot\cos(0.99\pi)$.

Solution Let $f(x, y) = \sin x \cos y$ and $x_0 = 0$, $y_0 = \pi$. Then if $x = 0.01$ and $y = 0.99\pi$,

$$\begin{aligned} f(x, y) &\approx f_x(0,\pi)(x - 0) + f_y(0,\pi)(y - \pi) + f(0,\pi) \\ &= -1(0.01) + 0 + 0 \\ &= -0.01. \end{aligned}$$

(The value of $f(x, y)$ computed on our calculator is -0.009994899.) ▲

Example 6 A multiplication problem is altered by taking a small amount from one factor and adding it to the other. How can you tell whether the product increases or decreases?

Solution Let $f(x, y) = xy$. If the amount moved from y to x is h, we must look at

$$f(x + h, y - h) - f(x, y),$$

which may be approximated by

$$hf_x(x, y) + (-h)f_y(x, y).$$

The partial derivatives are $f_x(x, y) = y$ and $f_y(x, y) = x$, so the linear approximation to the change in the product is $h(y - x)$. Thus, the product increases when the increment h is taken from the larger factor. ▲

Exercises for Section 15.2

Find equations for the planes tangent to the surfaces in Exercises 1–4 at the indicated points.

1. $z = x^3 + y^3 - 6xy$; $(1, 2, -3)$
2. $z = (\cos x)(\cos y)$; $(0, \pi/2, 0)$
3. $z = (\cos x)(\sin y)$; $(0, \pi/2, 1)$
4. $z = 1/xy$; $(1, 1, 1)$

Find the equation of the plane tangent to the graph of $z = f(x, y) = x^2 + 2y^3 + 1$ at the points in Exercises 5–8.

5. $(1, 1, 4)$
6. $(-1, -1, 0)$
7. $(0, 0, 1)$
8. $(1, -1, 0)$

Find the equation of the tangent plane of the graph of f at the point $(x_0, y_0, f(x_0, y_0))$ for the functions and points in Exercises 9–12.

9. $f(x, y) = x - y + 2$; $(x_0, y_0) = (1, 1)$.
10. $f(x, y) = x^2 + 4y^2$; $(x_0, y_0) = (2, -1)$.
11. $f(x, y) = xy$; $(x_0, y_0) = (1, 1)$.
12. $f(x, y) = x/(x + y)$; $(x_0, y_0) = (1, 0)$.

For each of the indicated functions and points in Exercises 13–16, find a unit normal vector to the graph at $(x_0, y_0, f(x_0, y_0))$.

13. f and (x_0, y_0) as in Exercise 9.
14. f and (x_0, y_0) as in Exercise 10.
15. f and (x_0, y_0) as in Exercise 11.
16. f and (x_0, y_0) as in Exercise 12.

Find an appropriate value for each of the quantities in Exercises 17–22 using the linear approximation.

17. $(1.01)^2[1-\sqrt{1.98}]$ [*Hint*: $1.96=(1.4)^2$.]

18. $\tan\left(\dfrac{\pi+0.01}{3.97}\right)$

19. $(0.99)^3+(2.01)^3-6(0.99)(2.01)$

20. $(0.98)\sin\left(\dfrac{0.99}{1.03}\right)$

21. $(0.98)(0.99)(1.03)$

22. $\sqrt{(4.01)^2+(3.98)^2+(2.02)^2}$

23. In the setup of Example 4, Section 15.1, at Dawson Creek, is the temperature increasing or decreasing as you proceed south? As you proceed east? Southeast?

24. Refer to Exercise 45, Section 15.1. If, in part (b), R_1 is increased by 1 ohm, R_2 is decreased by 2 ohms, and R_3 is increased by 4 ohms, use the linear approximation to calculate the change in R. Compare with a direct calculation on a calculator.

25. Let $f(a,v)$ be the length of a side of a cube whose surface area is a and whose volume is v. Find the linear approximation to $f(6+\Delta a, 1+\Delta v)$.

26. Let $g(u,v)$ be the gas mileage if you drive u miles and use v gallons of gasoline. How does $g(u,v)$ change if you go Δu extra miles on Δv extra gallons? (Use the linear approximation.)

27. Suppose that $z=f(x,y)=x^2+y^2$.
    (a) Find $\partial z/\partial y|_{(1,1)}$
    (b) Describe the curve obtained by intersecting the graph of f with the plane $x=1$.
    (c) Find a tangent vector to this curve at the point $(1,1,f(1,1))$.

28. Repeat Exercise 27 for $z=f(x,y)=e^{xy}$.

29. Let $f(x,y)=-(1-x^2-y^2)^{1/2}$ for (x,y) such that $x^2+y^2<1$. Show that the plane tangent to the graph of f at $(x_0,y_0,f(x_0,y_0))$ is orthogonal to the vector with components $(x_0,y_0,f(x_0,y_0))$. Interpret this geometrically.

★30. (a) Let k be a differentiable function of one variable, and let $f(x,y)=k(xy)$. Suppose that x and y are functions of t: $x=g(t)$, $y=h(t)$, and set $F(t)=f(g(t),h(t))$. Prove that

$$F'(t)=\frac{\partial f}{\partial x}\frac{dx}{dt}+\frac{\partial f}{\partial y}\frac{dy}{dt}.$$

(b) If $f(x,y)=k(x)l(y)$, show that the formula in (a) is still valid. (These are special cases of the chain rule, proved in the next section.)

15.3 The Chain Rule

The derivative of a composite function with several intermediate variables is a sum of products.

In Chapter 2 we developed the chain rule for functions of one variable: If y is a function of x and z is a function of y, then z also may be regarded as a function of x, and

$$\frac{dz}{dx}=\frac{dz}{dy}\cdot\frac{dy}{dx}.$$

For functions of several variables, the chain rule is more complicated. First we consider the case where z is a function of x and y, and x and y are functions of t; we can then regard z as a function of t. In this case the chain rule states that

$$\frac{dz}{dt}=\frac{\partial z}{\partial x}\frac{dx}{dt}+\frac{\partial z}{\partial y}\frac{dy}{dt}.$$

The chain rule applies when quantities in which we are interested depend in a known way upon other quantities which in turn depend upon a third set of quantities. Suppose, for example, that the temperature T on the surface of a pond is a function $f(x,y)$ of the position coordinates (x,y). If a duck swims on the pond according to the parametric equations $x=g(t)$, $y=h(t)$, it will feel the water temperature varying with time according to the function $T=F(t)=f(g(t),h(t))$. The rate at which this temperature changes with respect to time is the derivative dT/dt. By analogy with the chain rule in one variable, we may expect this derivative to depend upon the direction and magnitude of the duck's velocity, as given by the derivatives dx/dt and dy/dt, as well as upon the partial derivatives $\partial T/\partial x$ and $\partial T/\partial y$ of temperature with respect to position. The correct formula relating all these derivatives is given in the following box. The formula will be proved after we see how it works in an example.

The Chain Rule

To find dz/dt, when $z = f(x, y)$ has continuous partial derivatives and $x = g(t)$ and $y = h(t)$ are differentiable, multiplying the partial derivatives of z with respect to each of the intermediate variables x and y by the derivative of that intermediate variable with respect to t, and add the products: If $F(t) = f(g(t), h(t))$, then

$$F'(t) = f_x(g(t), h(t))\,g'(t) + f_y(g(t), h(t))h'(t).$$

In Leibniz notation,

$$\frac{dz}{dt} = \frac{\partial z}{\partial x}\frac{dx}{dt} + \frac{\partial z}{\partial y}\frac{dy}{dt}.$$

For three intermediate variables, if u depends on x, y, and z, and x, y, and z depend on t, then

$$\frac{du}{dt} = \frac{\partial u}{\partial x}\frac{dx}{dt} + \frac{\partial u}{\partial y}\frac{dy}{dt} + \frac{\partial u}{\partial z}\frac{dz}{dt}.$$

Example 1 Suppose that a duck is swimming in a circle, $x = \cos t$, $y = \sin t$, while the water temperature is given by the formula $T = x^2e^y - xy^3$. Find dT/dt: (a) by the chain rule; (b) by expressing T in terms of t and differentiating.

Solution (a) $\partial T/\partial x = 2xe^y - y^3$; $\partial T/\partial y = x^2e^y - 3xy^2$; $dx/dt = -\sin t$; $dy/dt = \cos t$. By the chain rule, $dT/dt = (\partial T/\partial x)(dx/dt) + (\partial T/\partial y)(dy/dt)$, so

$$\frac{dT}{dt} = (2xe^y - y^3)(-\sin t) + (x^2e^y - 3xy^2)\cos t$$

$$= (2\cos t\, e^{\sin t} - \sin^3 t)(-\sin t) + (\cos^2 t\, e^{\sin t} - 3\cos t\sin^2 t)\cos t$$

$$= -2\cos t\sin t\, e^{\sin t} + \sin^4 t + \cos^3 t\, e^{\sin t} - 3\cos^2 t\sin^2 t.$$

(b) Substituting for x and y in the formula for T gives

$$T = \cos^2 t e^{\sin t} - \cos t\sin^3 t,$$

and differentiating this gives

$$\frac{dT}{dt} = 2\cos t(-\sin t)e^{\sin t} + \cos^2 t\, e^{\sin t}\cos t + \sin t\sin^3 t - (\cos t)3\sin^2 t\cos t$$

$$= -2\cos t\sin t\, e^{\sin t} + \cos^3 t\, e^{\sin t} + \sin^4 t - 3\cos^2 t\sin^2 t,$$

which is the same as the answer in part (a). ▲

An intuitive argument for the chain rule is based on the linear approximation of Section 15.2. If the position of the duck changes from the point (x, y) to the point $(x + \Delta x, y + \Delta y)$, the temperature change ΔT is given approximately by $(\partial T/\partial x)\Delta x + (\partial T/\partial y)\Delta y$. On the other hand, the linear approximation for functions of one variable gives $\Delta x \approx (dx/dt)\Delta t$ and $\Delta y \approx (dy/dt)\Delta t$. Putting these two approximations together gives

$$\Delta T \approx \frac{\partial T}{\partial x}\frac{dx}{dt}\Delta t + \frac{\partial T}{\partial y}\frac{dy}{dt}\Delta t.$$

Hence

$$\frac{\Delta T}{\Delta t} \approx \frac{\partial T}{\partial x}\frac{dx}{dt} + \frac{\partial T}{\partial y}\frac{dy}{dt}. \tag{1}$$

As $\Delta t \to 0$, the approximations become more and more accurate and the ratio $\Delta T/\Delta t$ approaches dT/dt, so the approximation formula (1) becomes the chain rule. The argument for three variables is similar.

In trying to make a proof out of this intuitive argument, one discovers that more than differentiability is required; the partial derivatives of T should be continuous. The technical details are outlined in Exercise 20.

Example 2 Verify the chain rule for $u = xe^{yz}$ and $(x, y, z) = (e^t, t, \sin t)$.

Solution Substituting the formulas for x, y, and z in the formula for u gives

$$u = e^t \cdot e^{t\sin t} = e^{t(1+\sin t)},$$

$$\text{so}\quad \frac{du}{dt} = \left[t\cos t + (1+\sin t)\right]e^{t(1+\sin t)}.$$

The chain rule says that this should equal

$$\begin{aligned}\frac{\partial u}{\partial x}\frac{dx}{dt} + \frac{\partial u}{\partial y}\frac{dy}{dt} + \frac{\partial u}{\partial z}\frac{dz}{dt} &= e^{yz}e^t + xze^{yz}\cdot 1 + xye^{yz}\cos t\\ &= e^{t\sin t}e^t + e^t\sin t\, e^{t\sin t} + e^t\cdot t\cdot e^{t\sin t}\cos t\\ &= e^{t(1+\sin t)}(1+\sin t + t\cos t),\end{aligned}$$

which it does. ▲

Example 3 Suppose that $v = r\cos(st) - e^s\sin(rt)$ and that r, s, and t are functions of x. Find an expression for dv/dx.

Solution We use the chain rule with a change of notation. If $v = f(r,s,t)$ and r,s,t are functions of x, then

$$\frac{dv}{dx} = \frac{\partial v}{\partial r}\frac{dr}{dx} + \frac{\partial v}{\partial s}\frac{ds}{dx} + \frac{\partial v}{\partial t}\frac{dt}{dx}.$$

In this case we get

$$\begin{aligned}\frac{dv}{dx} &= \left[\cos(st) - te^s\cos(rt)\right]\frac{dr}{dx} - \left[tr\sin(st) + e^s\sin(rt)\right]\frac{ds}{dx}\\ &\quad - \left[rs\sin(st) + re^s\cos(rt)\right]\frac{dt}{dx}. \blacktriangle\end{aligned}$$

Example 4 What do you get if you apply the chain rule to the case $z = xy$, where x and y are arbitrary functions of t?

Solution If $z = xy$, then $\partial z/\partial x = y$ and $\partial z/\partial y = x$, so the chain rule gives $dz/dt = y(dx/dt) + x(dy/dt)$, which is precisely the product rule for functions of one variable. ▲

The chain rule for the case of two intermediate variables has a nice geometric interpretation involving the tangent plane. Recall from Section 15.2 that the tangent plane to the graph $z = f(x, y)$ at the point (x_0, y_0) is given by the linear equation $z = f(x_0, y_0) + f_x(x_0, y_0)(x - x_0) + f_y(x_0, y_0)(y - y_0)$. For this formula to be consistent with the definition of the tangent line to a curve, we would like the following statement to be true.

Tangents to Curves in Graphs

If $(x, y, z) = (g(t), h(t), k(t))$ is any curve on the surface $z = f(x, y)$ with $(g(t_0), h(t_0)) = (x_0, y_0)$, then the tangent line to the curve at t_0 lies in the tangent plane to the surface at (x_0, y_0). (In this statement, all derivatives are assumed to be continuous.)

To verify the above statement, we start with the fact that $(g(t), h(t), k(t))$ lies on the surface $z = f(x, y)$, i.e.,

$$k(t) = f(g(t), h(t)).$$

Differentiate both sides using the chain rule and then set $t = t_0$:

$$k'(t_0) = f_x(x_0, y_0)g'(t_0) + f_y(x_0, y_0)h'(t_0);$$

but this shows that the tangent line

$$z - z_0 = tk'(t_0), \qquad x - x_0 = tg'(t_0), \qquad y - y_0 = th'(t_0)$$

satisfies $z - z_0 = f_x(x_0, y_0)(x - x_0) + f_y(x_0, y_0)(y - y_0)$; that is, the tangent line lies in the tangent plane.

You may think of the preceding box as the "geometric statement" of the chain rule. It is illustrated in Fig. 15.3.1.

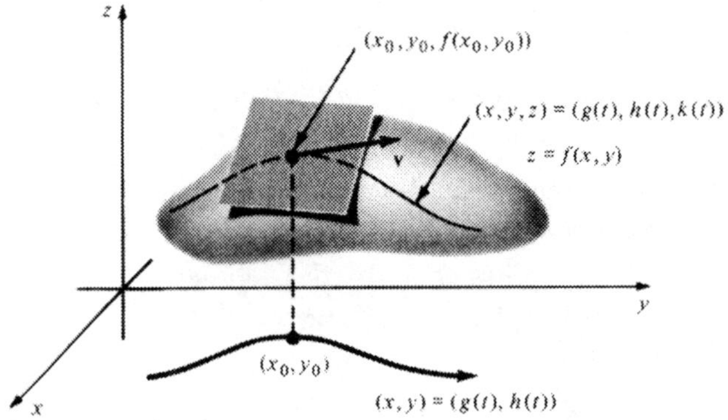

Figure 15.3.1. If a curve lies on the surface $z = f(x, y)$, then the tangent line (with direction vector $\mathbf{v}$) to the curve lies in the tangent plane of the surface.

Example 5 Show that for any curve $\boldsymbol{\sigma}(t)$ in the upper hemisphere $z = \sqrt{1 - x^2 - y^2}$, the velocity vector $\boldsymbol{\sigma}'(t)$ is perpendicular to $\boldsymbol{\sigma}(t)$.

Solution Let $(x, y, z) = \boldsymbol{\sigma}(t)$. By the preceding box, $\boldsymbol{\sigma}'(t)$ is perpendicular to the normal vector to the hemisphere at (x, y, z). From Example 1 of the previous section, this normal vector is just $x\mathbf{i} + y\mathbf{j} + z\mathbf{k} = \boldsymbol{\sigma}(t)$. Thus, $\boldsymbol{\sigma}'(t)$ is perpendicular to $\boldsymbol{\sigma}(t)$. ▲

Example 6 Show that the tangent plane at each point (x_0, y_0, z_0) of the cone $z = \sqrt{x^2 + y^2}$ $((x, y) \neq (0,0))$ contains the line passing through (x_0, y_0, z_0) and the origin.

Solution The line l through (x_0, y_0, z_0) and the origin has parametrization $(x, y, z) = (x_0t, y_0t, z_0t)$. Since this line lies in the cone for all $t > 0$, $(z^2 = z_0^2t^2 = (x_0^2 + y_0^2)t^2 = x^2 + y^2)$, the geometric interpretation of the chain rule implies that the tangent plane to the cone contains the tangent line to l; but the tangent line to l is l itself, so l is contained in the tangent plane. ▲

Exercises for Section 15.3

1. Suppose that a duck is swimming in a straight line $x = 3 + 8t$, $y = 3 - 2t$, while the water temperature is given by the formula $T = x^2\cos y - y^2\sin x$. Find dT/dt in two ways: (a) by the chain rule and (b) by expressing T in terms of t and differentiating.
2. Suppose that a duck is swimming along the curve $x = (3 + t)^2$, $y = 2 - t^2$, while the water temperature is given by the formula $T = e^x(y^2 + x^2)$. Find dT/dt in two ways: (a) by the chain rule and (b) by expressing T in terms of t and differentiating.

Verify the chain rule for the functions and curves in Exercises 3–6.

3. $f(x, y) = (x^2 + y^2)\ln(\sqrt{x^2 + y^2})$; $\boldsymbol{\sigma}(t) = (e^t, e^{-t})$.
4. $f(x, y) = xe^{x^2+y^2}$; $\boldsymbol{\sigma}(t) = (t, -t)$.
5. $f(x, y, z) = x + y^2 + z^3$; $\boldsymbol{\sigma}(t) = (\cos t, \sin t, t)$.
6. $f(x, y, z) = e^{x-z}(y^2 - x^2)$; $\boldsymbol{\sigma}(t) = (t, e^t, t^2)$.
7. Verify the chain rule for $u = x/y + y/z + z/x$, $x = e^t$, $y = e^{t^2}$, $z = e^{t^3}$.
8. Verify the chain rule for $u = \sin(xy)$, $x = t^2 + t$, $y = t^3$.
9. Show that applying the chain rule to $z = x/y$ (where x and y are arbitrary functions of t) gives the quotient rule for functions of one variable.
10. (a) Apply the chain rule to $u = xyz$, where x, y, and z are functions of t, to get a rule for differentiating a product of three functions of one variable.
    (b) Derive the rule in (a) by using one-variable calculus.
11. Let $z = \sqrt{x^2 + y^2} + 2xy^2$, where x and y are functions of u. Find an expression for dz/du.
12. If $u = \sin(a + \cos b)$, where a and b are functions of t, what is du/dt?
13. Describe the collection of vectors tangent to all possible curves on the paraboloid $z = x^2 + y^2$ through the point $(1, 2, 5)$.
14. Show that if a surface is defined by an equation $f(x, y, z) = 0$, and if $(x(t), y(t), z(t))$ is a curve in the surface which passes through the point (x_0, y_0, z_0) when $t = t_0$, then the two vectors $x'(t_0)\mathbf{i} + y'(t_0)\mathbf{j} + z'(t_0)\mathbf{k}$ and $f_x(x_0, y_0, z_0)\mathbf{i} + f_y(x_0, y_0, z_0)\mathbf{j} + f_z(x_0, y_0, z_0)\mathbf{k}$ are perpendicular.
15. (a) Use the chain rule to find $(d/dx)(x^x)$ by using the function $f(y, z) = y^z$.
    (b) Calculate $(d/dx)(x^x)$ by using one-variable calculus.
    (c) Which way do you prefer?
16. Suppose that the temperature at the point (x, y, z) in space is $T(x, y, z) = x^2 + y^2 + z^2$. Let a particle follow the right circular helix $\boldsymbol{\sigma}(t) = (\cos t, \sin t, t)$ and let $T(t)$ be its temperature at time t.
    (a) What is $T'(t)$?
    (b) Find an approximate value for the temperature at $t = (\pi/2) + 0.01$.
17. Use the chain rule to find a formula for the derivative $(d/dt)(f(t)g(t)/h(t))$.
18. Use the chain rule to differentiate $f(t)/[g(t)h(t)]$.

★19. A bug is swimming along the surface of a wave as in the Supplement to Section 15.1. Suppose that the motion of this wave is described by the function $f(t, y) = e^{-y}\cos t + \sin(y + t^2)$. At $t = 2$, the bug is at the position $y = 3$ and its horizontal velocity dy/dt is equal to 5. What is its vertical velocity dz/dt at that moment?

★20. Prove the chain rule by filling in the details in the following argument. Let $z = f(g(t), h(t))$, where g and h are differentiable and f has continuous partial derivatives.
(a) Show that

$$\frac{\Delta z}{\Delta t} = \frac{1}{\Delta t}\{[f(g(t+\Delta t), h(t+\Delta t)) - f(g(t), h(t+\Delta t))]$$
$$+ [f(g(t), h(t+\Delta t)) - f(g(t), h(t))]\}.$$

(b) Apply the mean value theorem for functions of one variable to each of the expressions in square brackets.
(c) Take the limit as $\Delta t \to 0$. (You will use the continuity of partial derivatives at this point.)

★21. Suppose that $z = f(x, y)$ is a surface with the property that if (x_0, y_0, z_0) lies on the surface, then so does the half-line from the origin through (x_0, y_0, z_0). Prove that this half-line also lies in the tangent plane to $z = f(x, y)$ at (x_0, y_0). Give an explicit example of such a surface.

★22. The differential equation $u_t + u_{xxx} + uu_x = 0$, called the *Korteweg–de Vries equation*, describes the motion of water waves in a shallow channel. Show that for any positive number c, the function

$$u(x, t) = 3c\,\mathrm{sech}^2\left[\tfrac{1}{2}(x - ct)\sqrt{c}\,\right]$$

is a solution of the Korteweg–de Vries equation. This solution represents a travelling "hump" of water in the channel and is called a *soliton*. How do the shape and speed of the soliton depend on c? (Solitons were first discovered by J. Scott Russell around 1840 in barge canals near Edinburgh. He reported his results in the Transactions of the Royal Society of Edinburgh, 1840, Vol. **14**, pp. 47–109.)

15.4 Matrix Multiplication and the Chain Rule

The derivative matrix of a composite function is the product of two matrices.

The chain rule of Section 2.2 enabled us to differentiate a function which depended on one independent variable through one intermediate variable. In Section 15.3, this result was extended to the case of two or three intermediate variables. When we allow the number of functions and independent variables to grow to two or three, the chain rule may be expressed in terms of matrix multiplication.

An $m \times n$ *matrix* is a rectangular array of mn numbers, called the *entries* of the matrix, arranged in m rows and n columns. The entry in the ith row and jth column is called the (i, j) entry. (See Fig. 15.4.1.)

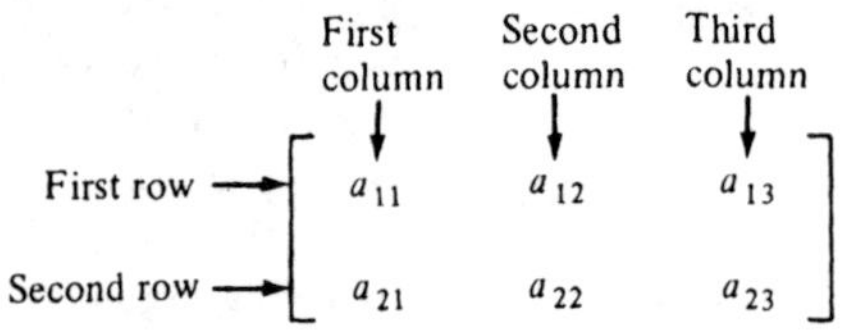

Figure 15.4.1. The (i, j) entry of this matrix is a_{ij}.

The chain rule of Section 15.3 involves both the partial derivatives of one function of three variables and the derivatives of three functions of one variable. We may assemble the derivatives of one function of three variables as a 1×3 matrix or row vector which we denote

$$\left[\frac{\partial u}{\partial x} \; \frac{\partial u}{\partial y} \; \frac{\partial u}{\partial z} \right] = \frac{\partial u}{\partial(x, y, z)} \tag{1}$$

and the velocity vector of three functions of one variable as a 3×1 matrix or column vector which we denote

$$\begin{bmatrix} \dfrac{dx}{dt} \\ \dfrac{dy}{dt} \\ \dfrac{dz}{dt} \end{bmatrix} = \frac{\partial(x, y, z)}{\partial t}. \tag{2}$$

(If there are only two intermediate variables, our row and column vectors will be 1×2 and 2×1 matrices.)

To express the chain rule in this new notation, we define a product between row and column vectors of the same length.

Multiplication of Row and Column Vectors

Let

$$A = [a_1 a_2 \ldots a_n] \quad \text{and} \quad B = \begin{bmatrix} b_1 \\ b_2 \\ \vdots \\ b_m \end{bmatrix}$$

be a row vector and a column vector, respectively. If $m = n$, we define the product AB to be the number $a_1b_1 + a_2b_2 + \cdots a_nb_n = \sum_{i=1}^{n} a_ib_i$. (If $m \neq n$, the product AB is not defined.)

In terms of this definition, the chain rule for three intermediate variables becomes

$$\frac{du}{dt} = \begin{bmatrix} \frac{\partial u}{\partial x} & \frac{\partial u}{\partial y} & \frac{\partial u}{\partial z} \end{bmatrix} \begin{bmatrix} \frac{dx}{dt} \\ \frac{dy}{dt} \\ \frac{dz}{dt} \end{bmatrix} = \frac{\partial u}{\partial(x,y,z)} \frac{\partial(x,y,z)}{\partial t}, \tag{3}$$

which looks very much like the chain rule of one-variable calculus.

The product of row and column vectors has many other applications. For example, every linear function $f(x, y, z) = ax + by + cz + d$ can be written as

$$f(x,y,z) = \begin{bmatrix} a & b & c \end{bmatrix} \begin{bmatrix} x \\ y \\ z \end{bmatrix} + d.$$

Your bill at the fruit market can be expressed as a product PQ, where

$$P = \begin{bmatrix} p_1 & p_2 & \cdots & p_n \end{bmatrix}$$

is the *price vector* whose ith entry is the price of the ith fruit in dollars per kilogram, and

$$Q = \begin{bmatrix} q_1 \\ \vdots \\ q_n \end{bmatrix}$$

is the *quantity vector* whose ith entry is the number of kilograms of the ith fruit purchased.

Example 1 Find AB if

$$A = \begin{bmatrix} 1 & 2 & 3 & 4 \end{bmatrix} \quad \text{and} \quad B = \begin{bmatrix} -1 \\ 1 \\ -1 \\ 1 \end{bmatrix}.$$

Solution $AB = (1)(-1) + (2)(1) + (3)(-1) + (4)(1) = -1 + 2 - 3 + 4 = 2.$ ▲

Having described the derivative of m functions of one variable by an $m \times 1$ matrix, and the derivative of one function of n variables by a $1 \times n$ matrix, it is natural for us to describe the derivative of m functions of n variables by an $m \times n$ matrix. For example, if $x = f(u,v)$, $y = g(u,v)$, and $z = h(u,v)$, we may put all six partial derivatives into a 3×2 matrix:

$$\begin{bmatrix} \frac{\partial x}{\partial u} & \frac{\partial x}{\partial v} \\ \frac{\partial y}{\partial u} & \frac{\partial y}{\partial v} \\ \frac{\partial z}{\partial u} & \frac{\partial z}{\partial v} \end{bmatrix} = \frac{\partial(x,y,z)}{\partial(u,v)}.$$

The rows of this matrix are the derivative vectors of f, g, and h. The columns are the "partial velocity vectors" with respect to u and v of the vector-valued function $\mathbf{r}(u,v) = (f(u,v), g(u,v), h(u,v))$.

In general, we may define the *derivative matrix* of m functions of n variables, as in the box on the following page.

Derivative Matrix

Let $u_1 = f_1(x_1, x_2, \ldots, x_n)$, $u_2 = f_2(x_1, x_2, \ldots, x_n), \ldots,$ and $u_m = f_m(x_1, x_2, \ldots, x_n)$ be m functions of the n variables $x_1, \ldots, x_n$. The *derivative matrix* of the u_i's with respect to the x_j's is the $m \times n$ matrix:

$$\frac{\partial(u_1, \ldots, u_m)}{\partial(x_1, \ldots, x_n)} = \begin{bmatrix} \dfrac{\partial u_1}{\partial x_1} & \dfrac{\partial u_1}{\partial x_2} & \cdots & \dfrac{\partial u_1}{\partial x_n} \\ \dfrac{\partial u_2}{\partial x_1} & \dfrac{\partial u_2}{\partial x_2} & \cdots & \dfrac{\partial u_2}{\partial x_n} \\ \vdots & \vdots & & \vdots \\ \dfrac{\partial u_m}{\partial x_1} & \dfrac{\partial u_m}{\partial x_2} & & \dfrac{\partial u_m}{\partial x_n} \end{bmatrix}$$

whose (i, j) entry is the partial derivative $\partial u_i / \partial x_j$.

The entries of the derivative matrix are functions of $(x_1, \ldots, x_n)$. If we fix values $(x_1^0, \ldots, x_n^0)$ for the independent variables, then the derivative matrix becomes a matrix of numbers and is denoted by

$$\left.\frac{\partial(u_1, \ldots, u_m)}{\partial(x_1, \ldots, x_n)}\right|_{(x_1^0, \ldots, x_n^0)}$$

Example 2 Let $u = x^2 + y^2$, $v = x^2 - y^2$, and $w = xy$. Find $\partial(u, v, w)/\partial(x, y)$ and evaluate

$$\left.\frac{\partial(u, v, w)}{\partial(x, y)}\right|_{(-2,3)}$$

Solution Applying the definition, with $m = 3$, $n = 2$, $u_1 = u$, $u_2 = v$, $u_3 = w$, $x_1 = x$, and $x_2 = y$, we get

$$\frac{\partial(u, v, w)}{\partial(x, y)} = \begin{bmatrix} 2x & 2y \\ 2x & -2y \\ y & x \end{bmatrix}.$$

Substituting $x = -2$ and $y = 3$, we get

$$\left.\frac{\partial(u, v, w)}{\partial(x, y)}\right|_{(-2,3)} = \begin{bmatrix} -4 & 6 \\ -4 & -6 \\ 3 & -2 \end{bmatrix}. \ \blacktriangle$$

Notice that the derivative matrix of one function $u = f(t)$ of one variable is a 1×1 matrix $\partial(u)/\partial(t)$ whose single entry is just the ordinary derivative du/dt. Thus the chain rule (3) can be rewritten as

$$\frac{\partial(u)}{\partial(t)} = \frac{\partial(u)}{\partial(x, y, z)} \frac{\partial(x, y, z)}{\partial(t)}. \tag{4}$$

In the remainder of this section, we will show how to multiply matrices of all sizes and, thereby, to generalize the chain rule (4) to several independent and dependent variables.

Example 3 (a) Suppose that $u = ax + by + cz + d$ and $v = ex + fy + gz + h$, where a, $b, \ldots, h$ are constants.

(i) Express u and v by using products of row and column vectors.
(ii) Find the derivative matrix $\partial(u,v)/\partial(x,y,z)$.

(b) Suppose that x, y, and z in (a) are linear functions of t:

$x = mt + n, y = pt + q$, *and* $z = rt + s$.

(i) Express u and v in terms of t and find the derivative matrix $\partial(u,v)/\partial(t)$.
(ii) Express the elements of $\partial(u,v)/\partial(t)$ as products of row and column vectors.

Solution (a) (i) We have $u = [a\ b\ c]\begin{bmatrix} x \\ y \\ z \end{bmatrix} + d$

and $v = [e\ f\ g]\begin{bmatrix} x \\ y \\ z \end{bmatrix} + h.$

(ii) The derivative matrix is

$$\frac{\partial(u,v)}{\partial(x,y,z)} = \begin{bmatrix} a & b & c \\ e & f & g \end{bmatrix},$$

all of whose entries are constants.

(b) (i) Substituting for x, y, and z their expressions in t, we get

$$u = a(mt+n) + b(pt+q) + c(rt+s) + d,$$

$$v = e(mt+n) + f(pt+q) + g(rt+s) + h.$$

We can find the derivative matrix without multiplying out:

$$\frac{\partial(u,v)}{\partial(t)} = \begin{bmatrix} \dfrac{du}{dt} \\ \dfrac{dv}{dt} \end{bmatrix} = \begin{bmatrix} am + bp + cr \\ em + fp + gr \end{bmatrix}.$$

(ii) The entries of $\partial(u,v)/\partial(t)$ are

$$\begin{bmatrix} a & b & c \end{bmatrix}\begin{bmatrix} m \\ p \\ r \end{bmatrix} \quad \text{and} \quad \begin{bmatrix} e & f & g \end{bmatrix}\begin{bmatrix} m \\ p \\ r \end{bmatrix}.$$

Notice that they are obtained by multiplying the rows of $\partial(u,v)/\partial(x,y,z)$ by the (single) column of $\partial(x,y,z)/\partial(t)$. ▲

The preceding example and the multiplication of row and column vectors suggest how we should multiply $m \times n$ matrices.

Matrix Multiplication

Let A and B be two matrices and assume that the number of columns of A equals the number of rows of B. To form $C = AB$:

1. Take the product of the first row of A and first column of B and let it be the $(1,1)$ entry of C.
2. Take the product of the first row and second column of B and let it be the $(1,2)$ entry of C.
3. Repeat. In general, the product of the ith row of A and jth column of B is the (i,j) entry of C.

Example 4 Let

$$A = \begin{bmatrix} 1 & -1 \\ 2 & 0 \\ 8 & -3 \end{bmatrix} \quad \text{and} \quad B = \begin{bmatrix} 2 & 0 \\ 1 & -1 \end{bmatrix}.$$

Find AB and BA.

Solution

$$\begin{bmatrix} 1 & -1 \\ 2 & 0 \\ 8 & -3 \end{bmatrix}\begin{bmatrix} 2 & 0 \\ 1 & -1 \end{bmatrix} = \begin{bmatrix} 1\cdot 2+(-1)\cdot 1 & \\ & \\ & \end{bmatrix} = \begin{bmatrix} 1 & \\ & \\ & \end{bmatrix}$$

(going across the first row of A and down the first column of B). Moving to the second column of B:

$$\begin{bmatrix} 1 & -1 \\ 2 & 0 \\ 8 & -3 \end{bmatrix}\begin{bmatrix} 2 & 0 \\ 1 & -1 \end{bmatrix} = \begin{bmatrix} 1\cdot 2+(-1)\cdot 1 & \\ & \\ & \end{bmatrix} = \begin{bmatrix} 1 & \\ & \\ & \end{bmatrix}$$

Moving to the second and third rows of A, we fill in the remaining entries:

$$\begin{bmatrix} 1 & -1 \\ 2 & 0 \\ 8 & -3 \end{bmatrix}\begin{bmatrix} 2 & 0 \\ 1 & -1 \end{bmatrix} = \begin{bmatrix} 1 & 1 \\ 4 & 0 \\ 13 & 3 \end{bmatrix}.$$

BA is not defined since the number of columns of B is not equal to the number of rows of A. ▲

Example 5 Find:

$$\begin{bmatrix} 1 & 1 \\ 2 & 0 \end{bmatrix}\begin{bmatrix} 1 & 2 \\ 0 & 1 \end{bmatrix} \quad \text{and} \quad \begin{bmatrix} 1 & 2 \\ 0 & 1 \end{bmatrix}\begin{bmatrix} 1 & 1 \\ 2 & 0 \end{bmatrix}.$$

Solution

$$\begin{bmatrix} 1 & 1 \\ 2 & 0 \end{bmatrix}\begin{bmatrix} 1 & 2 \\ 0 & 1 \end{bmatrix} = \begin{bmatrix} 1 & 3 \\ 2 & 4 \end{bmatrix} \quad \text{and} \quad \begin{bmatrix} 1 & 2 \\ 0 & 1 \end{bmatrix}\begin{bmatrix} 1 & 1 \\ 2 & 0 \end{bmatrix} = \begin{bmatrix} 5 & 1 \\ 2 & 0 \end{bmatrix}. \ ▲$$

Example 5 shows that even if AB and BA are defined, they may not be equal. In other words, matrix multiplication is *not commutative*.

Example 6 For 2×2 matrices A and B, verify that $|AB| = |A||B|$, where $|A|$ denotes the determinant of A (Section 13.6).

Solution Let

$$A = \begin{bmatrix} a & b \\ c & d \end{bmatrix} \quad \text{and} \quad B = \begin{bmatrix} e & f \\ g & h \end{bmatrix}.$$

Then $|A| = ad - bc$, $|B| = eh - gf$, and

$$\begin{aligned} |AB| &= \begin{vmatrix} ae + bg & af + bh \\ ce + dg & cf + dh \end{vmatrix} \\ &= (ae + bg)(cf + dh) - (ce + dg)(af + bh) \\ &= aecf + aedh + bgcf + bgdh - aecf - adgf - cebh - bgdh \\ &= aedh + bgcf - adgf - cebh \\ &= (ad - bc)(eh - gf) \\ &= |A| \cdot |B|. \ ▲ \end{aligned}$$

The result of Example 3(b) may be written in the following way in terms of derivative matrices:

$$\frac{\partial(u,v)}{\partial(t)} = \frac{\partial(u,v)}{\partial(x,y,z)}\frac{\partial(x,y,z)}{\partial(t)}.$$

This suggests a similar formula for the general chain rule.

The General Chain Rule

Let $u_1 = f_1(x_1, \dots, x_n), \dots, u_m = f_m(x_1, \dots, x_n)$ be m functions of n variables, and let $x_1 = g_1(t_1, \dots, t_k), \dots, x_n = g_n(t_1, \dots, t_k)$ be n functions of k variables, all with continuous partial derivatives.

Consider the u_i's as functions of the t_j's by

$$u_1 = f_1(g_1(t_1, \dots, t_k), \dots, g_n(t_1, \dots, t_k)).$$

Then

$$\frac{\partial(u_1, \dots, u_m)}{\partial(t_1, \dots, t_k)} = \frac{\partial(u_1, \dots, u_m)}{\partial(x_1, \dots, x_n)}\frac{\partial(x_1, \dots, x_n)}{\partial(t_1, \dots, t_k)}.$$

In other words,

$$\frac{\partial u_i}{\partial t_j} = \frac{\partial u_i}{\partial x_1}\frac{\partial x_1}{\partial t_j} + \frac{\partial u_i}{\partial x_2}\frac{\partial x_2}{\partial t_j} + \dots + \frac{\partial u_i}{\partial x_n}\frac{\partial x_n}{\partial t_j}.$$

(Note that there are as many terms in the sum as there are intermediate variables.)

We will carry out the proof for the "typical" case $m = 2$, $n = 3$, $k = 2$. We must prove that

$$\begin{bmatrix} \frac{\partial u_1}{\partial t_1} & \frac{\partial u_1}{\partial t_2} \\ \frac{\partial u_2}{\partial t_1} & \frac{\partial u_2}{\partial t_2} \end{bmatrix} = \begin{bmatrix} \frac{\partial u_1}{\partial x_1} & \frac{\partial u_1}{\partial x_2} & \frac{\partial u_1}{\partial x_3} \\ \frac{\partial u_2}{\partial x_1} & \frac{\partial u_2}{\partial x_2} & \frac{\partial u_2}{\partial x_3} \end{bmatrix} \begin{bmatrix} \frac{\partial x_1}{\partial t_1} & \frac{\partial x_1}{\partial t_2} \\ \frac{\partial x_2}{\partial t_1} & \frac{\partial x_2}{\partial t_2} \\ \frac{\partial x_3}{\partial t_1} & \frac{\partial x_3}{\partial t_2} \end{bmatrix}.$$

This matrix equation represents four ordinary equations. We will prove a typical one:

$$\frac{\partial u_2}{\partial t_1} = \frac{\partial u_2}{\partial x_1}\frac{\partial x_1}{\partial t_1} + \frac{\partial u_2}{\partial x_2}\frac{\partial x_2}{\partial t_1} + \frac{\partial u_2}{\partial x_3}\frac{\partial x_3}{\partial t_1}. \tag{5}$$

In taking the partial derivatives with respect to t_1, we hold t_2 fixed and take ordinary derivatives with respect to t_1. With this understood, we may rewrite (5) as

$$\frac{du_2}{dt_1} = \frac{\partial u_2}{\partial x_1}\frac{dx_1}{dt_1} + \frac{\partial u_2}{\partial x_2}\frac{dx_2}{dt_1} + \frac{\partial u_2}{\partial x_3}\frac{dx_3}{dt_1}. \tag{6}$$

We are now in the situation of Section 15.3—we have the independent variable t_1, the dependent variable u_2, and intermediate variables (x_1, x_2, x_3); but the chain rule for this case is just formula (6), so (6) is true and hence (5) is proved.

Example 7 Verify the chain rule for $\partial p/\partial x$, where

$$p = f(u,v,w) = u^2 + v^2 - w, \quad u = x^2y, \quad v = y^2, \quad \text{and} \quad w = e^{-xz}.$$

Solution $f(u,v,w) = (x^2y)^2 + y^4 - e^{-xz} = x^4y^2 + y^4 - e^{-xz}$. Thus

$$\frac{\partial p}{\partial x} = 4x^3y^2 + ze^{-xz}.$$

On the other hand,

$$\begin{aligned}\frac{\partial p}{\partial u}\frac{\partial u}{\partial x} + \frac{\partial p}{\partial v}\frac{\partial v}{\partial x} + \frac{\partial p}{\partial w}\frac{\partial w}{\partial x} &= 2u(2xy) + 2v\cdot 0 + ze^{-xz}\\ &= (2x^2y)(2xy) + ze^{-xz},\end{aligned}$$

which is the same. ▲

Example 8 Let (x, y) be cartesian coordinates in the plane and let (r,θ) be polar coordinates. (a) If $z = f(x, y)$ is a function on the plane, express the partial derivatives $\partial z/\partial r$ and $\partial z/\partial\theta$ in terms of $\partial z/\partial x$ and $\partial z/\partial y$. (b) Express $\partial^2 z/\partial r^2$ in cartesian coordinates.

Solution (a) By the general chain rule, using $x = r\cos\theta$ and $y = r\sin\theta$,

$$\begin{aligned}\begin{bmatrix}\dfrac{\partial z}{\partial r} & \dfrac{\partial z}{\partial\theta}\end{bmatrix} &= \begin{bmatrix}\dfrac{\partial z}{\partial x} & \dfrac{\partial z}{\partial y}\end{bmatrix}\begin{bmatrix}\dfrac{\partial x}{\partial r} & \dfrac{\partial x}{\partial\theta}\\[2mm] \dfrac{\partial y}{\partial r} & \dfrac{\partial y}{\partial\theta}\end{bmatrix}\\ &= \begin{bmatrix}\dfrac{\partial z}{\partial x} & \dfrac{\partial z}{\partial y}\end{bmatrix}\begin{bmatrix}\cos\theta & -r\sin\theta\\ \sin\theta & r\cos\theta\end{bmatrix}.\end{aligned}$$

Multiplying out,

$$\frac{\partial z}{\partial r} = \frac{\partial z}{\partial x}\cos\theta + \frac{\partial z}{\partial y}\sin\theta,$$

$$\frac{\partial z}{\partial\theta} = r\left[-\frac{\partial z}{\partial x}\sin\theta + \frac{\partial z}{\partial y}\cos\theta\right].$$

(b) By (a),

$$\frac{\partial z}{\partial r} = \frac{\partial z}{\partial x}\cos\theta + \frac{\partial z}{\partial y}\sin\theta. \tag{7}$$

Thus $\dfrac{\partial^2 z}{\partial r^2} = \dfrac{\partial}{\partial r}\left[\dfrac{\partial z}{\partial x}\right]\cos\theta + \dfrac{\partial}{\partial r}\left[\dfrac{\partial z}{\partial y}\right]\sin\theta$. Applying equation (7) with $\partial z/\partial x$ and $\partial z/\partial y$ replacing z, we get

$$\begin{aligned}\frac{\partial^2 z}{\partial r^2} &= \left(\frac{\partial^2 z}{\partial x^2}\cos\theta + \frac{\partial^2 z}{\partial y\,\partial x}\sin\theta\right)\cos\theta + \left(\frac{\partial^2 z}{\partial x\,\partial y}\cos\theta + \frac{\partial^2 z}{\partial y^2}\sin\theta\right)\sin\theta\\ &= \frac{\partial^2 z}{\partial x^2}\cos^2\theta + 2\frac{\partial^2 z}{\partial y\,\partial x}\sin\theta\cos\theta + \frac{\partial^2 z}{\partial y^2}\sin^2\theta\\ &= \frac{1}{x^2+y^2}\left[x^2\frac{\partial^2 z}{\partial x^2} + 2xy\frac{\partial^2 z}{\partial x\,\partial y} + y^2\frac{\partial^2 z}{\partial y^2}\right]. \ ▲\end{aligned}$$

Example 9 Suppose that $(t,s) = (f(x, y), g(x, y))$, $x = u - 2v$, and $y = u + 3v$. Express the derivative matrix $\partial(t,s)/\partial(u,v)$ in terms of $\partial(t,s)/\partial(x, y)$.

Solution By the chain rule,

$$\frac{\partial(t,s)}{\partial(u,v)} = \frac{\partial(t,s)}{\partial(x,y)}\,\frac{\partial(x,y)}{\partial(u,v)}.$$

In this example,

$$\frac{\partial(x,y)}{\partial(u,v)} = \begin{bmatrix} 1 & -2 \\ 1 & 3 \end{bmatrix},$$

so

$$\frac{\partial(t,s)}{\partial(u,v)} = \frac{\partial(t,s)}{\partial(x,y)}\begin{bmatrix} 1 & -2 \\ 1 & 3 \end{bmatrix} = \begin{bmatrix} \dfrac{\partial t}{\partial x} & \dfrac{\partial t}{\partial y} \\ \dfrac{\partial s}{\partial x} & \dfrac{\partial s}{\partial y} \end{bmatrix}\begin{bmatrix} 1 & -2 \\ 1 & 3 \end{bmatrix}$$

$$= \begin{bmatrix} \dfrac{\partial t}{\partial x} + \dfrac{\partial t}{\partial y} & -2\dfrac{\partial t}{\partial x} + 3\dfrac{\partial t}{\partial y} \\ \dfrac{\partial s}{\partial x} + \dfrac{\partial s}{\partial y} & -2\dfrac{\partial s}{\partial x} + 3\dfrac{\partial s}{\partial y} \end{bmatrix}. \blacktriangle$$

Exercises for Section 15.4

Find the matrix products in Exercises 1–4.

1. $[1 \quad 2 \quad 3]\begin{bmatrix} 4 \\ 5 \\ 6 \end{bmatrix}$

2. $\begin{bmatrix} \frac{1}{4} & \frac{1}{2} & \frac{1}{4} \end{bmatrix}\begin{bmatrix} 1 \\ 2 \\ 1 \end{bmatrix}$

3. $[2 \quad 2 \quad 6]\begin{bmatrix} 3 \\ 4 \\ 5 \end{bmatrix}$

4. $[1 \quad 2 \quad 3 \quad 4 \quad 5]\begin{bmatrix} 1 \\ 2 \\ 3 \\ 4 \\ 5 \end{bmatrix}$

Find the derivative matrices in Exercises 5–8 and evaluate at the given points.

5. $\partial(x,y)/\partial(u,v)$; $x = u\sin v$, $y = e^{uv}$; at $(0,1)$.
6. $\partial(x,y,z)/\partial(r,\theta,\phi)$; where $x = r\sin\phi\cos\theta$, $y = r\sin\phi\sin\theta$, $z = r\cos\phi$; at $(2,\pi/3,\pi/4)$.
7. $\partial(u,v)/\partial(x,y,z)$; $u = xyz$, $v = x+y+z$; at $(3,3,3)$.
8. $\partial(x,y)/\partial(r,\theta)$; $x = r\cos\theta$, $y = r\sin\theta$; at $(5,\pi/6)$.

Compute the matrix products in Exercises 9–20 or explain why they are not defined.

9. $\begin{bmatrix} 1 & 2 \\ 0 & 1 \end{bmatrix}\begin{bmatrix} 2 & 3 \\ 4 & 5 \end{bmatrix}$

10. $\begin{bmatrix} 1 & 0 \\ 0 & 0 \end{bmatrix}\begin{bmatrix} 0 & 0 \\ 0 & 1 \end{bmatrix}$

11. $\begin{bmatrix} 0 & 1 \\ 1 & 0 \end{bmatrix}\begin{bmatrix} a & b \\ c & d \end{bmatrix}$

12. $\begin{bmatrix} 1 & 0 \\ 0 & 1 \end{bmatrix}\begin{bmatrix} a & b \\ c & d \end{bmatrix}$

13. $\begin{bmatrix} 0 & 1 \\ 2 & 3 \end{bmatrix}\begin{bmatrix} 1 \\ 2 \\ 3 \end{bmatrix}$

14. $\begin{bmatrix} 1 \\ 2 \\ 3 \end{bmatrix}\begin{bmatrix} 4 \\ 5 \\ 6 \end{bmatrix}$

15. $\begin{bmatrix} 0 & 0 \\ 0 & 1 \end{bmatrix}\begin{bmatrix} a \\ b \end{bmatrix}$

16. $\begin{bmatrix} 1 & 2 \\ 3 & 4 \\ 5 & 6 \end{bmatrix}\begin{bmatrix} 0 & 0 & 0 \\ 3 & 2 & 1 \end{bmatrix}$

17. $\begin{bmatrix} 1 & 0 \\ 0 & 0 \end{bmatrix}\begin{bmatrix} a & b \\ c & d \end{bmatrix}$

18. $\begin{bmatrix} 1 & 0 & 0 \\ 0 & 1 & 0 \\ 0 & 0 & 1 \end{bmatrix}\begin{bmatrix} a & b & c \\ d & e & f \\ g & h & i \end{bmatrix}$

19. $\left(\begin{bmatrix} 1 & 0 \\ 2 & 3 \end{bmatrix}\begin{bmatrix} 2 & 4 \\ 1 & -1 \end{bmatrix}\right)\begin{bmatrix} 1 & 1 \\ 0 & 1 \end{bmatrix}$

20. $\begin{bmatrix} 1 & 0 \\ 2 & 3 \end{bmatrix}\left(\begin{bmatrix} 2 & 4 \\ 1 & -1 \end{bmatrix}\begin{bmatrix} 1 & 1 \\ 0 & 1 \end{bmatrix}\right)$

Compute $\partial z/\partial x$ and $\partial z/\partial y$ in Exercises 21–24 using matrix multiplication and by direct substitution.

21. $z = u^2 + v^2$; $u = 2x + 7$, $v = 3x + y + 7$.
22. $z = u^2 + 3uv - v^2$; $u = \sin x$, $v = -\cos x + \cos y$.
23. $z = \sin u \cos v$; $u = 3x^2 - 2y$, $v = x - 3y$.
24. $z = u/v^2$; $u = x + y$, $v = xy$.

25. (a) Compute derivative matrices $\partial(x,y)/\partial(t,s)$ and $\partial(u,v)/\partial(x,y)$ if

$$x = t + s, \qquad y = t - s,$$
$$u = x^2 + y^2, \qquad v = x^2 - y^2.$$

(b) Express (u,v) in terms of (t,s) and calculate $\partial(u,v)/\partial(t,s)$.

(c) Verify that the chain rule holds.

26. Do as in Exercise 25 for the functions $x = t^2 - s^2$, $y = ts$, $u = \sin(x+y)$, $v = \cos(x-y)$.
27. Do as in Exercise 25 for $x = ts$, $y = ts$; $u = x$, $v = -y$.
28. Do as in Exercise 25 for $x = t^2 + s^2$, $y = t^2 - s^2$, $z = 2ts$; $u = xy$, $v = xz$, $w = xz$.

29. Suppose that a function is given in terms of rectangular coordinates by $u = f(x,y,z)$. If

$$x = r\cos\theta\sin\phi,$$
$$y = r\sin\theta\sin\phi,$$
$$z = r\cos\phi,$$

express $\partial u/\partial r$, $\partial u/\partial\theta$, and $\partial u/\partial\phi$ in terms of $\partial u/\partial x$, $\partial u/\partial y$, and $\partial u/\partial z$.

30. Suppose that x, y, z are as in Exercise 29 and $u = x^2 + y^2 + z^2$. Find $\partial u/\partial r$, $\partial u/\partial\theta$, and $\partial u/\partial\phi$.
31. Express the polar coordinates r and θ in terms of the cartesian coordinates x and y, and find the derivative matrix $\partial(r,\theta)/\partial(x,y)$.

32. Let A be the derivative matrix $\partial(x, y)/\partial(r, \theta)$ for $x = r\cos\theta$, $y = r\sin\theta$. Let B be the derivative matrix $\partial(r, \theta)/\partial(x, y)$ of Exercise 31, with its entries expressed in terms of r and θ. Find AB and BA.

33. Let B be the $m \times 1$ column vector

$$\begin{bmatrix} \frac{1}{m} \\ \frac{1}{m} \\ \vdots \\ \frac{1}{m} \end{bmatrix}.$$

If $A = [a_1 \cdots a_m]$ is any row vector, what is AB?

Exercises 34–38 form a unit.

34. Let

$$I = \begin{bmatrix} 1 & 0 \\ 0 & 1 \end{bmatrix} \quad \text{and} \quad A = \begin{bmatrix} 1 & 2 \\ 0 & 1 \end{bmatrix}.$$

Find a matrix B such that $AB = I$.

35. In Exercise 34, show that we also have

$$BA = \begin{bmatrix} 1 & 0 \\ 0 & 1 \end{bmatrix}.$$

36. Show that the solution of the equation

$$A\begin{bmatrix} x \\ y \end{bmatrix} = \begin{bmatrix} e \\ f \end{bmatrix} \quad \text{is} \quad \begin{bmatrix} x \\ y \end{bmatrix} = B\begin{bmatrix} e \\ f \end{bmatrix},$$

where A and B are as in Exercises 34 and 35.

37. Find a matrix B such that

$$B\begin{bmatrix} 1 & 2 \\ 2 & 5 \end{bmatrix} = \begin{bmatrix} 1 & 0 \\ 0 & 1 \end{bmatrix}.$$

38. Using the results of Exercises 36 and 37, solve each of the following systems of equations:
(a) $x + 2y = 1$, $2x + 5y = 2$;
(b) $x + 2y = 0$, $2x + 5y = 0$.

Exercises 39–42 form a unit.

39. If $(f_1(x_1, \ldots, x_n), \ldots, f_n(x_1, \ldots, x_n)) = (u_1, \ldots, u_n)$ are n functions of n variables, then the (square) matrix of partial derivatives is called the *jacobian matrix*. Its determinant is called the *jacobian determinant* and is denoted by

$$\left|\frac{\partial(u_1, \ldots, u_n)}{\partial(x_1, \ldots, x_n)}\right|.$$

(a) Suppose that $n = 2$. Show that the absolute value of

$$\left|\left.\frac{\partial(u_1, u_2)}{\partial(x, y)}\right|_{(a,b)}\right|$$

is the area of the parallelogram spanned by

$$\left(\left.\frac{\partial u_1}{\partial x}\right|_{(a,b)}, \left.\frac{\partial u_2}{\partial x}\right|_{(a,b)}\right)$$

and

$$\left(\left.\frac{\partial u_1}{\partial y}\right|_{(a,b)}, \left.\frac{\partial u_2}{\partial y}\right|_{(a,b)}\right).$$

(b) Suppose that $n = 3$. Show that the absolute value of

$$\left|\left.\frac{\partial(u_1, u_2, u_3)}{\partial(x_1, x_2, x_3)}\right|_{(a,b,c)}\right|$$

is the volume of the parallelepiped spanned by the vectors

$$\left(\left.\frac{\partial u_1}{\partial x_i}\right|_{(a,b,c)}, \left.\frac{\partial u_2}{\partial x_i}\right|_{(a,b,c)}, \left.\frac{\partial u_3}{\partial x_i}\right|_{(a,b,c)}\right)$$

for $i = 1, 2, 3$.

40. Compute the following jacobian determinants:
(a) $(x, y) = (r\cos\theta, r\sin\theta)$. Find

$$\left|\frac{\partial(x, y)}{\partial(r, \theta)}\right|.$$

(b) Let $(x, y, z) = (r\cos\theta, r\sin\theta, z)$. Find

$$\left|\frac{\partial(x, y, z)}{\partial(r, \theta, z)}\right|.$$

(c) Let $(x, y, z) = (r\cos\theta\sin\phi, r\sin\theta\sin\phi, r\,c$
Find

$$\left|\frac{\partial(x, y, z)}{\partial(r, \theta, \phi)}\right|.$$

41. Compute the jacobian determinants (see Exercise 39) of the following functions at the indicated points:
(a) $(x, y) = (t^2 + s^2, t^2 - s^2)$; $(t, s) = (1, 2)$.
(b) $(u, v) = (x + y, xy)$; $(x, y) = (5, -3)$.
(c) Compute the jacobian determinant of (u, v) with respect to (t, s) from parts (a) and (b) at $(t, s) = (1, 2)$. Verify that your answer is the product of the answers in (a) and (b).

42. Prove the following equations (notation from Exercise 39) in light of the chain rule and the multiplicative property of determinants found in Example 6:

(a) $$\left|\frac{\partial(u, v)}{\partial(x, y)}\right|\left|\frac{\partial(x, y)}{\partial(t, s)}\right| = \left|\frac{\partial(u, v)}{\partial(t, s)}\right|;$$

(b) $$\left|\frac{\partial(x, y)}{\partial(t, s)}\right|\left|\frac{\partial(t, s)}{\partial(x, y)}\right| = 1.$$

43. Let v_1, v_2, v_3 be the components of a vector function $\mathbf{v}$, u a scalar function, a, b, ρ constants. Express in matrix notation the equations of elasticity:

$$\rho\left(\frac{\partial^2 v_1}{\partial t^2}\right) = (a + b)\left(\frac{\partial u}{\partial x}\right) + b\left(\frac{\partial^2 v_1}{\partial x^2} + \frac{\partial^2 v_1}{\partial y^2} + \frac{\partial^2 v_1}{\partial z^2}\right);$$

$$\rho\left(\frac{\partial^2 v_2}{\partial t^2}\right) = (a + b)\left(\frac{\partial u}{\partial y}\right) + b\left(\frac{\partial^2 v_2}{\partial x^2} + \frac{\partial^2 v_2}{\partial y^2} + \frac{\partial^2 v_2}{\partial z^2}\right);$$

$$\rho\left(\frac{\partial^2 v_3}{\partial t^2}\right) = (a + b)\left(\frac{\partial u}{\partial z}\right) + b\left(\frac{\partial^2 v_3}{\partial x^2} + \frac{\partial^2 v_3}{\partial y^2} + \frac{\partial^2 v_3}{\partial z^2}\right).$$

44. A rotation of points in the xy plane (relative to fixed axes) is given by
$$\begin{bmatrix} X \\ Y \end{bmatrix} = \begin{bmatrix} \cos\theta & -\sin\theta \\ \sin\theta & \cos\theta \end{bmatrix} \begin{bmatrix} x \\ y \end{bmatrix}$$
where θ is the angle of rotation of
$$\begin{bmatrix} x \\ y \end{bmatrix} \quad \text{into} \quad \begin{bmatrix} X \\ Y \end{bmatrix}$$
Show by means of matrix multiplication that a rotation of θ_1 followed by a rotation of θ_2 is the same as a rotation of θ_2 followed by a rotation of θ_1.

45. The coordinates u, φ, θ are defined by $x = au\sin\varphi\cos\theta$, $y = bu\sin\varphi\sin\theta$, $z = cu\cos\varphi$ for $u \geqslant 0$, $0 \leqslant \varphi \leqslant \pi$, $0 \leqslant \theta \leqslant 2\pi$.
(a) Show that the surfaces $u =$ constant are the *ellipsoids*
$$\left(\frac{x}{au}\right)^2 + \left(\frac{y}{bu}\right)^2 + \left(\frac{z}{cu}\right)^2 = 1.$$
(b) Show that the surfaces $\varphi =$ constant are *elliptical cones*.
(c) Show the surface $\theta =$ constant is a *plane*.
(d) Volume calculations involve the determinant of $\partial(x, y, z)/\partial(u, \varphi, \theta)$. Show that it equals $abcu^2\sin\varphi$.

46. The matrix equation
$$\begin{bmatrix} X \\ Y \\ 1 \end{bmatrix} = \begin{bmatrix} \cos\theta & -\sin\theta & 0 \\ \sin\theta & \cos\theta & 0 \\ 0 & 0 & 1 \end{bmatrix} \begin{bmatrix} x \\ y \\ 1 \end{bmatrix}$$
can be viewed as a *rotation* in the xy plane through the angle θ. (See Exercise 4.4). Similarly, the equation
$$\begin{bmatrix} X \\ Y \\ 1 \end{bmatrix} = \begin{bmatrix} 1 & 0 & -a \\ 0 & 1 & -b \\ 0 & 0 & 1 \end{bmatrix} \begin{bmatrix} x \\ y \\ 1 \end{bmatrix}$$
can be viewed as a *translation* in the xy plane.
(a) Use a matrix multiplication to find the matrix equation for a rotation followed by a translation.
(b) Is a rotation followed by a translation the same as a translation followed by a rotation?

★47. Verify the formula $|AB| = |A||B|$ for 3×3 matrices A and B, where $|A|$ denotes the determinant of A.

★48. Public health officials have located four persons, x_1, x_2, x_3, and x_4, known to be carrying a new strain of flu. Three persons, y_1, y_2, and y_3, report possible contact, and a first-order contact matrix A is set up whose i, jth entry is 1 if there was contact between x_i and y_j and zero otherwise. Five other people, z_1, z_2, z_3, z_4, and z_5, are questioned for possible contact with y_1, y_2, and y_3, and another first-order contact matrix B is set up whose i, jth entry is 1 if y_i has contacted z_j and zero otherwise.
(a) Show that the product matrix $C = AB$ counts the number of second-order contacts. That is, the i, jth entry of C is the number of possible paths of disease communication from x_i to z_j.
(b) Write down the three matrices for the situation shown in Fig. 15.4.2. Check the conclusion of part (a).

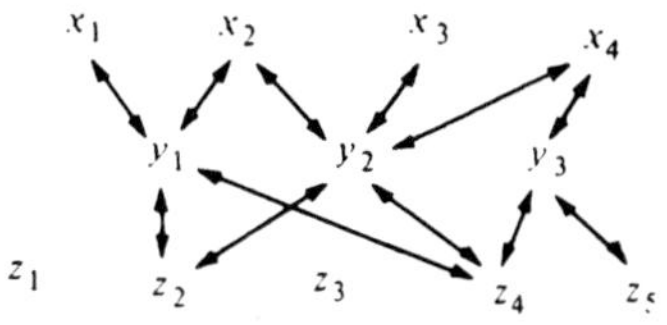

Figure 15.4.2. Contacts between three groups of people.

★49. Express Simpson's rule (Section 11.5) by using a product of row and column vectors.

★50. Suppose that f is a differentiable function of one variable and that a function $u = g(x, y)$ is defined by
$$u = g(x, y) = xyf\left(\frac{x + y}{xy}\right).$$
Show that u satisfies a (partial) differential equation of the form
$$x^2\frac{\partial u}{\partial x} - y^2\frac{\partial u}{\partial y} = G(x, y)u$$
and find the function $G(x, y)$.

Review Exercises for Chapter 15

Calculate all first partial derivatives for the functions in Exercises 1–10.

1. $u = g(x, y) = \dfrac{\sin(\pi x)}{1 + y^2}$.
2. $u = f(x, z) = \dfrac{x}{1 + \cos(2z)}$.
3. $u = k(x, z) = xz^2 - \cos(xz^3)$.
4. $u = m(y, z) = y^z$.
5. $u = h(x, y, z) = zx + y^2 + yz$.
6. $u = n(x, y, z) = x^{yz}$.
7. $u = f(x, y, z) = \ln[1 + e^{-x}\cos(xy)]$.
8. $u = h(x, y, z) = \cos(e^{-x^2 - 2y^2})$.
9. $u = g(x, y, z) = xz + e^z\left(\int_0^x t^2e^t\,dt\right)$.
10. $u = f(x, y) = \cos(xy^2) + \exp\left[\int_0^x \sqrt{t}\cos(ty)\,dt\right]$.

Check the equality of the given mixed partials for the indicated functions in Exercises 11–16.

11. $\partial^2u/\partial x\,\partial y = \partial^2u/\partial y\,\partial x$ for u in Exercise 1.
12. $\partial^2u/\partial x\,\partial z = \partial^2u/\partial z\,\partial x$ for u in Exercise 2.
13. $\partial^2u/\partial x\,\partial z = \partial^2u/\partial z\,\partial x$ for u in Exercise 3.
14. $\partial^2u/\partial y\,\partial z = \partial^2u/\partial z\,\partial y$ for u in Exercise 4.
15. $\partial^2u/\partial x\,\partial z = \partial^2u/\partial z\,\partial x$ for u in Exercise 5.
16. $\partial^2u/\partial x\,\partial z = \partial^2u/\partial z\,\partial x$ for u in Exercise 6.

17. Find $(\partial/\partial x)e^{x-\cos(yx)}|_{x=1,y=0}$.
18. Find $(\partial/\partial s)\exp(rs^3 - r^3s)|_{r=1,s=1}$.
19. Find $f_x(1,0)$ if $f(x, y) = \cos(x + e^{yx})$.
20. Find $f_s(-1,2)$ if $f(r,s) = (r + s^2)/(1 - r^2 - s^2)$.
21. The possible time T in minutes of a scuba dive is given by $T = 32V/(x + 32)$, where V is the volume of air in cubic feet at 15 psi (pounds per square inch) which is compressed into the air tanks, and x is the depth of the scuba dive in feet.
    (a) How long can a 27-foot dive last when $V = 65$?
    (b) Find $\partial T/\partial x$ and $\partial T/\partial V$ when $x = 27$, $V = 65$. Interpret.
22. The displacement of a certain violin string placed on the x axis is given by $u = \sin(x - 6t) + \sin(x + 6t)$. Calculate the velocity of the string at $x = 1$ when $t = \frac{1}{3}$.

Find the limits in Exercises 23–26, if they exist.

23. $\lim_{(x,y)\to(0,0)} (x^2 - 2xy + 4)$
24. $\lim_{(x,y)\to(0,0)} (x^3 - y^3 + 15)$
25. $\lim_{(x,y)\to(0,0)} \dfrac{x^3 - y^3}{x^2 + y^2}$
26. $\lim_{(x,y)\to(0,0)} \dfrac{xy + x^3 + x - 2}{\sqrt{x^2 + y^2}}$

Find the equation of the tangent plane to the given surface at the indicated point in Exercises 27–30.

27. $z = x^2 + y^2$; $x = 1, y = 1$.
28. $z = x\sin y$; $x = 2, y = \pi/4$.
29. $z = e^{xy}$; $x = 0, y = 0$.
30. $z = \sqrt{x^2 + y^2}$; $x = 3, y = 4$.

Use the linear approximation to find approximations for the quantities in Exercises 31–34.

31. $\sqrt{(1.01)^2 + (4.01)^2 + (8.002)^2}$
32. $(2.004)\ln(0.98)$
33. $(0.999)^{1.001}$
34. $(1.001)^{0.999}$

35. Find an approximate value for the hypotenuse of a right triangle whose legs are 3.98 and 3.03.
36. The capacitance per unit length of a parallel pair of wires of radii R and axis-to-axis separation D is given by

$$C = \frac{\pi\epsilon_0}{\ln\left(\dfrac{D + \sqrt{D^2 - 4R^2}}{2R}\right)}.$$

The capacitance between a wire and a plane parallel to it is

$$C^* = \frac{2\pi\epsilon_0}{\ln\left(\dfrac{h + \sqrt{h^2 - R^2}}{R}\right)},$$

where h = distance from the wire to the plane.

    (a) Find the expected change in capacitance for two parallel wires, separated by 2 centimeters, with radius 0.40 centimeter, due to a radius increase of 0.01 centimeter.
    (b) A wire of 0.57 centimeter radius has its central axis at a uniform distance of 3 centimeters from a conducting plane. Due to heating, the wire increases 0.02 centimeter in radius, but due to bowing of the wire, it can be assumed that the axis of the wire was raised to 3.15 centimeters above the plane. What is the expected change in capacitance?
37. At time $t = 0$, a particle is ejected from the surface $x^2 + 2y^2 + 3z^2 = 6$ at the point $(1,1,1)$ in a direction normal to the surface at a speed of 10 units per second. At what time does it cross the sphere $x^2 + y^2 + z^2 = 103$? [Hint: Solve for z].
38. At what point(s) on the surface in Exercise 37 is the normal vector parallel to the line $x = y = z$?
39. Verify the chain rule for the function $f(x, y, z) = \ln(1 + x^2 + 2z^2)/(1 + y^2)$ and the curve $\boldsymbol{\sigma}(t) = (t, 1 - t^2, \cos t)$.
40. Verify the chain rule for the function $f(x, y) = x^2/(2 + \cos y)$ and the curve $x = e^t$, $y = e^{-t}$.
41. (a) Let c be a constant. Show that, for every function $f(x)$, the function $u(x,t) = f(x - ct)$ satisfies the partial differential equation $u_t + cu_x = 0$.
    (b) With u as in (a), consider for each value of t the graph $z = u(x,t)$ in the xz plane. How does this change as t increases?
42. (a) Show that, if $u(x,t)$ is *any* solution of the equation $u_t + cu_x = 0$, then the function $g(y,t)$ defined by $g(y,t) = u(y + ct, t)$ is independent of t.
    (b) Conclude from (a) that u must be of the form $u(x,t) = f(x - ct)$ for some function f.
    (c) What kind of wave motion is described by the equation $u_t + cu_x = 0$?
43. A right circular cone of sand is gradually collapsing. At a certain moment, the cone has a height of 10 meters and a base radius of 3 meters. If the height of the cone is decreasing at a rate of 1 meter per hour, how is the radius changing, assuming that the volume remains constant?
44. A boat is sailing northeast at 20 kilometers per hour. Assuming that the temperature drops at a rate of 0.2°C per kilometer in the northerly direction and 0.3°C per kilometer in the easterly direction, what is the time rate of change of temperature as observed on the boat?
45. Use the chain rule to find a formula for $(d/dt)\exp[f(t)g(t)]$.
46. Use the chain rule to find a formula for $(d/dt)(f(t)^{g(t)})$.

47. If x and y are functions of t,

$$\left.\frac{dx}{dt}\right|_{t=0} = 1, \text{ and } \left.\frac{dy}{dt}\right|_{t=0} = -1,$$

find $\left.\frac{d}{dt}e^{x+2xy}\right|_{t=0}$ in terms of x and y.

48. If x, y, and z are functions of t and

$$\left.\frac{dx}{dt}\right|_{t=0} = 1, \quad \left.\frac{dy}{dt}\right|_{t=0} = 0$$

and $\left.\frac{dz}{dt}\right|_{t=0} = -1$, find $\left.\frac{d}{dt}\cos(xyz^2)\right|_{t=0}$ in terms of x, y, and z.

49. The tangent plane to $z = x^2 + 6y^2$ at $x = 1$, $y = 1$ meets the xy plane in a line. Find the equation of this line.

50. The tangent plane to $z = e^{x-y}$ at $x = 1$, $y = 2$ meets the line $x = t$, $y = 2t - 1$, $z = 5t$ in a point. Find it.

Find the products AB of the matrices in Exercises 51–60.

51. $A = [1 \quad 2 \quad 4]$, $B = \begin{bmatrix} 2 \\ -1 \\ 1 \end{bmatrix}$

52. $A = \begin{bmatrix} \frac{1}{3} & \frac{1}{3} & \frac{1}{2} & \frac{1}{2} \end{bmatrix}$, $B = \begin{bmatrix} 3/2 \\ 3/2 \\ 2 \\ 2 \end{bmatrix}$

53. $A = \begin{bmatrix} 0 & -1 \\ 1 & 0 \end{bmatrix}$, $B = \begin{bmatrix} 1 & 0 \\ 0 & -1 \end{bmatrix}$

54. $A = \begin{bmatrix} 1 & 2 \\ 2 & 1 \end{bmatrix}$, $B = \begin{bmatrix} -1 & 1 \\ 1 & 1 \end{bmatrix}$

55. $A = \begin{bmatrix} 1 & 2 \\ 2 & 4 \end{bmatrix}$, $B = \begin{bmatrix} 2 & -1 \\ 2 & -1 \end{bmatrix}$

56. $A = \begin{bmatrix} 2 & -1 \\ 2 & -1 \end{bmatrix}$, $B = \begin{bmatrix} 1 & 2 \\ 2 & 4 \end{bmatrix}$

57. $A = \begin{bmatrix} 1 & 2 \\ -1 & 1 \\ 2 & 1 \end{bmatrix}$, $B = \begin{bmatrix} 2 & 1 & 2 \\ -1 & -1 & 1 \end{bmatrix}$

58. $A = \begin{bmatrix} 1 & 2 & 3 & 4 & 5 \\ 6 & 7 & 8 & 9 & 10 \end{bmatrix}$, $B = \begin{bmatrix} 1 \\ -1 \\ 1 \\ -1 \\ 1 \end{bmatrix}$

59. $A = \begin{bmatrix} 3 & 0 & 1 \\ 1 & 2 & -1 \\ 1 & 0 & 1 \end{bmatrix}$, $B = \begin{bmatrix} 1 & 0 & -1 \\ 2 & 0 & 1 \\ 0 & 1 & 0 \end{bmatrix}$

60. $A = \begin{bmatrix} 1 & 0 & -1 \\ 2 & 0 & 1 \\ 0 & 1 & 0 \end{bmatrix}$, $B = \begin{bmatrix} 3 & 0 & 1 \\ 1 & 2 & -1 \\ 1 & 0 & 1 \end{bmatrix}$

61. Compute $\partial z/\partial x$ and $\partial z/\partial y$ if

$$z = \frac{u^2 + v^2}{u^2 - v^2}, \qquad u = e^{-x-y}, \quad v = e^{xy}$$

by (a) substitution and (b) the chain rule.

62. Do as in Exercise 61 if $z = uv$, $u = x + y$, and $v = x - y$.

63. Suppose that $z = f(x, y)$, $x = u + v$ and $y = u - v$. Express $\partial z/\partial u$ and $\partial z/\partial v$ in terms of $\partial z/\partial x$ and $\partial z/\partial y$.

64. In the situation of Exercise 63, express $\partial z/\partial x$ and $\partial z/\partial y$ in terms of $\partial z/\partial u$ and $\partial z/\partial v$.

65. The *ideal gas law* $PV = nRT$ involves a constant R, the number n of moles of the gas, the volume V, the Kelvin temperature T, and the pressure P.
    (a) Show that each of n, P, T, V is a function of the remaining variables, and determine explicitly the defining equations.
    (b) The quantity $\partial P/\partial T$ is a rate of change. Discuss this in detail, and illustrate with an example which identifies the variables held constant.
    (c) Calculate $\partial V/\partial T, \partial T/\partial P, \partial P/\partial V$ and show that their product equals -1.

66. The *potential temperature* θ is defined in terms of temperature T and pressure p by

$$\theta = T\left(\frac{1000}{p}\right)^{0.286}.$$

The temperature and pressure may be thought of as functions of position (x, y, z) in the atmosphere and also of time t.
    (a) Find formulas for $\partial\theta/\partial x, \partial\theta/\partial y, \partial\theta/\partial z, \partial\theta/\partial t$ in terms of partial derivatives of T and p.
    (b) The condition $\partial\theta/\partial z < 0$ is regarded as *unstable atmosphere*, for it leads to large vertical excursions of air parcels from a single upward or downward impetus. Meteorologists use the formula

$$\frac{\partial\theta}{\partial z} = \frac{\theta}{T}\left(\frac{\partial T}{\partial z} + \frac{g}{C_p}\right),$$

where $g = 32.2$. $C_p = \text{constant} > 0$. How does the temperature change in the upward direction for an unstable atmosphere?

67. The specific volume V, pressure P, and temperature T of a Van der Waals gas are related by $P = [RT/(V - \beta)] - \alpha/V^2$, where α, β, R are considered to be constants.
    (a) Explain why any two of V, P, or T can be considered independent variables which determine the third variable.
    (b) Find $\partial T/\partial P, \partial P/\partial V, \partial V/\partial T$. Identify which variables are constant, and interpret each partial derivative physically.
    (c) Verify that $(\partial T/\partial P)(\partial P/\partial V)(\partial V/\partial T) = -1$ (not $+1$!).

68. Dieterici's equation of state for a gas is

$$P(V - b)e^{a/RVT} = RT,$$

where a, b, and R are constants. Regard volume V as a function of temperature T and pressure P and show that

$$\frac{\partial V}{\partial T} = \left(R + \frac{a}{TV}\right)\left(\frac{RT}{V - b} - \frac{a}{V^2}\right)^{-1}.$$

69. What is wrong with the following argument? Suppose that $w = f(x, y)$ and $y = x^2$. By the chain rule,
$$\frac{\partial w}{\partial x} = \frac{\partial w}{\partial x}\frac{\partial x}{\partial x} + \frac{\partial w}{\partial y}\frac{\partial y}{\partial x} = \frac{\partial w}{\partial x} + 2x\frac{\partial w}{\partial y}.$$
Hence $0 = 2x(\partial w/\partial y)$, so $\partial w/\partial y = 0$.

70. What is wrong with the following argument? Suppose that $w = f(x, y, z)$ and $z = g(x, y)$. Then by the chain rule,
$$\frac{\partial w}{\partial x} = \frac{\partial w}{\partial x}\frac{\partial x}{\partial x} + \frac{\partial w}{\partial y}\frac{\partial y}{\partial x} + \frac{\partial w}{\partial z}\frac{\partial z}{\partial x}$$
$$= \frac{\partial w}{\partial x} + \frac{\partial w}{\partial z}\frac{\partial z}{\partial x}.$$
Hence
$$0 = \frac{\partial w}{\partial z}\frac{\partial z}{\partial x},$$
so $\partial w/\partial z = 0$ or $\partial z/\partial x = 0$, which is, in general, absurd.

71. For a function u of three variables (x, y, z), show that $\partial^3 u/\partial x\,\partial y\,\partial z = \partial^3 u/\partial y\,\partial z\,\partial x$.

72. For a function u of three variables (x, y, z), show that $\partial^3 u/\partial x\,\partial y\,\partial z = \partial^3 u/\partial z\,\partial x\,\partial y$.

73. Prove that the functions
(a) $f(x, y) = \ln(x^2 + y^2)$,
(b) $g(x, y, z) = \dfrac{1}{(x^2 + y^2 + z^2)^{1/2}}$,
(c) $h(x, y, z, w) = \dfrac{1}{x^2 + y^2 + z^2 + w^2}$,
satisfy the respective Laplace equations:
(a) $f_{xx} + f_{yy} = 0$,
(b) $g_{xx} + g_{yy} + g_{zz} = 0$,
(c) $h_{xx} + h_{yy} + h_{zz} + h_{ww} = 0$,
where $f_{xx} = \partial^2 f/\partial x^2$, etc.

74. If $z = f(x - y)/y$, show that
$$z + y(\partial z/\partial x) + y(\partial z/\partial y) = 0.$$

75. Given $w = f(x, y)$ with $x = u + v$, $y = u - v$, show that
$$\frac{\partial^2 w}{\partial u\,\partial v} = \frac{\partial^2 w}{\partial x^2} - \frac{\partial^2 w}{\partial y^2}.$$

★76. (a) A function $u = f(x_1, \ldots, x_m)$ is called *homogeneous of degree n* if
$$f(tx_1, \ldots, tx_m) = t^n f(x_1, \ldots, x_m).$$
Show that such a function satisfies *Euler's differential equation*
$$x_1\frac{\partial u}{\partial x_1} + x_2\frac{\partial u}{\partial x_2} + \cdots + x_m\frac{\partial u}{\partial x_m} = nf(x_1, \ldots, x_m).$$
(b) Show that each of the following functions satisfies a differential equation of the type in part (a), find n, and check directly that f is homogeneous of degree n.
(i) $f(x, y) = x^2 + xy + y^2$;
(ii) $f(x, y, z) = x + 3y - \sqrt{xz}$; $xz > 0$;
(iii) $f(x, y, z) = xyz + x^3 - x^2y$.

★77. In Exercise 77 on page 775 we saw that the mixed partial derivatives of
$$z = \frac{xy(x^2 - y^2)}{x^2 + y^2}$$
at $(0, 0)$ are not equal. Is this consistent with the graph in Fig. 15.R.1?

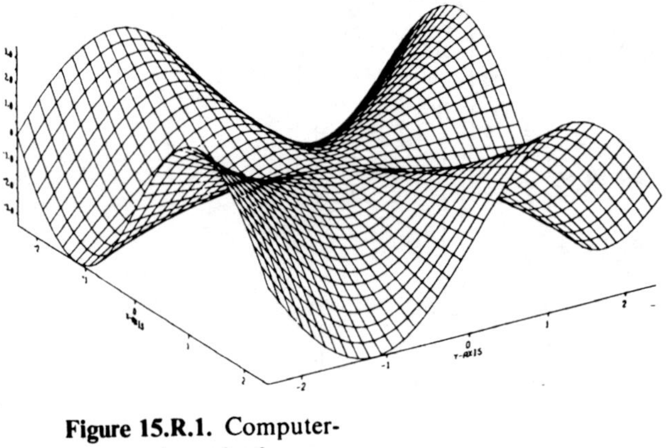

Figure 15.R.1. Computer-generated graph of
$$z = \frac{xy(x^2 - y^2)}{x^2 + y^2}.$$

Chapter 16

Gradients, Maxima, and Minima

The gradient of a function of several variables vanishes at a maximum or a minimum.

The gradient of a function f is a vector whose components are the partial derivatives of f. Derivatives in any direction can be found in terms of the gradient, using the chain rule. The gradient will be used to find the equations for tangent planes to level surfaces. The last two sections of the chapter extend our earlier studies of maxima and minima (Chapter 3) to functions of several variables.

16.1 Gradients and Directional Derivatives

The directional derivative is the dot product of the gradient and the direction vector.

The right-hand side of the chain rule

$$\frac{du}{dt} = \frac{\partial u}{\partial x}\frac{dx}{dt} + \frac{\partial u}{\partial y}\frac{dy}{dt} + \frac{\partial u}{\partial z}\frac{dz}{dt}$$

has the appearance of a dot product—in fact it is the dot product of the vectors

$$\frac{dx}{dt}\mathbf{i} + \frac{dy}{dt}\mathbf{j} + \frac{dz}{dt}\mathbf{k} \quad \text{and} \quad \frac{\partial u}{\partial x}\mathbf{i} + \frac{\partial u}{\partial y}\mathbf{j} + \frac{\partial u}{\partial z}\mathbf{k}.$$

We recognize the first vector as the *velocity vector* of a parametric curve; if $\boldsymbol{\sigma}(t)$ is the vector representation of the curve, it is just $\boldsymbol{\sigma}'(t)$. The second vector is something new: it depends upon the function $u = f(x, y, z)$ and contains in vector form all three partial derivatives of f. This is called the *gradient* of f and is denoted ∇f. Thus

$$\nabla f(x, y, z) = f_x(x, y, z)\mathbf{i} + f_y(x, y, z)\mathbf{j} + f_z(x, y, z)\mathbf{k}.$$

Example 1 (a) Find ∇f if $u = f(x, y, z) = xy - z^2$.
(b) Find ∇f for the function $f(x, y, z) = e^{xy} - x\cos(yz^2)$.

Solution (a) Substituting the partial derivatives of f into the formula for the gradient of f, we find $\nabla f(x, y, z) = y\mathbf{i} + x\mathbf{j} - 2z\mathbf{k}$.

(b) Here $f_x(x, y, z) = ye^{xy} - \cos(yz^2)$, $f_y(x, y, z) = xe^{xy} + xz^2\sin(yz^2)$, and $f_z(x, y, z) = 2xyz\sin(yz^2)$, so

$$\nabla f(x, y, z) = \left[ye^{xy} - \cos(yz^2)\right]\mathbf{i} + \left[xe^{xy} + xz^2\sin(yz^2)\right]\mathbf{j} + \left[2xyz\sin(yz^2)\right]\mathbf{k}. \blacktriangle$$

Notice that the vector $\nabla f(x, y, z)$ is a function of the point (x, y, z) in space; in other words, ∇f is a function of the point in space where the partial derivatives are evaluated. A rule $\mathbf{\Phi}$ which assigns a vector $\mathbf{\Phi}(x, y, z)$ in space to each point (x, y, z) of some domain in space is called a *vector field*. Thus, for a given function f, ∇f is a vector field. Similarly, a vector field in the xy plane is a rule $\mathbf{\Phi}$ which assigns to each point (x, y) a vector $\mathbf{\Phi}(x, y)$ in the plane.

The Gradient

If $z = f(x, y)$ is a function of two variables, its gradient vector field ∇f is defined by

$$\nabla f(x, y) = f_x(x, y)\mathbf{i} + f_y(x, y)\mathbf{j} = \frac{\partial z}{\partial x}\mathbf{i} + \frac{\partial z}{\partial y}\mathbf{j}.$$

If $u = f(x, y, z)$ is a function of three variables, its gradient vector field ∇f is defined by

$$\nabla f(x, y, z) = f_x(x, y, z)\mathbf{i} + f_y(x, y, z)\mathbf{j} + f_z(x, y, z)\mathbf{k} = \frac{\partial u}{\partial x}\mathbf{i} + \frac{\partial u}{\partial y}\mathbf{j} + \frac{\partial u}{\partial z}\mathbf{k}.$$

We may sketch a vector field $\mathbf{\Phi}(x, y)$ in the plane by choosing several values for (x, y), evaluating $\mathbf{\Phi}(x, y)$ at each point, and drawing the vector $\mathbf{\Phi}(x, y)$ with its tail at the point (x, y). The same thing may be done for vector fields in space, although they are more difficult to visualize.

Example 2 Sketch the gradient vector field of the function $f(x, y) = x^2/10 + y^2/6$.

Solution The partial derivatives are $f_x(x, y) = x/5$ and $f_y(x, y) = y/3$. Evaluating these for various values of x and y and plotting, we obtain the sketch in Fig. 16.1.1. For instance, $f_x(2, 2) = \frac{2}{5}$ and $f_y(2, 2) = \frac{2}{3}$; thus the vector $\frac{2}{5}\mathbf{i} + \frac{2}{3}\mathbf{j}$ is plotted at the point $(2, 2)$, as indicated in the figure. $\blacktriangle$

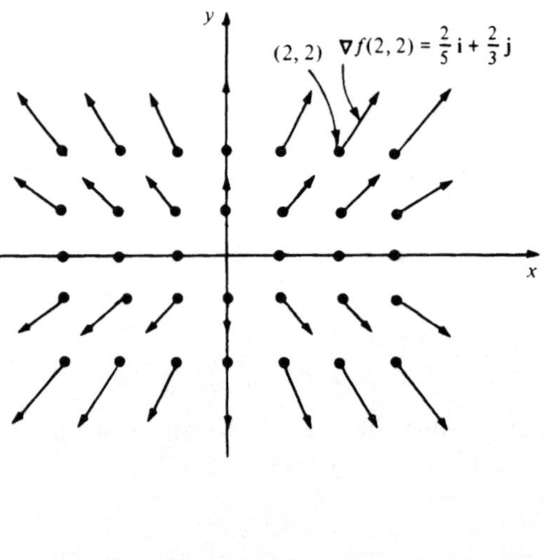

Figure 16.1.1. The gradient vector field ∇f, where $f(x, y) = (x^2/10) + (y^2/6)$.

In sketching a vector field $\mathbf{\Phi}(x, y)$, we sometimes find that the vectors are so long that they overlap one another, making the drawing confusing. In this case, it is better to sketch $\varepsilon\mathbf{\Phi}(x, y)$, where ε is a small positive number. This is illustrated in the next example.

Example 3 (a) Illustrate the vector field $\mathbf{\Phi}(x, y) = 3y\mathbf{i} - 3x\mathbf{j}$ by sketching $\frac{1}{6}\mathbf{\Phi}(x, y)$.
(b) Using the law of equality of mixed partial derivatives, show that the vector field in (a) is *not* the gradient vector field of any function.

Solution (a) If we sketched $\mathbf{\Phi}(x, y) = 3y\mathbf{i} - 3x\mathbf{j}$ itself, the vectors at different base points would overlap. Instead we sketch $\frac{1}{6}\mathbf{\Phi}(x, y) = \frac{1}{2}y\mathbf{i} - \frac{1}{2}x\mathbf{j}$ in Fig. 16.1.2.

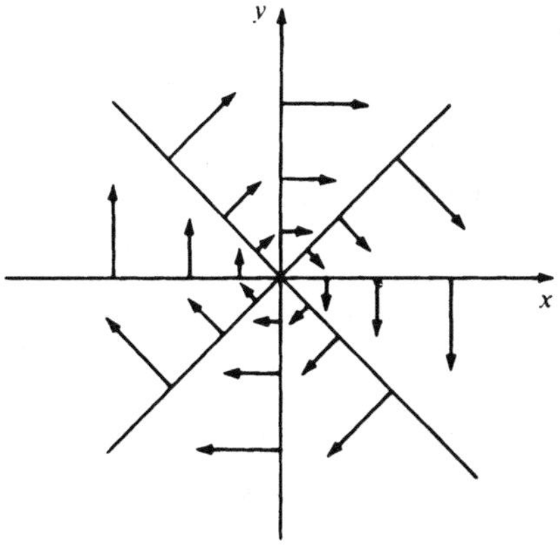

Figure 16.1.2. The vector field $3y\mathbf{i} - 3x\mathbf{j}$ is not the gradient of a function.

(b) If $\mathbf{\Phi}(x, y) = 3y\mathbf{i} - 3x\mathbf{j}$ were the gradient of a function $z = f(x, y)$, we would have $\partial z/\partial x = 3y$ and $\partial z/\partial y = -3x$. By the equality of mixed partial derivatives, $\partial^2 z/\partial x\,\partial y = -3$ and $\partial^2 z/\partial y\,\partial x = 3$ would have to be equal; but $3 \neq -3$, so our vector field cannot be a gradient. ▲

In a number of situations later in the book, the vector $\mathbf{r}$ from the origin to a point (x, y, z) plays a basic role. The next example illustrates its use.

Example 4 Let $\mathbf{r} = x\mathbf{i} + y\mathbf{j} + z\mathbf{k}$ and $r = \|\mathbf{r}\| = \sqrt{x^2 + y^2 + z^2}$. Show that

$$\nabla\left(\frac{1}{r}\right) = -\frac{\mathbf{r}}{r^3}, \qquad r \neq 0.$$

What is $\|\nabla(1/r)\|$?

Solution By definition of the gradient,

$$\nabla\left(\frac{1}{r}\right) = \frac{\partial}{\partial x}\left(\frac{1}{r}\right)\mathbf{i} + \frac{\partial}{\partial y}\left(\frac{1}{r}\right)\mathbf{j} + \frac{\partial}{\partial z}\left(\frac{1}{r}\right)\mathbf{k}.$$

Now

$$\frac{\partial}{\partial x}\left(\frac{1}{r}\right) = \frac{\partial}{\partial x}\left(\frac{1}{\sqrt{x^2 + y^2 + z^2}}\right) = -\frac{x}{(x^2 + y^2 + z^2)^{3/2}} = -\frac{x}{r^3},$$

and, similarly,

$$\frac{\partial}{\partial y}\left(\frac{1}{r}\right) = -\frac{y}{r^3}, \qquad \frac{\partial}{\partial z}\left(\frac{1}{r}\right) = -\frac{z}{r^3}.$$

Thus

$$\nabla\left(\frac{1}{r}\right) = -\frac{x}{r^3}\mathbf{i} - \frac{y}{r^3}\mathbf{j} - \frac{z}{r^3}\mathbf{k} = -\frac{1}{r^3}(x\mathbf{i} + y\mathbf{j} + z\mathbf{k}) = -\frac{\mathbf{r}}{r^3},$$

as required. Finally,

$$\left\|\nabla\left(\frac{1}{r}\right)\right\| = \left|-\frac{1}{r^3}\right|\|\mathbf{r}\| = \frac{r}{r^3} = \frac{1}{r^2} = \frac{1}{x^2+y^2+z^2}. \blacktriangle$$

In the next box we restate the chain rule from Section 15.3 in terms of gradients.

> ## The Chain Rule for Functions and Curves
>
> Let f be a function of two (three) variables, $\boldsymbol{\sigma}(t)$ a parametric curve in the plane (in space), and $h(t) = f(\boldsymbol{\sigma}(t))$ the composite function. Then
>
> $$h'(t) = \nabla f(\boldsymbol{\sigma}(t)) \cdot \boldsymbol{\sigma}'(t); \qquad \text{that is,} \quad \frac{d}{dt} f(\boldsymbol{\sigma}(t)) = \nabla f(\boldsymbol{\sigma}(t)) \cdot \boldsymbol{\sigma}'(t).$$

In this form, the chain rule looks more like it did for functions of one variable:

$$\frac{d}{dt}(f(g(t))) = f'(g(t))g'(t).$$

Example 5 Verify the chain rule for $u = f(x, y, z) = xy - z^2$ and $\boldsymbol{\sigma}(t) = (\sin t, \cos t, e^t)$.

Solution The gradient vector field of f is $y\mathbf{i} + x\mathbf{j} - 2z\mathbf{k}$; the velocity vector is given by $\boldsymbol{\sigma}'(t) = \cos t\,\mathbf{i} - \sin t\,\mathbf{j} + e^t\mathbf{k}$. By the chain rule,

$$\frac{du}{dt} = \nabla f \cdot \boldsymbol{\sigma}' = (y\mathbf{i} + x\mathbf{j} - 2z\mathbf{k}) \cdot (\cos t\,\mathbf{i} - \sin t\,\mathbf{j} + e^t\mathbf{k})$$

$$= y\cos t - x\sin t - 2ze^t = \cos^2 t - \sin^2 t - 2e^{2t}.$$

To verify this directly, we first compute the composition as $f(\boldsymbol{\sigma}(t)) = \sin t\cos t - e^{2t}$. Then by one-variable calculus, we find

$$\frac{d}{dt} f(\boldsymbol{\sigma}(t)) = -\sin^2 t + \cos^2 t - 2e^{2t}.$$

Thus the chain rule is verified in this case. ▲

Example 6 Suppose that f takes the value 2 at all points on a curve $\boldsymbol{\sigma}(t)$. What can you say about $\nabla f(\boldsymbol{\sigma}(t))$ and $\boldsymbol{\sigma}'(t)$?

Solution If $f(\boldsymbol{\sigma}(t))$ is always equal to 2, the derivative $(d/dt)f(\boldsymbol{\sigma}(t))$ is zero. By the chain rule, $0 = \nabla f(\boldsymbol{\sigma}(t)) \cdot \boldsymbol{\sigma}'(t)$, so the gradient vector $\nabla f(\boldsymbol{\sigma}(t))$ and the velocity vector $\boldsymbol{\sigma}'(t)$ are perpendicular at all points on the curve. ▲

Let $u = f(x, y, z)$ be a function (with continuous partial derivatives) and $\boldsymbol{\sigma}(t)$ a parametrized curve in space. The derivative with respect to t of the composite function $f(\boldsymbol{\sigma}(t))$ may be thought of as "the derivative of f *along the curve* $\boldsymbol{\sigma}(t)$." According to the chain rule, the value of this derivative at $t = t_0$ is $\nabla f(\boldsymbol{\sigma}(t_0)) \cdot \boldsymbol{\sigma}'(t_0)$. We may write this dot product as

$$\|\nabla f(\boldsymbol{\sigma}(t_0))\|\,\|\boldsymbol{\sigma}'(t_0)\|\cos\theta,$$

where θ is the angle between the gradient vector $\nabla f(\boldsymbol{\sigma}(t_0))$ and the velocity vector $\boldsymbol{\sigma}'(t_0)$ (Fig. 16.1.3). If we fix the function f and differentiate it along various curves through a given point $\mathbf{r}$ (here, as usual, we identify a point with the vector from the origin to the point), the derivative will be proportional to the *speed* $\|\boldsymbol{\sigma}'(t_0)\|$ and to the cosine of the angle between the gradient and

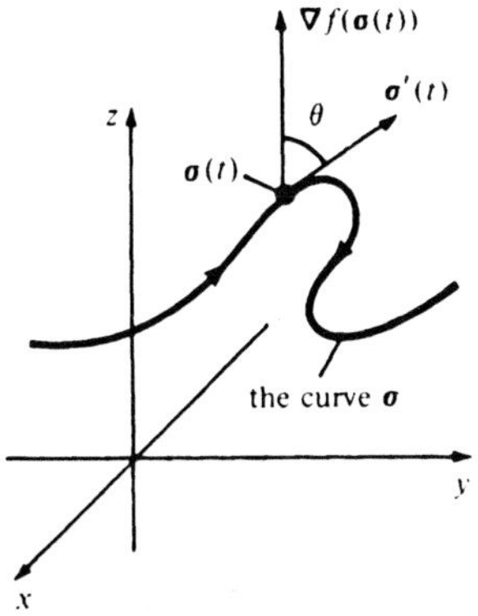

Figure 16.1.3. The derivative of f along the curve $\sigma(t)$ is

$\frac{d}{dt} f(\sigma(t)) = \nabla f(\sigma(t)) \cdot \sigma'(t) = \|\nabla f(\sigma(t))\| \, \|\sigma'(t)\| \cos\theta.$

velocity vectors. To describe how the derivative of f varies as we change the direction of the curve along which it is differentiated, we fix $\mathbf{r}$ and choose $\sigma(t) = \mathbf{r} + t\mathbf{d}$ for $\mathbf{d}$ a unit vector. (Note that since $\mathbf{d}$ is a unit vector, the speed of the curve $\sigma(t)$ is 1, so 1 unit of time corresponds to 1 unit of distance along the curve.)

We make the following definition: Let $f(x, y, z)$ be a function of three variables, $\mathbf{r}$ a point in its domain, and $\mathbf{d}$ a unit vector. Define the parametric curve $\sigma(t)$ by $\sigma(t) = \mathbf{r} + t\mathbf{d}$. The derivative $(d/dt) f(\sigma(t))|_{t=0}$ is called the *directional derivative* of f at $\mathbf{r}$ in the direction of $\mathbf{d}$.

Since $\sigma'(t) = \mathbf{d}$ and $\|\mathbf{d}\| = 1$, we see that if f has continuous partial derivatives, the directional derivative at $\mathbf{r}$ in the direction of $\mathbf{d}$ is

$$\nabla f(\mathbf{r}) \cdot \mathbf{d} = \|\nabla f(\mathbf{r})\| \cos\theta.$$

Notice that the directional derivatives in the directions of $\mathbf{i}$, $\mathbf{j}$, and $\mathbf{k}$ are just the partial derivatives. For instance, choosing $\mathbf{d} = \mathbf{i}$, $\nabla f \cdot \mathbf{i} = (f_x\mathbf{i} + f_y\mathbf{j} + f_z\mathbf{k}) \cdot \mathbf{i} = f_x$. Similarly, $\nabla f \cdot \mathbf{j} = f_y$ and $\nabla f \cdot \mathbf{k} = f_z$.

As we let $\mathbf{d}$ vary, the directional derivative takes its maximum value when $\cos\theta = 1$, that is, when $\mathbf{d}$ points in the direction of $\nabla f(\mathbf{r})$. The maximum value of the directional derivative is just the length $\|\nabla f(\mathbf{r})\|$.

The following box summarizes our findings.

Gradients and Directional Derivatives

The *directional derivative* at $\mathbf{r}$ in the direction of a unit vector $\mathbf{d}$ is the rate of change of f along the straight line through $\mathbf{r}$ in direction $\mathbf{d}$; i.e., along $\sigma(t) = \mathbf{r} + t\mathbf{d}$.

The directional derivative at $\mathbf{r}$ in the direction $\mathbf{d}$ equals $\nabla f(\mathbf{r}) \cdot \mathbf{d}$. It is greatest (for fixed $\mathbf{r}$) when $\mathbf{d}$ points in the direction of the gradient $\nabla f(\mathbf{r})$ and least when $\mathbf{d}$ points in the same direction as $-\nabla f(\mathbf{r})$.

Example 7 Compute the directional derivatives of the following functions at the indicated points in the given directions.

(a) $f(x, y) = x + 2x^2 - 3xy$; $(x_0, y_0) = (1, 1)$; $\mathbf{d} = (\frac{3}{5}, \frac{4}{5})$.
(b) $f(x, y) = \ln(\sqrt{x^2 + y^2})$; $(x_0, y_0) = (1, 0)$; $\mathbf{d} = (2\sqrt{5}/5, \sqrt{5}/5)$.
(c) $f(x, y, z) = xyz$; $(x_0, y_0, z_0) = (1, 1, 1)$; $\mathbf{d} = (1/\sqrt{2})\mathbf{i} + (1/\sqrt{2})\mathbf{k}$.
(d) $f(x, y, z) = e^x + yz$; $(x_0, y_0, z_0) = (1, 1, 1)$; $\mathbf{d} = (1/\sqrt{3})(\mathbf{i} - \mathbf{j} + \mathbf{k})$.

Solution (a) $\nabla f(x, y) = (1 + 4x - 3y, -3x)$. At $(1, 1)$ this is equal to $(2, -3)$. The directional derivative is $\nabla f(x_0, y_0) \cdot \mathbf{d} = (2, -3) \cdot (\frac{3}{5}, \frac{4}{5}) = -\frac{6}{5}$.

(b) $\nabla f(x, y) = (x/(x^2 + y^2), y/(x^2 + y^2))$, so $\nabla f(1, 0) = (1, 0)$. Thus, the directional derivative in direction $(2\sqrt{5}/5, \sqrt{5}/5)$ is $2\sqrt{5}/5$.

(c) $\nabla f(x, y, z) = (yz, xz, xy)$, which equals $(1, 1, 1)$ at $(1, 1, 1)$. For $\mathbf{d}$ equal to $(1/\sqrt{2}, 0, 1/\sqrt{2})$, the directional derivative is $1/\sqrt{2} + 0 + 1/\sqrt{2} = \sqrt{2}$.

(d) $\nabla f(x, y, z) = (e^x, z, y)$, which equals $(e, 1, 1)$ at $(1, 1, 1)$. For $\mathbf{d}$ equal to $(1/\sqrt{3}) \cdot (\mathbf{i} - \mathbf{j} + \mathbf{k})$, the directional derivative is $e(1/\sqrt{3}) + 1(-1\sqrt{3}) + 1(1/\sqrt{3}) = e/\sqrt{3}$. ▲

If one wishes to move from $\mathbf{r} = (x, y, z)$ in a direction in which f is *increasing* most quickly, one should move in the direction $\nabla f(\mathbf{r})$. This is because $\nabla f(\mathbf{r}) \cdot \mathbf{d} = \|\nabla f(\mathbf{r})\| \cos\theta$ is maximum when $\theta = 0$, i.e., $\cos\theta = 1$, so $\mathbf{d}$ is in the direction of $\nabla f(\mathbf{r})$. Likewise, $-\nabla f(\mathbf{r})$ is the direction in which f is *decreasing* at the fastest rate.

Example 8 Let $u = f(x, y, z) = (\sin xy)e^{-z^2}$. In what direction from $(1, \pi, 0)$ should one proceed to increase f most rapidly?

Solution We compute the gradient:

$$\nabla f = \frac{\partial u}{\partial x}\mathbf{i} + \frac{\partial u}{\partial y}\mathbf{j} + \frac{\partial u}{\partial z}\mathbf{k}$$

$$= y\cos(xy)e^{-z^2}\mathbf{i} + x\cos(xy)e^{-z^2}\mathbf{j} + (-2z\sin xy)e^{-z^2}\mathbf{k}.$$

At $(1, \pi, 0)$ this becomes

$$\pi\cos(\pi)\mathbf{i} + \cos(\pi)\mathbf{j} = -\pi\mathbf{i} - \mathbf{j}.$$

Thus one should proceed in the direction of the vector $-\pi\mathbf{i} - \mathbf{j}$. ▲

Example 9 Captain Astro is drifting in space near the sunny side of Mercury and notices that the hull of her ship is beginning to melt. The temperature in her vicinity is given by $T = e^{-x} + e^{-2y} + e^{3z}$. If she is at $(1, 1, 1)$, in what direction should she proceed in order to *cool* fastest?

Solution In order to cool the fastest, the captain should proceed in the direction in which T is decreasing the fastest; that is, in the direction $-\nabla T(1, 1, 1)$. However,

$$\nabla T = \frac{\partial T}{\partial x}\mathbf{i} + \frac{\partial T}{\partial y}\mathbf{j} + \frac{\partial T}{\partial z}\mathbf{k} = -e^{-x}\mathbf{i} - 2e^{-2y}\mathbf{j} + 3e^{3z}\mathbf{k}.$$

Thus,

$$-\nabla T(1, 1, 1) = e^{-1}\mathbf{i} + 2e^{-2}\mathbf{j} - 3e^{3}\mathbf{k}$$

is the direction required. ▲

Directional derivatives are also defined for functions of two variables. In this case, we have a geometric interpretation of the directional derivatives of $f(x, y)$ in terms of the graph $z = f(x, y)$. Given a point (x_0, y_0) in the plane and a unit vector $\mathbf{d} = a\mathbf{i} + b\mathbf{j}$, we can intersect the graph with the plane $\mathscr{P}$ in space which lies above the line through (x_0, y_0) with direction $\mathbf{d}$. (See Fig. 16.1.4.)

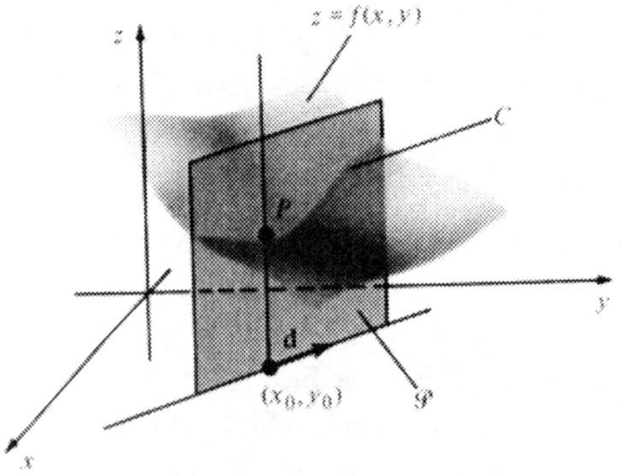

Figure 16.1.4. The slope at the point P of the curve C in the plane $\mathscr{P}$ is the directional derivative at (x_0, y_0) of f in the direction $\mathbf{d}$.

The result is a curve C which may be parametrized by the formula $(x, y, z) = (x_0 + at, y_0 + bt, f(x_0 + at, y_0 + bt))$. The tangent vector to this curve at $P = (x_0, y_0, f(x_0, y_0))$ is

$$\mathbf{v} = a\mathbf{i} + b\mathbf{j} + \frac{d}{dt}f(x_0 + at, y_0 + bt)|_{t=0}\mathbf{k}$$

$$= a\mathbf{i} + b\mathbf{j} + \left[af_x(x_0, y_0) + bf_y(x_0, y_0)\right]\mathbf{k}.$$

The slope of C in the plane $\mathscr{P}$ at P is the ratio of the vertical component

$af_x(x_0, y_0) + bf_y(x_0, y_0)$ of $\mathbf{v}$ to the length $\sqrt{a^2+b^2}$ of the horizontal component; but $\sqrt{a^2+b^2} = 1$, since $\mathbf{d} = a\mathbf{i} + b\mathbf{j}$ is a unit vector. Hence the slope of C in the plane $\mathscr{P}$ is just $af_x(x_0, y_0) + bf_y(x_0, y_0) = \mathbf{d} \cdot \nabla f(x_0, y_0)$, which is precisely the directional derivative of f at (x_0, y_0) in the direction of $\mathbf{d}$.

If we let the vector $\mathbf{d}$ rotate in the xy plane, then the plane $\mathscr{P}$ will rotate about the vertical line through (x_0, y_0) and the curve C will change. The slopes at P of all these curves are determined by the two numbers $f_x(x_0, y_0)$ and $f_y(x_0, y_0)$, and the tangent lines to all these curves lie in the tangent plane to $z = f(x, y)$ at P.

Example 10 Let $f(x, y) = x^2 - y^2$. In what direction from $(0, 1)$ should one proceed in order to increase f the fastest? Illustrate your answer with a sketch.

Solution The required direction is

$$\begin{aligned} \nabla f(0,1) &= \frac{\partial f}{\partial x}\mathbf{i} + \frac{\partial f}{\partial y}\mathbf{j} \quad \text{at } (0,1) \\ &= 2x\mathbf{i} - 2y\mathbf{j} \quad \text{at } (0,1) \\ &= -2\mathbf{j}. \end{aligned}$$

Thus one should head toward the origin along the y axis. The graph of f, sketched in Fig. 16.1.5, illustrates this. ▲

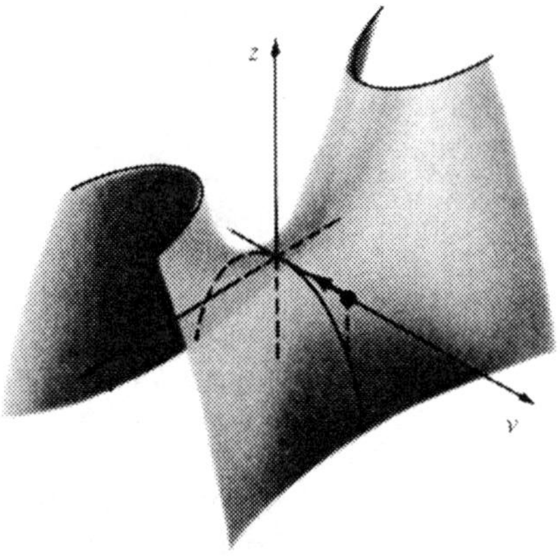

Figure 16.1.5. Starting from (0, 1), moving in along the y axis makes the graph rise the steepest.

Our final example concerns the "position vector" $\mathbf{r}$; see Example 4.

Example 11 Let $\mathbf{r} = x\mathbf{i} + y\mathbf{j} + z\mathbf{k}$ and $r = \|\mathbf{r}\|$. Compute ∇r. In what direction is r increasing the fastest? Interpret your answer geometrically.

Solution We know that $r = \sqrt{x^2+y^2+z^2}$, so

$$\nabla r = \frac{\partial r}{\partial x}\mathbf{i} + \frac{\partial r}{\partial y}\mathbf{j} + \frac{\partial r}{\partial z}\mathbf{k} = \frac{x}{r}\mathbf{i} + \frac{y}{r}\mathbf{j} + \frac{z}{r}\mathbf{k},$$

since $\partial r/\partial x = \frac{1}{2} \cdot 2x/\sqrt{x^2+y^2+z^2} = x/r$ and so forth. Thus

$$\nabla r = \frac{1}{r}(x\mathbf{i} + y\mathbf{j} + z\mathbf{k}) = \frac{\mathbf{r}}{r}.$$

Thus r is increasing fastest in the direction of $\mathbf{r}/r$, which is a unit vector pointing outward from the origin. This makes sense since r is the distance from the origin. ▲

Exercises for Section 16.1

Compute the gradients of the functions in Exercises 1–8.

1. $f(x, y, z) = \sqrt{x^2 + y^2 + z^2}$.
2. $f(x, y, z) = xy + yz + xz$.
3. $f(x, y, z) = x + y^2 + z^3$.
4. $f(x, y, z) = xy^2 + yz^2 + zx^2$.
5. $f(x, y) = \ln(\sqrt{x^2 + y^2})$.
6. $f(x, y) = (x^2 + y^2)\ln\sqrt{x^2 + y^2}$.
7. $f(x, y) = xe^{x^2+y^2}$.
8. $f(x, y) = x\exp(xy^3 + 3)$.

9. Sketch the gradient vector field of $f(x, y) = x^2/8 + y^2/12 + 6$.
10. Sketch the gradient vector field of $f(x, y) = x^2/8 - y^2/12$.
11. (a) Illustrate the vector field $\mathbf{\Phi}(x, y) = x\mathbf{j} - y\mathbf{i}$ by sketching $\frac{1}{3}\mathbf{\Phi}(x, y)$ instead. (b) Show that $\mathbf{\Phi}$ is not a gradient vector field.
12. (a) Sketch the vector field $\mathbf{\Phi}(x, y) = \frac{1}{4}\mathbf{i} + [1/(9 + x^2 + y^2)]\mathbf{j}$. (b) Explain why $\mathbf{\Phi}$ is or is not a gradient vector field.
13. Show that $\nabla(1/r^2) = -2\mathbf{r}/r^4$ $(r \neq 0)$.
14. Find $\nabla(1/r^3)$ $(r \neq 0)$.

Verify the chain rule for the functions and curves in Exercises 15–18.

15. $f(x, y, z) = xz + yz + xy$; $\boldsymbol{\sigma}(t) = (e^t, \cos t, \sin t)$.
16. $f(x, y, z) = e^{xyz}$; $\boldsymbol{\sigma}(t) = (6t, 3t^2, t^3)$.
17. $f(x, y, z) = \sqrt{x^2 + y^2 + z^2}$; $\boldsymbol{\sigma}(t) = (\sin t, \cos t, t)$.
18. $f(x, y, z) = xy + yz + xz$; $\boldsymbol{\sigma}(t) = (t, t, t)$.

19. Suppose that $f(\boldsymbol{\sigma}(t))$ is an increasing function of t. What can you say about the angle between the gradient ∇f and the velocity vector $\boldsymbol{\sigma}'$?
20. Suppose that $f(\boldsymbol{\sigma}(t))$ attains a minimum at the time t_0. What can you say about the angle between $\nabla f(\boldsymbol{\sigma}(t_0))$ and $\boldsymbol{\sigma}'(t_0)$?

In Exercises 21–28, compute the directional derivative of each function at the given point in the given direction.

21. $f(x, y) = x^2 + y^2 - 3xy^3$; $(x_0, y_0) = (1, 2)$; $\mathbf{d} = (1/2, \sqrt{3}/2)$.
22. $f(x, y) = e^x\cos y$; $(x_0, y_0) = (0, \pi/4)$; $\mathbf{d} = (\mathbf{i} + 3\mathbf{j})/\sqrt{10}$.
23. $f(x, y) = 17x^y$; $(x_0, y_0) = (1, 1)$; $\mathbf{d} = (\mathbf{i} + \mathbf{j})/\sqrt{2}$.
24. $f(x, y) = e^{x^2\cos y}$; $(x_0, y_0) = (1, \pi/2)$; $\mathbf{d} = (3\mathbf{i} + 4\mathbf{j})/5$.
25. $f(x, y, z) = x^2 - 2xy + 3z^2$; $(x_0, y_0, z_0) = (1, 1, 2)$; $\mathbf{d} = (\mathbf{i} + \mathbf{j} - \mathbf{k})/\sqrt{3}$.
26. $f(x, y, z) = e^{-(x^2+y^2+z^2)}$; $(x_0, y_0, z_0) = (1, 10, 100)$; $\mathbf{d} = (1, -1, -1)/\sqrt{3}$.
27. $f(x, y, z) = \sin(xyz)$; $(x_0, y_0, z_0) = (1, 1, \pi/4)$; $\mathbf{d} = (1/\sqrt{2}, 0, -1/\sqrt{2})$.
28. $f(x, y, z) = 1/(x^2 + y^2 + z^2)$; $(x_0, y_0, z_0) = (2, 3, 1)$; $\mathbf{d} = (\mathbf{j} - 2\mathbf{k} + \mathbf{i})/\sqrt{6}$.

In Exercises 29–32 determine the direction in which each of the functions is increasing fastest at (1, 1).

29. $f(x, y) = x^2 + 2y^2$
30. $g(x, y) = x^2 - 2y^2$
31. $h(x, y) = e^x\sin y$
32. $l(x, y) = e^x\sin y - e^{-x}\cos y$

33. Captain Astro is once again in trouble near the sunny side of Mercury. She is at location (1, 1, 1), and the temperature of the ship's hull when she is at location (x, y, z) will be given by $T(x, y, z) = e^{-x^2-2y^2-3z^2}$, where x, y and z are measured in meters.
    (a) In what direction should she proceed in order to decrease the temperature most rapidly?
    (b) If the ship travels at e^8 meters per second, how fast will be the temperature decrease if she proceeds in that direction?
    (c) Unfortunately, the metal of the hull will crack if cooled at a rate greater than $\sqrt{14}\,e^2$ degrees per second. Describe the set of possible directions in which she may proceed to bring the temperature down at no more than that rate.
34. Suppose that a mountain has the shape of an elliptic paraboloid $z = c - ax^2 - by^2$, where a, b, and c are positive constants, x and y are the east-west and north-south map coordinates, and z is the altitude above sea level (x, y, and z are all measured in meters). At the point (1, 1), in what direction is the altitude increasing most rapidly? If a marble were released at (1, 1), in what direction would it begin to roll?
35. An engineer wishes to build a railroad up the mountain of Exercise 34. Straight up the mountain is much too steep for the power of the engines. At the point (1, 1), in what directions may the track be laid so that it will be climbing with a 3% grade—that is, an angle whose tangent is 0.03. (There are two possibilities.) Make a sketch of the situation indicating the two possible directions for a 3% grade at (1, 1).
36. The height h of the Hawaiian volcano Mauna Loa is (roughly) described by the function $h(x, y) = 2.59 - 0.00024y^2 - 0.00065x^2$, where h is the height above sea level in miles and x and y measure east-west and north-south distances in miles from the top of the mountain.

    At $(x, y) = (-2, -4)$:
    (a) How fast is the height increasing in the direction (1, 1) (that is, northeastward)? Express your answer in miles of height per mile of horizontal distance travelled.
    (b) In what direction is the steepest upward path?

(c) In what direction is the steepest downward path?
(d) In what direction(s) is the path level?
(e) If you proceed south, are you ascending or descending? At what rate?
(f) If you move northwest, are you ascending or descending? At what rate?
(g) In what direction(s) may you proceed in order to be climbing with a grade of 3%?

37. In what direction from $(1,0)$ does the function $f(x, y) = x^2 - y^2$ increase the fastest? Illustrate with a sketch.
38. In what direction from $(-1,0)$ does the function $f(x, y) = x^2 - y^2$ increase fastest? Sketch.
39. In what direction is the length of $\mathbf{r} + \mathbf{j}$ increasing fastest at the point $(1,0,1)$? ($\mathbf{r} = x\mathbf{i} + y\mathbf{j} + z\mathbf{k}$).
40. In what direction should you travel from the point $(2,4,3)$ to make the length of $\mathbf{r} + \mathbf{k}$ decrease as fast as possible?
41. Suppose that f and g are real-valued functions (with continuous partial derivatives). Show that:
(a) $\nabla f = \mathbf{0}$ if f is constant;
(b) $\nabla(f + g) = \nabla f + \nabla g$;
(c) $\nabla(cf) = c\nabla f$ if c is a constant;
(d) $\nabla(fg) = f\nabla g + g\nabla f$;
(e) $\nabla(f/g) = (g\nabla f - f\nabla g)/g^2$ at points where $g \neq 0$.
42. What rate of change does $\nabla f(x, y, z) \cdot (-\mathbf{j})$ represent?
43. (a) In what direction is the directional derivative of $f(x, y) = (x^2 - y^2)/(x^2 + y^2)$ at $(1,1)$ equal to zero?
(b) How about at an arbitrary point (x_0, y_0) in the first quadrant?
(c) Describe the level curves of f. In particular, discuss them in terms of the result of (b).
44. Suppose that $f(x, y)$ is given (and has continuous partial derivatives). At $(1,1)$ the directional derivative in the direction toward $(2,4)$ is 2 and in the direction toward $(2,2)$ it is 3. Find the gradient of f at $(1,1)$ and the directional derivative there in the direction toward $(2,3)$.
45. A function $f(x, y)$ has, at the point $(1,3)$, directional derivatives of $+2$ in the direction toward $(2,3)$ and -2 in the direction toward $(1,4)$. Determine the gradient vector at $(1,3)$ and compute the directional derivative in the direction toward $(3,6)$.
46. In electrostatistics, the force $\mathbf{P}$ of attraction between two particles of opposite charge is given by $\mathbf{P} = k(\mathbf{r}/\|\mathbf{r}\|^3)$ (*Coulomb's law*), where k is a constant and $\mathbf{r} = x\mathbf{i} + y\mathbf{j} + z\mathbf{k}$. Show that $\mathbf{P}$ is the gradient of $f = -k/\|\mathbf{r}\|$.
47. The potential V due to two infinite parallel filaments of charge of linear densities λ and $-\lambda$ is $V = (\lambda/2\pi\varepsilon_0)\ln(r_2/r_1)$, where $r_1^2 = (x - x_0)^2 + y^2$ and $r_2^2 = (x + x_0)^2 + y^2$. We think of the filaments as being in the z direction, passing through the xy plane at $(-x_0, 0)$ and $(x_0, 0)$.
(a) Find $\nabla V(x, y)$, using the chain rule.
(b) Verify the *flux law* $\partial^2 V/\partial x^2 + \partial^2 V/\partial y^2 = 0$.
48. For each of the following find the maximum and minimum values attained by the function f along the curve $\boldsymbol{\sigma}(t)$:
(a) $f(x, y) = xy$; $\boldsymbol{\sigma}(t) = (\cos t, \sin t)$; $0 \leqslant t \leqslant 2\pi$.
(b) $f(x, y) = x^2 + y^2$; $\boldsymbol{\sigma}(t) = (\cos t, 2\sin t)$; $0 \leqslant t \leqslant 2\pi$.

★49. What conditions on the function $f(x, y)$ hold if the vector field $\mathbf{k} \times \nabla f$ is a gradient vector field?
★50. (a) Let F be a function of one variable and f a function of two variables. Show that the gradient vector of $g(x, y) = F(f(x, y))$ is parallel to the gradient vector of $f(x, y)$.
(b) Let $f(x, y)$ and $g(x, y)$ be functions such that $\nabla f = \lambda \nabla g$ for some function $\lambda(x, y)$. What is the relation between the level curves of f and g? Explain why there might be a function F such that $g(x, y) = F(f(x, y))$.

16.2 Gradients, Level Surfaces, and Implicit Differentiation

The gradient of a function of three variables is perpendicular to the surfaces on which the function is constant.

Recall that the tangent plane to a graph $z = f(x, y)$ was defined as the graph of the linear approximation to f. We found (Section 15.3) that the tangent plane at a point could also be characterized as the plane containing the tangent *lines* to all curves on the surface through the given point. For a general surface, we take this as a definition.

Definition: Tangent Plane to a Surface

Let S be a surface in space, $\mathbf{r}_0$ a point of S. If there is a plane which contains the tangent lines at $\mathbf{r}_0$ to all curves through $\mathbf{r}_0$ in S, then this plane is called *the tangent plane to S at* $\mathbf{r}_0$. A normal to the tangent plane is sometimes said to be *perpendicular to S*.

The next box tells how to find the tangent plane to a level surface.

Gradients and Tangent Planes

Let $\mathbf{r}_0$ lie on the level surface S defined by $f(x, y, z) = c$, and suppose that $\nabla f(\mathbf{r}_0) \neq \mathbf{0}$. Then $\nabla f(\mathbf{r}_0)$ is normal to the tangent plane to S at $\mathbf{r}_0$. (See Fig. 16.2.1.)

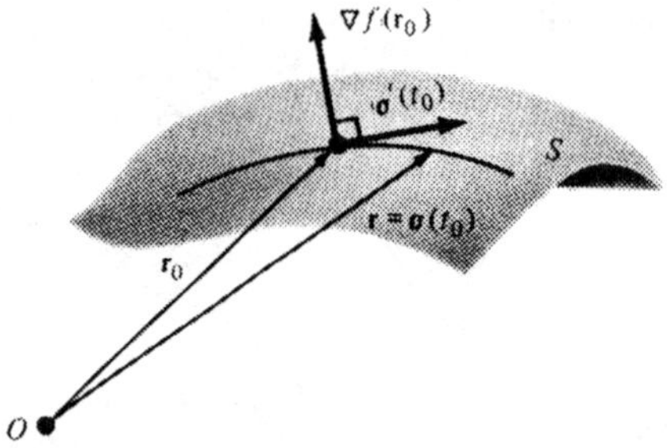

Figure 16.2.1. The gradient of f at $\mathbf{r}_0$ is perpendicular to the tangent vector of any curve in the level surface.

To prove this assertion, first observe that $f(\boldsymbol{\sigma}(t)) = c$ if the curve $\boldsymbol{\sigma}(t)$ lies in S. Hence

$$\frac{d}{dt} f(\boldsymbol{\sigma}(t)) = 0.$$

By the chain rule in terms of gradients, this gives

$$\nabla f(\boldsymbol{\sigma}(t)) \cdot \boldsymbol{\sigma}'(t) = 0.$$

Setting $t = t_0$, we have $\nabla f(\mathbf{r}_0) \cdot \boldsymbol{\sigma}'(t_0) = 0$ for every curve $\boldsymbol{\sigma}$ in S, so $\nabla f(\mathbf{r}_0)$ is normal to the tangent plane. (We required $\nabla f(\mathbf{r}_0) \neq \mathbf{0}$ so there would be a well-defined plane orthogonal to $\nabla f(\mathbf{r}_0)$.)

Example 1 Let $u = f(x, y, z) = x^2 + y^2 - z^2$. Find $\nabla f(0, 0, 1)$. Plot this on the level surface $f(x, y, z) = -1$.

Solution We have

$$\nabla f = \frac{\partial u}{\partial x}\mathbf{i} + \frac{\partial u}{\partial y}\mathbf{j} + \frac{\partial u}{\partial z}\mathbf{k} = 2x\mathbf{i} + 2y\mathbf{j} - 2z\mathbf{k}.$$

At $(0, 0, 1)$, $\nabla f(0, 0, 1) = -2\mathbf{k}$.

The level surface $x^2 + y^2 - z^2 = -1$ is a hyperboloid of two sheets (Section 14.4). If we plot $\nabla f(0, 0, 1)$ on it (Fig. 16.2.2), we see that it is indeed perpendicular to the surface. ▲

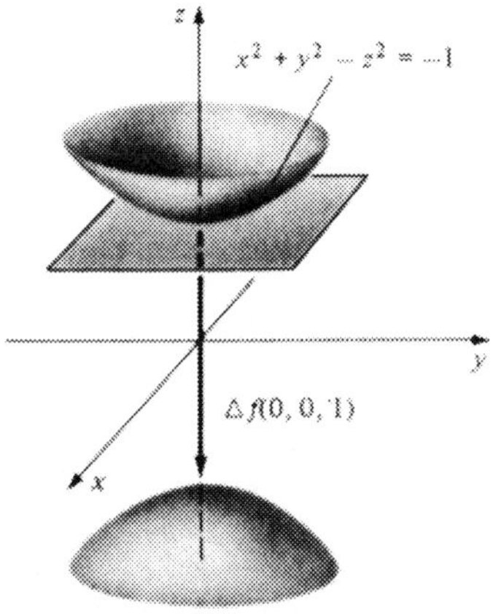

Figure 16.2.2. $\nabla f(0,0,1)$ is perpendicular to the surface.

Example 2 Find a unit normal to the surface $\sin(xy) = e^z$ at $(1, \pi/2, 0)$.

Solution Let $f(x, y, z) = \sin(xy) - e^z$, so the surface is $f(x, y, z) = 0$. A normal is $\nabla f = y\cos(xy)\mathbf{i} + x\cos(xy)\mathbf{j} - e^z\mathbf{k}$. At $(1, \pi/2, 0)$, we get $-\mathbf{k}$. Thus $-\mathbf{k}$ (or $\mathbf{k}$) is the required unit normal. (It already has length 1, so there is no need to normalize.) ▲

Example 3 The gravitational force exerted on a mass m at (x, y, z) by a mass M at the origin is, by Newton's law of gravitation,

$$\mathbf{F} = -\frac{GMm}{r^3}\mathbf{r}, \quad \text{where} \quad \mathbf{r} = x\mathbf{i} + y\mathbf{j} + z\mathbf{k} \quad \text{and} \quad r = \|\mathbf{r}\|.$$

Write $\mathbf{F}$ as the negative gradient of a function V (called the gravitational potential) and verify that F is orthogonal to the level surfaces of V.

Solution By Example 4, Section 16.1, $\nabla(1/r) = -(\mathbf{r}/r^3)$. Therefore we can choose $V = -GMm/r$ to give $\mathbf{F} = -\nabla V$. The vector $\mathbf{F}$ points toward the origin. The level surfaces of V are $1/r = c$—that is, $r = 1/c$, a sphere. Therefore, $\mathbf{F}$ is orthogonal to these surfaces. ▲

The gradient enables us to compute the equation of the tangent plane to the level surface S at $\mathbf{r}_0$. Indeed, $\nabla f(\mathbf{r}_0)$ will be a normal to this plane, which passes through $\mathbf{r}_0$. Therefore its equation can be read off immediately. (See Section 13.4.)

Example 4 Compute the equation of the plane tangent to the surface $3xy + z^2 = 4$ at $(1, 1, 1)$.

Solution Here $f(x, y, z) = 3xy + z^2$ and $\nabla f = (3y, 3x, 2z)$, which at $(1, 1, 1)$ is the vector $3\mathbf{i} + 3\mathbf{j} + 2\mathbf{k}$. Thus the tangent plane is

$$3(x - 1) + 3(y - 1) + 2(z - 1) = 0 \quad \text{or} \quad 3x + 3y + 2z = 8. \ ▲$$

Example 5
(a) Find a unit normal to the ellipsoid $x^2 + 2y^2 + 3z^2 = 10$ at each of the points $(\sqrt{10}, 0, 0)$, $(-\sqrt{10}, 0, 0)$, $(1, 0, \sqrt{3})$, and $(-1, 0, -\sqrt{3})$.
(b) Do the vectors you have found point to the inside or outside of the ellipsoid?
(c) Give equations for the tangent planes to the surface at the two points of the surface with $x_0 = y_0 = 1$.

Solution (a) Letting $f(x, y, z) = x^2 + 2y^2 + 3z^2 = 10$, we find $\nabla f(x, y, z) = (2x, 4y, 6z)$. At $(\sqrt{10}, 0, 0)$, a unit normal to the ellipsoid is

$$\frac{\nabla f(\sqrt{10}, 0, 0)}{\|\nabla f(\sqrt{10}, 0, 0)\|} = \frac{(2\sqrt{10}, 0, 0)}{\left((2\sqrt{10})^2 + 0^2 + 0^2\right)^{1/2}} = (1, 0, 0).$$

At $(-\sqrt{10}, 0, 0)$, it is $(-1, 0, 0)$. At $(1, 0, \sqrt{3})$, it is

$$\frac{\nabla f(1, 0, \sqrt{3})}{\|\nabla f(1, 0, \sqrt{3})\|} = \left(\frac{1}{\sqrt{28}}, 0, \frac{3\sqrt{3}}{\sqrt{28}}\right),$$

and at $(-1, 0, -\sqrt{3})$ it is $(-1/\sqrt{28}, 0, -3\sqrt{3}/\sqrt{28})$.

(b) The vectors are pointing to the outside of the ellipsoid.

(c) The two points are $(1, 1, \sqrt{7/3})$, and $(1, 1, -\sqrt{7/3})$. Evaluating the gradient, $\nabla f(1, 1, \sqrt{7/3}) = (2, 4, 2\sqrt{21})$ and $\nabla f(1, 1, -\sqrt{7/3}) = (2, 4, -2\sqrt{21})$, so the tangent planes to the surface at the points $(1, 1, \sqrt{7/3})$ and $(1, 1, -\sqrt{7/3})$ are given by $2(x-1) + 4(y-1) + 2\sqrt{21}(z - \sqrt{7/3}) = 0$ and $2(x-1) + 4(y-1) - 2\sqrt{21}(z + \sqrt{7/3}) = 0$, respectively. ▲

There is also a connection between gradients and tangents for functions of two variables: the tangent line to a level *curve* of a function $f(x, y)$ is perpendicular to the gradient of f at each point. Combining this fact with the box on p. 801, we see that the direction in which the function f is increasing or decreasing most rapidly is perpendicular to the level curves of f. For example, to get down most directly from the top of a hill, one should proceed in a direction perpendicular to the level contours. (See Fig. 16.2.3.)

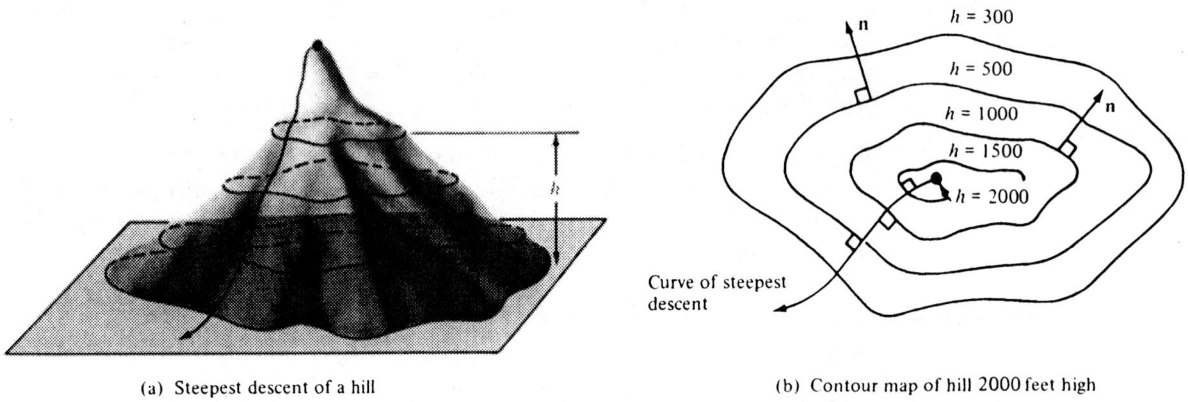

Figure 16.2.3. The curve of steepest descent is perpendicular to the level curves. (a) Steepest descent of a hill. (b) Contour map of hill 2000 feet high.

Gradients, Level Surfaces, and Level Curves

The normal to the tangent plane at $\mathbf{r}_0 = (x_0, y_0, z_0)$ of the level surface $f(x, y, z) = c$ is $\nabla f(\mathbf{r}_0)$. The equation of the plane is

$$f_x(x_0, y_0, z_0)(x - x_0) + f_y(x_0, y_0, z_0)(y - y_0) + f_z(x_0, y_0, z_0)(z - z_0) = 0.$$

The equation of the tangent line at (x_0, y_0) to the curve $f(x, y) = c$ is

$$f_x(x_0, y_0)(x - x_0) + f_y(x_0, y_0)(y - y_0) = 0.$$

Example 6 Find the equation of the tangent line to $xy = 6$ at $x = 1, y = 6$.

Solution With $f(x, y) = xy$, we have $f_x(x, y) = y$ and $f_y(x, y) = x$. Then $f_x(1, 6) = 6$ and $f_y(1, 6) = 1$, so from the preceding box, the equation of the tangent line through $(1, 6)$ is

$$6(x - 1) + 1(y - 6) = 0 \quad \text{or} \quad y = -6x + 12. \blacktriangle$$

In the next example we check that the equation given in Section 15.2 for the tangent plane to a graph is consistent with that given here.

Example 7 Let $z = g(x, y)$. The graph of g may be defined as the level surface $f(x, y, z) = 0$, where $f(x, y, z) = z - g(x, y)$. Compute the gradient of f and verify that it is perpendicular to the tangent plane of the graph $z = g(x, y)$ as defined in Section 15.2.

Solution With $f(x, y, z) = z - g(x, y)$,

$$\begin{aligned}\nabla f(x, y, z) &= f_x(x, y, z)\mathbf{i} + f_y(x, y, z)\mathbf{j} + f_z(x, y, z)\mathbf{k} \\ &= -g_x(x, y)\mathbf{i} - g_y(x, y)\mathbf{j} + \mathbf{k}.\end{aligned}$$

This is exactly the normal to the tangent plane at (x, y) to the graph of g. $\blacktriangle$

Many functions of several variables are built by combining functions of one variable. We actually found partial derivatives of such functions in our earlier work on implicit differentiation and related rates. For instance, suppose that $y = f(x)$ and that x and y satisfy the relation

$$x^3 + 8x \sin y = 0.$$

Then differentiating with respect to x, using the chain rule for functions of *one* variable, gives

$$3x^2 + 8 \sin y + 8x \cos y \frac{dy}{dx} = 0.$$

which we can solve for dy/dx to obtain

$$\frac{dy}{dx} = -\frac{3x^2 + 8 \sin y}{8x \cos y}.$$

From the point of view of multivariable calculus, we may say that the graph $y = f(x)$ lies on the level curve $F(x, y) = 0$, where

$$F(x, y) = x^3 + 8x \sin y.$$

A normal vector to this curve at (x, y) is

$$\nabla F = \frac{\partial F}{\partial x}\mathbf{i} + \frac{\partial F}{\partial y}\mathbf{j} = (3x^2 + 8 \sin y)\mathbf{i} + (8x \cos y)\mathbf{j},$$

so a tangent vector is given by any vector perpendicular to ∇F, such as

$$(-8x \cos y)\mathbf{i} + (3x^2 + 8 \sin y)\mathbf{j}.$$

Thus dy/dx, the slope of the tangent line, is

$$\frac{3x^2 + 8 \sin y}{-8x \cos y}.$$

The general procedure is indicated in the following box.

Implicit Differentiation and Partial Derivatives

If $y = f(x)$ is a function satisfying the relation $z = F(x, y) = 0$, then

$$\frac{dy}{dx} = -\frac{\partial z/\partial x}{\partial z/\partial y}, \tag{1}$$

i.e.

$$f'(x) = -\frac{F_x(x, f(x))}{F_y(x, f(x))}. \tag{1'}$$

Indeed, differentiating $F(x, y) = 0$ with respect to x using the chain rule gives

$$\frac{\partial F}{\partial x}\frac{dx}{dx} + \frac{\partial F}{\partial y}\frac{dy}{dx} = 0,$$

i.e.,

$$\frac{\partial F}{\partial x} + \frac{\partial F}{\partial y}\frac{dy}{dx} = 0.$$

Solving for dy/dx gives the result in the box. Notice that in (1) it is *incorrect* to "cancel the ∂z's," because the minus sign would be left.

Example 8 Suppose that y is defined implicitly in terms of x by $e^{x-y} + x^2 - y = 1$. Find dy/dx at $x = 0$, $y = 0$ using formula (1).

Solution Here $z = F(x, y) = e^{x-y} + x^2 - y - 1$, so

$$\frac{\partial z}{\partial x} = e^{x-y} + 2x \quad \text{and} \quad \left.\frac{\partial z}{\partial x}\right|_{\substack{x=0\\y=0}} = 1.$$

Likewise

$$\frac{\partial z}{\partial y} = -e^{x-y} - 1 \quad \text{and} \quad \left.\frac{\partial z}{\partial y}\right|_{\substack{x=0\\y=0}} = -2.$$

Therefore

$$-\frac{\partial z/\partial x}{\partial z/\partial y} = 1/2.$$

and so, by (1), $dy/dx = 1/2$. ▲

Formula (1) makes sense as long as $\partial z/\partial y \neq 0$. In fact there is a result called the *implicit function theorem*[1] which guarantees that $F(x, y) = 0$ does indeed define y as a function of x, provided that $\partial z/\partial y \neq 0$. The values of x and y may have to be restricted, as we found when studying implicit differentiation in Section 2.3 (see Figure 2.3.1).

Example 9 Discuss what happens to y as a function of x if $\partial z/\partial y = 0$ in (1) for the example $x - y^3 = 0$.

Solution The equation $z = F(x, y) = x - y^3 = 0$ implicitly defines the function $y = f(x) = \sqrt[3]{x}$. We have $\partial z/\partial x = 1$ and $\partial z/\partial y = -3y^2$; so $\partial z/\partial y$ vanishes

[1] For a proof based on the mean value and intermediate value theorems, see J. Marsden and A. Tromba, *Vector Calculus*, Second Edition, Freeman (1981), Section 4.4.

when $y = 0$; this is just the point on the graph $y = \sqrt[3]{x}$ where the cube-root function is not differentiable and the tangent line becomes vertical. ▲

In related rate problems, we have a parametric curve $(x, y) = (g(t), h(t))$ which lies on a level curve $F(x, y) = 0$. Differentiating with respect to t by the chain rule, we get

$$0 = F_x(x, y)\frac{dx}{dt} + F_y(x, y)\frac{dy}{dt} \tag{2}$$

which is a relation between the rates dx/dt and dy/dt. Such relations were obtained in Section 2.5 using one variable calculus.

Example 10 Suppose that $x = g(t)$ and $y = h(t)$ satisfy the relation $x^2 - y^2 = xy$. Find a relation between dx/dt and dy/dt:
(a) by one-variable calculus;
(b) by formula (2).

Solution (a) Differentiating the relation $x^2 - y^2 = xy$ with respect to t by one-variable calculus, we obtain $2x(dx/dt) - 2y(dy/dt) = y(dx/dt) + x(dy/dt)$ or, equivalently, $(2x - y)(dx/dt) - (2y + x)(dy/dt) = 0$.
(b) To apply formula (2), we set $F(x, y) = x^2 - y^2 - xy$. Then $F_x(x, y) = 2x - y$ and $F_y(x, y) = -(2y + x)$, so (2) gives the same relation between dx/dt and dy/dt: $(2x - y)(dx/dt) - (2y + x)(dy/dt) = 0$. ▲

Example 11 Suppose that $x = g(t)$ and $y = h(t)$ satisfy the relation $x^y = 2$. Find a relation between dx/dt and dy/dt.

Solution Let $z = F(x, y) = x^y - 2$. Then $\partial z/\partial x = yx^{y-1}$ and $\partial z/\partial y = x^y \ln x$, so the relation is

$$yx^{y-1}\frac{dx}{dt} + x^y \ln x\frac{dy}{dt} = 0.$$

Using the fact that $x^y = 2$, we can simplify this to

$$\frac{y}{x}\frac{dx}{dt} + \ln x\frac{dy}{dt} = 0. \ ▲$$

Exercises for Section 16.2

In Exercises 1–4, find $\nabla f(0, 0, 1)$ and plot it on the level surface $f(x, y, z) = c$ passing through $(0, 0, 1)$.

1. $f(x, y, z) = x^2 + y^2 + z^2$
2. $f(x, y, z) = z - x^2 - y^2$
3. $f(x, y, z) = z - x + y$
4. $f(x, y, z) = z^2 - x - y$

In Exercises 5–8, find a unit normal to the given surface at the given point.

5. $xyz = 8$; $(1, 1, 8)$.
6. $x^2y^2 + y - z + 1 = 0$ at $(0, 0, 1)$.
7. $\cos(xy) = e^z - 2$ at $(1, \pi, 0)$.
8. $e^{xyz} = e$ at $(1, 1, 1)$.

9. Coulomb's law states that the electric force on a charge q at (x, y, z) produced by a charge Q at the origin is $\mathbf{F} = Qq\mathbf{r}/r^3$. Find V so that $\mathbf{F} = -\nabla V$ and verify that $\mathbf{F}$ is orthogonal to the level surfaces of V.

10. Joe Perverse has invented a new law of gravitation. In this theory, the force exerted on a mass m at (x, y, z) by a mass M at the origin is $\mathbf{F} = -JMm\mathbf{r}/r^5$, where J is Joe's constant. Find V such that $\mathbf{F} = -\nabla V$ and verify that $\mathbf{F}$ is orthogonal to the level surfaces of V.

In Exercises 11–16, find the equation for the tangent plane to each surface at the indicated point.

11. $x^2 + 2y^2 + 3z^2 = 10$; $(1, \sqrt{3}, 1)$.
12. $xyz^2 = 1$; $(1, 1, 1)$.
13. $x^2 + 2y^2 + 3xz = 10$; $(1, 2, \frac{1}{3})$.
14. $y^2 - x^2 = 3$; $(1, 2, 8)$.
15. $xyz = 1$; $(1, 1, 1)$.
16. $xy/z = 1$; $(1, 1, 1)$.

Find the equation for the tangent line to each curve at the indicated point in Exercises 17–20.

17. $x^2 + 2y^2 = 3$; $(1, 1)$.

18. $xy = 17$; $(x_0, 17/x_0)$.
19. $\cos(x + y) = 1/2$; $x = \pi/2, y = 0$.
20. $e^{xy} = 2$; $(1, \ln 2)$.

Find the equation of the line normal to the given surface at the given point in Exercises 21–24.

21. $e^{-(x^2+y^2+z^2)} = e^{-3}$; $(1, 1, 1)$
22. $2x^2 + 3y^2 + z^2 = 9$; $(1, 1, 2)$
23. $x/yz = 1$; $(1, 1, 1)$
24. $xyz^2 = 4$; $(1, 1, 2)$

In Exercises 25–30, suppose that y is defined implicitly in terms of x by the given equation. Find dy/dx using formula (1).

25. $x^2 + 2y^2 = 3$
26. $x^2 - y^2 = 7$
27. $x/y = 10$
28. $y - \sin x^3 + x^2 - y^2 = 1$
29. $x^3 - \sin y + y^4 = 4$
30. $e^{x+y^2} + y^3 = 0$

In Exercises 31–34, find dy/dx at the indicated point using formula (1).

31. $3x^2 + y^2 - e^x = 0$; $x = 0, y = 1$.
32. $x^2 + y^4 = 1$; $x = 1, y = 1$.
33. $\cos(x + y) = x + 1/2$; $x = 0, y = \pi/3$.
34. $\cos(xy) = 1/2$; $x = 1, y = \pi/3$.

In Exercises 35–38, discuss what happens to y as a function of x if $\partial z/\partial y = 0$ in (1) for the given equation.

35. $x - y^2 = 0$
36. $x - \cos y = 0$
37. $x - y^5 = 0$
38. $x - \sin y = 0$

In Exercises 39–42, suppose that x and y are functions of t satisfying the given relation. Find a relation between dx/dt and dy/dt using formula (2).

39. $x \ln y = 1$
40. $\sin(xy) + \cos(xy) = 1$
41. $x^4 + y^4 = 1$
42. $x^2 + 3y^2 = 2$

43. (a) Derive a formula like (1) for dx/dy when x and y are related by $F(x, y) = 0$. (b) Use your result in (a) to find dx/dy for the functions in Exercises 29 and 30.
44. Let y be a function of x satisfying $F(x, y, x + y) = 0$, where $F(x, y, z)$ is a given function. Find a formula for dy/dx.

Suppose that $x = g(t)$ and $y = h(t)$ satisfy the equations in Exercises 45 and 46. Relate dx/dt and dy/dt.

45. $\ln(x \cos y) = x$
46. $\cos(x - 2y^2 + y^3) = y$

47. (a) Find the plane which is tangent to the surface $z = x^2 + y^2$ at the point $(1, -2, 5)$.
    ★(b) Letting $f(x, y) = x^2 + y^2$, define the "slope" of the tangent plane relative to the xy plane and show that it equals $\|\nabla f(1, -2)\|$.
48. (a) Show that the curve $x^2 - y^2 = c$, for *any* value of c, satisfies the differential equation $dy/dx = x/y$.
    (b) Draw in a few of the curves $x^2 - y^2 = c$, say for $c = \pm 1$. At several points (x, y) along each of these curves, draw a short segment of slope x/y; check that these segments appear to be tangent to the curve. What happens when $y = 0$? What happens when $c = 0$?
49. Suppose that a particle is ejected from the surface $x^2 + y^2 - z^2 = -1$ at the point $(1, 1, \sqrt{3})$ in a direction normal to the surface at time $t = 0$ with a speed of 10 units per second. When and where does it cross the xy plane?

★50. Let V be a function defined on a domain in space. The force field associated with V is $\mathbf{F} = \mathbf{\Phi}(x, y, z) = -\nabla V(x, y, z)$; we call V the *potential* of $\mathbf{\Phi}$. Let a point with mass m move on a parametric curve $\boldsymbol{\sigma}(t)$ and satisfy Newton's second law $m\mathbf{a} = \mathbf{F}$, where $\mathbf{a}$ is the acceleration of the curve. Use the chain rule to prove the law of conservation of energy: $E = \frac{1}{2}m\|\boldsymbol{\sigma}'(t)\|^2 + V[\boldsymbol{\sigma}(t)]$ is constant, where $\boldsymbol{\sigma}(t)$ is the position vector of the curve.

★51. The level surfaces of a potential function V are called *equipotential surfaces*.
    (a) What is the relation between the force vector and the equipotential surfaces?
    (b) Explain why "sea level" is approximately an equipotential surface for the earth's gravitational field. What spoils the approximation?

16.3 Maxima and Minima

First and second derivative tests are developed for locating maximum and minimum points for functions of two variables.

In studying maxima and minima for functions of one variable, we found that the basic tests involved the vanishing of the first derivative and the sign of the second derivative. In this section we develop tests involving first and second partial derivatives for locating maxima and minima of functions of two variables.

The definitions of maxima and minima for functions of two variables are similar to those in the one-variable case, except that we use disks instead o intervals. Recall that the disk of radius r about (x_0, y_0) consists of all (x, y) such that the distance $\sqrt{(x - x_0)^2 + (y - y_0)^2}$ is less than r. (See Fig. 15.1.2.)

Definition of Maxima and Minima

Let $f(x, y)$ be a function of two variables. We say that (x_0, y_0) is a *local minimum point* for f if there is a disk (of positive radius) about (x_0, y_0) such that $f(x, y) \geqslant f(x_0, y_0)$ for all (x, y) in the disk.

Similarly, if $f(x, y) \leqslant f(x_0, y_0)$ for all (x, y) in some disk (of positive radius) about (x_0, y_0), we call (x_0, y_0) a *local maximum point* for f.

A point which is either a local maximum or minimum point is called a *local extremum*.

We may also define *global* maximum and minimum points to be those at which a function attains the greatest and least values for all points in its domain.

Example 1 Refer to Fig. 16.3.1, a computer-drawn graph of $z = 2(x^2 + y^2)e^{-x^2-y^2}$. Where are the maximum and minimum points?

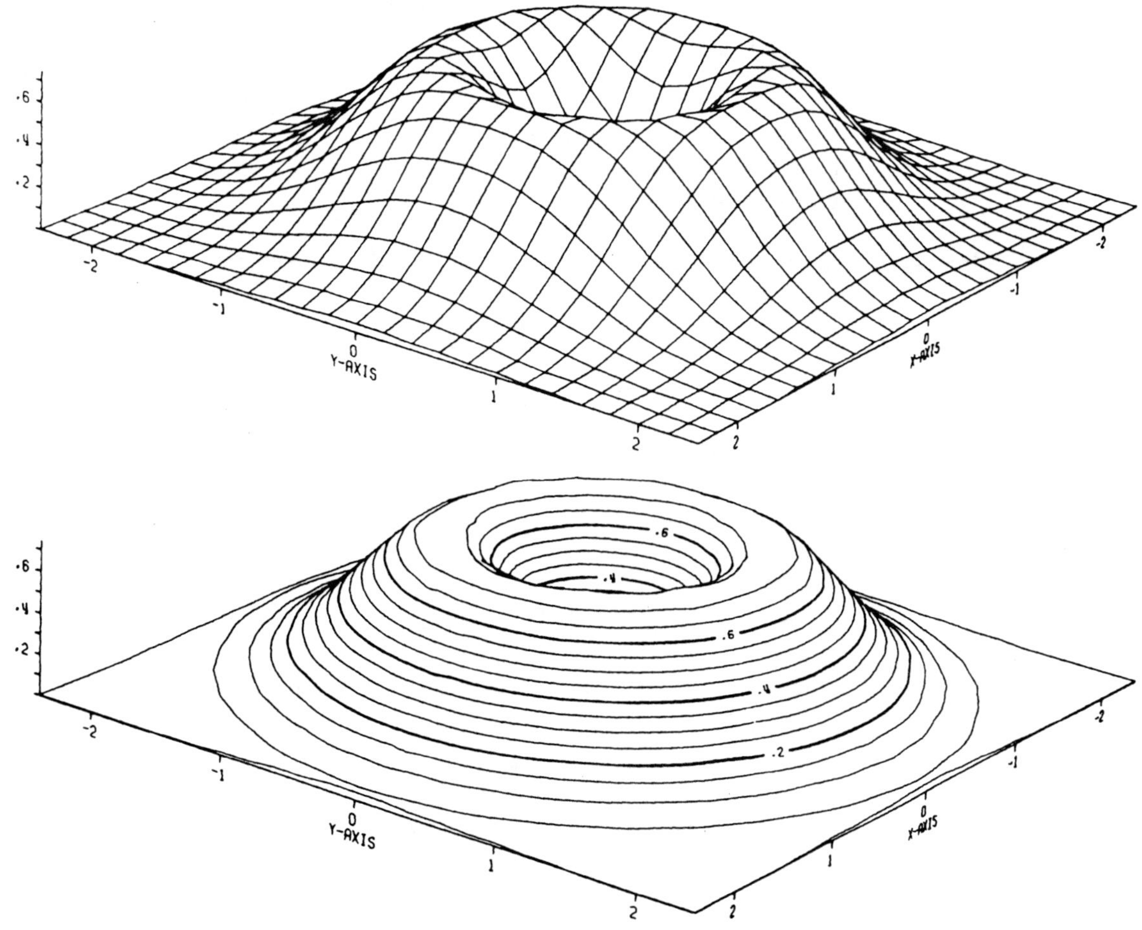

Figure 16.3.1. The volcano: $z = 2(x^2 + y^2)\exp(-x^2 - y^2)$. (a) Coordinate grid lifted to the surface. (b) Level curves lifted to the surface.

Solution There is a local (in fact, global) minimum at the volcano's center $(0,0)$, where $z=0$. There are maximum points all around the crater's rim (the circle $x^2+y^2=1$). ▲

The following is the analog in two variables of the first derivative test for one variable (see Section 3.2).

First Derivative Test

Suppose that (x_0, y_0) is a local extremum of f and that the partial derivatives of f exist at (x_0, y_0). Then $f_x(x_0, y_0) = f_y(x_0, y_0) = 0$.

We consider the case of a local minimum; the proof for a local maximum is essentially the same.

By assumption, there is a disk of radius r about (x_0, y_0) on which $f(x, y) \geqslant f(x_0, y_0)$. In particular, if $|x - x_0| < r$, then $f(x, y_0) \geqslant f(x_0, y_0)$, so the function $g(x) = f(x, y_0)$ has a local minimum at x_0. By the first derivative test of one-variable calculus, $g'(x_0) = 0$; but $g'(x_0)$ is just $f_x(x_0, y_0)$. Similarly, the function $h(y) = f(x_0, y)$ has a local minimum at y_0, so $f_y(x_0, y_0) = 0$.

The first derivative test has a simple geometric interpretation: at a local extremum of f, the tangent plane to the graph $z = f(x, y)$ is horizontal (that is, parallel to the xy plane.)

Points at which f_x and f_y both vanish are called *critical points* of f. As in one-variable calculus, finding critical points is only the first step in finding local extrema. A critical point could be a local maximum, local minimum, or neither. After looking at some examples, we will present the second derivative test for functions of two variables.

Example 2 Verify that the critical points of the function in Example 1 occur at $(0,0)$ and on the circle $x^2 + y^2 = 1$.

Solution Since $z = 2(x^2 + y^2)e^{-x^2-y^2}$, we have

$$\frac{\partial z}{\partial x} = 4x(e^{-x^2-y^2}) + 2(x^2+y^2)e^{-x^2-y^2}(-2x)$$

$$= 4x(e^{-x^2-y^2})(1 - x^2 - y^2)$$

and

$$\frac{\partial z}{\partial y} = 4y(e^{-x^2-y^2})(1 - x^2 - y^2).$$

These vanish when $x = y = 0$ or when $x^2 + y^2 = 1$. ▲

Example 3 Let $z = x^2 - y^2$. Show that $(0,0)$ is a critical point. Is it a local extremum?

Solution The partial derivatives $\partial z/\partial x = 2x$ and $\partial z/\partial y = -2y$ vanish at $(0,0)$, so the origin is a critical point. It is neither a local maximum nor minimum since $f(x, y) = x^2 - y^2$ is zero at $(0,0)$ and can be either positive (on the x axis) or negative (on the y axis) arbitrarily near the origin. This is also clear from the graph (see Fig. 16.1.5), which shows a saddle point at $(0,0)$. ▲

If we know in advance that a function has a minimum point, and that the partial derivatives exist there, then we can use the first derivative test to locate the point.

Example 4 (a) Find the minimum distance from the origin to a point on the plane $x + 3y - z = 6$.
(b) Find the minimum distance from $(1,2,0)$ to the cone $z^2 = x^2 + y^2$.

Solution (a) Geometric intuition tells us that any plane contains a point which is closest to the origin. To find that point, we must minimize the distance $d = \sqrt{x^2 + y^2 + z^2}$, where $z = x + 3y - 6$. It is equivalent but simpler to minimize $d^2 = x^2 + y^2 + (x + 3y - 6)^2$. By the first derivative test, we must have

$$\frac{\partial(d^2)}{\partial x} = 0 \qquad \text{that is,} \quad 2x + 2(x + 3y - 6) = 0$$

and

$$\frac{\partial(d^2)}{\partial y} = 0 \qquad \text{that is,} \quad 2y + 6(x + 3y - 6) = 0.$$

Solving these equations gives $y = \frac{18}{11}$, $x = \frac{6}{11}$. Thus $z = x + 3y - 6 = -\frac{6}{11}$, and so the minimum distance is $d = \sqrt{x^2 + y^2 + z^2} = 6\sqrt{11}/11$. (See Fig. 16.3.2.)

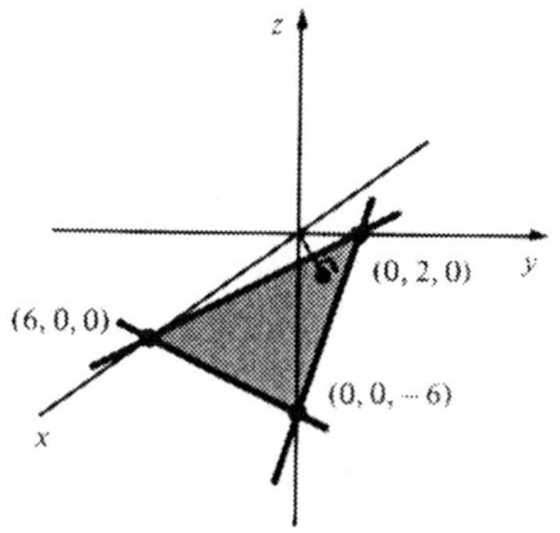

Figure 16.3.2. The point nearest to the origin on the plane $z = x + 3y - 6$ is $(6/11, 18/11, -6/11)$.

(b) We minimize the square of the distance: $d^2 = (x-1)^2 + (y-2)^2 + z^2$. Substituting $z^2 = x^2 + y^2$, we have the problem of minimizing

$$\begin{aligned} f(x, y) &= (x-1)^2 + (y-2)^2 + x^2 + y^2 \\ &= 2x^2 + 2y^2 - 2x - 4y + 5. \end{aligned}$$

Now

$$f_x(x, y) = 4x - 2 \quad \text{and} \quad f_y(x, y) = 4y - 4.$$

Thus the critical point, obtained by setting these equal to zero, is $x = \frac{1}{2}$, $y = 1$. This is the minimum point. The minimum distance is

$$\begin{aligned} d &= \sqrt{(1/2 - 1)^2 + (1-2)^2 + (1/2)^2 + 1} \\ &= \sqrt{1/4 + 1 + 1/4 + 1} = \sqrt{5/2} \approx 1.581. \ \blacktriangle \end{aligned}$$

Example 5 A rectangular box, open at the top, is to hold 256 cubic centimeters of cat food. Find the dimensions for which the surface area (bottom and four sides) is minimized.

Solution Let x and y be the lengths of the sides of the base. Since the volume of the box is to be 256, the height must be $256/xy$. Two of the sides have area $x(256/xy)$, two sides have area $y(256/xy)$, and the base has area xy, so the total surface area is $A = 2x(256/xy) + 2y(256/xy) + xy = 512/y + 512/x + xy$. To minimize A, we must have

$$0 = \frac{\partial A}{\partial x} = -\frac{512}{x^2} + y, \qquad 0 = \frac{\partial A}{\partial y} = -\frac{512}{y^2} + x.$$

The first equation gives $y = 512/x^2$; substituting this into the second equation gives $0 = -512(x^2/512)^2 + x = -x^4/512 + x$. Discarding the extraneous root $x = 0$, we have $x^3/512 = 1$, or $x = \sqrt[3]{512} = 8$. Thus $y = 512/x^2 = 8$, and the height is $256/xy = 4$, so the optimal box has a square base and is half as high as it is wide. (We have really shown only that the point $(8,8)$ is a critical point for f, but if there is any minimum point this must be it.) ▲

We now turn to the second derivative test for functions of two variables. Let us begin with an example.

Example 6 Captain Astro is being held captive by Jovians who are studying human intelligence. She is in a room where a loudspeaker emits a piercing noise. There are two knobs on the wall, whose positions, x and y, seem to affect the loudness of the noise. The knobs are initially at $x = 0$ and $y = 0$ and, when the first knob is turned, the noise gets even louder for $x < 0$ and for $x > 0$. So the captain leaves $x = 0$ and turns the second knob both ways, but, alas, the noise gets louder. Finally, she sees the formula $f(x, y) = x^2 + 3xy + y^2 + 16$ printed on the wall. What to do?

Solution First of all, she notices that $f(x,0) = x^2 + 16$ and $f(0, y) = y^2 + 16$, so the function f, like the loudness of the noise, increases if either x or y is moved away from zero. But look! If we set $y = -x$, then the "$3xy$" term becomes negative. In fact, $f(x, -x) = x^2 - 3x^2 + x^2 + 16 = -x^2 + 16$. Captain Astro rushes to the dials and turns them both at once, in opposite directions. (Why?) The noise subsides (and the Jovians cheer). ▲

The function $f(x, y) = x^2 + 3xy + y^2 + 16$ has a critical point at $(0,0)$, and the functions $g(x) = f(x,0)$ and $h(y) = f(0, y)$ both have zero as a local minimum point; but $(0,0)$ is not a local minimum point for f, because $f(x, -x) = -x^2 + 16$ is less than $f(0,0) = 16$ for arbitrarily small x. This example shows us that to tell whether a critical point (x_0, y_0) of a function $f(x, y)$ is a local extremum, we must look at the behavior of f along lines passing through (x_0, y_0) in all directions, not just those parallel to the axes.

The following test enables us to determine the nature of the critical point $(0,0)$ for any function of the form $Ax^2 + 2Bxy + Cy^2$.

Maximum–Minimum Test for Quadratic Functions

Let $g(x, y) = Ax^2 + 2Bxy + Cy^2$, where A, B, and C are constants.

1. If $AC - B^2 > 0$, and $A > 0$, [respectively $A < 0$], then $g(x, y)$ has a minimum [respectively maximum] at $(0,0)$.
2. If $AC - B^2 < 0$, then $g(x, y)$ takes both positive and negative values for (x, y) near $(0,0)$, so $(0,0)$ is not a local extremum for g.

To prove these assertions, we consider the two cases separately.

1. If $AC - B^2 > 0$, then A cannot be zero (why?), so we may write

$$g(x, y) = A\left(x^2 + \frac{2B}{A}xy + \frac{C}{A}y^2\right) = A\left(x^2 + \frac{2B}{A}xy + \frac{B^2y^2}{A^2} + \frac{C}{A}y^2 - \frac{B^2y^2}{A^2}\right)$$

$$= A\left(x + \frac{B}{A}y\right)^2 + \frac{1}{A}(AC - B^2)y^2. \qquad (1)$$

Both terms on the right-hand side of (1) have the same sign as A, and they are both zero only when $x + (B/A)y = 0$ and $y = 0$—that is, when $(x, y) = (0,0)$. Thus $(0,0)$ is a minimum point for g if $A > 0$ (since $g(x, y) > 0$ if $(x, y) \neq (0,0)$) and a maximum point if $A < 0$ (since $g(x, y) < 0$ if $(x, y) \neq (0,0)$).

2. If $AC - B^2 < 0$ and $A \neq 0$, then formula (1) still applies, but now the terms on the right-hand side have opposite signs. By suitable choices of x and y (see Exercise 49), we can make either term zero and the other nonzero. If $A = 0$, then $g(x, y) = y(2Bx + Cy)$, so we can again achieve both signs. ■

In case 2 of the preceding box, $(0,0)$ is called a *saddle point* for $g(x, y)$. (See Exercises 49 and 50 for a further discussion of this case and the case $AC - B^2 = 0$.)

Example 7 (a) Apply the maximum–minimum test to $f(x, y) = x^2 + 3xy + y^2 + 16$.
(b) Determine whether $(0,0)$ is a maximum point, a minimum point, or neither, of $g(x, y) = 3x^2 - 5xy + 3y^2$.

Solution (a) We may write this as $g(x, y) + 16$, where $g(x, y) = x^2 + 3xy + y^2$ has the form used in the test, with $A = 1$, $B = \frac{3}{2}$, and $C = 1$. Since $AC - B^2 = 1 - \frac{9}{4}$ is negative, there exist choices of x and y making $g(x, y)$ both positive and negative, so f has a saddle point at $(0,0)$. (Equation (1) gives $g(x, y) = (x + \frac{3}{2}y)^2 - \frac{9}{4}y^2$, so moving along the line $x = -\frac{3}{2}y$ makes g negative, while moving along $y = 0$ makes g positive.)
(b) $A = 3$, $B = -\frac{5}{2}$, and $C = 3$, so $A = 3 > 0$ and $AC - B^2 = 9 - \frac{25}{4} > 0$. Thus $(0,0)$ is a minimum point by part 1 of the maximum–minimum test. ▲

Note that for the quadratic function $g(x, y)$ in the preceding box, the constants A, B, and C can be recovered from g by the formulas

$$A = \frac{1}{2}\frac{\partial^2 g}{\partial x^2}, \qquad B = \frac{1}{2}\frac{\partial^2 g}{\partial x\,\partial y}, \qquad C = \frac{1}{2}\frac{\partial^2 g}{\partial y^2}$$

so that the signs of A and $AC - B^2$ are the same as those of $\partial^2 g/\partial x^2$ and $(\partial^2 g/\partial x^2)(\partial^2 g/\partial y^2) - (\partial^2 g/\partial x\,\partial y)^2$. The second derivative test for general functions involves just these combinations of partial derivatives.

Second Derivative Test

Let $f(x, y)$ have continuous second partial derivatives, and suppose that (x_0, y_0) is a critical point for f:

$$f_x(x_0, y_0) = 0 \quad \text{and} \quad f_y(x_0, y_0) = 0.$$

Let $A = f_{xx}(x_0, y_0)$, $B = f_{xy}(x_0, y_0)$, and $C = f_{yy}(x_0, y_0)$.

If:	then:
$A > 0$, $AC - B^2 > 0$	(x_0, y_0) is a local minimum;
$A < 0$, $AC - B^2 > 0$	(x_0, y_0) is a local maximum;
$AC - B^2 < 0$	(x_0, y_0) is a saddle point;
$AC - B^2 = 0$	the test is inconclusive.

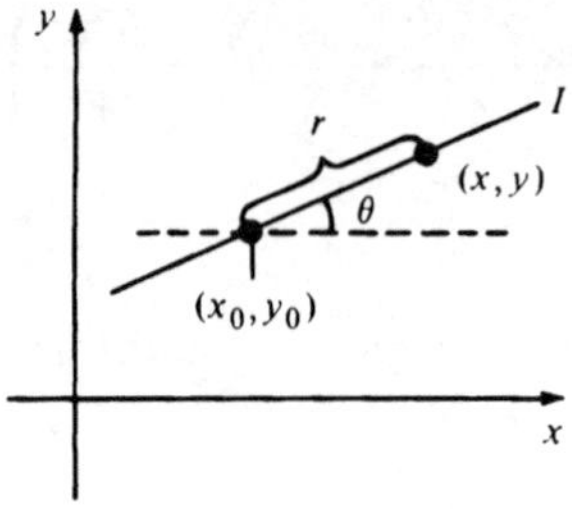

Figure 16.3.3. The point (x, y) is $(x_0 + r\cos\theta, y_0 + r\sin\theta)$.

To prove these assertions, we look at f along straight lines through (x_0, y_0). Specifically, for each fixed θ in $[0, 2\pi]$, we will consider the function $h(r) = f(x_0 + r\cos\theta, y_0 + r\sin\theta)$, which describes the behavior of f along the line through (x_0, y_0) in the direction of $\cos\theta\mathbf{i} + \sin\theta\mathbf{j}$. (See Fig. 16.3.3.)

For each θ, $h(r)$ is a function of one variable with a critical point at $r = 0$. To analyze the behavior of $h(r)$ near $r = 0$ by using the second derivative test for functions of one variable, we differentiate $h(r)$ using the chain rule of Section 15.3. Let $x = x_0 + r\cos\theta$ and $y = y_0 + r\sin\theta$; then

$$h'(r) = f_x(x, y)\frac{dx}{dr} + f_y(x, y)\frac{dy}{dr} = f_x(x, y)\cos\theta + f_y(x, y)\sin\theta.$$

We differentiate again, applying the chain rule to f_x and f_y:

$$h''(r) = f_{xx}(x, y)\cos\theta\frac{dx}{dr} + f_{xy}(x, y)\cos\theta\frac{dy}{dr}$$

$$+ f_{yx}(x, y)\sin\theta\frac{dx}{dr} + f_{yy}(x, y)\sin\theta\frac{dy}{dr}.$$

Since $f_{xy} = f_{yx}$ by equality of mixed partials, this becomes

$$h''(r) = f_{xx}(x, y)\cos^2\theta + 2f_{xy}(x, y)\cos\theta\sin\theta + f_{yy}(x, y)\sin^2\theta$$

or

$$h''(r) = f_{xx}(x_0 + r\cos\theta, y_0 + r\sin\theta)\cos^2\theta$$

$$+ 2f_{xy}(x_0 + r\cos\theta, y_0 + r\sin\theta)\cos\theta\sin\theta$$

$$+ f_{yy}(x_0 + r\cos\theta, y_0 + r\sin\theta)\sin^2\theta. \tag{2}$$

Setting $r = 0$, we get

$$h''(0) = f_{xx}(x_0, y_0)\cos^2\theta + 2f_{xy}(x_0, y_0)\cos\theta\sin\theta + f_{yy}(x_0, y_0)\sin^2\theta,$$

which has the form $Ax^2 + 2Bxy + Cy^2$, with $x = \cos\theta$, $y = \sin\theta$, and with $A = f_{xx}(x_0, y_0)$, $B = f_{xy}(x_0, y_0)$, $C = f_{yy}(x_0, y_0)$. Let $AC - B^2 = D$.

Now suppose that $D > 0$ and $f_{xx}(x_0, y_0) > 0$. By the maximum–minimum test for quadratic functions, $h''(0) > 0$, so h has a local minimum at $r = 0$. Since this is true for all values of θ, f has a local minimum along each line through (x_0, y_0). It is thus plausible that f has a local minimum at (x_0, y_0), so we will end the proof at this point. (Actually, more work is needed; for further details, see Exercises 51 and 52.)

If $D > 0$ and $f_{xx}(x_0, y_0) < 0$, then f has a local maximum along every line through (x_0, y_0); if $D < 0$, then f has a local minimum along some lines through (x_0, y_0) and a local maximum along others. ■

Example 8 Find the maxima, minima, and saddle points of $z = (x^2 - y^2)e^{(-x^2-y^2)/2}$.

Solution First we locate the critical points by setting $\partial z/\partial x = 0$ and $\partial z/\partial y = 0$. Here

$$\frac{\partial z}{\partial x} = \left[2x - x(x^2 - y^2)\right]e^{(-x^2-y^2)/2}$$

and

$$\frac{\partial z}{\partial y} = \left[-2y - y(x^2 - y^2)\right]e^{(-x^2-y^2)/2},$$

so the critical points are the solutions of

$$x\left[2-(x^2-y^2)\right]=0, \qquad y\left[-2-(x^2-y^2)\right]=0.$$

This has solutions $(0,0)$, $(\pm\sqrt{2},0)$, and $(0,\pm\sqrt{2})$.

The second derivatives are

$$\frac{\partial^2 z}{\partial x^2}=\left[2-5x^2+x^2(x^2-y^2)+y^2\right]e^{(-x^2-y^2)/2},$$

$$\frac{\partial^2 z}{\partial x\partial y}=xy(x^2-y^2)e^{(-x^2-y^2)/2},$$

$$\frac{\partial^2 z}{\partial y^2}=\left[5y^2-2+y^2(x^2-y^2)-x^2\right]e^{(-x^2-y^2)/2}.$$

Using the second derivative test results in the following data:

Point	A	B	C	$AC-B^2$	Type
$(0,0)$	2	0	-2	-4	saddle
$(\sqrt{2},0)$	$-4/e$	0	$-4/e$	$16/e^2$	maximum
$(-\sqrt{2},0)$	$-4/e$	0	$-4/e$	$16/e^2$	maximum
$(0,\sqrt{2})$	$4/e$	0	$4/e$	$16/e^2$	minimum
$(0,-\sqrt{2})$	$4/e$	0	$4/e$	$16/e^2$	minimum

The results of this example are confirmed by the computer-generated graph in Fig. 16.3.4. ▲

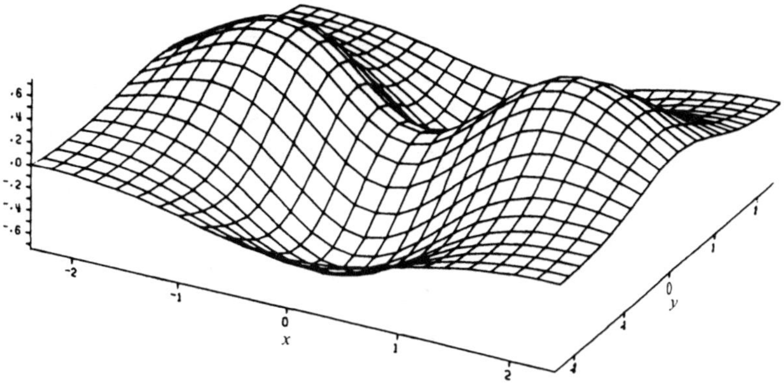

Figure 16.3.4. Computer-generated graph of $z=(x^2-y^2)e^{(-x^2-y^2)/2}$.

Example 9 Let $z=(x^2+y^2)\cos(x+2y)$. Show that $(0,0)$ is a critical point. Is it an extremum?

Solution We compute:

$$\frac{\partial z}{\partial x}=2x\cos(x+2y)-(x^2+y^2)\sin(x+2y),$$

$$\frac{\partial z}{\partial y}=2y\cos(x+2y)-2(x^2+y^2)\sin(x+2y).$$

These vanish at $(0,0)$, so $(0,0)$ is a critical point.

The second derivatives are:

$$\frac{\partial^2 z}{\partial x^2}=2\cos(x+2y)-4x\sin(x+2y)-(x^2+y^2)\cos(x+2y),$$

$$\frac{\partial^2 z}{\partial x\,\partial y} = -4x\sin(x+2y) - 2y\sin(x+2y) - 2(x^2+y^2)\cos(x+2y),$$

$$\frac{\partial^2 z}{\partial y^2} = 2\cos(x+2y) - 8y\sin(x+2y) - 4(x^2+y^2)\cos(x+2y).$$

Evaluating at $x=0$, $y=0$, we get $A=2$, $B=0$, and $C=2$, so $A>0$, $AC-B^2>0$, and thus $(0,0)$ is a local minimum. ▲

Example 10 Find the point or points on the elliptic paraboloid $z=4x^2+y^2$ closest to $(0,0,8)$.

Solution The typical point on the paraboloid is $(x,y,4x^2+y^2)$; its distance from $(0,0,8)$ is $\sqrt{x^2+y^2+(4x^2+y^2-8)^2}$. It is convenient to minimize the square of the distance:

$$f(x,y) = x^2+y^2+(4x^2+y^2-8)^2.$$

We begin by locating the critical points of f. The partial derivatives of f are

$$f_x(x,y) = 2x + 2(4x^2+y^2-8)\cdot 8x = 2x(32x^2+8y^2-63),$$

$$f_y(x,y) = 2y + 2(4x^2+y^2-8)\cdot 2y = 2y(8x^2+2y^2-15).$$

For $f_x(x,y)$ to be zero we must have $x=0$ or $32x^2+8y^2-63=0$. For $f_y(x,y)$ to be zero we must have $y=0$ or $8x^2+2y^2-15=0$. Thus there are four possibilities:

Case I. $x=0$ and $y=0$.

Case II. $x=0$ and $8x^2+2y^2-15=0$. Then $2y^2-15=0$ or $y=\pm\sqrt{15/2}$.

Case III. $32x^2+8y^2-63=0$ and $y=0$. Then $32x^2-63=0$ and so $x=\pm\sqrt{63/32}$.

Case IV. $32x^2+8y^2-63=0$ and $8x^2+2y^2-15=0$. Subtracting four times the second equation from the first gives $-3=0$, which is impossible, so case IV does not occur.

A simple way to see which of the points in cases I, II, and III minimizes the distance is to compute $f(x,y)$ in each case and choose the smallest value. We leave this method to the reader and, instead, use the second derivative test. The second derivatives are

$$f_{xx} = 2(32x^2+8y^2-63) + 2x\cdot 64x = 192x^2+16y^2-126,$$

$$f_{yy} = 2(8x^2+2y^2-15) + 2y\cdot 4y = 16x^2+12y^2-30,$$

$$f_{xy} = f_{yx} = 32xy.$$

Case I. $f_{xx}=-126$, $f_{yy}=-30$, $f_{xy}=0$. Thus $f_{xx}f_{yy}-f_{xy}^2 = 126\cdot 30>0$, so this point is *local maximum* for f.

Case II. $f_{xx}=16\cdot\frac{15}{2}-126<0$, $f_{yy}=12\cdot\frac{15}{2}-30>0$, $f_{xy}=0$. Therefore $f_{xx}f_{yy}-f_{xy}^2<0$, so these two points are *saddles* for f.

Case III. $f_{xx}=192\cdot\frac{63}{32}-126>0$, $f_{yy}=16\cdot\frac{63}{32}-30>0$, $f_{xy}=0$. Therefore $f_{xx}f_{yy}-f_{xy}^2>0$, so these two points are *local minima* for f. Thus the closest points to $(0,0,8)$ on the paraboloid are

$$\left(\tfrac{3}{4}\sqrt{\tfrac{7}{2}}\,,0,\tfrac{63}{8}\right) \quad\text{and}\quad \left(-\tfrac{3}{4}\sqrt{\tfrac{7}{2}}\,,0,\tfrac{63}{8}\right). \ ▲$$

Supplement to Section 16.3: Astigmatism

The visual problem called *astigmatism* results from a deviation from circular symmetry in the shape of the lens in your eye. Correcting astigmatism requires a compensating eyeglass or contact lens with the "opposite" deviation.

A piece of the lens surface may be described by a function $z = f(x, y) = Ax^2 + 2Bxy + Cy^2$, for x and y small. The lens is symmetric about the z axis when $B = 0$ and $A = C$. In general, if we slice the lens by a plane of the form $-x\sin\theta + y\cos\theta = 0$, which contains the z axis and the vector $\cos\theta\mathbf{i} + \sin\theta\mathbf{j}$, the slice is bounded by a curve through the origin whose curvature there is $2(A\cos^2\theta + 2B\sin\theta\cos\theta + C\sin^2\theta)$ (see Section 14.7 and Exercise 56). The maxima and minima of curvature occur when $\tan 2\theta = 2B/(A - C)$. Notice that the direction of maximum and minima curvature differ by 90°; this means that an optometrist must know only one of these directions in order to orient corrective lenses properly.

Exercises for Section 16.3

1. Refer to Fig. 16.3.5, a computer-generated graph of $z = (x^3 - 3x)/(1 + y^2)$. Where are the maximum and minimum points?

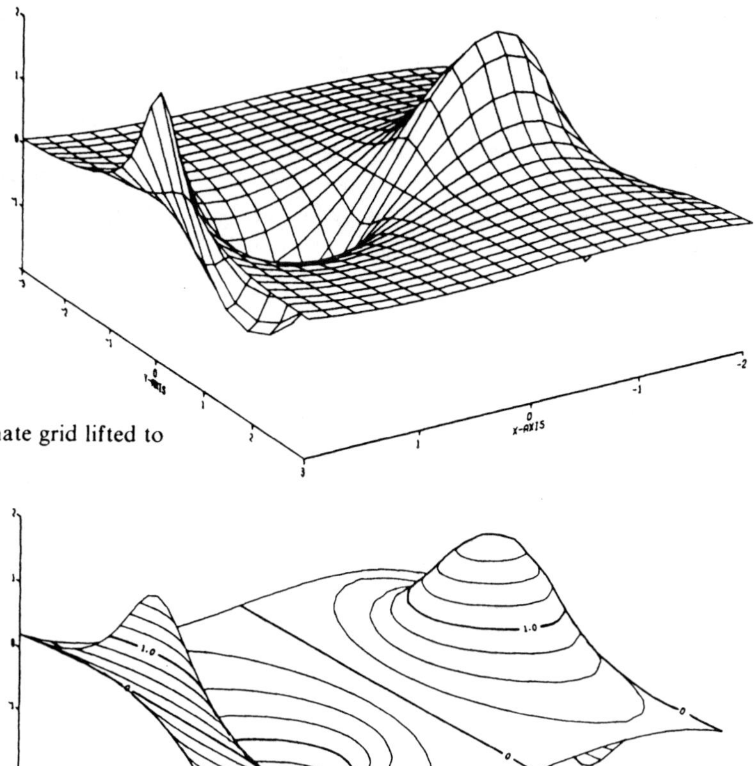

(a) Coordinate grid lifted to the graph.

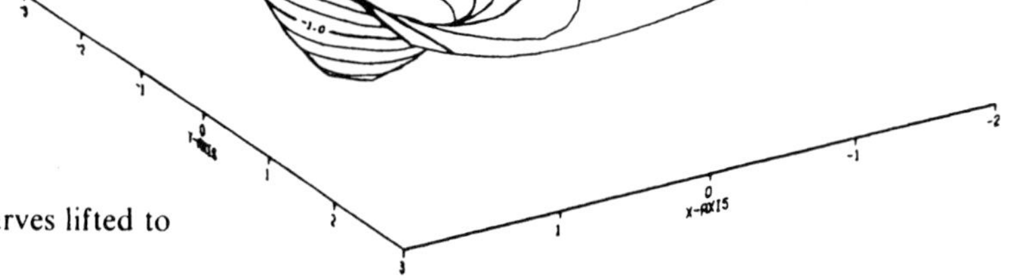

(b) Level curves lifted to the graph.

Figure 16.3.5. Computer-generated graph of $z = (x^3 - 3x)/(1 + y^2)$.

2. Refer to Fig. 16.3.6, a computer-generated graph of $z = \sin(\pi x)/(1 + y^2)$. Where are the maximum and minimum points?

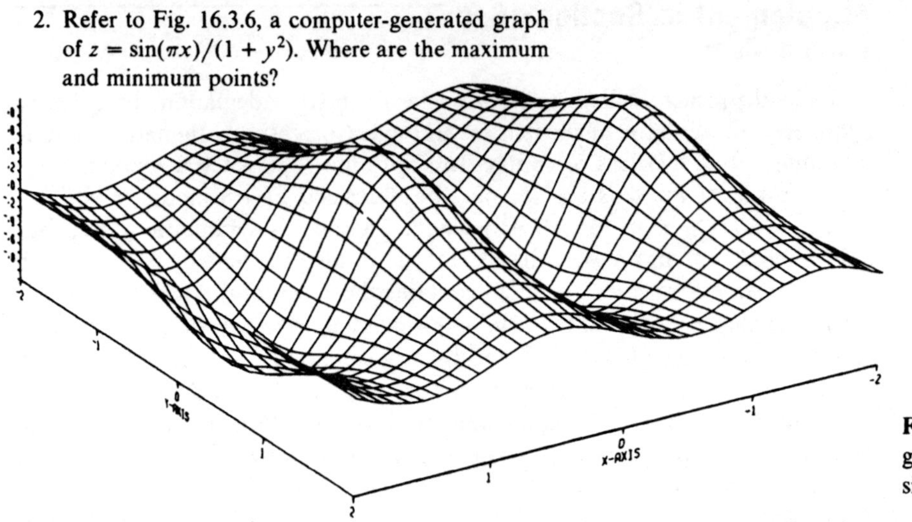

Figure 16.3.6. Computer-generated graph of $\sin(\pi x)/(1 + y^2)$.

Find the critical points of each of the functions in Exercises 3–6. Decide by inspection whether each of the critical points is a local maximum, minimum, or neither.

3. $f(x, y) = x^2 + 2y^2$
4. $f(x, y) = x^2 - 2y^2$
5. $f(x, y) = \exp(-x^2 - 7y^2 + 3)$
6. $f(x, y) = \exp(x^2 + 2y^2)$

7. Minimize the distance to the origin from the plane $x - y + 2z = 3$.
8. Find the distance from the plane given by $x + 2y + 3z - 10 = 0$: (a) To the origin. (b) To the point $(1, 1, 1)$.
9. Suppose that the material for the bottom of the box in Example 5 costs b cents per square centimeter, while that for the sides costs s cents per square centimeter. Find the dimensions which minimize the cost of the material.
10. Drug reactions can be measured by functions of the form $R(u, t) = u^2(c - u)t^2e^{-t}$, $0 \leqslant u \leqslant c$, $t \geqslant 0$. The symbols u and t are drug units and time in hours, respectively. Find the dosage u and time t at which R is a maximum.

In Exercises 11–16 use the maximum–minimum test for quadratic functions to decide whether $(0, 0)$ is a maximum, minimum, or saddle point.

11. $f(x, y) = x^2 + xy + y^2$.
12. $f(x, y) = x^2 - xy + y^2 + 1$.
13. $f(x, y) = y^2 - x^2 + 3xy$.
14. $f(x, y) = x^2 + y^2 - xy$.
15. $f(x, y) = y^2$.
16. $f(x, y) = 3 + 2x^2 - xy + y^2$.

Find the critical points of each of the functions in Exercises 17–30 and classify them as local maxima, minima, or neither.

17. $f(x, y) = x^2 + y^2 + 6x - 4y + 13$.
18. $f(x, y) = x^2 + y^2 + 3x - 2y + 1$.
19. $f(x, y) = x^2 - y^2 + xy - 7$.
20. $f(x, y) = x^2 + y^2 + 3xy + 10$.
21. $f(x, y) = x^2 + y^2 - 6x - 14y + 100$.
22. $f(x, y) = y^2 - x^2$.
23. $f(x, y) = 2x^2 - 2xy + y^2 - 2x + 1$.
24. $f(x, y) = x^2 - 3xy + 5x - 2y + 6y^2 + 8$.
25. $f(x, y) = 3x^2 + 2xy + 2y^2 - 3x + 2y + 10$.
26. $f(x, y) = x^2 + xy^2 + y^4$.
27. $f(x, y) = e^{1 + x^2 - y^2}$.
28. $f(x, y) = (x^2 + y^2)e^{x^2 - y^2}$.
29. $f(x, y) = \ln(2 + \sin xy)$. [Consider only the critical point $(0, 0)$.]
30. $f(x, y) = \sin(x^2 + y^2)$. [Consider only the critical point $(0, 0)$.]

31. Analyze the behavior of $z = x^5y + xy^5 + xy$ at its critical points.
32. Test for extrema: $z = \ln(x^2 + y^2 + 1)$.
33. Analyze the critical point at $(0, 0)$ for the function $f(x, y) = x^2 + y^3$. Make a sketch.
34. Locate any maxima, minima, or saddle points of $f(x, y) = \ln(ax^2 + by^2 + 1)$, $a, b > 0$.
35. A computer-generated graph of
$$z = (\sin \pi r)/\pi r, \qquad r = \sqrt{x^2 + y^2}\,,$$
is shown in Fig. 16.3.7. (a) Show, by calculation, that all critical points of the function lie on circles whose radius satisfies the equation $\pi r = \tan(\pi r)$. (b) Which points are maxima? Minima? (c) What symmetries does the graph have?
36. Show that $z = (x^3 - 3x)/(1 + y^2)$ has exactly one local maximum and one local minimum. What symmetries does the graph have? (It is computer drawn in Fig. 16.3.5.)
37. The work w done in a compressor with $k + 1$ compression cylinders is given by
$$w = c_1 y + c_2, \qquad c_1 \neq 0,$$
where
$$y = \sum_{i=0}^{k} T_i(p_i/p_{i+1})^{(n-1)/n}.$$

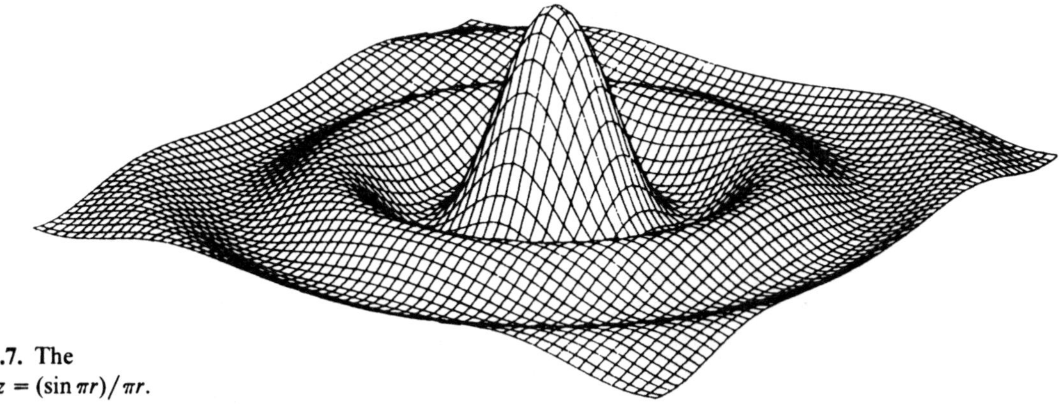

Figure 16.3.7. The sombrero: $z = (\sin \pi r)/\pi r$.

The symbols T_i and p_i stand for temperature and pressure in cylinder i, $1 \leqslant i \leqslant k+1$; the pressures $p_1, \dots, p_k$ are the independent variables, while $p_0, p_{k+1}, T_0, \dots, T_k$ and $n > 1$ are given.

(a) Find relations between the variables if w is a minimum.

★(b) Find p_1, p_2, p_3 explicitly for the case $k = 3$.

38. *Planck's law* gives the relationship of the energy E emitted by a blackbody to the wavelength λ and temperature T:

$$E = \frac{2\pi k^5 T^5}{h^4 c^3} \frac{x^5}{e^x - 1}, \qquad \text{where} \quad x = \frac{hc}{\lambda k T}.$$

The constants are $h = 6.6256 \times 10^{-34}$ joule seconds (Planck's constant), $k = 1.3805 \times 10^{-23}$ joule kilograms^{-1} (Boltzmann's constant), $c = 2.9979 \times 10^8$ meter second^{-1} (velocity of light). The plot of E versus λ for fixed T is called a *Planck curve*.

(a) The maximum along each Planck curve is obtained by setting $\partial E/\partial\lambda = 0$ and solving for $\lambda_{\max}$. The relationship so derived is called *Wien's displacement law*. Show that this law is just $\lambda_{\max} = hc/kTx_0$, where $5 - x_0 - 5e^{-x_0} = 0$.

(b) Clearly x_0 is close to 5. By examining the sign of $f(x) = 5 - x - 5e^{-x}$, use your calculator to complete the expansion $x_0 = 4.965\ldots$ to a full six digits.

(c) Improve upon the displacement law $\lambda_{\max} = 0.00289/T$ by giving a slightly better constant. The peak for the earth (288°K) is about 10 micrometers, the peak for the sun (6000°K) about 0.48 micrometers, so the maximum occurs in the infared and visible range, respectively.

39. Apply the second derivative test to the critical point in Example 5.

40. (a) Show that if (x_0, y_0, z_0) is a local minimum or maximum point of $w = f(x, y, z)$, then $\partial w/\partial x$, $\partial w/\partial y$, and $\partial w/\partial z$ are all zero at (x_0, y_0, z_0).

(b) Find the critical points of the function $\sin(x^2 + y^2 + z^2)$.

(c) Find the point in space which minimizes the sum of the squares of the distances from $(0,0,0)$, $(1,0,0)$, $(0,1,0)$, and $(0,0,1)$.

41. Analyze the behavior of the following functions at the indicated points:

(a) $f(x, y) = x^2 - y^2 + 3xy$; $(0,0)$.

(b) $f(x, y) = x^2 + y^2 + Cxy$; $(0,0)$. Determine what happens for various values of C. At what values of C does the behavior change qualitatively?

42. Find the local maxima and minima for $z = (x^2 + 3y^2)e^{1 - x^2 - y^2}$. (See Fig. 14.3.15.)

Exercises 43–48 deal with the *method of least squares*. It often happens that the theory behind an experiment indicates that the data should lie along a straight line of the form $y = mx + b$. The actual results, of course, will never exactly match with theory, so we are faced with the problem of finding the straight line which best fits some experimental data $(x_1, y_1), \dots, (x_n, y_n)$ as in Fig. 16.3.8. For the straight line $y = mx + b$, each point will deviate vertically from the line by an amount $d_i = y_i - (mx_i + b)$. We would like to choose m and b in such a way as to make the total effect of these deviations as small as possible. Since some deviations are negative and some are positive, however, a better measure of the total error is the sum of the squares of these deviations; so we are led to the problem of finding m and b to minimize the function

$$s = f(m, b) = d_1^2 + d_2^2 + \cdots + d_n^2$$
$$= \sum_{i=1}^{n} (y_i - mx_i - b)^2,$$

where $x_1, \dots, x_n$ and $y_1, \dots, y_n$ are given data.

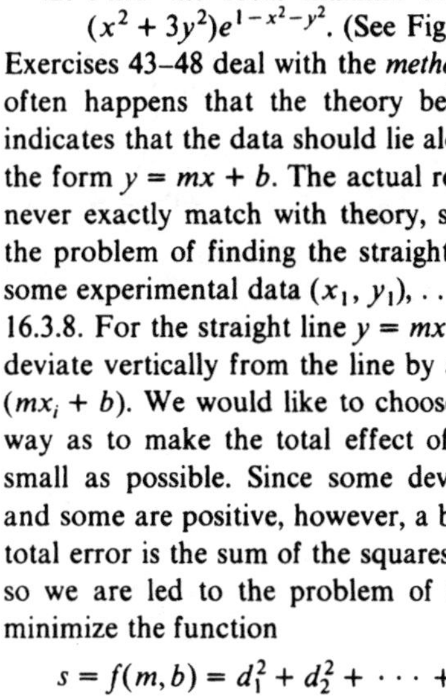

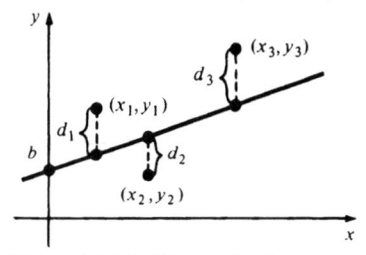

Figure 16.3.8. The method of least squares finds a straight line which "best" approximates a set of data.

43. For each set of three data points, plot the points, write down the function $f(m,b)$, find m and b to give the best straight-line fit according to the method of least squares, and plot the straight line.
    (a) $(x_1, y_1) = (1,1)$; $(x_2, y_2) = (2,3)$; $(x_3, y_3) = (4,3)$.
    (b) $(x_1, y_1) = (0,0)$; $(x_2, y_2) = (1,2)$; $(x_3, y_3) = (2,3)$.
44. Show that if only two data points (x_1, y_1) and (x_2, y_2) are given, then this method produces the line through (x_1, y_1) and (x_2, y_2).
45. Show that the equations for a critical point, $\partial s/\partial b = 0$ and $\partial s/\partial m = 0$, are equivalent to $m(\Sigma x_i) + nb = \Sigma y_i$ and $m(\Sigma x_i^2) + b(\Sigma x_i) = \Sigma x_i y_i$, where all the sums run from $i = 1$ to $i = n$.
46. If $y = mx + b$ is the best-fitting straight line to the data points $(x_1, y_1), \ldots, (x_n, y_n)$ according to the least squares method, show that

$$\sum_{i=1}^{n} (y_i - mx_i - b) = 0.$$

That is, show that the positive and negative deviations cancel (see Exercise 45).
47. Use the second derivative test to show that the critical point of f is actually a minimum.
48. Use the method of least squares to find the straight line that best fits the points $(0,1)$, $(1,3)$, $(2,2)$, $(3,4)$, and $(4,5)$. Plot your points and line.

★49. Complete the proof of the maximum–minimum test for quadratic functions by following these steps:
    (a) If $A \neq 0$ and $AC - B^2 < 0$, show that

$$g(x,y) = A\left[\left(x + \frac{B}{A}y\right) - ey\right]\left[\left(x + \frac{B}{A}y\right) + ey\right]$$

    for some number e. What is e?
    (b) Show that the set where $g(x,y) = 0$ consists of two intersecting lines. What are their equations?
    (c) Show that $g(x,y)$ is positive on two of the regions cut out by the lines in part (b) and negative on the other two.
    (d) If $A = 0$, $g(x,y) = 2Bxy + Cy^2$. B must be nonzero. (Why?) Write $g(x,y)$ as a product of linear functions and repeat parts (b) and (c).

★50. Discuss the function $Ax^2 + 2Bxy + Cy^2$ in the case where $AC - B^2 = 0$.
    (a) If $A \neq 0$, use formula (1) in the proof of the maximum–minimum test for quadratic functions.
    (b) Sketch a graph of the function $f(x,y) = x^2 + 2xy + y^2$.
    (c) What happens if $A = 0$?

★51. Let $f(x,y) = 3x^4 - 4x^2y + y^2$.
    (a) Show that $f(x,y)$ has a critical point at the origin.
    (b) Show that for all values of θ, the function $h(r) = f(r\cos\theta, r\sin\theta)$ has a local minimum at $r = 0$.
    (c) Show that, nevertheless, the origin is not a local minimum point for f.
    (d) Find the set of (x,y) for which $f(x,y) = 0$.
    (e) Sketch the regions in the plane where $f(x,y)$ is positive and negative.
    (f) Discuss why parts (b) and (c) do not contradict one another.

★52. Complete the proof of the second derivative test by following this outline:
    (a) We begin with the case in which $D > 0$ and $f_{xx}(x_0, y_0) > 0$. Using Exercise 78, Section 15.1, show that there is a number $\varepsilon > 0$ such that whenever (x,y) lies in the disk of radius ε about (x_0, y_0),

$$f_{xx}(x,y)f_{yy}(x,y) - f_{xy}(x,y)^2$$

    and $f_{xx}(x,y)$ are both positive.
    (b) Show that the function $h(r)$ is concave upward on the interval $(-\varepsilon, \varepsilon)$ for any choice of θ.
    (c) Conclude that $f(x,y) \geqslant f(x_0, y_0)$ for all (x,y) in the disk of radius ε about (x_0, y_0).
    (d) Complete the case in which $D > 0$ and $f_{xx}(x_0, y_0) < 0$.
    (e) Complete the case $D < 0$ by showing that f takes values near (x_0, y_0) which are greater and less than $f(_0, y_0)$.

★53. Find the point or points on the elliptic paraboloid $z = 4x^2 + y^2$ closest to $(0,0,a)$ for each a. (See Example 10.) How does the answer depend upon a?

★54. Let $f(x,y) > 0$ for all x and y. Show that $f(x,y)$ and $g(x,y) = [f(x,y)]^2$ have the same critical points, with the same "type" (maximum, minimum, or saddle).

★55. Consider the general problem of finding the points on a graph $z = k(x,y)$ closest to a point (a,b,c). Show that (x_0, y_0) is a critical point for the distance from $(x, y, k(x,y))$ to (a,b,c) if and only if the line from (a,b,c) to $(x_0, y_0, k(x_0, y_0))$ is orthogonal to the graph at $(x_0, y_0, k(x_0, y_0))$.

★56. (See the supplement to this section.)
    (a) Show that if the surface $z = Ax^2 + 2Bxy +$

Cy^2 is sliced by the plane $-x \sin\theta + y\cos\theta = 0$, then the curvature of the slice at the origin is twice the absolute value of $A\cos^2\theta + 2B\sin\theta\cos\theta + C\sin^2\theta$.

(b) Show that if $B^2 - AC < 0$ and $A > 0$, then the maximum and minima of curvature occur when $\tan 2\theta = 2B/(A - C)$.

16.4 Constrained Extrema and Lagrange Multipliers

The level surfaces of two functions must cross, except where the gradients of the functions are parallel.

In studying maximum–minimum problems for a function $f(x)$ defined on an interval $[a,b]$, we found in Section 3.5 that the maximum and minimum points could occur either at critical points (where $f'(x) = 0$) or at the endpoints a and b. For a function $f(x, y)$ in the plane, it is common to replace the interval $[a,b]$ by some region D; the role of the endpoints is now played by the boundary of D, which is a curve in the plane (possibly with corners).

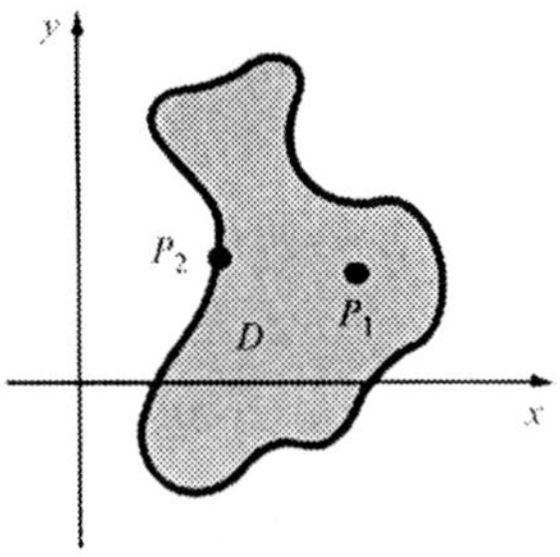

Figure 16.4.1. If the interior point P_1 is an extreme point of f on D, then the partial derivatives of f at P_1, if they exist, must be zero. If the boundary point P_2 is an extreme point, the partial derivatives there might not be zero.

The problem of finding extrema in several variables can be attacked in steps:

Step 1. Suppose that (x_0, y_0) is an extremum lying inside D, like the point P_1 in Fig. 16.4.1, and that the partial derivatives of f exist[2] at (x_0, y_0). Then our earlier analysis applies and (x_0, y_0) must be a critical point.

Step 2. The extreme point (x_0, y_0) may lie on the boundary of D, like P_2 in Fig. 16.4.1. At such a point, the partial derivatives of f might not be zero. Thus we must develop new techniques for finding candidates for the extreme points of f on the boundary.

Step 3. The function f should be evaluated at the points found in Steps 1 and 2, and the largest and smallest values should be identified.

If we can parametrize the boundary curve, say by $\boldsymbol{\sigma}(t)$ for t in $[a,b]$, then the restriction of f to the boundary[3] becomes a function of one variable, $h(t) = f(\boldsymbol{\sigma}(t))$, to which the methods of one-variable calculus apply, as in the following example.

Example 1 Find the extreme values of $z = f(x, y) = x^2 + 2y^2$ on the disk D consisting of points (x, y) satisfying $x^2 + y^2 \leqslant 1$.

Solution *Step 1.* At a critical point, $\partial z/\partial x = 2x = 0$ and $\partial z/\partial y = 4y = 0$. Thus, the only critical point is $(0,0)$. It is clearly a minimum point for f; we may also verify this by the second derivative test:

$$\frac{\partial^2 z}{\partial x^2} = 2 > 0$$

and

$$\left(\frac{\partial^2 z}{\partial x^2}\right)\left(\frac{\partial^2 z}{\partial y^2}\right) - \left(\frac{\partial^2 z}{\partial x\,\partial y}\right)^2 = 2 \cdot 4 - 0 > 0.$$

[2] As with functions of one variable, there may be points where the derivatives of f do not exist. If there are such points, they must be examined directly to see if they are maxima or minima.

[3] That is, the function which has the same values as f but whose domain consists only of the boundary points.

This result confirms that $(0,0)$ is a local minimum.

Step 2. The boundary of D is the unit circle, which we may parametrize by $(\cos t, \sin t)$. Along the boundary,

$$h(t) = f(\cos t, \sin t) = \cos^2 t + 2\sin^2 t = 1 + \sin^2 t.$$

Since $h'(t) = 2\sin t \cos t$, $h'(t) = 0$ at $t = 0$, $\pi/2$, π, and $3\pi/2$ (2π gives the same point as zero). Thus, the only boundary points which could possibly be local maxima and minima are $(\cos t, \sin t)$ for these values of t, i.e., $(1,0)$, $(0,1)$, $(-1,0)$, and $(0,-1)$.

Step 3. Evaluating f at the points found in Steps 1 and 2, we obtain:

$$f(0,0) = 0, \quad f(1,0) = 1, \quad f(0,1) = 2, \quad f(-1,0) = 1, \quad f(0,-1) = 2.$$

Thus, f has a minimum point at $(0,0)$ with value 0 and maximum points at $(0,1)$ and $(0,-1)$ with value 2. (See Fig. 16.4.2.) The points $(1,0)$ and $(-1,0)$ are neither maxima nor minima for f on D even though they are minima for f on the boundary. ▲

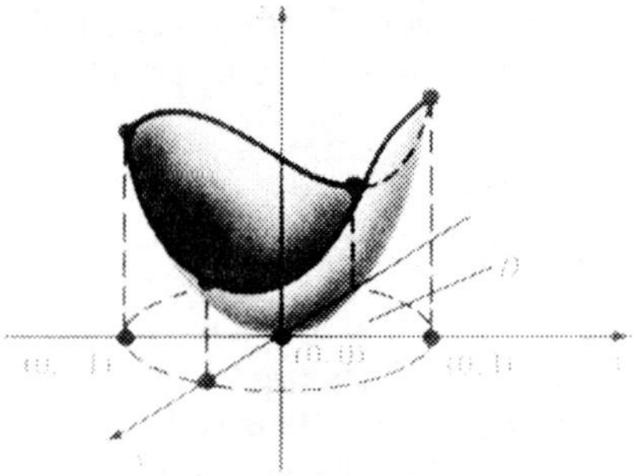

Figure 16.4.2. The function $f(x, y) = x^2 + 2y^2$ on the disk D has a minimum point at $(0,0)$ and maximum points at $(0,1)$ and $(0,-1)$.

Often it is inconvenient to find a parametrization for the curve C on which we are searching for extrema. Instead, the curve C may be given as a level curve of a function $g(x, y)$. In this case, we can still derive a first derivative test for local maxima and minima. The following result leads to the *method of Lagrange multipliers*.

First Derivative Test for Constrained Extrema

Let f and g be functions of two variables with continuous partial derivatives. Suppose that the function f, when restricted to the level curve C defined by $g(x, y) = c$, has a local extremum at (x_0, y_0) and that $\nabla g(x_0, y_0) \neq \mathbf{0}$. Then there is a number λ such that

$$\nabla f(x_0, y_0) = \lambda \nabla g(x_0, y_0).$$

If $\lambda \neq 0$, this formula says that the level curves of f and g through (x_0, y_0) have the same tangent line at (x_0, y_0).

To demonstrate the result in this box, choose a parametrization $(x, y) = \boldsymbol{\sigma}(t)$ for C near (x_0, y_0), with $\boldsymbol{\sigma}(0) = (x_0, y_0)$ and $\boldsymbol{\sigma}'(0) \neq 0$.[4] Since f has a local extremum at (x_0, y_0), the function $h(t) = f(\boldsymbol{\sigma}(t))$ has a local extremum at $t = 0$, so $h'(0) = 0$. According to the chain rule (Section 16.1), we get $h'(0) = \nabla f(x_0, y_0) \cdot \boldsymbol{\sigma}'(0)$, so $\nabla f(x_0, y_0)$ is perpendicular to $\boldsymbol{\sigma}'(0)$; but we already

[4] The implicit function theorem guarantees that such a parametrization exists; see J. Marsden and A. Tromba, *Vector Calculus*, Freeman (1981), p. 237. We will not need to know the explicit parametrization for the method to be effective.

know that the gradient $\nabla g(x_0, y_0)$ is perpendicular to the tangent vector $\sigma'(0)$ to the level curve C (Section 16.2). In the plane, any two vectors perpendicular to a given nonzero vector must be parallel, so $\nabla f(x_0, y_0) = \lambda \nabla g(x_0, y_0)$ for some number λ. If $\lambda \neq 0$, the tangent line to the level curve of f through (x_0, y_0), which is perpendicular to $\nabla f(x_0, y_0)$, is also perpendicular to the vector $\nabla g(x_0, y_0)$; the tangent line to C is also perpendicular to $\nabla g(x_0, y_0)$, so the level curves of f and g through (x_0, y_0) must have the same tangent line. This completes the demonstration. ■

There is a nice geometric way of seeing the result above. If the level curves of f and g had different tangent lines at (x_0, y_0), then the level curves would cross one another. It would follow that the level curve C of g would intersect level curves of f for both higher and lower values of f, so the point (x_0, y_0) would not be an extremum (see Fig. 16.4.3).

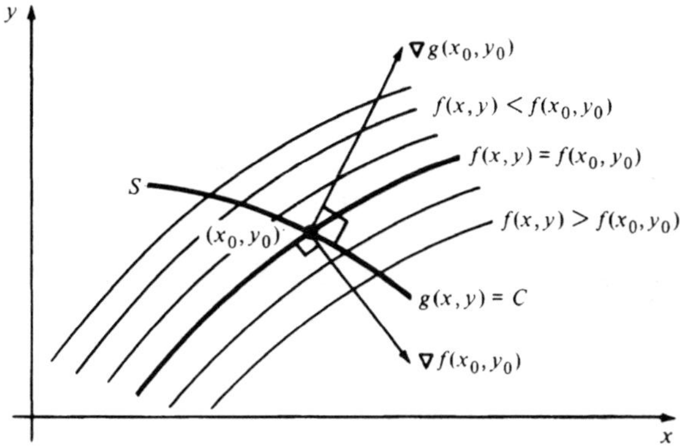

Figure 16.4.3. If $\nabla g(x_0, y_0)$ and $\nabla f(x_0, y_0)$ are not parallel, the level curve $g(x, y) = c$ cuts all nearby level curves of f.

In some problems, it is easiest to use the geometric condition of tangency directly. More often, however, we look for a point (x_0, y_0) on C and a constant λ, called a *Lagrange multiplier*, such that $\nabla f(x_0, y_0) = \lambda \nabla g(x_0, y_0)$. This means we wish to solve the three simultaneous equations

$$\begin{aligned} f_x(x, y) &= \lambda g_x(x, y), \\ f_y(x, y) &= \lambda g_y(x, y), \\ g(x, y) &= c \end{aligned} \tag{1}$$

for the three unknown quantities x, y, and λ. Another way of looking at equations (1) is that we seek the critical points of the auxiliary function $k(x, y, \lambda) = f(x, y) - \lambda[g(x, y) - c]$. (By a critical point of a function of three variables, we mean a point where all three of its partial derivatives vanish.) Here

$$\begin{aligned} k_x &= f_x - \lambda g_x, \\ k_y &= f_y - \lambda g_y, \\ k_\lambda &= c - g, \end{aligned}$$

and setting these equal to zero produces equations (1). We call this attack on the problem the *method of Lagrange multipliers*.

Method of Lagrange Multipliers

To find the extreme points of $f(x, y)$ subject to the constraint $g(x, y) = c$, seek points (x, y) and numbers λ such that (1) holds.

Example 2 Find the extreme values of $f(x, y) = x^2 - y^2$ along the circle S of radius 1 centered at the origin.

Solution The circle S is the level curve $g(x, y) = x^2 + y^2 = 1$, so we want x, y, and λ such that

$$f_x(x, y) = \lambda g_x(x, y),$$

$$f_y(x, y) = \lambda g_y(x, y),$$

$$g(x, y) = 1.$$

That is,

$$2x = \lambda 2x,$$

$$2y = -\lambda 2y,$$

$$x^2 + y^2 = 1.$$

From the first equation, either $x = 0$ or $\lambda = 1$. If $x = 0$, then from the third equation, $y = \pm 1$, and then from the second, $\lambda = -1$. If $\lambda = 1$, then $y = 0$ and $x = \pm 1$; so the eligible points are $(x, y) = (0, \pm 1)$ with $\lambda = -1$ and $(x, y) = (\pm 1, 0)$ with $\lambda = 1$. We must now check them to see if they really are extrema and, if so, what kind. To do this, we evaluate f:

$$f(0, 1) = f(0, -1) = -1,$$

$$f(1, 0) = f(-1, 0) = 1,$$

so the maximum and minimum values are 1 and -1. ▲

Example 3 Find the point(s) furthest from and closest to the origin on the curve $x^6 + y^6 = 1$.

Solution We extremize $f(x, y) = x^2 + y^2$ subject to the constraint $g(x, y) = x^6 + y^6 = 1$. The Lagrange multiplier equations (1) are

$$2x = 6\lambda x^5,$$

$$2y = 6\lambda y^5,$$

$$x^6 + y^6 = 1.$$

If we rewrite the first two of these equations as

$$x(6\lambda x^4 - 2) = 0,$$

$$y(6\lambda y^4 - 2) = 0,$$

we find the solutions $(0, \pm 1)$ and $(\pm 1, 0)$ with $\lambda = \frac{1}{3}$. If x and y are both nonzero, we have $x^4 = 1/3\lambda = y^4$, so $x = \pm y$, and we get the further solutions $(\pm \sqrt[6]{1/2}, \pm \sqrt[6]{1/2})$, with $\lambda = 2^{2/3}/3$.

To tell which points are maxima and minima, we compute; $f(0, \pm 1) = f(\pm 1, 0) = 1$, while $f(\pm \sqrt[6]{1/2}, \pm \sqrt[6]{1/2}) = 2\sqrt[3]{1/2} = 2^{2/3} > 1$, so the points $(0, \pm 1)$ and $(\pm 1, 0)$ are closest to the origin, while $(\pm \sqrt[6]{1/2}, \pm \sqrt[6]{1/2})$ are farthest (see Fig. 16.4.4). ▲

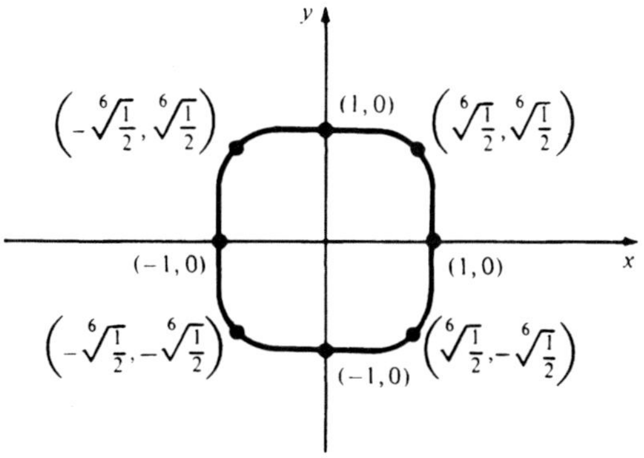

Figure 16.4.4. Extreme points of $x^2 + y^2$ on the curve $x^6 + y^6 = 1$.

For functions of three variables subject to a constraint, there is a similar method. (See Review Exercise 44 if there are two constraints.) Thus, if we are extremizing $f(x, y, z)$ subject to the constraint $g(x, y, z) = c$, we can proceed as follows (see Exercise 23).

Method 1. Find points (x_0, y_0, z_0) and a number λ such that

$$\nabla f(x_0, y_0, z_0) = \lambda \nabla g(x_0, y_0, z_0),$$

and

$$g(x_0, y_0, z_0) = c,$$

or

Method 2. Find critical points of the auxiliary function of four variables given by

$$k(x, y, z, \lambda) = f(x, y, z) - \lambda[g(x, y, z) - c].$$

Example 4 The density of a metallic spherical surface $x^2 + y^2 + z^2 = 4$ is given by $\rho(x, y, z) = 2 + xz + y^2$. Find the places where the density is highest and lowest.

Solution We want to extremize $\rho(x, y, z)$ subject to the constraint $g(x, y, z) = x^2 + y^2 + z^2 = 4$. Using either method 1 or 2 above gives the equations

$$\begin{aligned} \rho_x &= \lambda g_x \\ \rho_y &= \lambda g_y \\ \rho_z &= \lambda g_z \\ g &= 4 \end{aligned} \qquad \text{i.e.} \qquad \begin{aligned} z &= 2\lambda x \\ 2y &= 2\lambda y \\ x &= 2\lambda z \\ x^2 + y^2 + z^2 &= 4 \end{aligned}$$

If $y \neq 0$ then $\lambda = 1$ from the second equation, and so $z = 2x$ and $x = 2z$ which implies $x = z = 0$. From the last equation, $y = \pm 2$. If $y = 0$ then we have

$$z = 2\lambda x, \quad x = 2\lambda z \quad \text{and} \quad x^2 + z^2 = 4.$$

Thus $z = 4\lambda^2 z$, so if $z \neq 0$ then $\lambda = \pm 1/2$, so $x = \pm z$. If $x = z$, then from the last equation $x = z = \pm\sqrt{2}$. If $x = -z$, then $x = \pm\sqrt{2}$ and $z = \mp\sqrt{2}$. The case $y = 0$ and $z = 0$ cannot occur (why?).

Thus we have six possible extrema:

$$(0, \pm 2, 0), \qquad (\pm\sqrt{2}, 0, \pm\sqrt{2}), \qquad (\pm\sqrt{2}, 0, \mp\sqrt{2}).$$

Evaluating ρ at these six points, we find that ρ is a maximum at the two points $(0, \pm 2, 0)$ (where ρ is 6) and a minimum at the two points $(\pm\sqrt{2}, 0, \mp\sqrt{2})$ (where ρ is 0). ▲

The multiplier λ was introduced as an "artificial" device enabling us to find maxima and minima, but sometimes it represents something meaningful.

Example 5 Suppose that the output of a manufacturing firm is a quantity Q of product which is a function $f(K, L)$ of the amount K of capital equipment or investment and the amount L of labor used. If the price of labor is p, the price of capital is q, and the firm can spend no more than B dollars, how do you find the amount of capital and labor to maximize the output Q?

Solution It is useful to think about the problem before applying our machinery. We would expect that if the amount of capital or labor is increased, then the output Q should also increase; that is,

$$\frac{\partial Q}{\partial K} \geqslant 0, \quad \text{and} \quad \frac{\partial Q}{\partial L} \geqslant 0.$$

We also expect that as more and more labor is added to a given amount of capital equipment, we get less and less additional output for our effort; that is,

$$\frac{\partial^2 Q}{\partial L^2} < 0.$$

Similarly,

$$\frac{\partial^2 Q}{\partial K^2} < 0.$$

It is thus reasonable to expect the level curves of output (called *isoquants*) $Q = f(K, L) = c$ to look something like the curves sketched in Fig. 16.4.5, with $c_1 < c_2 < c_3$.

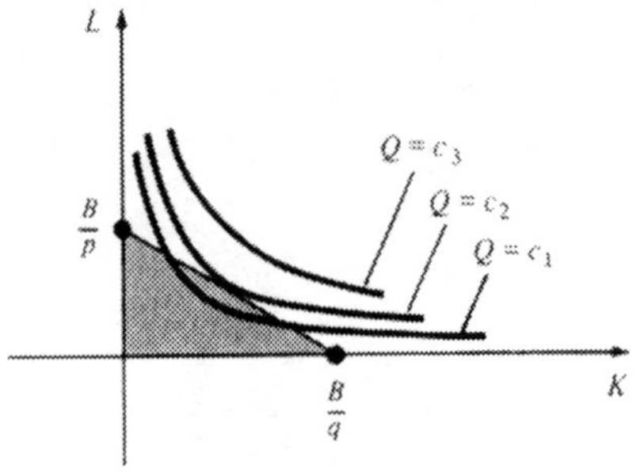

Figure 16.4.5. What is the largest value of Q in the shaded triangle?

We can interpret the convexity of the isoquants as follows. As you move to the right along a given isoquant, it takes more and more capital to replace a unit of labor and still produce the same output. The budget constraint means that we must stay inside the triangle bounded by the axes and the line $pL + qK = B$. Geometrically, it is clear that we produce the most by spending all our money in such a way as to pick the isoquant which just touches, but does not cross, the budget line.

Since the maximum point lies on the boundary of our domain, to find it we apply the method of Lagrange multipliers. To maximize $Q = f(K, L)$ subject to the constraint $pL + qK = B$, we look for critical points of the auxiliary function,

$$h(K, L, \lambda) = f(K, L) - \lambda(pL + qK - B);$$

so we want

$$\frac{\partial Q}{\partial K} = \lambda q, \qquad \frac{\partial Q}{\partial L} = \lambda p, \qquad pL + qK = B.$$

These are the conditions we must meet in order to maximize output. (We will work out a specific case in Example 6.)

In this example, λ does represent something interesting and useful. Let $k = qK$ and $l = pL$, so that k is the dollar value of the capital used and l is the dollar value of the labor used. Then the first two equations become

$$\frac{\partial Q}{\partial k} = \frac{1}{q}\frac{\partial Q}{\partial K} = \lambda = \frac{1}{p}\frac{\partial Q}{\partial L} = \frac{\partial Q}{\partial l}.$$

Thus, at the optimum production point, the marginal change in output per dollar's worth of additional capital investment is equal to the marginal change of output per dollar's worth of additional labor, and λ is this common value. At the optimum point, the exchange of a dollar's worth of capital for a dollar's worth of labor does not change the output. Away from the optimum point the marginal outputs are different, and one exchange or the other will increase the output.[5] ▲

Example 6 Carry out the analysis of Example 5 for the production function $Q(K,L) = AK^{\alpha}L^{1-\alpha}$, where A and α are positive constants and $\alpha < 1$. This *Cobb–Douglas production function* is sometimes used as a simple model for the national economy. Then Q is the output of the entire economy for a given input of capital and labor.

Solution The level curves of output are of the form $AK^{\alpha}L^{1-\alpha} = c$, or, solving for L,

$$L = \left(\frac{c}{A}\right)^{1/(1-\alpha)}K^{\alpha/(\alpha-1)}.$$

Since $\alpha/(\alpha - 1) < 0$, these curves do look like those in Fig. 16.4.5. The partial derivatives of Q are

$$\frac{\partial Q}{\partial K} = \alpha A K^{\alpha-1}L^{1-\alpha} \quad \text{and} \quad \frac{\partial Q}{\partial L} = (1-\alpha)AK^{\alpha}L^{-\alpha},$$

so there are no critical points except on the axes, where $Q = 0$. Thus the maximum must lie on the budget line $pL + qK = B$. The method of Lagrange multipliers gives the equations

$$\alpha A K^{\alpha-1}L^{1-\alpha} = \lambda q,$$
$$(1-\alpha)AK^{\alpha}L^{-\alpha} = \lambda p,$$
$$pL + qK = B.$$

Eliminating λ from the first two equations gives

$$\alpha p L = (1-\alpha)qK,$$

and from the third equation we obtain

$$K = \frac{\alpha B}{q} \quad \text{and} \quad L = \frac{(1-\alpha)B}{p}. \ \blacktriangle$$

[5] More of this type of mathematical analysis in economics can be found in *Microeconomic Theory*, by James Henderson and Richard Quandt, McGraw-Hill (1958). This reference discusses a second derivative test for the Lagrange multiplier method. [See also J. Marsden and A. Tromba, *Vector Calculus*, Second Edition, Freeman (1981).]

Exercises for Section 16.4

Use the method of Example 1 to find the extreme values of the functions in Exercises 1–4 on the disk $x^2 + y^2 \leqslant 1$.

1. $f(x, y) = 2x^2 + 3y^2$
2. $f(x, y) = xy + 5y$
3. $f(x, y) = 5x^2 - 2y^2 + 10$
4. $f(x, y) = 3xy - y + 5$

Find the extrema of f subject to the stated constraints in Exercises 5–12.

5. $f(x, y) = 3x + 2y$; $2x^2 + 3y^2 \leqslant 3$.
6. $f(x, y) = xy$; $2x + 3y \leqslant 10$, $0 \leqslant x$, $0 \leqslant y$.
7. $f(x, y) = x + y$; $x^2 + y^2 = 1$.
8. $f(x, y) = x - y$; $x^2 - y^2 = 2$.
9. $f(x, y) = xy$; $x + y = 1$.
10. $f(x, y) = \cos^2 x + \cos^2 y$; $x + y = \pi/4$.
11. $f(x, y) = x - 3y$; $x^2 + y^2 = 1$.
12. $f(x, y) = x^2 + y^2$; $x^4 + y^4 = 2$.

13. Cascade Container Company produces a cardboard shipping crate at three different plants in amounts x, y, z, respectively, producing an annual revenue of $R(x, y, z) = 8xyz^2 - 200{,}000(x + y + z)$. The company is to produce 100,000 units annually. How should production be handled to maximize the revenue?
14. The temperature T on the spherical surface $x^2 + y^2 + z^2 = 1$ satisfies the equation $T(x, y, z) = xz + yz$. Find all the hot spots.
15. A rectangular mirror with area A square feet is to have trim along the edges. If the trim along the horizontal edges costs p cents per foot and that for the vertical edges costs q cents per foot, find the dimensions which will minimize the total cost.
16. The Baraboo, Wisconsin, plant of International Widget Co. uses aluminum, iron, and magnesium to produce high-quality widgets. The quantity of widgets which may be produced using x tons of aluminum, y tons of iron, and z tons of magnesium is $Q(x, y, z) = xyz$. The cost of raw materials is aluminum, \$6 per ton; iron, \$4 per ton; and magnesium, \$8 per ton. How many tons each of aluminum, iron, and magnesium should be used to manufacture 1000 widgets at the lowest possible cost? [*Hint:* You want an extreme value for what function? Subject to what constraint?]
17. A water main consists of two sections of pipe of fixed lengths, l_1, l_2 carrying fixed amounts Q_1 and Q_2 liters per second. For a given total loss of head h, the (variable) diameters D_1, D_2 of the pipe will result in a minimum cost if
$$C = l_1(a + bD_1) + l_2(a + bD_2) = \text{minimum}$$
subject to the condition
$$h = \frac{cl_1 Q_1^2}{D_1^5} + \frac{cl_2 Q_2^2}{D_2^5}.$$
Find the ratio D_2/D_1.
18. A firm uses wool and cotton fiber to produce cloth. The amount of cloth produced is given by $Q(x, y) = xy - x - y + 1$, where x is the number of pounds of wool, y is the number of pounds of cotton, and $x > 1$ and $y > 1$. If wool costs p dollars per pound, cotton costs q dollars per pound, and the firm can spend B dollars on material, what should the mix of cotton and wool be to produce the most cloth?
19. Let $f(x, y) = x^2 + xy + y^2$.
    (a) Find the maximum and minimum points and values of f along the circle $x^2 + y^2 = 1$.
    (b) Moving counterclockwise along the circle $x^2 + y^2 = 1$, is the function increasing or decreasing at the points $(\pm 1, 0)$ and $(0, \pm 1)$?
    (c) Find extreme points and values for f in the disk D consisting of all (x, y) such that $x^2 + y^2 \leqslant 1$.
20. Locate extreme points and values for the function $f(x, y) = x^2 + y^2 - x - y + 1$ in the disk $x^2 + y^2 \leqslant 1$.
21. A transformer is built from wire of cross sections q_1 and q_2 wound with n_1 and n_2 turns onto the primary and secondary coils, respectively. The corresponding currents are i_1 and i_2. The thickness x of the primary winding and the thickness y of the secondary winding will result in minimum copper loss if
$$C = \frac{\rho n_1 \pi (D_1 + x) i_1^2}{q_1} + \frac{\rho n_2 \pi (D_2 - y) i_2^2}{q_2}$$
is a minimum. The resistivity ρ and iron core diameters D_1, D_2 are constants.
    (a) From transformer theory, $n_1 i_1 = n_2 i_2 =$ constant. By an argument involving insulation thickness, one can show that $q_1 = \alpha x h / n_1$ and $q_2 = \alpha y h / n_2$, where α and h are constants. Use these relations to simplify the expression for C.
    (b) Physical constraints give $x + y = \frac{1}{2}(D_2 - D_1)$. Apply the method of Lagrange multipliers to find x and y which minimize C subject to this condition.
22. The state of Megalomania occupies the region $x^4 + 2y^4 \leqslant 30{,}000$. The altitude at point (x, y) is $\frac{1}{8}xy + 200x$ meters above sea level. Where are the highest and lowest points in the state?
★23. Suppose that (x_0, y_0, z_0) is a critical point for the restriction of the function $f(x, y, z)$ to the surface $g(x, y, z) = c$. The method of Lagrange multipliers tells us that in this case the partial derivatives with respect to x, y, z, and λ of the function of four variables
$$k(x, y, z, \lambda) = f(x, y, z) - \lambda[g(x, y, z) - c]$$
are equal to zero.

(a) Interpret this fact as the statement about the gradient vectors of f and g at (x_0, y_0, z_0).
(b) Find the maxima and minima of xyz on the sphere $x^2 + y^2 + z^2 = 1$.
(c) Rework Example 5, Section 16.3 by minimizing a function of x, y, and h subject to the constraint $xyh = 256$.

Review Exercises for Chapter 16

Calculate the gradients of the functions in Exercises 1–4.

1. $f(x, y) = e^{xy} + \cos(xy)$
2. $f(x, y)\dfrac{x^2 - y^2}{x^2 + y^2}$
3. $f(x, y) = e^{x^2} - \cos(xy^2)$
4. $f(x, y) = \tan^{-1}(x^2 + y^2)$

In Exercises 5–8, calculate (a) the directional derivative of the function in the direction $\mathbf{d} = \mathbf{i}/\sqrt{2} - \mathbf{j}/\sqrt{2}$ and (b) the direction in which the function is increasing most rapidly at the given point.

5. $f(x, y) = \sin(x^3 - 2y^3)$; $(1, -1)$
6. $f(x, y) = \dfrac{x - y}{x + y}$; $(0, 1)$
7. $f(x, y) = \exp(x^2 - y^2 + 2)$; $(-1, 2)$
8. $f(x, y) = \sin^{-1}(x - 2y^2)$; $(0, 0)$

In Exercises 9–12, find the equation of the tangent plane to the surface at the indicated point.

9. $z = x^3 + 2y^2$; $(1, 1, 3)$
10. $z = \cos(x^2 + y^2)$; $(0, 0, 1)$
11. $x^2 + y^2 + z^2 = 1$; $\left(\dfrac{1}{\sqrt{3}}, \dfrac{1}{\sqrt{3}}, \dfrac{1}{\sqrt{3}}\right)$
12. $x^3 + y^3 + z^3 = 3$; $(1, 1, 1)$

Suppose that $x = f(t)$ and $y = g(t)$ satisfy the relations in Exercises 13–16. Relate dx/dt and dy/dt.

13. $x^2 + xy + y^2 = 1$
14. $\cos(x - y) = \frac{1}{2}$
15. $(x + y)^3 + (x - y)^3 = 42e^{-\pi}$
16. $\tan^{-1}(x - y) = \pi/4$

Suppose that x and y are related by the equations given in Exercises 17–20. Find dy/dx at the indicated points.

17. $x + \cos y = 1$, $x = 1$, $y = \pi/2$.
18. $x^4 + y^4 = 17$, $x = -1$, $y = 2$.
19. $\int_x^y u^2\,du = 5\frac{1}{3}$, $x = -2$, $y = 2$.
20. $\int_x^y f(t)\,dt = 7$, $x = 2$, $y = 4$; if $\int_2^4 f(t)\,dt = 7$, $f(2) = 3$, $f'(2) = 5$, $f(4) = 11$, $f'(4) = 13$.

Find and classify (as maxima, minima or saddles) the critical points of the functions in Exercises 21–24.

21. $f(x, y) = x^2 - 6xy - y^2$
22. $f(x, y) = 2x^2 - y^2 + 5xy$
23. $f(x, y) = \exp(x^2 - y^2)$
24. $f(x, y) = \sin(x^2 + y^2)$ (consider only $(0, 0)$).
25. Prove that

$$z = \frac{3x^4 - 4x^3 - 12x^2 + 18}{12(1 + 4y^2)}$$

has one local maximum, one local minimum, and one saddle point. (The computer-generated graph is shown in Fig. 16.R.1.)

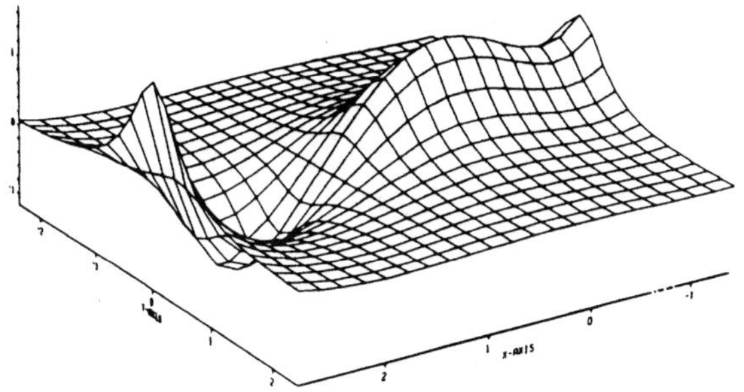

Figure 16.R.1. Computer-generated graph of $z = (3x^4 - 4x^3 - 12x^2 + 18)/12(1 + 4y^2)$.

26. Find the maxima, minima, and saddles of the function $z = (2 + \cos \pi x)(\sin \pi y)$, which is graphed in Fig. 16.R.2.

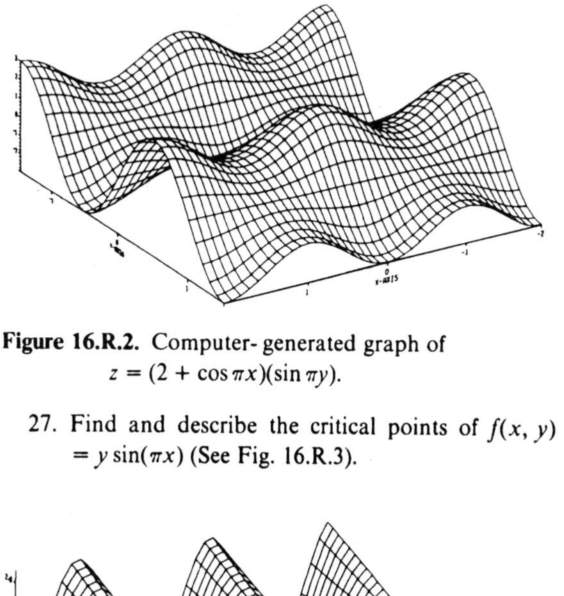

Figure 16.R.2. Computer-generated graph of $z = (2 + \cos \pi x)(\sin \pi y)$.

27. Find and describe the critical points of $f(x, y) = y \sin(\pi x)$ (See Fig. 16.R.3).

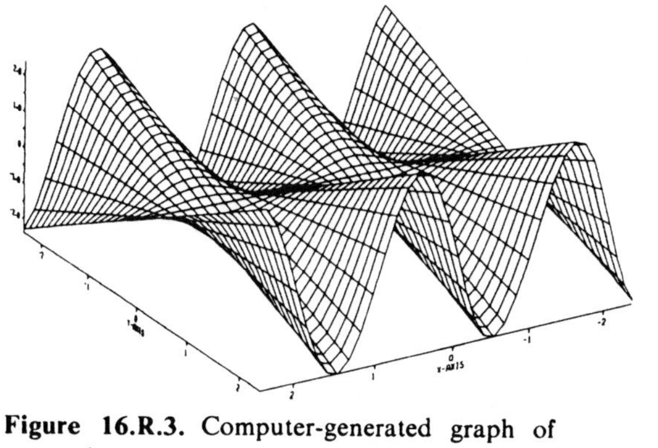

Figure 16.R.3. Computer-generated graph of $z = y \sin(\pi x)$.

28. A computer-generated graph of the function $z = \sin(\pi x)/(1 + y^2)$ is shown in Fig. 16.R.4. Verify that this function has alternating maxima and minima on the x axis with no other critical points.

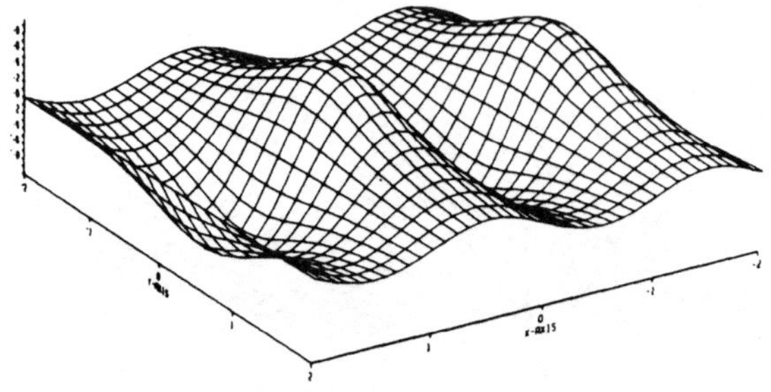

Figure 16.R.4. Computer-generated graph of $\sin(\pi x)/(1 + y^2)$.

29. In meteorology, the *pressure gradient* **G** is a vector quantity that points from regions of high pressure to regions of low pressure, normal to the lines of constant pressure (*isobars*).
   (a) In an xy coordinate system,
   $$\mathbf{G} = -\frac{\partial P}{\partial x}\mathbf{i} - \frac{\partial P}{\partial y}\mathbf{j}.$$
   Write a formula for the magnitude of the pressure gradient.
   (b) If the horizontal pressure gradient provided the only horizontal force acting on the air, the wind would blow directly across the isobars in the direction of **G**, and for a given air mass, with acceleration proportional to the magnitude of **G**. Explain, using Newton's second law.
   (c) *Buys–Ballot's law states*: "If in the Northern Hemisphere, you stand with your back to the wind, the high pressure is on your right and the low pressure on your left." Draw a figure and introduce xy coordinates so that **G** points in the proper direction.
   (d) State and graphically illustrate Buys–Ballot's law for the Southern Hemisphere, in which the orientation of high and low pressure is reversed.
30. A sphere of mass m, radius a, and uniform density has *potential* u and *gravitational force* **F**, at a distance r from the center $(0,0,0)$, given by
$$u = \frac{3m}{2a} - \frac{mr^2}{2a^3}, \qquad \mathbf{F} = -\frac{m}{a^3}\mathbf{r} \quad (r \leqslant a);$$
$$u = \frac{m}{r}, \qquad \mathbf{F} = -\frac{m}{r^3}\mathbf{r} \quad (r > a).$$
   Here, $r = \|\mathbf{r}\|$, $\mathbf{r} = x\mathbf{i} + y\mathbf{j} + z\mathbf{k}$.
   (a) Verify that $\mathbf{F} = \nabla u$ on the inside and outside of the sphere.
   (b) Check that u satisfies Poisson's equation: $\partial^2u/\partial x^2 + \partial^2u/\partial y^2 + \partial^2u/\partial z^2 = \text{constant}$ inside the sphere.
   (c) Show u satisfies Laplace's equation: $\partial^2u/\partial x^2 + \partial^2u/\partial y^2 + \partial^2u/\partial z^2 = 0$ outside the sphere.
31. Minimize the distance from $(0,0,0)$ to each of the following surfaces. [*Hint*: Write the square of the distance as a function of x and y.]
   (a) $z = \sqrt{x^2 - 1}$;
   (b) $z = 6xy + 7$;
   (c) $z = 1/xy$.
32. Suppose that $f(x, y) = x^2 + y$. Find the maximum and minimum values of f for (x, y) on a circle of radius 1 centered at the origin in two ways:
   (a) By parametrizing the circle.
   (b) By Lagrange multipliers.

In Exercises 33–36, find the extrema of the given functions subject to the given constraints.

33. $f(x, y) = x^2 - 2xy + 2y^2$; $x^2 + y^2 = 1$.
34. $f(x, y) = xy - y^2$; $x^2 + y^2 = 1$.
35. $f(x, y) = \cos(x^2 - y^2)$; $x^2 + y^2 = 1$.
36. $f(x, y) = \dfrac{x^2 - y^2}{x^2 + y^2}$; $x + y = 1$.
37. An irrigation canal in Arizona has concrete sides and bottom with trapezoidal cross section of area $A = y(x + y\tan\theta)$ and wetted perimeter $P = x + 2y/\cos\theta$, where x = bottom width, y = water depth, θ = side inclination, measured from vertical. The best design for fixed inclination θ is found by solving P = minimum subject to the condition A = constant. Show that $y^2 = A\cos\theta/(2 - \sin\theta)$.
38. The friction in an open-air aqueduct is proportional to the wetted perimeter of the cross section. Show that the best form of a rectangular cross section is one with the width x equal to twice the depth y, by solving the problem perimeter $= 2y + x$ = minimum, area $= xy$ = constant.
39. (a) Suppose that $z = f(x, y)$ is defined, has continuous second partial derivatives, and is harmonic:
$$\frac{\partial^2 z}{\partial x^2} + \frac{\partial^2 z}{\partial y^2} = 0.$$
   Assume that $(\partial^2 z/\partial x^2)(x_0, y_0) \neq 0$. Prove that f cannot have a local maximum or minimum at (x_0, y_0).
   (b) Conclude from (a) that if $f(x, y)$ is harmonic on the region $x^2 + y^2 < 1$ and is zero on $x^2 + y^2 = 1$, then f is zero everywhere on the unit disk. [*Hint:* Where are the maximum and minimum values of f?]

40. (a) Suppose that $u = f(x, y)$ and $v = g(x, y)$ have continuous partial derivatives which satisfy the *Cauchy–Riemann equations*:

$$\frac{\partial u}{\partial x} = \frac{\partial v}{\partial y} \quad \text{and} \quad \frac{\partial u}{\partial y} = -\frac{\partial v}{\partial x}$$

Show that the level curves of u are perpendicular to the level curves of v.

(b) Confirm this result for the functions $u = x^2 - y^2$ and $v = 2xy$. Sketch some of the level curves of these functions (all on the same set of axes).

41. Consider the two surfaces

$$S_1 : x^2 + y^2 + z^2 = f(x, y, z) = 6;$$

$$S_2 : 2x^2 + 3y^2 + z^2 = g(x, y, z) = 9.$$

(a) Find the normal vectors and tangent planes to S_1 and S_2 at $(1, 1, 2)$.
(b) Find the angle between the tangent planes.
(c) Find an expression for the line tangent at $(1, 1, 2)$ to the curve of intersection of S_1 and S_2. [*Hint:* It lies in both tangent planes.]

42. Repeat Exercise 41 for the surfaces $x^2 - y^2 + z^2 = 1$ and $2x^2 - y^2 + 5z^2 = 6$ at $(1, 1, -1)$.

43. (a) Consider the graph of a function $f(x, y)$ (Figure 16.R.5). Let (x_0, y_0) lie on a level curve C, so $\nabla f(x_0, y_0)$ is perpendicular to this curve. Show that the tangent plane to the graph is the plane that (i) contains the line perpendicular to $\nabla f(x_0, y_0)$ and lying in the horizontal plane $z = f(x_0, y_0)$, and (ii) has slope $\|\nabla f(x_0, y_0)\|$ relative to the xy plane. (By the slope of a plane P relative to the xy plane, we mean the tangent of the angle θ, $0 \leqslant \theta \leqslant \pi$, between the upward pointing normal $\mathbf{p}$ to P and the unit vector $\mathbf{k}$.)

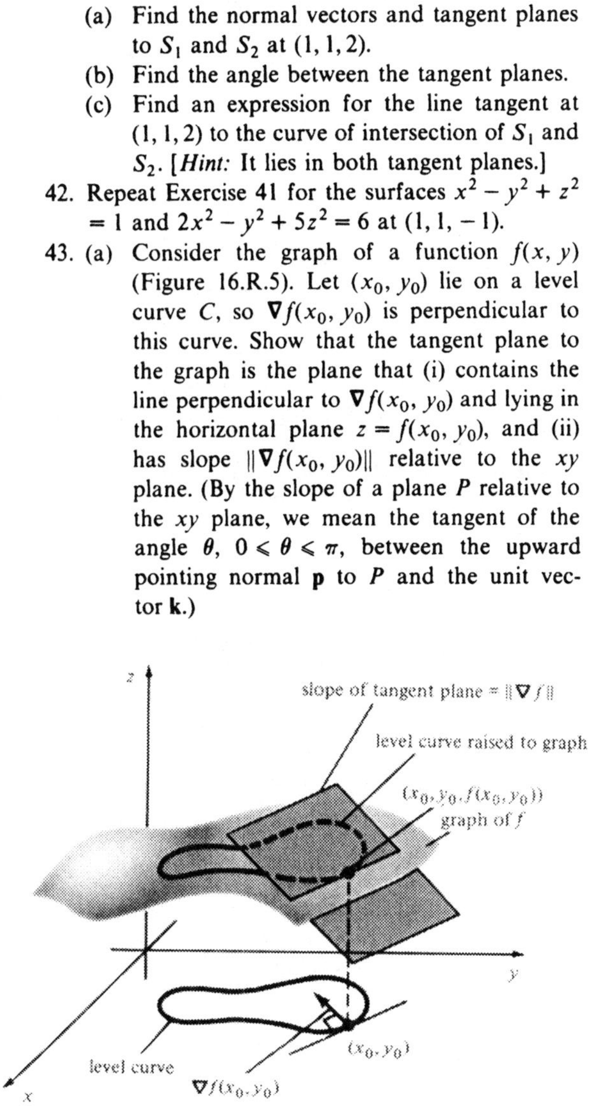

Figure 16.R.5. Relationship between the gradient of a function and the plane tangent to the function's graph (Exercise 43(a)).

(b) Use this method to show that the tangent plane of the graph of

$$f(x, y) = (x + \cos y)x^2$$

at $(1, 0, 2)$ is as sketched in Figure 16.R.6.

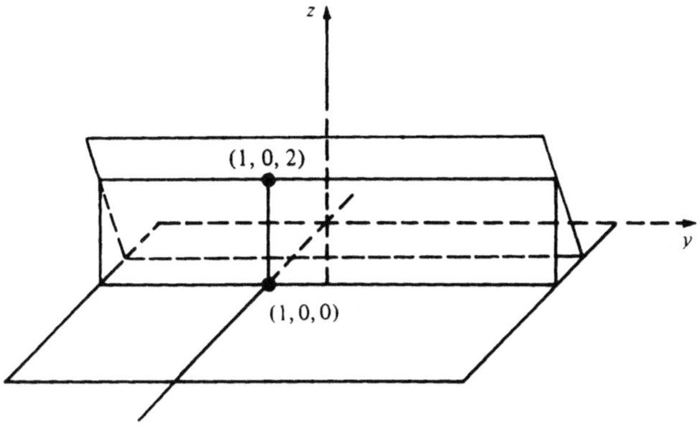

Figure 16.R.6. The plane referred to in Exercise 43(b).

44. (a) Use a geometric argument to demonstrate that if $f(x, y, z)$ is extremized at (x_0, y_0, z_0) subject to two constraints $g_1(x, y, z) = c_1$ and $g_2(x, y, z) = c_2$, then there should exist λ_1 and λ_2 such that

$$\nabla f(x_0, y_0, z_0) = \lambda_1 \nabla g_1(x_0, y_0, z_0) + \lambda_2 \nabla g_2(x_0, y_0, z_0).$$

(b) Extremize $f(x, y, z) = x - y + z$ subject to the constraints $x^2 + y^2 + z^2 = 1$ and $x + y + 2z = 1$.

45. A pipeline of length l is to be constructed from one pipe of length l_1 and diameter D_1 connected to another pipe of length l_2 and diameter D_2. The finished pipe must deliver Q liters per second at pressure loss h. The expense is reduced to a minimum by minimization of the cost $C = l_1(a + bD_1) + l_2(a + bD_2)$ $(a, b = \text{constants})$ subject to the conditions

$$l_1 + l_2 = l, \quad \text{and}$$

$$h = kQ^m \left\{ \frac{l_1}{D_1^{m_1}} + \frac{l_2}{D_2^{m_2}} \right\},$$

where k, m, m_1, m_2 are constants. Show that $D_1 = D_2$ in order to achieve the minimum cost. [*Hint*: As in Exercise 44, the partials of $C - \lambda_1 l - \lambda_2 h$ with respect to l_1, l_2, D_1 and D_2 must all be zero for suitable constants λ_1 and λ_2.]

46. Ammonia, NH_3, is to be produced at fixed temperature T and pressure p. The pressures of N_2, H_2, NH_3 are labeled as u, v, w, and are known to satisfy $u + v + w = p$, $w^2 = cuv^3$ (c = positive constant). Due to the nature of the reaction $N_2 + 3H_2 \rightleftharpoons 2NH_3$, the maximum ammonia production occurs for w = maximum. Find the maximum pressure $w_{\max}$.

For Exercises 47–50, consider the level curves for the function $f(x, y)$ shown in Figure 16.R.7. Find or estimate the maximum value of $f(x, y)$ for each of the given constraint conditions.

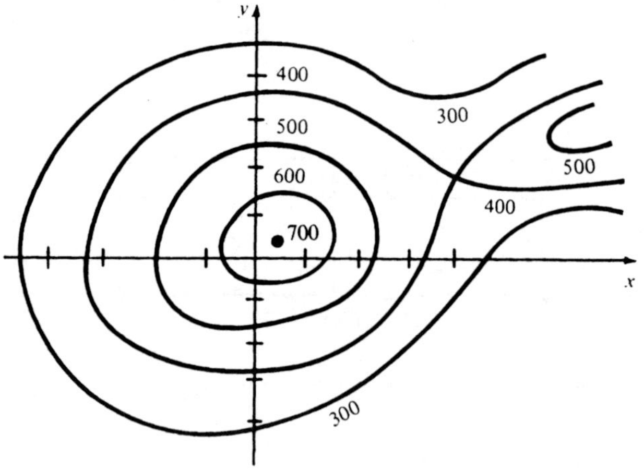

Figure 16.R.7. Level curves of a function f.

47. $x \geqslant 0, y \geqslant 0, y = -\frac{2}{3}x + 2$
48. $x^2 + y^2 \leqslant 4$
49. $x^2 + y^2 = 4$
50. $x = 4, 0 \leqslant y \leqslant 4$

51. Refer to Figure 16.R.7. The function f has exactly one saddle point. Find it.
52. Refer to Figure 16.R.7. There are two points on the graph $z = f(x, y)$ at which the tangent plane is horizontal. Give the equation of the tangent plane at each such point.
53. (a) Let y be defined implicitly by
$$x^2 + y^3 + e^y = 0.$$
Compute dy/dx in terms of x and y.
(b) Recall from p. 810 that
$$\frac{dy}{dx} = -\frac{\partial F/\partial x}{\partial F/\partial y} \quad \text{if} \quad \frac{\partial F}{\partial y} \neq 0.$$
Obtain a formula analogous to this if y_1, y_2 are defined implicitly by
$$F_1(x, y_1(x), y_2(x)) = 0,$$
$$F_2(x, y_1(x), y_2(x)) = 0.$$
(c) Let y_1 and y_2 be defined by
$$x^2 + y_1^2 = \cos x,$$
$$x^2 - y_2^2 = \sin x.$$
Compute dy_1/dx and dy_2/dx using (b).
54. Thermodynamics texts[6] use the relationship
$$\left(\frac{\partial y}{\partial x}\right)\left(\frac{\partial z}{\partial y}\right)\left(\frac{\partial x}{\partial z}\right) = -1.$$
Explain the meaning of this equation and prove that it is true. [*Hint:* Start with a relationship $F(x, y, z) = 0$ that implicitly defines $x = f(y, z)$, $y = g(x, z)$, and $z = h(x, y)$ and differentiate.]
55. (a) Suppose that $\mathbf{F}(x, y) = P(x, y)\mathbf{i} + Q(x, y)\mathbf{j}$. Show that if there is a function $f(x, y)$ with continuous second partial derivatives such that $\mathbf{F} = \nabla f$, then $P_y = Q_x$.
(b) Suppose that
$$\mathbf{F}(x, y, z) = P(x, y, z)\mathbf{i} + Q(x, y, z)\mathbf{j} + R(x, y, z)\mathbf{k}.$$
Show that if there is a function $f(x, y, z)$ with continuous second partial derivatives such that $\mathbf{F} = \nabla f$, then
$$P_y = Q_x, \qquad P_z = R_x, \qquad Q_z = R_y.$$
(c) Let $\mathbf{F} = 3xy\mathbf{i} - ye^x\mathbf{j}$. Is there an f such that $\mathbf{F} = \nabla f$?

★56. (Continuation of Exercise 55.) Suppose that P and Q have continuous partial derivatives everywhere in the xy plane and that $P_y = Q_x$. Follow the steps below to prove that there is a function f such that $\nabla f = P\mathbf{i} + Q\mathbf{j}$; that is, $f_x = P$ and $f_y = Q$.
(a) Let $g(x, y)$ be an antiderivative of Q with respect to y; that is, $g_y = Q$. Establish that $P - g_x$ is a function of x alone by showing that $(P - g_x)_y = 0$.
(b) If $P - g_x = 0$, then we may simply take $g = f$. Otherwise let $h(x)$ be an antiderivative of $P - g_x$; that is, $h'(x) = P - g_x$. Show that $f(x, y) = g(x, y) + y(x)$ satisfies $\nabla f = P\mathbf{i} + Q\mathbf{j}$.

★57. For each of the following vector fields $P\mathbf{i} + Q\mathbf{j}$, find a function f such that $f_x = P$ and $f_y = Q$ or show that no such function exists. (See Exercises 55 and 56.)
(a) $(x^2y^2 + 2x)e^{xy^2}\mathbf{i} + 2x^3ye^{xy^2}\mathbf{j}$;
(b) $(x^2y^2 + 2x)e^{xy^3}\mathbf{i} + 2x^3ye^{xy^3}\mathbf{j}$;
(c) $\dfrac{2x}{1 + x^2 + y^2}\mathbf{i} + \dfrac{2y}{1 + x^2 + y^2}\mathbf{j}$;
(d) $\dfrac{2y}{1 + x^2 + y^2}\mathbf{i} + \dfrac{2x}{1 + x^2 + y^2}\mathbf{j}$.

★58. (*The gradient and Laplacian in polar coordinates.*) Let r and θ be polar coordinates in the plane and let f be a given function of (x, y). Write
$$u = f(x, y) = f(r\cos\theta, r\sin\theta).$$
Let $\mathbf{i}_r = \cos\theta\mathbf{i} + \sin\theta\mathbf{j}$ and $\mathbf{i}_\theta = -\sin\theta\mathbf{i} + \cos\theta\mathbf{j}$.
(a) Show that when based at $\mathbf{v} = x\mathbf{i} + y\mathbf{j}$ the vectors $\mathbf{i}_r$ and $\mathbf{i}_\theta$ are orthogonal unit vectors in the directions of increasing r and θ, respectively.
(b) Show that
$$\nabla f = \left(\cos\theta\frac{\partial u}{\partial r} - \frac{\sin\theta}{r}\frac{\partial u}{\partial\theta}\right)\mathbf{i} + \left(\sin\theta\frac{\partial u}{\partial r} + \frac{\cos\theta}{r}\frac{\partial u}{\partial\theta}\right)\mathbf{j}$$

[6] See S. M. Binder, "Mathematical methods in elementary thermodynamics," *J. Chem. Educ.* **43** (1966): 85–92. A prope understanding of partial differentiation can be of significant use in applications; for example, see M. Feinberg, "Constitutiv equation for ideal gas mixtures and ideal solutions as consequences of simple postulates", *Chem. Eng. Sci.* **32** (1977): 75–78.

$$= \frac{\partial u}{\partial r}\mathbf{i}_r + \frac{1}{r}\frac{\partial u}{\partial \theta}\mathbf{i}_\theta .$$

(c) Show that

$$\frac{\partial^2 u}{\partial x^2} + \frac{\partial^2 u}{\partial y^2} = \frac{1}{r}\frac{\partial}{\partial r}\left(r\frac{\partial u}{\partial r}\right) + \frac{1}{r^2}\frac{\partial^2 u}{\partial \theta^2} .$$

★59. Find a family of curves orthogonal to the level curves of $f(x, y) = x^2 - y^2$ as follows:
(a) Find an expression for a vector normal to the level curve of f through (x_0, y_0) at the point (x_0, y_0).
(b) Use this expression to find a vector tangent to the level curve of f through (x_0, y_0) at (x_0, y_0).
(c) Find a function g which has these vectors as its gradient.
(d) Explain why the level curves of g should intersect those of f orthogonally.
(e) Draw a few of the level curves of f and g to illustrate this result.

★60. (a) Figure 16.R.8 shows the graph of the function $z = (x^2 - y^2)/(x^2 + y^2)$. Show that z has different limits if we come in along the x or y axis.

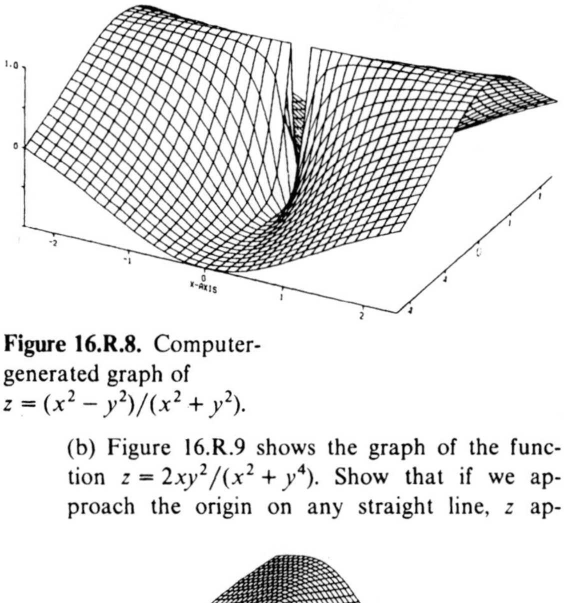

Figure 16.R.8. Computer-generated graph of $z = (x^2 - y^2)/(x^2 + y^2)$.

(b) Figure 16.R.9 shows the graph of the function $z = 2xy^2/(x^2 + y^4)$. Show that if we approach the origin on any straight line, z approaches zero, but z has different limits when $(0,0)$ is approached along the two parabolas $x = \pm y^2$.

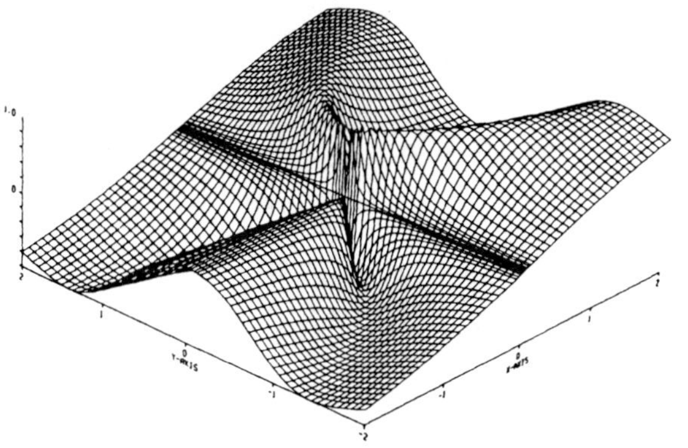

Figure 16.R.9. Computer-generated graph of $z = 2xy^2/(x^2 + y^4)$.

★61. Let $f(x, y) = y^3/(x^2 + y^2)$; $f(0,0) = 0$.
(a) Compute f_x, f_y, $f_x(0,0)$, and $f_y(0,0)$.
(b) Show that, for any θ, the directional derivative $(d/dr)\, f(r\cos\theta, r\sin\theta)|_{r=0}$ exists.
(c) Show that the directional derivatives are not all given by dotting the direction vector with the gradient vector (see Fig. 16.R.10). Why does this not contradict the chain rule?

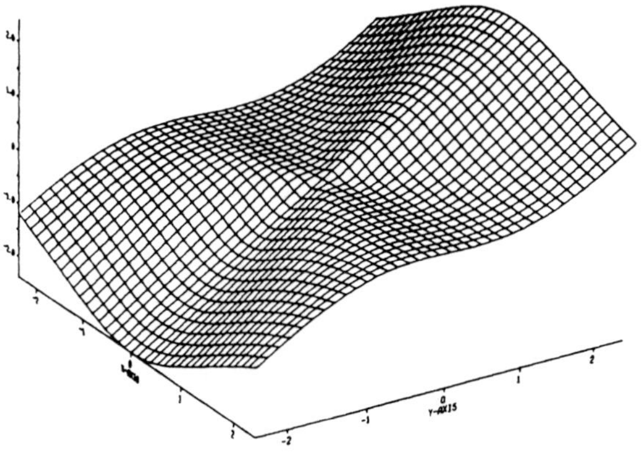

Figure 16.R.10. Computer-generated graph of $z = y^3/(x^2 + y^2)$.

★62. Do the same things as in Exercise 61 for $z = (x^3 - 3xy^2)/(x^2 + y^2)$, which is graphed in Fig. 16.R.11.

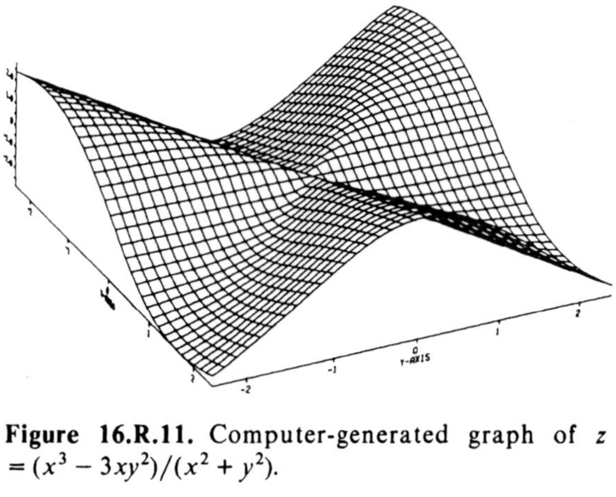

Figure 16.R.11. Computer-generated graph of $z = (x^3 - 3xy^2)/(x^2 + y^2)$.

Chapter 17

Multiple Integration

Functions can be integrated over regions in the plane and in space.

Double and triple integrals enable us to "sum" the values of real-valued functions of two or three variables; we evaluate them by integration with respect to one variable at a time, using the methods of one variable calculus. As in the previous chapters, we concentrate on the basic ideas and methods of calculation, leaving a few of the more theoretical points for a later course.

17.1 The Double Integral and Iterated Integral

The double integral of a non-negative function over a region in the plane is equal to the volume under its graph.

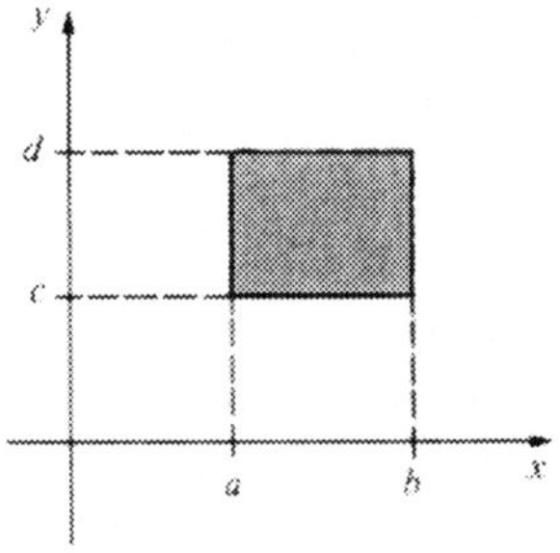

(a) The closed rectangle $a \leqslant x \leqslant b, c \leqslant y \leqslant d$

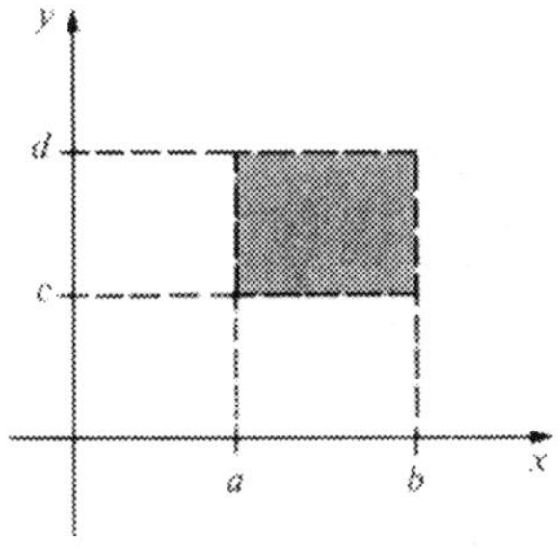

(b) The open rectangle $a < x < b, c < y < d$

Figure 17.1.1. Examples of closed and open rectangles.

The definite integral $\int_a^b f(x)\,dx$, defined in Chapter 4, represents a "sum" of the values of f at the (infinitely many) points of the interval $[a,b]$. To "sum" the values of a function $f(x, y)$ over the points of a region D in the plane, we will define the *double integral* $\iint_D f(x, y)\,dx\,dy$. We recommend a rapid review of Sections 4.1 to 4.5 as preparation for the present section.

Our development of double integrals will be similar to that of definite integrals in Chapter 4. We will give a formal definition first, but the actual calculation of double integrals will be done by reduction to repeated ordinary integrals as explained later in the section, rather than using the formal definition.

The sets in the plane which will play the role of "intervals" in double integration are the rectangles (see Fig. 17.1.1). A *closed rectangle D* consists of all x and y such that $a \leqslant x \leqslant b$ and $c \leqslant y \leqslant d$; it is denoted by $[a,b] \times [c,d]$. The *interior* of D consists of all x and y such that $a < x < b$ and $c < y < d$; it is called an *open rectangle* and is denoted by $(a,b) \times (c,d)$. The *area* of D is the product $(b-a)(d-c)$. Note that the rectangles considered here have their sides parallel to the coordinate axes.

We say that a function $g(x, y)$ defined in $[a,b] \times [c,d]$ is a *step function* provided there are partitions $a = t_0 < t_1 < t_2 < \cdots < t_n = b$ of $[a,b]$ and $c = s_0 < s_1 < s_2 < \cdots < s_m = d$ of $[c,d]$ such that, in each of the mn open rectangles $R_{ij} = (t_{i-1}, t_i) \times (s_{j-1}, s_j)$, $g(x, y)$ has a constant value k_{ij}. The graph of a step function is shown in Fig. 17.1.2.

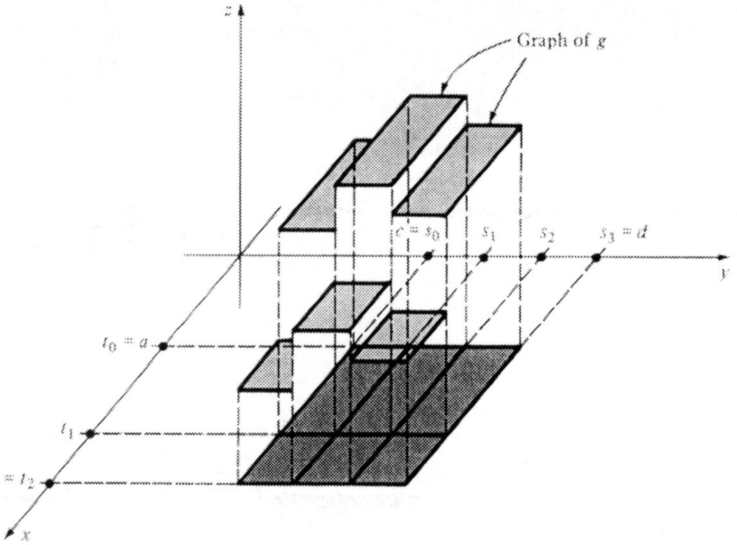

Figure 17.1.2. g is a step function since it is constant on each subrectangle.

By analogy with our definition for step functions of one variable, we define:

$$\iint_D g(x, y)\,dx\,dy = \sum_{i=1,j=1}^{n,m} (k_{ij})(\text{area } R_{ij}) = \sum_{i=1,j=1}^{n,m} (k_{ij})(\Delta t_i)(\Delta s_j),$$

where $\Delta t_i = t_i - t_{i-1}$ and $\Delta s_j = s_j - s_{j-1}$. The summation symbol means that we sum over all i and j, with i ranging from 1 to n and j from 1 to m; there are nm terms in the sum corresponding to the nm rectangles R_{ij}.

If $g(x, y) \geqslant 0$, the integral of g is exactly the volume under its graph. Indeed, the height of the box over the rectangle R_{ij} is k_{ij}, so the volume of the box is $k_{ij} \times$ area $(R_{ij}) = k_{ij}\,\Delta t_i\,\Delta s_j$; the integral of g is the sum of these and so it is the total volume.

Example 1 Let g take values on the rectangles as shown in Fig. 17.1.3. Calculate the integral of g over the rectangle $D = [0,5] \times [0,3]$.

Solution The integral of g is the sum of the values of g times the areas of the rectangles:

$$\iint_D g(x, y)\,dx\,dy = -8 \times 2 + 2 \times 3 + 6 \times 2 + 3 \times 3 + 4 \times 2 - 1 \times 3$$

$$= 16. \blacktriangle$$

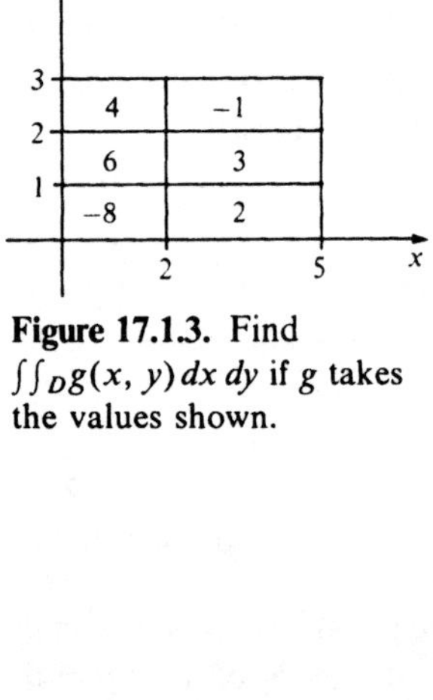

Figure 17.1.3. Find $\iint_D g(x, y)\,dx\,dy$ if g takes the values shown.

We proceed now to define the integral of a function over a closed rectangle in the same manner as we did in Section 4.3.

Upper and Lower Sums

The integral over D of a step function g such that $g(x, y) \leqslant f(x, y)$ on D is called a *lower sum* for f on D. The integral over D of a step function h such that $h(x, y) \geqslant f(x, y)$ on D is called an *upper sum* for f on D.

As in Section 4.3, every lower sum is less than or equal to every upper sum. The integral separates these sets of numbers.

The Double Integral

We say that f is *integrable* on D if there is a number S_0 such that every $S < S_0$ is a lower sum for f on D and every $S > S_0$ is an upper sum. The number S_0 is called the *integral* of f over D and is denoted by

$$\int\int_D f(x, y)\,dx\,dy.$$

Example 2 Let D be the rectangle $0 \leqslant x \leqslant 2$, $1 \leqslant y \leqslant 3$, and let $f(x, y) = x^2y$. Choose a step function $h(x, y) \geqslant f(x, y)$ to show that $\int\int_D f(x, y)\,dx\,dy \leqslant 25$.

Solution The constant function $h(x, y) = 12 \geqslant f(x, y)$ has integral $12 \times 4 = 48$, so we get only the crude estimate $\int\int_D f(x, y)\,dx\,dy \leqslant 48$. To get a better one, divide D into four pieces:

$$D_1 = [0,1] \times [1,2], \qquad D_3 = [1,2] \times [1,2],$$
$$D_2 = [0,1] \times [2,3], \qquad D_4 = [1,2] \times [2,3].$$

Let h be the step function given by taking the maximum value of f on each subrectangle (evaluated at the upper right-hand corner); that is,

$$h(x, y) = 2 \text{ on } D_1, 3 \text{ on } D_2, 8 \text{ on } D_3, \text{ and } 12 \text{ on } D_4.$$

The integral of h is $2 \times 1 + 3 \times 1 + 8 \times 1 + 12 \times 1 = 25$. Since $h \geqslant f$, we get

$$\int\int_D f(x, y)\,dx\,dy \leqslant 25. \blacktriangle$$

The basic properties of the double integral are similar to those of the ordinary integral:

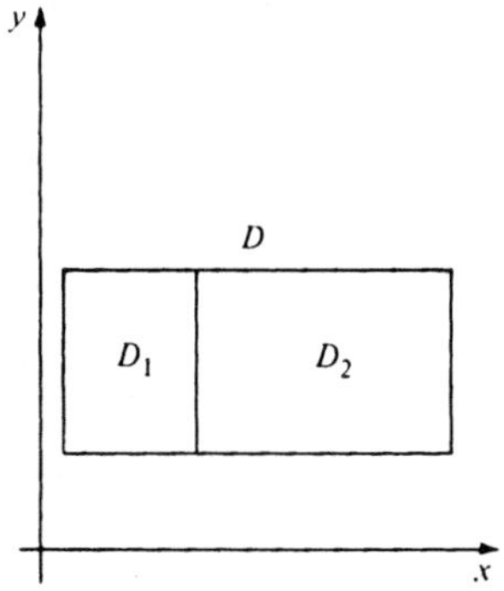

Figure 17.1.4. The rectangle D is divided into two smaller rectangles D_1 and D_2.

Properties of the Double Integral

1. Every continuous function is integrable.
2. If a rectangle D is divided by a line segment into two rectangles D_1 and D_2 (Fig. 17.1.4), and if $f(x, y)$ is integrable on D_1 and D_2, then f is integrable on D and
$$\int\int_D f(x, y)\,dx\,dy = \int\int_{D_1} f(x, y)\,dx\,dy + \int\int_{D_2} f(x, y)\,dx\,dy.$$
3. If f_1 and f_2 are integrable on D and if $f_1 \leqslant f_2$ on D, then
$$\int\int_D f_1(x, y)\,dx\,dy \leqslant \int\int_D f_2(x, y)\,dx\,dy.$$
4. If $f(x, y) = k$ on D,
$$\int\int_D f(x, y)\,dx\,dy = k(\text{area of } D).$$
5. $$\int\int_D [f_1(x, y) + f_2(x, y)]\,dx\,dy = \int\int_D f_1(x, y)\,dx\,dy + \int\int_D f_2(x, y)\,dx\,dy.$$
6. $$\int\int_D cf(x, y)\,dx\,dy = c\int\int_D f(x, y)\,dx\,dy.$$

We omit the proofs of these results since they are similar to the one-variable case. Choosing $k = 1$ in 4, note that $\iint_D dx\,dy =$ area of D.

We observed earlier that if $g(x, y) \geqslant 0$ and g is a step function, then the integral of g is the volume under its graph. If $f(x, y) \geqslant 0$ is any integrable function, then the volume under the graph of f lies between the volumes under the graphs of step functions $g \leqslant f$ and $h \geqslant f$; that is, between lower and upper sums. Since the integral has exactly this property, we conclude, as for functions of one variable, that *the integral of f over D equals the volume under the graph of f if* $f \geqslant 0$ (see Fig. 17.1.5).

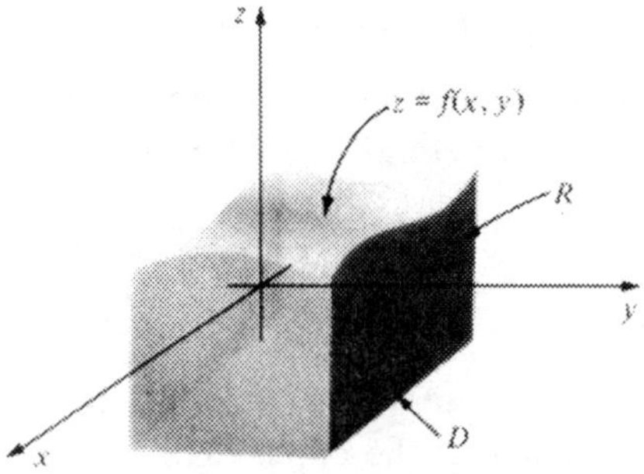

Figure 17.1.5. The volume of R, the region under the graph of f, equals $\iint_D f(x, y)\,dx\,dy$.

For the moment, we can evaluate integrals only approximately or by appealing to geometric formulas for volumes of special solids. Later in this section, we show how the fundamental theorem of calculus can be brought into play.

The double integral has other interpretations besides the volume of the region under the graph of the integrand. For example, suppose that a rectangular plate D has mass density $\rho(x, y)$ grams per square centimeter. Let us argue that $\iint_D \rho(x, y)\,dx\,dy$ is the mass of the plate. If ρ is constant, this is true since mass = density × area. Next, if ρ is a step function, then the integral of ρ over D is the mass of a plate with density ρ since the total mass is the sum of the masses of its parts. Now let ρ be arbitrary. If ρ_1 is a step function with $\rho_1 \leqslant \rho$, then $\iint_D \rho_1(x, y)\,dx\,dy \leqslant m$, where m is the mass of the plate with density ρ, since a lower density gives a smaller mass. Likewise, if ρ_2 is a step function with $\rho_2 \geqslant \rho$, then $m \leqslant \iint_D \rho_2(x, y)\,dx\,dy$. Thus the mass m lies between any pair of lower and upper sums for ρ, so it must equal the integral $\iint_D \rho(x, y)\,dx\,dy$.

Before establishing the fundamental result which will enable us to use one-variable techniques to evaluate double integrals, we must explain some notation. The *iterated integral*

$$\int_c^d \left[\int_a^b f(x, y)\,dx \right] dy$$

is evaluated, like most parenthesized expressions, from the inside out. One first holds y fixed and evaluates the integral $\int_a^b f(x, y)\,dx$ with respect to x; the result is a function of y which is then integrated from c to d.

The expression $\int_a^b [\int_c^d f(x, y)\,dy]\,dx$ is defined similarly; this time the integral with respect to y is evaluated first. Iterated integrals are often written without parentheses as

$$\int_c^d \int_a^b f(x, y)\,dx\,dy \quad \text{and} \quad \int_a^b \int_c^d f(x, y)\,dy\,dx.$$

Example 3 Evaluate $\int_0^2 \int_1^3 x^2 y \, dy \, dx$.

Solution

$$\int_0^2 \int_1^3 x^2 y \, dy \, dx = \int_0^2 \left(\int_1^3 x^2 y \, dy \right) dx = \int_0^2 \left(\frac{x^2 y^2}{2} \Bigg|_{y=1}^{3} \right) dx$$

$$= \int_0^2 x^2 \left(\frac{9}{2} - \frac{1}{2} \right) dx = 4 \int_0^2 x^2 \, dx = 4 \frac{x^3}{3} \Bigg|_0^2 = \frac{32}{3}.$$

Notice that the second step in this calculation is essentially the inverse of a partial differentiation. ▲

We claim, and shall prove below, that the double integral equals the iterated integral. That is, for $D = [a,b] \times [c,d]$,

$$\iint_D f(x, y) \, dx \, dy = \int_c^d \left[\int_a^b f(x, y) \, dx \right] dy = \int_a^b \left[\int_c^d f(x, y) \, dy \right] dx.$$

To see why this might be so, let us suppose that $f(x, y) \geqslant 0$, so that the integral $\iint_D f(x, y) \, dx \, dy$ represents the volume of the region R under the graph of f. If we take this volume and slice it by a plane parallel to the yz plane at a distance x from the origin, we get a two-dimensional region whose area is given by $A(x) = \int_c^d f(x, y) \, dy$ (see Fig. 17.1.6).

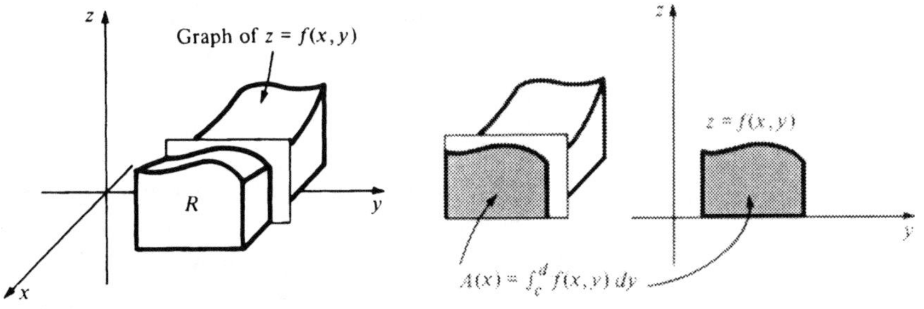

Figure 17.1.6. The area of the cross-section is the area under the graph of $z = f(x, y)$ from $y = c$ to $y = d$ (where x is fixed).

By Cavalieri's principle (Section 9.1), the total volume is the integral of the area function $A(x)$. Thus,

$$\iint_D f(x, y) \, dx \, dy = \text{volume of } R = \int_a^b A(x) \, dx = \int_a^b \left[\int_c^d f(x, y) \, dy \right] dx.$$

In the same way, if we use planes parallel to the xz plane, we get

$$\iint_D f(x, y) \, dx \, dy = \int_c^d \left[\int_a^b f(x, y) \, dx \right] dy,$$

which is what we claimed. We see that Cavalieri's principle gives a geometric "proof" of the reduction to iterated integrals; in fact, it is more appropriate to take our proof below as a justification for Cavalieri's principle.

Theorem: Reduction to Iterated Integrals

Assume that $f(x, y)$ is integrable on the rectangle $D = [a,b] \times [c,d]$. Then any iterated integral which exists is equal to the double integral; that is,

$$\iint_D f(x, y) \, dx \, dy = \int_c^d \left[\int_a^b f(x, y) \, dx \right] dy = \int_a^b \left[\int_c^d f(x, y) \, dy \right] dx.$$

Proof of the Reduction to Iterated Integrals To prove this theorem, we first show it is true for step functions. Let g be a step function, with $g(x, y) = k_{ij}$ on $(t_{i-1}, t_i) \times (s_{j-1}, s_j)$, so that

$$\iint_D g(x, y)\,dx\,dy = \sum_{i=1,j=1}^{n,m} k_{ij}\,\Delta t_i\,\Delta s_j.$$

If the summands $k_{ij}\,\Delta t_i\,\Delta s_j$ are laid out in a rectangular array, they may be added by first adding along rows and then adding up the subtotals, as follows:

$$\begin{array}{ccccccc}
k_{11}\Delta t_1\Delta s_1 & k_{21}\Delta t_2\Delta s_1 & \cdots & k_{n1}\Delta t_n\Delta s_1 & \longrightarrow & \left(\sum_{i=1}^{n} k_{i1}\Delta t_i\right)\Delta s_1 \\
k_{12}\Delta t_1\Delta s_2 & k_{22}\Delta t_2\Delta s_2 & \cdots & k_{n2}\Delta t_n\Delta s_2 & \longrightarrow & \left(\sum_{i=1}^{n} k_{i2}\Delta t_i\right)\Delta s_2 \\
\vdots & \vdots & \vdots & \vdots & & \vdots \\
k_{1m}\Delta t_1\Delta s_m & k_{2m}\Delta t_2\Delta s_m & \cdots & k_{nm}\Delta t_n\Delta s_m & \longrightarrow & \left(\sum_{i=1}^{n} k_{im}\Delta t_i\right)\Delta s_m \\
\hline
 & & & & & \sum_{j=1}^{m}\left(\sum_{i=1}^{n} k_{ij}\Delta t_i\right)\Delta s_j
\end{array}$$

The coefficient of Δs_j in the sum over the jth row, $\sum_{i=1}^{n} k_{ij}\,\Delta t_i$, is equal to $\int_a^b g(x, y)\,dx$ for any y with $s_{j-1} < y < s_j$, since, for y fixed, $g(x, y)$ is a step function of x. Thus the integral $\int_a^b g(x, y)\,dx$ is a step function of y, and its integral with respect to y is the sum:

$$\int_c^d\left[\int_a^b g(x, y)\,dx\right]dy = \sum_{j=1}^{m}\left(\sum_{i=1}^{n} k_{ij}\,\Delta t_i\right)\Delta s_j = \iint_D g(x, y)\,dx\,dy.$$

Similarly, by summing first over columns and then over rows, we obtain

$$\iint_D g(x, y)\,dx\,dy = \int_a^b\left[\int_c^d g(x, y)\,dy\right]dx.$$

The theorem is therefore true for step functions.

Now let f be integrable on $D = [a, b] \times [c, d]$ and assume that the iterated integral $\int_a^b[\int_c^d f(x, y)\,dy]\,dx$ exists. Denoting this integral by S_0, we will show that every lower sum for f on D is less than or equal to S_0, while every upper sum is greater than or equal to S_0, so S_0 must be the integral of f over D.

To carry out our program, let g be any step function such that

$$g(x, y) \leqslant f(x, y) \tag{1}$$

for all (x, y) in D. Integrating equation (1) with respect to y and using property 5 of the one-variable integral (see Section 4.5), we obtain

$$\int_c^d g(x, y)\,dy \leqslant \int_c^d f(x, y)\,dy \tag{2}$$

for all x in $[a, b]$. Integrating (2) with respect to x and applying property 5 once more gives

$$\int_a^b\left[\int_c^d g(x, y)\,dy\right]dx \leqslant \int_a^b\left[\int_c^d f(x, y)\,dy\right]dx. \tag{3}$$

Since g is a step function, it follows from the first part of this proof that the left-hand side of (3) is equal to the lower sum $\iint_D g(x, y)\,dx\,dy$; the right-hand side of (3) is just S_0, so we have shown that every lower sum is less than or equal to S_0. The proof that every upper sum is greater than or equal to S_0 is similar, and so we are done. ■

Example 4 Let $f(x, y) = e^{2x+y}$. Evaluate the integral of f over $D = [0, 1] \times [0, 3]$.

Solution

$$\iint_D f(x, y)\,dx\,dy = \int_0^3 \left(\int_0^1 e^{2x+y}\,dx \right) dy$$

$$= \int_0^3 \left(\frac{1}{2} e^{2x+y} \Big|_{x=0}^{1} \right) dy = \frac{1}{2} \int_0^3 (e^{2+y} - e^y)\,dy$$

$$= \frac{1}{2}(e^2 - 1) \int_0^3 e^y\,dy = \frac{(e^2 - 1)(e^3 - 1)}{2} \approx 60.9693.$$

(You should check that integrating with respect to y first gives the same answer.) ▲

Example 5 Evaluate $\int_1^3 \int_0^2 x^2 y\,dx\,dy$ and compare with Example 3.

Solution

$$\int_1^3 \int_0^2 x^2 y\,dx\,dy = \int_1^3 \left(\int_0^2 x^2 y\,dx \right) dy = \int_1^3 \left(\frac{x^3 y}{3} \Big|_{x=0}^{2} \right) dy$$

$$= \int_1^3 \frac{8}{3} y\,dy = \frac{4}{3} y^2 \Big|_1^3 = \frac{4}{3}(3^2 - 1^2) = \frac{32}{3}.$$

The answer is the same as that in Example 3, as predicted by the theorem above. (It is also consistent with Example 2.) ▲

Example 6 Compute $\iint_D \sin(x + y)\,dx\,dy$, where $D = [0, \pi] \times [0, 2\pi]$.

Solution

$$\iint_D \sin(x + y)\,dx\,dy = \int_0^{2\pi} \left[\int_0^{\pi} \sin(x + y)\,dx \right] dy$$

$$= \int_0^{2\pi} \left[-\cos(x + y)|_{x=0}^{\pi} \right] dy$$

$$= \int_0^{2\pi} [\cos y - \cos(y + \pi)]\,dy$$

$$= [\sin y - \sin(y + \pi)]|_{y=0}^{2\pi} = 0. \ ▲$$

Example 7 Find the volume under the graph of $f(x, y) = x^2 + y^2$ between the planes $x = 0$, $x = 3$, $y = -1$, and $y = 1$.

Solution The volume is

$$\int_{-1}^1 \int_0^3 (x^2 + y^2)\,dx\,dy = \int_{-1}^1 \left(\frac{x^3}{3} + y^2 x \Big|_{x=0}^{3} \right) dy = \int_{-1}^1 (9 + 3y^2)\,dy$$

$$= (9y + y^3)|_{-1}^{1} = 20. \ ▲$$

Example 8 If D is a plate defined by $1 \leqslant x \leqslant 2$, $0 \leqslant y \leqslant 1$, and the mass density is $\rho(x, y) = ye^{xy}$ grams per square centimeter, find the mass of the plate.

Solution The total mass is

$$\iint_D \rho(x, y)\,dx\,dy = \int_0^1 \int_1^2 ye^{xy}\,dx\,dy = \int_0^1 (e^{xy}|_{x=1}^{2})\,dy$$

$$= \int_0^1 (e^{2y} - e^y)\,dy = \left(\frac{e^{2y}}{2} - e^y\right)\bigg|_{y=0}^{1}$$

$$= \frac{e^2}{2} - e + \frac{1}{2} \approx 1.4762 \text{ grams. } \blacktriangle$$

Supplement for Section 17.1: Solar Energy and Double Integrals

To illustrate the process of summation which is represented by the double integral, we may use an example connected with solar energy. The intensity of solar radiation is a "local" quantity, which may be measured at any point on the earth's surface. Since the solar intensity is really a rate of power input (see the Supplement to Section 9.5), we can measure it in units of watts per square meter.

If the solar intensity is uniform over a region, the total power received is equal to the intensity times the area of the region. In practice, the intensity is a function of position (in particular, it is a function of latitude); so we cannot just multiply a value by the area of the region. Instead, we must *integrate* the intensity over the region. Thus, the method of this section would allow us to find (at least in principle) the total power received by the state of Colorado, which is a rectangle in longitude–latitude coordinates. The problem for Utah is also tractable, since that state is composed of two rectangles, but what happens if we are interested in Michigan or Florida? For this problem, we need to integrate over regions which are not rectangles: the method for doing this is presented in the next section.

Exercises for Section 17.1

In Exercises 1 and 2, the function g takes values on the rectangles as indicated in Fig. 17.1.7. Calculate the integral of g.

1. For the rectangle in Fig. 17.1.7(a).
2. For the rectangle in Fig. 17.1.7(b).

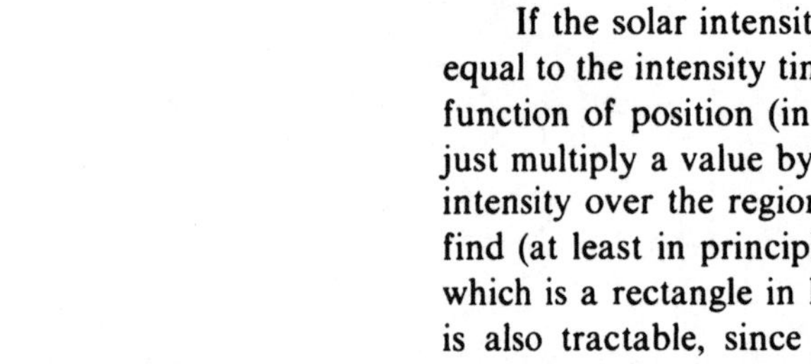

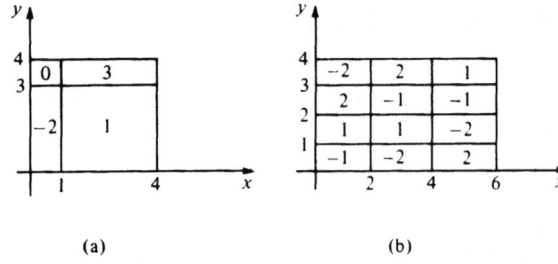

Figure 17.1.7. Find the integral of g.

3. (a) Let D be the rectangle determined by the inequalities $-1 \leqslant x \leqslant 1$ and $2 \leqslant y \leqslant 4$ and let $f(x, y) = x(1+y)$. Find a step function g satisfying $g(x, y) \leqslant f(x, y)$ to show that $\iint_D f(x, y)\,dx\,dy \geqslant -9$.
(b) Sketch the graph of $f(x, y)$ over D. Use symmetry to argue that the value of the integral in part (a) must in fact be zero.

4. Let D be the rectangle $1 \leqslant x \leqslant 4$, $0 \leqslant y \leqslant 2$. Suppose that D is a plate with mass density $\rho(x, y) = 2xy^2 + \cos \pi y + 1$ (grams per square centimeter). Find step functions to show that the mass m (in grams) of the plate satisfies the inequalities $30 \leqslant m \leqslant 62$.

Evaluate the iterated integrals in Exercises 5–10.

5. $\int_0^3 \int_0^2 x^3 y\,dx\,dy$
6. $\int_0^2 \int_1^4 x^3 y\,dy\,dx$
7. $\int_0^2 \int_{-1}^1 (yx)^2\,dy\,dx$
8. $\int_{-1}^1 \int_0^2 (yx)^2\,dx\,dy$
9. $\int_{-1}^1 \int_0^1 y e^x\,dy\,dx$
10. $\int_{-1}^1 \int_0^3 y^5 e^{xy^3}\,dx\,dy$

Evaluate $\iint_D f(x, y)\,dx\,dy$ for the indicated functions and rectangles in Exercises 11–16.

11. $f(x, y) = (x + 2y)^2$; $D = [-1, 2] \times [0, 2]$.
12. $f(x, y) = y^3 \cos^2 x$; $D = [-\pi/2, \pi] \times [1, 2]$.
13. $f(x, y) = x^2 + 2xy - y\sqrt{x}$; $D = [0, 1] \times [-2, 2]$.
14. $f(x, y) = xy^3 e^{x^2 y^2}$; $D = [1, 3] \times [1, 2]$.
15. $f(x, y) = xy + x/(y + 1)$; $D = [1, 4] \times [1, 2]$.
16. $f(x, y) = y^5 e^{y^3 \cos x} \sin x$; $D = [0, 1] \times [-1, 0]$.
17. Evaluate $\int_2^4 \int_{-1}^1 x(1 + y)\,dx\,dy$ and compare with Exercise 3.
18. Evaluate $\int_0^2 \int_1^4 (2xy^2 + \cos \pi y + 1)\,dx\,dy$ and compare with Exercise 4.

Sketch and find the volume under the graph of f between the planes $x = a$, $x = b$, $y = c$, and $y = d$ in Exercises 19 and 20.

19. $f(x, y) = x^3 + y^2 + 2$; $a = -1$, $b = 1$, $c = 1$, $d = 3$.
20. $f(x, y) = 2x + 3y^2 + 2$; $a = 0$, $b = 3$, $c = -2$, $d = 1$.
21. The density at each point of a 1 centimeter square (i.e., each side has length 1 centimeter) microchip is $4 + r^2$ grams per square centimeter, where r is the distance in centimeters from the point to the center of the chip. What is the mass of the chip?
22. Do as in Exercise 21, but now let r be the distance to the lower left-hand corner of the plate.

★23. Prove that the sum of two step functions defined on the same rectangle is again a step function.

★24. Prove that if $f(x, y) \leqslant g(x, y)$ for all (x, y) in the rectangle $[a, b] \times [c, d]$, then

$$\int_a^b \int_c^d f(x, y)\, dy\, dx \leqslant \int_c^d \int_a^b g(x, y)\, dx\, dy.$$

★25. The state of Colorado occupies the region between 33° and 41° latitude and 102° and 108° longitude. A degree of latitude is about 110 kilometers and a degree of longitude is about 83 kilometers. The intensity of solar radiation at time t on day T at latitude l is (in suitable units; see the Supplement to Section 9.5)

$$I = \cos l \sqrt{1 - \sin^2\alpha \cos^2\left(\frac{2\pi T}{365}\right)} \cos\left(\frac{2\pi t}{24}\right) + \sin l \sin \alpha \cos\left(\frac{2\pi T}{365}\right).$$

(a) What is the integrated solar energy over Colorado at time t on day T?

(b) Suppose that the result of part (a) is integrated with respect to t from t_1 to t_2. What does the integral represent?

17.2 The Double Integral Over General Regions

Double integrals over general regions become iterated integrals with variable endpoints.

Many applications involve double integrals $\iint_D f(x, y)\, dx\, dy$ over regions D which are not rectangles. For instance, the volume of a hemisphere, the mass of an elliptical plate, or the total solar power received by the state of Texas can be expressed as such integrals. We shall find that such integrals can be evaluated by iterated integration in a form slightly more complicated than that used for rectangles.

To begin, we must *define* what we mean by $\iint_D f(x, y)\, dx\, dy$ when D is not a rectangle. We shall assume that D is contained in some rectangle D^*. Let f^* be the function on D^* defined by

$$f^*(x, y) = \begin{cases} f(x, y) & \text{if } (x, y) \in D, \\ 0 & \text{if } (x, y) \notin D. \end{cases}$$

(See Fig. 17.2.1.)

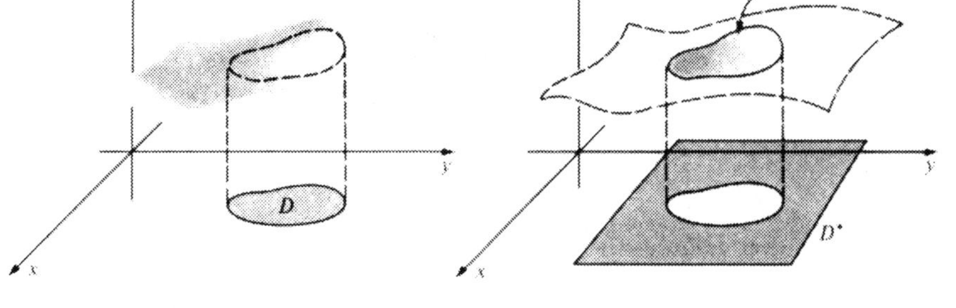

Figure 17.2.1. Given f and D, we construct f^* by setting $f^*(x, y)$ equal to zero outside D.

If D^{**} is another rectangle containing D, and f^{**} is the corresponding function defined as above, then $\iint_{D^*} f^*(x, y)\,dx\,dy = \iint_{D^{**}} f^{**}(x, y)\,dx\,dy$, since f^* and f^{**} are zero in the regions where D^* and D^{**} differ (see Fig. 17.2.2 and use the properties of the integral given in Section 17.1).

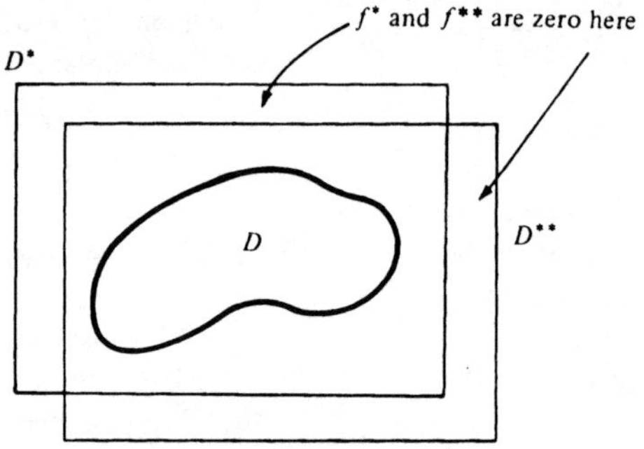

Figure 17.2.2. The choice of D^* does not matter.

We note that if $f(x, y) \geqslant 0$ on D, then both integrals above are equal to the volume of the region under the graph of f on D, i.e., the set of (x, y, z) such that $(x, y) \in D$ and $0 \leqslant z \leqslant f(x, y)$.

With these preliminaries, we can state the following definition.

The Double Integral Over a Region D

Extend f to a rectangle D^* containing D by letting f^* equal f on D and zero outside D. If f^* is integrable on D^*, then we say that f is *integrable* on D, and we define $\iint_D f(x, y)\,dx\,dy$ to be $\iint_{D^*} f^*(x, y)\,dx\,dy$. (By our preceding remarks, the choice of D^* does not affect the answer).

This definition serves the purpose of giving meaning to the double integral, but it is not very useful for computation. For this purpose, we need to choose D in a more specific way. We shall define two simple types of regions, which we will call *elementary regions*. Complicated regions can often be broken into elementary ones.

Suppose that we are given two continuous real-valued functions ϕ_1 and ϕ_2 on $[a,b]$ which satisfy $\phi_1(t) \leqslant \phi_2(t)$ for all t in $[a,b]$. Let D be the set of all points (x, y) such that

$$x \text{ is in } [a,b] \quad \text{and} \quad \phi_1(x) \leqslant y \leqslant \phi_2(x).$$

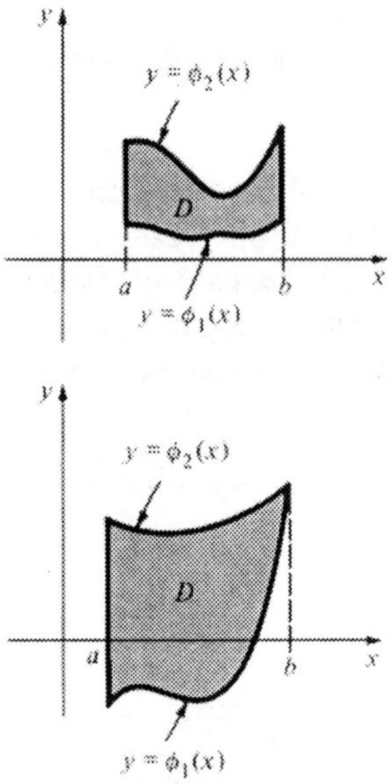

Figure 17.2.3. A region D is of type 1 if it is the region between the graphs of two functions, $y = \phi_1(x)$ and $y = \phi_2(x)$.

This region D is said to be of *type 1*. (See Fig. 17.2.3.) The curves and straight line segments that enclose the region constitute the *boundary* of D.

We say that a region D is of *type 2* if there are continuous functions ψ_1 and ψ_2 on $[c,d]$ such that D is the set of points (x, y) satisfying

$$y \text{ is in } [c,d] \quad \text{and} \quad \psi_1(y) \leqslant x \leqslant \psi_2(y),$$

where $\psi_1(t) \leqslant \psi_2(t)$ for t in $[c,d]$. (See Fig. 17.2.4.) Again the boundary of the region consists of the curves and line segments enclosing the region.

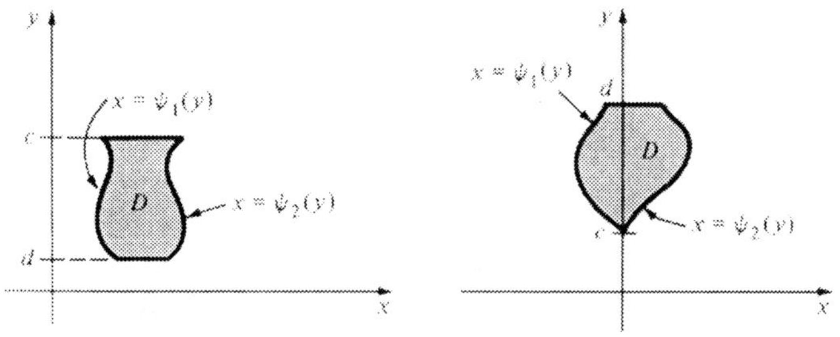

Figure 17.2.4. A region D is of type 2 if it is the region between the graphs of $x = \psi_1(y)$ and $x = \psi_2(y)$.

The following example shows that a given region may be of types 1 and 2 at the same time.

Example 1 Show that the region D defined by $x^2 + y^2 \leqslant 1$ (the unit disk) is a region of types 1 and 2.

Solution Descriptions of the disk, showing that it is of both types, are given in Fig. 17.2.5. ▲

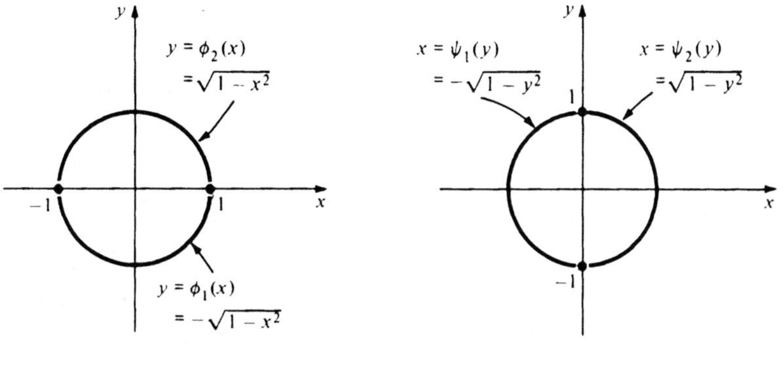

Figure 17.2.5. The unit disk as a type 1 region and a type 2 region.

We will use, without proof,[1] the following fact: If f is continuous on an elementary region D, then f is integrable on D. This fact, combined with the reduction to iterated integrals for rectangles, enables us to evaluate integrals over elementary regions by iterated integration. Indeed, if $D^* = [a,b] \times [c,d]$ is a rectangle containing D, then

$$\int\int_D f(x, y)\,dx\,dy = \int\int_{D^*} f^*(x, y)\,dx\,dy = \int_a^b \int_c^d f^*(x, y)\,dy\,dx \tag{1}$$

$$= \int_c^d \int_a^b f^*(x, y)\,dx\,dy, \tag{2}$$

where f^* equals f in D and is zero outside D. Assume that D is a region of type 1 determined by functions ϕ_1 and ϕ_2 on $[a,b]$. Consider the iterated integral

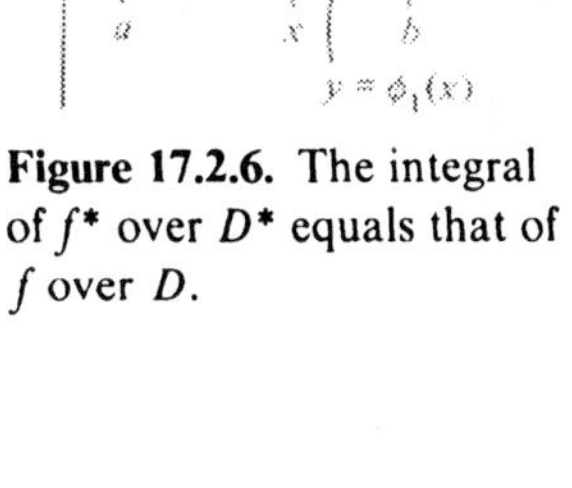

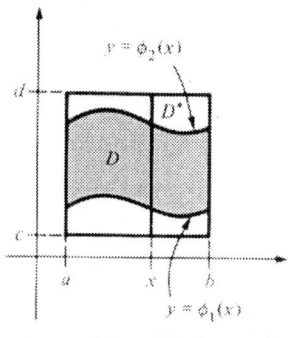

Figure 17.2.6. The integral of f^* over D^* equals that of f over D.

$$\int_a^b \int_c^d f^*(x, y)\,dy\,dx$$

and, in particular, the inner integral $\int_c^d f^*(x, y)\,dy$ for some fixed x (Fig. 17.2.6). By definition, $f^*(x, y) = 0$ if $y < \phi_1(x)$ or $y > \phi_2(x)$, so

$$\int_c^d f^*(x, y)\,dy = \int_{\phi_1(x)}^{\phi_2(x)} f^*(x, y)\,dy = \int_{\phi_1(x)}^{\phi_2(x)} f(x, y)\,dy. \tag{3}$$

[1] For a proof, see an advanced calculus text, such as J. Marsden, *Elementary Classical Analysis*, Freeman (1974), Chapter 8.

Substituting (3) into (1) gives

$$\iint_D f(x, y)\,dx\,dy = \int_a^b \left[\int_{\phi_1(x)}^{\phi_2(x)} f(x, y)\,dy \right] dx.$$

A similar construction works for type 2 regions.

Double Integrals

If D is a region of type 1 (Fig. 17.2.3),

$$\iint_D f(x, y)\,dx\,dy = \int_a^b \left[\int_{\phi_1(x)}^{\phi_2(x)} f(x, y)\,dy \right] dx. \tag{4}$$

If D is of type 2 (Fig. 17.2.4),

$$\iint_D f(x, y)\,dx\,dy = \int_c^d \left[\int_{\psi_1(y)}^{\psi_2(y)} f(x, y)\,dx \right] dy. \tag{5}$$

If D is of both types, either (4) or (5) is applicable.

If $f(x, y) \geqslant 0$ on D, we may understand the procedure of iterated integration as our old technique of finding volumes by slicing (Section 9.1). Suppose, for instance, that D is of type 1, determined by $\phi_1(x)$ and $\phi_2(x)$ on $[a, b]$. If we fix x and slice the volume under the graph of $f(x, y)$ on D by the plane which passes through the point $(x, 0, 0)$ and which is parallel to the yz plane, we obtain the region in the yz plane defined by the inequalities $\phi_1(x) \leqslant y \leqslant \phi_2(x)$ and $0 \leqslant z \leqslant f(x, y)$. The area $A(x)$ of this region is just $\int_{\phi_1(x)}^{\phi_2(x)} f(x, y)\,dy$. Now the double integral $\iint_D f(x, y)\,dx\,dy$, which is the volume of the entire solid, equals

$$\int_a^b A(x)\,dx = \int_a^b \left[\int_{\phi_1(x)}^{\phi_2(x)} f(x, y)\,dy \right] dx,$$

the iterated integral. Thus we get (4).

If D is of type 2, then slicing by planes parallel to the xz plane produces the corresponding result (5). The reader should draw figures similar to Fig. 17.1.6 to accompany this discussion.

Example 2 Find $\displaystyle\iint_D (x + y)\,dx\,dy$, where D is the shaded region in Fig. 17.2.7.

Solution D is a region of type 1, with $[a, b] = [0, \frac{1}{2}]$, $\varphi_1(x) = 0$, and $\varphi_2(x) = x^2$. By formula (4) in the preceding box,

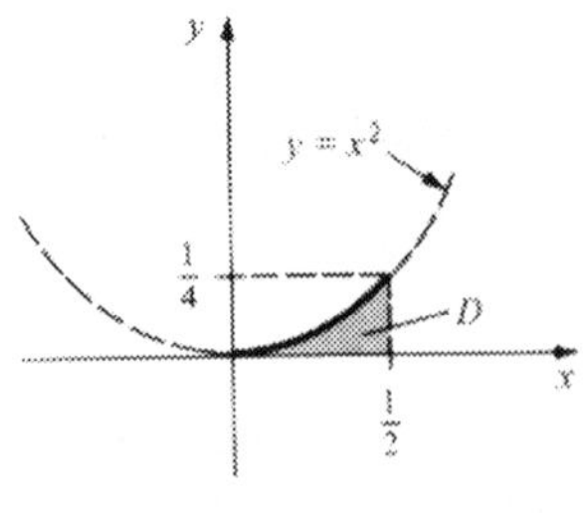

Figure 17.2.7. Find $\iint_D (x + y)\,dx\,dy$.

$$\iint_D (x + y)\,dx\,dy = \int_0^{1/2} \int_0^{x^2} (x + y)\,dy\,dx = \int_0^{1/2} \left[\left(xy + \frac{y^2}{2} \right) \Bigg|_{y=0}^{x^2} \right] dx$$

$$= \int_0^{1/2} \left(x^3 + \frac{x^4}{2} \right) dx = \left(\frac{x^4}{4} + \frac{x^5}{10} \right) \Bigg|_0^{1/2}$$

$$= \frac{1}{64} + \frac{1}{320} = \frac{3}{160}.$$

D is also a region of type 2, with $\psi_1(y) = \sqrt{y}$ and $\psi_2(y) = \frac{1}{2}$. We leave it to you to verify that the double integral calculated by formula (5) is also $\frac{3}{160}$. ▲

Example 3 Evaluate $\int_0^1 \int_{x^3}^{x^2} xy\,dy\,dx$. Sketch the region for the corresponding double integral.

Solution Here y ranges from x^3 to x^2, while x goes from 0 to 1. Hence the region is as shown in Fig. 17.2.8. The integral is

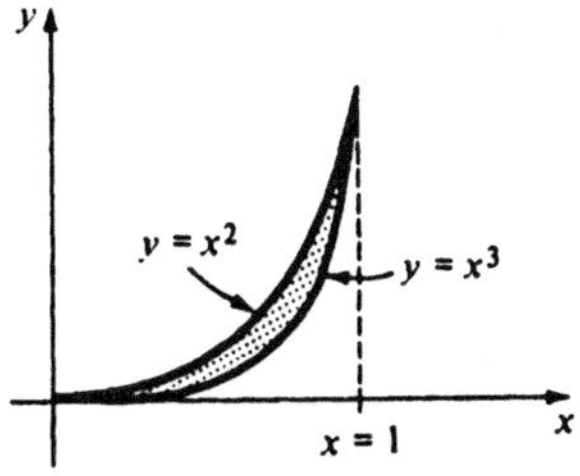

Figure 17.2.8. The region of integration for $\int_0^1 \int_{x^3}^{x^2} xy\,dy\,dx$.

$$\int_0^1 \left(\frac{xy^2}{2} \Big|_{y=x^3}^{x^2} \right) dx = \int_0^1 \left(\frac{x^5}{2} - \frac{x^7}{2} \right) dx = \left(\frac{x^6}{12} - \frac{x^8}{16} \right) \Big|_{x=0}^{1} = \frac{1}{12} - \frac{1}{16} = \frac{1}{48}. \blacktriangle$$

The next example shows that it sometimes saves labor to reverse the order of integration.

Example 4 Write $\int_0^1 \int_0^{\sqrt{1-x^2}} \sqrt{1-y^2}\,dy\,dx$ as an integral over a region. Sketch the region and show that it is of types 1 and 2. Reverse the order of integration and evaluate.

Solution The region D is shown in Fig. 17.2.9. D is a type 1 region with $\phi_1(x) = 0$, $\phi_2(x) = \sqrt{1-x^2}$ and a type 2 region with $\psi_1(y) = 0$, $\psi_2(y) = \sqrt{1-y^2}$. Thus

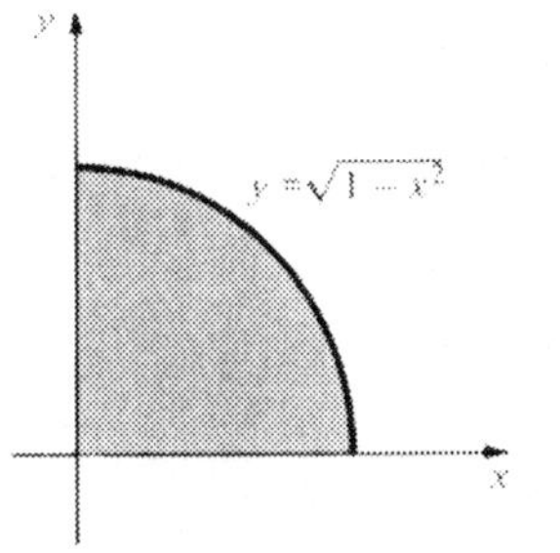

Figure 17.2.9. The region of integration for $\int_0^1 \int_0^{\sqrt{1-x^2}} \sqrt{1-y^2}\,dy\,dx$.

$$\int_0^1 \int_0^{\sqrt{1-x^2}} \sqrt{1-y^2}\,dy\,dx = \int_0^1 \int_0^{\sqrt{1-y^2}} \sqrt{1-y^2}\,dx\,dy$$
$$= \int_0^1 \left[\sqrt{1-y^2}\,x \Big|_{x=0}^{\sqrt{1-y^2}} \right] dy = \int_0^1 (1-y^2)\,dy$$
$$= \left(y - \frac{y^3}{3} \right) \Big|_0^1 = \frac{2}{3}.$$

Evaluating the integral in the original order requires considerably more computation! $\blacktriangle$

Example 5 Calculate the integral of $f(x, y) = (x+y)^2$ over the region shown in Fig. 17.2.10.

Solution In this example, there is a preferred order of integration for geometric reasons. The order $\int\int f(x, y)\,dx\,dy$, i.e., x first, requires us to break up the region into two parts by drawing the line $y = 1$; one then applies formula (5) to each part and adds the results. If we use the other order, we can cover the whole region at once:

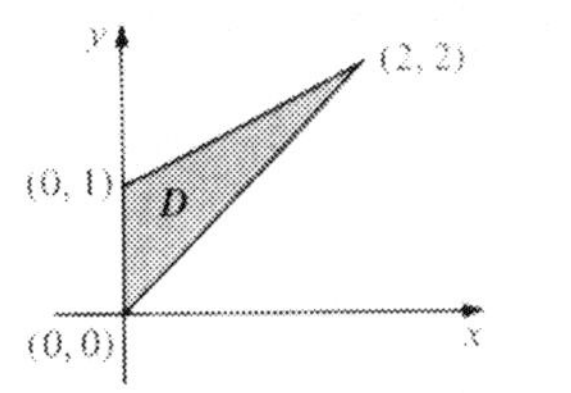

Figure 17.2.10. The region of integration for Example 5.

$$\int\int_D f(x, y)\,dx\,dy = \int_0^2 \int_x^{\frac{1}{2}x+1} f(x, y)\,dy\,dx.$$

(The lines bounding D on the bottom and top are $y = x$ and $y = \frac{1}{2}x + 1$.) The integral is thus

$$\int_0^2 \int_x^{\frac{1}{2}x+1} (x+y)^2\,dy\,dx = \int_0^2 \left[\frac{1}{3}(x+y)^3 \Big|_{y=x}^{\frac{1}{2}x+1} \right] dx$$
$$= \frac{1}{3} \int_0^2 \left[\left(\frac{3}{2}x + 1 \right)^3 - (2x)^3 \right] dx$$
$$= \frac{1}{3} \left[\frac{1}{6} \left(\frac{3}{2}x + 1 \right)^4 \Big|_0^2 - 2x^4 \Big|_0^2 \right]$$

$$= \frac{1}{3}\left[\frac{1}{6}(4^4 - 1) - 2 \cdot 16\right]$$
$$= \frac{1}{3}\left[\frac{21}{2}\right] = \frac{21}{6}. \blacktriangle$$

General regions can often be broken into elementary regions, and double integrals over these regions can be computed one piece at a time.

Example 6 Find $\int\int_R x^2\,dx\,dy$, where R is the shaded region in Fig. 17.2.11.

Solution Each of the regions R_1, R_2, R_3, R_4 is of types 1 and 2, so we may integrate over each one separately and sum the results. Using formula (4), we get

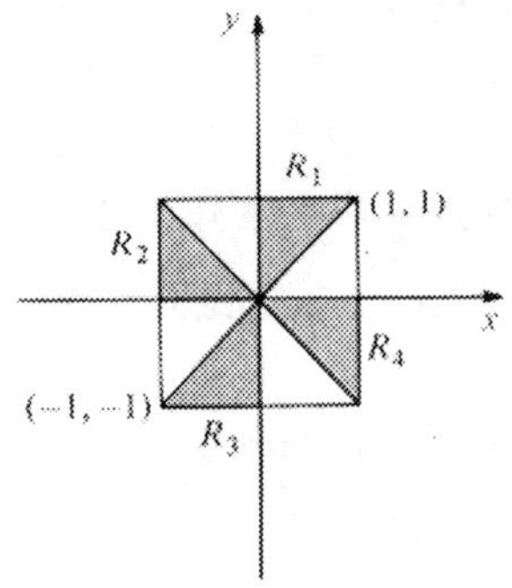

Figure 17.2.11. Integrate x^2 over the pinwheel.

$$\int\int_{R_1} x^2\,dx\,dy = \int_0^1\int_x^1 x^2\,dy\,dx = \int_0^1 (x^2y)\Big|_{y=x}^{1} dx$$
$$= \int_0^1 (x^2 - x^3)\,dx = \left(\frac{x^3}{3} - \frac{x^4}{4}\right)\Big|_0^1$$
$$= \frac{1}{3} - \frac{1}{4} = \frac{1}{12}.$$

Similarly, we find

$$\int\int_{R_2} x^2\,dx\,dy = \int_{-1}^{0}\int_0^{-x} x^2\,dy\,dx = \frac{1}{4}.$$

By symmetry,

$$\int\int_{R_3} x^2\,dx\,dy = \frac{1}{12} \quad \text{and} \quad \int\int_{R_4} x^2\,dx\,dy = \frac{1}{4},$$

so

$$\int\int_R x^2\,dx\,dy = \frac{1}{12} + \frac{1}{4} + \frac{1}{12} + \frac{1}{4} = \frac{2}{3}.$$

You may check that formula (5) gives the same answers. $\blacktriangle$

Exercises for Section 17.2

In Exercises 1–4, sketch each region and tell whether it is of type 1, type 2, both, or neither.

1. (x, y) such that $0 \leqslant y \leqslant 3x$, $0 \leqslant x \leqslant 1$.
2. (x, y) such that $y^2 \leqslant x \leqslant y$, $0 \leqslant y \leqslant 1$.
3. (x, y) such that $x^4 + y^4 \leqslant 1$.
4. (x, y) such that $\frac{1}{2} \leqslant x^4 + y^4 \leqslant 1$.
5. Find $\int\int_D (x + y)^2\,dx\,dy$, where D is the shaded region in Fig. 17.2.12.

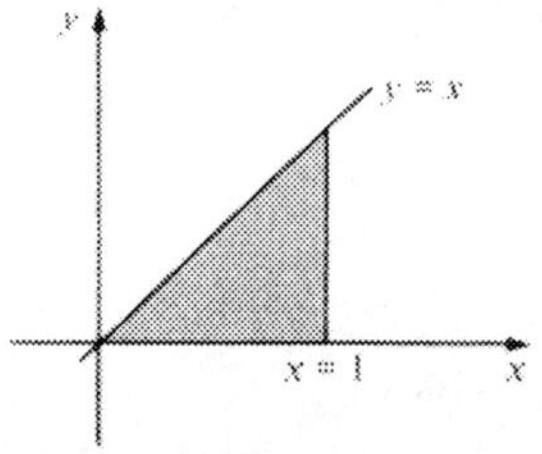

Figure 17.2.12. The region of integration for Exercises 5 and 6.

6. Find $\int\int_D (1 - \sin \pi x) y\,dx\,dy$, where D is the region in Figure 17.2.12.
7. Find $\int\int_D (x - y)^2 dx\,dy$, where D is the region in Figure 17.2.13.
8. Find $\int\int_D y(1 - \cos(\pi x/4))\,dx\,dy$, where D is the region in Fig. 17.2.13.

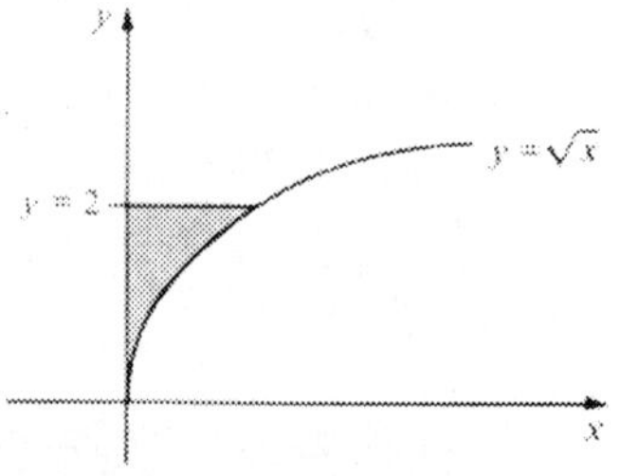

Figure 17.2.13. The region of integration for Exercises 7 and 8.

Evaluate the integrals in Exercises 9–16. Sketch and identify the type of the region (corresponding to the way the integral is written).

9. $\int_0^{\pi}\int_{\sin x}^{3\sin x} x(1+y)\,dy\,dx.$

10. $\int_0^1\int_{x-1}^{x\cos 2\pi x}(x^2+xy+1)\,dy\,dx.$

11. $\int_{-1}^1\int_{y^{2/3}}^{(2-y)^2}(y\sqrt{x}+y^3-2y)\,dx\,dy.$

12. $\int_0^2\int_{-3(\sqrt{4-x^2})/2}^{3(\sqrt{4-x^2})/2}\left(\frac{5}{\sqrt{2+x}}+y^3\right)dy\,dx.$

13. $\int_0^1\int_0^{x^2}(x^2+xy-y^2)\,dy\,dx.$

14. $\int_2^4\int_{y^2-1}^{y^3}3\,dx\,dy.$

15. $\int_0^1\int_{x^2}^{x}(x+y)^2\,dy\,dx.$

16. $\int_0^1\int_0^{3y}e^{x+y}\,dx\,dy.$

In Exercises 17–20, sketch the region of integration, interchange the order, and evaluate.

17. $\int_0^1\int_x^1 xy\,dy\,dx$

18. $\int_0^{\pi/2}\int_0^{\cos\theta}\cos\theta\,dr\,d\theta$

19. $\int_0^1\int_{1-y}^1(x+y^2)\,dx\,dy$

20. $\int_1^4\int_1^{\sqrt{x}}(x^2+y^2)\,dy\,dx$

In Exercises 21–24, integrate the given function f over the given region D.

21. $f(x,y)=x-y$; D is the triangle with vertices $(0,0)$, $(1,0)$, and $(2,1)$.

22. $f(x,y)=x^3y+\cos x$; D the triangle defined by $0\leqslant x\leqslant \pi/2$, $0\leqslant y\leqslant x$.

23. $f(x,y)=(x^2+2xy^2+2)$; D the region bounded by the graph of $y=-x^2+x$, the x axis, and the lines $x=0$ and $x=2$.

24. $f(x,y)=\sin x\cos y$; D the pinwheel in Fig. 17.2.11.

25. Show that evaluating $\iint_D dx\,dy$, where D is a region of type 1, simply reproduces the formula from Section 4.6 for the area between curves.

26. Let D be the region defined by $x^2+y^2\leqslant 1$.
 (a) Estimate $\iint_D dx\,dy$ (the area of D) within 0.1 by taking a rectangular grid in the plane and counting the number of rectangles: (i) contained entirely in D (lower sum); (ii) intersecting D (upper sum).
 (b) Compute $\iint_D dx\,dy$ exactly by using an iterated integral.

27. Which states in the United States are regions of type 1? Type 2? (Take $x=$ longitude, $y=$ latitude.)

★28. Prove: $\int_0^x\left[\int_0^t F(u)\,du\right]dt=\int_0^x(x-u)F(u)\,du.$

17.3 Applications of the Double Integral

Volumes, centers of mass, and surface areas can be calculated using double integrals.

We have observed in Section 17.1 that if $f(x,y)\geqslant 0$ on D, then the double integral $\iint_D f(x,y)\,dx\,dy$ represents the volume of the three-dimensional region R defined by (x,y) in D, $0\leqslant z\leqslant f(x,y)$. An "infinitesimal argument" for this result goes as follows. Consider R to be made of "infinitesimal rectangular prisms" with base dx and dy and height $f(x,y)$ (see Fig. 17.3.1). The total

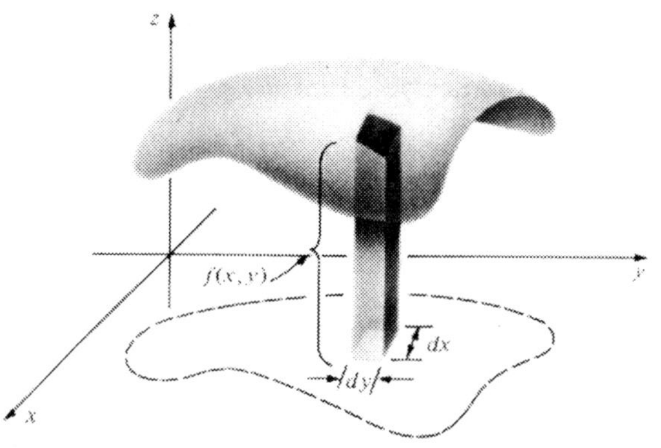

Figure 17.3.1. The region under the graph of f on D may be thought of as being composed of infinitesimal rectangular prisms.

volume is obtained by integrating (that is, "summing") the volumes of these cylinders. Notice, in particular, that if $f(x, y)$ is identically equal to 1, the volume of the region under the graph is just the area of D, so *the area of D is equal to $\iint_D dx\,dy$.*

Example 1 Compute the volume of the solid in space bounded by the four planes $x = 0$, $y = 0$, $z = 0$, and $3x + 4y = 10$, and the graph $z = x^2 + y^2$.

Solution The region is sketched in Fig. 17.3.2. Thus the volume is

$$\iint_D (x^2 + y^2)\,dx\,dy = \int_0^{5/2}\left[\int_0^{(10-4y)/3}(x^2 + y^2)\,dx\right]dy$$

$$= \int_0^{5/2}\left[\frac{(10-4y)^3}{3^4} + \frac{y^2(10-4y)}{3}\right]dy$$

$$= -\frac{(10-4y)^4}{3^4\cdot 4\cdot 4} + \frac{10y^3}{9} - \frac{y^4}{3}\Bigg|_0^{5/2}$$

$$= \frac{10^4}{3^4\cdot 2^4} + \frac{10\cdot 5^3}{3^2\cdot 2^3} - \frac{5^4}{3\cdot 2^4} = \frac{15625}{1296} \approx 12.056. \blacktriangle$$

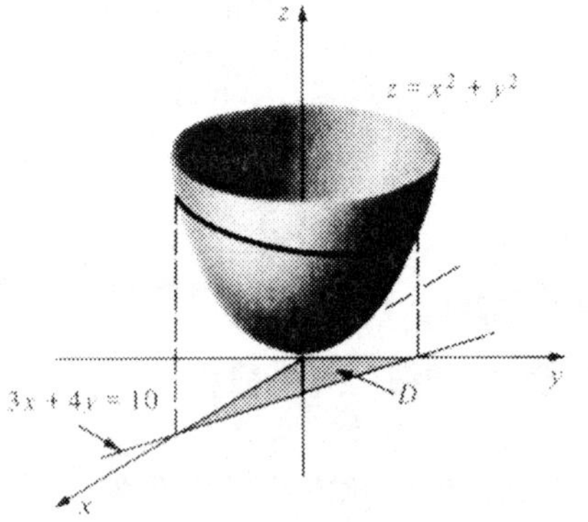

Figure 17.3.2. The volume of the region in space above D and below $z = x^2 + y^2$, is $\iint_D (x^2 + y^2)\,dx\,dy$.

(The volume in Example 1 can also be written as $5^6/(3^4\cdot 2^4)$. Can any reader explain this simple factorization?)

By reasoning similar to that for one-variable calculus (see Section 9.3), we are led to the following definition of the average value of a function on a plane region.

Average Value

If f is an integrable function on D, the ratio of the integral to the area of D,

$$\frac{\iint_D f(x, y)\,dx\,dy}{\iint_D dx\,dy},$$

is called the *average value* of f on D.

Example 2 Find the average value of $f(x, y) = x\sin^2(xy)$ on $D = [0, \pi]\times[0, \pi]$.

Solution First we compute

$$\iint_D f(x, y)\,dx\,dy = \int_0^{\pi}\int_0^{\pi} x\sin^2(xy)\,dx\,dy$$

$$= \int_0^{\pi}\left(\int_0^{\pi}\frac{1-\cos(2xy)}{2}\,x\,dy\right)dx = \int_0^{\pi}\left(\frac{y}{2}-\frac{\sin 2(xy)}{4x}\right)x\Big|_{y=0}^{\pi}dx$$

$$= \int_0^{\pi}\left(\frac{\pi x}{2}-\frac{\sin(2\pi x)}{4}\right)dx = \left(\frac{\pi x^2}{4}+\frac{\cos(2\pi x)}{8\pi}\right)\Big|_0^{\pi}$$

$$= \frac{\pi^3}{4}+\frac{\left[\cos(2\pi^2)-1\right]}{8\pi}.$$

Thus the average value of f is

$$\frac{\pi^3/4+\left[\cos(2\pi^2)-1\right]/8\pi}{\pi^2} = \frac{\pi}{4}+\frac{\cos(2\pi^2)-1}{8\pi^3} \approx 0.7839. \ \blacktriangle$$

Double integration also allows us to find the center of mass of a plate with variable density. Let D represent a plate with variable density $\rho(x, y)$. We can imagine breaking D into infinitesimal elements with mass $\rho(x, y)\,dx\,dy$; the total mass is thus $\iint_D \rho(x, y)\,dx\,dy$. Applying the consolidation principle (see Section 9.4) to the infinitesimal rectangles, one can derive the formulas in the following box.

Center of Mass

$$\bar{x} = \frac{\displaystyle\iint_D x\rho(x, y)\,dx\,dy}{\displaystyle\iint_D \rho(x, y)\,dx\,dy} \quad \text{and} \quad \bar{y} = \frac{\displaystyle\iint_D y\rho(x, y)\,dx\,dy}{\displaystyle\iint_D \rho(x, y)\,dx\,dy}.$$

Example 3 Find the center of mass of the rectangle $[0, 1] \times [0, 1]$ if the density is e^{x+y}.

Solution First we compute the mass:

$$\iint_D e^{x+y}\,dx\,dy = \int_0^1\int_0^1 e^{x+y}\,dx\,dy = \int_0^1\left(e^{x+y}\big|_{x=0}^1\right)dy$$

$$= \int_0^1 (e^{1+y}-e^y)\,dy$$

$$= e^{1+y}-e^y\big|_{y=0}^1 = e^2-e-(e-1) = e^2-2e+1.$$

The numerator in the formula for $\bar{x}$ is

$$\int_0^1\int_0^1 xe^{x+y}\,dx\,dy = \int_0^1 (xe^{x+y}-e^{x+y})\Big|_{x=0}^1\,dy$$

$$= \int_0^1\left[e^{1+y}-e^{1+y}-(0e^y-e^y)\right]dy$$

$$= \int_0^1 e^y\,dy = e^y\big|_{y=0}^1 = e-1,$$

so

$$\bar{x} = \frac{e-1}{e^2-2e+1} = \frac{e-1}{(e-1)^2} = \frac{1}{e-1} \approx 0.582.$$

The roles of x and y may be interchanged in all these calculations, so $\bar{y} = 1/(e-1) \approx 0.582$ as well. $\blacktriangle$

In Section 10.3 we used ordinary integration to determine the area of surfaces of revolution. Using the double integral, we can find the area of general curved surfaces. We confine ourselves here to an infinitesimal argument—the rigorous theory of surface area is quite subtle.[2]

To find the area of the graph $z = f(x, y)$ of a function f over the plane region D, we divide D into "infinitesimal rectangles" which are of the form $[x, x + dx] \times [y, y + dy]$. The image of this infinitesimal rectangle on the graph of f is approximately an "infinitesimal parallelogram" with vertices at

$$\begin{aligned}
P_1 &= (x, y, f(x, y)),\\
P_2 &= (x + dx, y, f(x + dx, y)) \approx (x + dx, y, f(x, y) + f_x(x, y)\,dx),\\
P_3 &= (x, y + dy, f(x, y + dy)) \approx (x, y + dy, f(x, y) + f_y(x, y)\,dy),\\
P_4 &= (x + dx, y + dy, f(x + dx, y + dy))\\
&\approx (x + dx, y + dy, f(x, y) + f_x(x, y)\,dx + f_y(x, y)\,dy).
\end{aligned}$$

(See Fig. 17.3.3.)

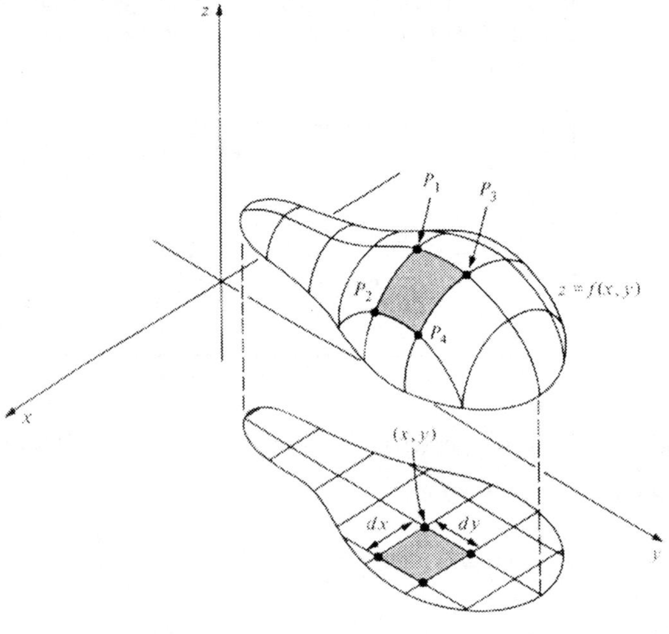

Figure 17.3.3. The "image" on the surface $z = f(x, y)$ of an infinitesimal rectangle in the plane is the infinitesimal parallelogram $P_1P_2P_4P_3$.

We compute the area dA of this parallelogram by taking the length of the cross product of the vectors from P_1 to P_2 and from P_1 to P_3 (see Section 13.5). The vectors in question are $dx\,\mathbf{i} + f_x(x, y)\,dx\,\mathbf{k}$ and $dy\,\mathbf{j} + f_y(x, y)\,dy\,\mathbf{k}$; their cross product is

$$\begin{vmatrix} \mathbf{i} & \mathbf{j} & \mathbf{k} \\ dx & 0 & f_x(x, y)\,dx \\ 0 & dy & f_y(x, y)\,dy \end{vmatrix} = -f_x(x, y)\,dx\,dy\,\mathbf{i} - f_y(x, y)\,dx\,dy\,\mathbf{j} + dx\,dy\,\mathbf{k},$$

and the length of this vector is $dA = \sqrt{1 + f_x(x, y)^2 + f_y(x, y)^2}\,dx\,dy$. To get the area of the surface, we "sum" the areas of the infinitesimal parallelograms by integrating over D.

[2] See T. Radó, *Length and Area*, American Mathematical Society Colloquium Publications. Volume 30 (1958).

Surface Area of a Graph

$$\text{Area} = \int\!\!\int_D dA = \int\!\!\int_D \sqrt{1 + f_x(x, y)^2 + f_y(x, y)^2}\, dx\, dy$$
$$= \int\!\!\int_D \sqrt{1 + \left(\frac{\partial z}{\partial x}\right)^2 + \left(\frac{\partial z}{\partial y}\right)^2}\, dx\, dy.$$

Note the similarity of this expression with the formula for the arc length of a graph (Section 10.3). As with arc length, the square root makes the analytic evaluation of surface area integrals difficult or even impossible in all but a few accidentally simple cases.

Example 4 Find the surface area of the part of the sphere $x^2 + y^2 + z^2 = 1$ lying above the ellipse $x^2 + (y^2/a^2) \leqslant 1$; ($a$ is a constant satisfying $0 < a \leqslant 1$).

Solution The region described by $x^2 + (y^2/a^2) \leqslant 1$ is type 1 with $\phi_1(x) = -a\sqrt{1 - x^2}$ and $\phi_2(x) = a\sqrt{1 - x^2}$; $-1 \leqslant x \leqslant 1$. The upper hemisphere may be described by the equation $z = f(x, y) = \sqrt{1 - x^2 - y^2}$. The partial derivatives of f are $\partial z/\partial x = -x/\sqrt{1 - x^2 - y^2}$ and $\partial z/\partial y = -y/\sqrt{1 - x^2 - y^2}$, so the area integrand is

$$\sqrt{1 + \frac{x^2}{1 - x^2 - y^2} + \frac{y^2}{1 - x^2 - y^2}} = \frac{1}{\sqrt{1 - x^2 - y^2}}$$

and the area is

$$A = \int\!\!\int_D \frac{dx\, dy}{\sqrt{1 - x^2 - y^2}} = \int_{-1}^{1} \left[\int_{-a\sqrt{1-x^2}}^{a\sqrt{1-x^2}} \frac{dy}{\sqrt{1 - x^2 - y^2}} \right] dx$$
$$= \int_{-1}^{1} \left(\sin^{-1} \frac{y}{\sqrt{1 - x^2}} \Bigg|_{-a\sqrt{1-x^2}}^{a\sqrt{1-x^2}} \right) dx = 2\int_{-1}^{1} \sin^{-1} a\, dx = 4\sin^{-1} a.$$

(See Fig. 17.3.4.) As a check on our answer, note that if $a = 1$, we get $4\sin^{-1} 1 = 4 \cdot \pi/2 = 2\pi$, the correct formula for the area of a hemisphere of radius 1 (the surface area of a full sphere of radius r is $4\pi r^2$). ▲

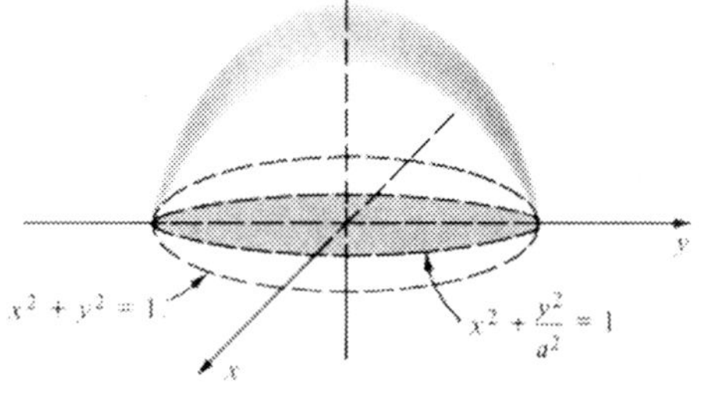

Figure 17.3.4. The area of the hemisphere above the ellipse $x^2 + y^2/a^2 \leqslant 1$ is $4\sin^{-1} a$.

Example 5 The equation in xyz space of the surface obtained by revolving the graph $y = f(x)$ about the x axis is $y^2 + z^2 = [f(x)]^2$. Express as a double integral the area of the part of this surface lying between the planes $x = a$ and $x = b$. Carry out the integration over y. Do you recognize the resulting integral over x?

Solution Writing z as a function of x and y, we have $z = g(x, y) = \pm\sqrt{f(x)^2 - y^2}$. The domain of this function consists of those (x, y) with $-f(x) \leqslant y \leqslant f(x)$, so the surface in question lies over the type 1 region D defined by $a \leqslant x \leqslant b$, $-f(x) \leqslant y \leqslant f(x)$. (See Fig. 17.3.5.) The partial derivatives of g are

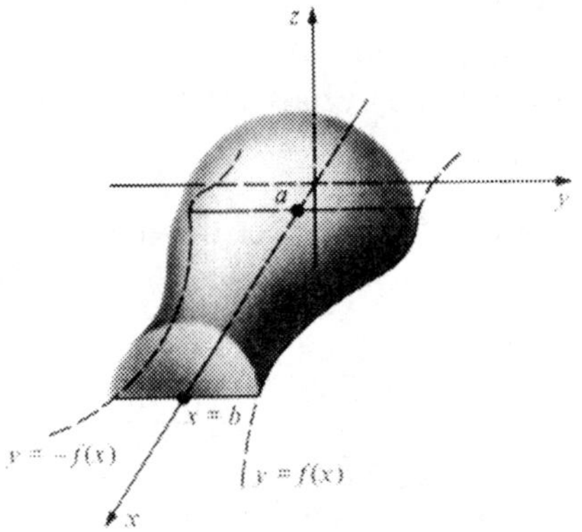

Figure 17.3.5. Finding the surface area when $y = f(x)$ is rotated about the x axis.

$$g_x(x, y) = f'(x)f(x)/\sqrt{f(x)^2 - y^2}\,, \qquad g_y(x, y) = -y/\sqrt{f(x)^2 - y^2}\,,$$

so the surface area integrand is

$$\sqrt{1 + \frac{f'(x)^2 f(x)^2}{f(x)^2 - y^2} + \frac{y^2}{f(x)^2 - y^2}} = \sqrt{\frac{f(x)^2 - y^2 + f'(x)^2 f(x)^2 + y^2}{f(x)^2 - y^2}}$$

$$= f(x)\sqrt{\frac{1 + f'(x)^2}{f(x)^2 - y^2}}$$

and the area is $A = 2\int_a^b \int_{-f(x)}^{f(x)} f(x)\sqrt{\dfrac{1 + f'(x)^2}{f(x)^2 - y^2}}\,dy\,dx.$

(The factor of 2 occurs since half the surface lies below the xy plane.) We can carry out the integration over y:

$$A = 2\int_a^b f(x)\sqrt{1 + f'(x)^2}\left[\int_{-f(x)}^{f(x)} \frac{1}{\sqrt{f(x)^2 - y^2}}\,dy\right]dx$$

$$= 2\int_a^b f(x)\sqrt{1 + f'(x)^2}\left[\sin^{-1}\left(\frac{y}{f(x)}\right)\Big|_{y = -f(x)}^{f(x)}\right]dx$$

$$= 2\int_a^b f(x)\sqrt{1 + f'(x)^2}\left(\frac{\pi}{2} + \frac{\pi}{2}\right)dx$$

$$= 2\pi\int_a^b f(x)\sqrt{1 + f'(x)^2}\,dx.$$

This is the formula for the area of a surface of revolution (Section 10.3). ▲

Exercises for Section 17.3

In Exercises 1–4, find the volume under the graph of $f(x, y)$ between the given planes $x = a$, $x = b$, $y = c$, and $y = d$,

1. $f(x, y) = x\sin y + 3$; $a = 0$, $b = 2$, $c = \pi$, $d = 3\pi$;
2. $f(x, y) = xy^5 + 2x^4 + 6$; $a = 0$, $b = 1$, $c = -1$, $d = 1$.
3. $f(x, y) = 2x^4 + 3y^{2/3}$; $a = -1$, $b = 1$, $c = 0$, $d = 2$;
4. $f(x, y) = xy\sqrt{x^2 + 3y^2}$; $a = 1$, $b = 2$, $c = 1$, $d = 2$.
5. Compute the volume under the graph of $f(x, y) = 1 + \sin(\pi y/2) + x$ on the parallelogram in the xy plane with vertices $(0,0), (1,2), (2,0), (3,2)$. Sketch.
6. Compute the volume under the graph of $f(x, y) = 4x^2 + 3y^2 + 27$ on the disk of radius 2 centered at $(0, 1)$. Sketch.
7. Compute the volume under the graph of $f(x, y) = (\cos y)e^{1-\cos 2x} + xy$ on the region bounded by the line $y = 2x$, the x axis, and the line $x = \pi/2$.
8. Compute the volume between the graphs of the functions $f(x, y) = 2x + 1$ and $g(x, y) = -x - 3y - 6$ on the region bounded by the y axis and the curve $x = 4 - y^2$. Sketch.
9. Find the volume of the region bounded by the planes $x = 0$ and $z = 0$ and the surfaces $x = -4y^2 + 3$, and $z = x^3y$.
10. Find the volume of the region bounded by the planes $x = 1$, $z = 0$, $y = x + 1$, $y = -x - 1$ and the surface $z = 2x^2 + y^4$.

In Exercises 11–14, find the average value of the given function on the given region.

11. $f(x, y) = y\sin xy$; $D = [0, \pi] \times [0, \pi]$.
12. $f(x, y) = x^2 + y^2$; $D =$ the ring between the circles $x^2 + y^2 = \frac{1}{2}$ and $x^2 + y^2 = 1$.
13. $f(x, y) = e^{x+y}$; $D =$ the triangle with vertices at $(0,0)$, $(0, 1)$, and $(1,0)$.
14. $f(x, y) = 1/(x + y)$; $D = [e, e^2] \times [e, e^2]$.

Find the average value of $x^2 + y^2$ over each of the regions in Exercises 15–18.

15. The square $[0, 1] \times [0, 1]$.
16. The square $[a, a + 1] \times [0, 1]$, where $a > 0$.
17. The square $[0, a] \times [0, a]$, where $a > 0$.
18. The set of (x, y) such that $x^2 + y^2 < a^2$.
19. Find the center of mass of the region between $y = x^2$ and $y = x$ if the density is $x + y$.
20. Find the center of mass of the region between $y = 0$, $y = x^2$, where $0 \leqslant x \leqslant \frac{1}{2}$.
21. Find the center of mass of the disk determined by $(x - 1)^2 + y^2 \leqslant 1$ if the density is x^2.
22. Repeat Exercise 21 if the density is y^2.
23. Find the area of the graph of the function $f(x, y) = \frac{2}{3}(x^{3/2} + y^{3/2})$ which lies over the domain $D = [0, 1] \times [0, 1]$.
24. Find the area cut out of the cylinder $x^2 + z^2 = 1$ by the cylinder $x^2 + y^2 = 1$.
25. Calculate the area of the part of the cone $z^2 = x^2 + y^2$ lying in the region of space defined by $x \geqslant 0$, $y \geqslant 0$, $z \leqslant 1$.
26. Find the area of the portion of the cylinder $x^2 + z^2 = 4$ which lies above the rectangle defined by $-1 \leqslant x \leqslant 1$, $0 \leqslant y \leqslant 2$.
27. Show that if a plate D has constant density, then the average values of x and y on D are the coordinates of the center of mass.
28. Find the center of mass of the region (composed of two pieces) bounded by $y = x^3$ and $y = \sqrt[3]{x}$ if the density is $(x - y)^2$. Try to minimize your work by exploiting some symmetry in the problem.
29. (a) Prove that the area on a sphere of radius r cut out by a cone of angle ϕ is $2\pi r^2(1 - \cos\phi)$ (Fig. 17.3.6).

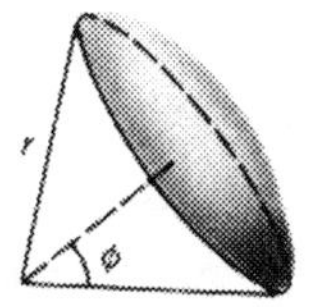

Figure 17.3.6. The area of the cap is $2\pi r^2(1 - \cos\phi)$.

(b) A sphere of radius 1 sits with its center on the surface of a sphere of radius $r > 1$. Show that the area of surface on the second sphere cut out by the first sphere is π. (Does something about this result surprise you?)

30. A uniform rectangular steel plate of sides a and b rotates about its center of gravity with constant angular velocity ω.
    (a) Kinetic energy equals $\frac{1}{2}$(mass)(velocity)^2. Argue that the kinetic energy of any element of mass $\rho\,dx\,dy$ ($\rho =$ constant) is given by $\rho(\omega^2/2)(x^2 + y^2)\,dx\,dy$, provided the origin $(0, 0)$ is placed at the center of gravity of the plate.
    (b) Justify the formula for kinetic energy:
$$\text{K.E.} = \int\int_{\text{plate}} \rho\frac{\omega^2}{2}(x^2 + y^2)\,dx\,dy.$$
    (c) Evaluate the integral, assuming that the plate is described by $-a/2 \leqslant x \leqslant a/2$, $-b/2 \leqslant y \leqslant b/2$.

31. A sculptured gold plate D is defined by $0 \leqslant x \leqslant 2\pi$ and $0 \leqslant y \leqslant \pi$ (centimeters) and has mass density $\rho(x, y) = y^2\sin^2 4x + 2$ (grams per square centimeter). If gold sells for \$7 per gram, how much is the gold in the plate worth?

★32. (a) Relate the integrand in the surface area formula to the angle between **k** and the normal to the surface $z = f(x, y)$.
(b) Express the ratio of the area of the graph of f over D to the area of D as the average value of some geometrically defined quantity.

★33. Express as a double integral the volume enclosed by the surface of revolution in Example 5. Carry out the integration over y and show that the resulting integral over x is a formula in Section 9.1.

★34. Let n right circular cylinders of radius r intersect such that their axes lie in a plane, meeting at one point with equal angles. Find the volume of their intersection.

17.4 Triple Integrals

Integrals over regions in three-dimensional space require the triple integral.

The basic ideas developed in Sections 17.1 and 17.2 can be readily extended from double to triple integrals. As with double integrals, one of the most powerful evaluation methods is reduction to iterated integrals. A second important technique, which we discuss in Section 17.5, is the method of changing variables.

If the temperature inside an oven is not uniform, determining the average temperature involves "summing" the values of the temperature function at all points in the solid region enclosed by the oven walls. Such a sum is expressed mathematically as a triple integral.

We formalize the ideas just as we did for double integrals. Suppose that W is a box (that is, rectangular parallelepiped) in space bounded by the planes $x = a$, $x = b$, $y = c$, $y = d$, and $z = p$, $z = q$, as in Fig. 17.4.1. We denote this

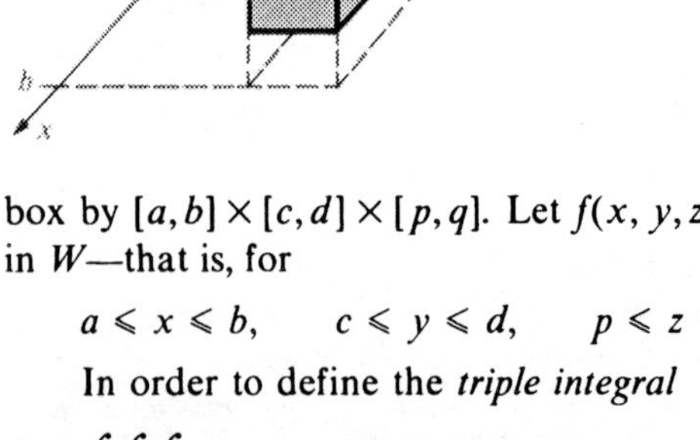

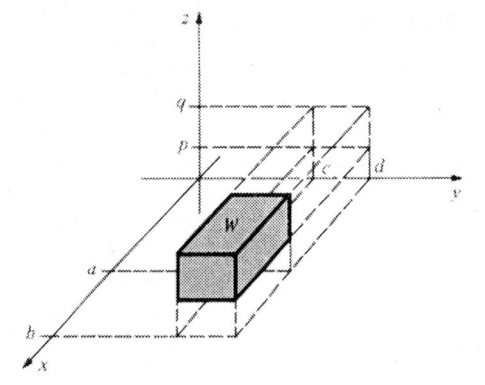

Figure 17.4.1. The box $W = [a,b] \times [c,d] \times [p,q]$ consists of points (x, y, z) satisfying $a \leqslant x \leqslant b$, $c \leqslant y \leqslant d$, and $p \leqslant z \leqslant q$.

box by $[a,b] \times [c,d] \times [p,q]$. Let $f(x, y, z)$ be a function defined for (x, y, z) in W—that is, for

$$a \leqslant x \leqslant b, \qquad c \leqslant y \leqslant d, \qquad p \leqslant z \leqslant q.$$

In order to define the *triple integral*

$$\iiint_W f(x, y, z)\, dx\, dy\, dz,$$

we first define the concept of a step function of three variables.

A function $g(x, y, z)$ defined on $[a,b] \times [c,d] \times [p,q]$ is called a *step function* if there are partitions

$$a = t_0 < t_1 < \cdots < t_n = b \quad \text{of} \quad [a,b],$$
$$c = s_0 < s_1 < \cdots < s_m = d \quad \text{of} \quad [c,d],$$
$$p = r_0 < r_1 < \cdots < r_l = q \quad \text{of} \quad [p,q]$$

such that $g(x, y, z)$ has the constant value k_{ijk} for (x, y, z) in the open box

$$W_{ijk} = (t_{i-1}, t_i) \times (s_{j-1}, s_j) \times (r_{k-1}, r_k).$$

We cannot draw the graphs of functions of three variables; however, we can indicate the value k_{ijk} associated with each box (see Fig. 17.4.2). The

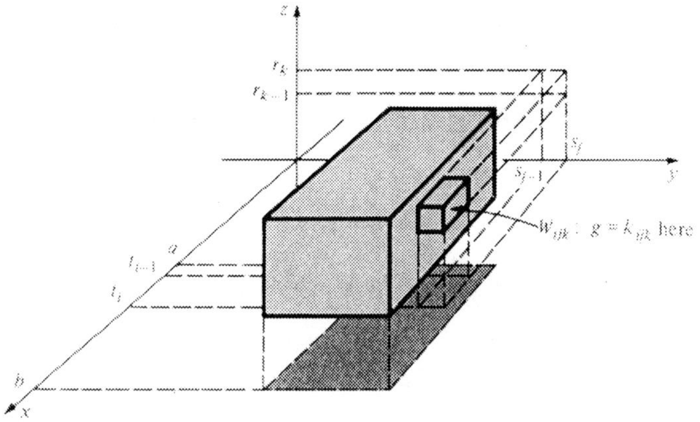

Figure 17.4.2. g has the value k_{ijk} on the small box W_{ijk}.

integral of g is defined as a sum of *nml* terms:

$$\iiint_W g(x, y, z)\,dx\,dy\,dz = \sum_{i=1,\, j=1,\, k=1}^{n,m,l} k_{ijk}(\text{volume } W_{ijk})$$
$$= \sum_{i=1,\, j=1,\, k=1}^{n,m,l} k_{ijk}\,(\Delta t_i)(\Delta s_j)(\Delta r_k).$$

If f is any function on W, its lower (respectively upper) sums are defined, as before, as the integrals of step functions g (respectively h) such that $g(x, y, z) \leqslant f(x, y, z)$ (respectively $h(x, y, z) \geqslant f(x, y, z)$) for all points (x, y, z) in W.

The Triple Integral

We say that f is *integrable* on W if there is a number S_0 such that every $S < S_0$ is a lower sum for f on W and every $S > S_0$ is an upper sum. The number S_0 is called the *integral* of f over W and is denoted by

$$\iiint_W f(x, y, z)\,dx\,dy\,dz.$$

At this point you should look back at the basic properties of the double integrals listed in Section 17.1. Similar properties hold for triple integrals. Furthermore, there is a similar reduction to iterated integrals.

Theorem: Reduction to Iterated Integrals

Let $f(x,y,z)$ be integrable on the box $W=[a,b]\times[c,d]\times[p,q]$. Then any iterated integral which exists is equal to the triple integral; that is,

$$\iiint_W f(x,y,z)\,dx\,dy\,dz = \int_p^q\int_c^d\int_a^b f(x,y,z)\,dx\,dy\,dz$$

$$= \int_p^q\int_a^b\int_c^d f(x,y,z)\,dy\,dx\,dz$$

$$= \int_a^b\int_p^q\int_c^d f(x,y,z)\,dy\,dz\,dx,$$

and so on. (There are six possible orders altogether.)

The proof of this result is just like the corresponding one in Section 17.1, so we omit it.

Example 1 (a) Let W be the box $[0,1]\times[-\frac{1}{2},0]\times[0,\frac{1}{3}]$. Evaluate

$$\iiint_W (x+2y+3z)^2\,dx\,dy\,dz.$$

(b) Verify that we get the same answer if the integration is done in the order y first, then z, and then x.

Solution (a) According to the reduction to iterated integrals, this integral may be evaluated as

$$\int_0^{1/3}\int_{-1/2}^0\int_0^1 (x+2y+3z)^2\,dx\,dy\,dz$$

$$= \int_0^{1/3}\int_{-1/2}^0\left[\left.\frac{(x+2y+3z)^3}{3}\right|_{x=0}^{1}\right]dy\,dz$$

$$= \int_0^{1/3}\int_{-1/2}^0 \frac{1}{3}\left[(1+2y+3z)^3-(2y+3z)^3\right]dy\,dz$$

$$= \int_0^{1/3}\frac{1}{24}\left[(1+2y+3z)^4-(2y+3z)^4\right]\Big|_{y=-1/2}^{0}\,dz$$

$$= \int_0^{1/3}\frac{1}{24}\left[(3z+1)^4-2(3z)^4+(3z-1)^4\right]dz$$

$$= \frac{1}{24\cdot 15}\left[(3z+1)^5-2(3z)^5+(3z-1)^5\right]\Big|_{z=0}^{1/3}$$

$$= \frac{1}{24\cdot 15}(2^5-2) = \frac{1}{12}.$$

(b) $$\iiint_W (x+2y+3z)^2\,dy\,dz\,dx$$

$$= \int_0^1\int_0^{1/3}\int_{-1/2}^0 (x+2y+3z)^2\,dy\,dz\,dx$$

$$= \int_0^1 \int_0^{1/3} \left[\frac{(x+2y+3z)^3}{6} \Bigg|_{y=-1/2}^{0} \right] dz\, dx$$

$$= \int_0^1 \int_0^{1/3} \frac{1}{6} \left[(x+3z)^3 - (x+3z-1)^3 \right] dz\, dx$$

$$= \int_0^1 \frac{1}{6} \left[\left(\frac{(x+3z)^4}{12} - \frac{(x+3z-1)^4}{12} \right) \Bigg|_{z=0}^{1/3} \right] dx$$

$$= \int_0^1 \frac{1}{72} \left[(x+1)^4 + (x-1)^4 - 2x^4 \right] dx$$

$$= \frac{1}{72} \frac{1}{5} \left[(x+1)^5 + (x-1)^5 - 2x^5 \right]_{x=0}^{1} = \frac{1}{12}. \ \blacktriangle$$

Example 2 Evaluate the integral of e^{x+y+z} over the box $[0,1] \times [0,1] \times [0,1]$.

Solution

$$\int_0^1 \int_0^1 \int_0^1 e^{x+y+z}\, dx\, dy\, dz = \int_0^1 \int_0^1 \left(e^{x+y+z} \big|_{x=0}^{1} \right) dy\, dz$$

$$= \int_0^1 \int_0^1 \left(e^{1+y+z} - e^{y+z} \right) dy\, dz = \int_0^1 \left[e^{1+y+z} - e^{y+z} \right]_{y=0}^{1} dz$$

$$= \int_0^1 \left[e^{2+z} - 2e^{1+z} + e^z \right] dz = \left[e^{2+z} - 2e^{1+z} + e^z \right]_0^1$$

$$= e^3 - 3e^2 + 3e - 1 = (e-1)^3. \ \blacktriangle$$

As in the two-variable case, we define the integral of a function f over a bounded region W by defining a new function f^*, equal to f on W and zero outside W, and then setting

$$\int\int\int_W f(x,y,z)\, dx\, dy\, dz = \int\int\int_{W^*} f^*(x,y,z)\, dx\, dy\, dz,$$

where W^* is any box containing the region W.

As before, we restrict our attention to particularly simple regions. A three-dimensional region W will be said to be of *type* I if there is an elementary region D in the xy plane and a pair of continuous functions, $\gamma_1(x,y)$ and $\gamma_2(x,y)$ defined on D, such that W consists of those triples (x,y,z) for which $(x,y) \in D$ and $\gamma_1(x,y) \leqslant z \leqslant \gamma_2(x,y)$. The region D may itself be of type 1 or type 2, so there are two possible descriptions of a type I region:

$$\begin{aligned} &a \leqslant x \leqslant b, \qquad \phi_1(x) \leqslant y \leqslant \phi_2(x), \\ &\gamma_1(x,y) \leqslant z \leqslant \gamma_2(x,y) \qquad \text{(if } D \text{ is of type 1)} \end{aligned} \tag{1}$$

or

$$\begin{aligned} &c \leqslant y \leqslant d, \qquad \psi_1(y) \leqslant x \leqslant \psi_2(y), \\ &\gamma_1(x,y) \leqslant z \leqslant \gamma_2(x,y) \qquad \text{(if } D \text{ is of type 2).} \end{aligned} \tag{2}$$

Figure 17.4.3 on the next page shows two regions of type I that are described by conditions (1) and (2), respectively.

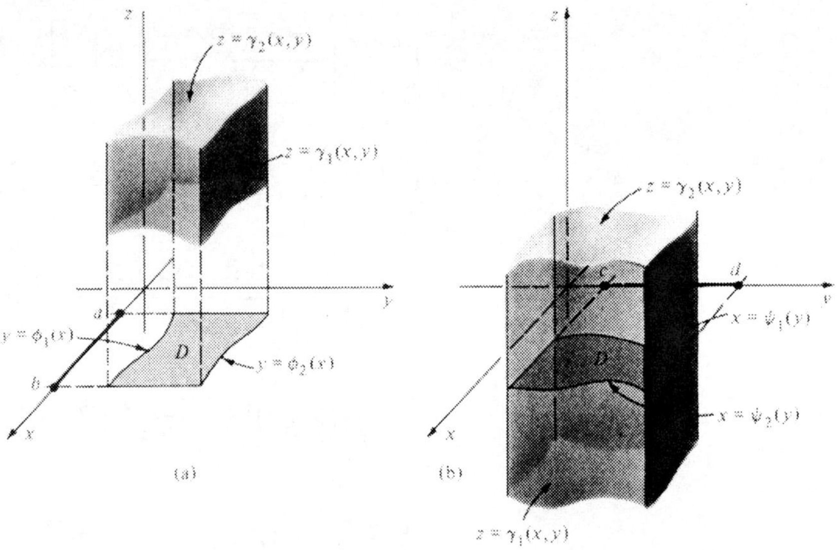

Figure 17.4.3. A region of type I lies between two graphs $z = \gamma_1(x, y)$ and $z = \gamma_2(x, y)$.

A region W is of *type II* if it can be expressed in form (1) or (2) with the roles of x and z interchanged, and W is of *type III* if it can be expressed in form (1) or (2) with y and z interchanged. See Fig. 17.4.4.

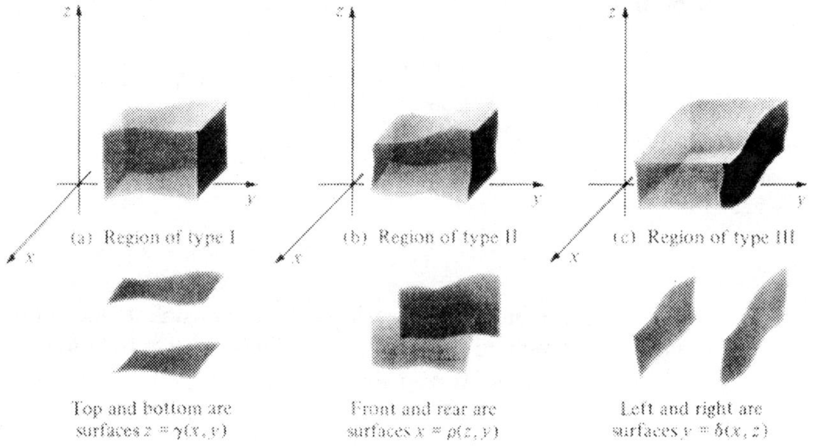

Figure 17.4.4. The three types of regions in space.

Notice that a given region may be of two or even three types at once. (See Fig. 17.4.5.) As with regions in the plane, we call a region of type I, II, or III in space an *elementary region*.

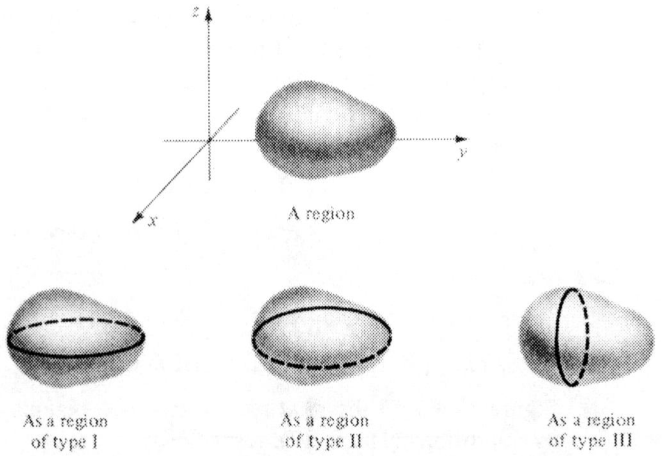

Figure 17.4.5. Regions in space can be of more than one type. This one is of all three types.

Example 3 Show that the unit ball $x^2 + y^2 + z^2 \leqslant 1$ is a region of all three types.

Solution As a type I region, we can express it as

$$-1 \leqslant x \leqslant 1,$$

$$-\sqrt{1 - x^2} \leqslant y \leqslant \sqrt{1 - x^2}\,,$$

$$-\sqrt{1 - x^2 - y^2} \leqslant z \leqslant \sqrt{1 - x^2 - y^2}\,.$$

In doing this, we first write the top and bottom hemispheres as $z = \sqrt{1 - x^2 - y^2}$ and $z = -\sqrt{1 - x^2 - y^2}$, where x and y vary over the unit disk (that is, $-\sqrt{1 - x^2} \leqslant y \leqslant \sqrt{1 - x^2}$ and x varies between -1 and 1). (See Fig. 17.4.6.) We write the region as a type II or III region in a similar manner by interchanging the roles of x, y, and z in the defining inequalities. ▲

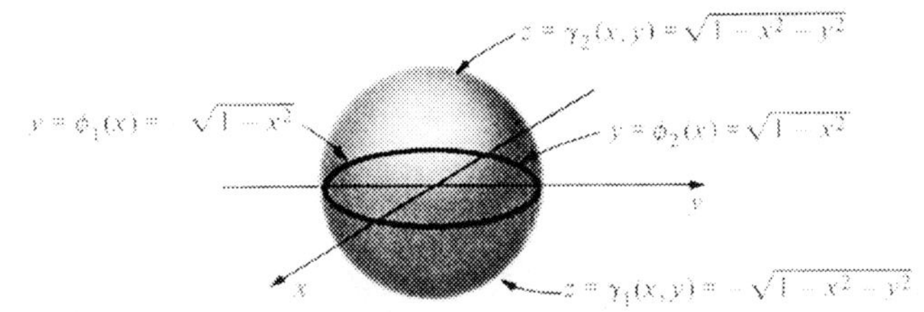

Figure 17.4.6. The unit ball described as a region of type I.

As with integrals in the plane, any function of three variables which is continuous over an elementary region is integrable on that region. An argument like that for double integrals shows that a triple integral over an elementary region can be rewritten as an iterated integral in which the limits of integration are functions. The formulas for such iterated integrals are given in the following display.

Triple Integrals

Suppose that W is of type I. Then either

$$\iiint_W f(x, y, z)\,dx\,dy\,dz = \int_a^b \int_{\phi_1(x)}^{\phi_2(x)} \int_{\gamma_1(x,y)}^{\gamma_2(x,y)} f(x, y, z)\,dz\,dy\,dx \qquad (3)$$

(see Fig. 17.4.3(a)) or

$$\iiint_W f(x, y, z)\,dx\,dy\,dz = \int_c^d \int_{\psi_1(y)}^{\psi_2(y)} \int_{\gamma_1(x,y)}^{\gamma_2(x,y)} f(x, y, z)\,dz\,dx\,dy \qquad (4)$$

(see Fig. 17.4.3(b)).

If W is of type II, it can be expressed as the set of all (x, y, z) such that

$$a \leqslant z \leqslant b, \qquad \phi_1(z) \leqslant y \leqslant \phi_2(z), \qquad \rho_1(z, y) \leqslant x \leqslant \rho_2(z, y).$$

Then

$$\iiint_W f(x,y,z)\,dx\,dy\,dz = \int_a^b \int_{\phi_1(z)}^{\phi_2(z)} \int_{\rho_1(z,y)}^{\rho_2(z,y)} f(x,y,z)\,dx\,dy\,dz. \qquad (5)$$

If W is expressed as the set of all (x, y, z) such that

$$c \leqslant y \leqslant d, \qquad \psi_1(y) \leqslant z \leqslant \psi_2(y), \qquad \rho_1(z,y) \leqslant x \leqslant \rho_2(z,y),$$

then

$$\iiint_W f(x,y,z)\,dx\,dy\,dz = \int_c^d \int_{\psi_1(y)}^{\psi_2(y)} \int_{\rho_1(z,y)}^{\rho_2(z,y)} f(x,y,z)\,dx\,dz\,dy. \qquad (6)$$

There are similar formulas for type III regions (Exercise 22).

Another way to write formula (3) is

$$\iiint_W f(x,y,z)\,dx\,dy\,dz = \iint_D \left[\int_{\gamma_1(x,y)}^{\gamma_2(x,y)} f(x,y,z)\,dz\right] dy\,dx,$$

and for formula (4),

$$\iiint_W f(x,y,z)\,dx\,dy\,dz = \iint_D \left[\int_{\gamma_1(x,y)}^{\gamma_2(x,y)} f(x,y,z)\,dz\right] dx\,dy.$$

Notice that the triple integral $\iiint_W dx\,dy\,dz$ is simply the volume of W.

Example 4 Verify the formula for the volume of a ball of radius 1: $\iiint_W dx\,dy\,dz = \frac{4}{3}\pi$, where W is the set of (x, y, z) with $x^2 + y^2 + z^2 \leqslant 1$.

Solution As explained in Example 3, the ball is a region of type I. By formula (3), the integral is

$$\int_{-1}^{1} \int_{-\sqrt{1-x^2}}^{\sqrt{1-x^2}} \int_{-\sqrt{1-x^2-y^2}}^{\sqrt{1-x^2-y^2}} dz\,dy\,dx.$$

Holding y and x fixed and integrating with respect to z yields

$$\int_{-1}^{1} \int_{-\sqrt{1-x^2}}^{\sqrt{1-x^2}} \left[z \Big|_{-\sqrt{1-x^2-y^2}}^{\sqrt{1-x^2-y^2}} \right] dy\,dx$$

$$= 2\int_{-1}^{1} \left[\int_{-\sqrt{1-x^2}}^{\sqrt{1-x^2}} (1 - x^2 - y^2)^{1/2}\,dy \right] dx.$$

Since x is fixed in the integral over y, this integral can be expressed as $\int_{-a}^{a}(a^2 - y^2)^{1/2}\,dy$, where $a = (1 - x^2)^{1/2}$. This integral represents the area of a semicircular region of radius a, so that

$$\int_{-a}^{a} (a^2 - y^2)^{1/2}\,dy = \frac{a^2}{2}\pi.$$

(We could also have used a trigonometric substitution.) Thus

$$\int_{-\sqrt{1-x^2}}^{\sqrt{1-x^2}} (1 - x^2 - y^2)^{1/2}\,dy = \frac{1 - x^2}{2}\pi,$$

and so

$$2\int_{-1}^{1} \int_{-\sqrt{1-x^2}}^{\sqrt{1-x^2}} (1 - x^2 - y^2)^{1/2}\,dy\,dx = 2\int_{-1}^{1} \pi \frac{1 - x^2}{2}\,dx$$

$$= \int_{-1}^{1} \pi(1 - x^2)\,dx = \pi\left(x - \frac{x^3}{3}\right)\Big|_{x=-1}^{1} = \frac{4}{3}\pi. \ \blacktriangle$$

Example 5 Let W be the region bounded by the planes $x = 0$, $y = 0$, and $z = 2$, and the surface $z = x^2 + y^2$. Compute $\iiint_W x\,dx\,dy\,dz$ and sketch the region.

Solution *Method* 1. The region W is sketched in Fig. 17.4.7. We may write this as a region of type I with $\gamma_1(x, y) = x^2 + y^2$, $\gamma_2(x, y) = 2$, $\phi_1(x) = 0$, $\phi_2(x) = \sqrt{2 - x^2}$, $a = 0$, and $b = \sqrt{2}$. By formula (3),

$$\iiint_W x\,dx\,dy\,dz = \int_0^{\sqrt{2}}\left[\int_0^{\sqrt{2-x^2}}\left(\int_{x^2+y^2}^{2} x\,dz\right)dy\right]dx$$

$$= \int_0^{\sqrt{2}}\int_0^{\sqrt{2-x^2}} x(2 - x^2 - y^2)\,dy\,dx$$

$$= \int_0^{\sqrt{2}} x\left[(2 - x^2)^{3/2} - \frac{(2 - x^2)^{3/2}}{3}\right]dx$$

$$= \int_0^{\sqrt{2}} \frac{2x}{3}(2 - x^2)^{3/2}\,dx = \left.\frac{-2(2 - x^2)^{5/2}}{15}\right|_0^{\sqrt{2}}$$

$$= 2\cdot\frac{2^{5/2}}{15} = \frac{8\sqrt{2}}{15}.$$

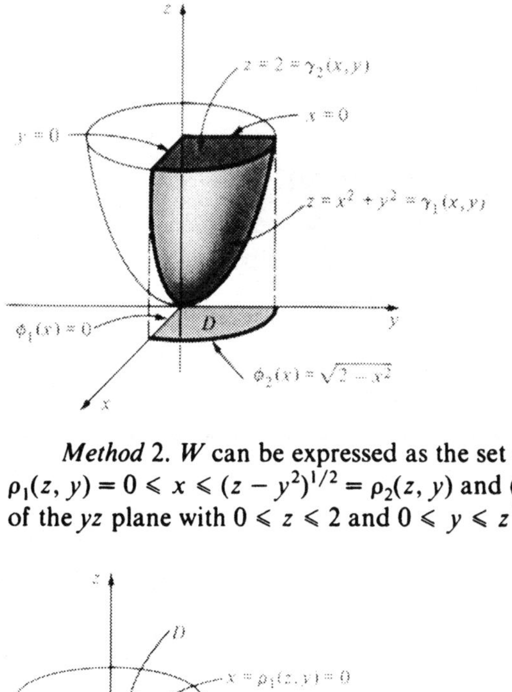

Figure 17.4.7. W is the region below the plane $z = 2$, above the paraboloid $z = x^2 + y^2$, and on the positive sides of the planes $x = 0$, $y = 0$.

Method 2. W can be expressed as the set of (x, y, z) with the property that $\rho_1(z, y) = 0 \leqslant x \leqslant (z - y^2)^{1/2} = \rho_2(z, y)$ and (z, y) in D, where D is the subset of the yz plane with $0 \leqslant z \leqslant 2$ and $0 \leqslant y \leqslant z^{1/2}$ (see Fig. 17.4.8). Therefore,

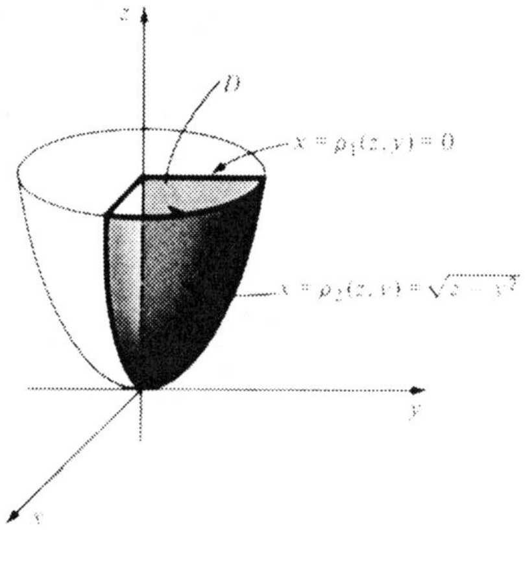

Figure 17.4.8. W as a type II region.

$$\iiint_W x\,dx\,dy\,dz = \iint_D\left(\int_{p_1(z,y)}^{p_2(z,y)} x\,dx\right)dy\,dz$$

$$= \int_0^2\left[\int_0^{z^{1/2}}\left(\int_0^{(z-y^2)^{1/2}} x\,dx\right)dy\right]dz = \int_0^2\int_0^{z^{1/2}}\left(\frac{z-y^2}{2}\right)dy\,dz$$

$$= \frac{1}{2}\int_0^2\left(z^{3/2} - \frac{z^{3/2}}{3}\right)dz = \frac{1}{2}\int_0^2 \frac{2}{3}z^{3/2}\,dz$$

$$= \left[\frac{2}{15}z^{5/2}\right]_0^2 = \frac{2}{15}2^{5/2} = \frac{8\sqrt{2}}{15},$$

which agrees with our other answer. ▲

Example 6 Evaluate $\int_0^1\int_0^x\int_{x^2+y^2}^1 dz\,dy\,dx$. Sketch the region W of integration and interpret.

Solution

$$\int_0^1\int_0^x\int_{x^2+y^2}^1 dz\,dy\,dx = \int_0^1\int_0^x(1 - x^2 - y^2)\,dy\,dx$$

$$= \int_0^1\left(x - x^3 - \frac{x^3}{3}\right)dx = \frac{1}{2} - \frac{1}{4} - \frac{1}{12} = \frac{1}{6}.$$

This is the volume of the region sketched in Fig. 17.4.9. ▲

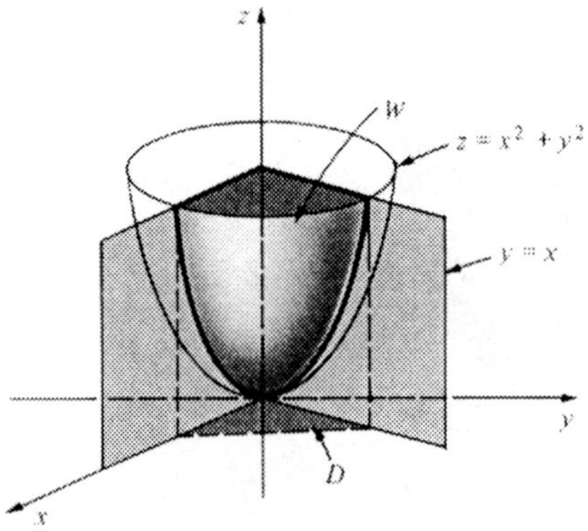

Figure 17.4.9. The region W (type I) for Example 6.

Exercises for Section 17.4

1. Evaluate $\iiint_W(2x + 3y + z)\,dx\,dy\,dz$, where $W = [1,2]\times[-1,1]\times[0,1]$, in at least two ways.
2. Evaluate the integral $\iiint_W x^2\,dx\,dy\,dz$, where $W = [0,1]\times[-1,1]\times[0,1]$, in at least two ways.
3. Integrate the function $\sin(x + y + z)$ over the box $[0,\pi]\times[0,\pi]\times[0,\pi]$.
4. Integrate ze^{x+y} over $[0,1]\times[0,1]\times[0,1]$.

Determine whether each of the regions in Exercises 5–8 is of type I, II, or III.

5. The region between the cone $z = \sqrt{x^2 + y^2}$ and the paraboloid $z = x^2 + y^2$.
6. The region cut out of the ball $x^2 + y^2 + z^2 \leqslant 4$ by the elliptic cylinder $2x^2 + z^2 = 1$ i.e. the region inside the cylinder and the ball.
7. The region inside the ellipsoid $x^2 + 2y^2 + z^2 = 1$ and above the plane $z = 0$.
8. The region bounded by the planes $x = 0$, $y = 0$, $z = 0$, $x + y = 4$, and $x = z - y - 1$.

Find the volumes of the regions in Exercises 9–12.

9. The region bounded by $z = x^2 + y^2$ and $z = 10 - x^2 - 2y^2$.
10. The solid bounded by $x^2 + 2y^2 = 2$, $z = 0$, and $x + y + 2z = 2$.
11. The solid bounded by $x = y$, $z = 0$, $y = 0$, $x = 1$ and $x + y + z = 0$.
12. The region common to the intersecting cylinders $x^2 + y^2 \leqslant a^2$ and $x^2 + z^2 \leqslant a^2$.

Evaluate the integrals in Exercises 13–20.

13. $\int_0^1\int_1^2\int_2^3 \cos[\pi(x+y+z)]\,dx\,dy\,dz$.

14. $\int_0^1\int_0^x\int_0^y (y+xz)\,dz\,dy\,dx$.

15. $\iiint_R (x^2+y^2+z^2)\,dx\,dy\,dz$; R is the region bounded by $x+y+z=a$ (where $a>0$), $x=0$, $y=0$, and $z=0$.

16. $\iiint_W z\,dx\,dy\,dz$; W is the region bounded by the planes $x=0$, $y=0$, $z=0$, $z=1$, and the cylinder $x^2+y^2=1$, with $x \geqslant 0$, $y \geqslant 0$.

17. $\iiint_W x^2\cos z\,dx\,dy\,dz$; W is the region bounded by $z=0$, $z=\pi$, $y=0$, $y=1$, $x=0$, and $x+y=1$.

18. $\int_0^2\int_0^x\int_0^{x+y} dz\,dy\,dx$.

19. $\iiint_W (1-z^2)\,dx\,dy\,dz$; W is the pyramid with top vertex at $(0,0,1)$ and base vertices at $(0,0)$, $(1,0)$, $(0,1)$, and $(1,1)$.

20. $\iiint_W (x^2+y^2)\,dx\,dy\,dz$; W is the same pyramid as in Exercise 19.

21. If $f(x,y,z)=F(x,y)$ for some function F—that is, if $f(x,y,z)$ does not depend on z—what is the triple integral of f over a box W?

22. Write general formulas analogous to (3) and (4) for the triple integral over a region of type III.

23. Do Example 4 by writing W as a region of type III.

24. Write out the property for triple integrals corresponding to property 2 of double integrals (Section 17.1, p. 841).

25. Show that the formula using triple integrals for the volume under the graph of a function $f(x,y)$, on an elementary region D in the plane, reduces to the double integral of f over D.

26. (a) Sketch the region for the integral

$$\int_0^1\int_0^x\int_0^y f(x,y,z)\,dz\,dy\,dx.$$

(b) Write the integral with the integration order $dx\,dy\,dz$.

27. (a) Show that the triple integral of a product over a box is the product of three ordinary integrals; that is, if $D=[a,b]\times[c,d]\times[p,q]$, then

$$\iiint_D f(x)g(y)h(z)\,dx\,dy\,dz$$

$$=\int_a^b f(x)\,dx\int_c^d g(y)\,dy\int_p^q h(z)\,dz.$$

(b) Use the result of part (a) to do Example 2.

17.5 Integrals in Polar, Cylindrical, and Spherical Coordinates

Problems with symmetry are often simplified by using coordinates that respect that symmetry.

We deal first with polar coordinates. Recall that a double integral

$$\iint_D f(x,y)\,dx\,dy$$

may be thought of as a "sum" of the values of f over infinitesimal rectangles with area $(dx)\cdot(dy)$. As in Section 10.5, however, we can also describe a region using polar coordinates and can use infinitesimal regions appropriate to those coordinates. The area of such a region is $r\,dr\,d\theta$, as is evident from Fig. 17.5.1. If $u=f(x,y)$, then u may be expressed in terms of r and θ by the formula $u=f(r\cos\theta, r\sin\theta)$.

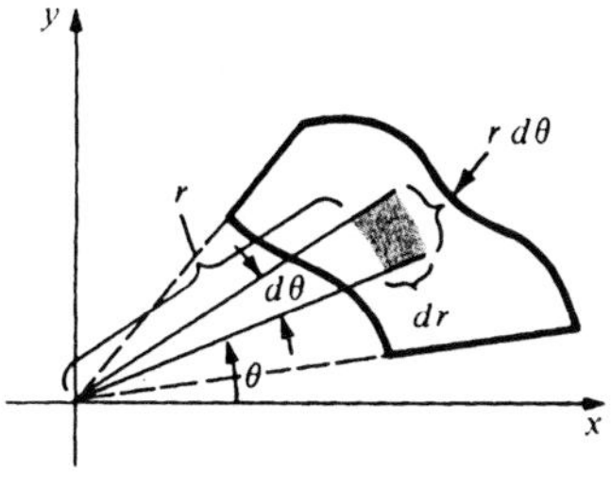

Figure 17.5.1. The area of the infinitesimal shaded region is $r\,dr\,d\theta$.

The preceding argument using infinitesimals suggests the formula in the following box (the rigorous proof is omitted).

Double Integrals in Polar Coordinates

$$\iint_D f(x, y)\,dx\,dy = \iint_{D'} f(r\cos\theta, r\sin\theta)\,r\,dr\,d\theta, \tag{1}$$

where D' is the region corresponding to D in the variables r and θ.

Example 1 Evaluate $\iint_{D_a} e^{-(x^2+y^2)}\,dx\,dy$, where D_a is the disk $x^2 + y^2 \leqslant a^2$.

Solution The presence of $r^2 = x^2 + y^2$ in the integrand and the symmetry of the disk suggest a change to polar coordinates. The disk is described by $0 \leqslant r \leqslant a$, $0 \leqslant \theta \leqslant 2\pi$, so by formula (1), we get

$$\iint_{D_a} e^{-(x^2+y^2)}\,dx\,dy = \int_0^{2\pi}\int_0^a e^{-r^2} r\,dr\,d\theta = \int_0^{2\pi}\left(-\frac{1}{2}e^{-r^2}\right)\Big|_0^a d\theta$$

$$= -\frac{1}{2}\int_0^{2\pi}(e^{-a^2} - 1)\,d\theta = \pi(1 - e^{-a^2}).$$

There is no direct way to evaluate this integral in xy coordinates! ▲

There is a remarkable application of the result of Example 1 to single-variable calculus: we will evaluate the Gaussian integral $\int_{-\infty}^{\infty} e^{-x^2}\,dx$, which is of basic importance in probability theory and quantum mechanics. There is no known way to evaluate this integral directly using only single-variable calculus. If we bring in two-variable calculus, however, the solution is surprisingly simple. Letting a go to ∞ in the formula $\iint_{D_a} e^{-(x^2+y^2)}\,dx\,dy = \pi(1 - e^{-a^2})$, we find that the limit $L = \lim_{a\to\infty}\iint_{D_a} e^{-(x^2+y^2)}\,dx\,dy$ exists and equals π. By analogy with the definition of improper integrals on the line, we may consider L as the (improper) integral of $e^{-(x^2+y^2)}$ over the entire plane, since the disks D_a grow to fill the whole plane as $a \to \infty$. The rectangles $R_a = [-a, a] \times [-a, a]$ grow to fill the whole plane, too, so we must have

$$\lim_{a\to\infty}\iint_{R_a} e^{-(x^2+y^2)}\,dx\,dy = \pi$$

as well[3]; but

[3] The technical details of the proof that

$$\lim_{a\to\infty}\iint_{D_a} e^{-(x^2+y^2)}\,dx\,dy = \lim_{a\to\infty}\iint_{R_a} e^{-(x^2+y^2)}\,dx\,dy$$

go as follows. We have already shown that

$$\lim_{a\to\infty}\iint_{D_a} e^{-(x^2+y^2)}\,dx\,dy$$

exists. Thus it suffices to show that

$$\lim_{a\to\infty}\left(\iint_{R_a} e^{-(x^2+y^2)}\,dx\,dy - \iint_{D_a} e^{-(x^2+y^2)}\,dx\,dy\right)$$

equals zero. The limit equals

$$\lim_{a\to\infty}\iint_{C_a} e^{-(x^2+y^2)}\,dx\,dy,$$

where C_a is the region between R_a and D_a (see Fig. 17.5.2). In the region C_a, $\sqrt{x^2+y^2} \geqslant a$ (the radius of D_a), so $e^{-(x^2+y^2)} \leqslant e^{-a^2}$. Thus

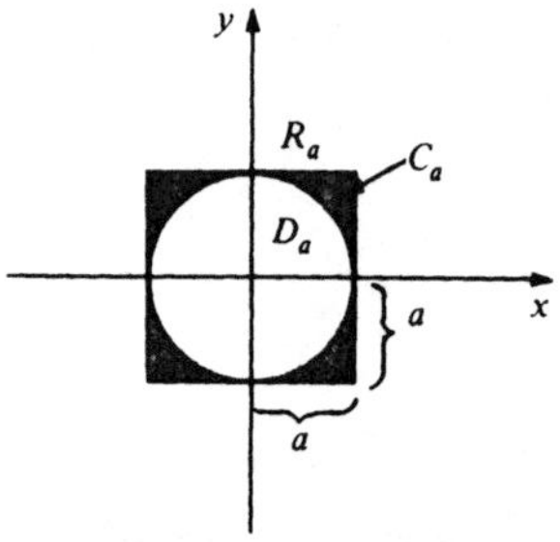

Figure 17.5.2. C_a is the shaded region between R_a and D_a.

$$\int\int_{R_a} e^{-(x^2+y^2)}\,dx\,dy = \int_{-a}^{a}\int_{-a}^{a} e^{-(x^2+y^2)}\,dx\,dy = \int_{-a}^{a}\int_{-a}^{a} e^{-x^2}e^{-y^2}\,dx\,dy$$

$$= \left(\int_{-a}^{a} e^{-x^2}\,dx\right)\left(\int_{-a}^{a} e^{-y^2}\,dy\right) = I_a^2,$$

where $I_a = \int_{-a}^{a} e^{-x^2}\,dx$ (see Exercise 27, Section 17.4). Thus,

$$\int_{-\infty}^{\infty} e^{-x^2}\,dx = \lim_{a\to\infty} I_a = \sqrt{\lim_{a\to\infty} I_a^2} = \sqrt{\lim_{a\to\infty}\int\int_{R_a} e^{-(x^2+y^2)}\,dx\,dy}$$

$$= \sqrt{\pi}\,.$$

The Gaussian Integral

$$\int_{-\infty}^{\infty} e^{-x^2}\,dx = \sqrt{\pi}\,.$$

Example 2 Find $\int_{-\infty}^{\infty} e^{-2x^2}\,dx$.

Solution We will use the change of variables $y = \sqrt{2}\,x$ to reduce the problem to the Gaussian integral just computed.

$$\int_{-\infty}^{\infty} e^{-2x^2}\,dx = \lim_{a\to\infty}\int_{-a}^{a} e^{-2x^2}\,dx = \lim_{a\to\infty}\int_{-\sqrt{2}a}^{\sqrt{2}a} e^{-y^2}\frac{dy}{\sqrt{2}}$$

$$= \frac{1}{\sqrt{2}}\int_{-\infty}^{\infty} e^{-y^2}\,dy = \frac{1}{\sqrt{2}}\sqrt{\pi} = \sqrt{\frac{\pi}{2}}\,. \blacktriangle$$

Example 3 Evaluate $\int\int_D \ln(x^2+y^2)\,dx\,dy$, where D is the region in the first quadrant lying between the circles $x^2+y^2=1$ and $x^2+y^2=4$.

Solution In polar coordinates, D is described by the set of points (r,θ) such that $1 \leqslant r \leqslant 2$, $0 \leqslant \theta \leqslant \pi/2$. Hence

$$\int\int_D \ln(x^2+y^2)\,dx\,dy = \int_{\theta=0}^{\pi/2}\int_{r=1}^{2} \ln(r^2)\,r\,dr\,d\theta$$

$$0 < \int\int_{C_a} e^{-(x^2+y^2)}\,dx\,dy < \int\int_{C_a} e^{-a^2}\,dx\,dy$$

$$= e^{-a^2}\,\text{area}\,(C_a) = e^{-a^2}(4a^2 - \pi a^2) = (4-\pi)a^2e^{-a^2}.$$

Thus it is enough to show that $\lim_{a\to\infty} a^2e^{-a^2} = 0$. But, by l'Hôpital's rule (see Section 11.2),

$$\lim_{a\to\infty} a^2e^{-a^2} = \lim_{a\to\infty}\left(\frac{a^2}{e^{a^2}}\right) = \lim_{a\to\infty}\left(\frac{2a}{2ae^{a^2}}\right) = \lim_{a\to\infty}\left(\frac{1}{e^{a^2}}\right) = 0,$$

as required.

$$= \int_0^{\pi/2}\int_1^2 2(\ln r)\cdot r\,dr\,d\theta$$

$$= \int_0^{\pi/2}\left(\frac{r^2}{2}(2\ln r - 1)\Big|_{r=1}^{2}\right)d\theta \qquad \text{(integration by parts)}$$

$$= \int_0^{\pi/2}\left(4\ln 2 - \frac{3}{2}\right)d\theta = \frac{\pi}{2}\left(4\ln 2 - \frac{3}{2}\right). \blacktriangle$$

We will now evaluate triple integrals in cylindrical and spherical coordinates. At this point you should review the basic features of these coordinates as discussed in Section 14.5.

Cylindrical coordinates consist of polar coordinates in the xy plane, together with the z coordinate. Therefore the infinitesimal "volume element" has volume $r\,dr\,d\theta\,dz$. See Fig. 17.5.3.

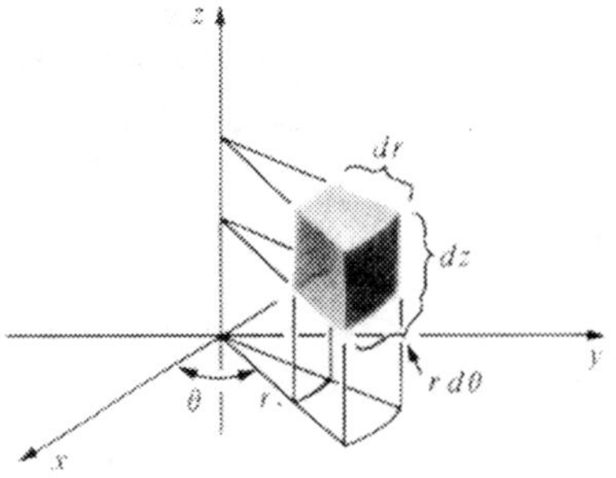

Figure 17.5.3. The infinitesimal shaded region has volume $r\,dr\,d\theta\,dz$.

As in the case of polar coordinates, this leads us to a formula for multiple integrals, presented in the next box.

Triple Integrals in Cylindrical Coordinates

$$\iiint_W f(x,y,z)\,dx\,dy\,dz = \iiint_{W'} f(r\cos\theta, r\sin\theta, z)\,r\,dr\,d\theta\,dz \qquad (2)$$

where W' is the region in r,θ,z coordinates corresponding to W.

Example 4 Evaluate $\iiint_W (z^2x^2 + z^2y^2)\,dx\,dy\,dz$, where W is the cylindrical region determined by $x^2+y^2 \leqslant 1$, $-1 \leqslant z \leqslant 1$.

Solution By Formula (2) we have

$$\iiint_W (z^2x^2+z^2y^2)\,dx\,dy\,dz = \int_{-1}^{1}\int_0^{2\pi}\int_0^1 (z^2r^2)r\,dr\,d\theta\,dz$$

$$= \int_{-1}^{1}\int_0^{2\pi} z^2\frac{r^4}{4}\Big|_{r=0}^{1}\,d\theta\,dz$$

$$= \int_{-1}^{1}\frac{2\pi}{4}z^2\,dz = \frac{\pi}{3}. \blacktriangle$$

Finally, we turn to spherical coordinates. The volume in space corresponding to infinitesimal changes $d\rho$, $d\theta$, and $d\phi$ is shown in Fig. 17.5.4. The sides of this "box" have lengths $d\rho$, $r\,d\theta(=\rho\sin\phi\,d\theta)$, and $\rho\,d\phi$ as shown. Therefore its volume is $\rho^2\sin\phi\,d\rho\,d\theta\,d\phi$. Hence we get the following:

Triple Integrals in Spherical Coordinates

$$\iiint_W f(x, y, z)\, dx\, dy\, dz$$

$$= \iiint_{W^*} f(\rho \sin\phi \cos\theta, \rho \sin\phi \sin\theta, \rho\cos\phi)\rho^2 \sin\phi\, d\rho\, d\theta\, d\phi, \quad (3)$$

where W^* is the region in ρ, θ, ϕ space corresponding to W; i.e., the limits on ρ, θ, ϕ are chosen so that the region in xyz coordinates is W.

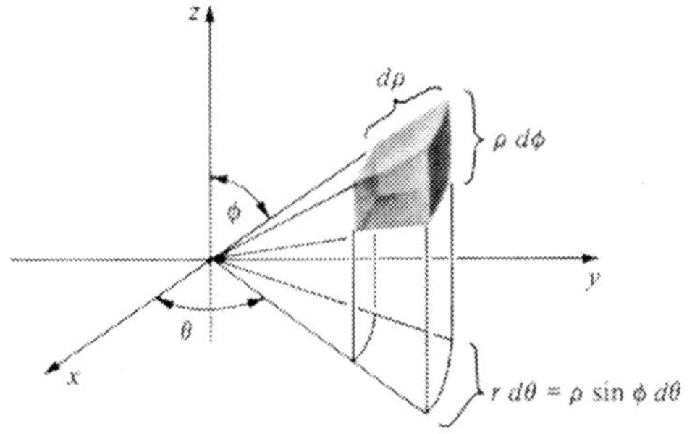

Figure 17.5.4. The infinitesimal shaded region has volume $\rho^2 \sin\phi\, d\rho\, d\theta\, d\phi$.

Example 5 Find the volume of the ball $x^2 + y^2 + z^2 \leqslant R^2$ by using spherical coordinates.

Solution The ball is described in spherical coordinates by $0 \leqslant \theta \leqslant 2\pi$, $0 \leqslant \phi \leqslant \pi$, and $0 \leqslant \rho \leqslant R$. Therefore, by formula (3),

$$\iiint_W dx\, dy\, dz = \int_0^\pi \int_0^{2\pi} \int_0^R \rho^2 \sin\phi\, d\rho\, d\theta\, d\phi = \frac{R^3}{3} \int_0^\pi \int_0^{2\pi} \sin\phi\, d\theta\, d\phi$$

$$= \frac{2\pi R^3}{3} \int_0^\pi \sin\phi\, d\phi = \frac{2\pi R^3}{3} \{-[\cos(\pi) - \cos(0)]\}$$

$$= \frac{4\pi R^3}{3},$$

which is the familiar formula for the volume of a ball. Compare the effort involved with Example 4, Section 17.4. ▲

Example 6 Evaluate $\iiint_W \exp[(x^2 + y^2 + z^2)^{3/2}]\, dx\, dy\, dz$, where W is the unit ball; i.e., the set of (x, y, z) satisfying $x^2 + y^2 + z^2 \leqslant 1$.

Solution In spherical coordinates, W is described by

$$0 \leqslant \rho \leqslant 1, \qquad 0 \leqslant \phi \leqslant \pi, \qquad 0 \leqslant \theta \leqslant 2\pi.$$

Hence

$$\iiint_W \exp\left[(x^2 + y^2 + z^2)^{3/2}\right] dx\, dy\, dz$$

$$= \int_0^{2\pi} \int_0^\pi \int_0^1 \exp(\rho^3) \cdot \rho^2 \sin\phi\, d\rho\, d\phi\, d\theta = \frac{1}{3} \int_0^{2\pi} \int_0^\pi (\exp(\rho^3)|_0^1) \sin\phi\, d\phi\, d\theta$$

$$= \frac{1}{3} \int_0^{2\pi} \int_0^\pi (e - 1)\sin\phi\, d\phi\, d\theta = \frac{1}{3}(e - 1) \int_0^{2\pi} ((-\cos\phi)|_{\phi=0}^\pi)\, d\theta$$

$$= \frac{1}{3}(e - 1) \int_0^{2\pi} 2\, d\theta = \frac{2}{3}(e - 1)(2\pi - 0) = \frac{4\pi}{3}(e - 1). \ ▲$$

It is important to spend a few moments of reflection with each integral to decide whether cylindrical, spherical, or rectangular coordinates are most useful; usually the symmetry of the problem provides the needed clue.

Example 7 Find the volumes of the following regions:

(a) The solid bounded by the circular cylinder $r = 2a\cos\theta$, the cone $z = r$, and the plane $z = 0$.

(b) The solid bounded by the cone $z = \sqrt{x^2 + y^2}$ and the paraboloid of revolution $z = x^2 + y^2$.

(c) The region bounded by $y = x^2$, $y = x + 2$, $4z = x^2 + y^2$, and $z = x + 3$.

Solution The formula for the volume of a region is $\iiint_W dx\,dy\,dz$.

(a) Since we can write $r = 2a\cos\theta$ as $r^2 = 2ax$ or $(x-a)^2 + y^2 = a^2$, we see that the base of the solid is a circle in the xy plane centered at $(a, 0)$ with radius a. (See Fig. 17.5.5.) The xz plane is a plane of symmetry, so the total volume is twice the volume over the shaded region. In cylindrical coordinates, the total volume is

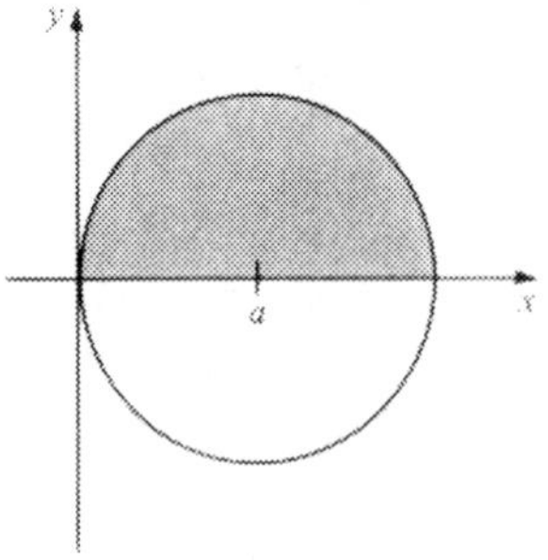

Figure 17.5.5. The base of W for Example 7(a).

$$2\int_0^{\pi/2}\int_0^{2a\cos\theta}\int_0^{r} r\,dz\,dr\,d\theta = 2\int_0^{\pi/2}\int_0^{2a\cos\theta}\left(rz\Big|_{z=0}^{r}\right)dr\,d\theta$$

$$= 2\int_0^{\pi/2}\int_0^{2a\cos\theta} r^2\,dr\,d\theta = 2\int_0^{\pi/2}\left(\frac{r^3}{3}\Bigg|_{r=0}^{2a\cos\theta}\right)d\theta$$

$$= 2\int_0^{\pi/2}\frac{8a^3\cos^3\theta}{3}\,d\theta = \left(\frac{16a^3}{3}\right)\int_0^{\pi/2}(1-\sin^2\theta)\cos\theta\,d$$

Let $u = \sin\theta$ to get $(16a^3/3)\int_0^1(1-u^2)\,du = (16a^3/3)(u - u^3/3)\big|_0^1 = 32a^3/9$.

(b) In cylindrical coordinates, the solid is bounded by $z = r$ and $z = r^2$ (Fig. 17.5.6.) The solid is obtained by rotating the shaded area around the z axis. Thus, the volume is

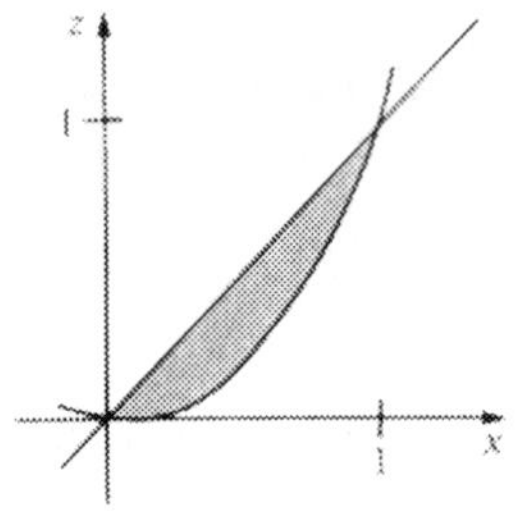

Figure 17.5.6. A cross section of W for Example 7(b).

$$\int_0^{2\pi}\int_0^1\int_{r^2}^{r} r\,dz\,dr\,d\theta = \int_0^{2\pi}\int_0^1\left(rz\Big|_{z=r^2}^{r}\right)dr\,d\theta = \int_0^{2\pi}\int_0^1 (r^2 - r^3)\,dr\,d\theta$$

$$= \int_0^{2\pi}\left(\frac{r^3}{3} - \frac{r^4}{4}\right)\Bigg|_{r=0}^{1} d\theta = \frac{1}{12}\int_0^{2\pi} d\theta = \frac{\pi}{6}.$$

(c) This part does *not* require cylindrical or spherical coordinates; $y = x^2 = x + 2$ has the solutions $x = -1$ and $x = 2$, so the volume is

$$\int_{-1}^{2}\int_{x^2}^{x+2}\int_{(x^2+y^2)/4}^{x+3} dz\,dy\,dx = \int_{-1}^{2}\int_{x^2}^{x+2}\left[(x+3) - \frac{x^2+y^2}{4}\right]dy\,dx$$

$$= \int_{-1}^{2}\left\{\left[\left(x + 3 - \frac{x^2}{4}\right)y - \frac{y^3}{12}\right]\Bigg|_{y=x^2}^{x+2}\right\}dx$$

$$= \int_{-1}^{2}\left(\frac{16}{3} + 4x - 3x^2 - \frac{4x^3}{3} + \frac{x^4}{4} + \frac{x^6}{12}\right)dx$$

$$= \left(\frac{16x}{3} + 2x^2 - x^3 - \frac{x^4}{3} + \frac{x^5}{20} + \frac{x^7}{84}\right)\Bigg|_{-1}^{2} =$$

Exercises for Section 17.5

1. Evaluate $\iint_D (x^2+y^2)^{3/2}\,dx\,dy$, where D is the disk $x^2+y^2 \leqslant 4$.
2. Evaluate $\iint_D (x^2+y^2)^{5/2}\,dx\,dy$; D is the disk $x^2+y^2 \leqslant 1$.
3. Evaluate $\int_{-\infty}^{\infty} e^{-10x^2}\,dx$.
4. Evaluate $\int_{-\infty}^{\infty} 3e^{-8x^2}\,dx$.
5. Integrate x^2+y^2 over the disk of radius 4 centered at the origin.
6. Find $\int_{-1}^{1}\int_{-\sqrt{1-x^2}}^{\sqrt{1-x^2}} \sin(x^2+y^2)\,dy\,dx$ by converting to polar coordinates.
7. Integrate $ze^{x^2+y^2}$ over the cylinder $x^2+y^2 \leqslant 4$, $2 \leqslant z \leqslant 3$.
8. Integrate $x^2+y^2+z^2$ over the cylinder given by $x^2+z^2 \leqslant 2$, $-2 \leqslant y \leqslant 3$.
9. Evaluate
$$\iiint_W \frac{dx\,dy\,dz}{\sqrt{1+x^2+y^2+z^2}},$$
where W is the ball $x^2+y^2+z^2 \leqslant 1$.
10. Evaluate $\iiint_W (x^2+y^2+z^2)^{5/2}\,dx\,dy\,dz$; W is the ball $x^2+y^2+z^2 \leqslant 1$.
11. Evaluate
$$\iiint_S \frac{dx\,dy\,dz}{(x^2+y^2+z^2)^{3/2}},$$
where S is the solid bounded by the spheres $x^2+y^2+z^2=a^2$ and $x^2+y^2+z^2=b^2$, where $a>b>0$.
12. Integrate $\sqrt{x^2+y^2+z^2}\,e^{-(x^2+y^2+z^2)}$ over the region in Exercise 11.
13. Find the volume of the region bounded by the surfaces $x^2+y^2+z^2=1$ and $x^2+y^2=\frac{1}{4}$.
14. Find the volume of the region enclosed by the cones $z=\sqrt{x^2+y^2}$ and $z=1-2\sqrt{x^2+y^2}$.
15. Find the volume inside the ellipsoid $x^2+y^2+4z^2=6$.
16. Find the volume of the intersection of the ellipsoid $x^2+2(y^2+z^2) \leqslant 10$ and the cylinder $y^2+z^2 \leqslant 1$.
17. Find the *normalizing constant* c, depending on σ, such that $\int_{-\infty}^{\infty} ce^{-x^2/\sigma}\,dx = 1$.
18. Integrate $(x^2+y^2)z^2$ over the part of the cylinder $x^2+y^2 \leqslant 1$ inside the sphere $x^2+y^2+z^2=4$.

★19. The general change of variables formula in two dimensions reads
$$\iint_D f(x,y)\,dx\,dy = \iint_{D^*} h(u,v)\left|\frac{\partial(x,y)}{\partial(u,v)}\right|du\,dv,$$
where $h(u,v)=f(x(u,v),y(u,v))$ and where $|\partial(x,y)/\partial(u,v)|$ is the absolute value of the determinant
$$\begin{vmatrix} \dfrac{\partial x}{\partial u} & \dfrac{\partial x}{\partial v} \\[2mm] \dfrac{\partial y}{\partial u} & \dfrac{\partial y}{\partial v} \end{vmatrix}.$$
Here $x(u,v)$ and $y(u,v)$ are the functions relating the variables (u,v) to the variables (x,y), and D^* is the region in the uv plane which corresponds to D.
   (a) Show that this formula is plausible by using the geometric interpretation of derivatives and determinants.
   (b) Show that the formula reduces to our earlier one when u and v are polar coordinates.

★20. Using the idea of Exercise 19, write down the general three-dimensional change of variables formula and show that it reduces to our earlier ones for cylindrical and spherical coordinates.

★21. By using the change of variables formula in Exercise 19 and $u=x+y$, $y=uv$, show that
$$\int_0^1\int_0^{1-x} e^{y/(x+y)}\,dy\,dx = \frac{e-1}{2}.$$
Also graph the region in the xy plane and the uv plane.

★22. Let D be the region bounded by $x+y=1$, $x=0$, $y=0$. Use the result of Exercise 19 to show that
$$\iint_D \cos\left(\frac{x-y}{x+y}\right)dx\,dy = \frac{\sin 1}{2},$$
and graph D on an xy plane and a uv plane, with $u=x-y$ and $v=x+y$.

17.6 Applications of Triple Integrals

The calculation of mass and center of mass of a region in space involves triple integrals.

Some of the applications in Section 17.3 carry over directly from double to triple integrals. We can compute the volume, mass, and center of mass of a region with variable density $\rho(x, y, z)$ by the formulas in the following box.

Volume, Mass, Center of Mass, and Average Value

$$\text{Volume} = \iiint_W dx\,dy\,dz,$$
$$\text{Mass} = \iiint_W \rho(x, y, z)\,dx\,dy\,dz. \tag{1}$$

Center of mass $= (\bar{x}, \bar{y}, \bar{z})$, where

$$\bar{x} = \frac{\iiint_W x\rho(x, y, z)\,dx\,dy\,dz}{\text{mass}},$$
$$\bar{y} = \frac{\iiint_W y\rho(x, y, z)\,dx\,dy\,dz}{\text{mass}}, \tag{2}$$
$$\bar{z} = \frac{\iiint_W z\rho(x, y, z)\,dx\,dy\,dz}{\text{mass}}.$$

The *average value* of a function f on a region W is defined by

$$\frac{\iiint_W f(x, y, z)\,dx\,dy\,dz}{\iiint_W dx\,dy\,dz}. \tag{3}$$

Example 1 The cube $[1,2] \times [1,2] \times [1,2]$ has mass density $\rho(x, y, z) = (1 + x)e^z y$. Find the mass of the box.

Solution The mass of the box is

$$\int_1^2 \int_1^2 \int_1^2 (1 + x)e^z y\,dx\,dy\,dz$$
$$= \int_1^2 \int_1^2 \left[\left(x + \frac{x^2}{2}\right)e^z y\right]_{x=1}^{x=2} dy\,dz = \int_1^2 \int_1^2 \frac{5}{2} e^z y\,dy\,dz$$
$$= \int_1^2 \frac{15}{4} e^z\,dz = \left[\frac{15}{4} e^z\right]_{z=1}^{z=2}$$
$$= \frac{15}{4}(e^2 - e). \ \blacktriangle$$

Example 2 Find the center of mass of the hemispherical region W defined by the inequalities $x^2 + y^2 + z^2 \leqslant 1$, $z \geqslant 0$. (Assume that the density is constant.)

Solution By symmetry, the center of mass must lie on the z axis, so $\bar{x} = \bar{y} = 0$. To find $\bar{z}$, we must compute, by formula (2), $I = \iiint_W z\,dx\,dy\,dz$. The hemisphere is of types I, II, and III; we will consider it to be of type III. Then the integral I becomes

$$I = \int_0^1 \int_{-\sqrt{1-z^2}}^{\sqrt{1-z^2}} \int_{-\sqrt{1-y^2-z^2}}^{\sqrt{1-y^2-z^2}} z\,dx\,dy\,dz.$$

Since z is a constant for the x and y integrations, we can bring it out to obtain

$$I = \int_0^1 z\left(\int_{-\sqrt{1-z^2}}^{\sqrt{1-z^2}} \int_{-\sqrt{1-y^2-z^2}}^{\sqrt{1-y^2-z^2}} dx\,dy\right) dz.$$

Instead of calculating the inner two integrals explicitly, we observe that they are simply the double integral $\iint_D dx\,dy$ over the disk $x^2 + y^2 \leqslant 1 - z^2$, considered as a type 2 region. The area of this disk is $\pi(1 - z^2)$, so

$$I = \pi\int_0^1 z(1 - z^2)\,dz = \pi\int_0^1 (z - z^3)\,dz = \pi\left[\frac{z^2}{2} - \frac{z^4}{4}\right]_0^1 = \frac{\pi}{4}.$$

The volume of the hemisphere is $(2/3)\pi$, so $\bar{z} = (\pi/4)/[(2/3)\pi] = 3/8$. ▲

Example 3 The temperature at points in the cube $W = [-1, 1] \times [-1, 1] \times [-1, 1]$ is proportional to the square of the distance from the origin.

(a) What is the average temperature?

(b) At which points of the cube is the temperature equal to the average temperature?

Solution (a) Let c be the constant of proportionality. Then $T = c(x^2 + y^2 + z^2)$ and the average temperature is $\bar{T} = \frac{1}{8}\iiint_W T\,dx\,dy\,dz$, since the volume of the cube is 8. Thus

$$\bar{T} = \frac{c}{8}\int_{-1}^1 \int_{-1}^1 \int_{-1}^1 (x^2 + y^2 + z^2)\,dx\,dy\,dz.$$

The triple integral is the sum of the integrals of x^2, y^2, and z^2. Since x, y, and z enter symmetrically into the description of the cube, the three integrals will be equal, so

$$\bar{T} = \frac{3c}{8}\int_{-1}^1 \int_{-1}^1 \int_{-1}^1 z^2\,dx\,dy\,dz = \frac{3c}{8}\int_{-1}^1 z^2\left(\int_{-1}^1 \int_{-1}^1 dx\,dy\right) dz.$$

The inner integral is equal to the area of the square $[-1, 1] \times [-1, 1]$. The area of that square is 4, so

$$\bar{T} = \frac{3c}{8}\int_{-1}^1 4z^2\,dz = \frac{3c}{2}\left(\frac{z^3}{3}\right)\Bigg|_{-1}^1 = c.$$

(b) The temperature is equal to the average temperature when $c(x^2 + y^2 + z^2) = c$; that is, on the sphere $x^2 + y^2 + z^2 = 1$, which is inscribed in the cube W. ▲

Example 4 The *moment of inertia* about the x axis of a solid S with uniform density ρ is defined by

$$I_x = \iiint_S \rho(y^2 + z^2)\, dx\, dy\, dz.$$

Similarly,

$$I_y = \iiint_S \rho(x^2 + z^2)\, dx\, dy\, dz, \qquad I_z = \iiint_S \rho(x^2 + y^2)\, dx\, dy\, dz.$$

For the following solid, compute I_z; assume that the density is a constant: The solid above the xy plane, bounded by the paraboloid $z = x^2 + y^2$ and the cylinder $x^2 + y^2 = a^2$.

Solution The paraboloid and cylinder intersect at the plane $z = a^2$. Using cylindrical coordinates, we find

$$I_z = \int_0^a \int_0^{2\pi} \int_0^{r^2} \rho r^2 \cdot r\, dz\, d\theta\, dr = \rho \int_0^a \int_0^{2\pi} \int_0^{r^2} r^3\, dz\, d\theta\, dr = \frac{\pi \rho a^6}{3}. \quad \blacktriangle$$

An interesting physical application of triple integration is the determination of the gravitational fields of solid objects. Example 3, Section 16.2, showed that the gravitational force field $\mathbf{F}(x, y, z)$ is the negative of the gradient of a function $V(x, y, z)$ called the *gravitational potential*. If there is a point mass m at (x, y, z), then the gravitational potential at (x_1, y_1, z_1) due to this mass is $Gm[(x - x_1)^2 + (y - y_1)^2 + (z - z_1)^2]^{-1/2}$, where G is the universal gravitational constant.

If our attracting object is an extended domain W with density $\rho(x, y, z)$, we may think of it as made of infinitesimal box-shaped regions with masses $dm = \rho(x, y, z)\, dx\, dy\, dz$ located at points (x, y, z). The total gravitational potential for W is then obtained by "summing" the potentials from the infinitesimal masses—that is, as a triple integral (see Fig. 17.6.1):

$$V(x_1, y_1, z_1) = G \iiint_W \frac{\rho(x, y, z)\, dx\, dy\, dz}{\sqrt{(x - x_1)^2 + (y - y_1)^2 + (z - z_1)^2}}. \tag{4}$$

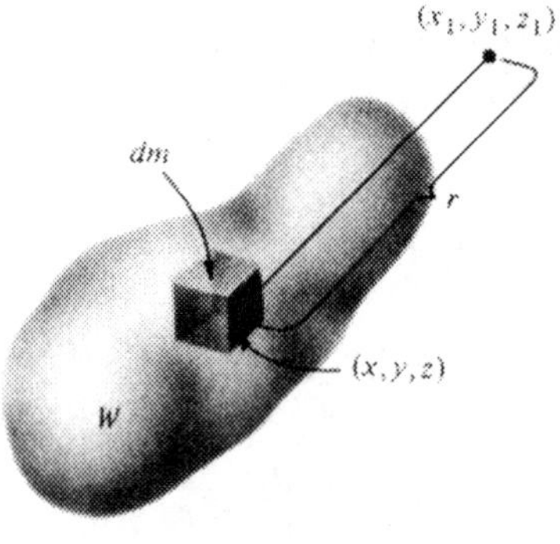

Figure 17.6.1. The gravitational potential at (x_1, y_1, z_1) arising from the mass $dm = \rho(x, y, z)\, dx\, dy\, dz$ at (x, y, z) is $[G\rho(x, y, z)\, dx\, dy\, dz]/r$.

The evaluation of the integral for the gravitational potential is usuall quite difficult. The few examples which can be carried out completely requir the use of cylindrical or spherical coordinate systems.

Historical Note Newton withheld publication of his gravitational theories for quite some tim until he could prove that a spherical planet has the same gravitational fiel that it would have if its mass were all concentrated at the planet's cente Using multiple integrals and spherical coordinates, we shall solve Newtor problem below; Newton's published solution used only euclidean geometry.

Example 5 Let W be a region of constant density and total mass M. Show that the gravitational potential is given by

$$V(x_1, y_1, z_1) = \overline{\left(\frac{1}{r}\right)} GM,$$

where $\overline{(1/r)}$ is the average over W of

$$f(x, y, z) = \frac{1}{\sqrt{(x - x_1)^2 + (y - y_1)^2 + (z - z_1)^2}}.$$

Solution According to formula (4),

$$\begin{aligned} V(x_1, y_1, z_1) &= G \iiint_W \frac{\rho(x, y, z)\, dx\, dy\, dz}{\sqrt{(x - x_1)^2 + (y - y_1)^2 + (z - z_1)^2}} \\ &= G\rho \iiint_W \frac{dx\, dy\, dz}{\sqrt{(x - x_1)^2 + (y - y_1)^2 + (z - z_1)^2}} \\ &= G[\rho \text{ volume } (W)] \frac{\displaystyle\iiint_W \frac{dx\, dy\, dz}{\sqrt{(x - x_1)^2 + (y - y_1)^2 + (z - z_1)^2}}}{\text{volume } (W)} \\ &= GM\overline{\left(\frac{1}{r}\right)}, \end{aligned}$$

as required. ▲

Let us now use formula (4) and spherical coordinates to find the gravitational potential $V(x_1, y_1, z_1)$ for the region W between the concentric spheres $\rho = \rho_1$ and $\rho = \rho_2$, assuming the density is constant. Before evaluating the integral in formula (4), we make some observations which will simplify the computation. Since G and the density are constants, we may ignore them at first. Since the attracting body W is symmetric with respect to all rotations about the origin, the potential $V(x_1, y_1, z_1)$ must itself be symmetric—thus $V(x_1, y_1, z_1)$ depends only on the distance $R = \sqrt{x_1^2 + y_1^2 + z_1^2}$ from the origin. Our computation will be simplest if we look at the point $(0, 0, R)$ on the z axis (see Fig. 17.6.2). Thus our integral is

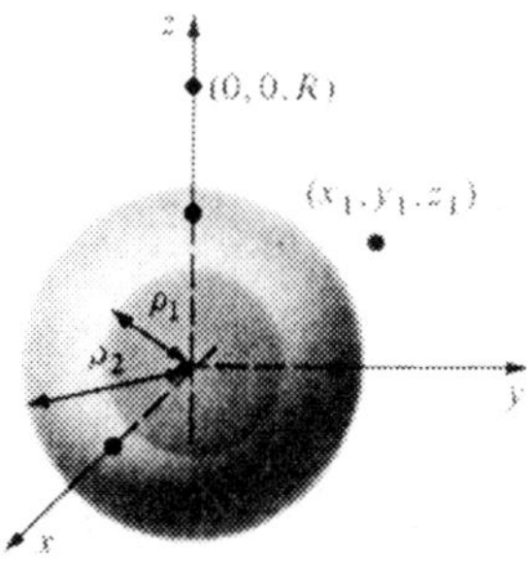

Figure 17.6.2. The gravitational potential at (x_1, y_1, z_1) is the same as at $(0, 0, R)$, where $R = \sqrt{x_1^2 + y_1^2 + z_1^2}$.

$$V(0, 0, R) = \iiint_W \frac{dx\, dy\, dz}{\sqrt{x^2 + y^2 + (z - R)^2}}.$$

In spherical coordinates, W is described by the inequalities $\rho_1 \leqslant \rho \leqslant \rho_2$, $0 \leqslant \theta \leqslant 2\pi$, and $0 \leqslant \phi \leqslant \pi$, so by formula (3) in Section 17.5,

$$V(0, 0, R) = \int_{\rho_1}^{\rho_2} \int_0^{\pi} \int_0^{2\pi} \frac{\rho^2 \sin\phi\, d\theta\, d\phi\, d\rho}{\sqrt{\rho^2 \sin^2\phi(\cos^2\theta + \sin^2\theta) + (\rho\cos\phi - R)^2}}.$$

Replacing $\cos^2\theta + \sin^2\theta$ by 1, so that the integrand no longer involves θ, we may integrate over θ to get

$$V(0,0,R)=2\pi\int_{\rho_1}^{\rho_2}\int_0^{\pi}\frac{\rho^2\sin\phi\,d\phi\,d\rho}{\sqrt{\rho^2\sin^2\phi(\rho\cos\phi-R)^2}}$$

$$=2\pi\int_{\rho_1}^{\rho_2}\rho^2\left[\int_0^{\pi}\frac{\sin\phi\,d\phi}{\sqrt{\rho^2-2R\rho\cos\phi+R^2}}\right]d\rho.$$

The inner integral is easily evaluated by the substitution $u=-2R\rho\cos\phi$: it becomes

$$\frac{1}{2R\rho}\int_{-2R\rho}^{2R\rho}(\rho^2+u+R^2)^{-1/2}du=\frac{2}{2R\rho}(\rho^2+u+R^2)^{1/2}\Big|_{-2R\rho}^{2R\rho}$$

$$=\frac{1}{R\rho}\left[(\rho^2+2R\rho+R^2)^{1/2}-(\rho^2-2R\rho+R^2)\right)$$

$$=\frac{1}{R\rho}\left\{\left[(\rho+R)^2\right]^{1/2}-\left[(\rho-R)^2\right]^{1/2}\right\}$$

$$=\frac{1}{R\rho}(\rho+R-|\rho-R|).$$

The expression $\rho+R$ is always positive, but $\rho-R$ may not be, so we must keep the absolute value sign. (Here we have used the formula $\sqrt{x^2}=|x|$.) Substituting into the formula for V, we get

$$V(0,0,R)=2\pi\int_{\rho_1}^{\rho_2}\frac{\rho^2}{R\rho}(\rho+R-|\rho-R|)\,d\rho$$

$$=\frac{2\pi}{R}\int_{\rho_1}^{\rho_2}\rho(\rho+R-|\rho-R|)\,d\rho.$$

We will now consider two possibilities for R, corresponding to the gravitational potential for objects outside and inside the hollow ball W.

If $R\geqslant\rho_2$ (that is, (x_1,y_1,z_1) is outside W), then $|\rho-R|=R-\rho$ for all ρ in the interval $[\rho_1,\rho_2]$, so

$$V(0,0,R)=\frac{2\pi}{R}\int_{\rho_1}^{\rho_2}\rho\left[\rho+R-(R-\rho)\right]d\rho$$

$$=\frac{4\pi}{R}\int_{\rho_1}^{\rho_2}\rho^2\,d\rho=\frac{1}{R}\frac{4\pi}{3}(\rho_2^3-\rho_1^3).$$

The factor $(4\pi/3)(\rho_2^3-\rho_1^3)$ is just the volume of W. Putting back the constants G and the mass density we find that *the gravitational potential is GM/R, where M is the mass of W. Thus V is just as it would be if all the mass of W were concentrated at the central point* (see Example 3, Section 16.2.)

If $R\leqslant\rho_1$ (that is, (x_1,y_1,z_1) is inside the hole), then $|\rho-R|=\rho-R$ for ρ in $[\rho_1,\rho_2]$, and

$$V(0,0,R)=\frac{2\pi}{R}\int_{\rho_1}^{\rho_2}\rho\left[\rho+R-(\rho-R)\right]d\rho=4\pi\int_{\rho_1}^{\rho_2}\rho\,d\rho=2\pi(\rho_2^2-\rho_1^2).$$

The result is independent of R, so the potential V is constant inside the hole. Since the gravitational force is minus the gradient of V, we conclude that *there is no gravitational force inside a uniform hollow planet*!

We leave it to you (Review Exercise 47) to compute $V(0,0,R)$ for the case $\rho_1<R<\rho_2$.

A similar argument shows that the gravitational potential outside any spherically symmetric body of mass M (even if the density is variable) is

$V = GM/R$, where R is the distance to its center (which is its center of mass) (Exercise 17).

Example 6 Find the gravitational potential of a spherical star with a mass $M = 3.02 \times 10^{30}$ kilograms at a distance of 2.25×10^{11} meters from its center ($G = 6.67 \times 10^{-11}$ Newton meter^2/kilogram^2).

Solution The potential is

$$V = \frac{GM}{R} = \frac{6.67 \times 10^{-11} \times 3.02 \times 10^{30}}{2.25 \times 10^{11}} = 8.95 \times 10^8 \text{ meters}^2/\text{sec}^2. \blacktriangle$$

Exercises for Section 17.6

1. (a) Find the mass of the box $[0, \frac{1}{2}] \times [0,1] \times [0,2]$, assuming the density to be uniform. (b) Same Exercise as part (a), but with a mass density $\rho(x, y, z) = x^2 + 3y^2 + z + 1$.
2. Find the mass of the solid bounded by the cylinder $x^2 + z^2 = 2x$ and the cone $z^2 = x^2 + y^2$ if the density is $\rho = \sqrt{x^2 + y^2}$.

Find the center of mass of the solids in Exercises 3 and 4, assuming them to have constant density.

3. S bounded by $x + y + z = 2$, $x = 0$, $y = 0$, $z = 0$.
4. S bounded by the parabolic cylinder $z = 4 - x^2$ and the planes $x = 0$, $y = 0$, $y = 6$, $z = 0$.
5. Evaluate the integral in Example 2 by considering the hemisphere as a region of type I.
6. Find the center of mass of the cylinder $x^2 + y^2 \leqslant 1$, $1 \leqslant z \leqslant 2$ if the density is $\rho = (x^2 + y^2)z^2$.
7. Redo Example 3 for the cube

$$W = [-c, c] \times [-c, c] \times [-c, c].$$

[*Hint*: Guess the answer to part (b) first.]

8. Find the average value of $x^2 + y^2$ over the conical region $0 \leqslant z \leqslant 2$, $x^2 + y^2 \leqslant z^2$.
9. Find the average value of $\sin^2 \pi z \cos^2 \pi x$ over the cube $[0,2] \times [0,4] \times [0,6]$.
10. Find the average value of e^{-z} over the ball $x^2 + y^2 + z^2 \leqslant 1$.
11. A solid with constant density is bounded above by the plane $z = a$ and below by the cone described in spherical coordinates by $\phi = k$, where k is a constant $0 < k < \pi/2$. Set up an integral for its moment of inertia about the z axis.
12. Find the moment of inertia around the y axis for the ball $x^2 + y^2 + z^2 \leqslant R^2$ if the mass density is a constant ρ.
13. Find the gravitational potential of a spherical planet with mass $M = 3 \times 10^{26}$ kilograms, at a distance of 2×10^8 meters from its center.
14. Find the gravitational force exerted on a 70-kilogram object at the position in Exercise 13. (See Example 3, Section 16.2.)
15. A body W in xyz coordinates is *symmetric with respect to a plane* if for every particle on one side of the plane there is a particle of equal mass located at its mirror image through the plane.
    (a) Discuss the planes of symmetry for the body of an automobile.
    (b) Let the plane of symmetry be the xy plane, and denote by $W+$ and $W-$ the portions of W above and below the plane, respectively. By our assumption, the mass density $\rho(x, y, z)$ satisfies $\rho(x, y, -z) = \rho(x, y, z)$. Justify these steps:

$$\bar{z} \cdot \iiint_W \rho(x, y, z)\, dx\, dy\, dz$$
$$= \iiint_W z\rho(x, y, z)\, dx\, dy\, dz$$
$$= \iiint_{W+} z\rho(x, y, z)\, dx\, dy\, dz + \iiint_{W-} z\rho(x, y, z)\, dx\, dy\, dz$$
$$= \iiint_{W+} z\rho(x, y, z)\, dx\, dy\, dz + \iiint_{W+} -w\rho(u, v, -w)\, du\, dv\, dw = 0.$$

    (c) Explain why part (b) proves that if a body is symmetrical with respect to a plane, then its center of mass lies in that plane.
    (d) Derive this law of mechanics: "If a body is symmetrical in two planes, then its center of mass lies on their line of intersection."

★16. If a body is composed of two or more parts whose centers of mass are known, then the center of mass of the composite body can be computed by regarding the component parts as single particles located at their respective centers of mass. Apply this consolidation principle below.
    (a) Find the center of mass of an aluminum block of constant density ρ, of base 4×6 centimeters, height 10 centimeters, with a hole drilled through. The cylinder removed is 2 centimeters in diameter and 6 centimeters long with its axis of symmetry 8 centi-

meters above the base and symmetrically placed.

(b) Repeat for a solid formed by pouring epoxy into a hemispherical form of radius 20 centimeters which contains a balloon of diameter 8 centimeters placed at the center of the circular base.

★17. Show that the gravitational potential outside of a spherically symmetric body whose density is a given function of the radius is $V(x_1, y_1, z_1) = GM/R$, where M is the mass of the body and $R = \sqrt{x_1^2 + y_1^2 + z_1^2}$ is the distance to the center of the body.

Review Exercises for Chapter 17

Evaluate the integrals in Exercises 1–10.

1. $\int_2^3 \int_4^8 [x^3 + \sin(x+y)]\,dx\,dy.$
2. $\int\int_D [x^3 + \sin(x+y)]\,dx\,dy$; D is the rectangle $[1,2] \times [-3,2]$.
3. $\int_1^2 \int_2^3 \int_3^4 (x + y + z)\,dx\,dy\,dz.$
4. $\int\int\int_W [e^x + (y+z)^5]\,dx\,dy\,dz$; W is the cube $[0,1] \times [0,1] \times [0,1]$.
5. $\int\int_D (x^3 + y^2 x)\,dx\,dy$; D = region under the graph of $y = x^2$ from $x = 0$ to $x = 2$.
6. $\int\int\int_W (x^2 + y^2 + z^2)\,dx\,dy\,dz$; W = solid hemisphere $x^2 + y^2 + z^2 \leqslant 1$, $z \geqslant 0$.
7. $\int\int_D \sec(x^2 + y^2)\,dx\,dy$, where D is the region defined by $x^2 + y^2 \leqslant 1$.
8. $\int\int\int_W (x^2 + 8yz)\,dx\,dy\,dz$, where W is the region bounded by the surfaces $z = x^2 - y^2$, $z = 0$, $y = \pm 1$, and $x = 0, 4$.
9. $\int_0^1 \int_0^x \int_0^y xyz\,dz\,dy\,dx.$
10. $\int_{-\pi}^{\pi} \int_{-\pi}^{\pi} \int_{-\pi}^{\pi} \cos(x + y + z)\,dx\,dy\,dz.$

Find the volume of each of the solids in Exercises 11–20.

11. The region bounded by the five planes $x = 0$, $x = 1$, $y = 0$, $y = 2$, $z = 0$, and the paraboloid $z = x^2 + y^2$.
12. The cone defined by $(x - \frac{1}{2}z)^2 + y^2 \leqslant 4z^2$, and $0 \leqslant z \leqslant 3$.
13. The ellipsoid $x^2/a^2 + y^2/b^2 + z^2/c^2 \leqslant 1$.
14. The intersection of a ball B of radius 1 with a ball of radius $\frac{1}{2}$ whose center is on the boundary of B.
15. The spherical sector $x^2 + y^2 + z^2 \leqslant 1$, $z \geqslant 0$, $x^2 + y^2 \leqslant az^2$.
16. The region between the graphs of $f(x, y) = \cos^2(y + x)$ and $g(x, y) = -\sin^2(y + x)$ on the domain $D = [0,1] \times [0,1]$.
17. The "ice cream cone" defined by $x^2 + y^2 \leqslant \frac{1}{5}z^2$, $0 \leqslant z \leqslant 5 + \sqrt{5 - x^2 - y^2}$.
18. The region below the plane $z + y = 1$ and inside the cylinder $x^2 + y^2 \leqslant 1$, $0 \leqslant z \leqslant 1$.
19. The solid bounded by $x^2 + y^2 + z^2 = 1$ and $z^2 \geqslant x^2 + y^2$.
20. The solid bounded by $x^2 + y^2 + z^2 = 1$ and $z \geqslant x^2 + y^2$.

Sketch and find the volume under the graph of f between the planes $x = a$, $x = b$, $y = c$, and $y = d$ in Exercises 21 and 22.

21. $f(x, y) = x^2 + \sin(2y) + 1$; $a = -3$, $b = 1$, $c = 0$, $d = \pi$.
22. $f(x, y) = 10 - x^2 - y^2$; $a = -2$, $b = 2$, $c = -1$, $d = 1$.

23. Find the average value of $f(x, y)$ on $D = [a,b] \times [c,d]$ for the function and region given in Review Exercise 21.
24. Find the average value of $f(x, y)$ on $D = [a,b] \times [c,d]$ for the function and region given in Review Exercise 22.
25. The *tetrahedron* defined by $x \geqslant 0$, $y \geqslant 0$, $z \geqslant 0$, $x + y + z \leqslant 1$ is to be sliced into n segments of equal volume by planes parallel to the plane $x + y + z = 1$. Where should the slices be made?
26. Show that the volume obtained by cutting the rectangular cylinder $a \leqslant x \leqslant b$, $c \leqslant y \leqslant d$ by the planes $z = 0$ and $z = px + qy + r$ is equal to the area of the base times the average of the heights of the four vertical edges. Assume that $px + qy + r \geqslant 0$ for all (x, y) in the rectangle $[a,b] \times [c,d]$. (See Fig. 17.R.1.)

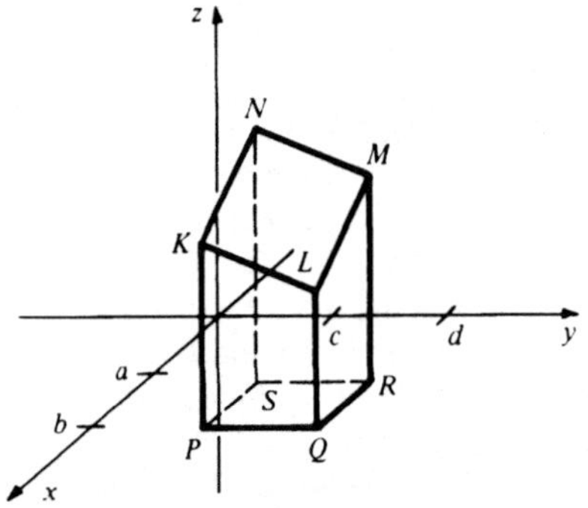

Figure 17.R.1. The volume of solid $PQRSKLMN$ is equal to the area of the base $PQRS$ times the average of the lengths of PK, QL, RM, and SN.

27. Show that the surface area of the part of the sphere $x^2 + y^2 + z^2 = 1$ lying above the rectangle $[-a, a] \times [-a, a]$ in the xy plane is

$$A = 2\int_{-a}^{a} \sin^{-1}\left(\frac{a}{\sqrt{1-x^2}}\right) dx \qquad (2a^2 < 1).$$

28. The sphere $x^2 + y^2 + z^2 = 1$ is to be cut into three pieces of equal surface area by two parallel planes. How should this be done?

29. Use cylindrical coordinates to find the center of mass of the "dish":

$$x^2 + y^2 \leqslant 1, \quad x^2 + y^2 + (z-2)^2 \leqslant 25/4, \quad z \leqslant 0.$$

30. Use cylindrical coordinates to find the center of mass of the region

$$y^2 + z^2 \leqslant 1/4, \quad (x-1)^2 + y^2 + z^2 \leqslant 1, \quad x \leqslant 1.$$

Using polar coordinates, find the surface area of the graph of each of the functions in Exercises 31–34 over the unit disk $x^2 + y^2 \leqslant 1$ (express your answer as an integral if necessary).

31. xy
32. $x^2 + y^2$
33. $x^2 - y^2$
34. $(x^2 + y^2)^2$

Evaluate the integrals in Exercises 35–38.

35. $\int_0^{\pi}\int_0^{\pi/2}\int_0^{\sin\phi} \rho^2 \sin\phi \, d\rho \, d\phi \, d\theta.$

36. $\int_0^1\int_0^{\pi/2}\int_0^{\cos\theta} rz \, dr \, d\theta \, dz.$

37. $\int\int_D dx\,dy/(x^2+y^2)$, where D is the region in the first quadrant bounded by the circles $x^2 + y^2 = 1$ and $x^2 + y^2 = 2$.

38. $\int\int_{[0,1]\times[0,1]} e^{yx}\, dy\, dx$. (Use a power series to evaluate to within 0.001.)

39. As is well known, the density of a typical planet is not constant throughout the planet. Assume that planet I.K.U. has a radius of 5×10^8 centimeters and a mass density (in grams per cubic centimeter)

$$\rho(x,y,z) = \begin{cases} \dfrac{3\times 10^4}{r}, & r \geqslant 10^4 \text{ centimeters}, \\ 3, & r \leqslant 10^4 \text{ centimeters}, \end{cases}$$

where $r = \sqrt{x^2 + y^2 + z^2}$. Find a formula for the gravitational potential outside I.K.U.

40. Derive the following three laws (see Exercise 15, Section 17.6).
   (a) If a body is symmetrical about an axis, then its center of mass lies on that axis.
   (b) If a body is symmetrical in three planes with a common point, then that point is its center of mass.
   (c) If a body has spherical symmetry about a point (that is, if the density depends only on the distance from that point), then that point is its center of mass.

41. The *flexural rigidity EI* of a uniform beam is the product of its Young's modulus of elasticity E and the moment of inertia I of the cross section of the beam at x with respect to a horizontal line l passing through the center of gravity of this cross section. Here

$$I = \int\int_R [D(x,y)]^2\, dx\, dy,$$

where $D(x, y)$ = the distance from (x, y) to l, R = the cross section of the beam being considered.
   (a) Assume the cross section R is the rectangle $-1 \leqslant x \leqslant 1$, $-1 \leqslant y \leqslant 2$ and l is the x axis. Find I.
   (b) Assume the cross section R is a circle of radius 4, and l is the x axis. Find I, using polar coordinates.

42. Justify the formula

$$\int\int_D f(x,y)\,dx\,dy = \int_a^b \left(\int_{\phi_1(x)}^{\phi_2(x)} f(x,y)\,dy\right) dx$$

for an integral over a region of type 1 by using the slice method.

43. Find $\int_{-\infty}^{\infty} e^{-5x^2}\,dx$.

44. Find $\int_{-\infty}^{\infty} (5 + x^2)e^{-2x^2}\,dx$.

45. (a) Interpret the result in Exercise 26 as a fact about the average value of a linear function on a rectangle.
   ★(b) What is the average value of a linear function on a parallelogram?

★46. If the world were two-dimensional,[4] the laws of physics would predict that the gravitational potential of a mass point is proportional to the logarithm of the distance from the point. Using polar coordinates, write an integral giving the gravitational potential of a disk of constant density.

★47. Find the gravitational potential in the situation of Fig. 17.6.2 when $\rho_1 < R < \rho_2$. [*Hint*: Break the integral over ρ into two parts.]

★48. There is an interesting application of double integrals to the problem of differentiation under the integral sign.[5] Study the following theorem:

Let $D = [a,b] \times [c,d]$ and let $g(x, y)$ be continuous, with $g_x(x, y)$ continuous on D. Then

$$\frac{d}{dx}\int_c^d g(x,y)\,dy = \int_c^d g_x(x,y)\,dy, \qquad a < x < b.$$

[4] See *Flatland* by Edwin A. Abbot (Barnes and Noble, 1963) for an amusing (but sexist) description of such a world.

[5] See "Fubini implies Leibniz implies $F_{yx} = F_{xy}$" by R. T. Seeley, American Mathematical Monthly **68**(1961), 56–57.

The idea of the proof is this: By the fundamental theorem of calculus,

$$\int_c^d g(x,y)\,dy = \int_c^d \left[\int_a^x g_x(s,y)\,ds + g(a,y) \right] dy.$$

Interchanging the order of integration, we get

$$\int_a^x \left[\int_c^d g_x(s,y)\,dy \right] ds + \int_c^d g(a,y)\,dy.$$

We can show that $\int_c^d g_x(s,y)\,dy$ is a continuous function of s, so we can use the fundamental theorem of calculus (alternative version) to get

$$\frac{d}{dx}\int_c^d g(x,y)\,dy = \frac{d}{dx}\int_a^x \left[\int_c^d g_x(s,y)\,dy \right] ds + \frac{d}{dx}\left[\int_c^d g(a,y)\,dy \right] = \int_c^d g_x(x,y)\,dy + 0$$

as asserted.

(a) Verify by direct integration that

$$\frac{d}{dx}\int_0^{\pi/2} \sin(xy)\,dy = \int_0^{\pi/2} \frac{\partial}{\partial x}\sin(xy)\,dy.$$

(b) Show that $F(k) = \int_0^{\pi/2} dx/\sqrt{1-k\cos^2 x}$, for $0 \leqslant k < 1$, is an increasing function of k. How fast is it increasing at $k = 0$?

★49. Use the discussion of the theorem in Exercise 48 to show that interchanging the order of integration allows one to prove that the mixed partials of a function are equal.

★50.[6] Show that each of the following functions has the curious property that the volume under its graph equals the surface area of its graph on any region:

(a) $f(x,y) = 1$,

(b) $f(x,y) = \cosh(x\cos\alpha + y\sin\alpha + c)$, ($\alpha, c$ constants) and

(c) $f(x,y) = \cosh(\sqrt{x^2+y^2} + c)$ (c constant). (Compare Review Exercise 85, Chapter 8).

★51. Suppose that $f(x,y) = \cosh[u(x,y)]$ has the property considered in Review Exercise 50. Find a partial differential equation satisfied by $u(x,y)$ if u is not identically zero (i.e., find a relation involving u, u_x and u_y).

[6] Suggested by Chris Fisher.

Chapter 18

Vector Analysis

The fundamental theorem of calculus is extended from the line to the plane and to space.

The major theorems of vector analysis relate double and triple integrals to integrals over curves and surfaces. These results have their origins in problems of fluid flow and electromagnetic theory; thus, it is not surprising that they are very important for physics as well as for their own mathematical beauty. Enough applications are given in this chapter so that the student can appreciate the physical meaning of the theorems.

18.1 Line Integrals

A vector field may be integrated along a curve to produce a number.

The integration of vector fields along curves is of fundamental importance in both mathematics and physics. We will use the concept of *work* to motivate the material in this section. In later sections, we establish Green's and Stokes' theorems, which relate line integrals, partial derivatives, and double integrals.

The motion of an object is described by a parametric curve—that is, by a vector function $\mathbf{r} = \boldsymbol{\sigma}(t)$. By differentiating this function, we obtain (see Section 14.6):

$$\mathbf{v} = \frac{d\mathbf{r}}{dt} = \boldsymbol{\sigma}'(t) = \text{velocity at time } t;$$

$$v = \|\mathbf{v}\| = \|\boldsymbol{\sigma}'(t)\| = \text{speed at time } t;$$

$$\mathbf{a} = \frac{d\mathbf{v}}{dt} = \boldsymbol{\sigma}''(t) = \text{acceleration at time } t.$$

According to Newton's second law,

$$\mathbf{F} = m\mathbf{a} = m\frac{d^2\mathbf{r}}{dt^2} = m\boldsymbol{\sigma}''(\mathbf{t}),$$

where $\mathbf{F}$ is the total force acting on the object. If the mass of the object is m, the kinetic energy K is defined by

$$K = \tfrac{1}{2}mv^2 = \tfrac{1}{2}m\mathbf{v}\cdot\mathbf{v}.$$

To investigate the relationship between force and kinetic energy, we differentiate K with respect to t, obtaining

$$\frac{dK}{dt} = \frac{1}{2}m\left(\frac{d\mathbf{v}}{dt}\cdot\mathbf{v} + \mathbf{v}\cdot\frac{d\mathbf{v}}{dt}\right) = m\frac{d\mathbf{v}}{dt}\cdot\mathbf{v} = m\mathbf{a}\cdot\mathbf{v} = \mathbf{F}\cdot\mathbf{v}.$$

The total change in kinetic energy from time t_1 to t_2 is the integral of dK/dt, so we get

$$\Delta K = \int_{t_1}^{t_2}\frac{dK}{dt}\,dt = \int_{t_1}^{t_2}\mathbf{F}\cdot\mathbf{v}\,dt = \int_{t_1}^{t_2}\mathbf{F}\cdot\frac{d\mathbf{r}}{dt}\,dt.$$

The integral

$$\int_{t_1}^{t_2}\mathbf{F}\cdot\frac{d\mathbf{r}}{dt}\,dt, \tag{1}$$

denoted W, is called the *work* done by the force $\mathbf{F}$ along the path $\mathbf{r} = \boldsymbol{\sigma}(t)$.

It is often possible to express the total force on an object as a sum of forces which are due to identifiable sources (such as gravity, friction, and fluid pressure). If $\mathbf{F}$ represents a force of a particular type, then the integral (1) is still called the work done by this particular force.

Let us now suppose that the force $\mathbf{F}$ at time t depends only on the position $\mathbf{r} = \boldsymbol{\sigma}(t)$. That is, we assume that there is a *vector field* $\boldsymbol{\Phi}(\mathbf{r})$ such that $\mathbf{F} = \boldsymbol{\Phi}(\boldsymbol{\sigma}(t))$. (Examples of such *position-dependent* forces are those caused by gravitational and electrostatic attraction; frictional and magnetic forces are *velocity dependent.*) Then we may write the integral (1) as

$$W = \int_{t_1}^{t_2}\boldsymbol{\Phi}(\boldsymbol{\sigma}(t))\cdot\boldsymbol{\sigma}'(t)\,dt. \tag{2}$$

In the one-dimensional case, (2) can be simplified, by a change of variables from time to position, to

$$W = \int_a^b F(x)\,dx,$$

where a and b are the starting and ending positions. This formula agrees with that found in Section 9.5. Notice that the work done depends only on F, a, and b and not on the details of the motion. We shall prove later in this section that, to a certain extent, the same situation remains true for motion in space: the work done by a force field as a particle moves along a path does not depend upon how the particle moves along the path; however, if *different paths* are taken between the same endpoints, the work may be different.

Example 1 Find the work done by the force field $\boldsymbol{\Phi}(x, y, z) = y\mathbf{i} - x\mathbf{j} + \mathbf{k}$ as a particle is moved from $(1,0,0)$ to $(1,0,1)$ along each of the following paths:

(a) $(x, y, z) = (\cos t, \sin t, t/2\pi)$; $0 \leqslant t \leqslant 2\pi$;
(b) $(x, y, z) = (\cos t^3, \sin t^3, t^3/2\pi)$; $0 \leqslant t \leqslant \sqrt[3]{2\pi}$;
(c) $(x, y, z) = (\cos t, -\sin t, t/2\pi)$; $0 \leqslant t \leqslant 2\pi$.

Solution The (helical) paths are sketched in Fig. 18.1.1.

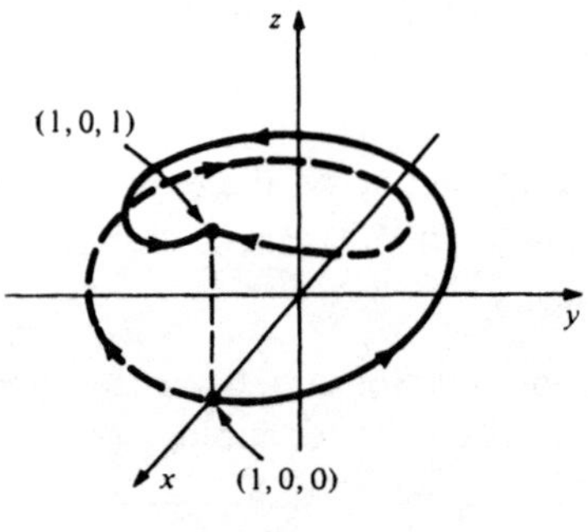

Figure 18.1.1. Paths (a) and (b) follow the solid line; path (c) follows the dashed line.

(a) By formula (2), with $t_1 = 0$, $t_2 = 2\pi$, $\boldsymbol{\sigma}(t) = \cos t\,\mathbf{i} + \sin t\,\mathbf{j} + (t/2\pi)\mathbf{k}$, and $\boldsymbol{\sigma}'(t) = -\sin t\,\mathbf{i} + \cos t\,\mathbf{j} + (1/2\pi)\mathbf{k}$, the work done by the force along path (a) is

$$\begin{aligned} W_a &= \int_0^{2\pi} (\sin t\,\mathbf{i} - \cos t\,\mathbf{j} + \mathbf{k}) \cdot \left(-\sin t\,\mathbf{i} + \cos t\,\mathbf{j} + \frac{1}{2\pi}\mathbf{k} \right) dt \\ &= \int_0^{2\pi} \left(-\sin^2 t - \cos^2 t + \frac{1}{2\pi} \right) dt \\ &= 2\pi\left(-1 + \frac{1}{2\pi} \right) = -2\pi + 1 \approx -5.28. \end{aligned}$$

(b) This time,

$$\begin{aligned} \boldsymbol{\sigma}'(t) &= -(\sin t^3)(3t^2)\mathbf{i} + (\cos t^3)(3t^2)\mathbf{j} + \frac{3t^2}{2\pi}\mathbf{k} \\ &= 3t^2\left[-\sin t^3\,\mathbf{i} + \cos t^3\,\mathbf{j} + \frac{1}{2\pi}\mathbf{k} \right], \end{aligned}$$

and

$$\begin{aligned} W_b &= \int_0^{\sqrt[3]{2\pi}} (\sin t^3\,\mathbf{i} - \cos t^3\,\mathbf{j} + \mathbf{k}) \cdot (3t^2)\left(-\sin t^3\,\mathbf{i} + \cos t^3\,\mathbf{j} + \frac{1}{2\pi}\mathbf{k} \right) dt \\ &= \int_0^{\sqrt[3]{2\pi}} \left(-\sin^2 t^3 - \cos^2 t^3 + \frac{1}{2\pi} \right) 3t^2\, dt \\ &= \int_0^{\sqrt[3]{2\pi}} \left(-1 + \frac{1}{2\pi} \right) 3t^2\, dt = \left(-1 + \frac{1}{2\pi} \right) t^3 \Big|_{t=0}^{\sqrt[3]{2\pi}} \\ &= \left(-1 + \frac{1}{2\pi} \right)(2\pi) = 1 - 2\pi, \end{aligned}$$

just as in part (a).
(c) Here

$$\boldsymbol{\sigma}'(t) = -\sin t\,\mathbf{i} - \cos t\,\mathbf{j} + \frac{1}{2\pi}\mathbf{k},$$

so

$$\begin{aligned} W_c &= \int_0^{2\pi} (-\sin t\,\mathbf{i} - \cos t\,\mathbf{j} + \mathbf{k}) \cdot \left(-\sin t\,\mathbf{i} - \cos t\,\mathbf{j} + \frac{1}{2\pi}\mathbf{k} \right) dt \\ &= \int_0^{2\pi} \left(\sin^2 t + \cos^2 t + \frac{1}{2\pi} \right) dt \\ &= 2\pi\left(1 + \frac{1}{2\pi} \right) = 2\pi + 1 \approx 7.28. \end{aligned}$$

In the case of path (c), the motion is "with the force," so the work is positive; for paths (a) and (b), the motion is "against the force" and the work is negative. ▲

The equality of work along paths (a) and (b) in Example 1 is no accident: it is a consequence of the fact that the two paths are simply two different parametrizations of the same curve in space. Shortly, we will prove that the work along a path is always independent of the parametrization, and in Section 18.2, we will identify those special force fields for which the work is independent of the path itself.

Example 2 Suppose that in the force field of Example 1, you pick up a unit mass at $(1,0,0)$, carry it along path (a), and leave it at $(1,0,1)$. How much work have *you* done?

Solution Since the kinetic energy is zero at the beginning and the end of the process, the change in energy is zero and the total work is zero. The two sources of work are you and the force field; since the work done by force field is -5.28, by part (a) of Example 1, the work done by you must be 5.28. ▲

Example 3 Consider the gravitational force field defined (for $(x, y, z) \neq (0,0,0)$) by

$$\mathbf{\Phi}(x, y, z) = \frac{-1}{(x^2+y^2+z^2)^{3/2}}(x\mathbf{i} + y\mathbf{j} + z\mathbf{k}).$$

Show that the work done by the gravitational force as a particle moves from (x_1, y_1, z_1) to (x_2, y_2, z_2) depends only on the radii $R_1 = \sqrt{x_1^2 + y_1^2 + z_1^2}$ and $R_2 = \sqrt{x_2^2 + y_2^2 + z_2^2}$.

Solution Let the path be given by $(x, y, z) = \boldsymbol{\sigma}(t)$, where $\boldsymbol{\sigma}(t_1) = (x_1, y_1, z_1)$ and $\boldsymbol{\sigma}(t_2) = (x_2, y_2, z_2)$. Think of x, y, and z as functions of t. Then $\boldsymbol{\sigma}'(t) = (dx/dt)\mathbf{i} + (dy/dt)\mathbf{j} + (dz/dt)\mathbf{k}$, and so

$$\begin{aligned}
W &= \int_{t_1}^{t_2} \frac{-1}{(x^2+y^2+z^2)^{3/2}}(x\mathbf{i} + y\mathbf{j} + z\mathbf{k}) \cdot \left(\frac{dx}{dt}\mathbf{i} + \frac{dy}{dt}\mathbf{j} + \frac{dz}{dt}\mathbf{k}\right) dt \\
&= -\int_{t_1}^{t_2} \frac{x\,(dx/dt) + y\,(dy/dt) + z\,(dz/dt)}{(x^2+y^2+z^2)^{3/2}}\, dt \\
&= \int_{t_1}^{t_2} - \frac{(1/2)(d/dt)(x^2+y^2+z^2)}{(x^2+y^2+z^2)^{3/2}}\, dt \\
&= \int_{t_1}^{t_2} \frac{d}{dt}(x^2+y^2+z^2)^{-1/2}\, dt = (x^2+y^2+z^2)^{-1/2}\Big|_{\boldsymbol{\sigma}(t_1)}^{\boldsymbol{\sigma}(t_2)} \\
&= (x_2^2+y_2^2+z_2^2)^{-1/2} - (x_1^2+y_1^2+z_1^2)^{-1/2} = \frac{1}{R_2} - \frac{1}{R_1}.
\end{aligned}$$

Thus, the work done by the gravitational field when a particle moves from (x_1, y_1, z_1) to (x_2, y_2, z_2) is $1/R_2 - 1/R_1$. Notice that, in this case, the work is independent of the path taken between the two points. ▲

To define the line integral, we use formula (2) abstracted from its physical interpretation.

The Line Integral

Let $\mathbf{\Phi}$ be a vector field defined in some region of space, and let $\mathbf{r} = \boldsymbol{\sigma}(t)$ be a parametric curve in that region defined for t in $[t_1, t_2]$. The integral

$$\int_{t_1}^{t_2} \mathbf{\Phi}(\boldsymbol{\sigma}(t)) \cdot \boldsymbol{\sigma}'(t)\, dt \tag{3}$$

is called the *line integral* of the vector field $\mathbf{\Phi}$ along this curve.

The work done by a force field on a moving particle is therefore the line integral of the force along the path travelled by the particle.

Example 4 Find the line integral of the vector field $e^y\mathbf{i} + e^x\mathbf{j} + e^z\mathbf{k}$ along the curve $(0, t, t^2)$, for $0 \leqslant t \leqslant \ln 2$.

Solution The velocity vector $\boldsymbol{\sigma}'(t)$ is $\mathbf{j} + 2t\mathbf{k}$, so the line integral is

$$\int_0^{\ln 2} (e^t\mathbf{i} + e^0\mathbf{j} + e^{t^2}\mathbf{k}) \cdot (\mathbf{j} + 2t\mathbf{k})\, dt = \int_0^{\ln 2} (1 + e^{t^2}2t)\, dt$$
$$= (t + e^{t^2})\Big|_0^{\ln 2} = \ln 2 + e^{(\ln 2)^2} - 1$$
$$= \ln 2 + 2^{\ln 2} - 1 \approx 1.31. \ \blacktriangle$$

The next theorem shows that the line integral does not depend upon how the path of integration is parametrized.

Theorem: Independence of Parametrization

Given $\boldsymbol{\Phi}$ and $\boldsymbol{\sigma}$ as in the preceding box, let $t = f(u)$ be a differentiable function defined on the interval $[u_1, u_2]$ such that $f(u_1) = t_1$ and $f(u_2) = t_2$. *Let* $\boldsymbol{\sigma}_1(u)$ *be the parametric curve defined by* $\boldsymbol{\sigma}_1(\mathrm{u}) = \boldsymbol{\sigma}(\mathrm{f(u)})$. *Then*

$$\int_{u_1}^{u_2} \boldsymbol{\Phi}(\boldsymbol{\sigma}_1(u)) \cdot \boldsymbol{\sigma}_1'(u)\, du = \int_{t_1}^{t_2} \boldsymbol{\Phi}(\boldsymbol{\sigma}(t)) \cdot \boldsymbol{\sigma}'(t)\, dt.$$

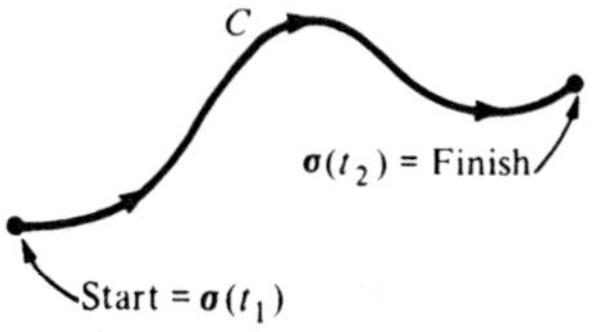

Figure 18.1.2. A geometric curve must be parametrized in a specified direction.

The basic idea of the proof was illustrated in Example 1. We apply the chain rule to $\boldsymbol{\sigma}_1(u) = \boldsymbol{\sigma}(f(u))$, obtaining $\boldsymbol{\sigma}_1'(u) = \boldsymbol{\sigma}'(f(u))f'(u)$. Hence

$$\int_{u_1}^{u_2} \boldsymbol{\Phi}(\boldsymbol{\sigma}_1(u)) \cdot \boldsymbol{\sigma}_1'(u)\, du = \int_{u_1}^{u_2} \boldsymbol{\Phi}(\boldsymbol{\sigma}(f(u))) \cdot \boldsymbol{\sigma}'(f(u))f'(u)\, du.$$

We next change variables from u to $t = f(u)$. Since $dt = f'(u)\, du$, the integral becomes $\int_{t_1}^{t_2} \boldsymbol{\Phi}(\boldsymbol{\sigma}(t)) \cdot \boldsymbol{\sigma}'(t)\, dt$ as required. ■

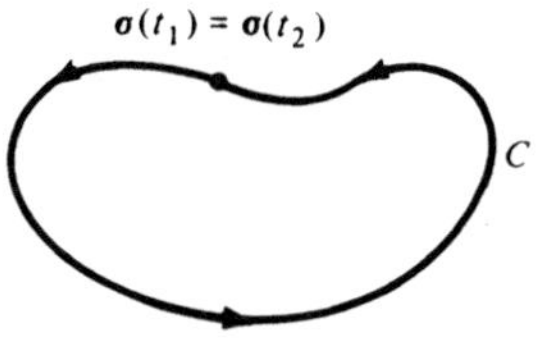

Figure 18.1.3. Any point may be taken as the starting point for integration around a closed curve.

A *geometric curve* C is a set of points in the plane which can be traversed by a parametrized curve; the direction of travel along C is specified, but not a specific parametrization (see Fig. 18.1.2). The theorem shows that the line integral of a vector field along a geometric curve is well defined.

A parametric curve $\boldsymbol{\sigma}(t)$ defined on $[t_1, t_2]$ is called *closed* if its endpoints coincide—that is, if $\boldsymbol{\sigma}(t_1) = \boldsymbol{\sigma}(t_2)$. A geometric curve is *closed* if it has a parametrization which is closed. When C is a closed curve, any point of C may be taken as the initial point for the parametrization, but we must be sure to go around C just once (see Fig. 18.1.3 and Example 5.)

In summary, there are two reservations which must be noted in choosing a parametrization of a geometric curve: *the parametrization must go in the correct direction and it must trace out the curve exactly once.*

Example 5 (a) Let C be the line segment joining $(0,0,0)$ to $(1,0,0)$ and let $\boldsymbol{\sigma}_1(t) = (t, 0, 0)$, $0 \leqslant t \leqslant 1$. Find the line integral of $\boldsymbol{\Phi}(x, y, z) = \mathbf{i}$ along this curve. If C is parametrized by $\boldsymbol{\sigma}_2(t) = (1 - t, 0, 0)$, $0 \leqslant t \leqslant 1$, find the line integral.
(b) Let C be the circle given by $x^2 + y^2 = 1$, $z = 0$. Let $\boldsymbol{\sigma}_1(t) = (\cos t, \sin t, 0)$, $0 \leqslant t \leqslant 2\pi$. Find the line integral of $\boldsymbol{\Phi}(x, y, z) = -y\mathbf{i} + x\mathbf{j}$ along this curve. If C is parametrized by $\boldsymbol{\sigma}_2(t) = (\cos t, \sin t, 0)$, $0 \leqslant t \leqslant 4\pi$, find the line integral.

Solution (a) Here $\sigma_1'(t) = \mathbf{i}$, $t_1 = 0$, $t_2 = 1$, and $\mathbf{\Phi}(\boldsymbol{\sigma}_1(t)) = \mathbf{i}$, so formula (3) gives

$$\int_{t_1}^{t_2} \mathbf{\Phi}(\boldsymbol{\sigma}_1(t)) \cdot \boldsymbol{\sigma}_1'(t)\,dt = \int_0^1 \mathbf{i} \cdot \mathbf{i}\,dt = \int_0^1 dt = 1.$$

For $\boldsymbol{\sigma}_2$, we similarly have $t_1 = 0$, $t_2 = 1$, $\boldsymbol{\sigma}_2'(t) = -\mathbf{i}$, and $\mathbf{\Phi}(\boldsymbol{\sigma}_2(t)) = \mathbf{i}$, so

$$\int_{t_1}^{t_2} \mathbf{\Phi}(\boldsymbol{\sigma}_2(t)) \cdot \boldsymbol{\sigma}_2'(t)\,dt = \int_0^1 -\mathbf{i} \cdot \mathbf{i}\,dt = -\int_0^1 dt = -1.$$

Here the geometric curve C is the same, but the two parametrizations, $\boldsymbol{\sigma}_1$ and $\boldsymbol{\sigma}_2$, traverse C in opposite directions. See Fig. 18.1.4.

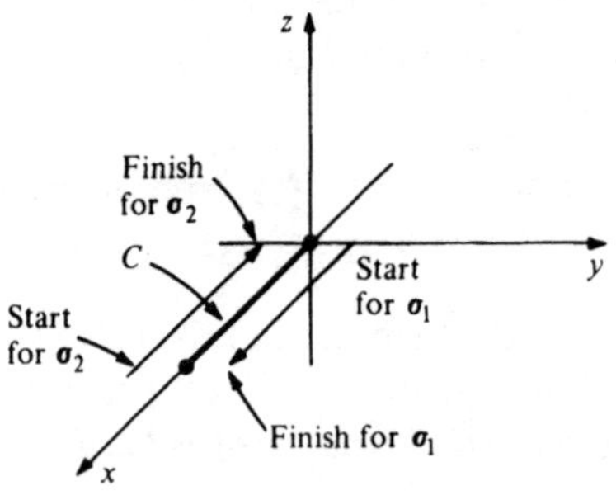

Figure 18.1.4. $\boldsymbol{\sigma}_1$ and $\boldsymbol{\sigma}_2$ traverse C in opposite directions.

(b) The line integral for $\boldsymbol{\sigma}_1$ is obtained from formula (3) by substituting $t_1 = 0$, $t_2 = 2\pi$, $\boldsymbol{\sigma}_1(t) = (\cos t, \sin t, 0)$, $\mathbf{\Phi}(x, y, z) = -y\mathbf{i} + x\mathbf{j}$, $\mathbf{\Phi}(\boldsymbol{\sigma}_1(t)) = -\sin t\,\mathbf{i} + \cos t\,\mathbf{j}$, $\boldsymbol{\sigma}_1(t) = \cos t\,\mathbf{i} + \sin t\,\mathbf{j}$, and $\boldsymbol{\sigma}_1'(t) = -\sin t\,\mathbf{i} + \cos t\,\mathbf{j}$, as follows:

$$\int_{t_1}^{t_2} \mathbf{\Phi}(\boldsymbol{\sigma}_1(t)) \cdot \boldsymbol{\sigma}_1'(t)\,dt = \int_0^{2\pi} (\sin^2 t + \cos^2 t)\,dt = 2\pi.$$

Notice that we get the same result if we choose any θ and parametrize C by the equation $\boldsymbol{\sigma}(t) = (\cos(t + \theta), \sin(t + \theta), 0)$, $0 \leqslant t \leqslant 2\pi$; this will start and finish at $(\cos\theta, \sin\theta, 0)$. If we go backwards along C, using the parametrization $\boldsymbol{\sigma}(t) = (\cos(-t), \sin(-t), 0)$, we will get the *negative* of our earlier answer.

If we use $\boldsymbol{\sigma}_2(t)$, the only change is that t_2 is changed to 4π, so we get

$$\int_{t_1}^{t_2} \mathbf{\Phi}(\boldsymbol{\sigma}_2(t)) \cdot \boldsymbol{\sigma}_2'(t)\,dt = 4\pi.$$

This is double our first answer, since $\boldsymbol{\sigma}_2$ traverses C twice. See Fig. 18.1.5. ▲

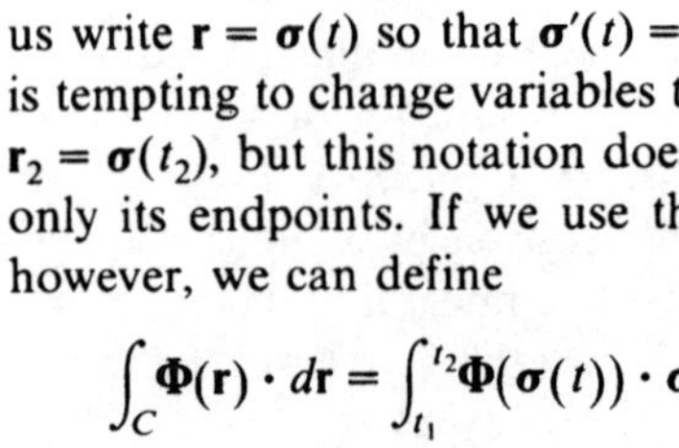

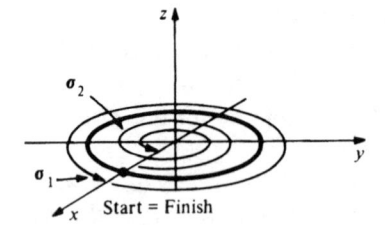

Figure 18.1.5. $\boldsymbol{\sigma}_1$ goes around C once while $\boldsymbol{\sigma}_2$ goes around twice.

A useful notation for the line integral is suggested by the Leibniz notation. Let us write $\mathbf{r} = \boldsymbol{\sigma}(t)$ so that $\boldsymbol{\sigma}'(t) = d\mathbf{r}/dt$ and (3) becomes $\int_{t_1}^{t_2} \mathbf{\Phi}(\mathbf{r}) \cdot (d\mathbf{r}/dt)\,dt$. It is tempting to change variables to $\mathbf{r}$ and write $\int_{\mathbf{r}_1}^{\mathbf{r}_2} \mathbf{\Phi}(\mathbf{r}) \cdot d\mathbf{r}$, where $\mathbf{r}_1 = \boldsymbol{\sigma}(t_1)$ and $\mathbf{r}_2 = \boldsymbol{\sigma}(t_2)$, but this notation does not display the curve traced out by $\mathbf{r} = \boldsymbol{\sigma}(t)$, only its endpoints. If we use the letter C to denote the path of integration, however, we can define

$$\int_C \mathbf{\Phi}(\mathbf{r}) \cdot d\mathbf{r} = \int_{t_1}^{t_2} \mathbf{\Phi}(\boldsymbol{\sigma}(t)) \cdot \boldsymbol{\sigma}'(t)\,dt, \tag{4}$$

where $\boldsymbol{\sigma}(t)$ is any parametrization of C (subject to the reservations discussed above). The notation $d\mathbf{r}$ for $\boldsymbol{\sigma}'(t)\,dt$ is consistent with our other uses of

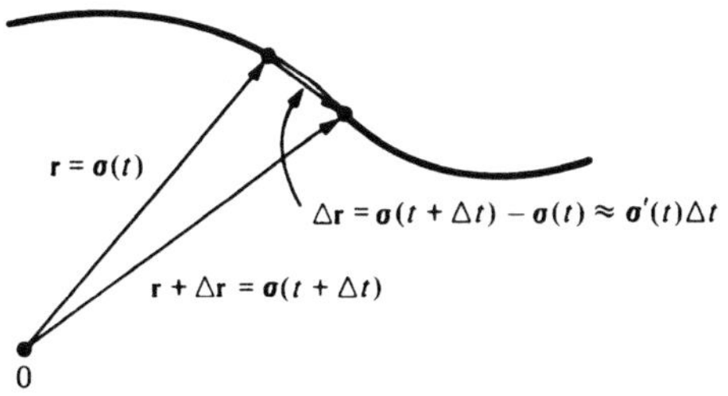

Figure 18.1.6. $\Delta\mathbf{r} \approx \boldsymbol{\sigma}'(t)\,\Delta t$.

infinitesimal notation. The change in $\mathbf{r}$ over a small time interval Δt is $\Delta\mathbf{r} \approx \boldsymbol{\sigma}'(t)\,\Delta t$ (see Fig. 18.1.6). As Δt becomes the infinitesimal dt, $\Delta\mathbf{r}$ passes over into $d\mathbf{r}$, and the approximate equality becomes exact.

Example 6 Let C be the straight line segment joining $(2,1,3)$ to $(-4,6,8)$. Find

$$\int_C \boldsymbol{\Phi}(\mathbf{r}) \cdot d\mathbf{r},$$

where $\boldsymbol{\Phi}(x, y, z) = x\mathbf{i} - y\mathbf{j} + xy\mathbf{k}$.

Solution We may choose any parametrization of C; the simplest is probably

$$\begin{aligned}\boldsymbol{\sigma}(t) &= (1-t)(2,1,3) + t(-4,6,8)\\ &= (2-6t, 1+5t, 3+5t), \qquad 0 \leqslant t \leqslant 1.\end{aligned}$$

As t varies from 0 to 1, $\boldsymbol{\sigma}(t)$ moves along C from $(2,1,3)$ to $(-4,6,8)$. By formula (4), we get

$$\begin{aligned}&\int_C \boldsymbol{\Phi}(\mathbf{r}) \cdot d\mathbf{r}\\ &\quad= \int_0^1 \big[(2-6t)\mathbf{i} - (1+5t)\mathbf{j} + (2-6t)(1+5t)\mathbf{k}\big] \cdot (-6\mathbf{i} + 5\mathbf{j} + 5\mathbf{k})\,dt\\ &\quad= \int_0^1 (-7 + 31t - 150t^2)\,dt = -\frac{83}{2}. \ \blacktriangle\end{aligned}$$

Example 7 Suppose that $\boldsymbol{\Phi}(\mathbf{r})$ is orthogonal to $\boldsymbol{\sigma}'(t)$ at each point of the curve $\boldsymbol{\sigma}(t)$. What can you say about the line integral $\int_C \boldsymbol{\Phi}(\mathbf{r}) \cdot d\mathbf{r}$?

Solution Since $\int_C \boldsymbol{\Phi}(\mathbf{r}) \cdot d\mathbf{r} = \int_{t_1}^{t_2} \boldsymbol{\Phi}(\boldsymbol{\sigma}(t)) \cdot \boldsymbol{\sigma}'(t)\,dt$, this integral will be zero because $\boldsymbol{\Phi}(\boldsymbol{\sigma}(t)) \cdot \boldsymbol{\sigma}'(t) = 0$, as $\boldsymbol{\Phi}$ and $\boldsymbol{\sigma}'$ are orthogonal. $\blacktriangle$

The formulas

$$\int_a^b f(x)\,dx + \int_b^c f(x)\,dx = \int_a^c f(x)\,dx \tag{5}$$

and

$$\int_a^b f(x)\,dx = -\int_b^a f(x)\,dx \tag{6}$$

for ordinary integrals have their counterparts in line integration. If we choose a point on a curve C, it divides C into two curves, C_1 and C_2 (see Fig. 18.1.7(a)). We write $C = C_1 + C_2$. Then (4) and (5) give

$$\int_{C_1 + C_2} \boldsymbol{\Phi}(\mathbf{r}) \cdot d\mathbf{r} = \int_{C_1} \boldsymbol{\Phi}(\mathbf{r}) \cdot d\mathbf{r} + \int_{C_2} \boldsymbol{\Phi}(\mathbf{r}) \cdot d\mathbf{r}. \tag{7}$$

Let $-C$ be the curve C traversed in the reverse direction (see Fig. 18.1.7(b)). Then (4) and (6) give

$$\int_{-C} \boldsymbol{\Phi}(\mathbf{r}) \cdot d\mathbf{r} = -\int_C \boldsymbol{\Phi}(\mathbf{r}) \cdot d\mathbf{r}. \tag{8}$$

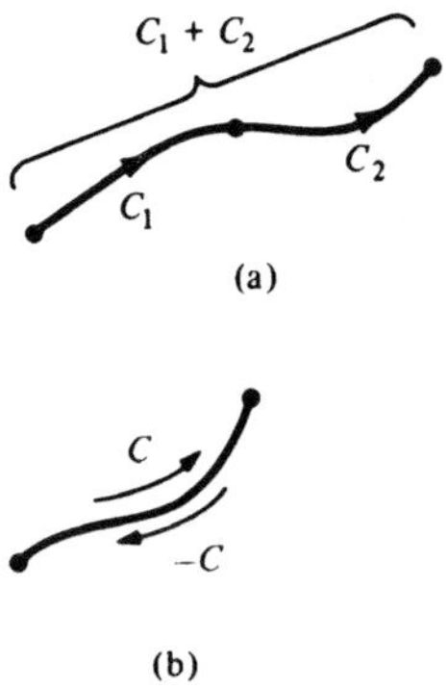

Figure 18.1.7. "Algebraic" operations on curves.

Addition formula (7) suggests a way to define line integrals over curves with "corners"—that is, continuous curves on which the tangent vector is undefined at certain points. If C is such a curve, we write $C = C_1 + C_2 + C_3 + \cdots + C_n$ by dividing it at the corner points, and we *define* $\int_C \mathbf{\Phi}(\mathbf{r}) \cdot d\mathbf{r}$ to be

$$\sum_{i=1}^{n} \int_{C_i} \mathbf{\Phi}(\mathbf{r}) \cdot d\mathbf{r} \qquad \text{(see Fig. 18.1.8).}$$

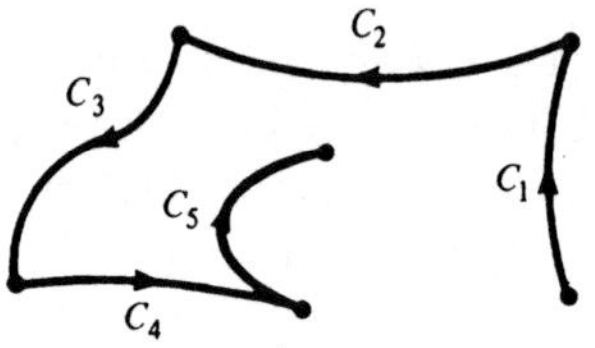

Figure 18.1.8. If $C = C_1 + \cdots + C_n$, then $\int_C \mathbf{\Phi}(\mathbf{r}) \cdot d\mathbf{r} = \sum_{i=1}^{n} \int_{C_i} \mathbf{\Phi}(\mathbf{r}) \cdot d\mathbf{r}$.

Example 8 Let C be the perimeter of the unit square $[0, 1] \times [0, 1]$ in the plane, traversed in the counterclockwise direction (see Fig. 18.1.9). Evaluate the line integral $\int_C \mathbf{\Phi}(\mathbf{r}) \cdot d\mathbf{r}$, where $\mathbf{\Phi}(x, y) = x^2\mathbf{i} + xy\mathbf{j}$.

Solution We do this problem by integrating along each of the sides C_1, C_2, C_3, C_4 separately and adding the results (see Fig. 18.1.9). The parametrizations are

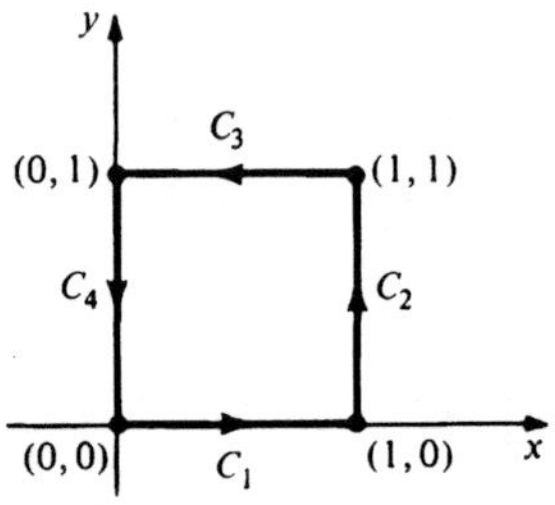

Figure 18.1.9. The perimeter of the unit square broken into four pieces.

C_1: $(t, 0)$, $0 \leqslant t \leqslant 1$; $\boldsymbol{\sigma}_1(t) = t\mathbf{i}$.
C_2: $(1, t)$, $0 \leqslant t \leqslant 1$; $\boldsymbol{\sigma}_2(t) = \mathbf{i} + t\mathbf{j}$.
C_3: $(1 - t, 1)$, $0 \leqslant t \leqslant 1$; $\boldsymbol{\sigma}_3(t) = (1 - t)\mathbf{i} + \mathbf{j}$.
C_4: $(0, 1 - t)$, $0 \leqslant t \leqslant 1$; $\boldsymbol{\sigma}_4(t) = (1 - t)\mathbf{j}$.

Thus, by (4),

$$\int_{C_1} \mathbf{\Phi}(\mathbf{r}) \cdot d\mathbf{r} = \int_0^1 t^2\, dt = \frac{1}{3},$$

$$\int_{C_2} \mathbf{\Phi}(\mathbf{r}) \cdot d\mathbf{r} = \int_0^1 t\, dt = \frac{1}{2},$$

$$\int_{C_3} \mathbf{\Phi}(\mathbf{r}) \cdot d\mathbf{r} = \int_0^1 (1 - t)^2(-1)\, dt = -\frac{1}{3},$$

$$\int_{C_4} \mathbf{\Phi}(\mathbf{r}) \cdot d\mathbf{r} = \int_0^1 0\, dt = 0.$$

Adding, we get

$$\int_C \mathbf{\Phi}(\mathbf{r}) \cdot d\mathbf{r} = \frac{1}{3} + \frac{1}{2} - \frac{1}{3} + 0 = \frac{1}{2}. \ \blacktriangle$$

Another notation for the line integral arises if we write our vectors in components. Suppose that $\mathbf{\Phi}(x, y, z) = a(x, y, z)\mathbf{i} + b(x, y, z)\mathbf{j} + c(x, y, z)\mathbf{k}$.

The expression for the derivative,

$$\frac{d\mathbf{r}}{dt} = \frac{dx}{dt}\mathbf{i} + \frac{dy}{dt}\mathbf{j} + \frac{dz}{dt}\mathbf{k}$$

can be written formally as

$$d\mathbf{r} = dx\,\mathbf{i} + dy\,\mathbf{j} + dz\,\mathbf{k},$$

and so

$$\begin{aligned} \int_C \mathbf{\Phi}(\mathbf{r}) \cdot d\mathbf{r} &= \int_C \big[a(x, y, z)\mathbf{i} + b(x, y, z)\mathbf{j} + c(x, y, z)\mathbf{k}\big] \cdot (dx\,\mathbf{i} + dy\,\mathbf{j} + dz\,\mathbf{k}) \\ &= \int_C \big[a(x, y, z)\, dx + b(x, y, z)\, dy + c(x, y, z)\, dz\big]. \end{aligned} \tag{9}$$

The expression inside the last integral is called a *differential form.* To evaluate the integral, we choose a parametrization of C. Then x, y, and z become functions of t; dx, dy, dz are expressed in terms of t and dt, and the integrand becomes an ordinary integrand in t which may be integrated over the parametrization interval.

Example 9 Evaluate $\int_C \cos z\,dx + e^x\,dy + e^y\,dz$, where C is parametrized by $(x, y, z) = (1, t, e^t)$, $0 \leqslant t \leqslant 2$.

Solution We compute $dx = 0$, $dy = dt$, $dz = e^t dt$, and so

$$\int_C [\cos z\,dx + e^x\,dy + e^y\,dz] = \int_0^2 (0 + e^1 + e^{2t})\,dt$$

$$= \left[et + \frac{1}{2}e^{2t} \right]_0^2 = 2e + \frac{1}{2}e^4 - \frac{1}{2}. \blacktriangle$$

The following box summarizes the notations and definitions developed so far.

Line Integrals

The line integral of a vector function (x, y, z) along a curve $\boldsymbol{\sigma}(t)$, $t_1 \leqslant t \leqslant t_2$, is

$$\int_{t_1}^{t_2} \boldsymbol{\Phi}(\boldsymbol{\sigma}(t)) \cdot \boldsymbol{\sigma}'(t)\,dt.$$

If $\boldsymbol{\Phi}(x, y, z) = a(x, y, z)\mathbf{i} + b(x, y, z)\mathbf{j} + c(x, y, z)\mathbf{k}$, and $\boldsymbol{\sigma}(t) = x(t)\mathbf{i} + y(t)\mathbf{j} + z(t)\mathbf{k}$, then

$$\int_{t_1}^{t_2} \boldsymbol{\Phi}(\boldsymbol{\sigma}(t)) \cdot \boldsymbol{\sigma}'(t)\,dt = \int_{t_1}^{t_2} a(x(t), y(t), z(t))\frac{dx}{dt}\,dt$$

$$+ \int_{t_1}^{t_2} b(x(t), y(t), z(t))\frac{dy}{dt}\,dt + \int_{t_1}^{t_2} c(x(t), y(t), z(t))\frac{dz}{dt}\,dt,$$

which is also written

$$\int_C a\,dx + b\,dy + c\,dz \quad \text{or} \quad \int_C \boldsymbol{\Phi}(\mathbf{r}) \cdot d\mathbf{r} \quad \text{or simply} \quad \int_C \boldsymbol{\Phi}.$$

Example 10 Find $\int_C \sin \pi x\,dy - \cos \pi y\,dz$, where C is the triangle with vertices $(1, 0, 0)$, $(0, 1, 0)$, and $(0, 0, 1)$ in that order.

Solution We write $C = C_1 + C_2 + C_3$, where C_1 is the line segment from $(1, 0, 0)$ to $(0, 1, 0)$, C_2 is the line segment from $(0, 1, 0)$ to $(0, 0, 1)$, and C_3 is the line segment from $(0, 0, 1)$ to $(1, 0, 0)$. Parametrizations on $[0, 1]$ for these segments are

C_1: $(x, y, z) = (1 - t, t, 0)$ so $dx = -dt$, $dy = dt$, $dz = 0$;
C_2: $(x, y, z) = (0, 1 - t, t)$ so $dx = 0$, $dy = -dt$, $dz = dt$;
C_3: $(x, y, z) = (t, 0, 1 - t)$ so $dx = dt$, $dy = 0$, $dz = -dt$.

Then

$$\int_C \sin(\pi x)\,dy - \cos(\pi y)\,dz = \sum_{i=1}^{3} \int_{C_i} \sin(\pi x)\,dy - \cos(\pi y)\,dz$$

$$= \int_0^1 \sin[\pi(1-t)]\,dt - \cos(\pi t)\cdot 0$$

$$+ \int_0^1 \sin(\pi\cdot 0)(-dt) - \cos[\pi(1-t)]\,dt$$

$$+ \int_0^1 \sin \pi t\cdot 0 - \cos(\pi\cdot 0)(-dt)$$

$$= \int_0^1 \sin[\pi(1-t)]\,dt - \int_0^1 \cos[\pi(1-t)]\,dt + \int_0^1 dt$$

$$= -\frac{1}{\pi}\{-\cos[\pi(1-t)]\}\Big|_0^1 - \left(-\frac{1}{\pi}\right)\{\sin[\pi(1-t)]\}\Big|_0^1 + 1$$

$$= \frac{1}{\pi}(\cos 0 - \cos\pi) + \frac{1}{\pi}(\sin 0 - \sin\pi) + 1$$

$$= \frac{1}{\pi}[1-(-1)] + \frac{1}{\pi}(0-0) + 1 = \frac{2}{\pi} + 1. \blacktriangle$$

Exercises for Section 18.1

1. Calculate the work which is done by the force field $\mathbf{\Phi}(x, y, z) = x\mathbf{i} + y\mathbf{j}$ when a particle is moved along the path $(3t^2, t, 1)$, $0 \leqslant t \leqslant 1$.
2. Find the work done by the force field in Exercise 1 when a particle is moved along the straight line segment from $(0,0,1)$ to $(3,1,1)$.
3. Find the work which is done by the force field $\mathbf{\Phi}(x, y) = (x^2 + y^2)(\mathbf{i} + \mathbf{j})$ around the loop $(x, y) = (\cos t, \sin t)$, $0 \leqslant t \leqslant 2\pi$.
4. Find the work done by the force field in Exercise 3 around the loop $(x, y) = (1 + \cos t, 1 + \sin t)$, $0 \leqslant t \leqslant 2\pi$.
5. Suppose that you pick up a unit mass which was at rest at $(1,0,0)$ and carry it to $(1,0,1)$ along the path $(1,0,t)$ under the field $xy\mathbf{i} + (x+y)\mathbf{k}$. If you leave the particle with velocity vector $\mathbf{i} + 2\mathbf{j}$ at the end of the trip, how much work have you done?
6. Do as in Exercise 5, except that the particle is left at rest at the end of the trip.
7. Show that if a particle is moved along the closed curve $(\cos t, \sin t, 0)$, $0 \leqslant t \leqslant 2\pi$, then the force field in Example 1 does a nonzero amount of work on the particle. How much is the work?
8. Show that if a particle is moved along a closed curve (that is, $\boldsymbol{\sigma}(t_1) = \boldsymbol{\sigma}(t_2)$), then the work done on it by the gravitational field in Example 3 is zero.

Let $\mathbf{\Phi}(x, y) = [1/(x^2 + y^2)](-y\mathbf{i} + x\mathbf{j})$ be a force field in the plane (minus the origin). Compute the work done by this force along each of the paths in Exercises 9–12.

9. $(\cos t, \sin t)$; $0 \leqslant t \leqslant \pi$
10. $(\cos t, -\sin t)$; $0 \leqslant t \leqslant \pi$
11. $(\cos t, \sin t)$; $0 \leqslant t \leqslant 2\pi$
12. $(-\cos t, \sin t)$; $0 \leqslant t \leqslant 2\pi$

In Exercises 13–20, evaluate the integral of the given vector field $\mathbf{\Phi}$ along the given path.

13. $\boldsymbol{\sigma}(t) = (\sin t, \cos t, t)$, $0 \leqslant t \leqslant 2\pi$, $\mathbf{\Phi}(x, y, z) = x\mathbf{i} + y\mathbf{j} + z\mathbf{k}$.
14. $\boldsymbol{\sigma}(t) = (t, t, t)$; $0 \leqslant t \leqslant 1$, $\mathbf{\Phi}(x, y, z) = x\mathbf{i} - y\mathbf{j} + z\mathbf{k}$.
15. $\boldsymbol{\sigma}(t) = (\cos t, \sin t, 0)$; $0 \leqslant t \leqslant \pi/2$, $\mathbf{\Phi}(x, y, z) = x\mathbf{i} - y\mathbf{j} + z\mathbf{k}$.
16. $\boldsymbol{\sigma}(t) = (\cos t, \sin t, 0)$; $0 \leqslant t \leqslant \pi/2$, $\mathbf{\Phi}(x, y, z) = x\mathbf{i} - y\mathbf{j} + 2\mathbf{k}$.
17. $\boldsymbol{\sigma}(t) = (\sin t, t^2, t)$; $0 \leqslant t \leqslant 2\pi$, $\mathbf{\Phi}(x, y, z) = \sin z\mathbf{i} + \cos\sqrt{y}\,\mathbf{j} + x^3\mathbf{k}$.
18. $\boldsymbol{\sigma}(t) = (\cos t, \sec t, \tan t)$; $-\pi/4 \leqslant t \leqslant \pi/4$, $\mathbf{\Phi}(x, y, z) = xz\mathbf{i} + xy\mathbf{j} + yz\mathbf{k}$.
19. $\boldsymbol{\sigma}(t) = ((1 + t^2)^2, 1, t)$; $0 \leqslant t \leqslant 1$, $\mathbf{\Phi}(x, y, z) = [1/(z^2 + 1)]\mathbf{i} + x(1 + y^2)\mathbf{j} + e^y\mathbf{k}$.
20. $\boldsymbol{\sigma}(t) = 3t\mathbf{i} + (t - 1)\mathbf{j} + t^2\mathbf{k}$; $0 \leqslant t \leqslant 1$, $\mathbf{\Phi}(x, y, z) = (x^2 + x)\mathbf{i} + \dfrac{x - y}{x + y}\mathbf{j} + (z - z^3)\mathbf{k}$.

Let $\mathbf{\Phi}(x, y, z) = x^2\mathbf{i} - xy\mathbf{j} + \mathbf{k}$. Evaluate the line integral of $\mathbf{\Phi}$ along each of the curves in Exercises 21–24.

21. The straight line joining $(0,0,0)$ to $(1,1,1)$.
22. The circle of radius 1, center at the origin and lying in the yz plane, traversed counterclockwise as viewed from the positive x axis.
23. The parabola $z = x^2$, $y = 0$, between $(-1,0,1)$ and $(1,0,1)$.
24. The straight line between $(-1,0,1)$ and $(1,0,1)$.
25. Let C be parametrized by $x = \cos^3\theta$, $y = \sin^3\theta$, $z = \theta$, $0 \leqslant \theta \leqslant 7\pi/2$. Evaluate the integral $\int_C \sin z\,dx + \cos z\,dy - (xy)^{1/3}\,dz$.

26. Evaluate $\int_C x^2\,dx + xy\,dy + dz$, where C is parametrized by $\sigma(t) = (t, t^2, 1)$, $0 \leqslant t \leqslant 1$.
27. Evaluate $\int_C \mathbf{\Phi}(\mathbf{r}) \cdot d\mathbf{r}$, where $\mathbf{\Phi}(x, y, z) = \sin z\mathbf{i} + \cos\sqrt{y}\,\mathbf{j} + x^3\mathbf{k}$ and C is the line segment from $(1,0,0)$ to $(0,0,3)$.
28. Evaluate $\int_C e^{x+y-z}(\mathbf{i} + \mathbf{j} - \mathbf{k}) \cdot d\mathbf{r}$, where C is the path $(\ln t, t, t)$ for $2 \leqslant t \leqslant 4$.

The *line integral of a scalar function* f along a parametric curve $\sigma(t)$, $t_1 \leqslant t \leqslant t_2$, is defined by

$$\int_{t_1}^{t_2} f(\sigma(t))\|\sigma'(t)\|\,dt.$$

Note that if $f = 1$, this is just the arc length of the curve. Evaluate the line integrals of the functions along the indicated curves in Exercises 29–34.

29. $f(x, y, z) = x + y + yz$, where $\sigma(t) = (\sin t, \cos t, t)$; $0 \leqslant t \leqslant 2\pi$.
30. $f(x, y, z) = x + \cos^2 z$, σ as in Exercise 29.
31. $f(x, y, z) = x \cos z$, $\sigma(t) = t\mathbf{i} + t^2\mathbf{j}$; $0 \leqslant t \leqslant 1$.
32. $f(x, y, z) = \exp\sqrt{z}$, $\sigma(t) = (1, 2, t^2)$; $0 \leqslant t \leqslant 1$.
33. $f(x, y, z) = yz$, $\sigma(t) = (t, 3t, 2t)$; $1 \leqslant t \leqslant 3$.
34. $f(x, y, z) = (x + y)/(y + z)$, $\sigma(t) = (t, \frac{2}{3}t^{3/2}, t)$; $1 \leqslant t \leqslant 2$.

★35. Show that the value of the line integral of a scalar function over a parametric curve, defined after Exercise 28, is unchanged if the curve is reparametrized.

18.2 Path Independence

The line integral of a gradient vector field depends only on the endpoints of the curve.

We saw in the last section that the line integral of a vector field along a curve from a point A to a point B depends not just on A and B, but on the path of integration itself. There is, however, an important class of vector fields for which line integrals are path independent. In this section and the next, we shall give several different ways to recognize and use such vector fields.

A vector field $\mathbf{\Phi}(x, y, z)$ defined on some domain D in space (or a vector field $\mathbf{\Phi}(x, y)$ defined on a domain in the plane) is called *conservative* if, whenever C_1 and C_2 are curves in D with the same endpoints, the line integrals $\int_{C_1}\mathbf{\Phi}(\mathbf{r}) \cdot d\mathbf{r}$ and $\int_{C_2}\mathbf{\Phi}(\mathbf{r}) \cdot d\mathbf{r}$ are equal.

Our first observation is that a vector field on D is conservative if and only if its integral around every *closed* curve in D is zero. (A conservative force field is thus one in which no net work is done, i.e., energy is conserved if a particle goes around a closed path.)

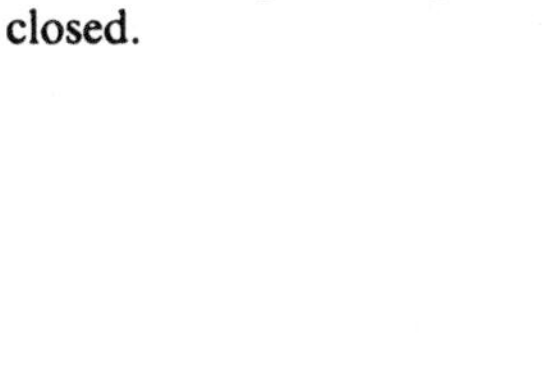

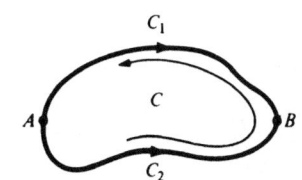

Figure 18.2.1. C_1 and C_2 have the same endpoints when $C = C_2 + (-C_1)$ is closed.

To justify this observation, we consider Figure 18.2.1, which can be interpreted in two different ways. First of all, if C_1 and C_2 are given curves from A to B, then $C = C_2 + (-C_1)$ is a closed curve (from A to A). If $\mathbf{\Phi}$ has the property that its integral around every closed curve is zero, then by formulas (7) and (8) in Section 18.1,

$$\int_{C_2}\mathbf{\Phi}(\mathbf{r}) \cdot d\mathbf{r} - \int_{C_1}\mathbf{\Phi}(\mathbf{r}) \cdot d\mathbf{r} = \int_{C_2+(-C_1)}\mathbf{\Phi}(\mathbf{r}) \cdot d\mathbf{r} = \int_C\mathbf{\Phi}(\mathbf{r}) \cdot d\mathbf{r} = 0;$$

so $\int_{C_1}\mathbf{\Phi}(\mathbf{r}) \cdot d\mathbf{r} = \int_{C_2}\mathbf{\Phi}(\mathbf{r}) \cdot d\mathbf{r}$. Since this is true for all curves C_1 and C_2 with common endpoints, $\mathbf{\Phi}$ is conservative.

The second way to look at Figure 18.2.1 is to consider the closed curve C as given; the pieces C_1 and C_2 are then manufactured by choosing points A and B on C so that $C = C_1 + (-C_2)$. Now if $\mathbf{\Phi}$ is conservative, then

$$0 = \int_{C_2}\mathbf{\Phi}(\mathbf{r}) \cdot d\mathbf{r} - \int_{C_1}\mathbf{\Phi}(\mathbf{r}) \cdot d\mathbf{r} = \int_{C_2+(-C_1)}\mathbf{\Phi}(\mathbf{r}) \cdot d\mathbf{r} = \int_C\mathbf{\Phi}(\mathbf{r}) \cdot d\mathbf{r};$$

so the integral of $\mathbf{\Phi}$ around a closed curve C is zero.

The argument just given to connect path independence with integrals around closed curves has many applications in mathematics. A related geometric example is given in Exercise 37.

Conservative Fields

A vector field $\mathbf{\Phi}$ is conservative when the line integral of $\mathbf{\Phi}$ around any closed curve is zero, which is the same as the line integrals of $\mathbf{\Phi}$ being path independent.

Example 1 Let $\mathbf{\Phi}$ be a conservative vector field in the plane. If the line integral of $\mathbf{\Phi}$ along the curve AOB in Figure 18.2.2 equals 3.5, find the integral of $\mathbf{\Phi}$ along the broken lines: (a) ACB, (b) BDA, (c) $ACBDA$.

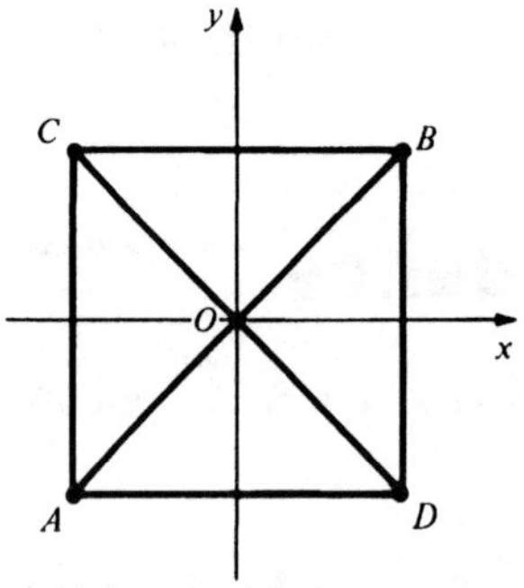

Figure 18.2.2. Paths for Example 1.

Solution (a) Since AOB and ACB have the same endpoints, and $\mathbf{\Phi}$ is conservative, the integral of $\mathbf{\Phi}$ along ACB is 3.5.
(b) BDA has the same endpoints as BOA, which is $-AOB$, so the integral is -3.5.
(c) $ACBDA$ is a closed curve, so the line integral around it is zero. ▲

Example 2 Using examples from the previous section, show that neither of the vector fields $y\mathbf{i} - x\mathbf{j} + \mathbf{k}$ and $-y\mathbf{i} + x\mathbf{j}$ is conservative.

Solution In Example 1, Section 18.1, the line integrals of $y\mathbf{i} - x\mathbf{j} + \mathbf{k}$ along the paths (a) and (c) are different, although the curves have the same endpoints. In Example 5(b), the line integral of $-y\mathbf{i} + x\mathbf{j}$ around a closed curve is not zero, so it cannot be conservative. ▲

Conservative vector fields are easy to integrate because we can replace complicated paths by simple ones, but we still do not know how to recognize these fields. A first step in this direction is the following result.

Theorem: Gradient Vector Fields Are Conservative

If $\mathbf{\Phi} = \nabla f$, then $\mathbf{\Phi}$ is conservative. In fact, if $\mathbf{\Phi} = \nabla f$, then for any curve C from A to B,

$$\int_C \mathbf{\Phi}(\mathbf{r}) \cdot d\mathbf{r} = f(B) - f(A).$$

To prove this theorem, we let $\boldsymbol{\sigma}(t)$ be a parametrized curve on $[t_1, t_2]$ representing C and going from A to B, so that

$$\int_{C_1} \mathbf{\Phi}(\mathbf{r}) \cdot d\mathbf{r} = \int_{t_1}^{t_2} \mathbf{\Phi}(\boldsymbol{\sigma}(t)) \cdot \boldsymbol{\sigma}'(t) dt = \int_{t_1}^{t_2} \nabla f(\boldsymbol{\sigma}(t)) \cdot \boldsymbol{\sigma}'(t)\, dt$$

By the chain rule for gradients and curves (Section 16.1), $\nabla f(\boldsymbol{\sigma}(t)) \cdot \boldsymbol{\sigma}'(t) = (d/dt)[f(\boldsymbol{\sigma}(t))]$, so

$$\int_C \boldsymbol{\Phi}(\mathbf{r}) \cdot d\mathbf{r} = \int_{t_1}^{t_2} \frac{d}{dt}\left[f(\boldsymbol{\sigma}(t)) \right] dt = f(\boldsymbol{\sigma}(t_2)) - f(\boldsymbol{\sigma}(t_1)) = f(B) - f(A).$$

In the next-to-last equality, we used the fundamental theorem of calculus (Section 4.4). ■

Example 3 Evaluate $\int_C y\,dx + x\,dy$ if C is parametrized by $(t^9, \sin^9(\pi t/2))$, $0 \leqslant t \leqslant 1$.

Solution We recognize the vector field $\boldsymbol{\Phi} = y\mathbf{i} + x\mathbf{j}$ (by guesswork) as the gradient of $f(x, y) = xy$, since

$$f_x = y \quad \text{and} \quad f_y = x.$$

The path of integration goes from $(0,0)$ at $t = 0$ to $(1,1)$ at $t = 1$, so the theorem above gives

$$\int_C \boldsymbol{\Phi}(\mathbf{r}) \cdot d\mathbf{r} = f(1,1) - f(0,0) = 1. \blacktriangle$$

Note the resemblance of the formula

$$\int_C \nabla f(\mathbf{r}) \cdot d\mathbf{r} = f(B) - f(A),$$

in the box above to the fundamental theorem of calculus. By analogy with the one-dimensional case, we call a function f such that $\nabla f = \boldsymbol{\Phi}$ an *antiderivative* for f (the term *primitive* is sometimes used). The theorem also shows a big difference between the one-dimensional and multi-dimensional cases: whereas every continuous function of one variable has an antiderivative (see Section 4.5), in several variables, only conservative vector fields *can* have antiderivatives. In fact, every conservative vector field *does* have an antiderivative.

Theorem: Every Conservative Vector Field is a Gradient

If $\boldsymbol{\Phi}(x, y)$ is a vector field defined in a region D, and $\boldsymbol{\Phi}$ is conservative, then there is a function f defined on D such that $\boldsymbol{\Phi} = \nabla f$.

To prove this theorem, we must construct an antiderivative for any given conservative vector field $\boldsymbol{\Phi}$. We do so by "integrating $\boldsymbol{\Phi}$" as follows. Arbitrarily choose a point O in the domain D of the vector field. For every point A in D, we define $f(A)$ by integrating $\boldsymbol{\Phi}$ along some path from O to A. (The word "path" is synonymous with "curve.") Since $\boldsymbol{\Phi}$ is conservative, we get the same result no matter what path we choose. (We are implicitly assuming that D is *connected*, i.e., that it consists of just one piece, so that every point may be joined to O by some path in D. If D has several pieces, the argument is still valid if we work with one piece at a time.)

To show that $\nabla f = \boldsymbol{\Phi}$, we shall show that $\int_C \nabla f(\mathbf{r}) \cdot d\mathbf{r} = \int_C \boldsymbol{\Phi}(\mathbf{r}) \cdot d\mathbf{r}$ for every curve C. Once we have done this, it follows that

$$\int_C \left[\boldsymbol{\Phi}(\mathbf{r}) - \nabla f(\mathbf{r}) \right] \cdot d\mathbf{r} = 0$$

for all C; but this implies that $\boldsymbol{\Phi}(\mathbf{r}) - \nabla f(\mathbf{r})$ is identically zero (if not, its integral over a suitably chosen short straight path would be nonzero—see Exercise 36), so $\boldsymbol{\Phi} = \nabla f$.

To prove that $\int_C \nabla f(\mathbf{r}) \cdot d\mathbf{r} = \int_C \mathbf{\Phi}(\mathbf{r}) \cdot d\mathbf{r}$ for any path C, we refer to Figure 18.2.3. Since we know that $\int_C \nabla f(\mathbf{r}) \cdot d\mathbf{r} = f(B) - f(A)$ by the previous theorem, we need to show that $\int_C \mathbf{\Phi}(\mathbf{r}) \cdot d\mathbf{r} = f(B) - f(A)$. Now $f(B) = \int_{C_2} \mathbf{\Phi}(\mathbf{r}) \cdot d\mathbf{r}$ and $f(A) = \int_{C_1} \mathbf{\Phi}(\mathbf{r}) \cdot d\mathbf{r}$. Thus $f(B) - f(A) = \int_{(-C_1)+C_2} \mathbf{\Phi}(\mathbf{r}) \cdot d\mathbf{r} = \int_C \mathbf{\Phi}(\mathbf{r}) \cdot d\mathbf{r}$ since $(-C_1) + C_2$ and C have the same endpoints and $\mathbf{\Phi}$ is conservative. ■

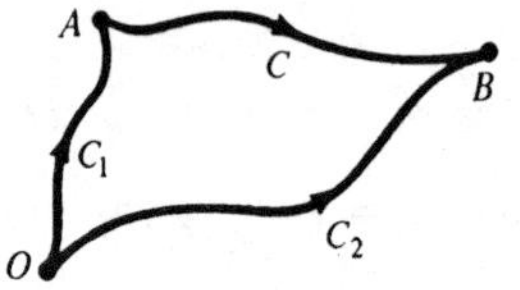

Figure 18.2.3. C and $(-C_1) + C_2$ are both paths from A to B.

Example 4 Calculate the work done in moving a mass m from a distance r_1 out to a distance r_2 in the gravitational field of mass M which produces the force field $\mathbf{F} = -(GMm/r^3)\mathbf{r}$, where $\mathbf{r} = x\mathbf{i} + y\mathbf{j} + z\mathbf{k}$ and the mass M is located at the origin.

Solution From Example 4, Section 16.1, $\mathbf{F} = -\nabla V$, where $V = -GMm/r$. Let the curve C join the points A and B at distances r_1 and r_2 from the origin. The work done by $\mathbf{F}$ is

$$W = \int_C \mathbf{F} \cdot d\mathbf{r} = -\int_C \nabla V \cdot d\mathbf{r} = -(V(B) - V(A))$$

$$= V(A) - V(B) = GMm\left(\frac{1}{r_2} - \frac{1}{r_1}\right).$$

The work done in doing this move is therefore $GMm(1/r_1 - 1/r_2)$. If the signs confuse you, note that a spacecraft moving a payload from a distance r_1 to a distance $r_2 > r_1$ does *positive* work. ▲

We still do not know how to tell whether a vector field is conservative just by looking at a formula like $\mathbf{\Phi}(x, y, z) = \cos yz\mathbf{i} + e^x\mathbf{j} + \mathbf{k}$, nor do we have a computationally efficient way of finding antiderivatives. The following theorem provides the method we need. For simplicity, we present it for two variables; the analogous result for three variables is given in Exercise 30.

Theorem: The Cross-Derivative Test

A vector field $\mathbf{\Phi}(x, y) = a(x, y)\mathbf{i} + b(x, y)\mathbf{j}$ defined on the whole plane is conservative if and only if $a_y = b_x$.

The following proof gives an explicit method for finding an antiderivative of $\mathbf{\Phi}$. First, we note from pages 896 and 897 that $\mathbf{\Phi}$ is conservative if and only if it is of the form ∇f, i.e., if and only if we can solve the simultaneous equations[1]

$$f_x = a, \tag{1}$$

$$f_y = b. \tag{2}$$

We notice that if (1) and (2) have a solution f, then (1) implies $f_{xy} = a_y$ while (a) implies $f_{yx} = b_x$. By the equality of mixed partial derivatives (Section 15.1), $f_{xy} = f_{yx}$, so we must have $a_y = b_x$ whenever the vector field $a(x, y)\mathbf{i} + b(x, y)\mathbf{j}$ is conservative.

[1] These are called *partial differential equations*, since they involve partial derivatives of the unknown function f.

To prove that $a_y = b_x$ implies that $a(x, y)\mathbf{i} + b(x, y)\mathbf{j}$ is a gradient, we begin by solving equation (1). We do this by integrating (1) with respect to x, thinking of y as being held fixed just as we do in partial differentiation, to get a trial solution $\bar{f}(x, y)$ of our equations. (You may wish to skip ahead and read Example 5 before finishing this proof.) For instance, we may take

$$\bar{f}(x, y) = \int_0^x a(t, y)\,dt.$$

The choice of 0 as the starting x value for integration was arbitrary, and indeed we get other solutions of equation (1) by adding any function $g(y)$, since the x derivative of $f(x, y) = \bar{f}(x, y) + g(y)$ is still $a(x, y)$. The function $g(y)$ plays the role of the "arbitrary constant" in this partial indefinite integration; what we must do next is to choose it so that equation (2) as well as (1) will be satisfied. With $f(x, y) = \bar{f}(x, y) + g(y)$, equation (2) becomes

$$\bar{f}_y(x, y) + g'(y) = b(x, y),$$

i.e., $g'(y) = b(x, y) - \bar{f}_y(x, y)$. (3)

We have written $g'(y)$ rather than $g_y(y)$ since g is a function of *one* variable.

Equation (3) can be solved by ordinary integration with respect to y provided that the right-hand side is a function of y alone, i.e., if the x derivative of $b - \bar{f}_y$ is zero; but $(b - \bar{f}_y)_x = b_x - \bar{f}_{yx} = b_x - \bar{f}_{xy} = b_x - a_y$, which is zero by the hypothesis $a_y = b_x$. ■

Example 5 Show that the vector field $(2x + 3y^3)\mathbf{i} + (9xy^2 + 2y)\mathbf{j}$ is conservative, and find an antiderivative.

Solution With $a(x, y) = 2x + 3y^3$ and $b(x, y) = 9xy^2 + 2y$, we have $a_y(x, y) = 9y^2$ and $b_x(x, y) = 9y^2$, so the cross-derivative test shows that the given field is conservative. To find an antiderivative f, we first solve the equation

$$f_x(x, y) = 2x + 3y^3$$

by integrating with respect to x to obtain the trial solution $x^2 + 3xy^3$. To this we may add an arbitrary function $g(y)$ without destroying the property that its x derivative is $2x + 3y^3$. Thus, our trial solution is $f(x, y) = x^2 + 3xy^3 + g(y)$. Now the equation $f_y = b$ becomes

$$9xy^2 + g'(y) = 9xy^2 + 2y,$$

so $g'(y) = 2y$. (The fact that the equation for $g'(y)$ does not involve x is a consequence of the condition $a_y = b_x$.) A solution of $g'(y) = 2y$ is $g(y) = y^2$, so we may take $f(x, y) = x^2 + 3xy^3 + y^2$ as our antiderivative. ▲

Example 6 Show that the vector field $x^2\mathbf{i} + xy\mathbf{j}$ is not conservative.

Solution With $a(x, y) = x^2$ and $b(x, y) = xy$, we have $a_y = 0$ and $b_x = y$. Since these cross-derivatives are unequal, the field cannot be conservative. ▲

The next example illustrates the importance of the condition in the previous theorem that the vector field be defined in the *entire* plane.

Example 7 (a) Show that the vector field

$$\mathbf{\Phi}(x, y) = \frac{-y}{x^2 + y^2}\mathbf{i} + \frac{x}{x^2 + y^2}\mathbf{j}$$

is not conservative by integrating it around the circle $x^2 + y^2 = 1$.

(b) Verify that the cross-derivative condition $a_y = b_x$ is nevertheless satisfied for this vector field.

(c) What is going on here?

Solution (a) Parametrizing the unit circle C by $\boldsymbol{\sigma}(t) = (\cos t)\mathbf{i} + (\sin t)\mathbf{j}$ gives

$$\int_C \boldsymbol{\Phi}(\mathbf{r}) \cdot d\mathbf{r} = \int_0^{2\pi} \left(\frac{-\sin t}{\sin^2 t + \cos^2 t} \mathbf{i} + \frac{\cos t}{\sin^2 t + \cos^2 t} \mathbf{j} \right) \cdot ((-\sin t)\mathbf{i} + (\cos t)\mathbf{j})\, dt$$

$$= \int_0^{2\pi} \frac{\sin^2 t + \cos^2 t}{\sin^2 t + \cos^2 t}\, dt = 2\pi.$$

Since this is not zero, $\boldsymbol{\Phi}$ is not conservative.

(b) With $a(x, y) = -y/(x^2 + y^2)$ and $b(x, y) = x/(x^2 + y^2)$, we have

$$a_y(x, y) = \frac{(x^2 + y^2)(-1) - (-y)(2y)}{(x^2 + y^2)^2} = \frac{y^2 - x^2}{(x^2 + y^2)^2}$$

and

$$b_x(x, y) = \frac{(x^2 + y^2)(1) - (x)(2x)}{(x^2 + y^2)^2} = \frac{y^2 - x^2}{(x^2 + y^2)^2},$$

so the cross-derivative condition is satisfied.

(c) The vector field $\boldsymbol{\Phi}(x, y)$ is not defined at the origin, so the cross-derivative test does not apply. ▲

Exercises for Section 18.2

Let $\boldsymbol{\Phi}$ be a conservative vector field in the plane. Suppose that the integral of $\boldsymbol{\Phi}$ along AOF is 3, along OF is 2, and along AB is -5. Compute the integral of $\boldsymbol{\Phi}$ along the paths in Exercises 1–4.

1. $AODEF$
2. $FEDO$
3. $BOEF$
4. $BAODEF$

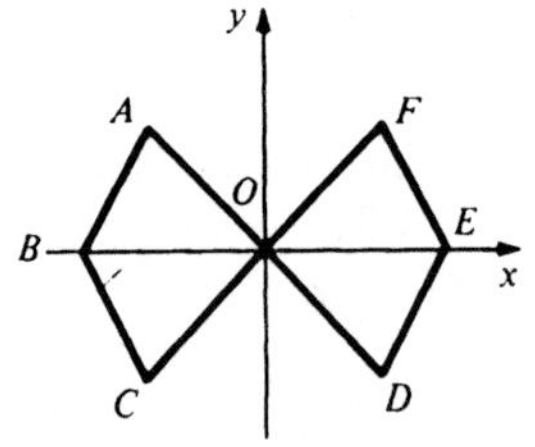

Figure 18.2.4. Paths for Exercise 4.

In Exercises 5–8, evaluate the integral of the given vector field around the given closed curve to demonstrate that it is not conservative.

5. $\boldsymbol{\Phi} = y\mathbf{i} + y\mathbf{j} + \mathbf{k}$, C the path consisting of straight lines joining $(0,0,0)$, $(0,1,0)$, $(1,1,0)$ and $(0,0,0)$.
6. $\boldsymbol{\Phi} = 2\mathbf{i} + x\mathbf{j}$, C the path in the plane consisting of straight lines from $(0,0)$ to $(1,1)$ to $(-1,1)$ to $(0,0)$.
7. $\boldsymbol{\Phi} = 3\mathbf{i} + x\mathbf{j}$, C the unit circle $x^2 + y^2 = 1$.
8. $\boldsymbol{\Phi} = y\mathbf{i} - xy\mathbf{j}$, C the unit circle $x^2 + y^2 = 1$.

In Exercises 9–12, recognize the vector field as a gradient and use this to evaluate the given integral.

9. $\int_C 2xy\,dx + x^2\,dy$, C parametrized by $x = \cos 8t$, $y = 5 \sin 16t$, $0 \leqslant t \leqslant \pi/4$.
10. $\int_C ye^{xy}\,dx + xe^{xy}\,dy$, C parametrized by $x = 5t^3$, $y = -t^3$, $-1 \leqslant t \leqslant 1$.
11. $\int_C 3x^2y^2\,dx + 2x^3y\,dy$, C parametrized by $x = 3t^2 + 1$, $y = 2t$, $0 \leqslant t \leqslant 1$.
12. $\int_C y \sin(xy)\,dx + x \sin(xy)\,dy$, C parametrized by $x = \cos 2t$, $y = 3 \sin 2t$, $0 \leqslant t \leqslant \pi/2$.

13. A certain force field exerted on a mass m is given by $\mathbf{F} = -(JMm/r^5)\mathbf{r}$. Find the work done in moving the mass m from a distance r_1 out to a distance $r_2 > r_1$.
14. The mass of the earth is approximately 6×10^{27} grams; the mass of the sun is 330,000 times as much. The gravitational constant (in units of grams, seconds, and centimeters) is 6.7×10^{-8}. The distance of the earth from the sun is about 1.5×10^{12} centimeters. Compute, approximately, the work necessary to increase the distance of the earth from the sun by 1 centimeter.
15. Suppose that the kinetic energy of a particle which moves in a circular path increases after it makes one circuit. Can the force field governing the particle's motion be conservative?
16. In the earth's gravitational field, show that if a mass is taken on a journey, however complicated, the net amount of work done against the

gravitational field is zero, provided the mass ends up in the same spot it began, with the same speed.

17. Is $2xy\mathbf{i} + (x^2 + \cos y)\mathbf{j}$ conservative? If so, find an antiderivative.
18. Is $x^2y\mathbf{i} + (\frac{1}{3}x^3 + ye^y)\mathbf{j}$ conservative? If so, find an antiderivative.
19. Is $4x\cos^2(y/2)\mathbf{i} - x^2\sin y\mathbf{j}$ conservative? If so, find an antiderivative.
20. Is $2xy\sin(x^2y)\mathbf{i} + (e^y + x^2\sin(x^2y))\mathbf{j}$ conservative? If so, find an antiderivative.
21. Consider the vector field in Example 7. Show that if we restrict its domain to be those (x, y) with $y > 0$, then it is conservative. (Show that $f(x, y) = \tan^{-1}(y/x)$ is an antiderivative.)
22. In Exercise 21, what prevents f from being an antiderivative for $\mathbf{\Phi}$ on the whole plane minus the origin?

In Exercises 23–26, recognize $\mathbf{\Phi}$ as a gradient and compute the work done in moving a particle along the given path with the given force.

23. $\boldsymbol{\sigma}(t) = (\cos t, \sin t, t)$; $0 \leqslant t \leqslant 2\pi$, $\mathbf{\Phi}(x, y, z) = x\mathbf{i} + y\mathbf{j} + z\mathbf{k}$.
24. $\boldsymbol{\sigma}(t) = (\cos t, \sin t, 0)$; $0 \leqslant t \leqslant 2\pi$, $\mathbf{\Phi}(x, y, z) = x\mathbf{i} + y\mathbf{j} + z\mathbf{k}$.
25. Same as Exercise 24 but with $0 \leqslant t \leqslant \pi$.
26. $\boldsymbol{\sigma}(t) = (a\cos t, 0, b\sin t)$; $0 \leqslant t \leqslant 2\pi$, $\mathbf{\Phi}(r) = -\mathbf{r}/\|\mathbf{r}\|^3$.

27. Let $\mathbf{\Phi}(x, y, z) = (z^3 + 2xy)\mathbf{i} + x^2\mathbf{j} + 3xz^2\mathbf{k}$. Show that the integral of $\mathbf{\Phi}$ around the circumference of the unit square $[0, 1] \times [0, 1]$ in the xy plane is zero by:
    (a) Evaluating directly.
    (b) Demonstrating that $\mathbf{\Phi}$ is the gradient of some function f.
28. Let
$$f(x, y, z) = e^x\cos(yz + x^8) + [\sin(yz)]\ln(1 + x^2).$$
Show that
$$\int_{C_1} \nabla f(\mathbf{r}) \cdot d\mathbf{r} = \int_{C_2} \nabla f(\mathbf{r}) \cdot d\mathbf{r},$$
where C_1 is the straight line joining $(0, 1, 0)$ to $(1, 1, 0)$ and C_2 is the curve parametrized by $(\sin t, \cos 4t, \sin 4t)$; $0 \leqslant t \leqslant \pi/2$.
29. If $f(x, y, z) = x^3 - y^3 + \sin(\pi yz/2)$, evaluate $\int_C \nabla f(\mathbf{r}) \cdot d\mathbf{r}$ along the curve
$$\boldsymbol{\sigma}(t) = \left(\frac{t^3}{1 + t^2}, \sin\frac{\pi}{4}t^5, t^2 + 2\right); \qquad 0 \leqslant t \leqslant 1.$$

★30. Extend the cross-derivative test to vector fields in space: show that $\mathbf{\Phi}(x, y, z) = a(x, y, z)\mathbf{i} + b(x, y, z)\mathbf{j} + c(x, y, z)\mathbf{k}$ is conservative if and only if $c_y = b_z$, $c_x = a_z$ and $a_y = b_x$.

Use the cross-derivative test in Exercise 30 to determine whether each of the vector fields in Exercises 31–34 is conservative. If it is, find an antiderivative.

★31. $2xy\mathbf{i} + (x^2 + z^2)\mathbf{j} + y\mathbf{k}$
★32. $xy\mathbf{i} + yz\mathbf{j} + xz\mathbf{k}$
★33. $e^{yz}\mathbf{i} + xze^{yz}\mathbf{j} + xye^{yz}\mathbf{k}$
★34. $\cos(xy)\mathbf{i} + yx\mathbf{j} - \sin(yz)\mathbf{k}$

★35. Show that any two antiderivatives of a vector field in the plane or space differ by a constant.
★36. Show that only the zero vector field is *totally path independent* in the sense that its integrals over all paths, even with different endpoints, are equal.
★37. *Two-color problem*. Several intersecting circles are drawn in the plane. Show that the resulting "map" can be colored with two colors in such a way that adjacent regions have different colors (as in Fig. 18.2.5). [*Hint*: First show that every closed curve crosses the union of the circles an even number of times. Then divide the regions into two classes according to whether an arc from the region to a fixed point crosses the circles an even or odd number of times. Compare your argument with the proof of the basic properties of conservative vector fields.]

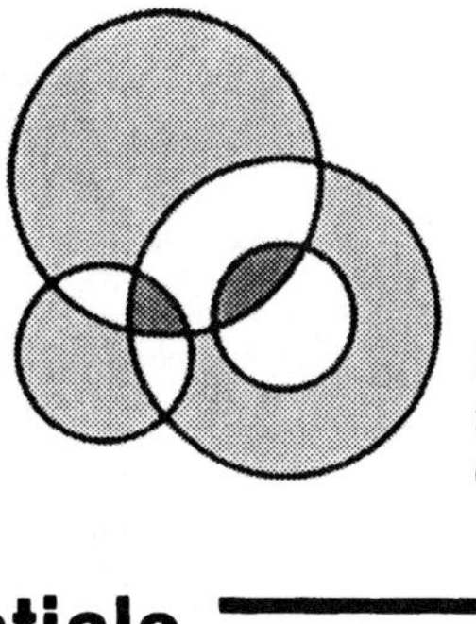

Figure 18.2.5. Adjacent regions have opposite colors.

18.3 Exact Differentials

Gradient vector fields correspond to exact differentials.

In the preceding section, we established the cross-derivative test: a vector field $\mathbf{\Phi}(x, y) = a(x, y)\mathbf{i} + b(x, y)\mathbf{j}$ defined on the whole plane is the gradient of some function if and only if $a_y = b_x$. In this section, we shall use this result to solve a class of differential equations called exact equations. In doing this, it is convenient to use the notation of differential forms, so we begin by summarizing this notation.

Just as the differential notation $\int f(x)\,dx$ was convenient for functions of one variable, we have seen that the notation

$$\int_C \mathbf{\Phi}(x, y) \cdot d\mathbf{r} = \int_C a(x, y)\,dx + b(x, y)\,dy$$

is a good one for the line integral of a vector field $\mathbf{\Phi}(x, y) = a(x, y)\mathbf{i} + b(x, y)\mathbf{j}$. The expression $a(x, y)\,dx + b(x, y)\,dy$ is called a *differential form*; such an expression is often written as $P\,dx + Q\,dy$, where $P = a(x, y)$ and $Q = b(x, y)$. The differential form $P\,dx + Q\,dy$ is called *exact* if there is a function $u = f(x, y)$ such that $P = \partial u/\partial x$ and $Q = \partial u/\partial y$, so $P\,dx + Q\,dy = (\partial u/\partial x)\,dx + (\partial u/\partial y)\,dy$. We call $u = f(x, y)$ an *antiderivative* of the differential form. Using this notation, we can rewrite the cross-derivative test as follows.

Cross-Derivative Test for Differential Forms

Let $P\,dx + Q\,dy$ be a differential form defined in the plane. If P and Q have continuous partials and $\partial P/\partial y = \partial Q/\partial x$, then $P\,dx + Q\,dy$ is exact—that is, there is a function $u = f(x, y)$ such that $P = \partial u/\partial x$ and $Q = \partial u/\partial y$. (In other words, the vector field $P\mathbf{i} + Q\mathbf{j}$ is the gradient of f.)

Given a differential form $P\,dx + Q\,dy$ satisfying $\partial P/\partial y = \partial Q/\partial x$, there are two ways of finding $u = f(x, y)$. The first is the method we used in Example 5, Section 18.2.

Method 1. If $P = \partial u/\partial x$, integrate to give $u = \int P\,dx + g(y)$, where $g(y)$ is a function of y to be determined. Differentiate u with respect to y and equate your answer to Q. This will give an equation for $g'(y)$ which may be integrated to yield $g(y)$ and hence u.

Method 2. If C is a curve from (x_0, y_0) to (x, y) and if $\mathbf{\Phi}(x, y) = \nabla f(x, y)$ then from p. 896,

$$f(x, y) - f(x_0, y_0) = \int_C \mathbf{\Phi} \cdot d\mathbf{r}.$$

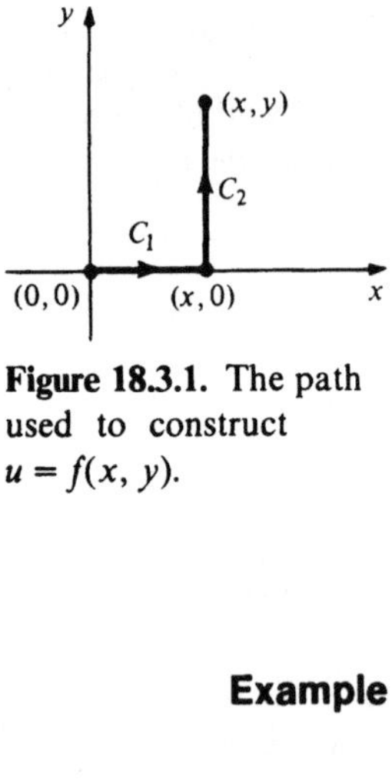

Figure 18.3.1. The path used to construct $u = f(x, y)$.

Choose C to be the curve from $(0,0)$ to (x, y) shown in Figure 18.3.1. We can also adjust f by a constant so that $f(0, 0) = 0$. This leads to the explicit formula

$$f(x, y) = \int_0^x a(t, 0)\,dt + \int_0^y b(x, t)\,dt \tag{1}$$

for $u = f(x, y)$, where $P = a(x, y)$ and $Q = b(x, y)$. In (1), $\int_0^x a(t, 0)\,dt$ is the integral of $P\,dx + Q\,dy$ along C_1, while $\int_0^y b(x, t)\,dt$ is the integral along C_2.

Example 1 Is $(2xy\cos y + x^2)\,dx + (x^2\cos y - x^2y\sin y)\,dy$ an exact differential? If so, find an antiderivative; that is, find u such that $\partial u/\partial x = 2xy\cos y + x^2$ and $\partial u/\partial y = x^2\cos y - x^2y\sin y$.

Solution To test for exactness, we let $P = 2xy\cos y + x^2$ and $Q = x^2\cos y - x^2y\sin y$ and compute that

$$\frac{\partial P}{\partial y} = -2xy\sin y + 2x\cos y = \frac{\partial Q}{\partial x},$$

and so a u exists. To find u, we can use either of the two methods above.

Method 1. We must have

$$\frac{\partial u}{\partial x} = P \quad \text{and} \quad \frac{\partial u}{\partial y} = Q,$$

Hence

$$\frac{\partial u}{\partial x} = 2xy\cos y + x^2 \quad \text{and} \quad \frac{\partial u}{\partial y} = x^2\cos y - x^2y\sin y.$$

Integrating the first equation with respect to x gives $u = x^2y\cos y + x^3/3 + g(y)$; the "constant" of integration g must be a function of y alone. Differentiating with respect to y gives $\partial u/\partial y = x^2\cos y - x^2y\sin y + g'(y)$. Since $\partial u/\partial y = x^2\cos y - x^2y\sin y$, $g'(y) = 0$, and so we may take $g(y) = 0$. Thus $u = x^3/3 + x^2y\cos y$.

Method 2. We use formula (1) above:

$$f(x, y) = \int_0^x a(t, 0)\,dt + \int_0^y b(x, t)\,dt,$$

where $P = a(x, y)$ and $Q = b(x, y)$. In this case, $a(x, y) = 2xy\cos y + x^2$ and $b(x, y) = x^2\cos y - x^2y\sin y$, so we get

$$f(x, y) = \int_0^x t^2\,dt + \int_0^y (x^2\cos t - x^2t\sin t)\,dt.$$

Evaluating the last integral by integrating by parts gives

$$\begin{aligned} u = f(x, y) &= \frac{x^3}{3} + \int_0^y x^2\cos t\,dt - \int_0^y x^2\cos t\,dt + x^2t\cos t\Big|_0^y \\ &= \frac{x^3}{3} + x^2y\cos y, \end{aligned}$$

which agrees with the answer using Method 1. We check that $\partial u/\partial x = P$ and $\partial u/\partial y = Q$, as required. ▲

Exact Differentials

To determine whether there is a function $u = f(x, y)$ such that $\partial u/\partial x = P = a(x, y)$ and $\partial u/\partial y = Q = b(x, y)$, check whether $\partial P/\partial y = \partial Q/\partial x$.

If so, f may be constructed by formula (1) or by integrating: $u = \int P\,dx + g(y)$; equating $\partial u/\partial y$ and Q determines $g'(y)$ and hence, by integration, $g(y)$.

If $\partial P/\partial y \neq \partial Q/\partial x$, no such f can exist.

Example 2 Show that, in place of formula (1), we can also choose

$$f(x, y) = \int_0^1 \left[xa(tx, ty) + yb(tx, ty) \right] dt.$$

Solution The function f is the integral of $P\,dx + Q\,dy$ along any path from $(0,0)$ to (x, y). (In the proof, we used paths parallel to the axes.) The expression $\int_0^1[xa(tx, ty) + yb(tx, ty)]\,dt$ is just the integral of $P\,dx + Q\,dy$ along the path (tx, ty), $0 \leqslant t \leqslant 1$; that is, the straight line segment joining $(0,0)$ to (x, y). ▲

In Chapters 8 and 12, we studied a number of useful classes of differential equations. A new class may be solved by the methods of this section. A differential equation of the form

$$P(x, y) + Q(x, y)\frac{dy}{dx} = 0 \tag{2}$$

will be called *exact* if the corresponding differential form $P\,dx + Q\,dy$ is exact.

These equations may be solved as follows: there is a function $f(x, y)$ such that $P = f_x$ and $Q = f_y$. The curves

$$f(x, y) = C \tag{3}$$

for C a constant are solutions of (2) assuming that they define y as a function of x. Indeed, differentiating (3) implicitly by the chain rule, we get

$$f_x + f_y \frac{dy}{dx} = 0, \quad \text{i.e.,} \quad P + Q\frac{dy}{dx} = 0.$$

Exact Differential Equations

To test if the equation

$$P + Q\frac{dy}{dx} = 0$$

is exact, see if

$$\frac{\partial P}{\partial y} = \frac{\partial Q}{\partial x}.$$

To solve an exact equation $P + Q(dy/dx) = 0$, find $f(x, y)$ such that

$$P = f_x \quad \text{and} \quad Q = f_y.$$

(See the preceding box.) The solutions are $f(x, y) = C$ for any constant C.

Example 3 Find the solution of the differential equation

$$2xy\cos y + x^2 + (x^2\cos y - x^2y\sin y)\frac{dy}{dx} = 0$$

that passes through $(1,0)$.

Solution By Example 1, this equation is exact with

$$f(x, y) = \frac{x^3}{3} + x^2y\cos y.$$

The solution is thus $(x^3/3) + x^2y\cos y = C$. Since $y = 0$ when $x = 1$, $C = \frac{1}{3}$. Thus our solution is given implicitly by

$$\frac{x^3}{3} + x^2y\cos y = \frac{1}{3}. \ \blacktriangle$$

Example 4 (a) Let $(x, y) = (e^{t-1}, \sin(\pi/t))$, $1 \leqslant t \leqslant 2$, be a parametrization of the curve C. Calculate $\int_C 2x\cos y\,dx - x^2\sin y\,dy$.

(b) Find the solution of $2x\cos y = x^2\sin y(dy/dx)$ that satisfies $y(3) = 0$.

Solution (a) The curve C goes from $(1,0)$ to $(e, 1)$. Since $\partial(2x\cos y)/\partial y = -2x\sin y = \partial(-x^2\sin y)/\partial x$, the integrand is exact. Thus we can replace C by any curve having the same endpoints, in particular by the polygonal path from $(1,0)$ to $(e,0)$ to $(e,1)$. Thus the line integral must be equal to

$$\int_1^e 2t\cos 0\,dt + \int_0^1 -e^2\sin t\,dt = (e^2 - 1) + e^2(\cos 1 - 1)$$
$$= e^2\cos 1 - 1.$$

Alternatively, using the antiderivative $u = f(x, y) = x^2\cos y$, which may be found by the methods of Example 1,

$$\int_C 2x\cos y\,dx - x^2\sin y\,dy = \int_C \nabla f(\mathbf{r})\cdot d\mathbf{r} = f(e,1) - f(1,0)$$

$$= e^2\cos 1 - 1.$$

(b) The general solution is $x^2\cos y = C$. Since $y = 0$ when $x = 3$ we have $x^2\cos y = 9$; that is, $y = \cos^{-1}(9/x^2)$. ▲

Example 5 (a) Verify that $\partial P/\partial y = \partial Q/\partial x$ and find $u = f(x, y)$ such that $P = \partial u/\partial x$ and $Q = \partial u/\partial y$ in each case:

(i) $P = x^2 + y^3$; $Q = 3xy^2 + 1$.

(ii) $P = y\cos x$; $Q = y + \sin x$.

(b) Solve the following differential equations with the given conditions:

(i) $x^2 + y^3 + (3xy^2 + 1)dy/dx = 0$; $y(0) = 1$.

(ii) $y\cos x + (y + \sin x)dy/dx = 0$; $y(\pi/4) = 1$.

Solution (a) (i) We have $P_y = 3y^2$ and $Q_x = 3y^2$, and so $P_y = Q_x$. Let $f(x, y)$ be such that $f_x = x^2 + y^3$ and $f_y = 3xy^2 + 1$. Then $f(x, y) = \int f_x\,dx = x^3/3 + xy^3 + g(y)$. Then $f_y(x, y) = 3xy^2 + g'(y) = 3xy^2 + 1$ implies $g'(y) = 1$, so $g(y) = y + C$. Thus, $f(x, y) = x^3/3 + xy^3 + y + C$.

(ii) We have $\partial P/\partial y = \cos x$ and $\partial Q/\partial x = \cos x$; thus $\partial P/\partial y = \partial Q/\partial x$. Integrating, $f(x, y) = \int f_x\,dx = y\sin x + g(y)$; $f_y = \sin x + g'(y) = \sin x + y$ implies $g'(y) = y$ or $g(y) = y^2/2 + C$. We thus have $f(x, y) = y\sin x + y^2/2 + C$.

(b) (i) The differential equation $x^2 + y^3 + (3xy^2 + 1)(dy/dx) = 0$ is exact by our calculation in part (a)(i). From that calculation we find that $f(x, y) = x^3/3 + y^3x + y = K$ is a solution. Applying the condition $y(0) = 1$ gives $K = 1$. Thus the solution is $x^3/3 + y^3x + y = 1$.

(ii) Similarly, using the result of part (a)(ii), we get the solution $f(x, y) = y\sin x + y^2/2 = K$. Applying the condition that $y(\pi/4) = 1$ gives $K = (\sqrt{2} + 1)/2$. Thus the solution is $y\sin x + y^2/2 = (\sqrt{2} + 1)/2$. ▲

A differential equation

$$M + N\frac{dy}{dx} = 0 \tag{4}$$

which is not exact may sometimes be made exact if we multiply it through by a function $\mu(x, y)$. The equivalent equation is $\mu M + \mu N\,dy/dx = 0$, which is exact if $\partial(\mu M)/\partial y = \partial(\mu N)/\partial x$. This is a condition on μ, which can be written as

$$\mu_y M + \mu M_y = \mu_x N + \mu N_x\,. \tag{5}$$

Any function $\mu(x, y)$ satisfying equation (5) is called an *integrating factor* for the original equation (4).

In general, it is not easy to solve (5), but occasionally it is possible to solve it with a μ which is a function of x alone, i.e., $\mu(x, y) = \mu(x)$, in which case $\mu_x = \mu'$ and $\mu_y = 0$. In this case (5) becomes $\mu M_y = \mu' N + \mu N_x$ or

$$\frac{M_y - N_x}{N} = \frac{\mu'}{\mu} = (\ln\mu)'.$$

This can be solved for μ if $(M_y - N_x)/N$ is a function of x alone, in which case $\ln \mu = \int[(M_y - N_x)/N]\,dx$, and so the integrating factor is

$$\mu = \exp\left[\int \frac{M_y - N_x}{N}\,dx\right].$$

Having found an integrating factor, one may now solve the exact equation $\mu M + \mu N\,dy/dx = 0$ by the methods described earlier in this section. The following example shows how linear equations may be solved by this method (compare this approach with the methods of Section 8.6).

Example 6 (a) Solve the equation $x\,dy/dx - x^5 + x^3y - y = 0$.
(b) Solve the linear equation $dy/dx + P(x)y + Q(x) = 0$ using the method of integrating factors.

Solution (a) Comparing the equation with (4), we see that $M = -x^5 + x^3y - y$ and $N = x$. Thus $M_y = x^3 - 1$ and $N_x = 1$. Hence $(M_y - N_x)/N = (x^3 - 1 - 1)/x = x^2 - 2/x$. Therefore, the integrating factor is $\mu = \exp[\int(x^2 - 2/x)\,dx] = \exp(x^3/3 - 2\ln x) = \exp(x^3/3)/x^2$. According to formula (1) applied to $\mu M\,dx + \mu N\,dy$,

$$\begin{aligned} F(x,y) &= \int_0^x a(t,0)\,dt + \int_0^y b(x,t)\,dt \\ &= \int_0^x \frac{\exp(t^3/3)}{t^2}(-t^5)\,dt + \int_0^y \frac{\exp(x^3/3)}{x^2}\,x\,dt \\ &= -\int_0^x t^3\exp\left(\frac{t^3}{3}\right)dt + \frac{y}{x}\exp(x^3/3) = C. \end{aligned}$$

If we rearrange and use the indefinite integral notation

$$\int_0^x t^3\exp\left(\frac{t^3}{3}\right)dt + C = \int x^3\exp\left(\frac{x^3}{3}\right)dx,$$

we get

$$y = x\exp\left(\frac{-x^3}{3}\right)\int x^3\exp\left(\frac{x^3}{3}\right)dx.$$

(b) Here, $M = P(x)y + Q(x)$, and so $M_y = P(x)$; $N = 1$, so $N_x = 0$. Thus $(M_y - N_x)/N = P(x)$. Therefore, the integrating factor is $\mu = \exp(\int P(x)\,dx)$. Applying (1) to $\mu M\,dx + \mu N\,dy$, we find that the solution is

$$\begin{aligned} f(x,y) &= \int_0^x a(t,0)\,dt + \int_0^y b(x,t)\,dt \\ &= \int_0^x \exp\left(\int P(t)\,dt\right)Q(t)\,dt + \int_0^y \exp\left(\int P(x)\,dx\right)dt \\ &= \int \exp\left(\int P(x)\,dx\right)Q(x)\,dx + y\exp\left(\int P(x)\,dx\right) = C. \end{aligned}$$

Absorbing the constant into the first integration and solving for y gives

$$y = -e^{-\int P(x)\,dx}\left[\int Q(x)e^{\int P(x)\,dx}\,dx\right].$$

(The methods of Section 8.6 yield the same answer.) ▲

Exercises for Section 18.3

1. Is $[2x + e^{xy}(xy + 1)]dx + x^2e^{xy}\,dy$ an exact differential? If so, find an antiderivative.
2. Is $(\cos xy - xy\sin xy)\,dx - (x^2\sin xy)\,dy$ an exact differential? If so, find an antiderivative.
3. Is there a function $u = f(x, y)$ such that $\partial u/\partial x = x\sqrt{x^2y^2+1}$ and $\partial u/\partial y = y\sqrt{x^2y^2+1}$? If so, find it.
4. Is there a function u of (x, y) such that $\partial u/\partial x = 2x\cos y + \cos y$ and $\partial u/\partial y = -x^2\sin y - x\sin y$? If so, find it.

In Exercises 5–8, determine which differentials are exact.

5. $x^2y\,dx + (x^3y/3)\,dy$
6. $x^2\,dx + (x^3/3)\,dy$
7. $x^2y\,dx + (x^3/3)\,dy$
8. $xy^2\,dx + (y^3/3)\,dy$
9. Show that we can use
$$f(x, y) = \int_0^y b(0, t)\,dt + \int_0^x a(t, y)\,dt$$
in place of formula (1).
10. Use the path $(\sqrt{t}\,x, tx)$, $0 \leqslant t \leqslant 1$ to find another formula (other than those in Example 2 and Exercise 9) that may be used in place of (1).
11. Let C_1 be parametrized by $(t^2+1, t\sin(\pi t/8))$, $0 \leqslant t \leqslant 4$, and C_2 by $((t^2+1)\cos\pi t, t)$ $0 \leqslant t \leqslant 4$. Determine whether $\int_{C_1} P\,dx + Q\,dy = \int_{C_2} P\,dx + Q\,dy$, where $P\,dx + Q\,dy = \frac{1}{2}(x^4 + y^2)\,dx + (xy + e^y)\,dy$.
12. Let $P = \ln(x^2+1) - 2xe^{-y}$ and $Q = x^2e^{-y} - \ln(y^2+1)$. Determine whether $\int_{C_1} P\,dx + Q\,dy = \int_{C_2} P\,dx + Q\,dy$, where C_1 and C_2 are two curves with the same endpoints.

Solve the differential equations satisfying the given conditions in Exercises 13–16.

13. $ye^x + e^y + (xe^y + e^x)\,dy/dx = 0$, $y(0) = 2$.
14. $e^y + (xe^y + 2y)\,dy/dx = 0$, $y(0) = 1$.
15. $3x^2 + 2xy + (x^2 + y^2)\,dy/dx = 0$, $y(1) = 2$.
16. $\cos y\sin x + \sin y\cos x\,dy/dx = 0$, $y(\pi/2) = 1$.

Determine which of the equations in Exercises 17–20 are exact and solve the ones that are.

17. $x\,dy/dx + y + x^2y^2 - 1 = 0$.
18. $y - x^3 + (x + y^3)\,dy/dx = 0$.
19. $xy^2 + 3x^2y + (x + y)x^2\,dy/dx = 0$.
20. $e^x\sin y - e^y\sin x\,dy/dx = 0$.
21. (a) Let $(x, y) = (t^3 - 1, t^6 - t)$, $0 \leqslant t \leqslant 1$ parametrize the curve C. Calculate
$$\int_C\left(\frac{2x}{y^2+1}\,dx - \frac{2y(x^2+1)}{(y^2+1)^2}\,dy\right).$$
(b) Find the solution of
$$\frac{x}{y} = \frac{x^2+1}{y^2+1}\frac{dy}{dx}$$
that satisfies $y(1) = 1$.
22. (a) Let $(x, y) = (e^t, e^{t+1})$, $-1 \leqslant t \leqslant 0$ be a parametrization of the curve C. Calculate
$$\int_C\left[\cos(xy^2) - xy^2\sin(xy^2)\right]dx - 2x^2y\sin(xy^2)\,dy.$$
(b) Find the solution of
$$\cos(xy^2) - xy^2\sin(xy^2) = 2x^2y\sin(xy^2)\frac{dy}{dx}$$
that satisfies $y(1) = 0$.
23. Solve the equation in Exercise 21(b) using the method of separation of variables (Section 8.5).
24. Solve the equation $dy/dx = (1 - y)/(x + 1)$, $y(0) = 2$ using (a) the method of exact equations and (b) separation of variables. Verify that the two answers agree.
25. Find an integrating factor $\mu(x)$ for the equation
$$2y\cos y + x + (x\cos y - xy\sin y)\frac{dy}{dx} = 0.$$
26. Solve the equation $dy/dx = 3x^2y + x$ by finding an integrating factor. (Leave your answer in the form of an integral.)
27. Solve the equation $x\,dy/dx = xy^2 + y$ by using the integrating factor $1/y^2$.
28. Use the integrating factor xy to solve
$$\frac{1}{x} + \frac{1}{y} + \left(\frac{y}{x} + \frac{1}{y}\right)\frac{dy}{dx} = 0.$$
29. Find the equation which must be satisfied by a function μ of y alone if it is to be an integrating factor for $M + N\,dy/dx = 0$.
30. (a) Find a solution of the equation $3x^2y^2 + 2(x^3y + x^3y^3)\,dy/dx = 0$ in the form $u(x, y) = \text{constant}$.
(b) If $y = f(x)$ is a solution of the equation with $f(1) = 1$, what is $f(2)$?
★31. If $P\,dx + Q\,dy$ satisfies $\partial P/\partial y = \partial Q/\partial x$, show that functions $g(y)$ and $h(x)$ can always be found such that $\int P\,dx + g(y) = \int Q\,dy + h(x)$.
★32. Generalize the assertion in Exercise 31 to $P\,dx + Q\,dy + R\,dz$.

18.4 Green's Theorem

A line integral along a closed curve can be converted to a double integral over a region.

Green's[2] theorem relates the line integral of a vector field (or differential form) around a closed curve to the double integral of a certain function over the region bounded by the curve. Among the many applications of this theorem is a formula for the area of a region in terms of a line integral around its boundary.

Green's theorem states that

$$\int_C P\,dx + Q\,dy = \int\int_D \left(\frac{\partial Q}{\partial x} - \frac{\partial P}{\partial y}\right) dx\,dy, \tag{1}$$

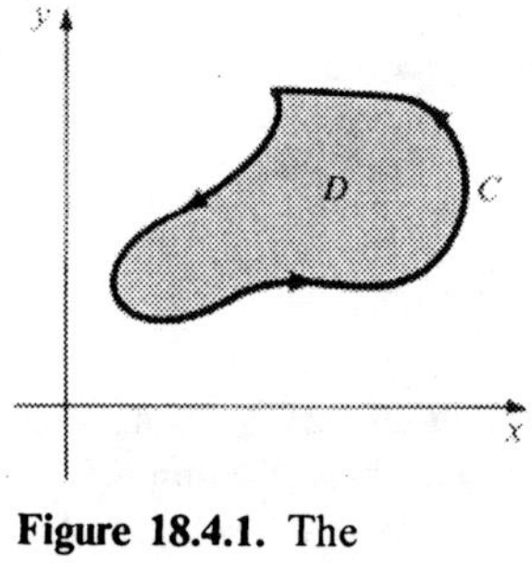

Figure 18.4.1. The boundary C of a region D must be oriented counterclockwise in Green's theorem.

where C is the curve bounding a region D, C is traversed counterclockwise (see Fig. 18.4.1), and P and Q have continuous partial derivatives. We have seen that $P\,dx + Q\,dy$ is exact when $\partial Q/\partial x = \partial P/\partial y$; in this case, Green's theorem is obviously true because each side of (1) is zero.

We shall prove Green's theorem for regions which are of type 1 and 2 and then indicate through examples how the result may be proved for more general regions.

Consider a region of type 1 in the plane, as shown in Fig. 18.4.2. The *boundary* of this region is defined to be the closed curve C which goes once

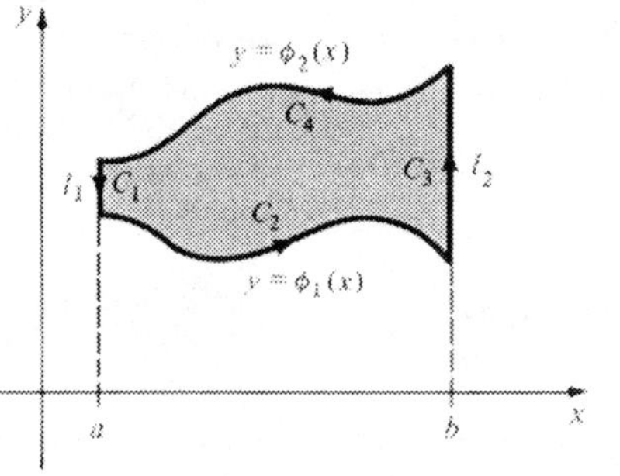

Figure 18.4.2. The boundary of this type 1 region consists of the graphs $y = \phi_1(x)$ and $y = \phi_2(x)$ and segments of the lines l_1 and l_2.

around the region in the counterclockwise direction. If we start at the upper left-hand corner, the boundary curve first traverses the vertical line l_1 (call this C_1), then goes along the graph of ϕ_1 (call this C_2), then up l_2 (call this C_3), and, finally, backwards along the graph of ϕ_2 (call this C_4). The following lemma is a preliminary form of Green's theorem.

Lemma *Let D be a type 1 region as above, with C its bounding curve. Let $P = f(x, y)$ have continuous partial derivatives in D and on C. Then*

$$\int_C P\,dx = -\int\int_D \frac{\partial P}{\partial y}\,dx\,dy. \tag{2}$$

[2] George Green (1793–1841), an English mathematician and physicist, was one of the early investigators of electricity and magnetism. His work on potentials led to what is commonly called Green's theorem (although it may also be due to Cauchy). This theorem, as generalized to three dimensions by Lord Kelvin, Stokes, Gauss, and Ostrogradsky, was crucial to later developments in electromagnetic, gravitational, and other physical theories. Some of these applications are discussed later in this chapter. For more history, see "The history of Stokes' theorem" by V. J. Katz, Mathematics Magazine, **52** (1979) 146–156.

Proof Since the double integral may be evaluated as an iterated integral (see p. 850), we have

$$\int\int_D \frac{\partial P}{\partial y}\,dx\,dy = \int_a^b \int_{\phi_1(x)}^{\phi_2(x)} f_y(x, y)\,dy\,dx$$

$$= \int_a^b \left[f(x, \phi_2(x)) - f(x, \phi_1(x)) \right] dx.$$

The latter equality uses the fundamental theorem of calculus.

To compute the line integral, we parametrize each of the segments of C:

C_1: $(x, y) = (a, -t)$; t in $[-\phi_2(a), -\phi_1(a)]$; $dx = 0$, $dy = -dt$.
C_2: $(x, y) = (t, \phi_1(t))$; t in $[a, b]$; $dx = dt$, $dy = \phi_1'(t)\,dt$.
C_3: $(x, y) = (b, t)$; t in $[\phi_1(b), \phi_2(b)]$; $dx = 0$, $dy = dt$.
C_4: $(x, y) = (-t, \phi_2(-t))$; t in $[-b, -a]$; $dx = -dt$, $dy = -\phi_2'(-t)\,dt$.

Now $\int_C P\,dx$ is the sum of the line integrals over the four C_i's. The integrals over C_1 and C_3 are zero, since $dx = 0$ on those curves (x is constant). The integrals over C_2 and C_4 are given by

$$\int_{C_2} P\,dx = \int_a^b f(t, \phi_1(t))\,dt$$

and

$$\int_{C_4} P\,dx = \int_{-b}^{-a} f(-t, \phi_2(-t))(-dt)$$

$$= \int_b^a f(t, \phi_2(t))\,dt \qquad \text{(substituting } -t \text{ for } t\text{)}$$

$$= -\int_a^b f(t, \phi_2(t))\,dt.$$

Thus

$$\int_C P\,dx = \int_a^b f(t, \phi_1(t))\,dt - \int_a^b f(t, \phi_2(t))\,dt$$

$$= -\int_a^b \left[f(t, \phi_2(t)) - f(t, \phi_1(t)) \right] dt$$

$$= -\int\int_D \frac{\partial P}{\partial y}\,dx\,dy. \blacksquare$$

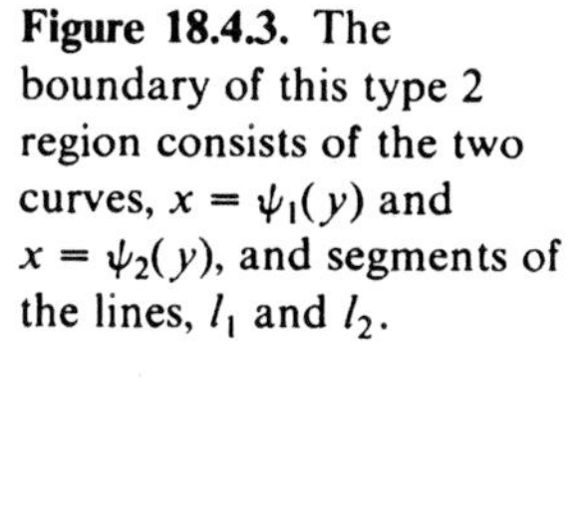

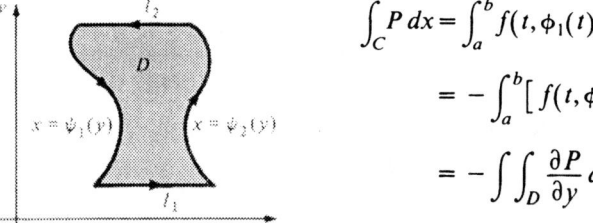

Figure 18.4.3. The boundary of this type 2 region consists of the two curves, $x = \psi_1(y)$ and $x = \psi_2(y)$, and segments of the lines, l_1 and l_2.

In exactly the same way, we can prove that if D is a type 2 region with boundary curve C traversed counterclockwise (Fig. 18.4.3), then for $Q = g(x, y)$ with continuous partial derivatives,

$$\int_C Q\,dy = \int\int_D \frac{\partial Q}{\partial x}\,dx\,dy. \tag{3}$$

If we have a region which is of both types 1 and 2, then equations (2) and (3) are both valid. Adding them yields formula (1).

Green's Theorem *If D is a region of types 1 and 2 with boundary curve C traversed counterclockwise, and if $P = f(x, y)$ and $Q = g(x, y)$ have continuous partial derivatives in D and on C, then*

$$\int_C P\,dx + Q\,dy = \int\int_D \left(\frac{\partial Q}{\partial x} - \frac{\partial P}{\partial y} \right) dx\,dy.$$

Example 1 Verify Green's theorem for $P = x$ and $Q = xy$, where D is the unit disk $x^2 + y^2 \leqslant 1$.

Solution We can evaluate both sides in Green's theorem directly. The boundary C of D is the unit circle parametrized by $x = \cos t$, $y = \sin t$, $0 \leqslant t \leqslant 2\pi$, so

$$\int_C P\,dx + Q\,dy = \int_0^{2\pi}\left[(\cos t)(-\sin t) + \cos t \sin t \cos t\right] dt$$

$$= \left[\frac{\cos^2 t}{2}\right]_0^{2\pi} + \left[-\frac{\cos^3 t}{3}\right]_0^{2\pi} = 0.$$

On the other hand,

$$\int\int_D\left(\frac{\partial Q}{\partial x} - \frac{\partial P}{\partial y}\right) dx\,dy = \int\int_D y\,dx\,dy,$$

which is zero also, since the contributions from the upper and lower half-circles cancel one another. Thus Green's theorem is verified in this case. ▲

Green's theorem applies as well to many regions other than just those of types 1 and 2. Often one can show this by dividing up the region, as in the following example.

Example 2 Show that Green's theorem is valid for the region D shown in Fig. 18.4.4.

Solution Figure 18.4.5 shows how to divide up D into three regions, D_1, D_2, D_3, each of which is of types 1 and 2. Let C_1, C_2, C_3 be the boundary curves of these regions. Then

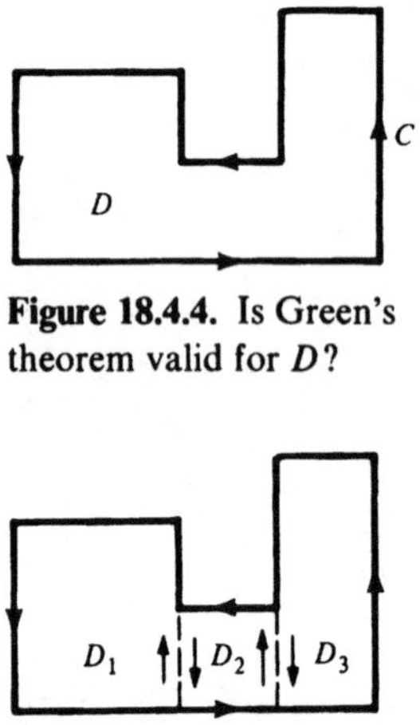

Figure 18.4.4. Is Green's theorem valid for D?

$$\int\int_{D_i}\left(\frac{\partial Q}{\partial x} - \frac{\partial P}{\partial y}\right) dx\,dy = \int_{C_i} P\,dx + Q\,dy.$$

The double integral over the D_i's adds up to the double integral over D, so

$$\int\int_D\left(\frac{\partial Q}{\partial x} - \frac{\partial P}{\partial y}\right) dx\,dy = \int_{C_1} P\,dx + Q\,dy + \int_{C_2} P\,dx + Q\,dy + \int_{C_3} P\,dx + Q\,dy.$$

However, the dotted portions of the boundaries shown in Fig. 18.4.5 are traversed twice in opposite directions; these cancel in the line integrals. Thus we are left with

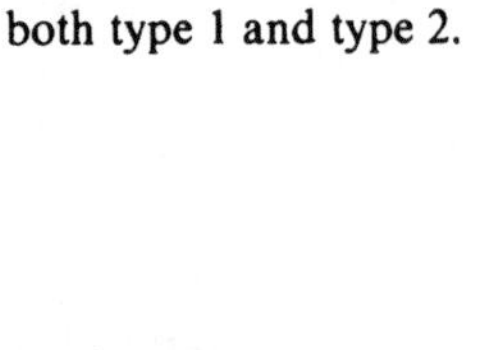

Figure 18.4.5. Breaking a region up into smaller regions, each of which is both type 1 and type 2.

$$\int\int_D\left(\frac{\partial Q}{\partial x} - \frac{\partial P}{\partial y}\right) dx\,dy = \int_C P\,dx + Q\,dy,$$

and so Green's theorem is valid. ▲

Example 2 illustrates a special case of the following procedure for a region D:

(a) Break up D into smaller regions, $D_1, D_2, \ldots, D_n$, each of which is of types 1 and 2.
(b) Apply Green's theorem as proven above to each of $D_1, \ldots, D_n$, and add the resulting integrals.
(c) The line integrals along interior boundaries cancel, leaving the line integral around the boundary of D.

This procedure yields Green's theorem for D. It is plausible that this method applies to any region bounded by piecewise smooth curves, and so we may expect a general form of Green's theorem.

Green's Theorem

If D is a region and C is the boundary of D, oriented as in Fig. 18.4.1, then

$$\int_C P\,dx + Q\,dy = \int\int_D \left(\frac{\partial Q}{\partial x} - \frac{\partial P}{\partial y} \right) dx\,dy.$$

Example 3 Let $\mathbf{\Phi}(x, y) = y\mathbf{i} - x\mathbf{j}$ and let C be a circle of radius r traversed counterclockwise. Write $\int_C \mathbf{\Phi}(\mathbf{r}) \cdot d\mathbf{r}$ as a double integral using Green's theorem. Evaluate.

Solution If $\mathbf{\Phi}(x, y) = P\mathbf{i} + Q\mathbf{j}$, then $\mathbf{\Phi}(\mathbf{r}) \cdot d\mathbf{r} = P\,dx + Q\,dy$. Now we apply Green's theorem to the case where D is the disk of radius r, $Q = -x$, and $P = y$, so $\partial Q/\partial x - \partial P/\partial y = -2$. Thus $\int_C \mathbf{\Phi}(\mathbf{r}) \cdot d\mathbf{r} = \int\int_D (-2)\,dx\,dy = (-2)(\text{area of } D) = -2\pi r^2$. ▲

Example 4 Let C be the boundary of the square $[0, 1] \times [0, 1]$ oriented counterclockwise. Evaluate

$$\int_C (y^4 + x^3)\,dx + 2x^6\,dy.$$

Solution We could, of course, evaluate the integral directly, but it is easier to use Green's theorem. Let D be the unit square (bounded by the lines $x = 0$, $y = 0$, $x = 1$, and $y = 1$). Then

$$\begin{aligned}\int_C (y^4 + x^3)\,dx + 2x^6\,dy &= \int\int_D \left[\frac{\partial}{\partial x} 2x^6 - \frac{\partial}{\partial y}(y^4 + x^3) \right] dx\,dy \\ &= \int\int_D (12x^5 - 4y^3)\,dx\,dy \\ &= \int_0^1 \left[\int_0^1 (12x^5 - 4y^3)\,dx \right] dy \\ &= \int_0^1 (2 - 4y^3)\,dy = 1. \text{ ▲}\end{aligned}$$

Example 5 Show that if C is the boundary of D, then

$$\int_C PQ\,dx + PQ\,dy = \int\int_D \left[Q\left(\frac{\partial P}{\partial x} - \frac{\partial P}{\partial y} \right) + P\left(\frac{\partial Q}{\partial x} - \frac{\partial Q}{\partial y} \right) \right] dx\,dy.$$

Solution By Green's theorem,

$$\begin{aligned}\int_C PQ\,dx + PQ\,dy &= \int\int_D \left[\frac{\partial}{\partial x}(PQ) - \frac{\partial}{\partial y}(PQ) \right] dx\,dy \\ &= \int\int_D \left(\frac{\partial P}{\partial x} Q + P\frac{\partial Q}{\partial x} - \frac{\partial P}{\partial y} Q - P\frac{\partial Q}{\partial y} \right) dx\,dy \\ &= \int\int_D \left[Q\left(\frac{\partial P}{\partial x} - \frac{\partial P}{\partial y} \right) + P\left(\frac{\partial Q}{\partial x} - \frac{\partial Q}{\partial y} \right) \right] dx\,dy. \text{ ▲}\end{aligned}$$

We can use Green's theorem to obtain a formula for the *area of a region bounded by a curve C*.

Corollary: Area of a Region

If C is a curve that bounds a region D, then the area of D is

$$A = \frac{1}{2}\int_C x\,dy - y\,dx. \tag{4}$$

The proof is as follows. Let $P = -y$, $Q = x$; then by Green's theorem we have

$$\frac{1}{2}\int_C x\,dy - y\,dx = \frac{1}{2}\int\int_D\left(\frac{\partial x}{\partial x} - \frac{\partial(-y)}{\partial y}\right)dx\,dy = \int\int_D dx\,dy$$

which is the area of D. ■

Example 6 Verify formula (4) in the case where D is the disk $x^2 + y^2 \leqslant r^2$.

Solution The area is πr^2. Formula (4) with $x = r\cos t$, $y = r\sin t$, $0 \leqslant t \leqslant 2\pi$, gives

$$A = \frac{1}{2}\int_C x\,dy - y\,dx = \frac{1}{2}\int_0^{2\pi}(r\cos t)(r\cos t)\,dt - (r\sin t)(-r\sin t)\,dt$$

$$= \frac{1}{2}\int_0^{2\pi} r^2\,dt = \pi r^2,$$

so formula (4) checks. ▲

Example 7 Use formula (4) to find the area bounded by the ellipse C: $x^2/a^2 + y^2/b^2 = 1$.

Solution Parametrize C by $(a\cos t, b\sin t)$, $0 \leqslant t \leqslant 2\pi$. Then (4) gives

$$A = \frac{1}{2}\int_C x\,dy - y\,dx = \frac{1}{2}\int_0^{2\pi}(a\cos t)(b\cos t)\,dt - (b\sin t)(-a\sin t)\,dt$$

$$= \frac{1}{2}\int_0^{2\pi} ab\,dt = ab\pi. \text{ ▲}$$

Exercises for Section 18.4

1. Check the validity of Green's theorem for the region between the curves $y = x^2$ and $y = x$ between $x = 0$ and $x = 1$, with $P = xy$ and $Q = x$.
2. Verify Green's theorem when D is the disk of radius r, center $(0,0)$, and $P = xy^2$, $Q = -yx^2$.
3. Verify Green's theorem for $P = 2y$, $Q = x$, and D the unit disk $x^2 + y^2 \leqslant 1$.
4. Verify Green's Theorem for $P = \cos(xy^2) - xy^2\sin(xy^2)$, $Q = -2x^2y\sin(xy^2)$, and D the ellipse $x^2/4 + y^2/9 \leqslant 1$. [Hint: do not evaluate the line integral directly.]

Exercises 5–8 refer to Fig. 18.4.6. Show how to decompose each region into subregions, each of which is of types 1 and 2.

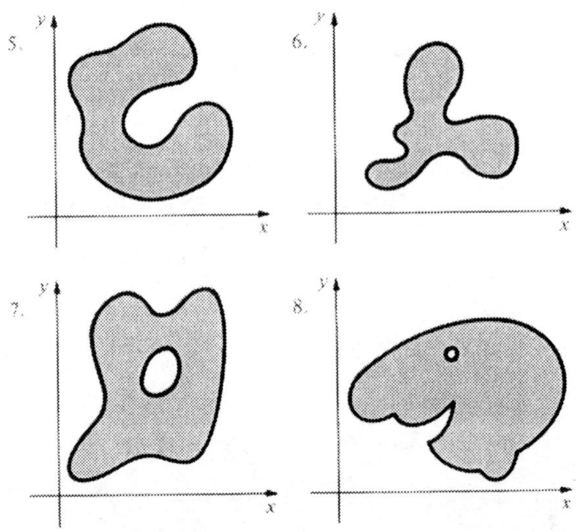

Figure 18.4.6. Subdivide each of these regions into subregions of types 1 and 2.

9. Let C be the ellipse $x^2/a^2 + y^2/b^2 = 1$, and let $\mathbf{\Phi}(x, y) = xy^2\mathbf{i} - yx^2\mathbf{j}$. Write $\int_C \mathbf{\Phi}(\mathbf{r}) \cdot d\mathbf{r}$ as a double integral using Green's theorem. Evaluate.
10. Let $\mathbf{\Phi}(x, y) = (2y + e^x)\mathbf{i} + (x + \sin(y^2))\mathbf{j}$ and C be the circle $x^2 + y^2 = 1$. Write $\int_C \mathbf{\Phi}(\mathbf{r}) \cdot d\mathbf{r}$ as a double integral and evaluate.
11. Let C be the boundary of the rectangle $[1,2] \times [1,2]$. Evaluate $\int_C x^2y\,dx + 3yx^2\,dy$ by using Green's theorem.
12. Evaluate $\int_C (x^5 - 2xy^3)\,dx - 3x^2y^2\,dy$, where C is parametrized by (t^8, t^{10}), $0 \leqslant t \leqslant 1$.

Let C be the boundary of the rectangle with sides $x = 1$, $y = 2$, $x = 3$, and $y = 3$. Evaluate the integrals in Exercises 13–16.

13. $\displaystyle\int_C (2y^2 + x^5)\,dx + 3y^6\,dy.$
14. $\displaystyle\int_C (xy^2 - y^3)\,dx + (-5x^2 + y^3)\,dy.$
15. $\displaystyle\int_C (3x^4 + 5)\,dx + (y^5 + 3y^2 - 1)\,dy.$
16. $\displaystyle\int_C \left[\frac{2y + \sin x}{1 + x^2}\right] dx + \left[\frac{x + e^y}{1 + y^2}\right] dy.$
17. Suppose that $P\mathbf{i} + Q\mathbf{j}$ is parallel to the tangent vector of a closed curve C.
    (a) Show that $Q\mathbf{i} - P\mathbf{j}$ is perpendicular to the tangent vector.
    (b) Show that $\int\int_D (\partial P/\partial x + \partial Q/\partial y)\,dx\,dy = 0$, where D is the region whose boundary is C.

We call $\nabla^2 u = \partial^2 u/\partial x^2 + \partial^2 u/\partial y^2$ the *Laplacian* of $u = f(x, y)$. Prove the identities in Exercises 18 and 19.

18. $\displaystyle\int\int_D u\,\nabla^2 v\,dx\,dy = -\int\int_D \nabla u \cdot \nabla v\,dx\,dy$
$\displaystyle + \int_C u\frac{\partial v}{\partial x}\,dy - u\frac{\partial v}{\partial y}\,dx$, (Green's first identity).
19. $\displaystyle\int\int_D (u\,\nabla^2 v - v\,\nabla^2 u)\,dx\,dy = \int_C \left(u\frac{\partial v}{\partial x} - v\frac{\partial u}{\partial x}\right) dy$
$\displaystyle -\left(u\frac{\partial v}{\partial y} - v\frac{\partial u}{\partial y}\right) dx$ (Green's second identity).
[*Hint:* Write down Green's first identity again with u and v interchanged and subtract.]
20. Suppose that $\partial^2 u/\partial x^2 + \partial^2 u/\partial y^2 = 0$ on D. Show that
$$\int_C \frac{\partial u}{\partial y}\,dx - \frac{\partial u}{\partial x}\,dy = 0.$$

Use formula (4) to determine the area of the regions in the plane bounded by the figures in Exercises 21 and 22.

21. The triangle with vertices $(1,0)$, $(3,4)$, and $(5,-1)$.
22. The rhombus with vertices $(0,-1)$, $(3,0)$, $(1,2)$, and $(4,3)$.
23. Show that the area enclosed by the hypocycloid $x = a\cos^3\theta$, $y = a\sin^3\theta$, $0 \leqslant \theta \leqslant 2\pi$ is $\frac{3}{8}\pi a^2$. (Use Green's theorem.)
24. Find the area bounded by one arc of the cycloid $x = a(\theta - \sin\theta)$, $y = a(1 - \cos\theta)$, where $a > 0$, $0 \leqslant \theta \leqslant 2\pi$.
25. Find the area between the curves $y = x^3$ and $y = \sqrt{x}$ by using Green's theorem.
26. Use formula (4) to recover the formula $A = \frac{1}{2}\int_a^b r^2\,d\theta$ for a region in polar coordinates (see Section 10.5).
27. Sketch the proof of Green's theorem for the region shown in Fig. 18.4.7.

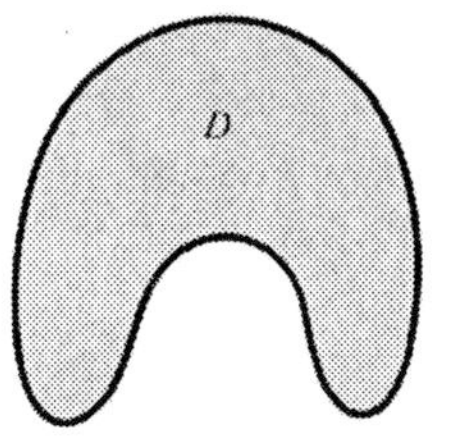

Figure 18.4.7. Prove Green's theorem for this region.

28. Find the work which is done by the force field $(3x + 4y)\mathbf{i} + (8x + 9y)\mathbf{j}$ on a particle which moves once around the ellipse $4x^2 + 9y^2 = 36$ by (a) directly evaluating the line integral and by (b) using Green's theorem.
29. Let $P = y$ and $Q = x$. What is $\int_C P\,dx + Q\,dy$ if C is a closed curve?
30. Use Green's first identity (Exercise 18) with $v = u$ to prove that if $\nabla^2 u = 0$ on D and $u = 0$ on C, then $\nabla u = \mathbf{0}$ on D, and hence $u = 0$ on D.

★31. Green's theorem can be used to give another proof that a differential form $P\,dx + Q\,dy$ defined on the plane is exact if $\partial P/\partial y = \partial Q/\partial x$. (In fact, the argument in Section 18.3 or the one outlined here also works in other regions, such as a disk.) Define f by
$$f(x, y) = \int_0^x a(t,0)\,dt + \int_0^y b(x,t)\,dt,$$
where $P = a(x, y)$ and $Q = b(x, y)$ and set $u = f(x, y)$. Show that $\partial u/\partial y = Q$. The function f is the line integral of $P\,dx + Q\,dy$ along a horizontal segment C_1 and a vertical segment C_2. Define another function $\hat{u} = \hat{f}(x, y)$ by letting $\hat{C}$ be the path which consists of the vertical segment $\hat{C}_1$ from $(0,0)$ to $(0, y)$, followed by the horizontal segment $\hat{C}_2$ from $(0, y)$ to (x, y) and setting $\hat{f}(x, y) = \int_{\hat{C}_1 + \hat{C}_2} P\,dx + Q\,dy$. Show that $\partial\hat{u}/\partial x = P$. Apply Green's theorem to the rectangle D bounded by $C_1 + C_2 + (-\hat{C}_2) + (-\hat{C}_1)$ to get
$$\int\int_D \left(\frac{\partial P}{\partial x} - \frac{\partial Q}{\partial y}\right) dx\,dy$$

$$= \int_{C_1+C_2+(-\hat{C}_2)+(-\hat{C}_1)} (P\,dx + Q\,dy)$$
$$= \int_{C_1+C_2} (P\,dx + Q\,dy) - \int_{\hat{C}_1+\hat{C}_2} (P\,dx + Q\,dy)$$
$$= u - \hat{u};$$

but the double integral over D is zero, since $\partial P/\partial x = \partial Q/\partial y$ by assumption. Conclude that $u = \hat{u}$, to complete the proof.

★32. If $P\,dx + Q\,dy$ satisfies $\partial P/\partial y = \partial Q/\partial x$, use Green's theorem to show that $P\,dx + Q\,dy$ is conservative by showing its integral around "every" closed curve in zero.

★33. *Project:* The formula $A = \frac{1}{2}\int_C x\,dy - y\,dx$ is the basis for the operation of the *planimeter*, a mechanical device for measuring areas. Find out about planimeters from an encyclopedia or the American Mathematical Monthly, Vol. **88**, No. 9, November (1981), p. 701, and relate their operation to this formula for area.

★34. (a) If D is a region to which Green's theorem applies, write the identity of Example 5 this way:

$$\int\int_D \left(P\frac{\partial Q}{\partial x} - P\frac{\partial Q}{\partial y} \right) dx\,dy$$
$$= \int_C PQ\,dx + PQ\,dy$$
$$- \int\int_D \left(Q\frac{\partial P}{\partial x} - Q\frac{\partial P}{\partial y} \right) dx\,dy.$$

How is this like integration by parts?
(b) Elaborate the following statement: Green's theorem is "the fundamental theorem of calculus" in the plane since it relates a double integral to an integral around the boundary, just as the fundamental theorem relates an integral over an interval to a sum over the boundary of the interval (that is, the two endpoints).

18.5 Circulation and Stokes' Theorem

The line integral of a vector field around the boundary of a surface in space equals the surface integral of the curl of the vector field.

If $\mathbf{\Phi}$ is a vector field defined at points in the plane, we can write $\mathbf{\Phi} = P\mathbf{i} + Q\mathbf{j}$. The line integral of $\mathbf{\Phi}$ around a curve C, namely

$$\int_C \mathbf{\Phi}(\mathbf{r}) \cdot d\mathbf{r} = \int_C P\,dx + Q\,dy,$$

occurs on the left-hand side of the equation in the statement of Green's theorem. The expression "*circulation* of $\mathbf{\Phi}$ around C" is often used for the number $\int_C \mathbf{\Phi}(\mathbf{r}) \cdot d\mathbf{r}$. This terminology arose through the application of Green's theorem to fluid mechanics; we shall now briefly discuss this application.

Imagine a fluid moving in the plane. Each particle of the fluid (or piece of dust suspended in the fluid) has a well-defined velocity. If, at a particular time, we assign to each point (x, y) of the plane the velocity $\mathbf{V}(x, y)$ of the fluid particle moving through (x, y) at that time, we obtain a vector field $\mathbf{V}$ on the plane. See Fig. 18.5.1. The integral $\int_C \mathbf{V}(\mathbf{r}) \cdot d\mathbf{r}$ of $\mathbf{V}$ around a closed curve

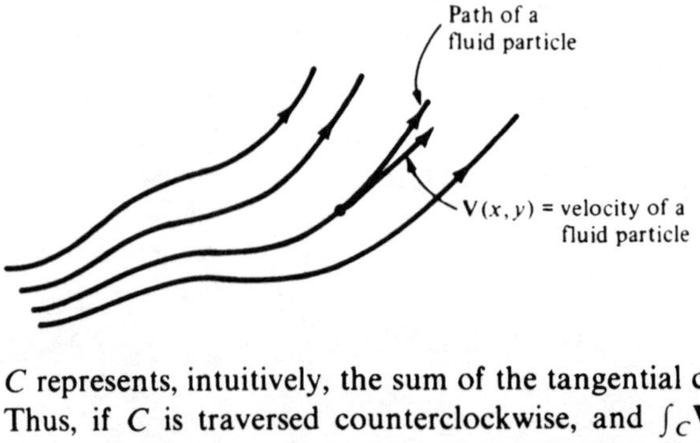

Figure 18.5.1. The velocity field of a fluid.

C represents, intuitively, the sum of the tangential components of $\mathbf{V}$ around C. Thus, if C is traversed counterclockwise, and $\int_C \mathbf{V}(\mathbf{r}) \cdot d\mathbf{r} > 0$, there is a net counterclockwise motion of the fluid. Likewise, if $\int_C \mathbf{V}(\mathbf{r}) \cdot d\mathbf{r} < 0$, the fluid is circulating clockwise. This explains the origin of the term "circulation" and is illustrated in Fig. 18.5.2.

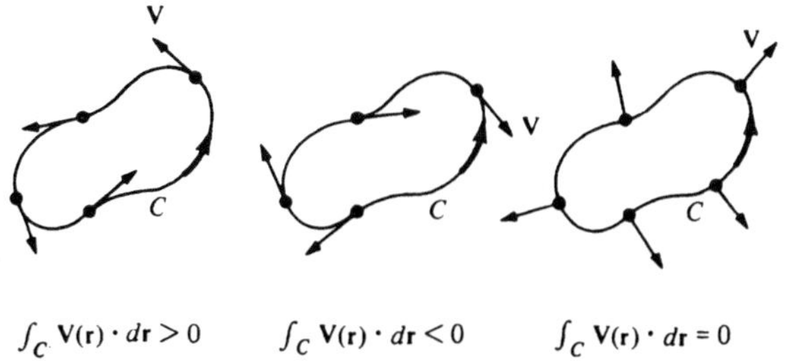

Figure 18.5.2. The intuitive meaning of the possible signs of $\int_C \mathbf{V}(\mathbf{r}) \cdot d\mathbf{r}$.

In Section 18.1, we interpreted the line integral of a force vector field along a curve as the work done by the force on a particle traversing the curve. Notice that the single mathematical concept of a line integral is subject to different interpretations, depending on what physical quantity is represented by the vector field.

The integrand on the right-hand side of Green's theorem,

$$\frac{\partial Q}{\partial x} - \frac{\partial P}{\partial y},$$

is important because, when integrated over the region whose boundary is C, it produces the circulation of $\mathbf{\Phi}$ around C according to Green's theorem.

The Scalar Curl

If $\mathbf{\Phi} = P\mathbf{i} + Q\mathbf{j}$ is a vector field in the plane,

$$\frac{\partial Q}{\partial x} - \frac{\partial P}{\partial y}$$

is called the *scalar curl* of $\mathbf{\Phi}$.

Example 1 In the plane, the vector field

$$\mathbf{V}(x, y) = \frac{y\mathbf{i}}{x^2 + y^2} - \frac{x\mathbf{j}}{x^2 + y^2}$$

approximates (the horizontal part of) the velocity field of water flowing down a drain (see Fig. 18.5.3). (a) Calculate its scalar curl. (b) Is Green's theorem valid for this vector field on the unit disk D, defined by $x^2 + y^2 \leqslant 1$?

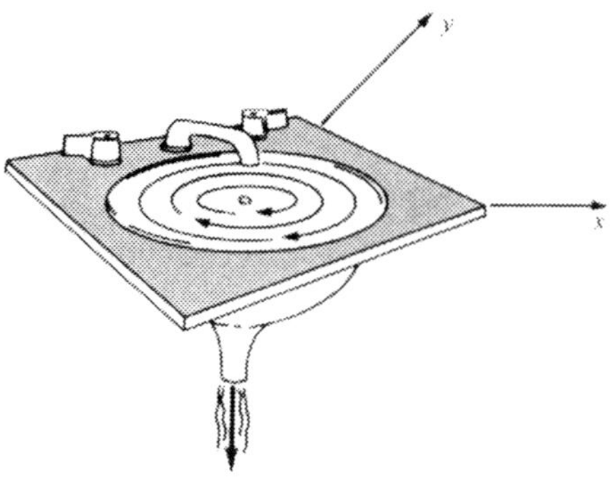

Figure 18.5.3. The velocity field near a drain.

Solution Here $P(x, y) = y/(x^2 + y^2)$ and $Q(x, y) = -x/(x^2 + y^2)$, so the scalar curl is

$$\frac{\partial Q}{\partial x} - \frac{\partial P}{\partial y} = -\frac{\partial}{\partial x}\left(\frac{x}{x^2+y^2}\right) - \frac{\partial}{\partial y}\left(\frac{y}{x^2+y^2}\right)$$

$$= \frac{-(x^2+y^2)+2x^2}{(x^2+y^2)^2} + \frac{-(x^2+y^2)+2y^2}{(x^2+y^2)^2} = 0.$$

The circulation of $\mathbf{V}$ about the circle $x^2 + y^2 = 1$ is

$$\int_C P\,dx + Q\,dy = \int_C y\,dx - x\,dy = 2 \cdot (\text{area of disk}) = -2\pi.$$

(See formula (4), Section 18.4.) This is an apparent contradiction to Green's theorem! The explanation is the fact that the hypotheses of Green's theorem are not satisfied; $\mathbf{V}$ is not defined at (0,0). ▲

If we write (scalar curl $\mathbf{\Phi}$) $= \partial Q/\partial x - \partial P/\partial y$, then Green's theorem becomes

$$\int_C \mathbf{\Phi}(\mathbf{r}) \cdot d\mathbf{r} = \int\int_D (\text{scalar curl } \mathbf{\Phi})\,dx\,dy,$$

where C is the boundary of D. Now choose a point P_0 in the plane and let D_ε be the disk of radius ε about P_0 and C_ε the circle of radius ε. By the mean value theorem,

$$\int\int_{D_\varepsilon} (\text{scalar curl } \mathbf{\Phi})\,dx\,dy = [\text{scalar curl } \mathbf{\Phi}(P_\varepsilon)][\text{area } D_\varepsilon]$$

for some point P_ε in D_ε. Dividing by (area D_ε) and letting $\varepsilon \to 0$ gives

$$(\text{scalar curl } \mathbf{\Phi})(P_0) = \lim_{\varepsilon \to 0}\left[\frac{1}{\text{area } D_\varepsilon}\int_{C_\varepsilon} \mathbf{\Phi}(\mathbf{r}) \cdot d\mathbf{r}\right],$$

i.e., *the scalar curl may be thought of as the circulation per unit area.*

A fluid moving in space is represented by a vector field $\mathbf{\Phi}(x, y, z)$ in three variables. The generalization of Green's theorem to this case is called *Stokes' theorem*. We shall next prove a special case of this theorem.

Consider a region D in the xy plane and a function f defined in D. The equation $z = f(x, y)$ defines a surface over D. If we refer to Section 15.2, we see that a normal vector to this surface is given by $-f_x\mathbf{i} - f_y\mathbf{j} + \mathbf{k}$, and so a unit normal is

$$\mathbf{n} = \frac{-f_x\mathbf{i} - f_y\mathbf{j} + \mathbf{k}}{\sqrt{f_x^2 + f_y^2 + 1}}.$$

The area element on the surface is given by

$$dA = \sqrt{1 + f_x^2 + f_y^2}\;dx\,dy$$

as was shown in Section 17.3. Thus $\mathbf{n}\,dA = (-f_x\mathbf{i} - f_y\mathbf{j} + \mathbf{k})\,dx\,dy$.

The Surface Integral

If $\mathbf{\Phi} = P\mathbf{i} + Q\mathbf{j} + R\mathbf{k}$ is a vector field in space and S is the surface $z = f(x, y)$, the *surface integral* of $\mathbf{\Phi}$ over S is the integral of the normal component of $\mathbf{\Phi}$ over S:

$$\int\int_S \mathbf{\Phi} \cdot \mathbf{n}\,dA = \int\int_D (-Pf_x - Qf_y + R)\,dx\,dy. \tag{1}$$

We shall discuss a physical interpretation of this definition in Section 18.6.

Example 2 Let $\mathbf{\Phi} = x^2\mathbf{i} + y^2\mathbf{j} + z\mathbf{k}$. Evaluate $\int\int_S \mathbf{\Phi}\cdot\mathbf{n}\,dA$, where S is the graph of the function $z = x + y + 1$ over the rectangle $0 \leqslant x \leqslant 1$, $0 \leqslant y \leqslant 1$.

Solution By formula (1), with $P(x, y, z) = x^2$, $Q(x, y, z) = y^2$, $R(x, y, z) = z$, and $z = f(x, y) = x + y + 1$,

$$\int\int_S \mathbf{\Phi}\cdot\mathbf{n}\,dA = \int\int_D \left[-x^2\cdot 1 - y^2\cdot 1 + (x + y + 1)\right]dx\,dy$$

$$= \int_0^1\int_0^1 (x + y + 1 - x^2 - y^2)\,dx\,dy$$

$$= \frac{1}{2} + \frac{1}{2} + 1 - \frac{1}{3} - \frac{1}{3} = \frac{4}{3}. \blacktriangle$$

Stokes' theorem will involve the concept of *curl* defined as follows.

The Curl of a Vector Field

Let $\mathbf{\Phi} = P\mathbf{i} + Q\mathbf{j} + R\mathbf{k}$ be a vector field in space. Its *curl* is defined by

$$\operatorname{curl}\mathbf{\Phi} = (R_y - Q_z)\mathbf{i} + (P_z - R_x)\mathbf{j} + (Q_x - P_y)\mathbf{k}.$$

If $\mathbf{\Phi}$ is a vector field in the plane, but regarded as being in space with $R = 0$ and P, Q independent of z, then the curl of $\mathbf{\Phi}$ is just the scalar curl of $\mathbf{\Phi}$ times $\mathbf{k}$.

Formally, we can write $\nabla = \frac{\partial}{\partial x}\mathbf{i} + \frac{\partial}{\partial y}\mathbf{j} + \frac{\partial}{\partial z}\mathbf{k}$, treating it as if it were a vector; then

$$\operatorname{curl}\mathbf{\Phi} = \nabla\times\mathbf{\Phi} = \begin{vmatrix} \mathbf{i} & \mathbf{j} & \mathbf{k} \\ \dfrac{\partial}{\partial x} & \dfrac{\partial}{\partial y} & \dfrac{\partial}{\partial z} \\ P & Q & R \end{vmatrix}$$

which helps one remember the formula.

Example 3 Find the curl of $xy\mathbf{i} - \sin z\,\mathbf{j} + \mathbf{k}$.

Solution

$$\operatorname{curl}\mathbf{\Phi} = \nabla\times\mathbf{\Phi} = \begin{vmatrix} \mathbf{i} & \mathbf{j} & \mathbf{k} \\ \dfrac{\partial}{\partial x} & \dfrac{\partial}{\partial y} & \dfrac{\partial}{\partial z} \\ xy & -\sin z & 1 \end{vmatrix}$$

$$= \begin{vmatrix} \dfrac{\partial}{\partial y} & \dfrac{\partial}{\partial z} \\ -\sin z & 1 \end{vmatrix}\mathbf{i} - \begin{vmatrix} \dfrac{\partial}{\partial x} & \dfrac{\partial}{\partial z} \\ xy & 1 \end{vmatrix}\mathbf{j} + \begin{vmatrix} \dfrac{\partial}{\partial x} & \dfrac{\partial}{\partial y} \\ xy & -\sin z \end{vmatrix}\mathbf{k}$$

$$= \cos z\,\mathbf{i} - x\mathbf{k}. \blacktriangle$$

Example 4 If f is a twice differentiable function in space, prove that $\nabla\times(\nabla f) = \mathbf{0}$.

Solution Let us write out the components. Since $\nabla f = (\partial f/\partial x, \partial f/\partial y, \partial f/\partial z)$, we have

$$\nabla\times\nabla f=\begin{vmatrix}\mathbf{i} & \mathbf{j} & \mathbf{k}\\ \dfrac{\partial}{\partial x} & \dfrac{\partial}{\partial y} & \dfrac{\partial}{\partial z}\\ \dfrac{\partial f}{\partial x} & \dfrac{\partial f}{\partial y} & \dfrac{\partial f}{\partial z}\end{vmatrix}$$

$$=\left(\frac{\partial^2 f}{\partial y\,\partial z}-\frac{\partial^2 f}{\partial z\,\partial y}\right)\mathbf{i}+\left(\frac{\partial^2 f}{\partial z\,\partial x}-\frac{\partial^2 f}{\partial x\,\partial z}\right)\mathbf{j}+\left(\frac{\partial^2 f}{\partial x\,\partial y}-\frac{\partial^2 f}{\partial y\,\partial x}\right)\mathbf{k}.$$

Each component is zero because of the symmetry property of mixed partial derivatives. ▲

Now we are ready to state Stokes' theorem. Like Green's theorem, it relates an integral over a surface with an integral around a curve.

Stokes' Theorem

Let D be a region in the plane (to which Green's theorem applies) and S the surface $z=f(x,y)$, where f is twice continuously differentiable. Let ∂D be the boundary of D traversed counterclockwise and ∂S the corresponding boundary of S (see Fig. 18.5.4). If $\mathbf{\Phi}$ is a continuously differentiable vector field in space, then

$$\int_{\partial S}\mathbf{\Phi}(\mathbf{r})\cdot d\mathbf{r}=\int\!\!\int_S(\nabla\times\mathbf{\Phi})\cdot\mathbf{n}\,dA.$$

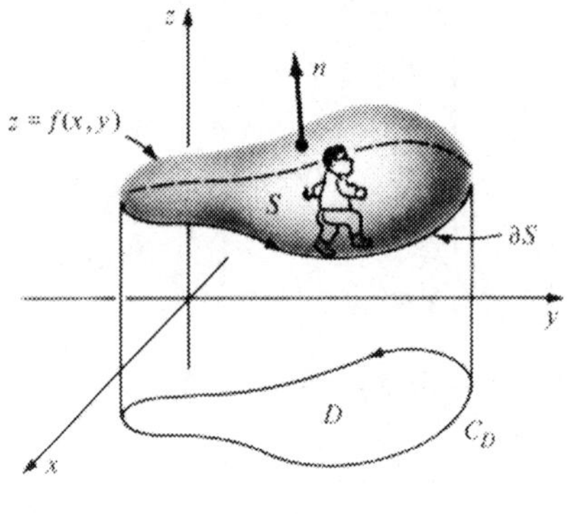

Figure 18.5.4. As you traverse ∂S counterclockwise, the surface is on your left.

Proof of Stokes Theorem

Let $\mathbf{\Phi}=P\mathbf{i}+Q\mathbf{j}+R\mathbf{k}$, so that

$$\nabla\times\mathbf{\Phi}=(R_y-Q_z)\mathbf{i}+(P_z-R_x)\mathbf{j}+(Q_x-P_y)\mathbf{k}.$$

We may use formula (1) to write

$$\int\!\!\int_S\operatorname{curl}\mathbf{\Phi}\cdot\mathbf{n}\,d\mathbf{A}=\int\!\!\int_D\left[\left(\frac{\partial R}{\partial y}-\frac{\partial Q}{\partial z}\right)\left(-\frac{\partial z}{\partial x}\right)\right.$$

$$\left.+\left(\frac{\partial P}{\partial z}-\frac{\partial R}{\partial x}\right)\left(-\frac{\partial z}{\partial y}\right)+\left(\frac{\partial Q}{\partial x}-\frac{\partial P}{\partial y}\right)\right]dA.\quad(2)$$

On the other hand, if $\boldsymbol{\sigma}(t)=x(t)\mathbf{i}+y(t)\mathbf{j}$ parametrizes ∂D, then $\boldsymbol{\eta}(t)=x(t)\mathbf{i}+y(t)\mathbf{j}+f(x(t),y(t))\mathbf{k}$ is an orientation-preserving parameterization of the oriented simple closed curve ∂S. Thus

$$\int_{\partial S}\mathbf{\Phi}(\mathbf{r})\cdot d\mathbf{r}=\int_a^b\left(P\frac{dx}{dt}+Q\frac{dy}{dt}+R\frac{dz}{dt}\right)dt;\quad(3)$$

but, by the chain rule,

$$\frac{dz}{dt} = \frac{\partial z}{\partial x}\frac{dx}{dt} + \frac{\partial z}{\partial y}\frac{dy}{dt}.$$

Substituting this expression into (3), we obtain

$$\int_{\partial S} \mathbf{\Phi}(\mathbf{r}) \cdot d\mathbf{r} = \int_a^b \left[\left(P + R\frac{\partial z}{\partial x} \right) \frac{dx}{dt} + \left(Q + R\frac{\partial z}{\partial y} \right) \frac{dy}{dt} \right] dt$$

$$= \int_{\partial D} \left(P + R\frac{\partial z}{\partial x} \right) dx + \left(Q + R\frac{\partial z}{\partial y} \right) dy. \qquad (4)$$

Applying Green's theorem to (4) yields

$$\iint_D \left[\frac{\partial(Q + R\,\partial z/\partial y)}{\partial x} - \frac{\partial(P + R\,\partial z/\partial x)}{\partial y} \right] dA.$$

Now we use the chain rule, remembering that P, Q, and R are functions of x, y, and z, and z is a function of x and y, to obtain

$$\int_{\partial S} \mathbf{\Phi}(\mathbf{r}) \cdot d\mathbf{r}$$

$$= \iint_D \left[\left(\frac{\partial Q}{\partial x} + \frac{\partial Q}{\partial z} \cdot \frac{\partial z}{\partial x} + \frac{\partial R}{\partial x} \cdot \frac{\partial z}{\partial y} + \frac{\partial R}{\partial z} \cdot \frac{\partial z}{\partial x} \cdot \frac{\partial z}{\partial y} + R \cdot \frac{\partial^2 z}{\partial x\,\partial y} \right) \right.$$

$$\left. - \left(\frac{\partial P}{\partial y} + \frac{\partial P}{\partial z} \cdot \frac{\partial z}{\partial y} + \frac{\partial R}{\partial y} \cdot \frac{\partial z}{\partial x} + \frac{\partial R}{\partial z} \cdot \frac{\partial z}{\partial y} \cdot \frac{\partial z}{\partial x} + R\frac{\partial^2 z}{\partial y\,\partial x} \right) \right] dA.$$

The last two terms in each set of parentheses cancel each other, and we can rearrange terms to obtain the integral (2). ■

As with Green's theorem, Stokes' theorem is valid for a much wider class of surfaces than graphs, but for simplicity we have treated only this case.

Example 5 Let $\mathbf{\Phi} = ye^z\mathbf{i} + xe^z\mathbf{j} + xye^z\mathbf{k}$. Show that the integral of $\mathbf{\Phi}$ around an oriented simple closed curve C that is the boundary of a surface S is 0. (Assume S to be the graph of a function.)

Solution By Stokes' theorem,

$$\int_{\partial S} \mathbf{\Phi}(\mathbf{r}) \cdot d\mathbf{r} = \iint_S (\nabla \times \mathbf{\Phi}) \cdot \mathbf{n}\, dA.$$

However,

$$\nabla \times \mathbf{\Phi} = \begin{vmatrix} \mathbf{i} & \mathbf{j} & \mathbf{k} \\ \dfrac{\partial}{\partial x} & \dfrac{\partial}{\partial y} & \dfrac{\partial}{\partial z} \\ ye^z & xe^z & xye^z \end{vmatrix} = \mathbf{0},$$

so $\int_C \mathbf{\Phi}(\mathbf{r}) \cdot d\mathbf{r} = \iint_S \nabla \times \mathbf{\Phi} \cdot \mathbf{n}\, dA$
$= \iint_S \mathbf{0} \cdot \mathbf{n}\, dA = 0.$ ▲

Example 6 Find the integral of $\mathbf{F}(x, y, z) = x^2\mathbf{i} + y^2\mathbf{j} - z\mathbf{k}$ around the triangle with vertices (0, 0, 0), (0, 2, 0) and (0, 0, 2), using Stokes' theorem.

Solution Refer to Fig. 18.5.5. C is the triangle in question and S is a surface it bounds. By Stokes' theorem,

$$\int_C \mathbf{F}\cdot d\mathbf{r} = \int\int_S (\nabla\times\mathbf{F})\cdot\mathbf{n}\,dA.$$

Now

$$\nabla\times\mathbf{F} = \begin{vmatrix} \mathbf{i} & \mathbf{j} & \mathbf{k} \\ \frac{\partial}{\partial x} & \frac{\partial}{\partial y} & \frac{\partial}{\partial z} \\ x^2 & y^2 & -z \end{vmatrix} = 0\mathbf{i} + 0\mathbf{j} + 0\mathbf{k} = \mathbf{0}.$$

Therefore the integral of $\mathbf{F}$ around C is zero. ▲

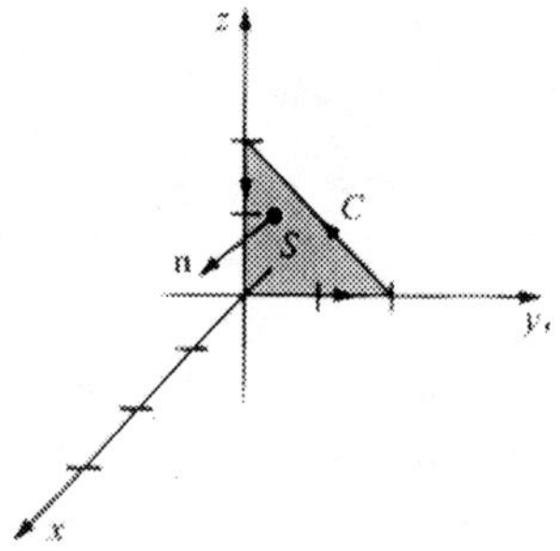

Figure 18.5.5. The curve C of integration for Example 6.

Example 7 Evaluate the integral $\int\int_S(\nabla\times\mathbf{F})\cdot\mathbf{n}\,dA$, where S is the portion of the surface of a sphere defined by $x^2+y^2+z^2=1$ and $x+y+z\geqslant 1$, and where $\mathbf{F} = \mathbf{r}\times(\mathbf{i}+\mathbf{j}+\mathbf{k})$, $\mathbf{r} = x\mathbf{i}+y\mathbf{j}+z\mathbf{k}$.

Solution The surface S and its boundary ∂S are shown in Fig. 18.5.6. We choose the orientation shown.

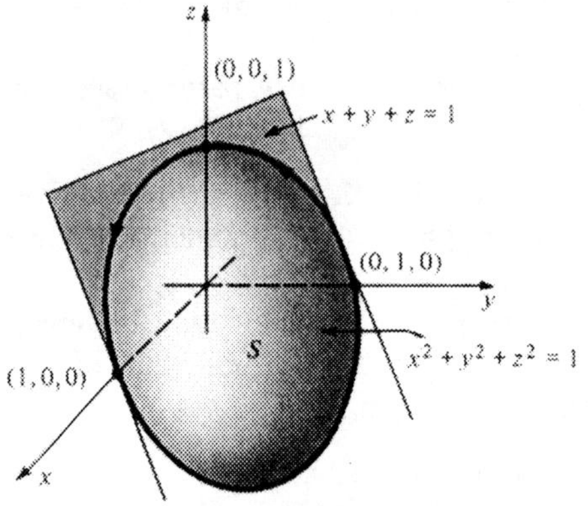

Figure 18.5.6. The surface S in Example 7.

Method 1. Using Stokes' theorem directly. We need to parametrize the boundary circle ∂S, which consists of all unit vectors $\mathbf{v} = x\mathbf{i}+y\mathbf{j}+z\mathbf{k}$ satisfying the equation $x+y+z=1$. The vectors $\mathbf{i}$, $\mathbf{j}$, and $\mathbf{k}$ point from the origin to points on ∂S, and $\mathbf{m} = \frac{1}{3}(\mathbf{i}+\mathbf{j}+\mathbf{k})$ points from the origin to the center of the circle ∂S. The radius of the circle ∂S is $\|\frac{1}{3}(\mathbf{i}+\mathbf{j}+\mathbf{k})-\mathbf{i}\| = (\frac{4}{9}+\frac{1}{9}+\frac{1}{9})^{1/2} = \frac{1}{3}\sqrt{6} = \sqrt{2/3}$. To describe the general point on ∂S, we choose orthogonal unit vectors parallel to the plane $x+y+z=1$, say $\mathbf{u} = (1/\sqrt{2})(\mathbf{i}-\mathbf{j})$ and $\mathbf{v} = (1/\sqrt{6})(\mathbf{i}+\mathbf{j}-2\mathbf{k})$. The general point on ∂S is $\mathbf{m}+\sqrt{2/3}\,[(\cos t)\mathbf{u}+(\sin t)\mathbf{v}]$. As t goes from 0 to 2π, the circle is traversed once. The orientation is correct if the triple product $(\mathbf{u}\times\mathbf{v})\cdot\mathbf{m}$ is positive[3] (Figure 18.5.7). Up to positive factors, this triple product equals the determinant

$$\begin{vmatrix} 1 & 1 & 1 \\ 1 & -1 & 0 \\ 1 & 1 & -2 \end{vmatrix} = \begin{vmatrix} 0 & 0 & 3 \\ 1 & -1 & 0 \\ 1 & 1 & -2 \end{vmatrix} = 3\begin{vmatrix} 1 & -1 \\ 1 & 1 \end{vmatrix} = 3\cdot 2 = 6.$$

[3]Our original computation of $(\mathbf{u}\times\mathbf{v})\cdot\mathbf{m}$ when writing this solution came out negative, so we interchanged $\mathbf{u}$ and $\mathbf{v}$.

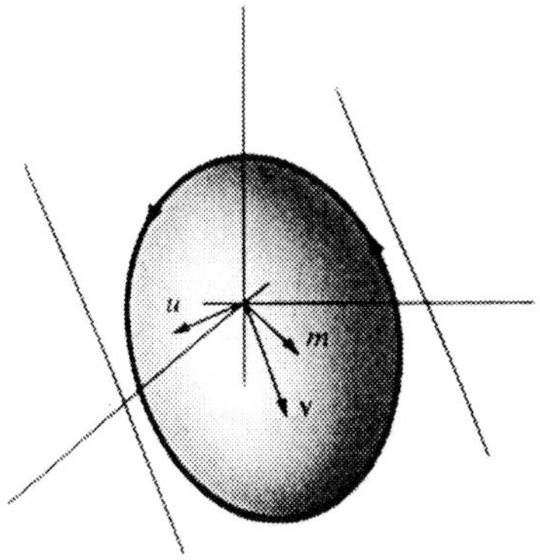

Figure 18.5.7. **u**, **v**, **m** are oriented so that rotations about **m** counterclockwise in the **uv** plane correspond to the correct orientation for Stokes' theorem.

Now $\mathbf{r} = \mathbf{m} + \sqrt{2/3}\,(\cos t\,\mathbf{u} + \sin t\,\mathbf{v})$ and $d\mathbf{r} = \sqrt{2/3}\,(-\sin t\,\mathbf{u} + \cos t\,\mathbf{v})\,dt$, so by Stokes' theorem,

$$\int\int_S (\nabla \times \mathbf{F}) \cdot \mathbf{n}\,dA$$
$$= \int_{\partial S} \mathbf{F} \cdot d\mathbf{r} = \int_{\partial S} [\mathbf{r} \times (\mathbf{i} + \mathbf{j} + \mathbf{k})] \cdot d\mathbf{r} = \int_{\partial S} (d\mathbf{r} \times \mathbf{r}) \cdot (\mathbf{i} + \mathbf{j} + \mathbf{k})$$
$$= \int_0^{2\pi} \left[\left(\sqrt{\frac{2}{3}}\,(-\sin t\mathbf{u} + \cos t\mathbf{v}) \right) \times \left(\mathbf{m} + \sqrt{\frac{2}{3}}\,(\cos t\,\mathbf{u} + \sin t\,\mathbf{v}) \right) \right] \cdot (\mathbf{i} + \mathbf{j} + \mathbf{k})\,dt.$$

Now $\mathbf{i} + \mathbf{j} + \mathbf{k} = 3\mathbf{m}$, so the term involving $\mathbf{m}$ drops out, and the integral becomes

$$\frac{2}{3} \int_0^{2\pi} -(\sin^2 t + \cos^2 t)(\mathbf{u} \times \mathbf{v}) \cdot (\mathbf{i} + \mathbf{j} + \mathbf{k})\,dt = -\frac{2}{3} \int_0^{2\pi} \frac{1}{\sqrt{6}} \cdot \frac{1}{\sqrt{2}} \cdot 6\,dt$$
$$= -\frac{4\pi}{\sqrt{3}}.$$

Method 2. Simplifying the Surface. We compute $\nabla \times \mathbf{F} = -2\mathbf{i} - 2\mathbf{j} - 2\mathbf{k}$. By Stokes' theorem,

$$\int\int_S (\nabla \times \mathbf{F}) \cdot \mathbf{n}\,dA = \int_{\partial S} \mathbf{F} \cdot d\mathbf{r} = \int\int_P (\nabla \times \mathbf{F}) \cdot \mathbf{n}\,dA,$$

where P is *any* surface having ∂S as its boundary. We take for P the portion of the plane $x + y + z = 1$ inside the circle in Fig. 18.5.6; $\mathbf{n} = (\mathbf{i} + \mathbf{j} + \mathbf{k})/\sqrt{3}$ is a unit vector orthogonal to the plane, so $(\nabla \times \mathbf{F}) \cdot \mathbf{n}$ is constant and equal to $-6/\sqrt{3}$. In method 1, we found the radius of P to be $\sqrt{2/3}$. Thus

$$\int\int_S (\nabla \times \mathbf{F}) \cdot \mathbf{n}\,dA = \int\int_P -\frac{6}{\sqrt{3}}\,dA = \frac{6}{\sqrt{3}}\,(\text{area of } P)$$
$$= -\frac{6}{\sqrt{3}} \cdot \pi \cdot \left(\frac{2}{3}\right) = -\frac{4\pi}{\sqrt{3}}.$$

which agrees with the answer in Method 1. ▲

Just as with the scalar curl, we can show that $\mathbf{n} \cdot \operatorname{curl} \Phi(P_0)$ *is the circulation per unit area at* P_0 *in the plane through* P *orthogonal to* $\mathbf{n}$.

Indeed, let D_ε be the disk centered at P_0 with radius ε and lying in the plane orthogonal to **n**, and let ∂D_ε be its boundary. See Fig. 18.5.8. By Stokes' theorem,

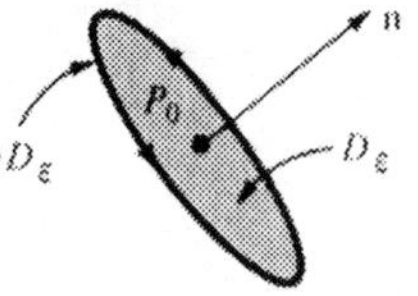

Figure 18.5.8. The curl gives the circulation per unit area.

$$\iint_{D_\varepsilon} (\nabla \times \mathbf{\Phi}) \cdot \mathbf{n}\, dA = \int_{\partial D_\varepsilon} \mathbf{\Phi} \cdot d\mathbf{r}.$$

In one variable calculus we have seen that a mean value theorem holds for integrals (see p. 435). There is a similar result for double and triple integrals; thus there is a point P_ε in D_ε such that

$$\iint_{D_\varepsilon} (\nabla \times \mathbf{\Phi}) \cdot \mathbf{n}\, dA = \left[(\nabla \times \mathbf{\Phi})(P_\varepsilon) \cdot \mathbf{n}\right](\text{area } D_\varepsilon).$$

Thus

$$\mathbf{n} \cdot (\nabla \times \mathbf{\Phi})(P_\varepsilon) = \frac{1}{\pi\varepsilon^2} \int_{\partial D_\varepsilon} \mathbf{\Phi} \cdot d\mathbf{r}, \quad \text{and so} \quad \mathbf{n} \cdot (\nabla \times \mathbf{\Phi})(P_0) = \lim_{\varepsilon \to 0} \frac{1}{\pi\varepsilon^2} \int_{\partial D_\varepsilon} \mathbf{\Phi} \cdot d\mathbf{r},$$

as we wanted to show.

Example 8 Let **E** and **H** be time-dependent electric and magnetic fields in space. Let S be a surface with boundary C. We define

$$\int_C \mathbf{E} \cdot d\mathbf{r} = \text{voltage drop around } C,$$

$$\iint_S \mathbf{H} \cdot \mathbf{n}\, dA = \text{magnetic flux across } S.$$

Faraday's law (see Fig. 18.5.9) states that the voltage around C equals the negative rate of change of magnetic flux through S.

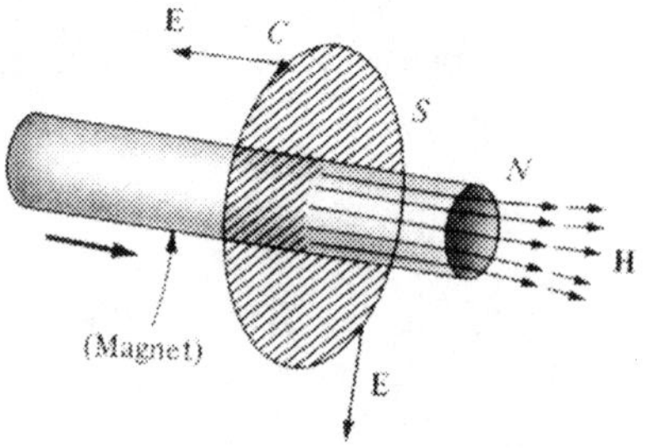

Figure 18.5.9. Faraday's law.

Show that Faraday's law follows from the following differential equation (one of the Maxwell equations):

$$\nabla \times \mathbf{E} = -\frac{\partial \mathbf{H}}{\partial t}.$$

Solution In symbols, Faraday's law states $\int_C \mathbf{E} \cdot d\mathbf{r} = -(\partial/\partial t)\iint_S \mathbf{H} \cdot \mathbf{n}\, dA$. By Stokes' theorem, $\int_C \mathbf{E} \cdot d\mathbf{r} = \iint_S (\nabla \times \mathbf{E}) \cdot \mathbf{n}\, dA$. Assuming that we can move $\partial/\partial t$ under the integral sign (see Review Exercise 48, Chapter 17), we get

$$-\frac{\partial}{\partial t} \iint_S \mathbf{H} \cdot \mathbf{n}\, dA = \iint_S -\frac{\partial \mathbf{H}}{\partial t} \cdot \mathbf{n}\, dA.$$

Since the two integrals

$$\iint_S (-(\partial \mathbf{H}/\partial t) \cdot \mathbf{n})\, dA \quad \text{and} \quad \iint_S (\nabla \times \mathbf{E}) \cdot \mathbf{n}\, dA$$

are equal for all S, it must be the case that $\nabla \times \mathbf{E} = -\partial \mathbf{H}/\partial t$ (compare the proof of the theorem following Example 3 in Section 18.2). ▲

Exercises for Section 18.5

Calculate the scalar curl of the plane vector fields in Exercises 1–4.

1. $\mathbf{V}(x, y) = y\mathbf{i} - x\mathbf{j}$.
2. $\mathbf{V}(x, y) = xy\mathbf{i} - e^{xy}\mathbf{j}$.
3. $\mathbf{V}(x, y) = \dfrac{x\mathbf{i}}{x^2 + y^2} - \dfrac{y\mathbf{j}}{x^2 + y^2}$.
4. $\mathbf{V}(x, y) = e^{xy}\mathbf{i} - \dfrac{x}{x^2 + y^2}\mathbf{j}$.

Evaluate the surface integrals of the vector fields over the surfaces given in Exercises 5–8.

5. $\mathbf{\Phi} = 3x^2\mathbf{i} - 2yx\mathbf{j} + 8\mathbf{k}$; S is the graph of $z = 2x - y$ over the rectangle $[0, 2] \times [0, 2]$.
6. $\mathbf{\Phi} = x\mathbf{i} - 2y\mathbf{j} + xz\mathbf{k}$, S is the graph of $z = -x - y - 1$ over the rectangle $[0, 1] \times [0, 1]$.
7. $\mathbf{\Phi} = x\mathbf{k}$, S is the disk $x^2 + y^2 \leqslant 1$ in the xy plane.
8. $\mathbf{\Phi} = \mathbf{j}$, S is the disk $x^2 + z^2 \leqslant 1$ in the xz plane.

Calculate the curl of the vector fields in Exercises 9–12.

9. $\mathbf{F}(x, y, z) = e^z\mathbf{i} - \cos(xy)\mathbf{j} + z^3y\mathbf{k}$.
10. $\mathbf{\Phi}(x, y, z) = xz\cos x\mathbf{i} - yz\sin x\mathbf{j} - xy\tan y\mathbf{k}$.
11. $\mathbf{\Phi}(x, y, z) = \dfrac{yz}{x^2 + y^2 + z^2}\mathbf{i} - \dfrac{xz}{x^2 + y^2 + z^2}\mathbf{j} + \dfrac{xy}{x^2 + y^2 + z^2}\mathbf{k}$.
12. $\mathbf{F}(x, y, z) = (\nabla \times \mathbf{\Phi})(x, y, z)$, where $\mathbf{\Phi}$ is given in Exercise 10.

13. Prove the identity $\operatorname{curl}(f\mathbf{\Phi}) = f\operatorname{curl}\mathbf{\Phi} + \nabla f \times \mathbf{\Phi}$.
14. If $\mathbf{r} = x\mathbf{i} + y\mathbf{j} + z\mathbf{k}$, prove that $\nabla \times \mathbf{r} = \mathbf{0}$.
15. Show that Stokes' theorem reduces to $0 = 0$ for $\mathbf{\Phi} = \nabla f$, by evaluating each side directly.
16. Prove the identity
$$\int\int_S (\nabla f \times \nabla g) \cdot \mathbf{n}\, dA = \int_{\partial S} f\nabla g \cdot d\mathbf{r} = -\int_{\partial S} g\nabla f \cdot d\mathbf{r}.$$
17. Let
$$\mathbf{\Phi} = \frac{\mathbf{i}}{y + z} - \frac{x\mathbf{j}}{(y + z)^2} - \frac{x\mathbf{k}}{(y + z)^2}.$$
Show that the integral of $\mathbf{\Phi}$ around an oriented simple curve C that is the boundary of a surface S is zero.
18. Let $\mathbf{\Phi} = (yze^x + xyze^x)\mathbf{i} + xze^x\mathbf{j} + xye^x\mathbf{k}$. Repeat Exercise 17.
19. Let $\mathbf{\Phi} = 2x\mathbf{i} - y\mathbf{j} + (x + z)\mathbf{k}$. Evaluate the integral of $\mathbf{\Phi}$ around the curve consisting of straight lines joining $(1, 0, 1)$, $(0, 1, 0)$, and $(0, 0, 1)$, using Stokes' theorem.
20. Let C consist of straight lines joining $(2, 0, 0)$, $(0, 1, 0)$, and $(0, 0, 3)$. Evaluate the integral of $\mathbf{\Phi}(x, y, z) = xy\mathbf{i} + yz\mathbf{j} + xz\mathbf{k}$ around C by using Stokes' theorem.
21. Let $\mathbf{\Phi}$ be perpendicular to the tangent vector of the boundary ∂S of a surface S. Show that $\int\int_S (\nabla \times \mathbf{\Phi}) \cdot \mathbf{n}\, dA = 0$.
22. Let $\mathbf{\Phi} = ax\mathbf{i} + by\mathbf{j} + cz\mathbf{k}$ and let C be a curve in a plane with normal $\mathbf{n}$ and enclosing the area A. Find an expression for $\int_C \mathbf{\Phi} \cdot d\mathbf{r}$ using Stokes' theorem.
23. Evaluate $\int\int_S (\nabla \times \mathbf{F}) \cdot \mathbf{n}\, dA$, where S is the portion of the sphere $x^2 + y^2 + z^2 = 9$ defined by $x + y \geqslant 1$, and where $\mathbf{F} = \mathbf{r} \times (\mathbf{i} + \mathbf{j})$.
24. Evaluate $\int\int_S (\nabla \times \mathbf{F}) \cdot \mathbf{n}\, dA$, where S is the portion of the surface of a sphere $x^2 + y^2 + z^2 = 4$ and $3x + 2y - z \geqslant 1$, and where $\mathbf{F}$ is the vector field $\mathbf{r} \times (3\mathbf{i} + 2\mathbf{j} - \mathbf{k})$.
25. *Ampere's law* states that if the electric current density is described by a vector field $\mathbf{J}$ and the induced magnetic field is $\mathbf{H}$, then the circulation of $\mathbf{H}$ around the boundary C of a surface S equals the integral of $\mathbf{J}$ over S (i.e., the total current crossing S). See Fig. 18.5.10. Show that this is implied by the steady-state *Maxwell equation* $\nabla \times \mathbf{H} = \mathbf{J}$.

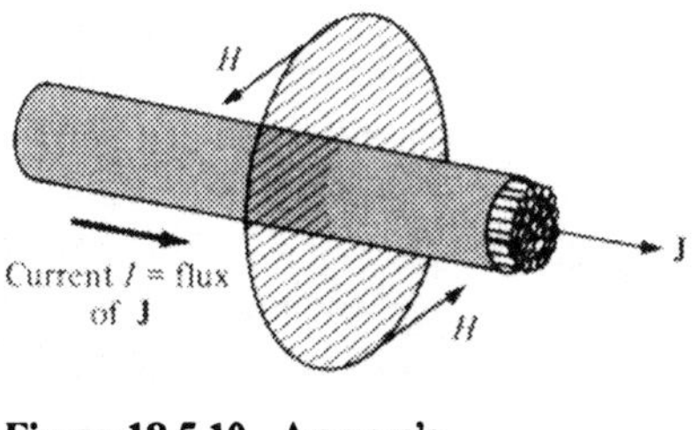

Figure 18.5.10. Ampere's law.

26. Let $\mathbf{F} = x^2\mathbf{i} + (2xy + x)\mathbf{j} + z\mathbf{k}$. Let C be the circle $x^2 + y^2 = 1$ oriented counterclockwise and S the disk $x^2 + y^2 \leqslant 1$. Determine:
    (a) The integral of $\mathbf{F}$ over S.
    (b) The circulation of $\mathbf{F}$ around C.
    (c) Find the integral of $\nabla \times \mathbf{F}$ over S. Verify Stokes' theorem directly in this case.

★27. Imagine a fluid moving in space. Take a paddle wheel on the end of a stick and put it in the fluid. Move the stick around until the paddle wheel rotates the fastest in a counter-clockwise direction. Your stick now points in the direction of the curl. Justify.

★28. Generalize Example 7 by replacing $x + y + z \geqslant 1$ by $ax + by + cz \geqslant d$ and $\mathbf{F} = \mathbf{r} \times (\mathbf{i} + \mathbf{j} + \mathbf{k})$ by $\mathbf{F} = \mathbf{r} \times (a\mathbf{i} + b\mathbf{j} + c\mathbf{k})$. (What inequality is needed to ensure that the plane $ax + by + cz = d$ intersects the sphere $x^2 + y^2 + z^2 = 1$?)

★29. Prove that Stokes' theorem holds for the surface $x^2 + y^2 + z^2 = 1$, $z \geqslant -1/2$.

18.6 Flux and the Divergence Theorem

The integral of the normal component of a vector field over a closed surface equals the integral of its divergence over the enclosed volume.

Let **V** be the velocity field of a fluid moving in the plane. In the previous section, we explained why the line integral of **V** around a closed curve C is called the circulation of **V** around C. The line integral is the integral of the tangential component of **V**. The integral around C of the normal component of **V** also has physical meaning.

Imagine first that **V** is constant and C is a line segment; see Fig. 18.6.1.

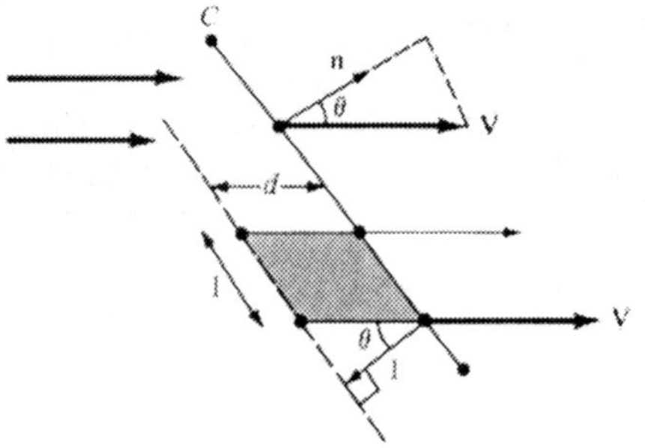

Figure 18.6.1. The amount of fluid crossing C per unit time is the normal component of **V** times the length of C.

Suppose we consider a parallelogram consisting of a unit area of fluid, the shaded area in Fig. 18.6.1. The parallelogram's base is one unit in length along C and has its other side parallel to **V**. Since its area is one, the other side has length $d = 1/\cos\theta$, where θ is the angle between **n** and **V**. It takes this parallelogram $t = d/\|\mathbf{V}\| = 1/[\cos\theta\|\mathbf{V}\|]$ units of time to cross C. Thus $\cos\theta\|\mathbf{V}\|$ square units of fluid cross each unit length of C per unit time. Since **n** has unit length, this rate equals $\mathbf{V}\cdot\mathbf{n}$.

If we now imagine C to consist of straight line segments and **V** to be constant across each one, we are led to interpret the integral of the normal component of **V** along C, that is,

$$\int_C \mathbf{V}\cdot\mathbf{n}\,ds,$$

as the amount of fluid crossing C per unit of time. This integral is the *flux* of **V** across C.

Let C be parametrized by $\boldsymbol{\sigma}(t) = x(t)\mathbf{i} + y(t)\mathbf{j}$. Then a unit tangent vector is

$$\mathbf{t} = \frac{x'\mathbf{i} + y'\mathbf{j}}{\sqrt{(x')^2 + (y')^2}}, \qquad \text{where } x' = dx/dt \text{ and } y' = dy/dt.$$

The element of length is

$$ds = \sqrt{(x')^2 + (y')^2}\,dt$$

and a unit normal is

$$\mathbf{n} = \frac{y'\mathbf{i} - x'\mathbf{j}}{\sqrt{(x')^2 + (y')^2}}.$$

This **n** has length 1 and is perpendicular to **t**, as is easily checked. We chose this $\mathbf{n}$ and not its negative so that if C is a closed curve traversed counterclockwise, $\mathbf{n}$ will be the unit *outward* normal, as in Fig. 18.6.2. If $\mathbf{V} = P\mathbf{i} + Q\mathbf{j}$, then substitution of the above formulas for $\mathbf{n}$ and ds gives

$$\mathbf{V} \cdot \mathbf{n}\, ds = (Py' - Qx')\, dt.$$

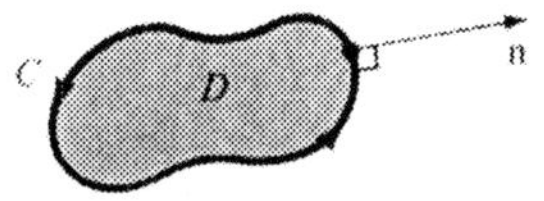

Figure 18.6.2. $\mathbf{n} = \dfrac{y'\mathbf{i} - x'\mathbf{j}}{\sqrt{(x')^2 + (y')^2}}$ is the unit outward normal.

This leads to the following definition.

The Flux of a Vector Field

The *flux* of $\mathbf{V}$ across a curve C is defined to be

$$\int_C \mathbf{V} \cdot \mathbf{n}\, ds = \int_C P\, dy - Q\, dx.$$

If C is parametrized by $\boldsymbol{\sigma}(t) = x(t)\mathbf{i} + y(t)\mathbf{j}$, $a \leqslant t \leqslant b$, the flux equals

$$\int_a^b \left(P(x(t), y(t)) \frac{dy}{dt} - Q(x(t), y(t)) \frac{dx}{dt} \right) dt,$$

as our calculations above show.

The divergence theorem relates the flux of a vector field $\mathbf{V}$ across C to the integral of the *divergence* of $\mathbf{V}$ over D defined as follows.

Divergence in the Plane

The *divergence* of a vector field $\mathbf{V} = P\mathbf{i} + Q\mathbf{j}$ is the function given by

$$\operatorname{div} \mathbf{V} = \nabla \cdot \mathbf{V} = \frac{\partial P}{\partial x} + \frac{\partial Q}{\partial y}.$$

We can remember the formula for $\operatorname{div} \mathbf{V}$ in the plane by writing the "dot product" of $\nabla = (\partial/\partial x)\mathbf{i} + (\partial/\partial y)\mathbf{j}$ with $\mathbf{V}$, where instead of multiplying ∂ ∂x and P, we let $\partial/\partial x$ *operate on* P. This is similar to the way we regarded curl $\boldsymbol{\Phi} = \nabla \times \boldsymbol{\Phi}$ as the cross product of $\nabla = (\partial/\partial x)\mathbf{i} + (\partial/\partial y)\mathbf{j} + (\partial/\partial z)\mathbf{k}$ with $\boldsymbol{\Phi}$ (see Example 3, Section 18.5).

Gauss' Divergence Theorem in the Plane

Let D be a region in the plane to which Green's theorem applies and let C be its boundary traversed in a counterclockwise direction. Then

$$\int_C \mathbf{V} \cdot \mathbf{n}\, ds = \iint_D (\operatorname{div} \mathbf{V})\, dx\, dy.$$

The proof is as follows. The left-hand side equals

$$\int_C P\, dy - Q\, dx.$$

By Green's theorem, this equals (with P replaced by $-Q$ and Q by P in the statement of Green's theorem)

$$\iint_D \left(\frac{\partial P}{\partial x} - \frac{\partial(-Q)}{\partial y} \right) dx\, dy = \iint_D (\operatorname{div} \mathbf{V})\, dx\, dy. \ \blacksquare$$

Example 1 Calculate the flux of $\mathbf{V} = x\cos y\,\mathbf{i} - \sin y\,\mathbf{j}$ across the boundary of the unit square in the plane with vertices $(0,0)$, $(1,0)$, $(1,1)$, and $(0,1)$.

Solution The divergence of $\mathbf{V}$ is

$$\operatorname{div}\mathbf{V} = \frac{\partial}{\partial x}(x\cos y) + \frac{\partial}{\partial y}(-\sin y) = \cos y - \cos y = 0,$$

so by the divergence theorem, the flux across any closed curve is zero. Thus the flux across the given boundary is zero. (Notice that directly computing the flux across C is possible, but more tedious.) ▲

A vector field in the plane is called *incompressible* or *divergence free* if $\operatorname{div}\mathbf{V} = 0$. This terminology arises from the divergence theorem and the example in which $\mathbf{V}$ is the velocity of a fluid. In fact, the divergence theorem implies that the flux across closed curves is zero, that is, the net area of fluid entering and leaving the region enclosed by C is zero. For a compressible fluid, it can happen that the fluid inside C is squeezed so that the net flux across C is negative. In this case, $\operatorname{div}\mathbf{V}$ would be negative. Likewise, if $\operatorname{div}\mathbf{V} > 0$ in a region, the fluid is expanding. See Fig. 18.6.3.

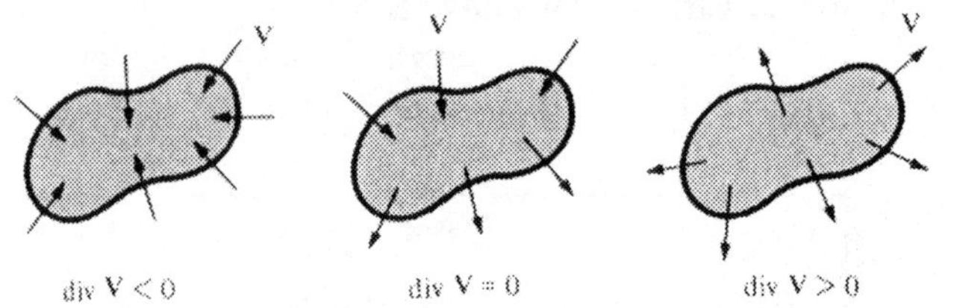

Figure 18.6.3. A compressing, incompressible and expanding fluid.

Example 2 Figure 18.6.4 shows some flow lines for a fluid moving in the plane with velocity field $\mathbf{V}$. What would you guess[4] the sign of $\operatorname{div}\mathbf{V}$ to be at points A, B, C, and D?

Figure 18.6.4. Find the sign of div **V**.

Solution The fluid appears to be emerging from small regions near A, B, and C, so at these points it is reasonable to suppose $\operatorname{div}\mathbf{V} > 0$. At D the fluid appears to be converging, so there $\operatorname{div}\mathbf{V} < 0$. ▲

We saw that a generalization of Green's theorem to three dimensions which relies on the idea of circulation is given by Stokes' theorem. It is natural to also seek a generalization of the divergence theorem to three dimensions.

Let $\mathbf{V}$ be a vector field defined in space. Reasoning as we did in the plane, we see that if $\mathbf{V}$ represents the velocity field of a fluid and S is a surface, then the surface integral $\iint_S \mathbf{V}\cdot\mathbf{n}\,dA$ is the volume of fluid crossing the surface S in the direction of the normal $\mathbf{n}$ per unit time. Thus we call $\iint_S \mathbf{V}\cdot\mathbf{n}\,dA$ the *flux* of $\mathbf{V}$ across S.

If $\mathbf{V} = P\mathbf{i} + Q\mathbf{j} + R\mathbf{k}$, we can generalize our definition of two-dimensional divergence by setting

$$\operatorname{div}\mathbf{V} = \frac{\partial P}{\partial x} + \frac{\partial Q}{\partial y} + \frac{\partial R}{\partial z},$$

again called the *divergence* of the three-dimensional vector field $\mathbf{V}$.

[4]One has to guess, because one cannot really tell from the picture without knowing how fast the fluid is moving.

Gauss' Divergence Theorem in Space

Let W be a region in space which is of type I, II, and III (see Section 17.4), and let ∂W be the surface of W with $\mathbf{n}$ the outward pointing unit normal. Let $\mathbf{V}$ be a vector field defined on W. Then

$$\iiint_W (\operatorname{div}\mathbf{V})\,dx\,dy\,dz = \iint_{\partial W}(\mathbf{V}\cdot\mathbf{n})\,dA.$$

In words, the total flux across the boundary of W equals the total divergence in W.

Proof of Gauss' Theorem It is sufficient to prove the three equalities

$$\iint_{\partial W} P\mathbf{i}\cdot\mathbf{n}\,dA = \iiint_W \frac{\partial P}{\partial x}\,dx\,dy\,dz, \tag{1}$$

$$\iint_{\partial W} Q\mathbf{j}\cdot\mathbf{n}\,dA = \iiint_W \frac{\partial Q}{\partial y}\,dx\,dy\,dz, \tag{2}$$

$$\iint_{\partial W} R\mathbf{k}\cdot\mathbf{n}\,dA = \iiint_W \frac{\partial R}{\partial z}\,dx\,dy\,dz. \tag{3}$$

This is because

$$\iiint_W (\operatorname{div}\mathbf{V})\,dx\,dy\,dz = \iiint_W \frac{\partial P}{\partial x}\,dx\,dy\,dz + \iiint_W \frac{\partial Q}{\partial y}\,dy\,dy\,dz + \iiint_W \frac{\partial R}{\partial z}\,dx\,dy\,dz,$$

and

$$\begin{aligned}\iint_{\partial W}\mathbf{V}\cdot\mathbf{n}\,dA &= \iint_{\partial W}(P\mathbf{i}+Q\mathbf{j}+R\mathbf{k})\cdot\mathbf{n}\,dA\\ &= \iint_{\partial W}P\mathbf{i}\cdot\mathbf{n}\,dA + \iint_{\partial W}Q\mathbf{j}\cdot\mathbf{n}\,dA + \iint_{\partial W}R\mathbf{k}\cdot\mathbf{n}\,dA.\end{aligned}$$

The equality (3) will be proved here; the other two are proved in an analogous fashion.

Express W by the inequalities

$$f_2(x,y) \leqslant z \leqslant f_1(x,y), \qquad (x,y) \text{ in } D$$

for functions f_1 and f_2 on a domain D in the xy plane. (See Fig. 18.6.5.) The boundary of W is closed surface whose top S_1 is the graph of $z = f_1(x,y)$, (x,y) in D, and whose bottom S_2 is the graph of $z = f_2(x,y)$, (x,y) in D. The four other sides of ∂W (if they are not reduced to curves) consist of surfaces S_3, S_4, S_5, and S_6 whose normals are always perpendicular to the z axis. We claim that

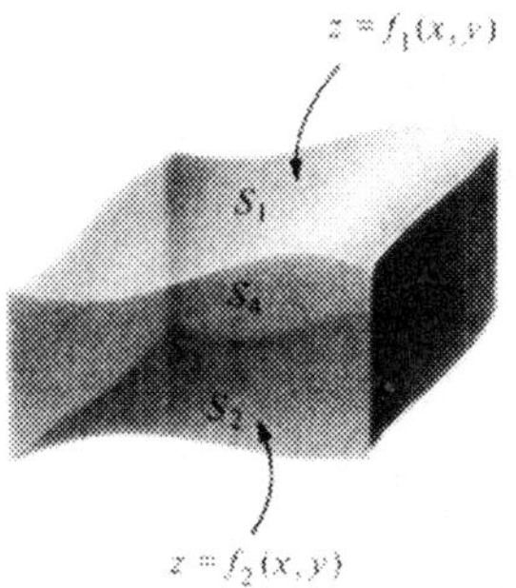

Figure 18.6.5. A region W of type I. The four sides of ∂W, namely S_3, S_4, S_5, S_6 have normals perpendicular to the z axis.

$$\iiint_W \frac{\partial R}{\partial z}\,dx\,dy\,dz = \iint_D \left[R(x,y,f_1(x,y)) - R(x,y,f_2(x,y))\right]dx\,dy. \tag{4}$$

Indeed, by the fundamental theorem of calculus and the reduction to iterated integrals,

$$\begin{aligned}\iiint_W \frac{\partial R}{\partial z}\,dz\,dy\,dx &= \iint_D \left[R(x,y,z)\Big|_{z=f_2(x,y)}^{f_1(x,y)}\right]dy\,dx\\ &= \iint_D \left[R(x,y,f_1(x,y)) - R(x,y,f_2(x,y))\right]dx\,dy.\end{aligned}$$

Next we break up the left side of (3) into the sum of six terms:

$$\iint_{\partial W} R\mathbf{k}\cdot\mathbf{n}\,dA = \iint_{S_1} R\mathbf{k}\cdot\mathbf{n}_1\,dA + \iint_{S_2} R\mathbf{k}\cdot\mathbf{n}_2\,dA + \sum_{i=3}^{6}\iint_{S_i} R\mathbf{k}\cdot\mathbf{n}_i\,dA. \quad (5)$$

Since on each of S_3, S_4, S_5, and S_6, the normal $\mathbf{n}_i$ is perpendicular to $\mathbf{k}$, we have $\mathbf{k}\cdot\mathbf{n}_i = 0$ along these faces, and so the integral (5) reduces to

$$\iint_{\partial W} R\mathbf{k}\cdot\mathbf{n}\,dA = \iint_{S_1} R\mathbf{k}\cdot\mathbf{n}_1\,dA + \iint_{S_2} R\mathbf{k}\cdot\mathbf{n}_2\,dA. \quad (6)$$

The surface S_2 is defined by $z = f_2(x, y)$, so

$$\mathbf{n}_2 = \frac{(\partial f_2/\partial x)\mathbf{i} + (\partial f_2/\partial y)\mathbf{j} - \mathbf{k}}{\sqrt{(\partial f_2/\partial x)^2 + (\partial f_2/\partial y)^2 + 1}}. \quad (7)$$

Thus

$$\mathbf{n}_2\cdot\mathbf{k} = \frac{-1}{\sqrt{(\partial f_2/\partial x)^2 + (\partial f_2/\partial y)^2 + 1}},$$

and so

$$\iint_{S_2} R(\mathbf{k}\cdot\mathbf{n}_2)\,dA$$

$$= \iint_D R(x, y, f_2(x, y))\left[\frac{-1}{\sqrt{\left(\frac{\partial f_2}{\partial x}\right)^2 + \left(\frac{\partial f_2}{\partial y}\right)^2 + 1}}\right]\sqrt{\left(\frac{\partial f_2}{\partial x}\right)^2 + \left(\frac{\partial f_2}{\partial y}\right)^2 + 1}$$

$$= -\iint_D R(x, y, f_2(x, y))\,dx\,dy. \quad (8)$$

The formula for $\mathbf{n}_1$ is similar to formula (7) for $\mathbf{n}_2$. However, $\mathbf{n}_1$ points upward, so the numerator of $\mathbf{n}_1$ is $-(\partial f_1/\partial x)\mathbf{i} - (\partial f_1/\partial y)\mathbf{j} + \mathbf{k}$ (Note the positive $\mathbf{k}$ component). Thus

$$\mathbf{n}_1\cdot\mathbf{k} = \left[\left(\frac{\partial f_1}{\partial x}\right)^2 + \left(\frac{\partial f_1}{\partial y}\right)^2 + 1\right]^{-1/2},$$

and so

$$\iint_{S_1} R(\mathbf{k}\cdot\mathbf{n}_1)\,dA = \iint_D R(x, y, f_1(x, y))\,dx\,dy. \quad (9)$$

Substituting (8) and (9) in (6) gives

$$\iint_{\partial W} R(\mathbf{k}\cdot\mathbf{n})\,dA$$

$$= \iint_D R(x, y, f_1(x, y))\,dx\,dy - \iint_D R(x, y, f_2(x, y))\,dx\,dy$$

$$= \iint_D \left[R(x, y, f_1(x, y)) - R(x, y, f_2(x, y))\right]dx\,dy$$

$$= \iiint_W \frac{\partial R}{\partial z}\,dx\,dy\,dz, \qquad \text{by formula (4).}$$

This proves that

$$\iint_{\partial W} R\mathbf{k}\cdot\mathbf{n}\,dA = \iiint_W \frac{\partial R}{\partial z}\,dx\,dy\,dz$$

which is the identity (3) which we wanted to show. ■

As with Green's and Stokes' theorems, Gauss' theorem holds for regions more general than those of types I, II, and III. This assertion may be established by breaking up a region W into smaller subregions $W_1, \ldots, W_n$ each of which is of types I, II, and III. We apply Gauss' theorem to each of these smaller regions and add the results. The surface integrals along common boundaries interior to W cancel, and we are left with Gauss' theorem for the original region W.

Example 3 Evaluate

$$\iint_S \mathbf{F}\cdot\mathbf{n}\,dA, \qquad \text{where} \quad \mathbf{F}(x,y,z) = xy^2\mathbf{i} + x^2y\mathbf{j} + y\mathbf{k}$$

and S is the surface of the cylinder W defined by $x^2+y^2 \leqslant 1$, $-1 \leqslant z \leqslant 1$.

Solution One can compute the integral directly, but it is easier to use the divergence theorem.

Since S is the boundary of the region W, the divergence theorem gives $\iint_S \mathbf{F}\cdot\mathbf{n}\,dA = \iiint_W \operatorname{div}\mathbf{F}\,dx\,dy\,dz$. Now

$$\begin{aligned}\operatorname{div}\mathbf{F} &= \frac{\partial}{\partial x}(xy^2) + \frac{\partial}{\partial y}(x^2y) + \frac{\partial}{\partial z}(y)\\ &= x^2+y^2,\end{aligned}$$

and so

$$\begin{aligned}\iiint_W \operatorname{div}\mathbf{F}\,dx\,dy\,dz &= \iiint_W (x^2+y^2)\,dx\,dy\,dz\\ &= \int_{-1}^{1}\left(\iint_{x^2+y^2\leqslant 1}(x^2+y^2)\,dx\,dy\right)dz\\ &= 2\iint_{x^2+y^2\leqslant 1}(x^2+y^2)\,dx\,dy.\end{aligned}$$

We change variables to polar coordinates to evaluate the double integral:

$$x = r\cos\theta, \qquad y = r\sin\theta, \qquad 0\leqslant r\leqslant 1, \qquad 0\leqslant\theta\leqslant 2\pi.$$

Replacing x^2+y^2 by r^2 and $dx\,dy$ by $r\,dr\,d\theta$, we have

$$\iint_{x^2+y^2\leqslant 1}(x^2+y^2)\,dx\,dy = \int_0^{2\pi}\left(\int_0^1 r^3\,dr\right)d\theta = \frac{1}{2}\pi.$$

Therefore

$$\iint_S \mathbf{F}\cdot\mathbf{n}\,dA = \iiint_W \operatorname{div}\mathbf{F}\,dx\,dy\,dz = \pi. \;\blacktriangle$$

Example 4 Let $\mathbf{V} = 2x\mathbf{i} + 2y\mathbf{j} + 2z\mathbf{k}$ and let S be the unit sphere $x^2+y^2+z^2=1$. Calculate the flux of $\mathbf{V}$ across S.

Solution By the divergence theorem, the flux of $\mathbf{V}$ across S equals

$$\iiint_W \operatorname{div}\mathbf{V}\,dx\,dy\,dz,$$

where W is the ball $x^2+y^2+z^2\leqslant 1$. However,

$$\operatorname{div}\mathbf{V}(x,y,z) = \frac{\partial}{\partial x}(2x) + \frac{\partial}{\partial y}(2y) + \frac{\partial}{\partial z}(2z) = 6.$$

Thus the flux is

$$6\times\text{volume}(W) = 6\cdot\frac{4}{3}\pi = 8\pi. \;\blacktriangle$$

Example 5 Calculate the flux of $\mathbf{V}(x, y, z) = x^3\mathbf{i} + y^3\mathbf{j} + z^3\mathbf{k}$ across the surface of the unit sphere $x^2 + y^2 + z^2 = 1$.

Solution For $\mathbf{V}(x, y, z) = x^3\mathbf{i} + y^3\mathbf{j} + z^3\mathbf{k}$, $\operatorname{div}\mathbf{V} = 3x^2 + 3y^2 + 3z^2$. By Gauss' theorem in space, the flux equals

$$\iiint_W (\operatorname{div}\mathbf{V})\,dx\,dy\,dz.$$

Using spherical coordinates (Section 17.5), this becomes

$$\int_0^{2\pi}\int_0^{\pi}\int_0^1 3\rho^4 \sin\phi\,d\rho\,d\phi\,d\theta = \frac{12}{5}\pi. \ \blacktriangle$$

Example 6 Prove the vector identities (a) $\nabla\cdot(f\mathbf{\Phi}) = (\nabla f)\cdot\mathbf{\Phi} + f\nabla\cdot\mathbf{\Phi}$ (b) $\nabla\cdot(\nabla\times\mathbf{\Phi}) = 0$.

Solution Suppose that $\mathbf{\Phi} = a\mathbf{i} + b\mathbf{j} + c\mathbf{k}$.

(a)

$$\begin{aligned}\nabla\cdot(f\mathbf{\Phi}) &= \frac{\partial(fa)}{\partial x} + \frac{\partial(fb)}{\partial y} + \frac{\partial(fc)}{\partial z}\\ &= \frac{\partial f}{\partial x}a + \frac{\partial f}{\partial y}b + \frac{\partial f}{\partial z}c + f\left(\frac{\partial a}{\partial x} + \frac{\partial b}{\partial y} + \frac{\partial c}{\partial z}\right) = (\nabla f)\cdot\mathbf{\Phi} + f\nabla\cdot\mathbf{\Phi}.\end{aligned}$$

(b)

$$\nabla\cdot(\nabla\times\mathbf{\Phi}) = \frac{\partial}{\partial x}\left(\frac{\partial b}{\partial z} - \frac{\partial c}{\partial y}\right) + \frac{\partial}{\partial y}\left(\frac{\partial c}{\partial x} - \frac{\partial a}{\partial z}\right) + \frac{\partial}{\partial z}\left(\frac{\partial a}{\partial y} - \frac{\partial b}{\partial x}\right).$$

By commutativity of partial derivatives, all terms cancel to give 0 as the result. ▲

Example 7 A basic law of electrostatics is that an electric field $\mathbf{E}$ in space satisfies $\operatorname{div}\mathbf{E} = \rho$, where ρ is the charge density. Show that the flux of $\mathbf{E}$ across a closed surface equals the total charge *inside* the surface.

Solution Let W be a region in space with boundary surface S. By the divergence theorem,

$$\begin{aligned}\left\{\begin{matrix}\text{flux of } \mathbf{E}\\ \text{across } S\end{matrix}\right\} &= \iint_S \mathbf{E}\cdot\mathbf{n}\,dA\\ &= \iiint_W \operatorname{div}\mathbf{E}\,dx\,dy\,dz\\ &= \iiint_W \rho(x, y, z)\,dx\,dy\,dz,\end{aligned}$$

since $\operatorname{div}\mathbf{E} = \rho$ by assumption; but since ρ is the charge per unit volume,

$$Q = \iiint_W \rho\,dx\,dy\,dz$$

is the total charge inside S. ▲

Exercises for Section 18.6

Calculate the divergence of the vector fields in Exercises 1–4.

1. $\mathbf{\Phi}(x, y) = x^3\mathbf{i} - x\sin(xy)\mathbf{j}$.
2. $\mathbf{\Phi}(x, y) = y\mathbf{i} - x\mathbf{j}$.
3. $\mathbf{F}(x, y) = \sin(xy)\mathbf{i} - \cos(x^2y)\mathbf{j}$.
4. $\mathbf{F}(x, y) = xe^y\mathbf{i} - [y/(x + y)]\mathbf{j}$.
5. Calculate the flux of $\mathbf{\Phi}(x, y) = x^2\mathbf{i} - y^3\mathbf{j}$ across the perimeter of the square whose vertices are $(-1, -1), (-1, 1), (1, 1), (1, -1)$.
6. Evaluate the flux of $\mathbf{\Phi}(x, y, z) = 3xy^2\mathbf{i} + 3x^2y\mathbf{j}$ out of the unit circle $x^2 + y^2 = 1$ in the plane.

7. Calculate the flux of $\mathbf{\Phi}(x, y) = y\mathbf{i} + e^x\mathbf{j}$ across the boundary of the square with vertices $(0,0)$, $(1,0)$, $(0,1)$ and $(1,1)$.
8. Calculate the flux of $x^3\mathbf{i} + y^3\mathbf{j}$ out of the unit circle $x^2 + y^2 = 1$.
9. Fig. 18.6.6 shows some flow lines for a fluid moving in the plane. Let $\mathbf{V}$ be the velocity field. At which of the indicated points A, B, C, D can one reasonably expect that (a) $\operatorname{div}\mathbf{V} > 0$? (b) $\operatorname{div}\mathbf{V} < 0$?

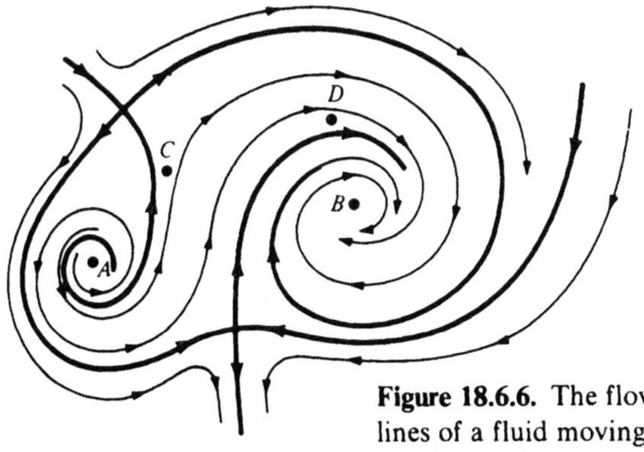

Figure 18.6.6. The flow lines of a fluid moving in the plane.

10. Fig. 18.6.7 shows some flow lines for a fluid moving in the plane. Let $\mathbf{V}$ be the velocity field. At which of the indicated points A, B, C, D can one reasonably expect that (a) $\operatorname{div}\mathbf{V} > 0$? (b) $\operatorname{div}\mathbf{V} < 0$?

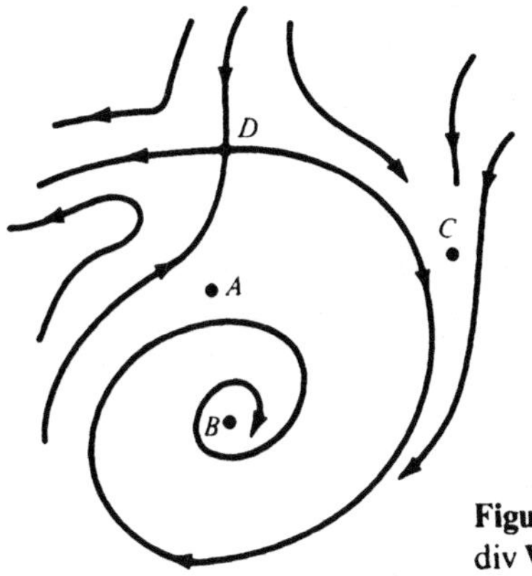

Figure 18.6.7. Where is $\operatorname{div}\mathbf{V} > 0$? < 0?

Find the divergence of the vector fields in Exercises 11–14.

11. $\mathbf{V}(x, y, z) = e^{xy}\mathbf{i} - e^{xy}\mathbf{j} + e^{yz}\mathbf{k}$.
12. $\mathbf{V}(x, y, z) = yz\mathbf{i} + xz\mathbf{j} + xy\mathbf{k}$.
13. $\mathbf{V}(x, y, z) = x\mathbf{i} + (y + \cos x)\mathbf{j} + (z + e^{xy})\mathbf{k}$.
14. $\mathbf{V}(x, y, z) = x^2\mathbf{i} + (x + y)^2\mathbf{j} + (x + y + z)^2\mathbf{k}$.

15. Find the flux of $\mathbf{\Phi}(x, y, z) = 3xy^2\mathbf{i} + 3x^2y\mathbf{j} + z^3\mathbf{k}$ out of the unit sphere.
16. Evaluate the flux of $\mathbf{\Phi}(x, y, z) = x\mathbf{i} + y\mathbf{j} + z\mathbf{k}$ out of the unit sphere.
17. Evaluate $\iint_{\partial W}\mathbf{F}\cdot\mathbf{n}\,dA$, where $\mathbf{F}(x, y, z) = x\mathbf{i} + y\mathbf{j} - z\mathbf{k}$ and W is the unit cube in the first octant. Perform the calculation directly and check by using the divergence theorem.
18. Evaluate the surface integral $\iint_{\partial S}\mathbf{F}\cdot\mathbf{n}\,dA$, where $\mathbf{F}(x, y, z) = \mathbf{i} + \mathbf{j} + z(x^2 + y^2)^2\mathbf{k}$ and ∂S is the surface of the cylinder $x^2 + y^2 \leqslant 1$, $0 \leqslant z \leqslant 1$.
19. Suppose a vector field $\mathbf{V}$ is tangent to the boundary of a region W in space. Prove that $\iiint_W(\operatorname{div}\mathbf{V})\,dx\,dy\,dz = 0$.
20. Prove the identity
$$\nabla\cdot(\mathbf{F}\times\mathbf{\Phi}) = \mathbf{\Phi}\cdot(\nabla\times\mathbf{F}) - \mathbf{F}\cdot(\nabla\times\mathbf{\Phi}).$$
21. Prove that
$$\iiint_W(\nabla f)\cdot\mathbf{\Phi}\,dx\,dy\,dz$$
$$= \iint_{\partial W} f\mathbf{\Phi}\cdot\mathbf{n}\,dA - \iiint_W f\nabla\cdot\mathbf{\Phi}\,dx\,dy\,dz.$$
22. Prove that $(\partial/\partial t)(\nabla\cdot\mathbf{H}) = 0$ from the Maxwell equation $\nabla\times\mathbf{E} = -\partial\mathbf{H}/\partial t$ (Example 8, Section 18.5).
23. i(i) Prove that $\nabla\cdot\mathbf{J} = 0$ from the steady state Maxwell equation $\nabla\times\mathbf{H} = \mathbf{J}$ (Exercise 25, Section 18.5).
    (ii) Argue physically that the flux of $\mathbf{J}$ through any closed surface is zero (conservation of charge). Use this to deduce $\nabla\cdot\mathbf{J} = 0$ from Gauss' theorem.
24. (a) Use Gauss' theorem to show that
$$\iint_{S_1}\nabla\times\mathbf{F}\cdot\mathbf{n}\,dA = \iint_{S_2}\nabla\times\mathbf{F}\cdot\mathbf{n}\,dA,$$
    where S_1 and S_2 have a common boundary.
    (b) Prove the same assertion using Stokes' Theorem.
    (c) Where was this used in Example 7, Section 18.5?

★25. Let ρ be a continuous function of $\mathbf{q} = (x, y, z)$ such that $\rho(\mathbf{q}) = 0$ except for $\mathbf{q}$ in some region Ω. The *potential* of ρ is defined as the function
$$\phi(\mathbf{p}) = \iiint_\Omega \frac{\rho(\mathbf{q})}{4\pi\|\mathbf{p} - \mathbf{q}\|}\,dx\,dy\,dz,$$
where $\|\mathbf{p} - \mathbf{q}\|$ is the distance between $\mathbf{p}$ and $\mathbf{q}$.
    (a) Show that for a region D in space $\iint_{\partial D}\nabla\phi\cdot\mathbf{n}\,dA = \iiint_D\rho\,dx\,dy\,dz$.
    (b) Show that ϕ satisfies *Poisson's equation* $\nabla^2\phi = \rho$.

Review Exercises for Chapter 18

Evaluate the line integrals in Exercises 1–10.

1. $\int_C xy\,dx + x\sin y\,dy$, where C is the straight line segment joining $(0,1,1)$ to $(2,2,-3)$.
2. $\int_C x\,dy$, where C is the unit circle $x^2+y^2=1$ traversed counterclockwise.
3. $\int_C xe^y\,dx - ye^x\,dy$ where C is the straight line segment joining $(0,1,0)$ to $(2,1,1)$.
4. $\int_C x^2\,dx + y^2\,dy$, where C is the circle $x^2+y^2=1$, traversed counterclockwise.
5. $\int_C \nabla(xy^2\cos z)\cdot d\mathbf{r}$ where C is a curve in space joining $(0,1,0)$ to $(8,2,\pi)$.
6. $\int_C \nabla(\cos(xyz))\cdot d\mathbf{r}$, where C is a curve in space joining $(0,0,0)$ to $(1,\pi/2,1)$.
7. $\int_C \frac{\partial}{\partial x}(e^{xyz})\,dx + \frac{\partial}{\partial y}(e^{xyz})\,dy + \frac{\partial}{\partial z}(e^{xyz})\,dz$, where C is a curve in space joining $(1,1,1)$ to $(0,2,0)$.
8. $\int_C \frac{\partial}{\partial x}(x\cos yz)\,dx + \frac{\partial}{\partial y}(x\cos yz)\,dy + \frac{\partial}{\partial z}(x\cos yz)\,dz$, where C is a curve in space joining $(0,0,0)$ to $(1,\pi,1)$.
9. $\int_C \sin x\,dx - \ln z\,dy + xy\,dz$, where C is parametrized by $(2t+1, \ln t, t^2)$, $1 \leqslant t \leqslant 2$.
10. $\int_C xy^2\,dx + (y+z)\,dy + [\sin(e^y)]\,dz$, where C is parametrized by $(\sin 2t, \cos t, 5)$, $0 \leqslant t \leqslant 2\pi$.
11. Let $f(x,y) = x^2\sin y - xy^2$. Evaluate the integral $\int_C(\partial f/\partial x)\,dx + (\partial f/\partial y)\,dy$, where C is parametrized by:
    (a) $(\tan \pi t, \ln(t+1))$; $-\frac{1}{4} \leqslant t \leqslant \frac{1}{4}$.
    (b) $(t\cos^2(\pi t/4), e^{1+t})$; $0 \leqslant t \leqslant 2$.
    (c) $(t^3 - 2t^2 + 1, \sin^5(\pi t/2) - 2t^{3/2})$; $0 \leqslant t \leqslant 1$.
12. If $f(x,y,z) = e^{x^2+y} - \ln(z^2+1)$, evaluate the line integral $\int_C \nabla f(\mathbf{r})\cdot d\mathbf{r}$ along the curve given by $\boldsymbol{\sigma}(t) = (-t^2, t^3+1, t\sin\frac{15}{2}\pi t)$, $0 \leqslant t \leqslant 1$.

In Exercises 13–16, let $f(x,y,z) = xze^y - z^3/(1+y^2)$. Evaluate the line integral $\int_C(\partial f/\partial x)\,dx + (\partial f/\partial y)\,dy + (\partial f/\partial z)\,dz$, where C has the given parametrization.

13. $\boldsymbol{\sigma}(t) = \left(\sqrt{t}\sin\frac{\pi}{4}(1+t), t^2-1, \frac{2-t^2}{2+t^2}\right)$; $0 \leqslant t \leqslant 1$.
14. $\boldsymbol{\sigma}(t) = (\cos^4 t, e^{\sin(\pi t/2)}, 2-t)$; $0 \leqslant t \leqslant 1$.
15. $\boldsymbol{\sigma}(t) = \left(\cos\frac{\pi}{2}t, \sin\frac{\pi}{2}t, t\right)$; $-1 \leqslant t \leqslant 1$.
16. $\boldsymbol{\sigma}(t) = (e^t, e^{-t}, t^2)$, $0 \leqslant t \leqslant 1$.
17. Find the work done moving a particle along the path $\boldsymbol{\sigma}(t) = (t, t^2)$, $0 \leqslant t \leqslant 1$ subject to the force $\boldsymbol{\Phi}(x,y) = e^x\mathbf{i} - xe^{xy}\mathbf{j}$.
18. Find the work done moving a particle from $(0,0,0)$ to $(1,1,1)$ along the path
    $$(x,y,z) = (t^3, t, t^2)$$
    subject to the force $\boldsymbol{\Phi}(x,y,z) = z\cos x\mathbf{j} + 8yz\mathbf{k}$.

In Exercises 19–22, determine if the given vector field $\boldsymbol{\Phi}$ is conservative; if it is, find an f such that $\boldsymbol{\Phi} = \nabla f$.

19. $\boldsymbol{\Phi}(x,y) = x\tan y\mathbf{i} + x\sec^2 y\mathbf{j}$.
20. $\boldsymbol{\Phi}(x,y) = \tan y\mathbf{i} + x\sec^2 y\mathbf{j}$.
21. $\boldsymbol{\Phi}(x,y) = 3x^2y^2\mathbf{i} + 2x^3y^2\mathbf{j}$.
22. $\boldsymbol{\Phi}(x,y) = 3x^2y^2\mathbf{i} + 2x^3y\mathbf{j}$.
23. In Exercise 30, Section 18.2, it was shown that a vector field
    $$\boldsymbol{\Phi}(x,y,z) = a(x,y,z)\mathbf{i} + b(x,y,z)\mathbf{j} + c(x,y,z)\mathbf{k}$$
    defined in all of space is conservative if and only if $a_y = b_x$, $a_z = c_x$, and $b_z = c_y$. Is $\boldsymbol{\Phi}(x,y,z) = 2xy\mathbf{i} + (3z^3 + x^2)\mathbf{j} + 9yz^2\mathbf{k}$ conservative? If so, find a function f such that $\nabla f = \boldsymbol{\Phi}$.
24. (a) If $\boldsymbol{\Phi}$ is conservative, prove that $\nabla\times\boldsymbol{\Phi} = \mathbf{0}$.
    (b) Find an f such that
    $$\nabla f = \tan^{-1}(yz)\mathbf{i} + \left(1 + \frac{xz}{y^2z^2+1}\right)\mathbf{j} + \frac{xy}{y^2z^2+1}\mathbf{k}.$$

Which of the differential forms in Exercises 25–28 are exact? Find antiderivatives for those that are.

25. $(e^y\sin x + xe^y\cos x)\,dx + xe^y\sin x\,dy$.
26. $(y\cos x + \sin z)\,dx + (z\cos y + \sin x)\,dy + (x\cos z + \sin y)\,dz$. (You may wish to study Review Exercise 23 first.)
27. $xe^y\,dx + ye^x\,dy$.
28. $\exp(x^2+y^2)(x\,dx + y\,dy)$.

Solve the differential equations in Exercises 29–32.

29. $y\cos x + 2xe^y + (\sin x + x^2e^y + 2)\frac{dy}{dx} = 0$, $y(\pi/2) = 0$.
30. $3x^2 - 2y + e^xe^y + (e^xe^y - 2x + 4)\frac{dy}{dx} = 0$, $y(0) = 0$.
31. $2xy + (x^2+1)\frac{dy}{dx} = 1$, $y(1) = 1$.
32. $2xy - 2x + (x^2+1)\frac{dy}{dx} = 0$.

Test each of the equations in Exercises 33–36 for exactness and solve the exact ones.

33. $xy - (x^2+3y^2)\frac{dy}{dx} = 0$.
34. $\sin x\sin y - xe^y + (e^y + \cos x\cos y)\frac{dy}{dx} = 0$.
35. $5x^3y^4 - 2y + (3x^2y^5 + x)\frac{dy}{dx} = 0$.
36. $9x^2 + y = (4y - x)\frac{dy}{dx} + 1$.

37. If D is a region in the plane with boundary curve C traversed counterclockwise, express the following three integrals in terms of the area of D: (a) $\int_C x\,dy$, (b) $\int_C y\,dx$, (c) $\int_C x\,dx$.
38. Use Green's theorem to calculate the line integral $\int_C (x^3 + y^3)\,dy - (x^3 + y)\,dx$, where C is the circle $x^2 + y^2 = 1$ traversed counterclockwise.

Calculate the curl and the divergence of the vector fields in Exercises 39–42.

39. $\mathbf{\Phi}(x, y, z) = x\mathbf{i} + [y/(x + z)]\mathbf{j} - z\mathbf{k}$.
40. $\mathbf{\Phi}(x, y, z) = 2xe^{y}\mathbf{i} - y^2e^{z}\mathbf{j} + ze^{x-y}\mathbf{k}$.
41. $\mathbf{\Phi}(x, y, z) = \mathbf{r} \times (x\mathbf{i} - y\mathbf{j} - z\mathbf{k})$, where $\mathbf{r} = x\mathbf{i} + y\mathbf{j} + z\mathbf{k}$.
42. $\mathbf{\Phi} = \nabla \times \mathbf{F}$, where $\mathbf{F}(x, y, z) = 3x^2\mathbf{i} + \cos(yz)\mathbf{j} - \sin(xy)\mathbf{k}$.
43. (a) Let $\mathbf{F} = y\mathbf{i} - x\mathbf{j} + zx^3y^2\mathbf{k}$. Calculate $\nabla \times \mathbf{F}$ and $\nabla \cdot \mathbf{F}$.
    (b) Evaluate $\int\int_{S_1} (\nabla \times \mathbf{F}) \cdot \mathbf{n}\,dA$ where S_1 is the surface $x^2 + y^2 + z^2 = 1$, $z \leqslant 0$.
    (c) Evaluate $\int\int_{S_2} \mathbf{F} \cdot \mathbf{n}\,dA$ where S_2 is the surface of the unit cube in the first octant.
44. (a) Let $\mathbf{F} = (x^2 + y - 4)\mathbf{i} + 3xy\mathbf{j} + (2xz + z^2)\mathbf{k}$. Calculate the divergence and curl of $\mathbf{F}$.
    (b) Find the flux of the curl of $\mathbf{F}$ across the surface $x^2 + y^2 + z^2 = 16$, $z \geqslant 0$.
    (c) Find the flux of $\mathbf{F}$ across the surface of the unit cube in the first quadrant.
45. Express as a surface integral the work done by a force field $\mathbf{F}$ going around a closed curve C in space.
46. Suppose that $\operatorname{div} \mathbf{F} > 0$ inside the unit ball $x^2 + y^2 + z^2 \leqslant 1$. Show that $\mathbf{F}$ cannot be everywhere tangent to the surface of the sphere. Give a physical interpretation of this result.
47. Calculate the surface integral $\int\int_S (\nabla \times \mathbf{F}) \cdot \mathbf{n}\,dA$, where S is the hemisphere $x^2 + y^2 + z^2 = 1$, $z \geqslant 0$ and $\mathbf{F} = x^3\mathbf{i} - y^3\mathbf{j}$.
48. Calculate the integral of the vector field in Exercise 47 over the hemisphere $x^2 + y^2 + z^2 = 1$, $z \leqslant 0$.
49. Calculate the integral $\int\int_S \mathbf{F} \cdot \mathbf{n}\,dA$, where S is the surface of the half ball $x^2 + y^2 + z^2 \leqslant 1$, $z \geqslant 0$, and $\mathbf{F} = (x + 3y^5)\mathbf{i} + (y + 10xz)\mathbf{j} + (z - xy)\mathbf{k}$.
50. Find $\int\int_S (\nabla \times \mathbf{\Phi}) \cdot \mathbf{n}\,dA$, where S is the ellipsoid $x^2 + y^2 + 2z^2 = 10$ and $\mathbf{\Phi} = \sin xy\mathbf{i} + e^x\mathbf{j} - yz\mathbf{k}$.

For a region W in space with boundary ∂W, unit outward normal $\mathbf{n}$ and functions f and g defined on W and ∂W, prove *Green's identities* in Exercises 51 and 52, where $\nabla^2 f = \dfrac{\partial^2 f}{\partial x^2} + \dfrac{\partial^2 f}{\partial y^2} + \dfrac{\partial^2 f}{\partial z^2}$ is the Laplacian of f.

51. $$\int\int_{\partial W} f(\nabla g) \cdot \mathbf{n}\,dA = \int\int\int_W (f\nabla^2 g + \nabla f \cdot \nabla g)\,dx\,dy\,dz.$$
52. $$\int\int_{\partial W} (f\nabla g - g\nabla f) \cdot \mathbf{n}\,dA = \int\int\int_W (f\nabla^2 g - g\nabla^2 f)\,dx\,dy\,dz.$$
53. Show that $\operatorname{div} \mathbf{\Phi}$ at a point P_0 in space is the "flux of $\mathbf{\Phi}$ per unit volume" at P_0.
54. In Section 18.4, we gave an example of a region to which Green's theorem as stated did not apply, but which could be cut up into smaller regions to which it did apply. In this way, Green's theorem was extended. Give an example of a similar procedure for Stokes' theorem.
55. Surface integrals apply to the study of heat flow. Let $T(x, y, z)$ be the temperature at a point (x, y, z) in W where W is some region in space and T is a function with continuous partial derivatives. Then
$$\nabla T = \frac{\partial T}{\partial x}\mathbf{i} + \frac{\partial T}{\partial y}\mathbf{j} + \frac{\partial T}{\partial z}\mathbf{k}$$
represents the temperature gradient, and heat "flows" with the vector field $-k\nabla T = \mathbf{F}$, where k is the positive constant. Therefore $\int\int_S \mathbf{F} \cdot \mathbf{n}\,dA$ is the total rate of heat flow or flux across the surface S. ($\mathbf{n}$ is the unit outward normal.)
    Suppose a temperature function is given as $T(x, y, z) = x^2 + y^2 + z^2$, and let S be the unit sphere $x^2 + y^2 + z^2 = 1$. Find the heat flux across the surface S if $k = 1$.

★56. (a) Express conservation of thermal energy by means of the statement that for any volume W in space
$$\frac{d}{dt}\int\int\int_W e\,dx\,dy\,dz = -\int\int_{\partial W} \mathbf{F} \cdot \mathbf{n}\,dA,$$
where $\mathbf{F} = -k\nabla T$, as in Exercise 55 and $e = c\rho_0 T$, where c is the specific heat (a constant) and ρ_0 is the mass density (another constant). Use the divergence theorem to show that this statement of conservation of energy is equivalent to the statement
$$\frac{\partial T}{\partial t} = \frac{k}{c\rho_0}\nabla^2 T \qquad \text{(heat equation)},$$
where $\nabla^2 T = \operatorname{div} \operatorname{grad} T = \partial^2 T/\partial x^2 + \partial^2 T/\partial y^2 + \partial^2 T/\partial z^2$ is the *Laplacian* of T.
(b) Make up an integral statement of conservation of mass for fluids that is equivalent to the *continuity equation*
$$\frac{\partial \rho}{\partial t} + \operatorname{div}(\rho \mathbf{V}) = 0,$$
where ρ is the mass density of a fluid and $\mathbf{V}$ is the fluid's velocity field.

★57. (a) Let $\mathbf{\Phi}$ be a vector field in space. Follow the pattern of Exercise 31, Section 18.4, replacing Green's theorem by Stokes' theorem to show that $\mathbf{\Phi} = \nabla f$ for some f (that is, $\mathbf{\Phi}$ is conservative) if and only if $\nabla \times \mathbf{\Phi} = \mathbf{0}$.
(b) Is $\mathbf{F} = (2xyz + \sin x)\mathbf{i} + x^2z\mathbf{j} + x^2y\mathbf{k}$ a gradient? If so, find f.

★58. (a) Use Green's theorem to find a formula for the area of the triangle with vertices (x_1, y_1), (x_2, y_2), and (x_3, y_3).
(b) Use Green's theorem to find a formula for the area of the n-sided polygon whose consecutive vertices are $(x_1, y_1), (x_2, y_2), \ldots, (x_n, y_n)$.

★59. Let f be a function on the region W in space such that: (i) $\dfrac{\partial^2 f}{\partial x^2} + \dfrac{\partial^2 f}{\partial y^2} + \dfrac{\partial^2 f}{\partial z^2} = 0$ everywhere on W and (ii) ∇f is tangent to the boundary ∂W at each point of ∂W. Use the identity in Exercise 51 to prove that f is constant.

★60. Show that the result in Exercise 59 is true if the condition (ii) is replaced by (ii′) f is constant on ∂W.

★61. Considering a closed curve in space as the boundary of two different surfaces, discuss the relation between:
(a) the divergence theorem;
(b) Stokes' theorem;
(c) the identity $\nabla \cdot (\nabla \times \Phi) = 0$.

Chapter 13 Answers

13.1 Vectors in the Plane

1. (4, 9)
3. $(-15, 3)$
5. $y = 1$
7. No solution
9. No solution
11. No solution
13. $a = 4$, $b = -1$
15. $a = 0$, $b = 1$
17. $(x_1, y_1) + (0, 0) = (x_1 + 0, y_1 + 0) = (x_1, y_1)$
19. $[(x_1, y_1) + (x_2, y_2)] + (x_3, y_3)$
 $= (x_1 + x_2 + x_3, y_1 + y_2 + y_3)$
 $= (x_1, y_1) + [(x_2, y_2) + (x_3, y_3)]$
21. $a(b(x, y)) = a(bx, by) = (abx, aby) = ab(x, y)$
23.

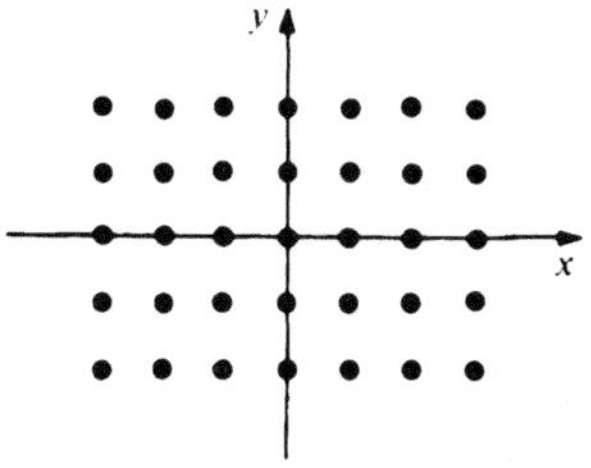

25. (a) $k(1, 3) + l(2, 0) = m(1, 2)$
 (b) $k + 2l = m$ and $3k + 0 = 2m$
 (c) $k = 4$, $l = 1$ and $m = 6$; i.e. $4SO_3 + S_2 = 6SO_2$
27. (a) **d**
 (b) **e**
29. (a) $\mathbf{c} + \mathbf{d} = (6, 2)$

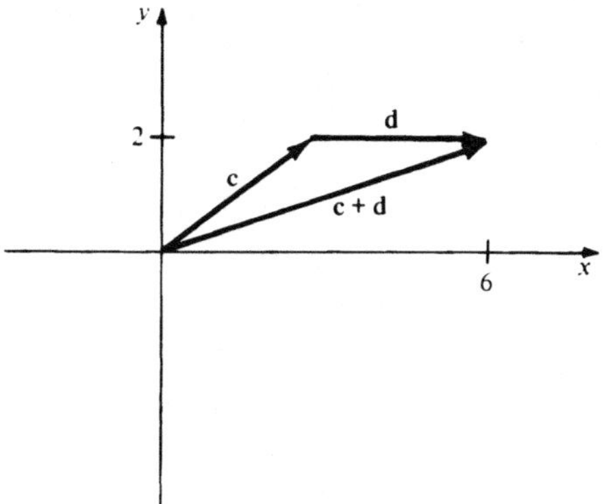

 (b) $-2\mathbf{e} + \mathbf{a} = (-1, 0)$

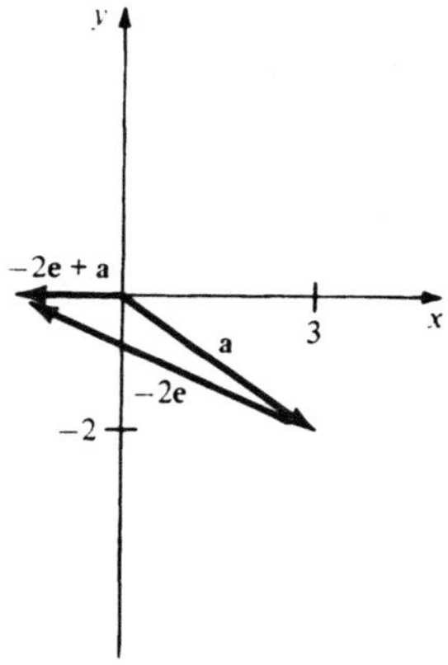

31.

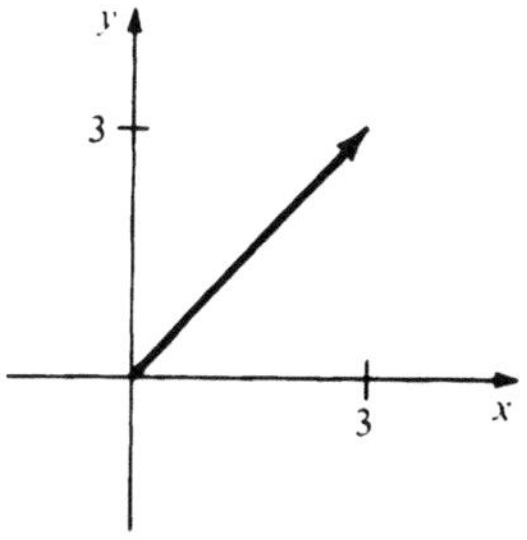

33.

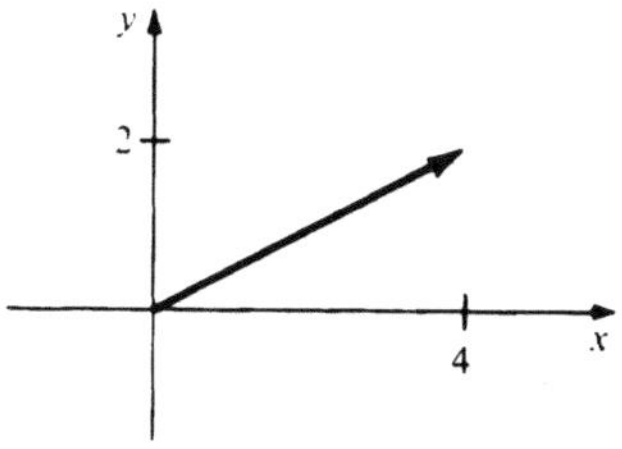

35.

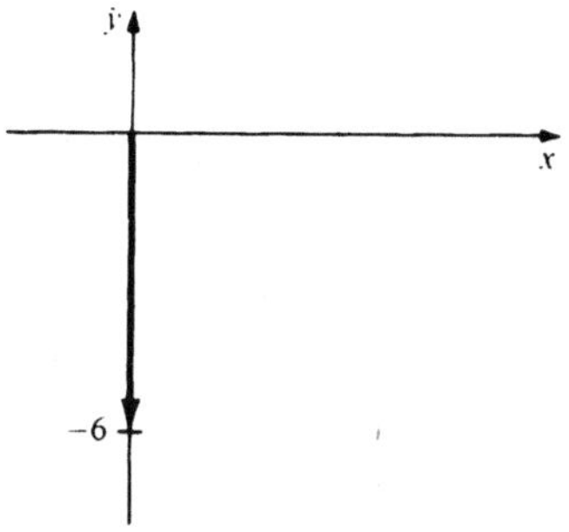

37. (a)

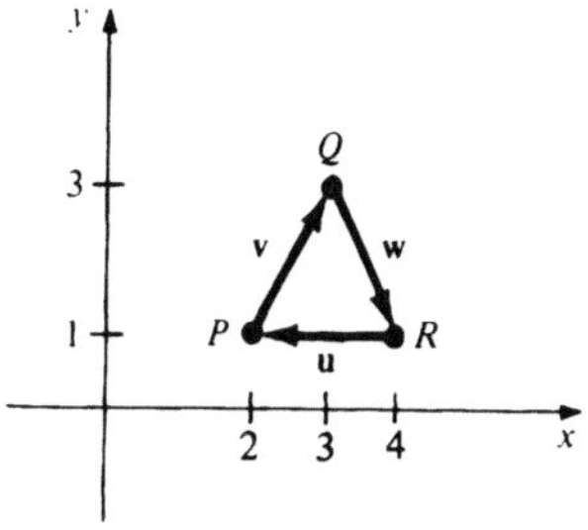

 (b) $\mathbf{v} = (1, 2)$; $\mathbf{w} = (1, -2)$; $\mathbf{u} = (-2, 0)$
 (c) **0**
39. (a)

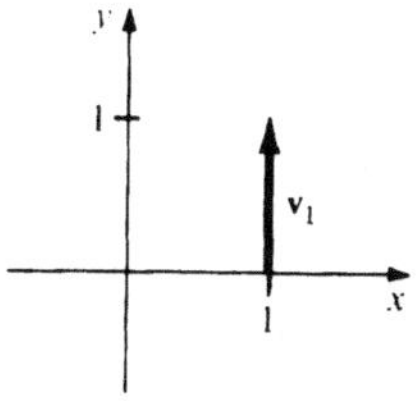

 (b) (0, 1)

(c) (0, 5/2)

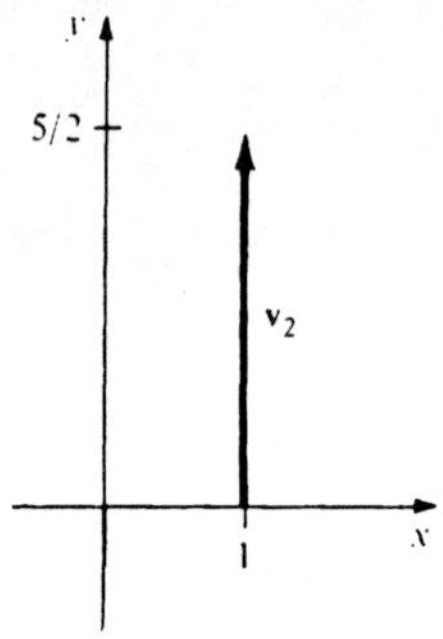

(d) (0, −2)

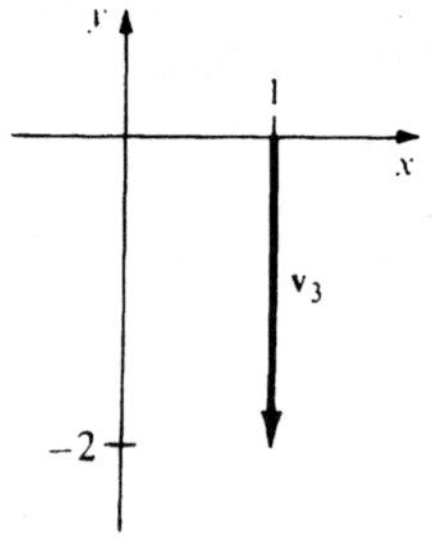

(e) (1, y)
(f) **v** = (0, y)

41. (a) Yes
(b) **v** = −(s/r)**w**
(c) Eliminate r and s.
(d) Solve linear equations.

13.2 Vectors in Space

1.

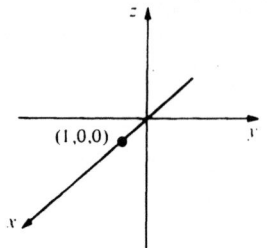

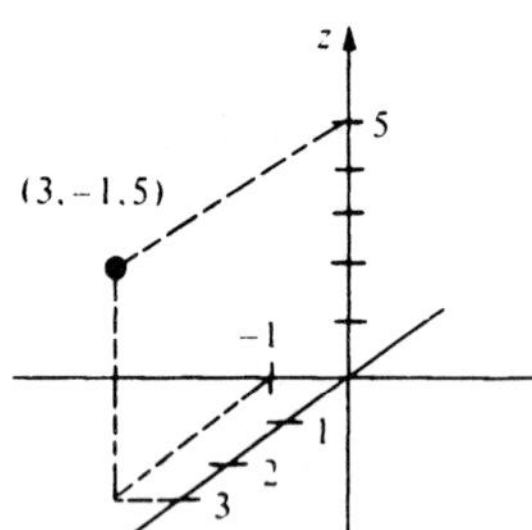

3.

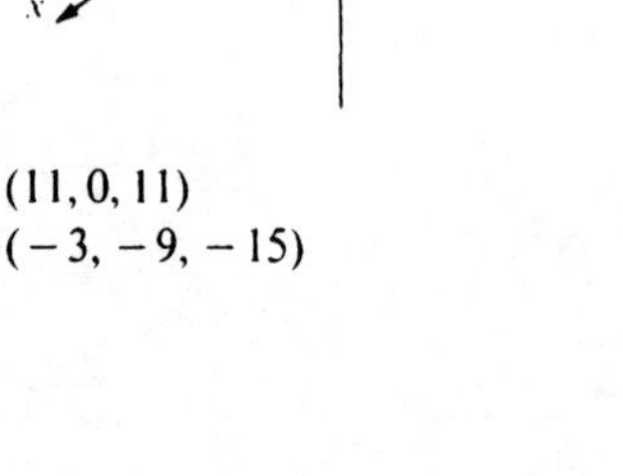

5. (11, 0, 11)

7. (−3, −9, −15)

9.

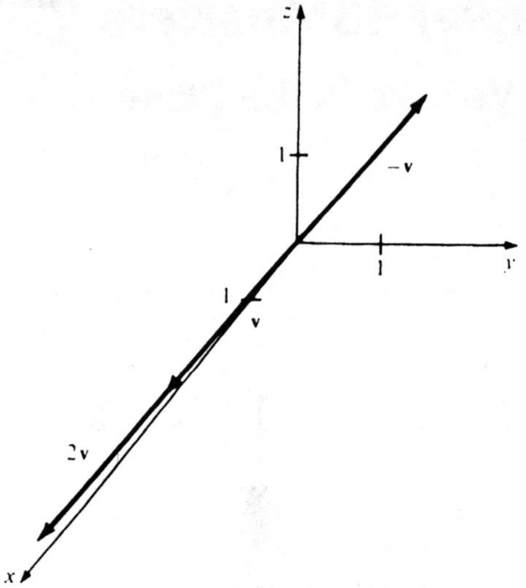

11.

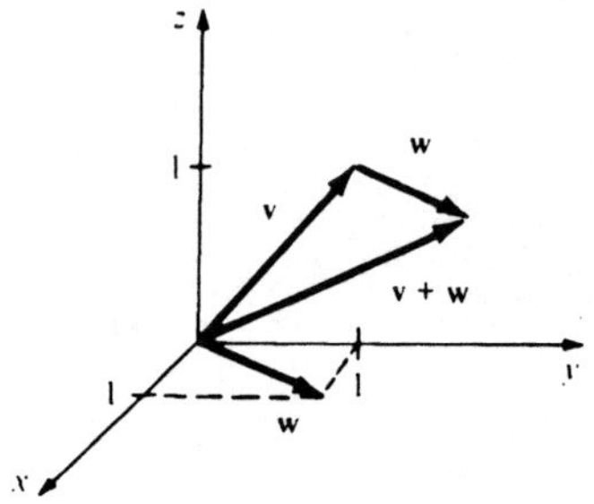

13. −**i** + 2**j** + 3**k**

15. 7**i** + 2**j** + 3**k**

17. **i** − **k**

19. **i** − **j** + **k**

21. **i** + 4**j**, $\theta \approx 0.24$ radians east of north

23. (a) 12:03 P.M.
(b) 4.95 kilometers

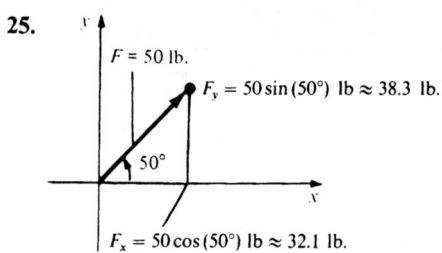

25.

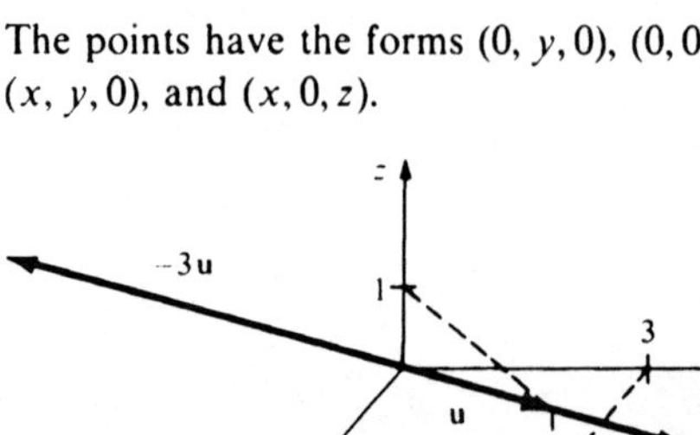

27. The points have the forms (0, y, 0), (0, 0, z), (x, y, 0), and (x, 0, z).

29.

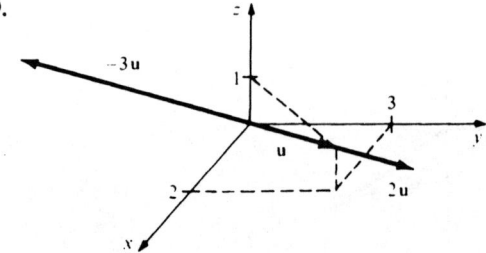

31. $9\mathbf{i} + 12\mathbf{j} + 15\mathbf{k}$ **33.** $26\mathbf{i} + 16\mathbf{j} + 38\mathbf{k}$

35. $a = \frac{1}{2}, b = -\frac{1}{2}$ **37.** $a = 5, b = 2$

39. $(4.9, 4.9, 4.9)$ and $(-4.9, -4.9, 4.9)$ newtons.

41. (a) Letting x, y, and z coordinates be the number of atoms of C, H, and O respectively, we get $p(3,4,3) + q(0,0,2) = r(1,0,2) + s(0,2,1)$.

(b) $p = 2$, $r = 6$, $s = 4$, $q = 5$

(c)

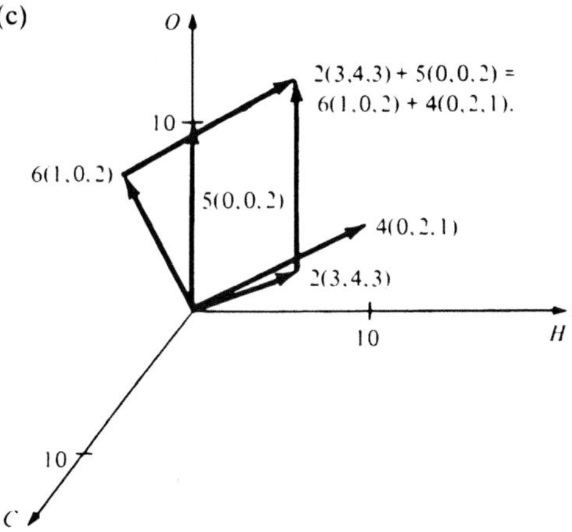

43. (a) $P_{-1} = (-1,-1,-1)$, $P_0 = (1,0,0)$, $P_1 = (3,1,1)$, $P_2 = (5,2,2)$

(b)

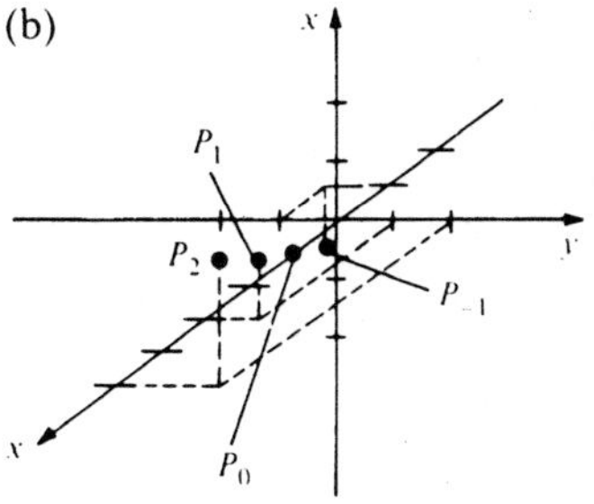

(c) The line through $(1,0,0)$ parallel to the vector $(2,1,1)$.

13.3 Lines and Distances

1. Use vectors with tails at the vertex containing the two sides.

3. Use the distributive law for scalar multiplication.

5. $(1, -\frac{1}{3})$

7. $x = 1 - t, y = 1 - t, z = t$

9. $x = t, y = t, z = t$

11. $x = 1 - t, y = 1 - t, z = t$

13. $x = -1 + 3t, y = -2 - 2t$

15. $(-2, -1, 0)$ **17.** No

19. $\sqrt{3}$ **21.** $\sqrt{2}$

23. $2\sqrt{2}$ **25.** $\pm\sqrt{246}$

27. $\|2\mathbf{i} + \mathbf{j} + 2\mathbf{k}\| = 3$, which is less than $\sqrt{3} + \sqrt{2}$.

29. One solution is $\mathbf{u} = \mathbf{i}$, $\mathbf{v} = -\mathbf{i}$, $\mathbf{w} = \mathbf{i}$.

31. Each side has length $\sqrt{2}$.

33. $(1/\sqrt{3})\mathbf{i} + (1/\sqrt{3})\mathbf{j} + (1/\sqrt{3})\mathbf{k}$, $(1/\sqrt{2})\mathbf{i} + (1/\sqrt{2})\mathbf{k}$

35. $\sqrt{3}$

37. $\sqrt{2}$

39. (i) $\sqrt{14t^2 - 12t + 4}$

(ii) $t = 3/7$

(iii) $\sqrt{10/7}$

41. 13 knots

43. Solve one equation for t and substitute. The line is vertical when $x_1 = x_2$.

45. When the angle between the vectors is 0.

13.4 The Dot Product

1. 4

3. 0

5. ≈ 0.34 radian

7. $\pi/2$ radians

9. $(1/\sqrt{5})\mathbf{i} + (2/\sqrt{5})\mathbf{j}$

11. 0.955 radians

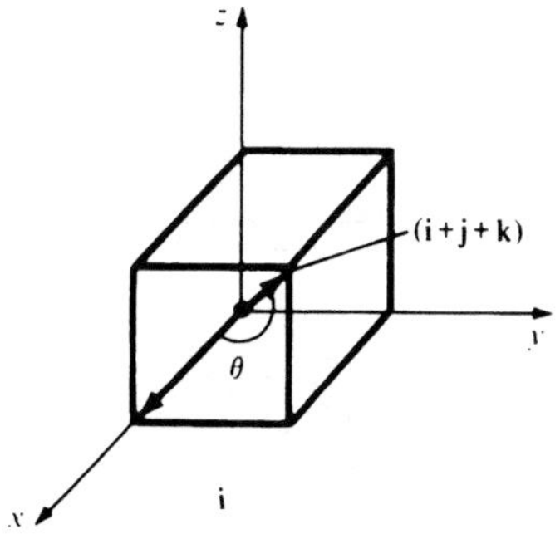

13. Use Figure 13.4.2. **15.** $\sqrt{42}$

17. $\sqrt{50/11}$ **19.** $x + y + z = 0$

21. $x = 0$ **23.** $-x + y + z - 1 = 0$

25. $(2/\sqrt{14})\mathbf{i} + (3/\sqrt{14})\mathbf{j} + (1/\sqrt{14})\mathbf{k}$

27. $(1/\sqrt{2})\mathbf{i} + (1/\sqrt{2})\mathbf{j}$

29. $x + y + z - 1 = 0$

31. $x = 1 + t, y = 1 + t, z = 1 + t$

33. $(1, -1/2, 3/2)$, $\sqrt{14}/2$

35. $x = (22 - 9t)/7, y = (-6 - 2t)/7, z = t$

37. $3\sqrt{3}$.

39. $\sqrt{2}/2$.

41. Letting (a,b) and (c,d) be the given points, the equation of the line is $(a - c)x + (b - d)y = (1/2)(a^2 - c^2 + b^2 - d^2)$. Use this to show that the two points are equidistant from points on the line.

43. (a) 3

(b) -2

(c) $2\sqrt{3}$

(d) 3

45. Letting $P_1 = (p,q)$ and $P_2 = (r,s)$, we have $(r - p)x + (s - q)y = (r^2 + s^2 - p^2 - q^2)/2$.

47. To show that $\mathbf{v}$ and $\mathbf{w} = (\mathbf{v}\cdot\mathbf{e}_1)\mathbf{e}_1 + (\mathbf{v}\cdot\mathbf{e}_2)\mathbf{e}_2$ are equal, show that $\mathbf{v} - \mathbf{w}$ is orthogonal to both $\mathbf{e}_1$ and $\mathbf{e}_2$.

49. $\mathbf{F}_1 = -(F/2)(\mathbf{i}+\mathbf{j})$ and $\mathbf{F}_2 = (F/2)(\mathbf{i}-\mathbf{j})$

51. (a) $\mathbf{F} = (3\sqrt{2}\,\mathbf{i} + 3\sqrt{2}\,\mathbf{j})$
(b) ≈ 0.322 radians
(c) $18\sqrt{2}$

53. Use the component formula for the dot product.

55. (a) $[12.5)^2 + (16.7)^2 - (20.9)^2]/[(12.5)(16.7)]$ is close to 0.
(b) 0.54%

57. (a) Let $s = t\sqrt{a^2+b^2+c^2}$
(b) Use the fact that $\|\mathbf{u}\| = 1$.
(c) Use $\|\mathbf{u}\|^2 = 1$.
(d) For L_1 and L_2,
$\cos\alpha = 1/\sqrt{3}$ so $\alpha = \cos^{-1}(1/\sqrt{3})$,
$\cos\beta = 1/\sqrt{3}$ so $\beta = \cos^{-1}(1/\sqrt{3})$,
$\cos\gamma = 1/\sqrt{3}$ so $\gamma = \cos^{-1}(1/\sqrt{3})$.
For L_3 and L_4,
$\cos\alpha = 1/\sqrt{83}$ so $\alpha = \cos^{-1}(1/\sqrt{83})$,
$\cos\beta = 1/\sqrt{83}$ so $\beta = \cos^{-1}(1/\sqrt{83})$
and $\cos\gamma = 9/\sqrt{83}$ so $\gamma = \cos^{-1}(9/\sqrt{83})$.
(e) Only the line $t(1, 1, 1)$.

13.5 The Cross Product

1. $\mathbf{j}+\mathbf{k}$
3. $2\mathbf{i} - 2\mathbf{j} + 4\mathbf{k}$
5. $9\mathbf{i} + 18\mathbf{j}$
7. $6\mathbf{i} - 2\mathbf{k}$
9. $-\mathbf{i}+\mathbf{k}$
11. $3\sqrt{2}$
13. 2
15. $-(1/\sqrt{2})\mathbf{j} + (1/\sqrt{2})\mathbf{k}$
17. $-(1/\sqrt{2})\mathbf{i} - (1/\sqrt{2})\mathbf{j}$
19. $(\sqrt{2}/6)\mathbf{i} - (\sqrt{2}/6)\mathbf{j} + (2\sqrt{2}/3)\mathbf{k}$
21. $2x + 3y + 4z = 0$
23. $x - 3y + 2z = 0$
25. $3\sqrt{2}/2$
27. The points are collinear, so the area is zero.
29. Substitute component expressions for $\mathbf{v}_1$ and $\mathbf{v}_2$.
31. The angle between the vectors is $\theta - \psi$. Now use property 1 in the box on p. 679.
33. Use the result of Exercise 32.
35. Show that $\mathbf{M}$ satisfies the defining properties of $\mathbf{R}\times\mathbf{F}$.
37. Show that $n_1(\mathbf{N}\times\mathbf{a})$ and $n_2(\mathbf{N}\times\mathbf{b})$ have the same magnitude and direction.
39. (a) Draw a figure showing the two lines and the plane in the hint.
(b) $\sqrt{2}$
41. If $\mathbf{F}$ is the gravitational force, the gyroscope rotates to the left (viewed from above).

13.6 Matrices and Determinants

1. 2
3. 0
5. -2
7. 25
9. ac
11. Compute the two determinants.
13. Compute the two determinants.
15. 0
17. 4
19. -6
21. 9
23. abc

25. $\begin{vmatrix} \mathbf{i} & \mathbf{j} & \mathbf{k} \\ 3 & -1 & 0 \\ 0 & 1 & 1 \end{vmatrix} = -\mathbf{i} - 3\mathbf{j} + 3\mathbf{k}$

27. $\begin{vmatrix} \mathbf{i} & \mathbf{j} & \mathbf{k} \\ 1 & 1 & 0 \\ 0 & 1 & 1 \end{vmatrix} = \mathbf{i} - \mathbf{j} + \mathbf{k}$

29. $\begin{vmatrix} \mathbf{i} & \mathbf{j} & \mathbf{k} \\ 1 & 0 & -1 \\ 1 & 0 & 1 \end{vmatrix} = -2\mathbf{j}$

31. 6
33. 12
35. Compute and simplify.
37. Compute both determinants and compare.
39. Use Example 8 after renumbering the vectors.
41.

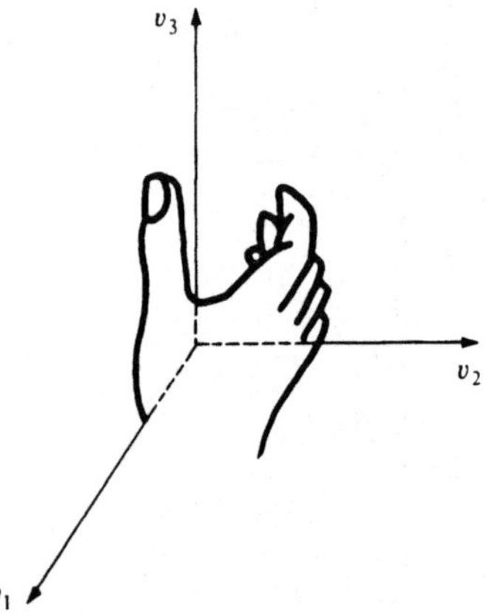

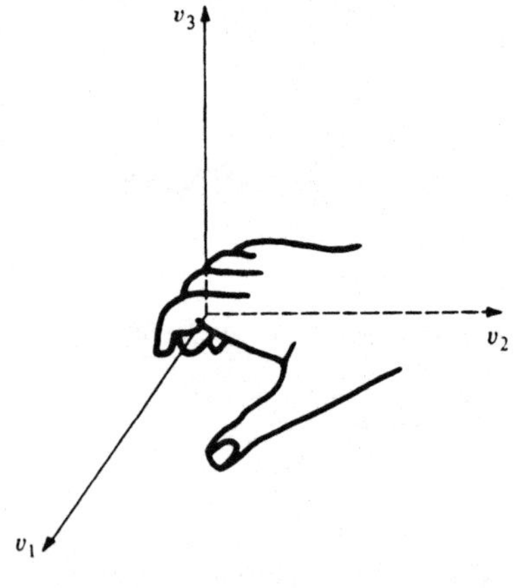

43. Substitute the expressions for x and y in the equations.
45. Substitute the given expressions for x, y, and z in the equations.
47. $x = 37/13$, $y = -3/13$
49. Compute both determinants and compare.

51. Subtract four times row 1 from row 2, subtract seven times row 1 from row 3, expand by column 1 and then evaluate the 2×2 determinant.

53. 3, -6

Review Exercises for Chapter 13

1. $(2, 8)$
3. $(-1, -2, 17)$
5. $11\mathbf{i} + \mathbf{j} - \mathbf{k}$
7. $-4\mathbf{i} + 7\mathbf{j} - 11\mathbf{k}$
9. 6
11. $-2\mathbf{k}$
13. $\mathbf{i} - 2\mathbf{j}$
15. $2\mathbf{i} + \mathbf{j} - 3\mathbf{k}$
17. $(-2/\sqrt{22})\mathbf{i} + (3/\sqrt{22})\mathbf{j} + (3/\sqrt{22})\mathbf{k}$
19. 0 (the three vectors lie in a plane).
21. (a) $(6, 6)$

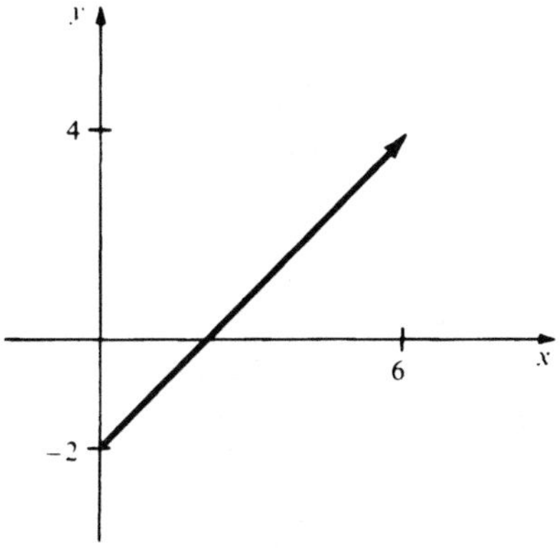

(b) $(9, 7)$

23. Put the triangle in the xy-plane; use cross products with $\mathbf{k}$.

25. $(1825 - 600\sqrt{2})^{1/2} \approx 31.25$ km/hr.

27. (a) $70\cos\theta + 20\sin\theta$
(b) $(21\sqrt{3} + 6)$ ft.-lbs.

29. $x = 1 + t, y = 1 + t, z = 2 + t$
31. $x = 1 + t, y = 1 - t, z = 1 - t$
33. $-x + y = 0$
35. $x - y - z - 1 = 0$
37. $x = -t, y = t, z = 3$
39. $x = 2 + t, y = 3 - t, z = 1 - t$
41. $(1/\sqrt{38})\mathbf{i} - (6/\sqrt{38})\mathbf{j} + (1/\sqrt{38})\mathbf{k}$
43. $(2/\sqrt{5})\mathbf{i} - (1/\sqrt{5})\mathbf{j}$
45. $(\sqrt{3}/2)\mathbf{i} + (1/2\sqrt{2})\mathbf{j} + (1/2\sqrt{2})\mathbf{k}$
47. It is parallel to the z-axis.

49. This is a (double) cone with vertex at the origin.

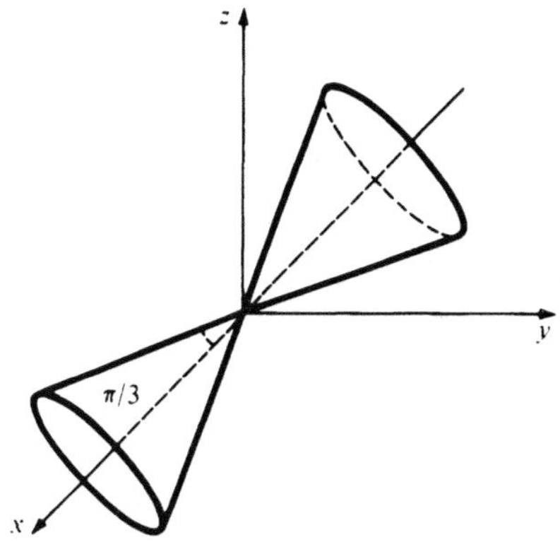

51. (a) Draw a vector diagram. (b) Use $\mathbf{c} \times \mathbf{c} = \mathbf{0}$.
(c) Use part (b).

53. Use the dot product to show that the vectors $\mathbf{a} - \mathbf{b}$ and $-\mathbf{a} - \mathbf{b}$ are perpendicular.

55. 3
57. 1
59. -2
61. 0
63. $\sqrt{381}$
65. $29/2$

67. (a) $\dfrac{1}{6}\begin{vmatrix} a_1 & a_2 & a_3 \\ b_1 & b_2 & b_3 \\ c_1 & c_2 & c_3 \end{vmatrix}$
(b) $1/3$

69. Use the fact that $\|\mathbf{a}\|^2 = \mathbf{a} \cdot \mathbf{a}$, expand both sides and use the definition of $\mathbf{c}$.

71. $x = 3/7, y = -29/21, z = 23/21$

73. -162

75. Each side equals
$2xy - 7yz + 5z^2 - 48x + 54y - 5z - 96.$
(Or switch the first two columns and then subtract the first row from the second.)

77. $\mathbf{v}$ is orthogonal to $\mathbf{i}$

79. (a) $4\mathbf{k}$
(b) $20\sqrt{2}\,\mathbf{i} + 20\sqrt{2}\,\mathbf{j}$

81. (a) Substitute $\mathbf{i}$, $\mathbf{j}$, and $\mathbf{k}$ for $\mathbf{w}$.
(b) $(\mathbf{u} - \mathbf{v}) \cdot \mathbf{w} = 0$.
(c) Repeat the reasoning in (a)
(d) Apply (c) to $\mathbf{u} - \mathbf{v}$.

83. (a)

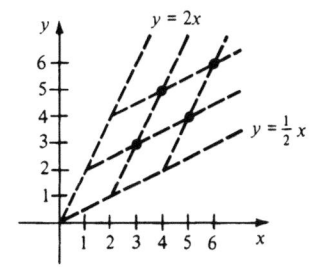

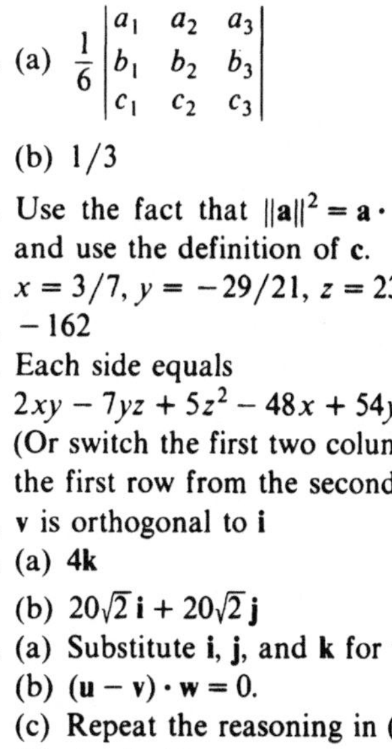

(b)

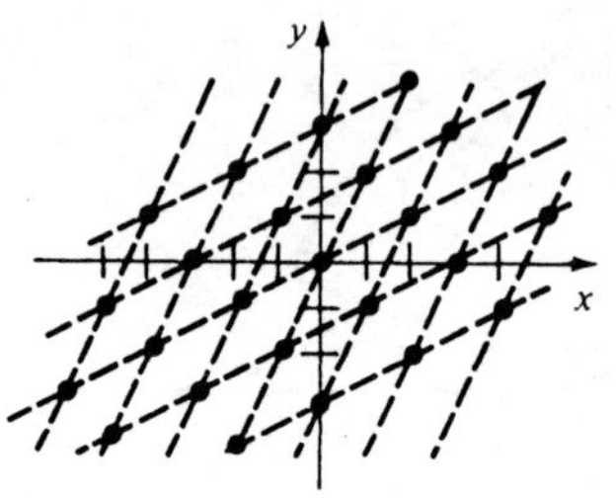

(c)

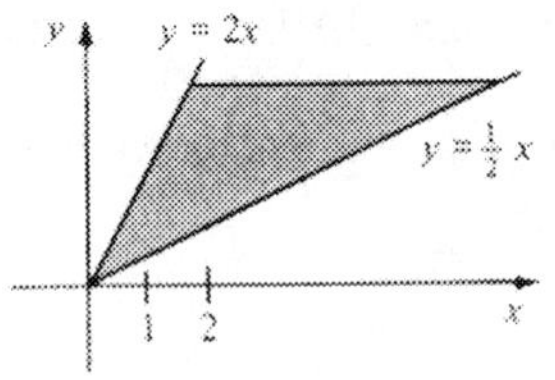

(d) The set is the entire plane.

85. (a)

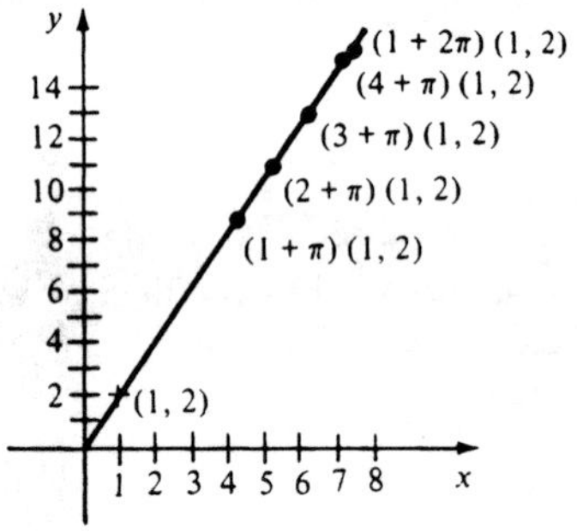

(b)

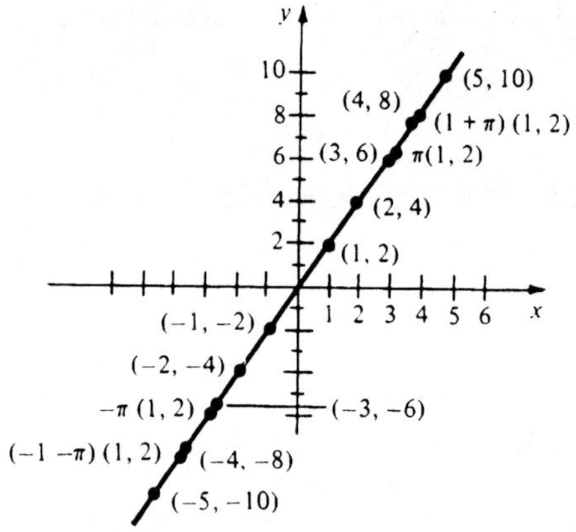

(c)

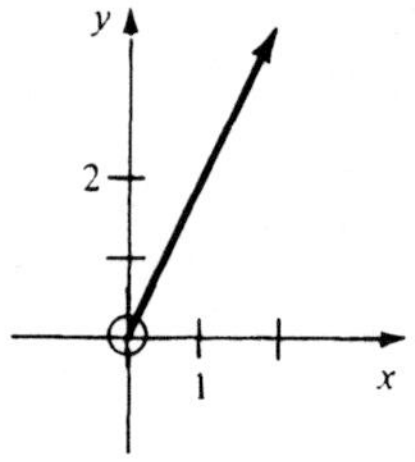

(d)

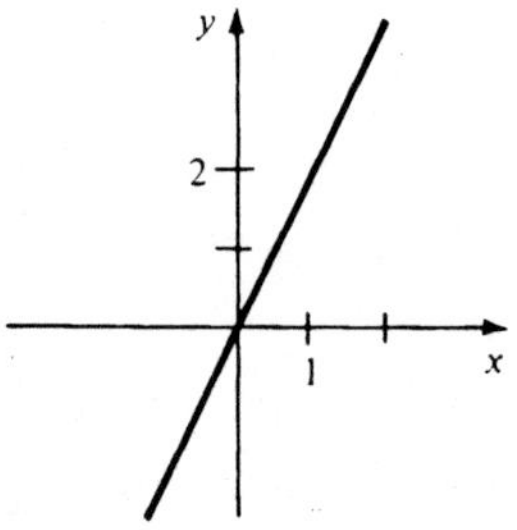

87. $\theta = \sin^{-1}(\sqrt{8}/3)$

Chapter 14 Answers

14.1 The Conic Sections

1. Foci at $(\pm 4\sqrt{2}, 0)$.

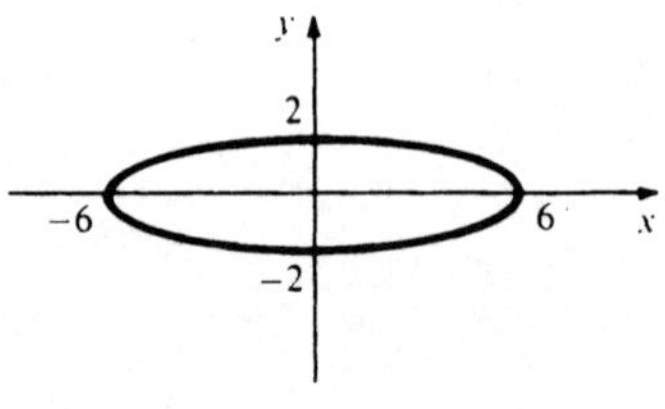

3.

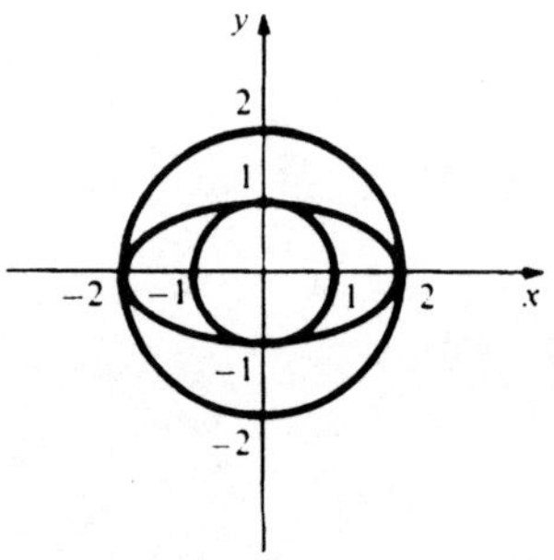

5.

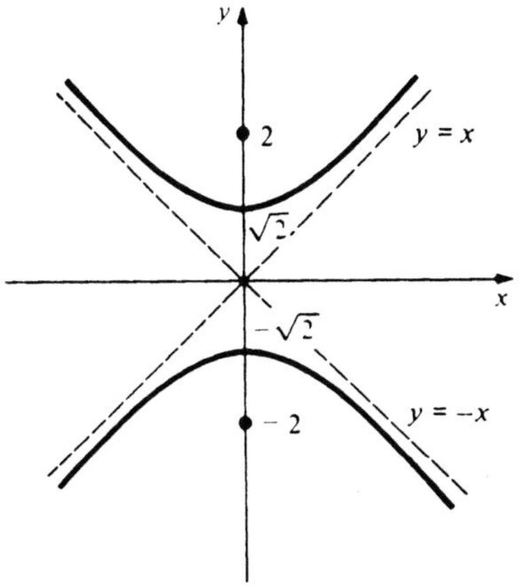

7.

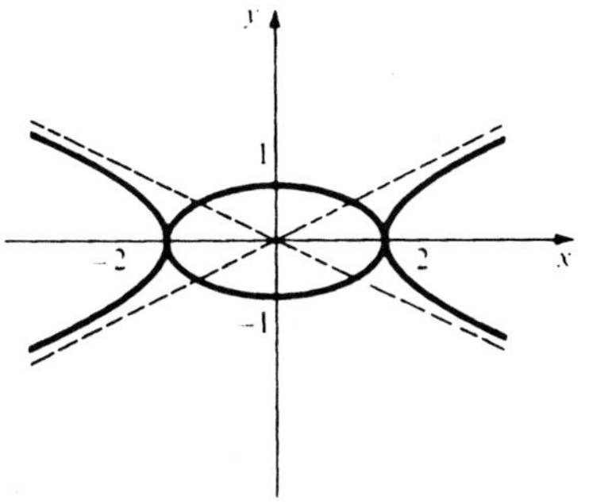

9. $y = x^2/16$
11. $(0, 1/4)$, $y = -1/4$
13. $(1/4, 0)$, $x = -1/4$
15. $x^2 + y^2 = 25$
17. $y = x^2/4$
19. $y^2 - \frac{x^2}{3} = 1$
21. On the axis, 2/15 meters from the mirror.
23. $4/\sqrt{15}$
25. Use the dot product to find an expression for the cosine of the incident and reflected angles.

14.2 Translation and Rotation of Axes

1.

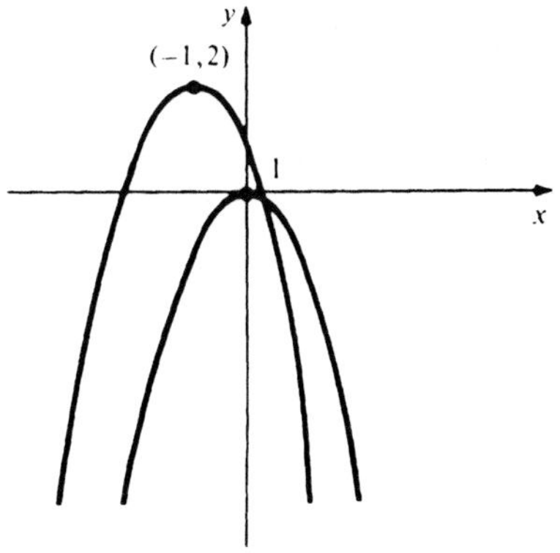

3.

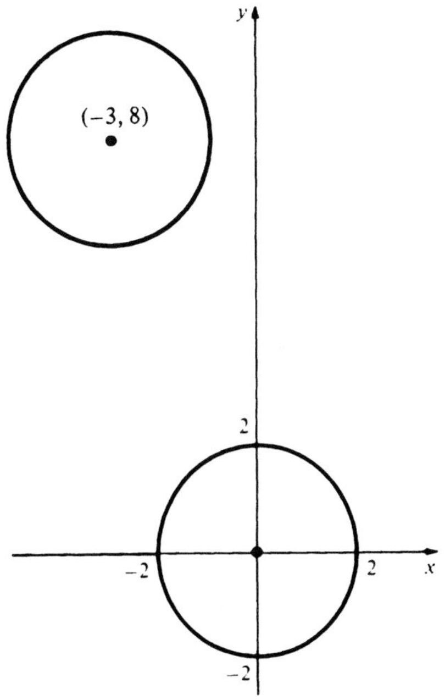

5. A circle of radius 1 centered at (1, 0)

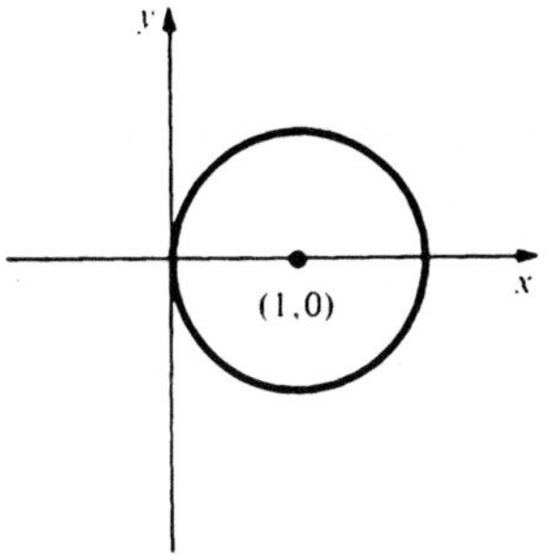

7. An ellipse shifted to (0, 3/4).

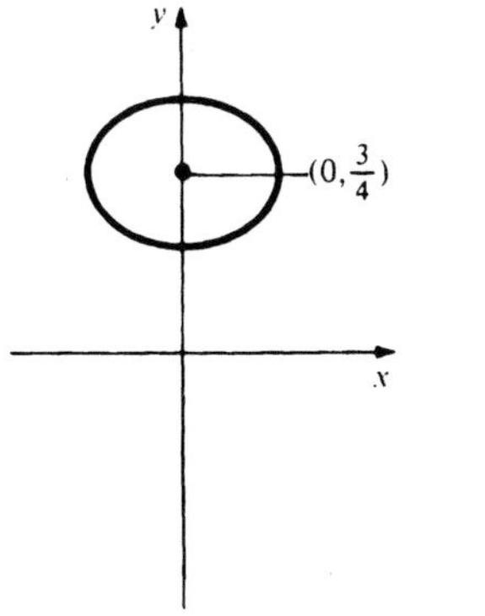

9. A hyperbola with asymptotes $y = \pm x$, shifted to $(-1, -1)$.

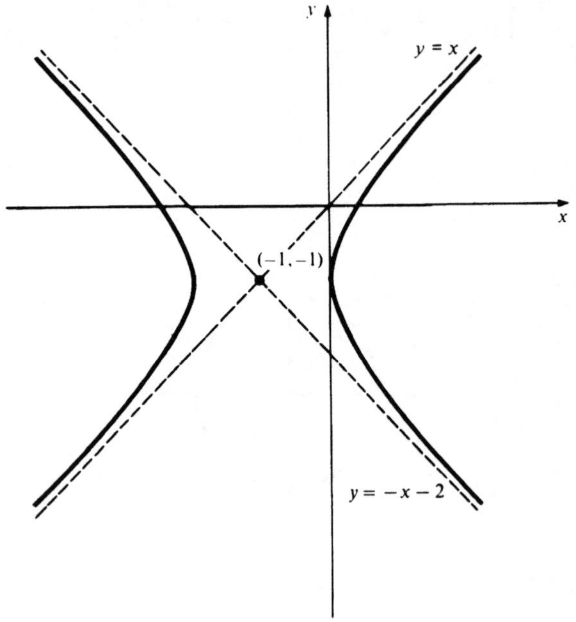

11. $x = X/2 - \sqrt{3}\,Y/2,\ y = \sqrt{3}\,X/2 + Y/2,$
$X = x/2 + \sqrt{3}\,y/2,\ Y = -\sqrt{3}\,x/2 + y/2.$

13. $x = 0.97X - 0.26Y,\ y = 0.26X + 0.97Y,$
$X = 0.97x + 0.26y,\ Y = -0.26x + 0.97y.$

15. Hyperbola

17. Ellipse

19.

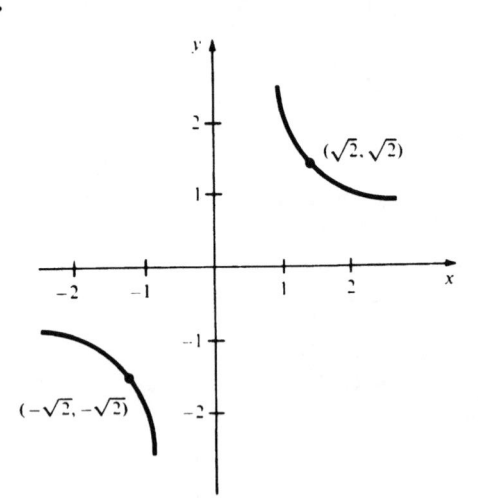

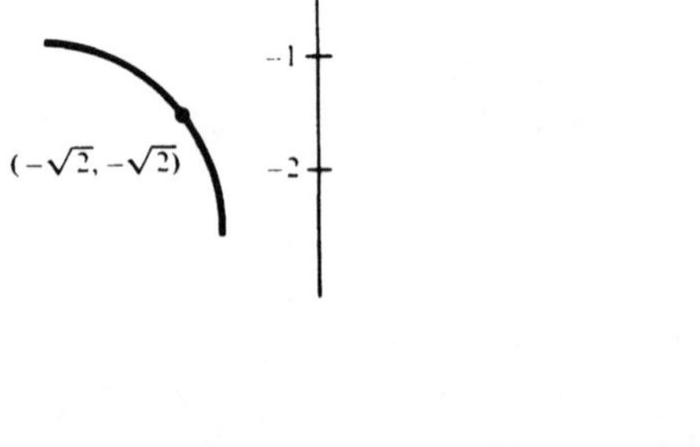

21.

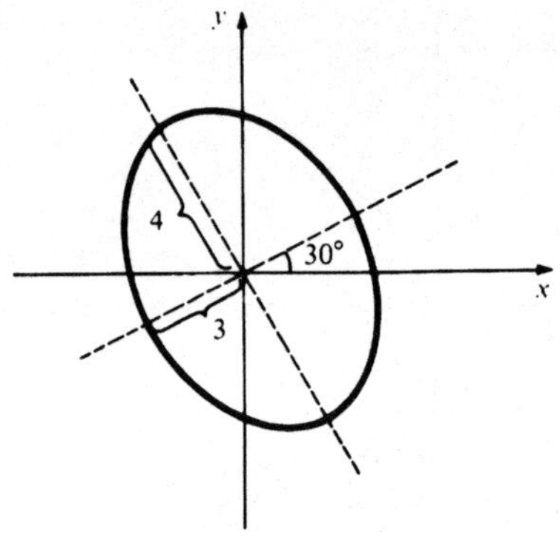

23. $x^2 + y^2 - 4x - 6y = 12$

25. $y = (x-1)^2$ **27.** $(y-1)^2 - \dfrac{x^2}{3} = 1.$

29. $X^2 + Y^2 + (1 - \sqrt{3})X + (-1 - \sqrt{3})Y = 2.$

31. For translations, $A = \bar{A}$ and $C = \bar{C}$. For rotations use equations (9) to compute $A + C$.

33. The area of the rotated ellipse is $\pi\sqrt{\dfrac{1}{\bar{A}}}\cdot\sqrt{\dfrac{1}{\bar{C}}}$.

14.3 Functions, Graphs and Level Surfaces

1. All (x, y) with $x \neq 0$; 0, 1.

3. All (x, y) with $x^2 + y^2 \neq 1$; 2, 0.

5. All (x, y, z) with $x^2 + y^2 + z^2 \neq 1$; 1, $-2/3$

7. All (x, y) with $x \neq \pi + 2n\pi$, n an integer; $-\sqrt{2}/4$, $\pi(2 - \sqrt{2})/2$.

9.

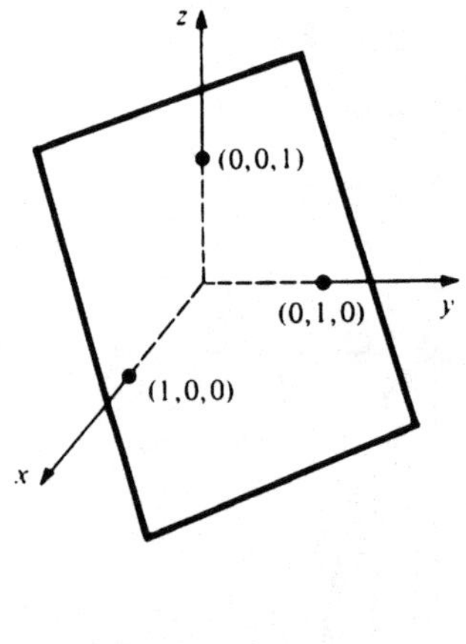

11.

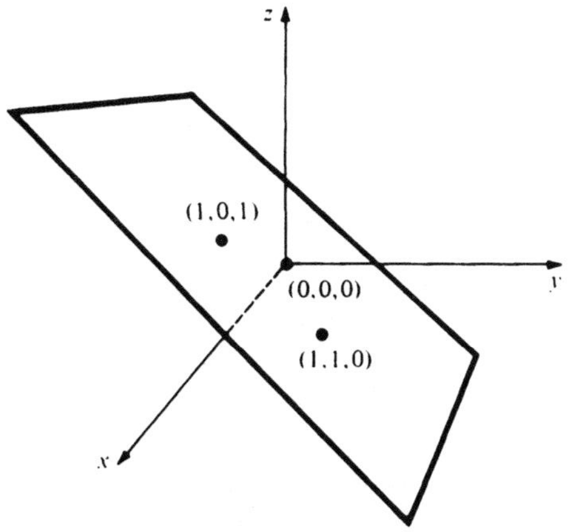

13.

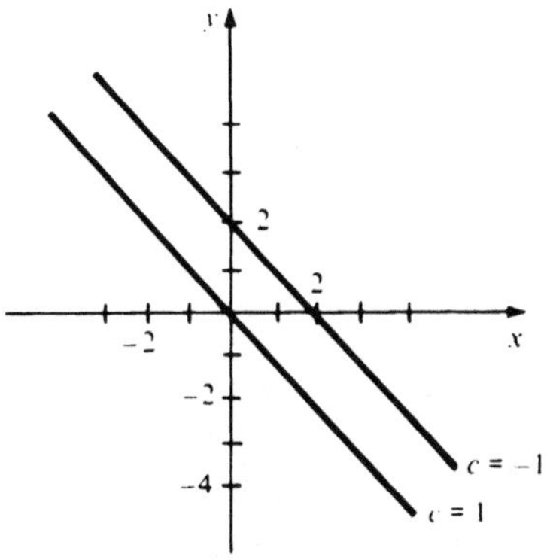

15. Circles with centers on the line $y = x$ and passing through the points $(\pm\sqrt{2}/2, \mp\sqrt{2}/2)$, excluding points on the circle $x^2 + y^2 = 1$.

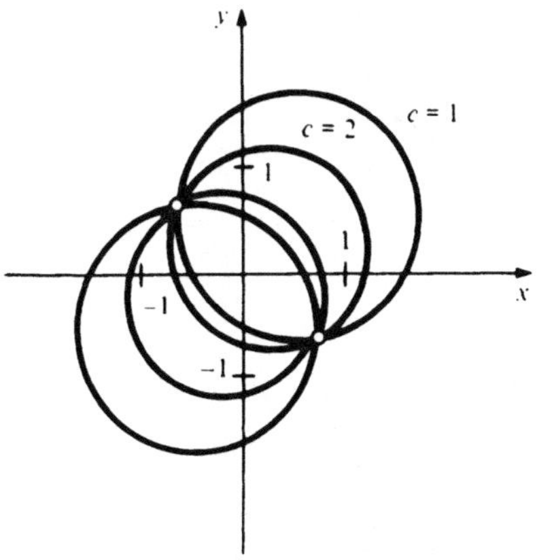

17. Lines through the origin, excluding points on the line $x = y$.

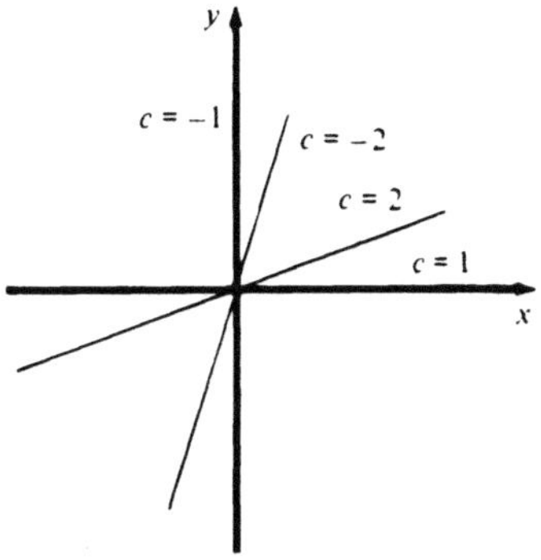

19.

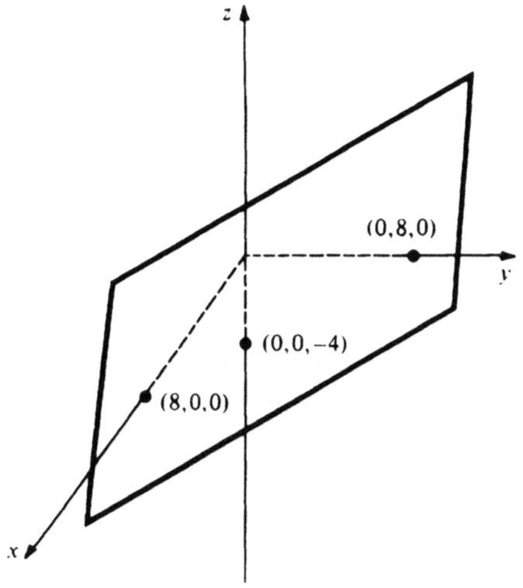

21.

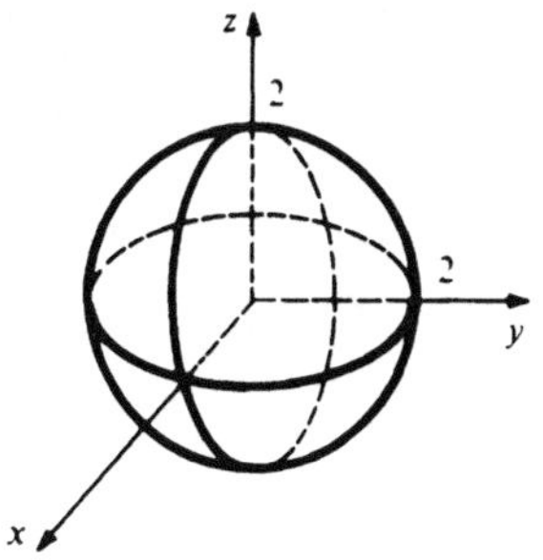

23.

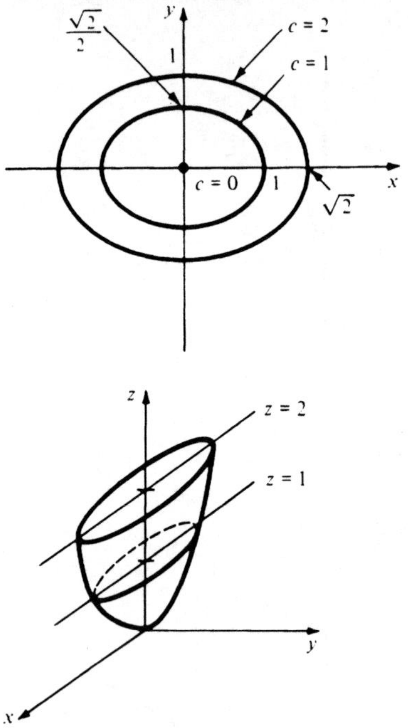

25.

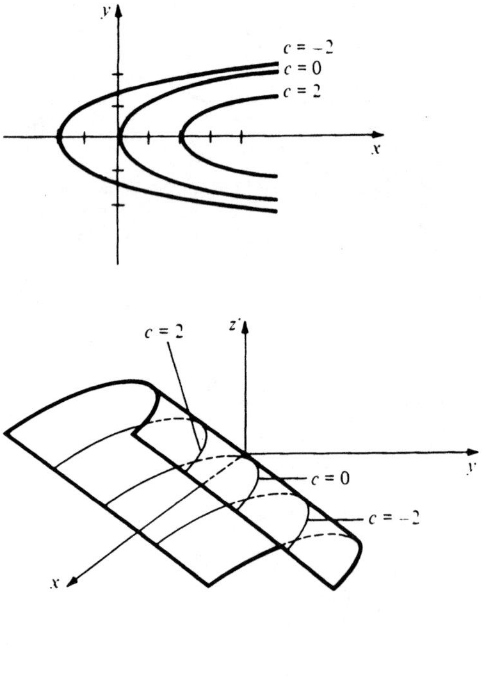

27.

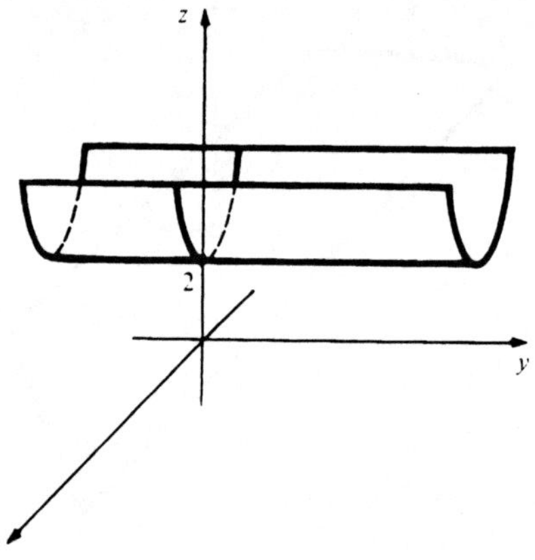

29.

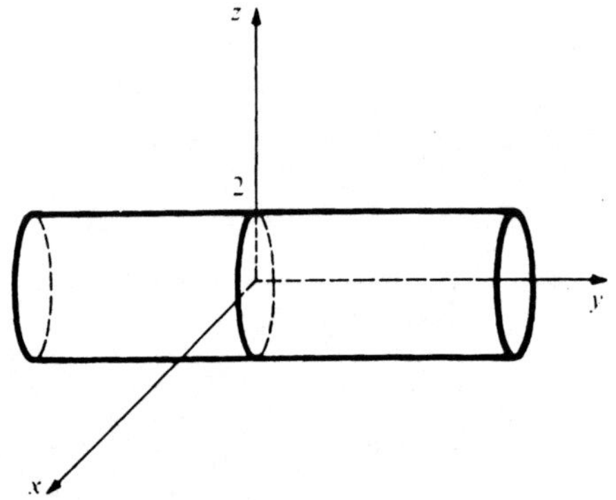

31.

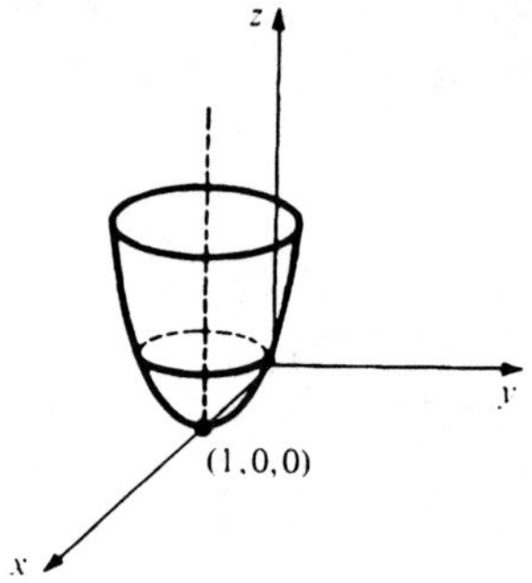

33.

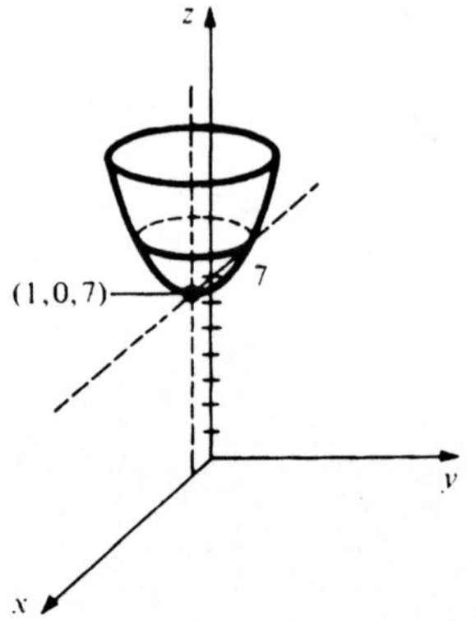

35.

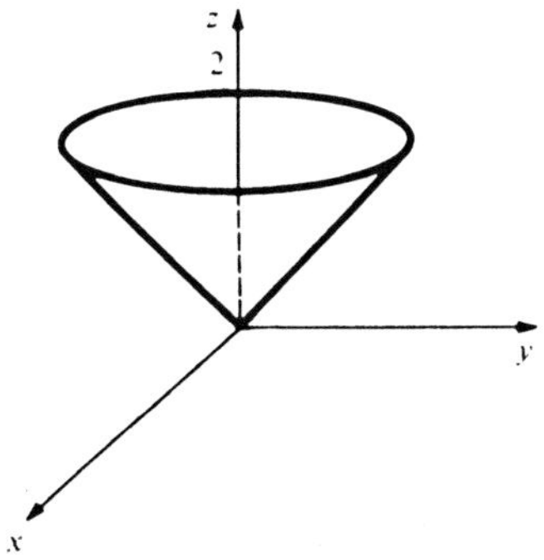

37.

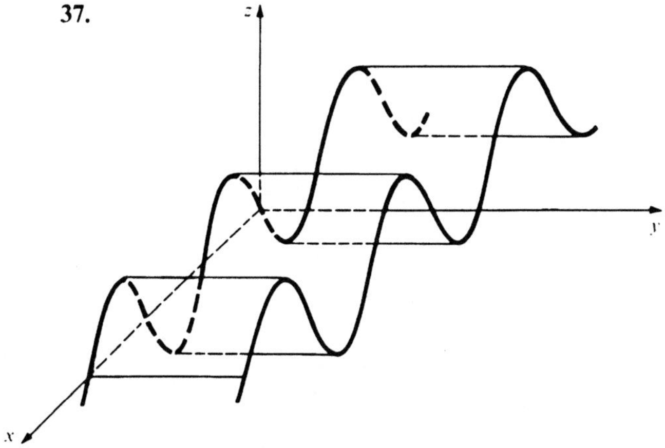

39.

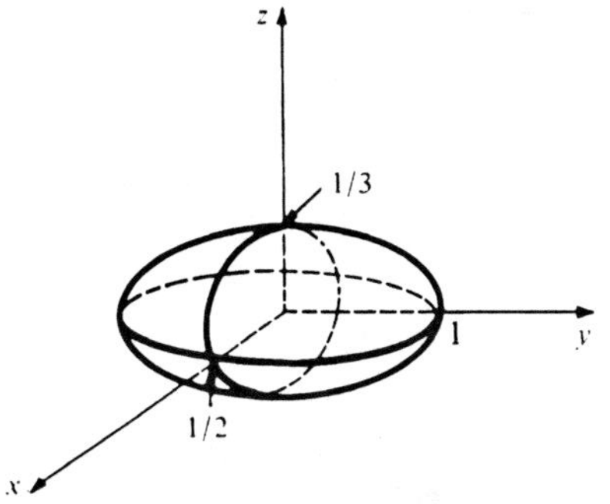

41. (a)

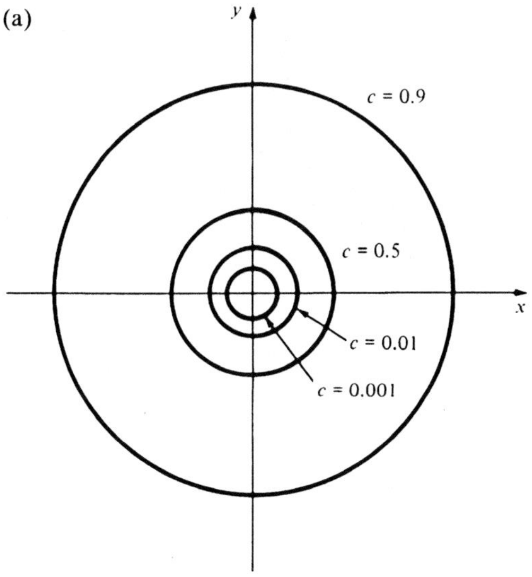

(b) No level curve.

(c)

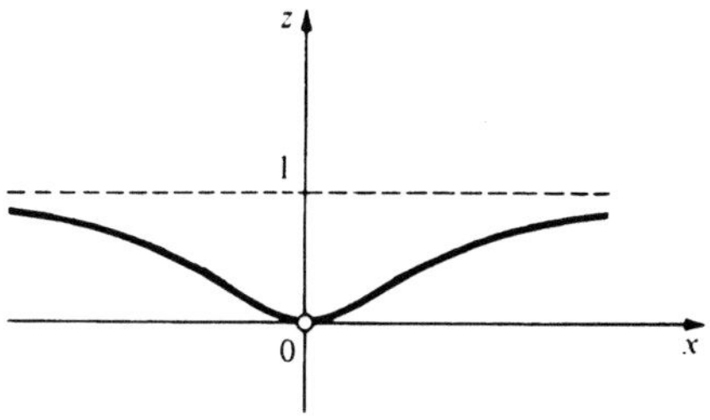

(d) In polar coordinates, the equation is $f(r,\theta) = e^{-1/r^2}$. This is independent of θ.

(e) The graph looks like a plane gradually sloping down to a pit in the center.

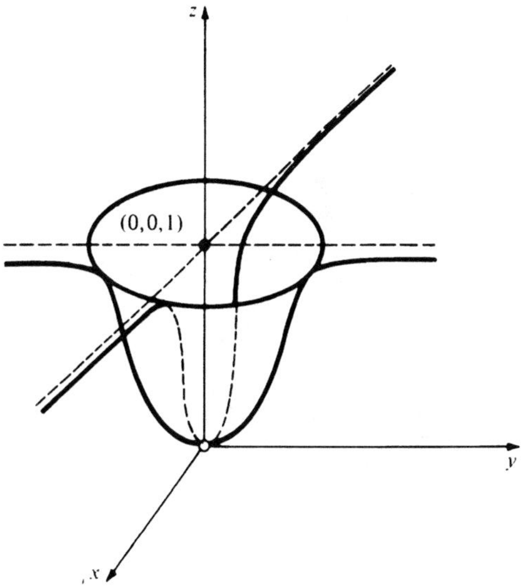

43. (a) $z = \exp[(-12)(x+y)/25(x-y)]$

(b)

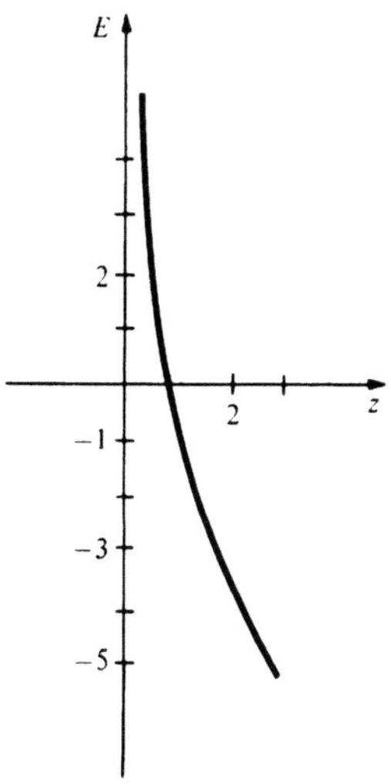

14.4 Quadric Surfaces

1.

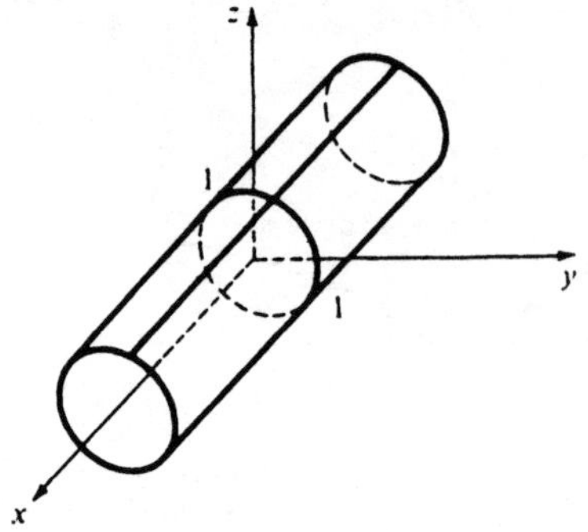

3.

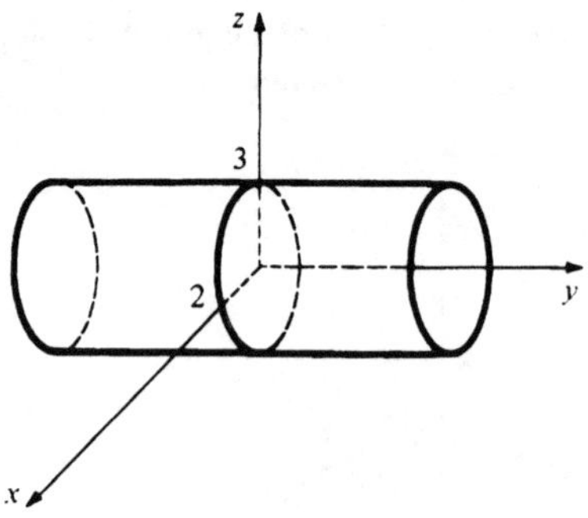

5.

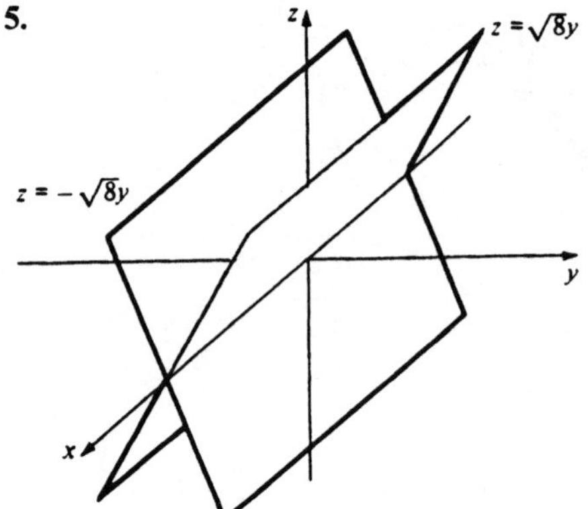

7.

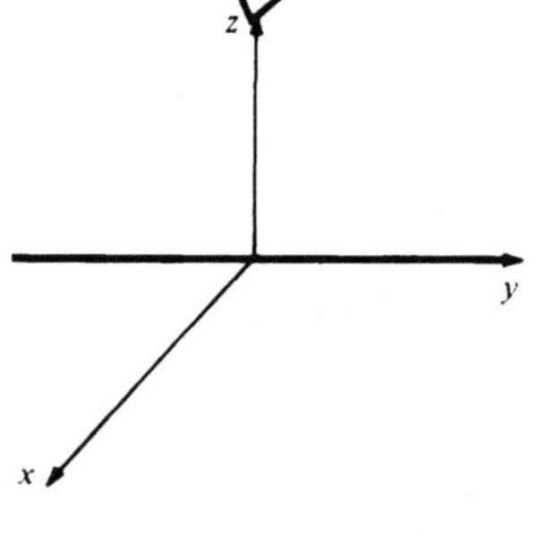

9.

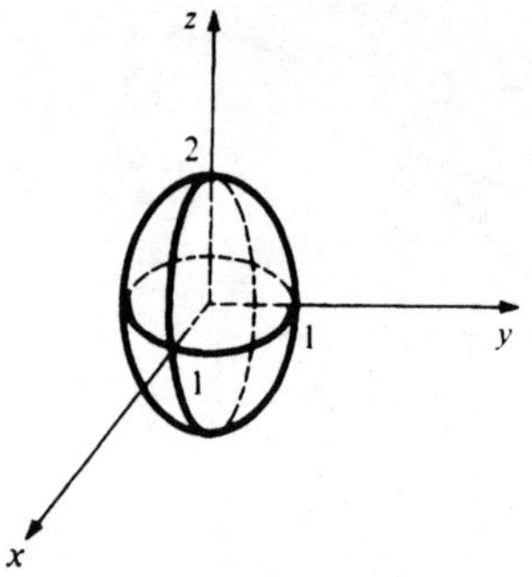

11.

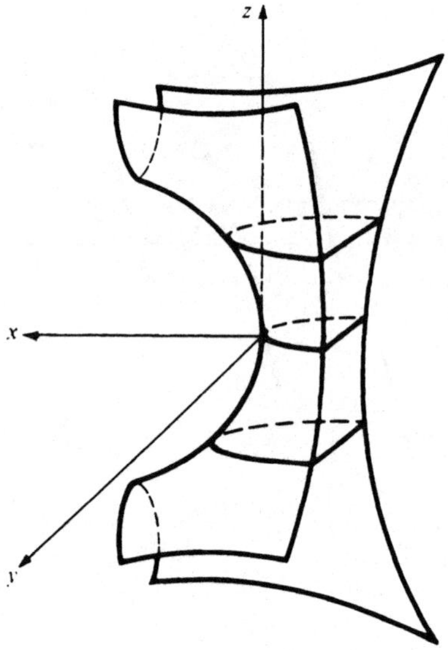

13.

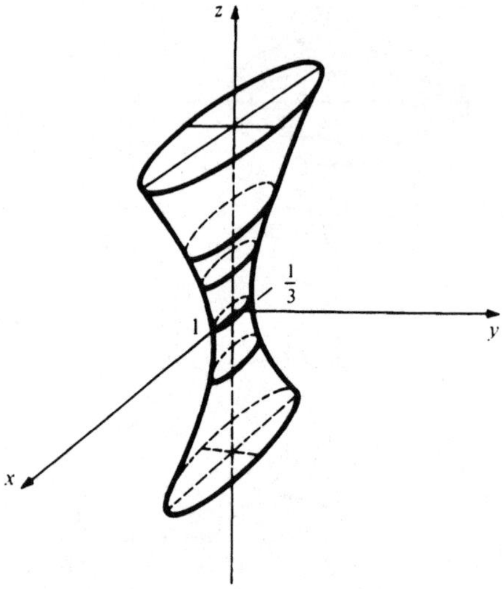

15.

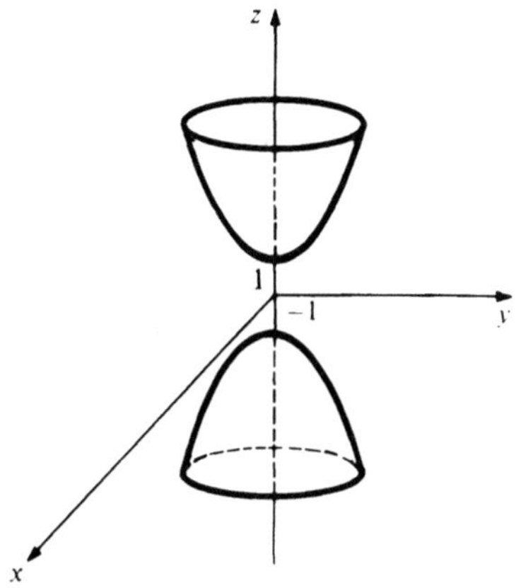

17. (a)

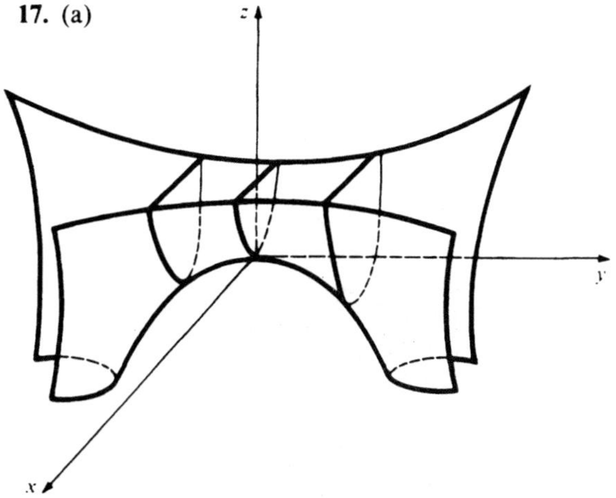

(b) Rotate the x and y axes by 45°.

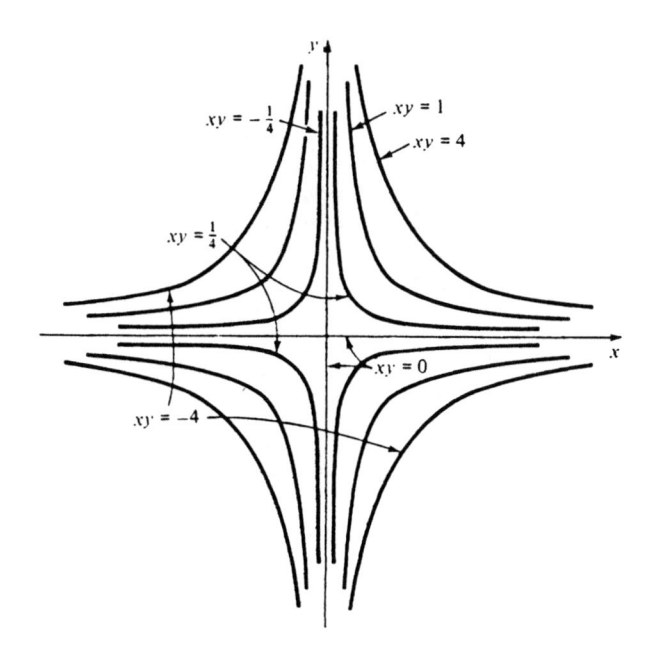

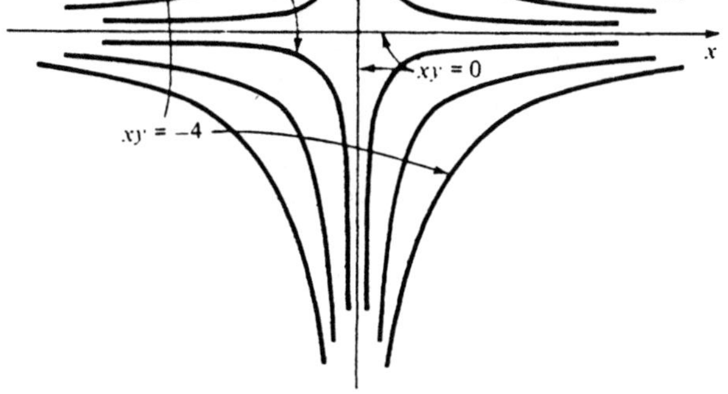

19.

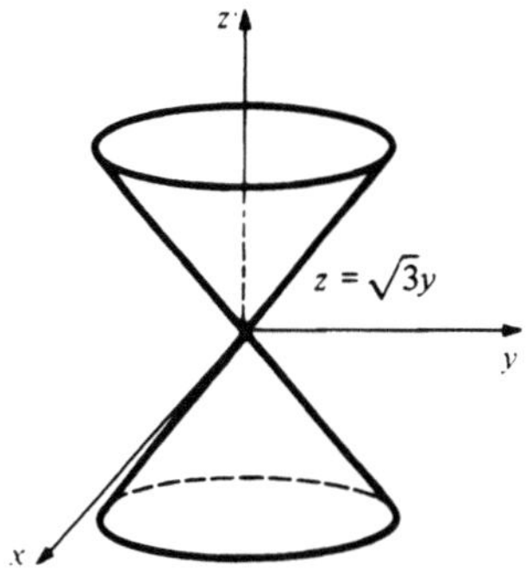

21.

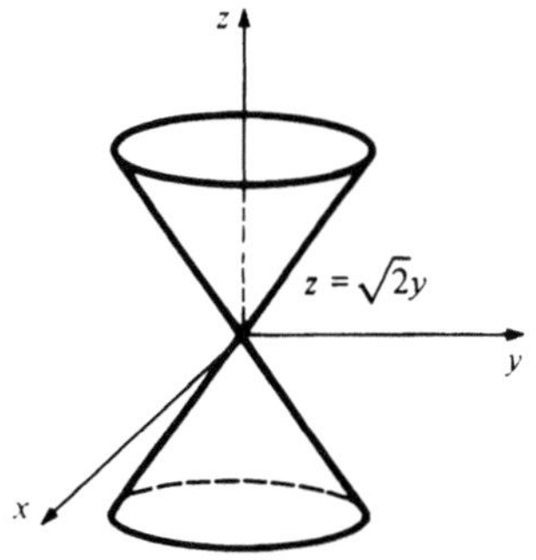

23. Substituting $z = 1$ gives $x^2 + y^2 = 1$, a circle.

25. (a) They are ellipses.

(b) In each case the cross section is two straight lines.

(c) If (x_0, y_0, z_0) satisfies the equation, so does (tx_0, ty_0, tz_0).

27. Substitute $x = x_0 + \xi u_1 + \eta v_1$, $y = y_0 + \xi u_2 + \eta v_2$, $z = z_0 + \xi u_3 + \eta v_3$ into $x^2 + y^2 = z^2$, where $\mathbf{u} = (u_1, u_2, u_3)$ and $\mathbf{v} = (v_1, v_2, v_3)$.

14.5 Cylindrical and Spherical Coordinates

1. $(\sqrt{2}, -\pi/4, 0)$

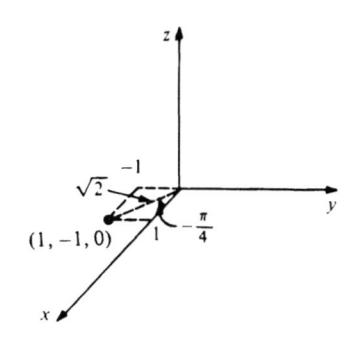

3. $(\sqrt{13}, -0.588, 1)$

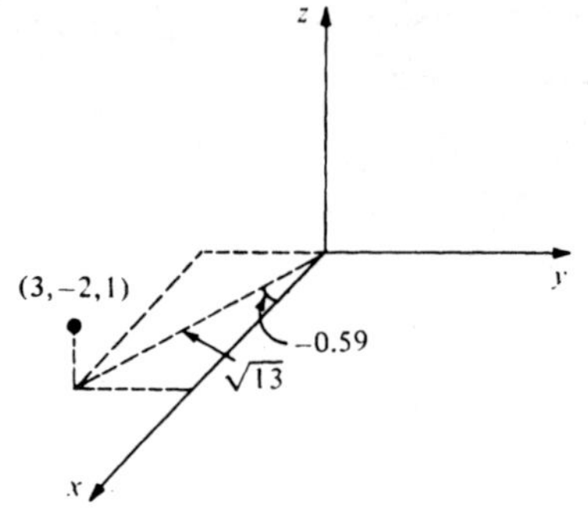

5. $(6, 0, -2)$

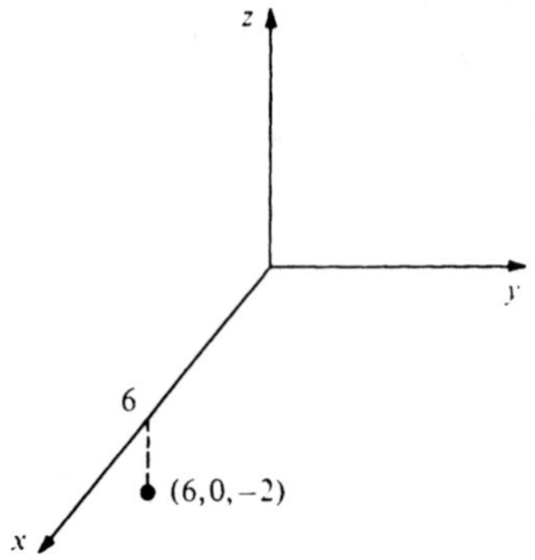

7. $(0, 1, 0)$

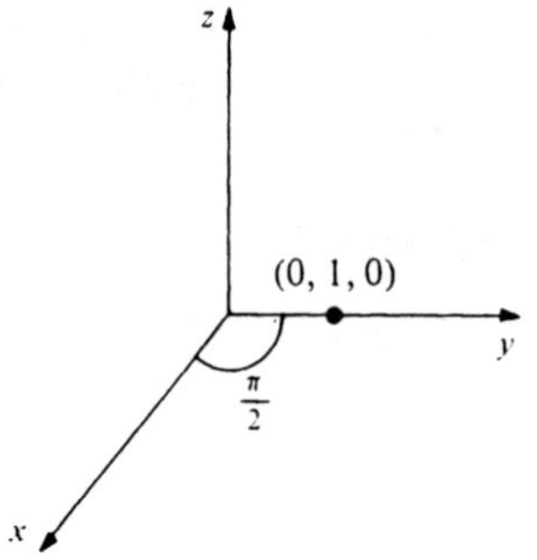

9. $(-\sqrt{3}/2, -1/2, 4)$

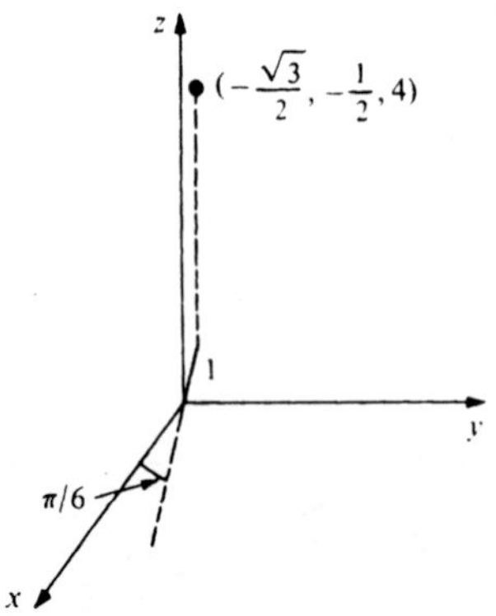

11. $(0, 0, 6)$

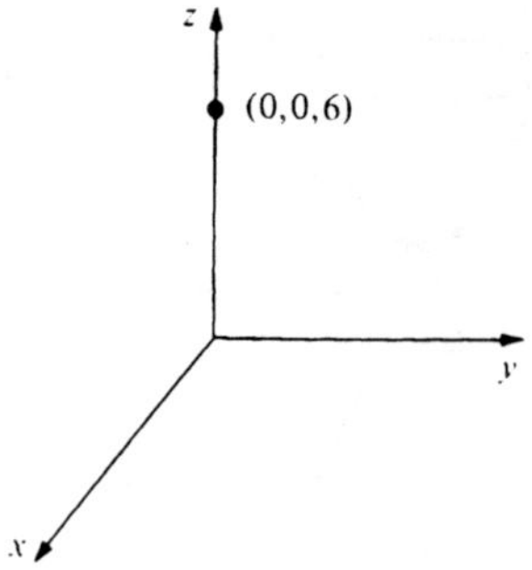

13.

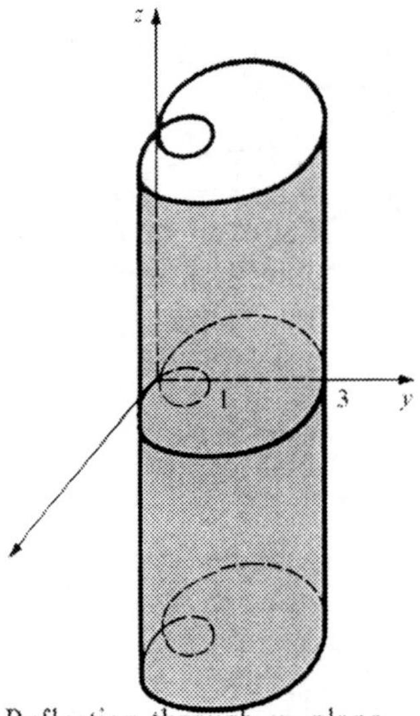

15. Reflection through xy-plane.

17. Stretching by a factor of 2 away from the z-axis, and a reflection through the xy-plane.

19. Right circular cylinder with radius r; vertical plane making an angle θ with the xz-plane; horizontal plane containing $(0, 0, z)$.

21. $(\sqrt{2}, \pi/2, \pi/4)$

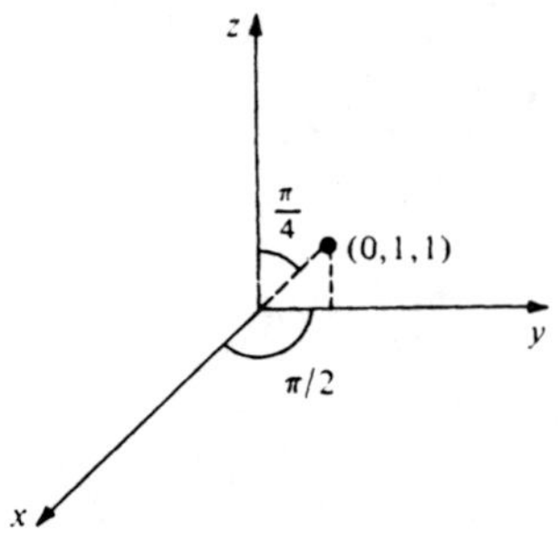

23. $(\sqrt{14}, 2.68, 2.50)$

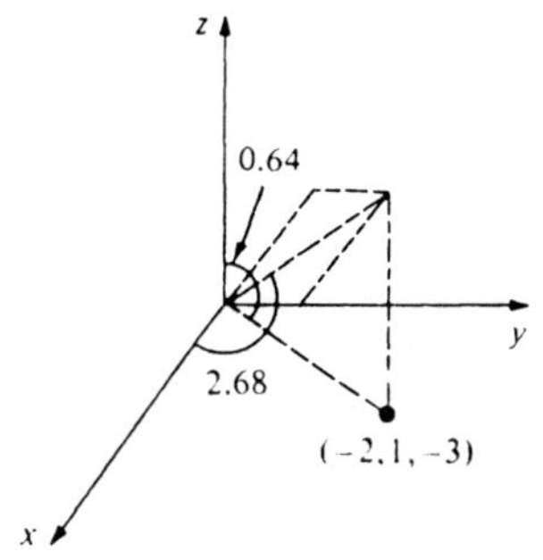

25. $(\sqrt{29}, 3.73, 2.41)$

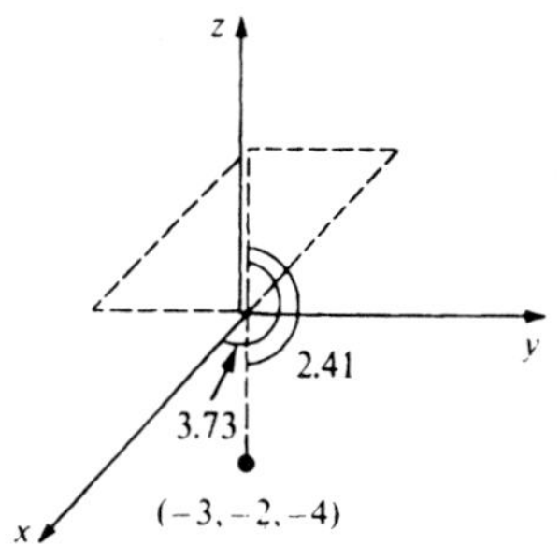

27. $(0, 0, -3)$

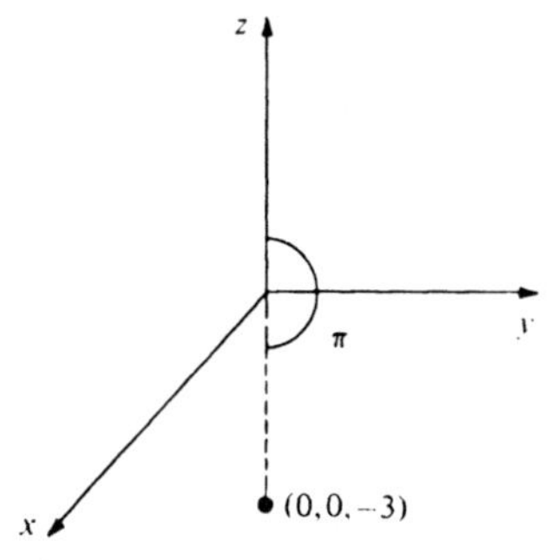

29. $(0, 0, 3)$

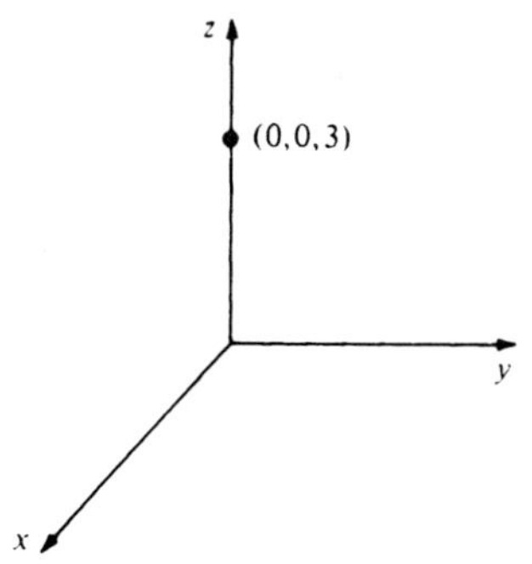

31. $(0, 0, -8)$

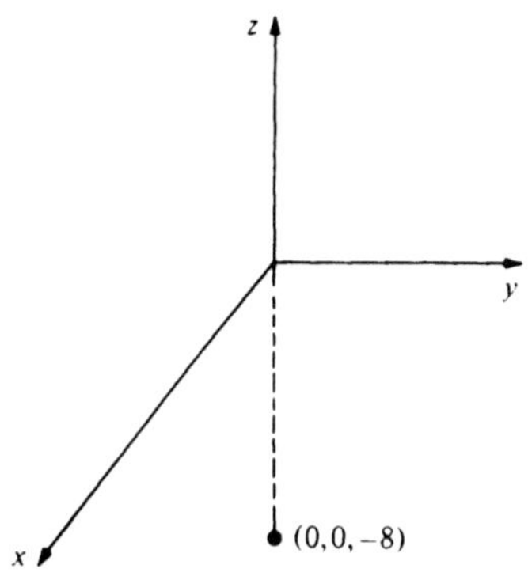

33. $\rho^2 = 2/\sin 2\phi \cos\theta$

35. The vertical half plane with positive y-coordinates and making a 45° angle with the xz-plane.

37. It moves each point twice as far from the origin along the same line through the origin.

39. The unit circle in the xy-plane.

41. $(3, \pi/2, 4)$, $(5, \pi/2, 0.64)$

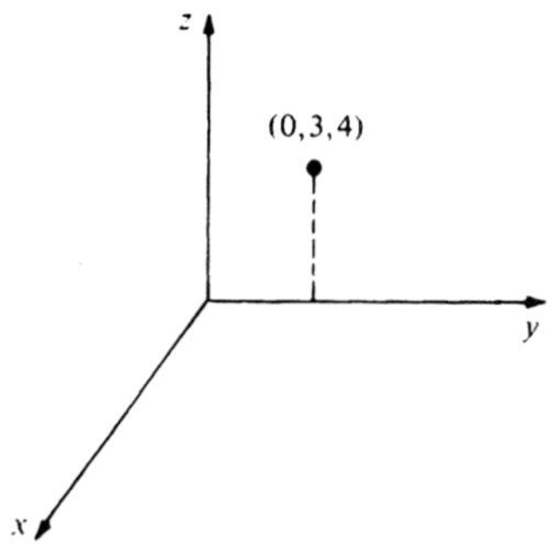

43. $(0, \theta, 0)$, $(0, \theta, \phi)$ for any θ, ϕ.

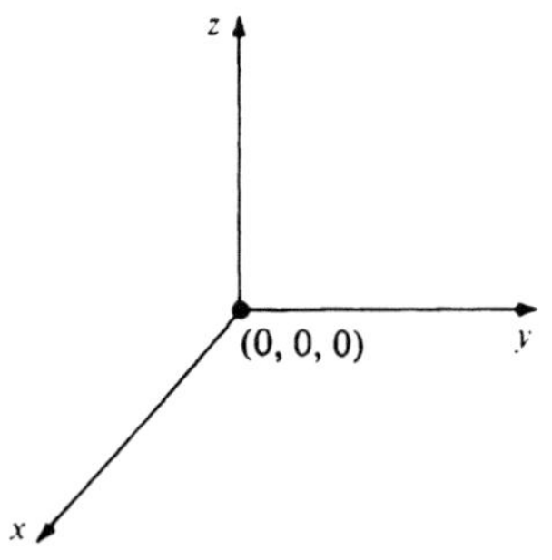

45. $(4, 7\pi/6, 3)$, $(5, 7\pi/6, 0.93)$

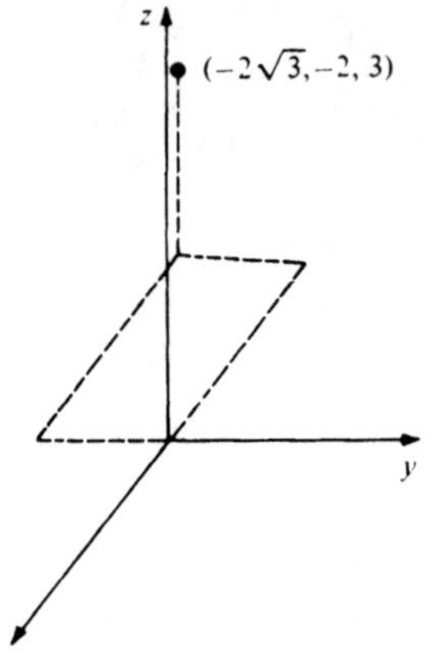

47. $(\sqrt{2}/2, \sqrt{2}/2, 1)$, $(\sqrt{2}, \pi/4, \pi/4)$
49. $(0, 0, 1)$, $(1, \pi/4, 0)$
51. $(0, 2, 1)$, $(\sqrt{5}, \pi/2, 1.11)$
53. $(0, 0, -1)$, $(0, \theta, -1)$ for any θ

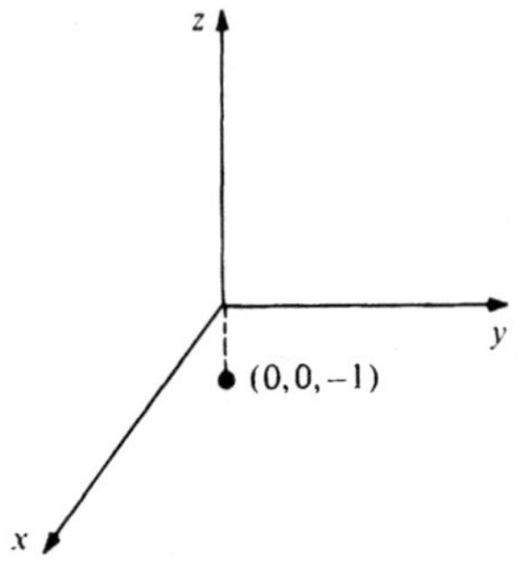

55. $(0, 0, 0)$, $(0, \theta, 0)$ for any θ

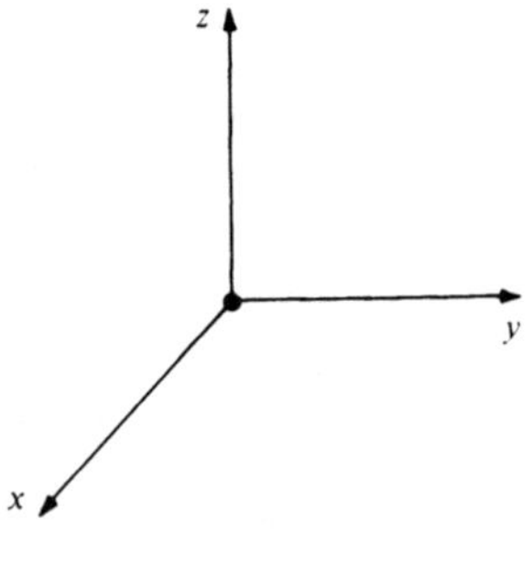

57. $(1/2, 0, -\sqrt{3}/2)$, $(1/2, 0, -\sqrt{3}/2)$

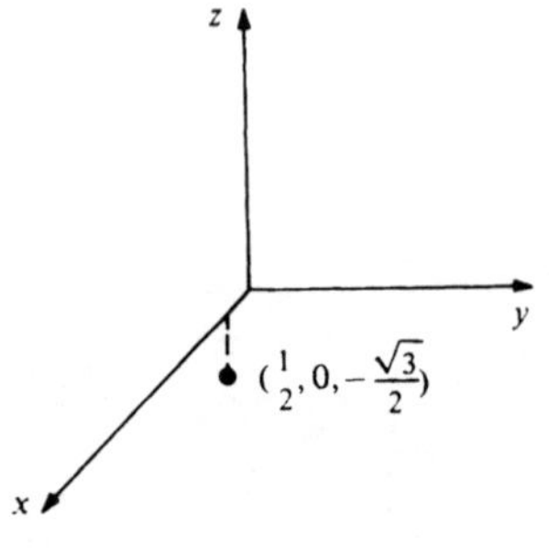

59. (a) $z = r^2\cos 2\theta$
(b) $1 = \rho \tan\phi \sin\phi \cos 2\theta$

61. (a) The length of $x\mathbf{i} + y\mathbf{j} + z\mathbf{k}$ is $(x^2 + y^2 + z^2)^{1/2} = \rho$
(b) $\cos\phi = z/(x^2 + y^2 + z^2)^{1/2}$
(c) $\cos\theta = x/(x^2 + y^2)^{1/2}$

63. $0 \leqslant r \leqslant a$, $0 \leqslant \theta \leqslant 2\pi$ means that (r, θ, z) is inside the cylinder with radius a centered on the z-axis, and $|z| \leqslant b$ means that it is no more than a distance b from the xy plane.

65. $-(d/6)\cos\phi \leqslant \rho \leqslant d/2$, $0 \leqslant \theta \leqslant 2\pi$, and $\pi - \cos^{-1}(1/3) \leqslant \phi \leqslant \pi$.

67. This is a surface whose cross-section with each surface $z = c$ is a four-petaled rose. The petals shrink to zero as $|c|$ changes from 0 to 1.

14.6 Curves in Space

1.

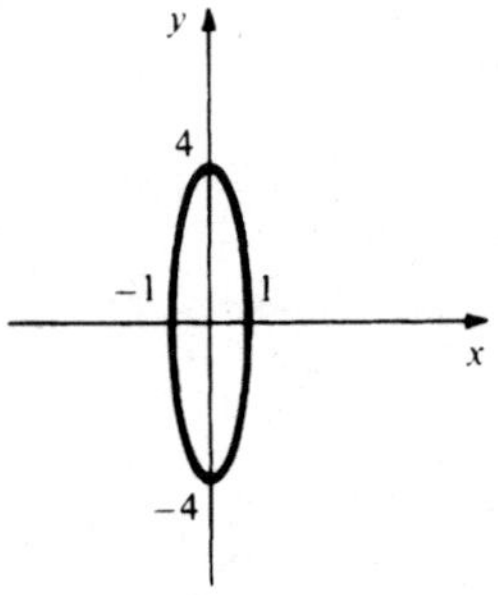

3.

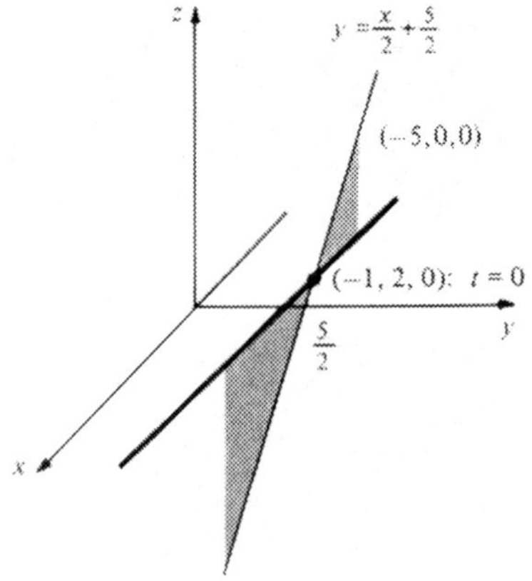

5.

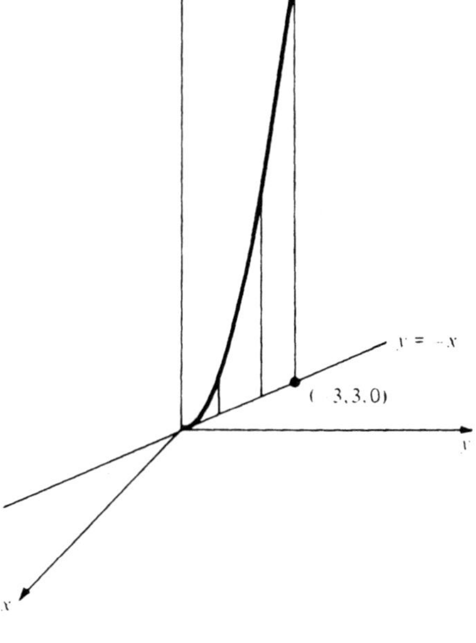

7.

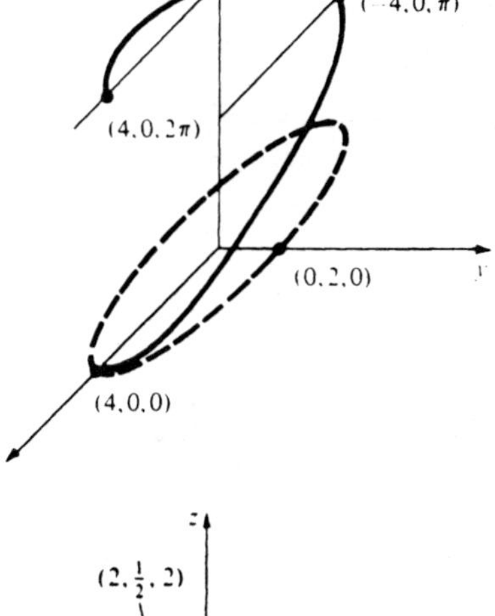

9.

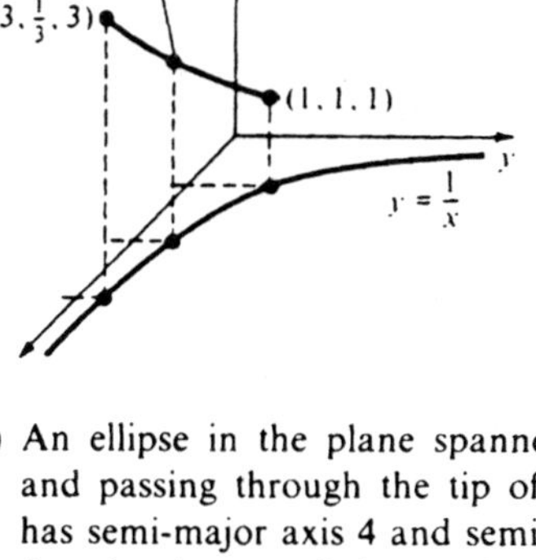

11. (a) An ellipse in the plane spanned by **v** and **w** and passing through the tip of **u**. The ellipse has semi-major axis 4 and semi-minor axis 2.
(b) $1, -1 + 4\cos t + 8\sin t$, and $4\cos t - 8\sin t$.

13. $\boldsymbol{\sigma}'(t) = -3\sin t\mathbf{i} - 8\cos t\mathbf{j} + e^t\mathbf{k}$;
$\boldsymbol{\sigma}''(t) = -3\cos t\mathbf{i} + 8\sin t\mathbf{j} + e^t\mathbf{k}$.

15. $(e^t - e^{-t})\mathbf{i} + (\cos t - \csc t\cot t)\mathbf{j} - 3t^2\mathbf{k}$.

17. $[t^3\cos t(-2 + \csc^2 t) - 3t^2\sin t(2 + \csc^2 t)]\mathbf{i} + [t^2e^{-t}(3 - t) + 2t^2e^t(t + 3)]\mathbf{j} + [e^t\csc t(1 - \cot t) - e^{-t}(\cos t + \sin t)]\mathbf{k}$

19. $e^t[2e^t\mathbf{i} + (\sin t + \cos t)\mathbf{j} + t^2(3 + t)\mathbf{k}]$

21. $\dfrac{d}{dt}\|\boldsymbol{\sigma}'(t)\|^2 = 2\boldsymbol{\sigma}'(t)\cdot\boldsymbol{\sigma}''(t) = 0.$

23. (a) $\cos t\mathbf{i} - 4\sin t\mathbf{j}$
(b) $-\sin t\mathbf{i} - 4\cos t\mathbf{j}$
(c) $\sqrt{\cos^2 t + 16\sin^2 t}$

25. (a) $2\mathbf{i} + \mathbf{j} + \mathbf{k}$ (b) 0 (c) $\sqrt{6}$

27. (a) $-\mathbf{i} + \mathbf{j} + 2t\mathbf{k}$ (b) $2\mathbf{k}$ (c) $\sqrt{2 + 4t^2}$

29. (a) $-4\sin t\mathbf{i} + 2\cos t\mathbf{j} + \mathbf{k}$ (b) $-4\cos t\mathbf{i} - 2\sin t\mathbf{j}$
(c) $\sqrt{5 + 12\sin^2 t}$

31. (a) $\mathbf{i} - (1/t^2)\mathbf{j} + \mathbf{k}$ (b) $(2/t^3)\mathbf{j}$ (c) $\sqrt{2t^4 + 1}\,/t^2$

33. $(6, 6t, 3t^2)$; $(0, 6, 6t)$; $(x, y, z) = t(6, 0, 0)$

35. $(-2\sin t\cos t, 3 - 3t^2, 1)$; $(-2\cos 2t, -6t, 0)$; $(x, y, z) = (1, 0, 0) + t(0, 3, 1)$

37. $(\sqrt{2}, e^t, -e^{-t})$; $(0, e^t, e^{-t})$;
$(x, y, z) = (0, 1, 1) + t(\sqrt{2}, 1, -1)$

39. $(2e, 0, \cos 1 - \sin 1)$

41. (a) $t(1,0,1)$

(a) (b)

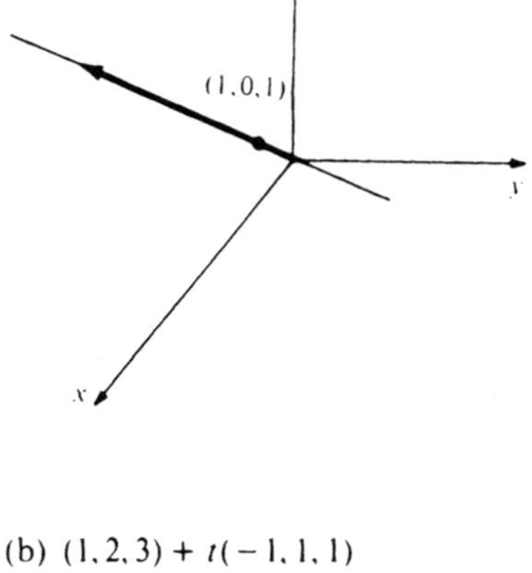

(b) $(1, 2, 3) + t(-1, 1, 1)$

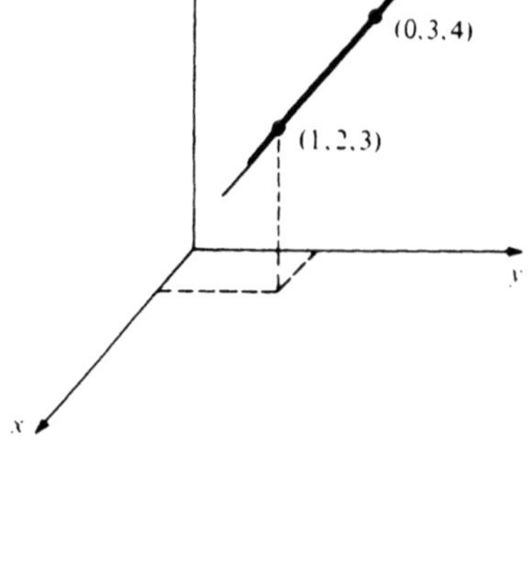

(c) $t(-1,1,1)$

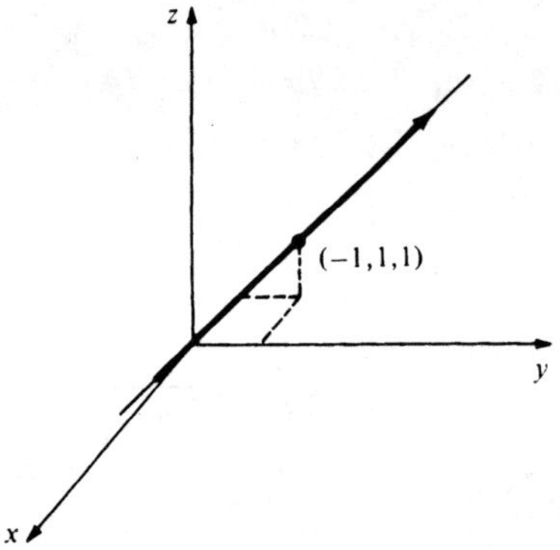

43. (a) The curve is a right circular helix with axis parallel to the z-axis.
(b)

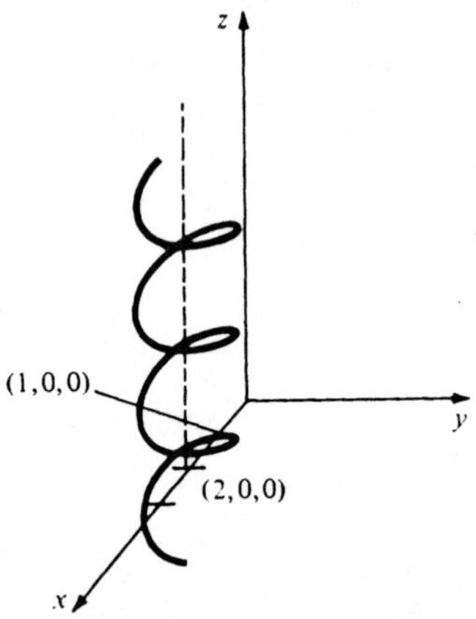

(c) The curve becomes a circle in the xy plane with center $(2,0,0)$ and radius 1.

45. (a) Substitute
(b) $(A_1\cos t + B_1\sin t, A_2\cos t + B_2\sin t, A_3\cos t + B_3\sin t)$ where A_1, A_2, A_3, B_1, B_2 and B_3 are constants.

47. (a) (i)

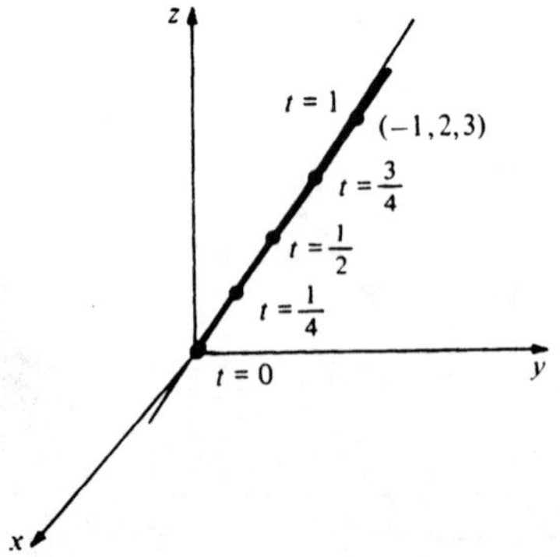

(ii)

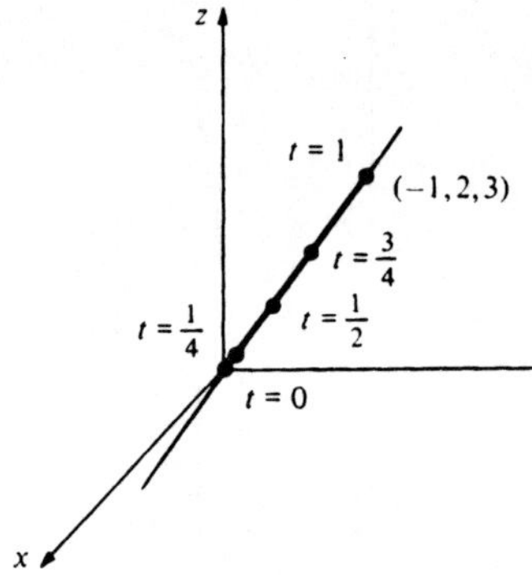

(iii)

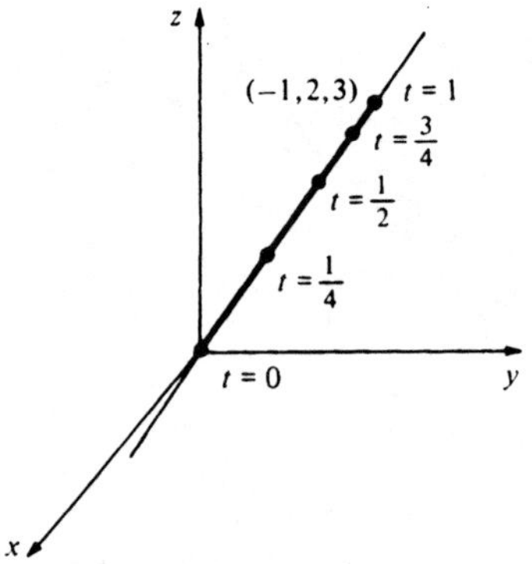

(b) Each curve is the line segment joining $(0,0,0)$ to $(-1,2,3)$. It is covered once by (i) and (iii) and twice by (ii). The velocity is constant in (i), variable in (ii) and (iii).

49. In each case, verify that $x^2 + y^2 + z^2 = 1$, so the curve lies on the sphere.

(a)

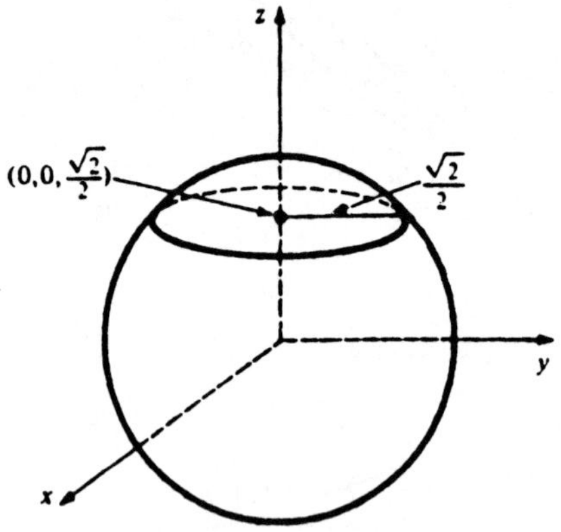

(b)

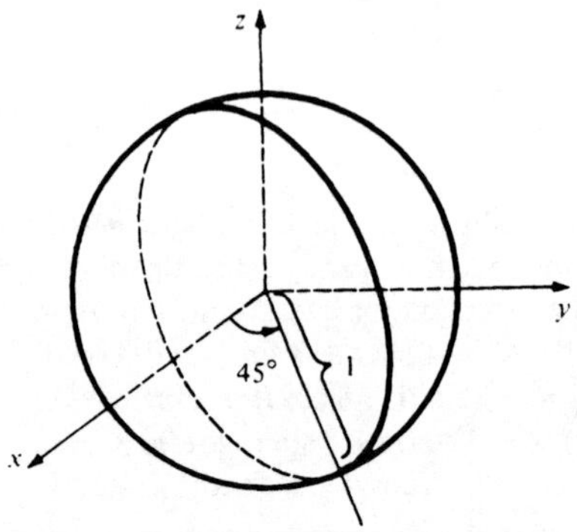

51. The set of points above or below P_0 have coordinates (x_0, y_0, z) where $z = \cos^{-1}x_0 + 2n\pi$ if $x_0 \geqslant 0$ or $z = -\cos^{-1}x_0 + 2n\pi$ if $x_0 \leqslant 0$, n an integer. The vertical distance is 2π.

53. Let $\boldsymbol{\sigma}_1 = f_1\mathbf{i} + g_1\mathbf{j} + h_1\mathbf{k}$, $\boldsymbol{\sigma}_2 = f_2\mathbf{i} + g_2\mathbf{j} + h_2\mathbf{k}$. Form $\boldsymbol{\sigma}_1 + \boldsymbol{\sigma}_2$ and differentiate using the sum rule for scalar functions.

55. Using notation in the answer to Exercise 53, form $\boldsymbol{\sigma}_1 \times \boldsymbol{\sigma}_2$ and differentiate using the product rule for scalar functions.

57. (a) $\boldsymbol{\sigma}(t)$ describes a curve in the plane through the origin perpendicular to $\mathbf{u}$.
(b) Same as (a), except that the plane need not go through the origin.
(c) $\boldsymbol{\sigma}(t)$ describes a curve lying in the cone with $\mathbf{u}$ as its axis and vertex angle $2\cos^{-1}b$.

14.7 The Geometry and Physics of Space Curves

1. $2\pi\sqrt{5}$
3. $4\sqrt{2} - 2$
5. ≈ 3.326
7. $-0.32\pi^2\mathbf{r}$, where $\mathbf{r}$ is the vector from the center to the particle.
9. (6.05×10^3) seconds.
11. (a) From $ma = GmM/R^2$, $g = GM/R^2 = (6.67 \times 10^{-11})(5.98 \times 10^{24})/(6.37 \times 10^6)^2 = 9.83\ \text{m/sec}^2$.
(b) The acceleration is $-9.8\mathbf{k}$ if $\mathbf{k}$ points upward.
13. (a) $x'' = (qb/cm)y'$; $y'' = (-qb/cm)x'$; $z'' = 0$.
(b) $x = -(amc/qb)\cos(qbt/mc) + (amc/qb) + 1$, $y = (amc/qb)\sin(qbt/mc)$, $z = ct$.
(c) $r = amc/qb$, the axis is the line parallel to the z-axis through $(amc/qb + 1, 0, 0)$.
15. The circle parametrized by arc length is $\boldsymbol{\sigma}(s) = (r\cos(s/r), r\sin(s/r))$. Calculate $\mathbf{T} = d\boldsymbol{\sigma}/ds$ and $d\mathbf{T}/ds$.
17. $k = 1/\sqrt{2}\,(2y^2 + x^2/2)^{3/2}$.
19. Assume that the curve is parametrized by arc length and show that $\mathbf{v}$ is constant.
21. Force magnitude = (mass) × (speed)^2 × (curvature).
23. (a) $\mathbf{n}$ is the normal to the plane. Since $\boldsymbol{\sigma}'$, $\boldsymbol{\sigma}''$, $\boldsymbol{\sigma}'''$ are perpendicular to $\mathbf{n}$, their triple product is zero. By Exercise 22(e), $\tau = 0$.
(b) By Exercise 22(e), $d\mathbf{B}/dt = \mathbf{0}$. By Exercise 22(a), $\mathbf{B}$ lies in the direction of $\mathbf{v} \times \mathbf{a}$.
25. Use the hint for the second equation and Exercise 22 (a) for the third.

14.S Rotations and the Sunshine Formula

1. (a) $\mathbf{m}_0 = (1/\sqrt{6})(\mathbf{i} + \mathbf{j} + 2\mathbf{k})$,
$\mathbf{n}_0 = (1/2\sqrt{3})(\mathbf{i} + \mathbf{j} - \mathbf{k})$
(b) $\mathbf{r} = [\cos(\pi t/12)/2\sqrt{2} + \sin(\pi t/12)/4]\mathbf{i} + [-1/2\sqrt{2} + \cos(\pi t/12)/2\sqrt{2} + \sin(\pi t/12)/4]\mathbf{j} + [-1/2\sqrt{2} + \cos(\pi t/12)/\sqrt{2} - \sin(\pi t/12)/4]\mathbf{k}$
(c) $(x, y, z) = (-1/2\sqrt{2})(\mathbf{i} + 2\mathbf{j} + 3\mathbf{k}) + (-\pi/48)(\mathbf{i} + \mathbf{j} + \mathbf{k})(t - 12)$
3. T_d would be longer.
5. The "exact" formula is $-\tan l \sin\alpha = \cos(2\pi t/T_d)[\tan(2\pi t/T_y)\tan(2\pi t/T_d) - \cos\alpha]$.
7. $A = 9.4°$
9. The equator would receive approximately six times as much solar energy as Paris.

Review Exercises for Chapter 14

1.

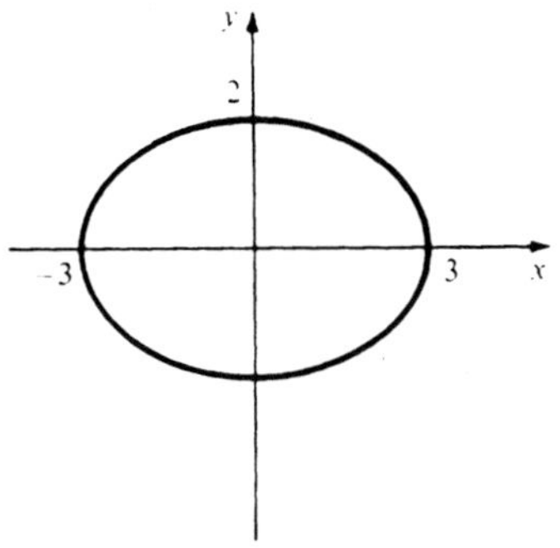

3.

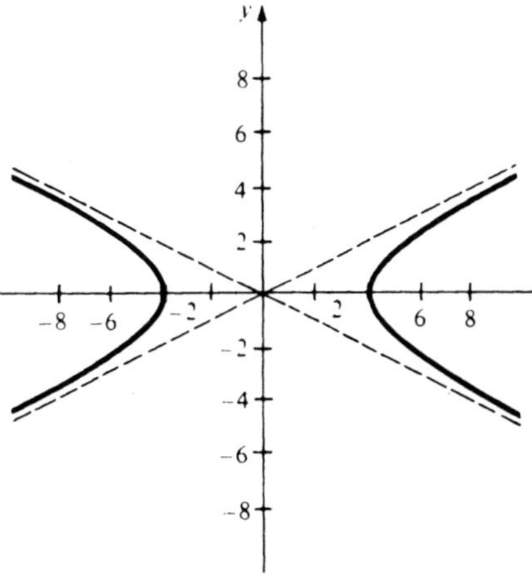

5.

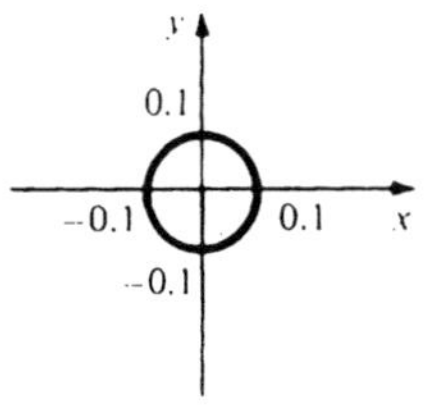

7.

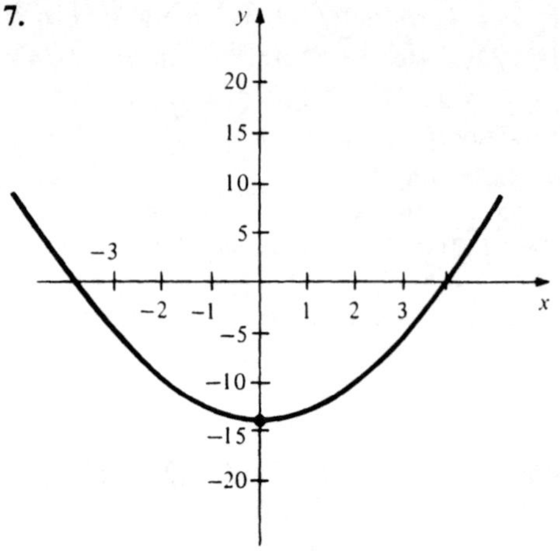

9.

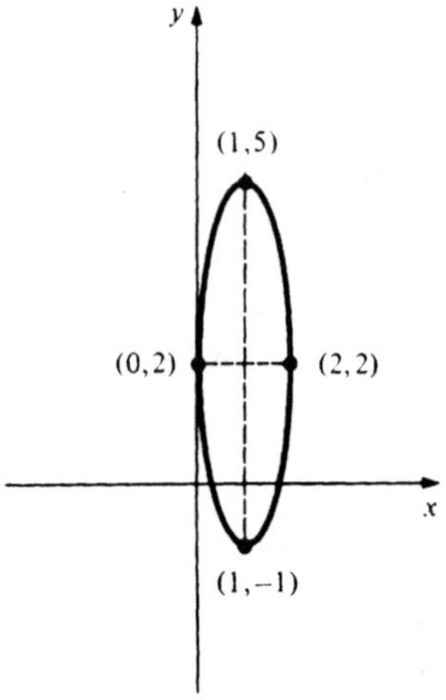

11.

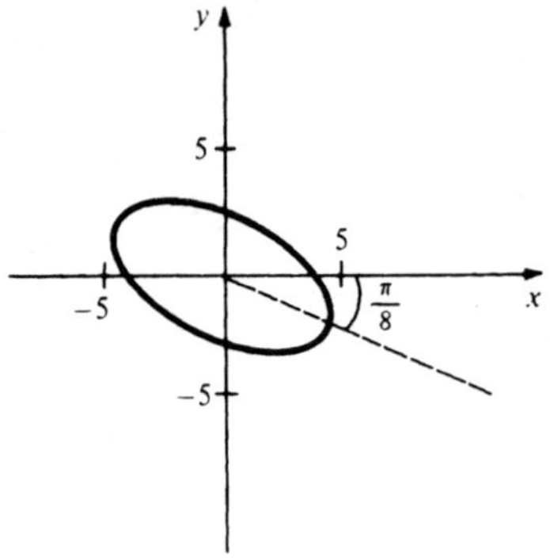

13.

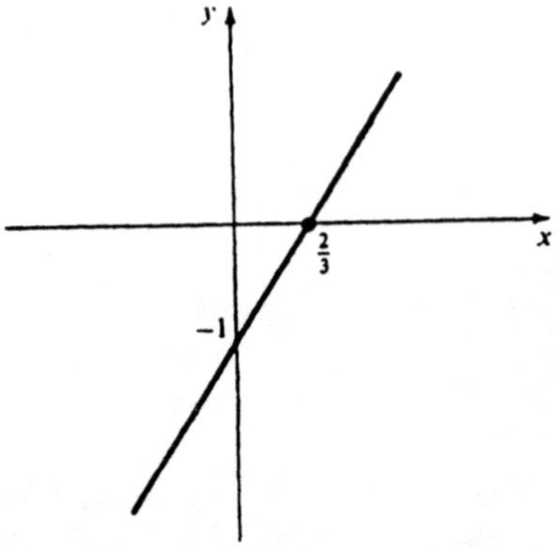

15.

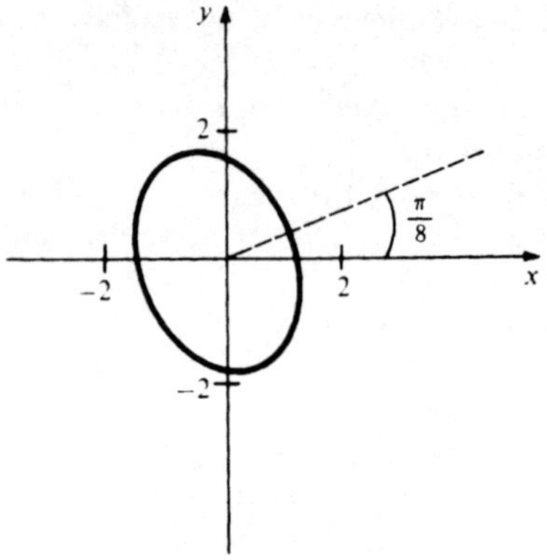

17. The level surfaces are parallel planes.

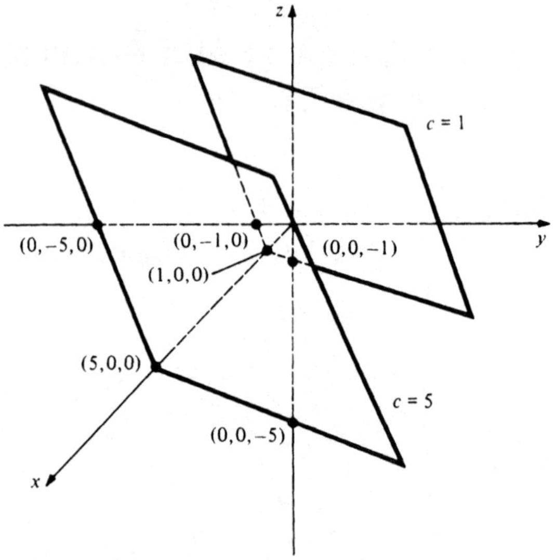

19. The level surfaces are spheres of radius $\sqrt{c-1}$.

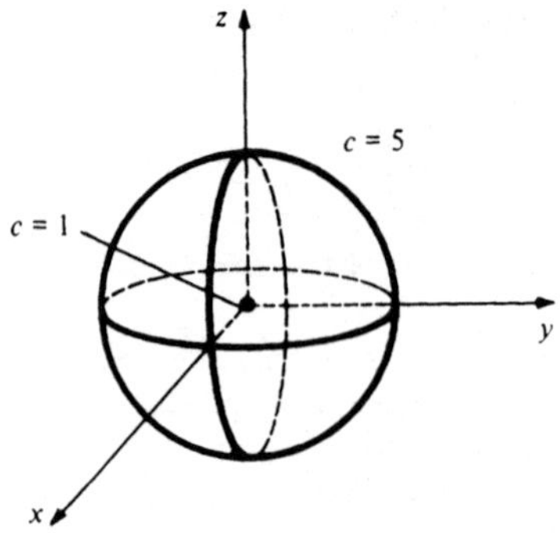

21. Ellipsoid with intercepts $(\pm 1, 0, 0)$, $(0, \pm 1/2, 0)$ and $(0, 0, \pm 1)$.

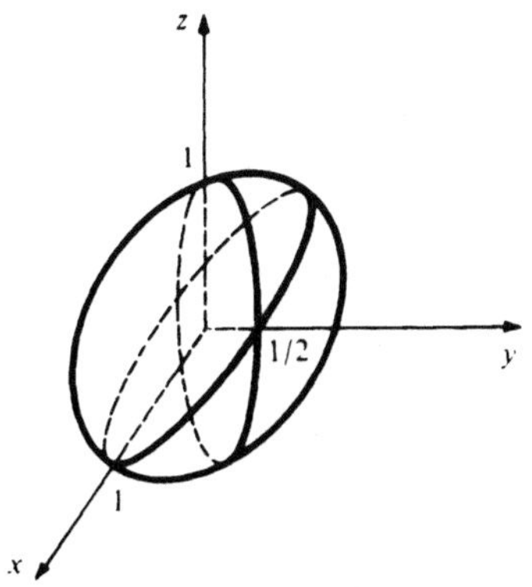

23. Elliptic hyperboloid with intercepts $(\pm 1, 0, 0)$ and $(0, \pm 1/2, 0)$.

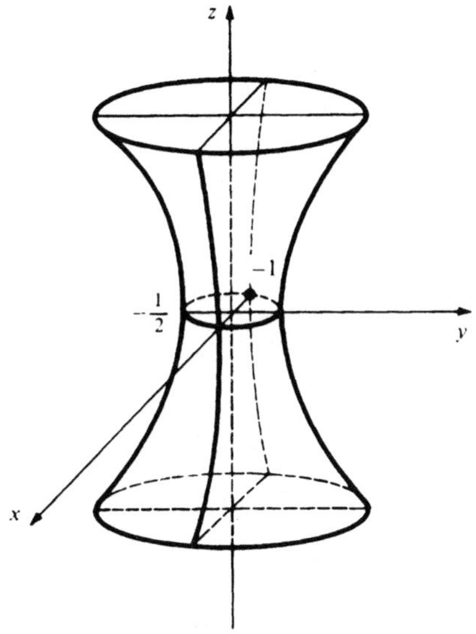

25. Elliptic paraboloid with intercepts $(\pm 1, 0, 0)$, $(0, \pm 1/2, 0)$ and $(0, 0, -1)$.

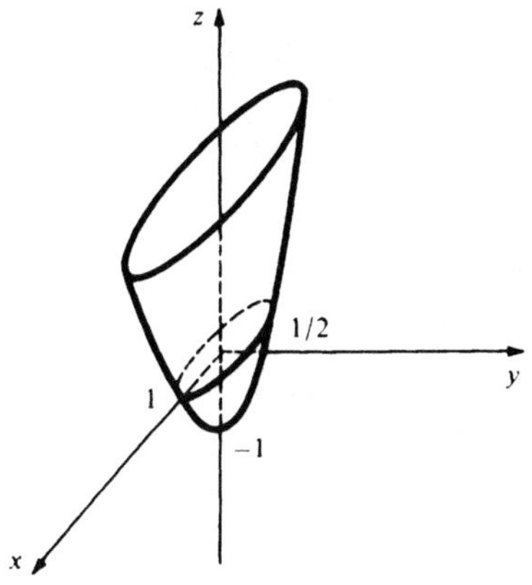

27. Elliptic paraboloid with intercepts $(\pm 1, 0, 0)$, $(0, \pm 1/2, 0)$ and $(0, 0, 1)$.

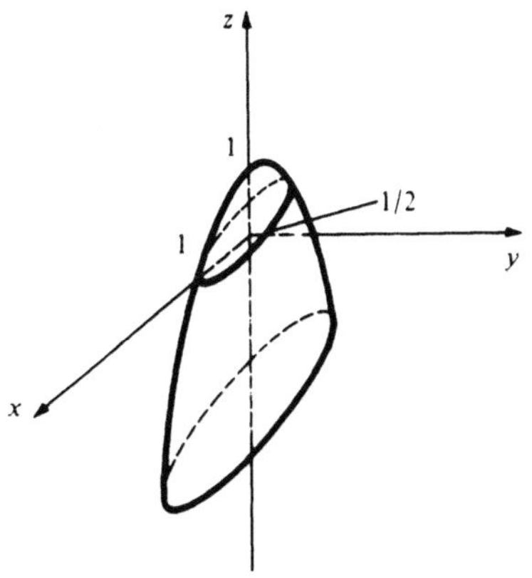

29. (a) $(x/a)^2 + (y/b)^2 = 1 + (z/c)^2$, which are ellipses.

(b) x constant; $|x| < a$ gives a hyperbola opening along the y-direction, $|x| = a$ gives two lines, and $|x| > a$ gives a hyperbola opening along z-direction.

(c)

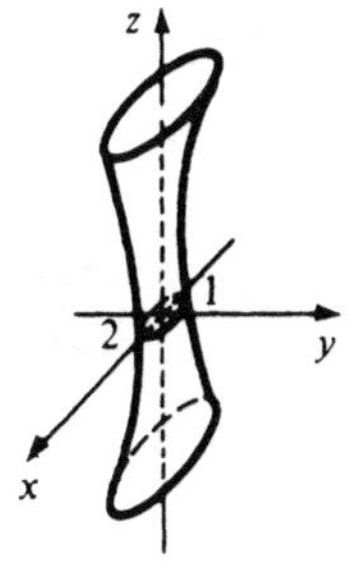

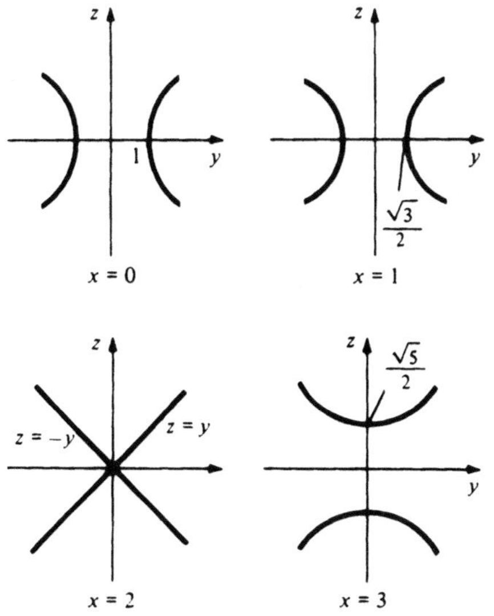

31. (a)

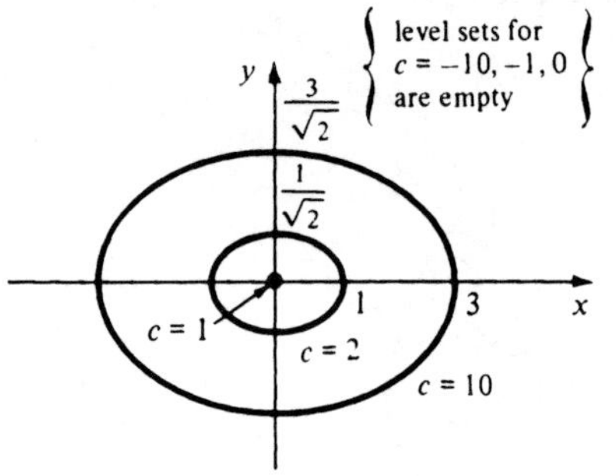

(b) For $x = \pm 1$ and $x = 2$, the equations are $z = 2y^2 + 2$ and $z = 2y^2 + 5$ which give parabolas opening upward in planes parallel to the yz-plane. For $y = \pm 1$ and $y = 2$, the equations are $z = x^2 + 3$ and $z = x^2 + 9$ which give parabolas opening upward in planes parallel to the xz-plane.

31. (c)

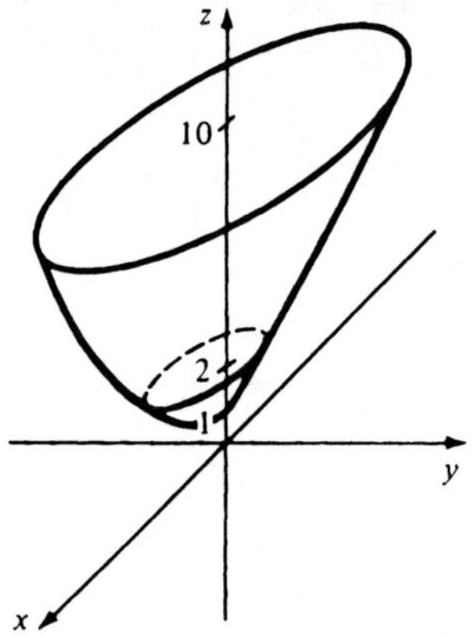

Rectangular Coordinates	Cylindrical Coordinates	Spherical Coordinates
33. $(1, -1, 1)$	$(\sqrt{2}, -\pi/4, 1)$	$(\sqrt{3}, -\pi/4, \cos^{-1}(1/\sqrt{3}))$
35. $(5\cos(\pi/12), 5\sin(\pi/12), 4)$	$(5, \pi/12, 4)$	$(\sqrt{41}, \pi/12, \cos^{-1}(4/\sqrt{41}))$
37. $(3\sqrt{6}/4, -3\sqrt{2}/4, 3\sqrt{2}/2)$	$(3\sqrt{2}/2, -\pi/6, 3\sqrt{2}/2)$	$(3, -\pi/6, \pi/4)$

33.

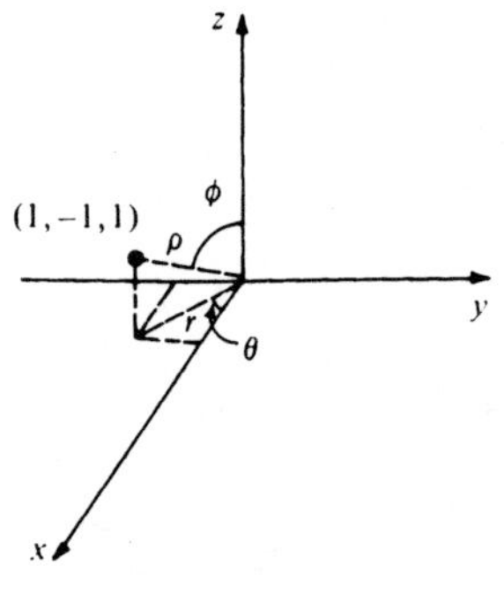

35.

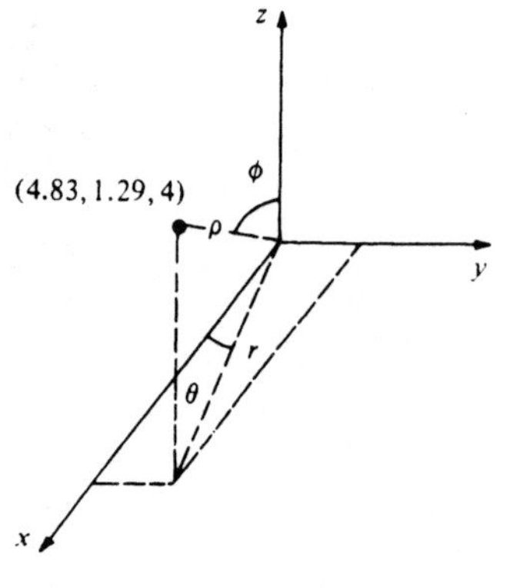

37.

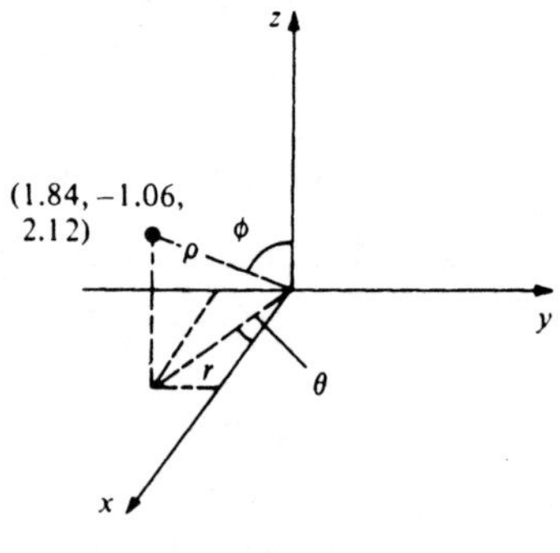

39. $3x^2 + 3y^2 = z^2 + 1$,

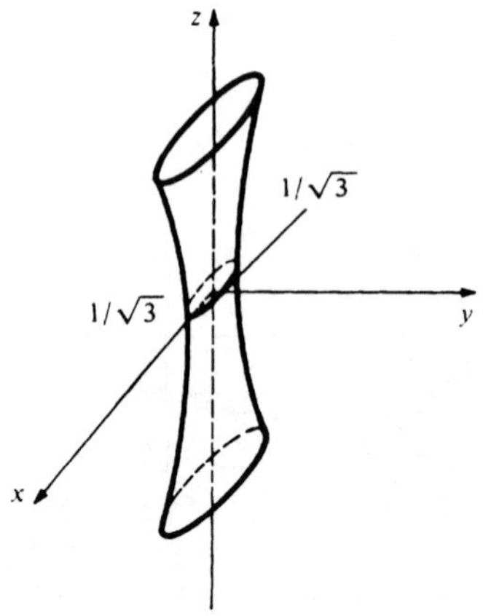

41. Rotate 180° around the z-axis and 90° away from the positive z-axis.

43. The rod is described by $0 \leqslant r \leqslant 5$, $0 \leqslant \theta \leqslant 2\pi$, and $0 \leqslant z \leqslant 15$. The winding is described by $5 \leqslant r \leqslant 6.2$, $0 \leqslant \theta \leqslant 2\pi$, and $0 \leqslant z \leqslant 15$.

45.

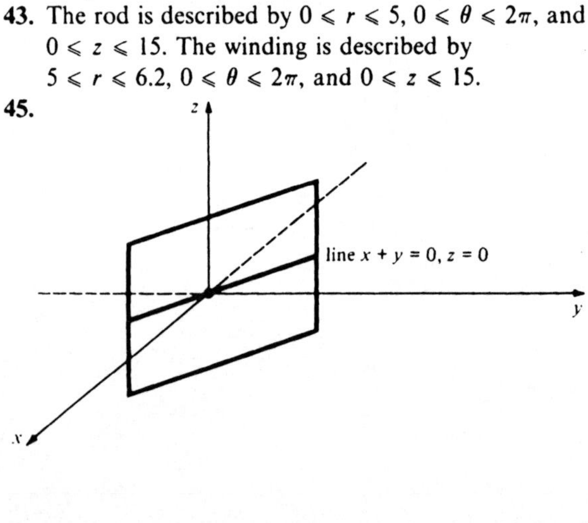

47.

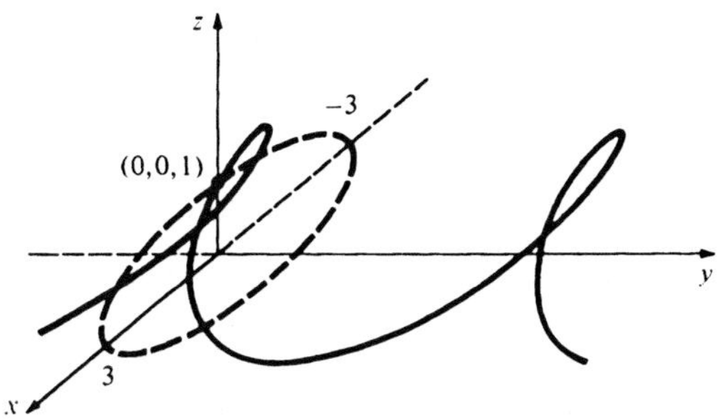

49.

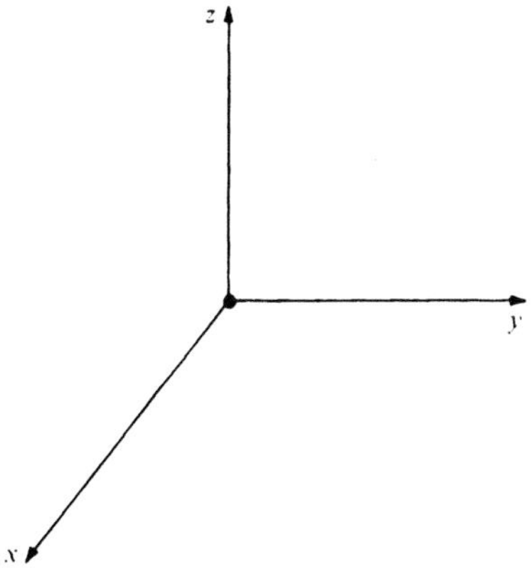

51.

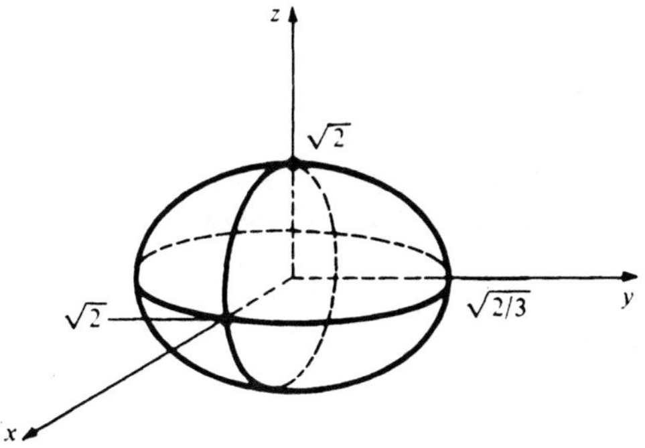

53. $(x, y, z) = (2, 1/e, 0) + (t-1)(3, -1/e, -\pi/2)$

55. $e^t\mathbf{i} + \cos t\mathbf{j} - \sin t\mathbf{k}$; $e^t\mathbf{i} - \sin t\mathbf{j} - \cos t\mathbf{k}$

57. $[e^t + 2t/(1+t^2)^2]\mathbf{i} + (\cos t + 1)\mathbf{j} - \sin t\mathbf{k}$;
$[e^t + (2-6t^2)/(1+t^2)^3]\mathbf{i} - \sin t\mathbf{j} - \cos t\mathbf{k}$

59. $x = t+1$, $y = (t-1)/2$, $z = (t-2)/3$.

61. $1 + \ln 2$

63. $(9/4, -\sin(1/2) - (1/2)\cos(1/2), -2e^{1/2})$

65. (a) $x''(t) = -(k/m)x(t)$, $y''(t) = -(k/m)y(t)$, and $z''(t) = -(k/m)z(t)$.
(b) $x(t) = 0$, $y(t) = (2m/k)\sin(kt/m)$, and $z(t) = (m/k)\sin(kt/m)$.

67. $(8/81)\cos 2t/(20\sin^2 t + 16)^{3/2}$, where $x = 2\cos t$, $y = 4/3 \sin t$

69. $k = |f''(x)|/[1 + (f'(x))^2]^{3/2}$

71. $(t^4 + 3t^2 + 8)^{1/2}/(t^2+2)^{3/2}$

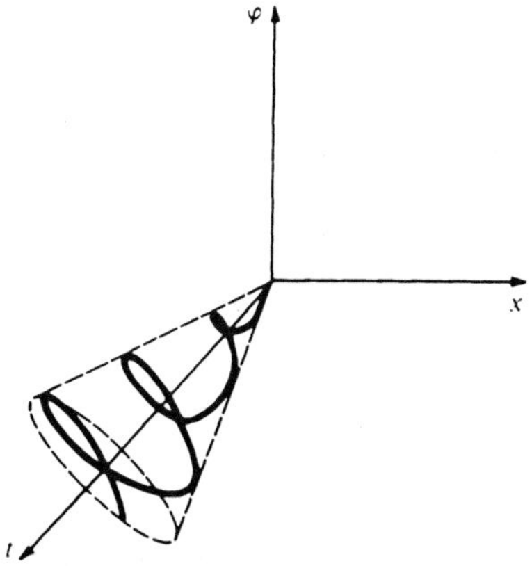

73. (a) The curve is $x^2 + z^2 = 2$ which can be expressed as $x = \sqrt{2}\cos t$, $y = 1$, $z = \sqrt{2}\sin t$.
(b) $(x, y, z) = (1, 1, 1) + t(-1, 0, 1)$
(c) $\int_0^{2\pi}\sqrt{(-\sqrt{2}\sin t)^2 + (0)^2 + (\sqrt{2}\cos t)^2}\,dt$
$= 2\sqrt{2}\,\pi.$

75. (a) $\boldsymbol{\sigma}(0) = \mathbf{u}_1$, $\boldsymbol{\sigma}(\pi/2) = \mathbf{u}_2$
(b) $\boldsymbol{\sigma}(t)$ lies on the unit sphere and in the plane determined by $\mathbf{u}_1$ and $\mathbf{u}_2$
(c) $\cos^{-1}(\mathbf{u}_1 \cdot \mathbf{u}_2)$
(d) $\int_0^{\pi/2}\sqrt{1 - 2\mathbf{u}_1 \cdot \mathbf{u}_2 \sin t \cos t}\,dt$
(e) Let $\mathbf{w}$ be the unit vector in the direction of $(\mathbf{u}_1 \times \mathbf{u}_2) \times \mathbf{u}_1$. Let $\omega = \frac{2}{\pi}\cos^{-1}(\mathbf{u}_1 \cdot \mathbf{u}_2)$. Then $\boldsymbol{\sigma}_1(t) = \mathbf{u}_1\cos(\omega t) + \mathbf{w}\sin(\omega t)$.

Chapter 15 Answers

15.1 Introduction to Partial Derivatives

1. $f_x = y$, $f_x(1,1) = 1$; $f_y = x$, $f_y(1,1) = 1$

3. $f_x = 1/[1 + (x - 3y^2)^2]$, $f_x(1,0) = 1/2$;
$f_y = -6y/[1 + (x-3y^2)^2]$, $f_y(1,0) = 0$.

5. $f_x = ye^{xy}\sin(x+y) + e^{xy}\cos(x+y)$, $f_x(0,0) = 1$;
$f_y = xe^{xy}\sin(x+y) + e^{xy}\cos(x+y)$, $f_y(0,0) = 1$.

7. $f_x = -3x^2/(x^3+y^3)^2$, $f_x(-1,2) = -3/49$;
$f_y = -3y^2/(x^3+y^3)^2$, $f_y(-1,2) = -12/49$.

9. $f_x = yz$, $f_x(1,1,1) = 1$; $f_y = xz$, $f_y(1,1,1) = 1$;
$f_z = xy$, $f_z(1,1,1) = 1$.

11. $f_x = -y^2\sin(xy^2) + 3yze^{3xyz}$, $f_x(\pi,1,1) = 3e^{3\pi}$;
$f_y = -2xy\sin(xy^2) + 3xze^{3xyz}$, $f_y(\pi,1,1) = 3\pi e^{3\pi}$;
$f_z = 3xye^{3xyz}$, $f_z(\pi,1,1) = 3\pi e^{3\pi}$.

13. $\partial z/\partial x = 6x$; $\partial z/\partial y = 4y$.

15. $\partial z/\partial x = 2/3y + 7/3$; $\partial z/\partial y = -2x/3y^2$.

17. $\partial u/\partial x = e^{-xyz}[-yz(xy + xz + yz) + (y+z)]$;
$\partial u/\partial y = e^{-xyz}[-xz(xy + xz + yz) + (x+z)]$;
$\partial u/\partial z = e^{-xyz}[-xy(xy + xz + yz) + (x+y)]$.

19. $\partial u/\partial x = e^x\cos(yz^2)$; $\partial u/\partial y = -z^2e^x\sin(yz^2)$; $\partial u/\partial z = -2yze^x\sin(yz^2)$

21. $(xye^xe^y - xe^xe^y + xe^y + e^x)/(ye^x + 1)^2$

23. $16b(mx + b^2)^7$

25. $12 + (2/9)\cos(2/9) - 27e^2$.

27. $-4\cos(1) + 3 - 3e$

29. (a) $-x(\sin x)e^{-xy}$

(b) $0, 0, -\pi/2, -(\pi/2)e^{-\pi^2/4}$

31. 1

33. $1/6 - 3\sec^2(6)$

35. $1/7 + 3\sec^2(-15)$

37. $(tu^2)e^{stu^2}$

39. $\dfrac{(-\mu\sin\lambda\mu)(1+\lambda^2+\mu^2) - 2\lambda\cos\lambda\mu}{(1+\lambda^2+\mu^2)^2}$

41. $f_z = \lim_{\Delta z\to 0}\{[f(x,y,z+\Delta z) - f(x,y,z)]/\Delta z\}$.

43. The rate of change is approximately zero.

45. (a) $1/(1 + R_1/R_2 + R_1/R_3)^2$

(b) 36/121 times as fast.

47. $\partial^2 z/\partial x^2 = 6$, $\partial^2 z/\partial y^2 = 4$, $\partial^2 z/\partial x\partial y = \partial^2 z/\partial y\partial x = 0$

49. $\partial^2 z/\partial x^2 = 0$, $\partial^2 z/\partial y^2 = 4x/3y^3$, $\partial^2 z/\partial x\partial y = \partial^2 z/\partial y\partial x = -2/3y^2$

51. $f_{xy} = 2x + 2y$, $f_{yz} = 2z$, $f_{zx} = 0$, $f_{xyz} = 0$

53. $\partial^2 u/\partial x^2 = 24xy(x^2 - y^2)/(x^2+y^2)^4$, $\partial^2 u/\partial y\partial x = \partial^2 u/\partial x\partial y = -6(x^4 - 6x^2y^2 + y^4)/(x^2+y^2)^4$, $\partial^2 u/\partial y^2 = -24xy(x^2-y^2)/(x^2+y^2)^4$.

55. $\partial^2 u/\partial x^2 = y^4e^{-xy^2} + 12x^2y^3$, $\partial^2 u/\partial y\partial x = \partial^2 u/\partial x\partial y = e^{-xy^2}(-2y + 2xy^3) + 12x^3y^2$, $\partial^2 u/\partial y^2 = 2xe^{-xy^2}(2xy^2 - 1) + 6x^4y$.

57. Take $\delta = \varepsilon$.

59. 0

61. $52/\sqrt{13} = 4\sqrt{13}$

63. $-e$

65. 0

67. $g'(t_0) = -2\cos t_0\sin t_0 + 2e^{2t_0}$

69. Evaluate the derivatives and add.

71. (a) Evaluate the derivatives and compare.

(b)

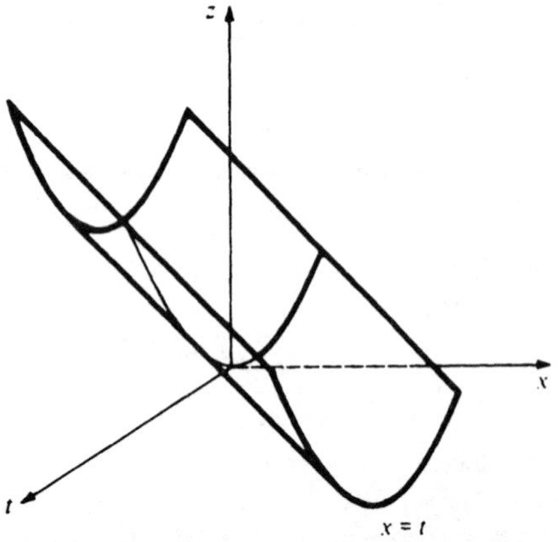

73. (a) 170 units

(b) 276 units. This is the marginal productivity of capital per million dollars invested, with a labor force of 5 people and investment level of three million dollars.

75. (a) Look at the function restricted to the x-, y-, and z-axes.

75. (a) Look at the function restricted to the x-, y-, and z-axes.

77. (a) Substitute $x = 0$ into f_x.

(b) Substitute $y = 0$ into f_y to get $f_y(x, 0) = x$

(c) $f_{yx}(0,0) = \lim_{y\to 0}[(f_x(0,y) - f_x(0,0))/y]$, etc.

(d) Notice that f_x and f_y are not continuous at $(0, 0)$.

15.2 Linear Approximations and Tangent Planes

1. $z = -9x + 6y - 6$.

3. $z = 1$

5. $z = 2x + 6y - 4$

7. $z = 1$

9. $z = x - y + 2$

11. $z = x + y - 1$

13. $-(1/\sqrt{3})(\mathbf{i} - \mathbf{j} - \mathbf{k})$

15. $-(1/\sqrt{3})(\mathbf{i} + \mathbf{j} - \mathbf{k})$

17. -0.415

19. -2.85

21. 1.00

23. Increasing, decreasing, increasing.

25. $1 - \Delta a/6 + \Delta v$

27. (a) 2

(b) A parabola in the yz-plane, opening upward with vertex at $(1, 0, 1)$.

(c) $(0, 1, 2)$

29. See Example 1; in this case we are dealing with the lower hemisphere.

15.3 The Chain Rule

1. $(48 + 128t)\cos(3 - 2t) - 8(3 - 2t)^2\cos(3 + 8t) + 2(3 + 8t)^2\sin(3 - 2t) + (12 - 8t)\sin(3 + 8t)$

3. $(e^{2t} - e^{-2t})(\ln(e^{2t} + e^{-2t}) + 1)$.

5. $-\sin t + 2\cos t\sin t + 3t^2$.

7. $e^{t-t^2}(1 - 2t) + e^{t^2-t^3}(2t - 3t^2) + e^{t^3-t}(3t^2 - 1)$.

9. Let $f(x, y) = x/y$.

11. $(x/\sqrt{x^2+y^2} + 2y^2)(dx/du) + (y/\sqrt{x^2+y^2} + 4xy)(dy/du)$.

13. $a\mathbf{i} + b\mathbf{j} + c\mathbf{k}$ where $-2a - 4b + c = 0$.

15. (a) $x^x(1 + \ln x)$

(b) $x^x(1 + \ln x)$

(c) One author prefers (a), the other (b).

17. $\dfrac{f'(t)g(t)h(t) + f(t)g'(t)h(t) - f(t)g(t)h'(t)}{[h(t)]^2}$

19. 6.843

21. The half-line lies in its own tangent line. The cone in Example 6 is such a surface, as is any other surface obtained by drawing rays from the origin to the points of a space curve.

15.4 Matrix Multiplication and the Chain Rule

1. [32]

3. [44]

5. $\begin{bmatrix}\sin v & u\cos v\\ ve^{uv} & ue^{uv}\end{bmatrix}$; $\begin{bmatrix}\sin 1 & 0\\ 1 & 0\end{bmatrix}$

9. $\begin{bmatrix}10 & 13\\ 4 & 5\end{bmatrix}$

7. $\begin{bmatrix}yz & xz & xy\\ 1 & 1 & 1\end{bmatrix}$; $\begin{bmatrix}9 & 9 & 9\\ 1 & 1 & 1\end{bmatrix}$

11. $\begin{bmatrix}c & d\\ a & b\end{bmatrix}$

13. Undefined, the first matrix has two columns and the second matrix has three rows.

15. $\begin{bmatrix} 0 \\ b \end{bmatrix}$ 17. $\begin{bmatrix} a & b \\ 0 & 0 \end{bmatrix}$ 19. $\begin{bmatrix} 2 & 6 \\ 7 & 12 \end{bmatrix}$

21. $\partial z/\partial x = 26x + 6y + 70$; $\partial z/\partial y = 6x + 2y + 14$.

23. $\partial z/\partial x = \cos(3x^2 - 2y)(6x)\cos(x - 3y) + \sin(3x^2 - 2y)[-\sin(x - 3y)]$;
$\partial z/\partial y = -2\cos(3x^2 - 2y)\cos(x - 3y) + 3\sin(3x^2 - 2y)\sin(x - 3y)$.

25. (a) $\begin{bmatrix} 1 & 1 \\ 1 & -1 \end{bmatrix}, \begin{bmatrix} 2x & 2y \\ 2x & -2y \end{bmatrix}$

(b) $u = (t + s)^2 + (t - s)^2$, $v = (t + s)^2 - (t - s)^2$,
$\partial(u, v)/\partial(t, s) = \begin{bmatrix} 4t & 4s \\ 4s & 4t \end{bmatrix}$

(c) Multiply the matrices in (a) and express in terms of s and t.

27. (a) $\begin{bmatrix} s & t \\ s & t \end{bmatrix}, \begin{bmatrix} 1 & 0 \\ 0 & -1 \end{bmatrix}$

(b) $u = ts$, $v = -ts$, $\partial(u, v)/\partial(t, s) = \begin{bmatrix} s & t \\ -s & -t \end{bmatrix}$

(c) Multiply the matrices in (a).

29. $\partial u/\partial r = \cos\theta \sin\phi(\partial u/\partial x) + \sin\theta\sin\phi(\partial u/\partial y) + \cos\phi(\partial u/\partial z)$, $\partial u/\partial\theta = -r\sin\theta\sin\phi(\partial u/\partial x) + r\cos\theta\sin\phi(\partial u/\partial y)$,
$(\partial u/\partial\phi) = r\cos\theta\cos\phi(\partial u/\partial x) + r\sin\theta\cos\phi(\partial u/\partial y) - r\sin\phi(\partial u/\partial z)$

31. $r = \sqrt{x^2 + y^2}$, $\theta = \tan^{-1}(y/x)$.

$$\partial(r, \theta)/\partial(x, y) = \begin{bmatrix} \dfrac{x}{\sqrt{x^2 + y^2}} & \dfrac{y}{\sqrt{x^2 + y^2}} \\ \dfrac{-y}{x^2 + y^2} & \dfrac{x}{x^2 + y^2} \end{bmatrix}$$

33. $AB = \sum_{i=1}^{m} a_i \cdot (1/m) = \frac{1}{m}\sum_{i=1}^{m} a_i$, the average of the entries of A.

35. Multiply B (found in Exercise 34) by A.

37. $\begin{bmatrix} 5 & -2 \\ -2 & 1 \end{bmatrix}$

39. Use the relations between areas, volumes, and determinants in Section 13.6.

41. (a) -16
(b) 8
(c) $-128 = -16 \cdot 8$

43. $$\rho\begin{bmatrix} \partial^2 v_1/\partial t^2 \\ \partial^2 v_2/\partial t^2 \\ \partial^2 v_3/\partial t^2 \end{bmatrix} = (a + b)\begin{bmatrix} \partial u/\partial x \\ \partial u/\partial y \\ \partial u/\partial z \end{bmatrix} + b\begin{bmatrix} \partial^2 v_1/\partial x^2 + \partial^2 v_1/\partial y^2 + \partial^2 v_1/\partial z^2 \\ \partial^2 v_2/\partial x^2 + \partial^2 v_2/\partial y^2 + \partial^2 v_2/\partial z^2 \\ \partial^2 v_3/\partial x^2 + \partial^2 v_3/\partial y^2 + \partial^2 v_3/\partial z^2 \end{bmatrix}$$

45. (a) Substitute and use $\cos^2 + \sin^2 = 1$.
(b) Eliminate θ to find a relation between x, y, z, and φ.
(c) Look at the ratio y/x.
(d) Find $\partial(x, y, z)/\partial(u, \varphi, \theta)$ and evaluate its determinant.

47. Express $|A||B|$ as a sum of 36 terms.

49. $$\int_a^b f(x)\,dx \approx \frac{b-a}{3n}[f(x_0)\, f(x_1) \ldots f(x_n)]\begin{bmatrix} 1 \\ 4 \\ 2 \\ 4 \\ \vdots \\ 4 \\ 2 \\ 4 \\ 1 \end{bmatrix}$$

Review Exercises for Chapter 15

1. $g_x = \pi\cos(\pi x)/(1 + y^2)$;
$g_y = -2y\sin(\pi x)/(1 + y^2)^2$

3. $k_x = z^2 + z^3\sin(xz^3)$; $k_z = 2xz + 3xz^2\sin(xz^3)$.

5. $h_x = z$; $h_y = 2y + z$; $h_z = x + y$

7. $f_x = -[\cos(xy) + y\sin(xy)]/[e^x + \cos(xy)]$;
$f_y = -x\sin(xy)/[e^x + \cos(xy)]$; $f_z = 0$.

9. $g_x = z + x^2e^{x+z}$; $g_y = 0$; $g_z = x + e^z\int_0^x t^2e^t\,dt$

11. $g_{xy} = g_{yx} = -2\pi y\cos(\pi x)/(1 + y^2)^2$

13. $k_{xz} = k_{zx} = 2z + 3z^2\sin(xz^3) + 3xz^5\cos(xz^3)$

15. $h_{xz} = h_{zx} = 1$ 17. 1 19. $-\sin(2)$

21. (a) 35.25 minutes
(b) $\partial T/\partial x|_{(27,65)} = -0.598$ minutes/foot; this means that in diving from 27 to 28 feet, your time decreases about 36 seconds. $\partial T/\partial V|_{(27)} = 0.542$ minutes/cubic foot; this means that bringing an extra cubic foot of air will give you about 33 seconds more diving time.

23. 4

25. 0

27. $z = 2x + 2y - 2$

29. $z = 1$

31. 9.00733

33. 0.999

35. 5.002

37. $t = \sqrt{14}(-3 + 2\sqrt{709})/70$

39. $d[f(\sigma(t))]/dt = 2t/[(1 + t^2 + 2\cos^2 t)(2 - 2t^2 + t^4)] - 4t(t^2 - 1)\ln(1 + t^2 + 2\cos^2 t)/(2 - 2t^2 + t^4)^2 - 4\cos t\sin t/[(1 + t^2 + 2\cos^2 t)(2 - 2t^2 + t^4)]$

41. (a) Use the chain rule with $x - ct$ as intermediate variable.
(b) It shifts with velocity c along the x axis, without changing its shape.

43. The radius is increasing by 15 cm/hr.

45. $[f'(t)g(t) + f(t)g'(t)]\exp[f(t)g(t)]$

47. $(1 + 2y - 2x)\exp(x + 2xy)$

49. $y = -x/6 + 7/12$

51. [4]

53. $\begin{bmatrix} 0 & 1 \\ 1 & 0 \end{bmatrix}$ 55. $\begin{bmatrix} 6 & -3 \\ 12 & -6 \end{bmatrix}$

57. $\begin{bmatrix} 0 & -1 & 4 \\ -3 & -2 & -1 \\ 3 & 1 & 5 \end{bmatrix}$

59. $\begin{bmatrix} 3 & 1 & -3 \\ 5 & -1 & 1 \\ 1 & 1 & -1 \end{bmatrix}$

61. $\partial z/\partial x = 4(e^{-2x-2y+2xy})(1+y)/(e^{-2x-2y}-e^{2xy})^2$, $\partial z/\partial y = 4(e^{-2y-2x+2xy})(1+x)/(e^{-2x-2y}-e^{2xy})^2$

63. $\partial z/\partial u = \partial z/\partial x + \partial z/\partial y$, $\partial z/\partial v = \partial z/\partial x - \partial z/\partial y$.

65. (a) $n = PV/RT$; $P = nRT/V$; $T = PV/nR$; $V = nRT/P$
 (b) $\partial P/\partial T$ represents the ratio between the change ΔP in pressure and the change ΔT in temperature when the volume and number of moles of gas are held fixed.
 (c) $\partial V/\partial T = nR/P$; $\partial T/\partial P = V/nR$; $\partial P/\partial V = -nRT/V^2$. Multiply, remembering that $PV = nRT$.

67. (a) One may solve for any of the variables in terms of the other two.
 (b) $\partial T/\partial P = (V-\beta)/R$; $\partial P/\partial V = -RT/(V-\beta)^2 + 2\alpha/V^3$; $\partial V/\partial T = R/[(V-\beta)(RT/(V-\beta)^2 - 2\alpha/V^3)]$
 (c) Multiply and cancel factors.

69. Notice that $y = x^2$, so if y is constant, x cannot be a variable.

71. $\partial^3 u/\partial x\partial y\partial z = \partial^3 u/\partial y\partial x\partial z = \partial^3 u/\partial y\partial z\partial x$

73. Differentiate and substitute.

75. Use the chain rule.

77. Yes. The second partial derivatives are not continuous at the origin; the graph has a 'crinkle' at the origin.

Chapter 16 Answers

16.1 Gradients and Directional Derivatives

1. $(x/\sqrt{x^2+y^2+z^2})\mathbf{i} + (y/\sqrt{x^2+y^2+z^2})\mathbf{j} + (z/\sqrt{x^2+y^2+z^2})\mathbf{k}$

3. $\mathbf{i} + 2y\mathbf{j} + 3z^2\mathbf{k}$

5. $[x/(x^2+y^2)]\mathbf{i} + [y/(x^2+y^2)]\mathbf{j}$

7. $(1+2x^2)\exp(x^2+y^2)\mathbf{i} + 2xy\exp(x^2+y^2)\mathbf{j}$.

9.

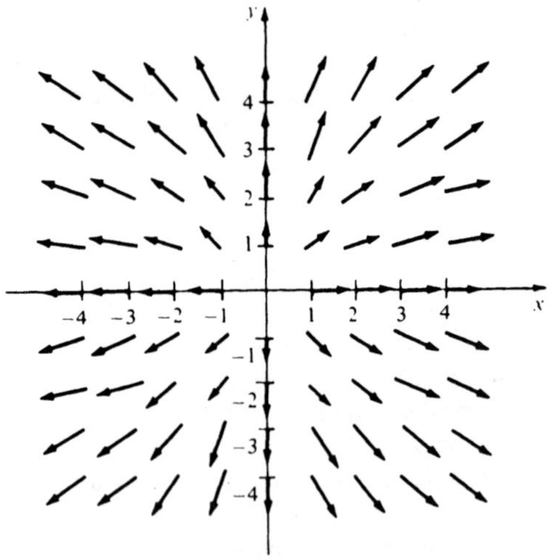

11. (a)

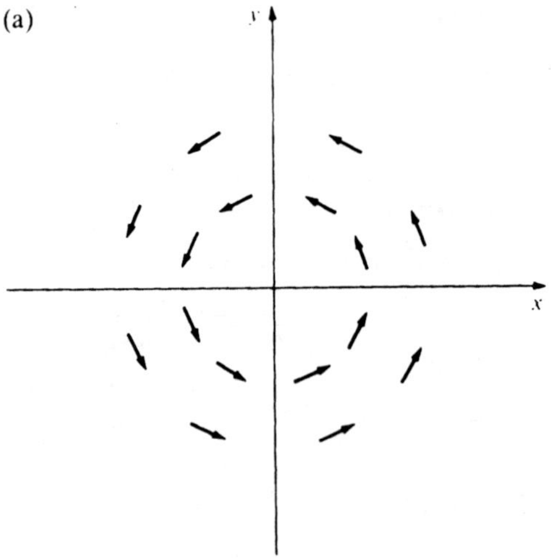

 (b) $\partial(-y)/\partial y \neq \partial(x)/\partial x$

13. $\dfrac{\partial}{\partial x}\left(\dfrac{1}{x^2+y^2+z^2}\right) = \dfrac{-2x}{r^4}$, etc.

15. $2e^t\cos t + \cos^2 t - \sin^2 t$

17. $t/\sqrt{1+t^2}$

19. The angle between the gradient and the velocity vector is between 0 and $\pi/2$.

21. $-11 - 16\sqrt{3}$

23. $17/\sqrt{2}$

25. $-14/\sqrt{3}$

27. $\pi/8 - 1/2$

29. $(\mathbf{i} + 2\mathbf{j})/\sqrt{5}$

31. $e[(\sin 1)\mathbf{i} + (\cos 1)\mathbf{j}]$

33. (a) $(1,2,3)$
 (b) $-2\sqrt{14}\,e^2$ degrees per second
 (c) She should fly outside the cone with vertex at $(1,1,1)$, axis along $(1,2,3)$ and sides at an angle of $\pi/3$ from the axis.

35. $\mathbf{d}_1 = [-(0.03 + 2by_1)/2a]\mathbf{i} + y_1\mathbf{j}$, $\mathbf{d}_2 = [-(0.03 + 2by_2)/2a]\mathbf{i} + y_2\mathbf{j}$ where y_1 and y_2 are the solutions of $(a^2 + b^2)y^2 + 0.03by + \left(\frac{0.03^2}{4} - a^2\right) = 0$.

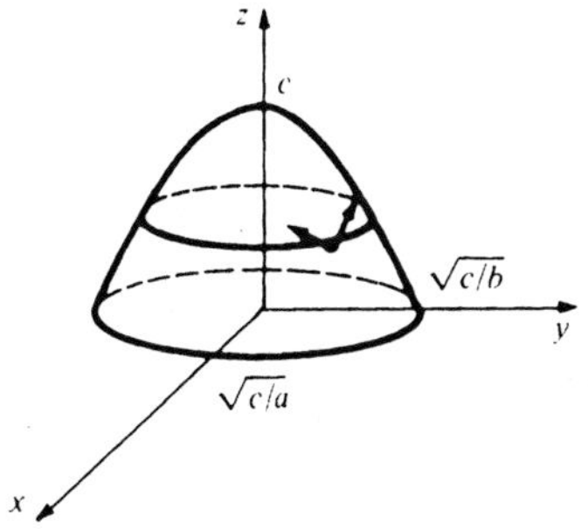

37. $2\mathbf{i}$

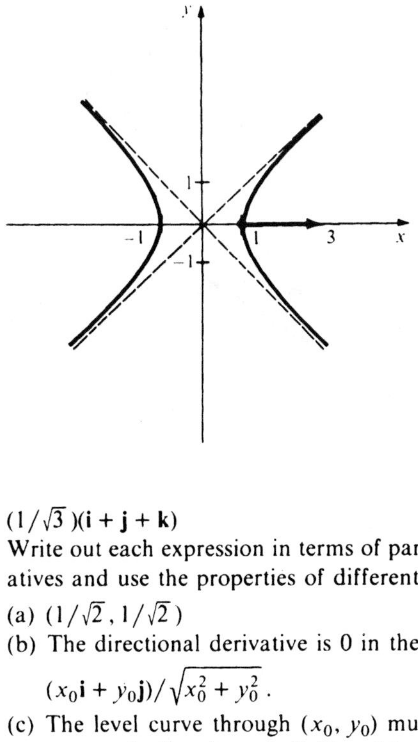

39. $(1/\sqrt{3})(\mathbf{i} + \mathbf{j} + \mathbf{k})$

41. Write out each expression in terms of partial derivatives and use the properties of differentiation.

43. (a) $(1/\sqrt{2}, 1/\sqrt{2})$
 (b) The directional derivative is 0 in the direction $(x_0\mathbf{i} + y_0\mathbf{j})/\sqrt{x_0^2 + y_0^2}$.
 (c) The level curve through (x_0, y_0) must be tangent to the line through (0, 0) and (x_0, y_0). The level curves are lines or half lines emanating from the origin.

45. $\nabla f(1, 3) = (2, -2)$; $-2/\sqrt{13}$

47. (a) $\dfrac{\lambda}{2\pi\varepsilon_0}\left\{\left(\dfrac{x + x_0}{r_1^2} - \dfrac{x - x_0}{r_2^2}\right)\mathbf{i} + 2y\left(\dfrac{1}{r_1^2} - \dfrac{1}{r_2^2}\right)\mathbf{j}\right\}$
 (b) Compute the indicated partial derivatives.

49. The function f must satisfy Laplace's equation: $f_{xx} + f_{yy} = 0$.

16.2 Gradients, Level Surfaces, and Implicit Differentiation

1. $\nabla f(0, 0, 1) = 2\mathbf{k}$

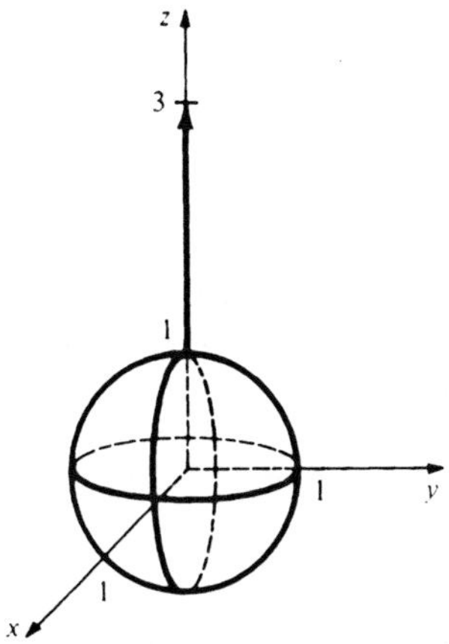

3. $\nabla f(0, 0, 1) = -\mathbf{i} + \mathbf{j} + \mathbf{k}$

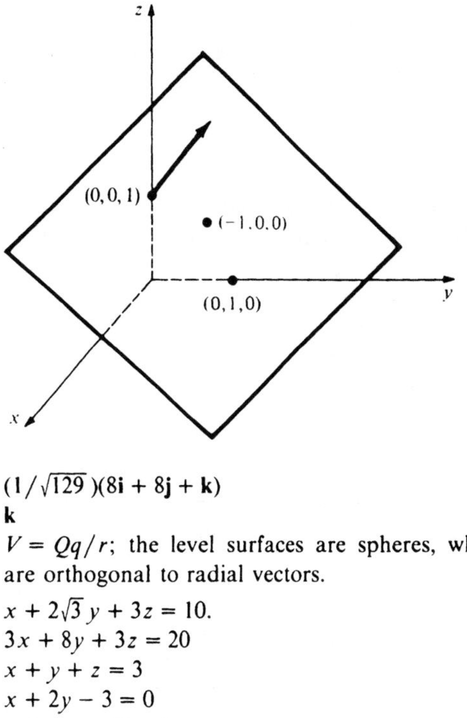

5. $(1/\sqrt{129})(8\mathbf{i} + 8\mathbf{j} + \mathbf{k})$

7. $\mathbf{k}$

9. $V = Qq/r$; the level surfaces are spheres, which are orthogonal to radial vectors.

11. $x + 2\sqrt{3}\,y + 3z = 10$.

13. $3x + 8y + 3z = 20$

15. $x + y + z = 3$

17. $x + 2y - 3 = 0$

19. $x + y - \pi/2 = 0$

21. $(1, 1, 1) + t(1, 1, 1)$

23. $(1, 1, 1) + t(1, -1, -1)$

25. $-x/2y$

27. $y/x = 1/10$

29. $3x^2/(\cos y - 4y^3)$

31. $1/2$

33. $-1 - 2\sqrt{3}/3$

35. At (0, 0), the slope of $y = \sqrt{x}$ is infinite.

37. At (0, 0), the slope of $y = x^{1/5}$ is infinite.

39. $dx/dt = (-1/y)(dy/dt)$

41. $x^3(dx/dt) + y^3(dy/dt) = 0$

43. (a) $dx/dy = -(\partial z/\partial y)/(\partial z/\partial x)$

(b) $(\cos y - 4y^3)/3x^2$; $-y(2e^{x+y^2} + 3y)/e^{x+y^2}$

45. $(1/x - 1)(dx/dt) + (-\tan y)(dy/dt) = 0$

47. (a) $z = 2x - 4y - 5$

(b) The slope is the tangent of the angle between the upward pointing unit normal vector and the z-axis. The slope in this case is $2\sqrt{5}$.

49. Crosses at $(2,2,0)$, $\sqrt{5}/10$ seconds later.

51. (a) They are perpendicular.

(b) If it were not equipotential, there would be places where the force of gravity is not perpendicular to the surface and the water would flow to correct this. The rotation of the earth and tides (among other things) spoil the approximation.

16.3 Maxima and Minima

1. Local minimum at $(1,0)$, local maximum at $(-1,0)$

3. $(0,0)$ is a local minimum.

5. $(0,0)$ is a local maximum.

7. $\sqrt{3/2}$

9. The height is $4b^{2/3}/s^{2/3}$.

11. Minimum

13. Saddle point

15. Minimum (although the test per se is inconclusive)

17. $(-3,2)$, minimum

19. $(0,0)$, neither

21. $(3,7)$, minimum

23. $(1,1)$, minimum

25. $(4/5, -9/10)$, minimum

27. $(0,0)$, neither

29. $(0,0)$, neither

31. $(0,0)$ is a saddle point.

33. The second derivative test fails, but from the accompanying graph, we can see that $(0,0)$ is neither a local maximum nor minimum.

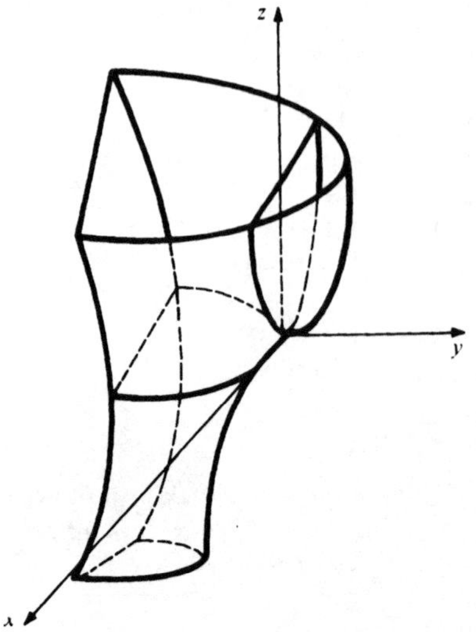

35. (a) Calculate $\partial z/\partial x$ and $\partial z/\partial y$ and set them equal to zero.

(b) The maximum is at $(0,0)$ and local maxima [resp. minima] occur on circles of radius $r_2, r_4, \ldots$ [resp. $r_1, r_3, \ldots$] where $0 < r_1 < r_2 < r_3 < \cdots$ are the solutions of $\pi r = \tan(\pi r)$.

(c) Symmetric in every vertical plane through the origin and under any rotation about the z-axis.

37. (a) Set $\partial w/\partial p_i = 0$. This occurs when $T_{i-1}/T_i = (p_i^2/(p_{i-1}p_{i+1}))^{1-(1/n)}$.

(b) $$p_1 = \left[\left(\frac{T_0^3}{T_1T_2T_3}\right)^{n/(n-1)} p_0^3 p_4\right]^{1/4}$$

$$p_2 = \left[\left(\frac{T_0T_1}{T_2T_3}\right)^{n/(n-1)} p_0 p_4\right]^{1/2}$$

$$p_3 = \left[\left(\frac{T_0T_1T_2}{T_3^3}\right)^{n/(n-1)} p_0 p_4^3\right]^{1/4}$$

39. $A = 2$, $B = 1$, $C = 2$ so $A > 0$ and $AC - B^2 = 3 > 0$. Thus the point is a local minimum.

41. (a) $(0,0)$ is a saddle point.

(b) The behavior changes qualitatively at $C = \pm 2$. For $-2 < C < 2$, $(0,0)$ is a strict minimum; for $C < -2$ or $C > 2$, $(0,0)$ is a saddle point. For $C = \pm 2$, $(0,0)$ is a minimum.

43. (a) $b = 1$, $m = 4/7$

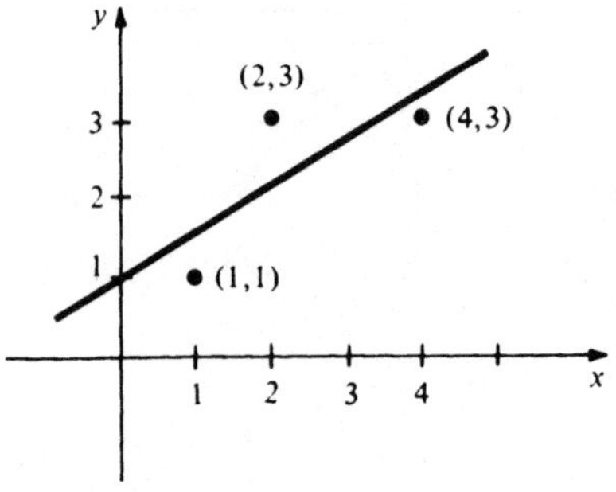

(b) $b = 1/6$, $m = 3/2$

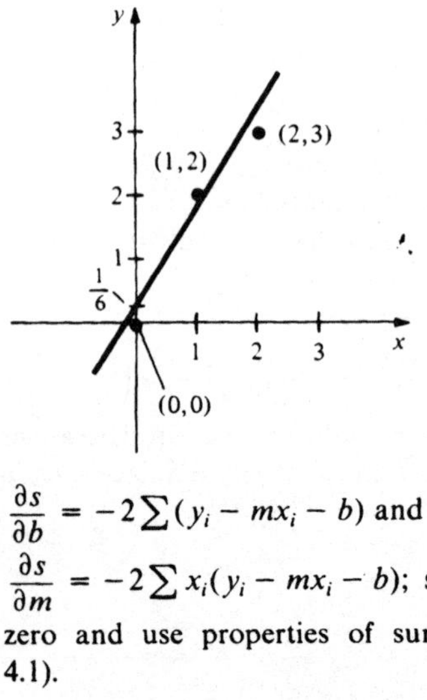

45. $\dfrac{\partial s}{\partial b} = -2\sum(y_i - mx_i - b)$ and $\dfrac{\partial s}{\partial m} = -2\sum x_i(y_i - mx_i - b)$; set these equal t zero and use properties of summation (Sectio 4.1).

47. Compute $\partial^2 s/\partial m^2$, $\partial^2 s/\partial m \partial b$, and $\partial^2 s/\partial b^2$ directly.

49. (a) $e = \sqrt{B^2 - CA}\,/A$
(b) $y = Ax/(Ae - B)$, $y = -Ax/(B + Ae)$
(c) $g(x, y)$ is positive when (i) $y > Ax/(Ae - B)$ and $y > -Ax/(B + Ae)$ or
(ii) $y < Ax/(Ae - B)$ and $y < -Ax/(B + Ae)$
(d) If $A = 0$ and $B = 0$, then $AC - B^2 = 0$. Thus B cannot be zero if $A = 0$ and $AC - B^2 < 0$. Rewrite g as $g(x, y) = y(2Bx + Cy)$. Note that $y = 0$ and $2Bx + Cy = 0$ are two lines intersecting at the origin. Thus $g(x) > 0$ in the region above $2Bx + Cy = 0$ and the negative x-axis, and the region below $2Bx + Cy = 0$ and the positive x-axis.

51. (a) $f_x(0,0) = 0$, $f_y(0,0) = 0$.
(b) Use one variable calculus on h
(c) $f > 0$ if $y > 3x^2$ or $y < x^2$ and $f < 0$ if $x^2 < y < 3x^2$.
(d) $f(x, y) = 0$ if $y = x^2$ or $y = 3x^2$.

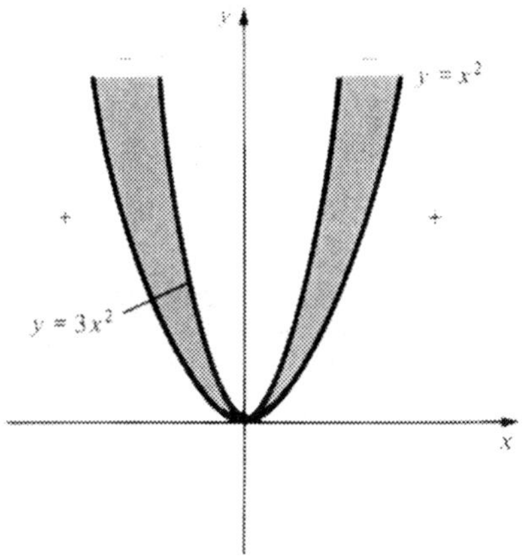

(f) The segment on which h is positive shrinks to 0 as $\theta \to 0$.

53. $(0,0,0)$ is closest for $a \le 1/8$;
$(\pm\sqrt{(8a-1)/32},\ 0,\ (8a-1)/8)$ are closest for $a > 1/8$.

55. Let $d^2 = (x-a)^2 + (y-b)^2 + (k(x,y) - c)^2$ and set $\partial(d^2)/\partial x$ and $\partial(d^2)/\partial y$ equal to zero.

16.4 Constrained Extrema and Lagrange Multipliers

1. The minimum value is 0 (occurs at $(0,0)$), maximum value is 3 (occurs at $(0,1)$ and $(0,-1)$).

3. The minimum value is 8 (occurs at $(0,1)$ and $(0,-1)$) maximum value is 15 (occurs at $(1,0)$ and $(-1,0)$).

5. $\sqrt{35/2}$ is the maximum value, $-\sqrt{35/2}$ is the minimum value. There are no interior critical points.

7. $\sqrt{2}$ is the maximum value, $-\sqrt{2}$ is the minimum value. $(\pm\sqrt{1/2}, \pm\sqrt{1/2})$ are the critical points.

9. $1/4$ is the maximum value. $(1/2, 1/2)$ is the critical point.

11. $\sqrt{10}$ is the maximum value, $-\sqrt{10}$ is the minimum value. $(\pm\sqrt{10}/10, \mp 3\sqrt{10}/10)$ are the critical points.

13. $x = y = 25{,}000$; $z = 50{,}000$.

15. Horizontal length is $\sqrt{qA/p}$, vertical length is $\sqrt{pA/q}$.

17. $(Q_2/Q_1)^{1/3}$

19. (a) $(\sqrt{2}/2, \sqrt{2}/2, 3/2)$ and $(-\sqrt{2}/2, -\sqrt{2}/2, 3/2)$ are maxima, while $(-\sqrt{2}/2, \sqrt{2}/2, 1/2)$, and $(\sqrt{2}/2, -\sqrt{2}/2, 1/2)$ are minima.
(b) h is increasing at $(\pm 1, 0)$ and decreasing at $(0, \pm 1)$.
(c) $(\sqrt{2}/2, \sqrt{2}/2, 3/2)$, $(-\sqrt{2}/2, -\sqrt{2}/2, 3/2)$ are maxima, $(0,0,0)$ is the minimum.

21. (a) $C = \dfrac{\rho\pi n_1^2 i_1^2}{\alpha h}\left(\dfrac{D_1}{x} + \dfrac{D_2}{y}\right)$

(b) $x = \dfrac{D_2 - D_1}{2\left(1 + \sqrt{D_2/D_1}\right)}$,

$y = \dfrac{D_2 - D_1}{2\left(\sqrt{D_1/D_2} + 1\right)}$

23. (a) ∇f is parallel to ∇g.
(b) The maximum value of $\sqrt{3}/9$ occurs at $(\sqrt{3}/3, \sqrt{3}/3, \sqrt{3}/3)$, $(\sqrt{3}/3, -\sqrt{3}/3, -\sqrt{3}/3)$, $(-\sqrt{3}/3, -\sqrt{3}/3, \sqrt{3}/3)$ and $(-\sqrt{3}/3, \sqrt{3}/3, -\sqrt{3}/3)$. The minimum value of $-\sqrt{3}/9$ occurs at $(-\sqrt{3}/3, \sqrt{3}/3, \sqrt{3}/3)$, $(\sqrt{3}/3, -\sqrt{3}/3, \sqrt{3}/3)$, $(\sqrt{3}/3, \sqrt{3}/3, -\sqrt{3}/3)$ and $(-\sqrt{3}/3, -\sqrt{3}/3, -\sqrt{3}/3)$.
(c) $x = y = 8$, $h = 4$.

Review Exercises for Chapter 16

1. $[y\exp(xy) - y\sin(xy)]\mathbf{i} + [x\exp(xy) - x\sin(xy)]\mathbf{j}$

3. $[2x\exp(x^2) + y^2\sin(xy^2)]\mathbf{i} + 2xy\sin(xy^2)\mathbf{j}$

5. (a) $(9/\sqrt{2})\cos(3)$
(b) $(\mathbf{i} - 2\mathbf{j})/\sqrt{5}$

7. (a) $\sqrt{2}/e$
(b) $(-\mathbf{i} - 2\mathbf{j})/\sqrt{5}$

9. $3x + 4y - z = 4$

11. $x + y + z = \sqrt{3}$

13. $(2x + y)(dx/dt) + (x + 2y)(dy/dt) = 0$

15. $(x^2 + y^2)(dx/dt) + 2xy(dy/dt) = 0$

17. 1

19. 1

21. $(0,0)$ is a saddle point.

23. $(0,0)$ is a saddle point.

25. $(-1,0)$ is a saddle point, $(0,0)$ a local maximum, and $(2,0)$ a local minimum.

27. $(n, 0)$, n an integer, are saddle points.

29. (a) $\|\mathbf{G}\| = ((\partial P/\partial x)^2 + (\partial P/\partial y)^2)^{1/2}$

(b) According to Newton's second law of motion, **G** creates a force on the air mass which produces a proportionate acceleration in the direction of **G**.

(c)

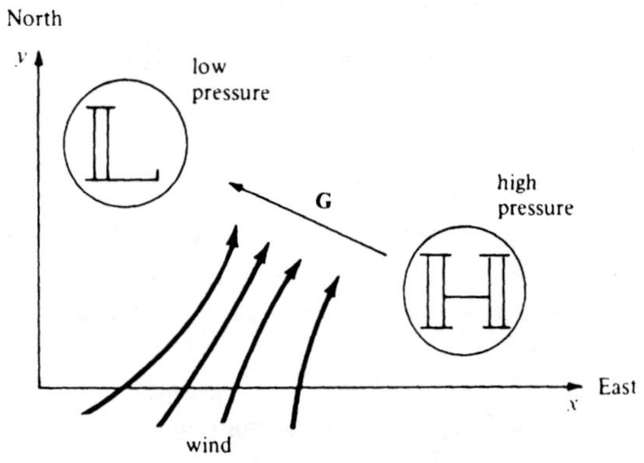

(d) If, in the Southern Hemisphere, you stand with your back to the wind, the high pressure is on your left and the low pressure is on your right.

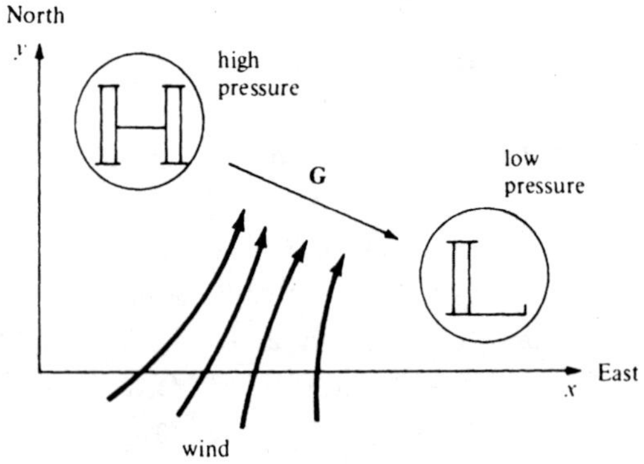

31. (a) $(\pm 1, 0, 0)$ are closest, distance is 1.
(b) $(-\sqrt{41}/6, \sqrt{41}/6, 1/6)$, $(\sqrt{41}/6, -\sqrt{41}/6, 1/6)$ are closest, the distance is $\sqrt{83}/6$.
(c) $(-1,-1,1)$, $(-1,1,-1)$, $(1,-1,-1)$ and $(1,1,1)$ are closest, the distance is $\sqrt{3}$.

33. 0.382 is the minimum value, 2.618 is the maximum value.

35. 0.540 is the minimum value, 1 is the maximum value.

37. The partials of $k(x, y, \lambda) = x + 2y \sec\theta + \lambda(xy + y^2 \tan\theta - A)$ must all vanish

39. (a) Use the second derivative test.
(b) Since the maximum and minimum must occur on the boundary (by (a)), both are zero. Hence f is everywhere zero.

41. (a) $(1, 1, 2)$, $x + y + 2z = 6$; $(2, 3, 2)$, $2x + 3y + 2z = 9$.
(b) $\theta = 0.47$
(c) $(1, 1, 2) + t(-4, 2, 1)$

43. (a) A normal vector to the tangent plane is $f_x\mathbf{i} + f_y\mathbf{j} - \mathbf{k}$.
(b) The slope of the plane relative to the xy plane is 5; the plane contains the line through the point $(1, 0, 2)$ parallel to the y-axis.

45. Equate the four partial derivatives equal to zero. Eliminate λ_1 by subtracting two equations and λ_2 by dividing two equations.

47. 600

49. Approximately 570

51. $(4, 2)$

53. (a) $dy/dx = -(2x)/(3y^2 + e^y)$
(b)

$$dy_1/dx = \frac{(\partial F_1/\partial y_2)(\partial F_2/\partial x) - (\partial F_2/\partial y_2)(\partial F_1/\partial x}{(\partial F_2/\partial y_2)(\partial F_1/\partial y_1) - (\partial F_2/\partial y_1)(\partial F_1/\partial y}$$

$$dy_2/dx = \frac{(\partial F_1/\partial y_1)(\partial F_2/\partial x) - (\partial F_1/\partial x)(\partial F_2/\partial y_1}{(\partial F_1/\partial y_2)(\partial F_2/\partial y_1) - (\partial F_2/\partial y_2)(\partial F_1/\partial y}$$

(c) $dy_1/dx = -(2x + \sin x)/2y_1$, $dy_2/dx = (\cos x - 2x)/y_2$.

55. (a) Use equality of mixed partials of $f(x, y)$
(b) Use equality of mixed partials of $f(x, y, z)$
(c) No

57. (a) $x^2\exp(xy^2) + C$
(b) No function exists.
(c) $\ln(1 + x^2 + y^2) + C$.
(d) No function exists.

59. (a) $(x_0, -y_0)$
(b) (y_0, x_0)
(c) $g(x, y) = xy$
(d) Since ∇g is normal to the level curves of g, the tangent to these curves is normal to level curves of f.
(e)

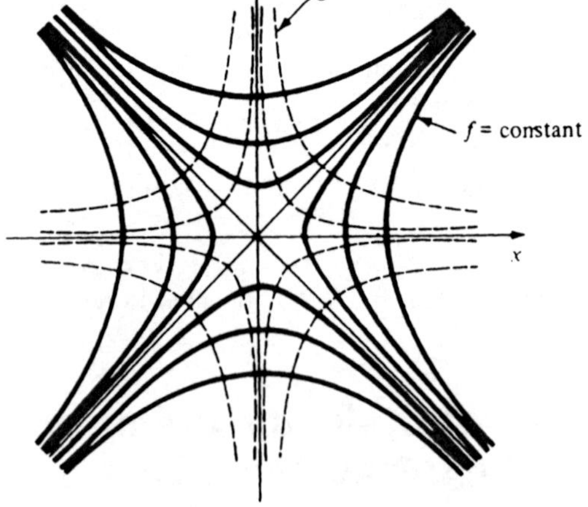

61. (a) $f_x = -2xy^3/(x^2 + y^2)^2$, $f_y = y^2(3x^2 + y^2)/(x^2 + y^2)^2$. $f_x(0,0) = 0$ and $f_y(0,0) = 1$ (compute the latter two using limits).
(b) $(\partial/\partial r)f(r\cos\theta, r\sin\theta) = \cos^2\theta(-2\sin^3\theta + 3\sin^2\theta + 1)$ is defined for all θ.
(c) $\nabla f(0,0) \cdot (\cos\theta, \sin\theta) = \sin\theta$ disagrees with the formula in (b). There is no contradiction because the partial derivatives are not continuous at $(0, 0)$ (see Exercise 20, p. 783).

Chapter 17 Answers

17.1 The Double Integral and Iterated Integral

1. 12

3. (a) Divide D into $D_1 = [-1, 0] \times [2, 3]$, $D_2 = [-1, 0] \times (3, 4]$, *and* $D_3 = (0, 1] \times [2, 4]$. Let $g(x, y)$ be -4 on D_1, -5 on D_2, and 0 on D_3.

(b) The part of the integral for $x \leqslant 0$ is the negative of the part for $x \geqslant 0$.

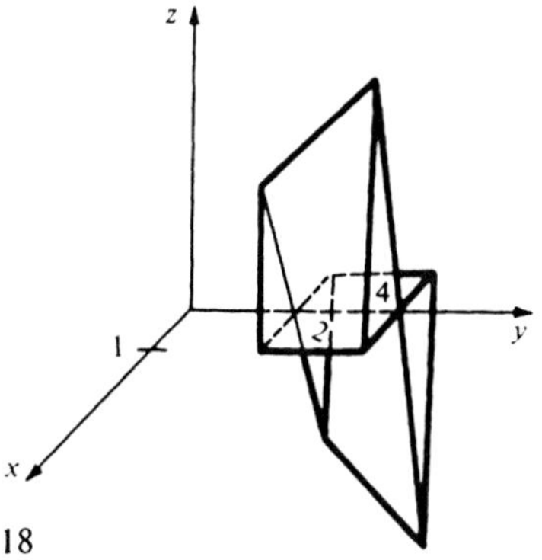

5. 18

7. 16/9

9. $e/2 - 1/2e$

11. 50

13. 4/3

15. $45/4 + (15/2)\ln(3/2)$

17. 0; this agrees with the answer in Exercise 3(b).

19. 76/3

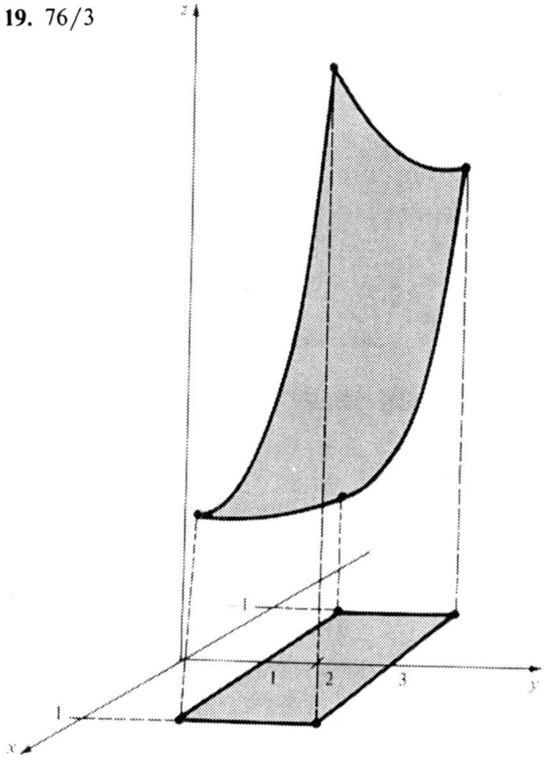

21. 25/6 grams.

23. Use partition points obtained from subrectangles for both the functions being added.

25. (a) $0.88[1 - \sin^2\alpha \cos^2(2\pi T/365)]^{1/2}\cos(2\pi t/24)] + 0.67 \sin\alpha \cos(2\pi T/365)$, where $\alpha = 23.5°$

(b) This is the total solar energy received in the state between times t_1 and t_2 on day T.

17.2 The Double Integral Over General Regions

1. Both a type 1 and a type 2 region.

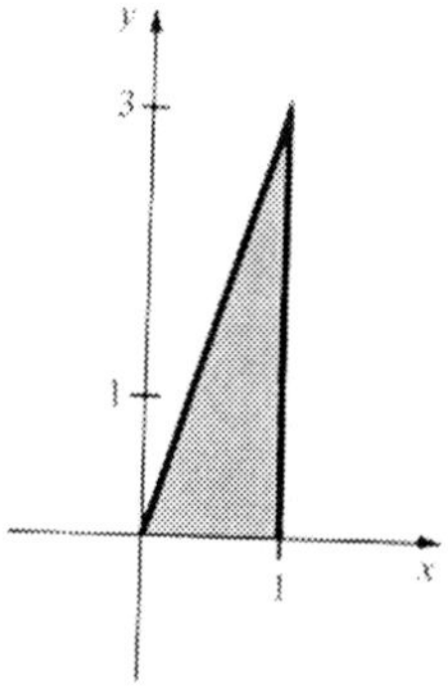

3. Both a type 1 and a type 2 region.

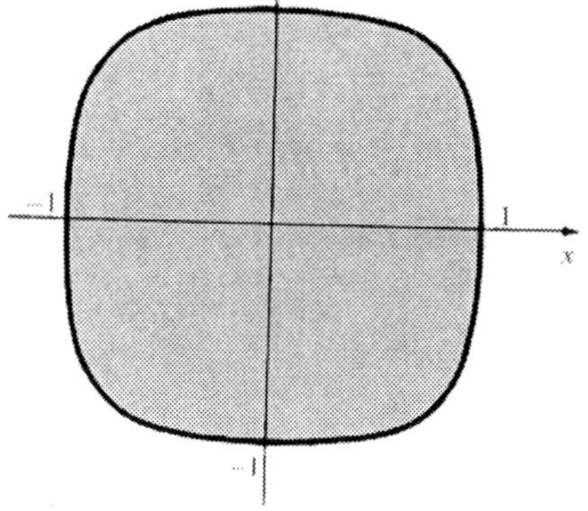

5. 7/12

7. 64/35

9. Type 1; $2\pi + \pi^2$

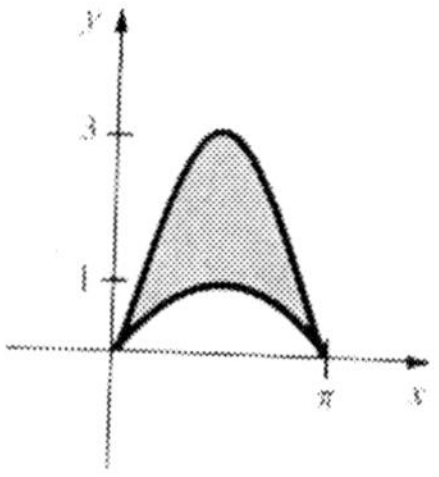

11. Type 2; 104/45

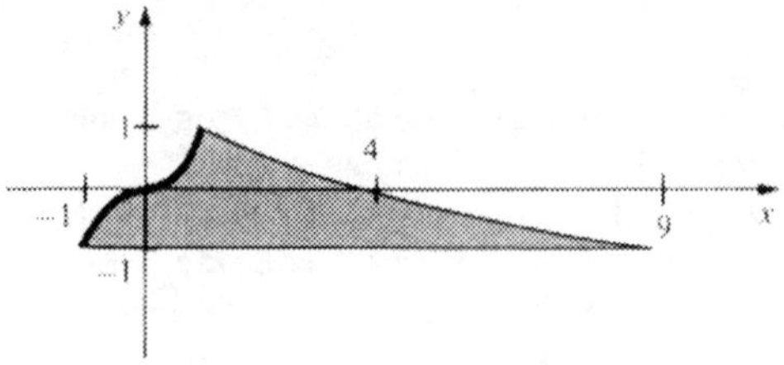

13. Type 1; 33/140

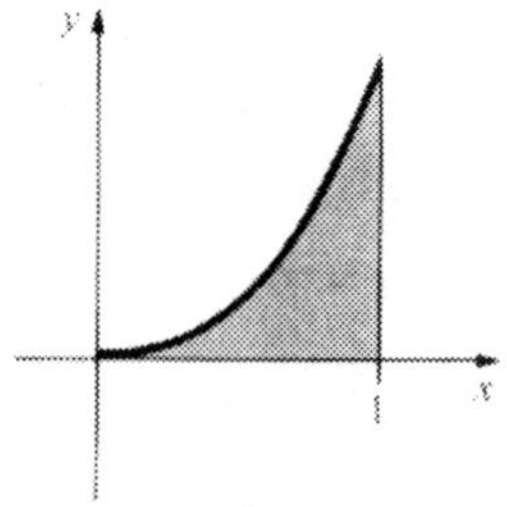

15. Type 1; 71/420

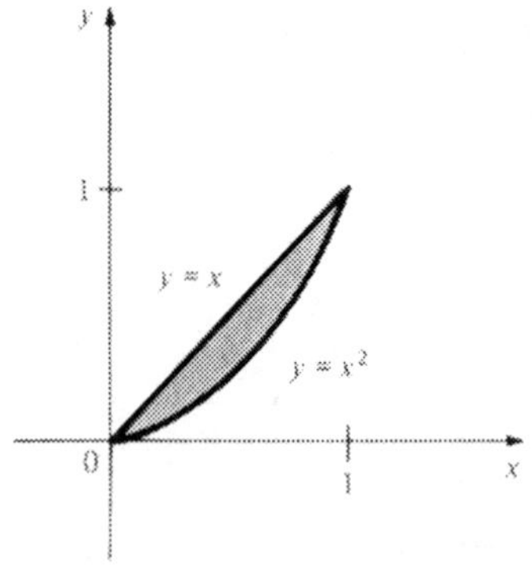

17. 1/8

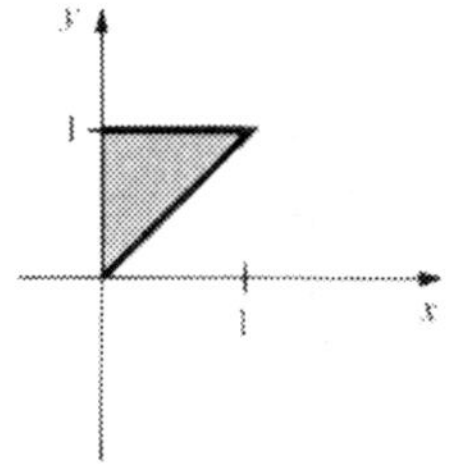

19. 7/12

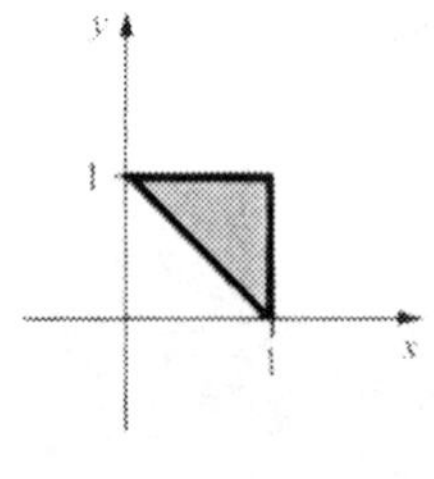

21. 1/3

23. 19/3

25. The result of the first integration is the length of a section of the region.

27. Type 1 states include: Wyoming, Colorado, Nevada, New Mexico, Kansas, and Ohio. Type 2 states include: Wyoming, Colorado, Kansas, North Dakota, and Vermont.

17.3 Applications of the Double Integral

1. 12π

3. $(4/5)(2 + 9^3\sqrt{4})$

5. $10 + 8/\pi$

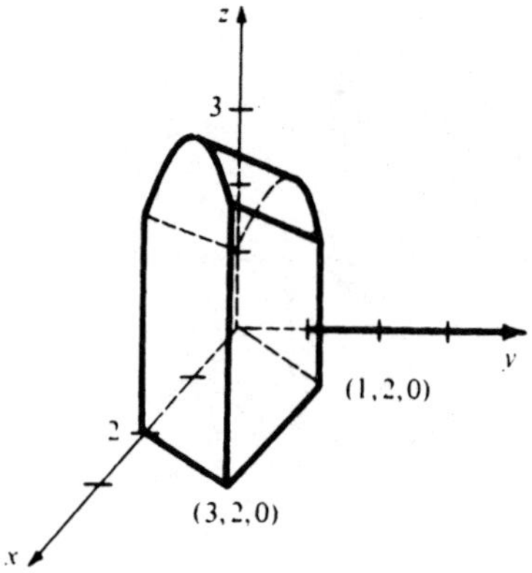

7. $(16e^2 - 16 + \pi^4)/32$ **9.** $243/80$

11. $[\pi^2 - \sin(\pi^2)]/\pi^3$ **13.** 2

15. $2/3$ **17.** $2a^2/3$

19. $(11/18, 65/126)$ **21.** $(7/5, 0)$

23. $(4/15)(9\sqrt{3} - 8\sqrt{2} + 1)$ **25.** $\sqrt{2}\,\pi/4$

27. Compare the formulas for average value and center of mass.

29. (a) Write $z = \pm\sqrt{r^2 - x^2 - y^2}$ over the region $0 \leqslant x \leqslant r,\ -\sqrt{r^2 - x^2} \leqslant y \leqslant \sqrt{r^2 - x^2}$.

(b) The result is independent of r.

31. \$503.64

33. $2\int_a^b \int_{-f(x)}^{f(x)} \sqrt{[f(x)]^2 - y^2}\, dy\, dx = \pi \int_a^b [f(x)]^2\, dx.$

(This is the disk method; see p. 423.)

17.4 Triple Integrals

1. 7 **3.** -8 **5.** Type I

7. Type I **9.** $25\sqrt{2/3}\,\pi$ **11.** $1/2$

13. 0 **15.** $a^5/20$ **17.** 0

19. 3/10

21. The double integral of F over the base of the box times the height of the box.

23. Interchange x and z in Example 4.

25. The region under the graph is of type I.

27. (a) Use iterated integrals and the constant multiple rule for definite integrals.

(b) $e^{x+y+z} = e^x e^y e^z$.

17.5 Integrals in Polar, Cylindrical, and Spherical Coordinates

1. $64\pi/5$
3. $\sqrt{\pi}/10$
5. 128π
7. $5\pi(e^4 - 1)/2$
9. $2\pi[\sqrt{2} - \ln(1 + \sqrt{2})]$
11. $4\pi \ln(a/b)$
13. $(\pi/6)(8 - 3\sqrt{3})$
15. $4\pi\sqrt{6}$
17. $1/\sqrt{\pi\sigma}$
19. (a) Use the linear approximations

$$\Delta x \approx \frac{\partial x}{\partial u}\Delta u + \frac{\partial x}{\partial v}\Delta v \text{ and } \Delta y \approx \frac{\partial y}{\partial u}\Delta u + \frac{\partial y}{\partial v}\Delta v.$$

(b) Use the fact that $\left|\frac{\partial(x, y)}{\partial(r, \theta)}\right| = r.$

21. The conditions $0 \leqslant y \leqslant 1 - x$ and $0 \leqslant x \leqslant 1$ are equivalent to $0 \leqslant x + y \leqslant 1$ and $0 \leqslant \frac{y}{x + y} \leqslant 1$.

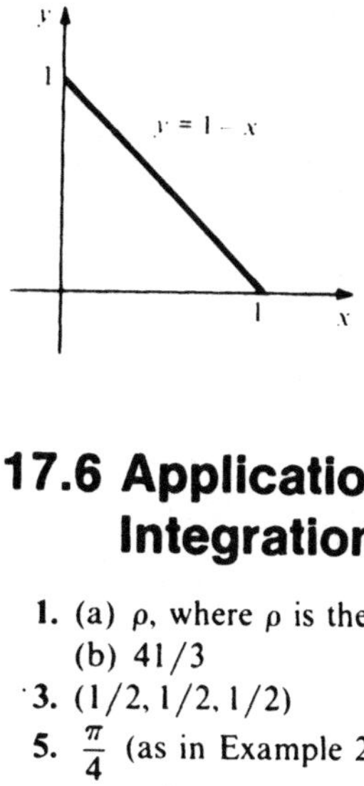

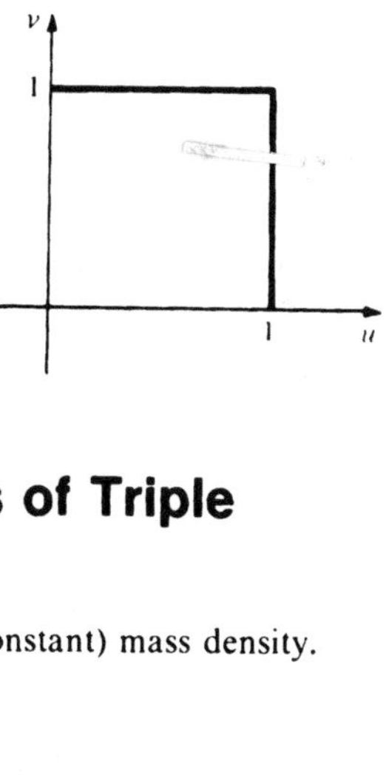

17.6 Applications of Triple Integration

1. (a) ρ, where ρ is the (constant) mass density.
(b) $41/3$
3. $(1/2, 1/2, 1/2)$
5. $\frac{\pi}{4}$ (as in Example 2)
7. (a) kc^2
(b) Along the sphere $x^2 + y^2 + z^2 = c^2$
9. $1/4$
11. Letting d be density, the moment of inertia is $d\int_0^k \int_0^{2\pi} \int_0^{a \sec \phi} \rho^4 \sin^3\phi \, d\rho \, d\theta \, d\phi$
13. $1.00 \times 10^8 (\mathrm{m/s})^2$
15. (a) The only plane of symmetry for the body of an automobile is the one dividing the left and right sides of the car.
(b) $\bar{z} \cdot \int\int\int_W \rho(x, y, z)\, dx\, dy\, dz$ is the z-coordinate of the center of mass times the mass of W. Rearrangement of the formula for $\bar{z}$ gives the first line of the equation. The next step is justified by the additivity property of integrals (see rule 2 on p. 841). By symmetry, we can replace z by $-z$ and integrate in the region above the xy-plane. Finally, we can factor the minus sign outside the second integral and since $\rho(x, y, z) = \rho(u, v, -w)$, we are subtracting the second integral from itself. Thus, the answer is 0.
(c) In part (b), we showed that $\bar{z}$ times the mass of W is 0. Since the mass must be positive, $\bar{z}$ must be 0.
(d) By part (c), the center of mass must lie in both planes.
17. Follow the pattern on p. 880 for the case of constant density. Be sure to use different symbols for the density and the spherical coordinate $\sqrt{x^2 + y^2 + z^2}$.

Review Exercises for Chapter 17

1. $960 - \sin(11) + \sin(10) + \sin(7) - \sin(6) \approx 961.4$
3. $15/2$
5. $64/3$
7. $\pi \ln(\sec 1 + \tan 1)$
9. $1/48$
11. $10/3$
13. $4\pi abc/3$
15. $2\pi(1 - 1/\sqrt{a + 1})/3$
17. $(25 + 10\sqrt{5})\pi/3$
19. $(4\pi/3)(1 - \sqrt{2}/2)$
21. $40\pi/3$

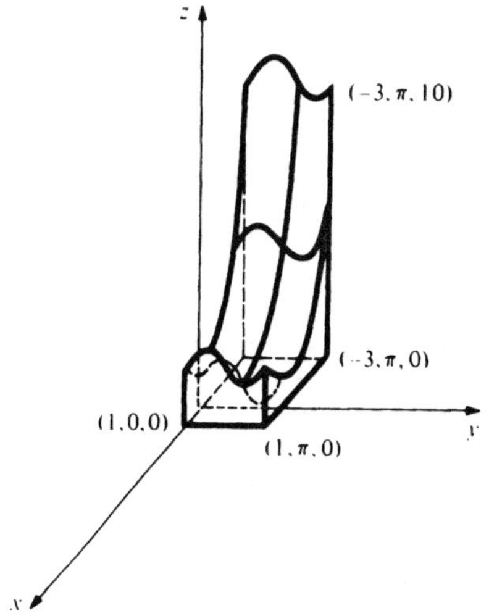

23. $10/3$
25. Cut with the planes $x + y + z = \sqrt[3]{k/n}$, $1 \leqslant k \leqslant n - 1$, k an integer.
27. Use the formula for surface area.

29. $(0, 0, -0.203)$
31. $2\pi(2\sqrt{2} - 1)/3$
33. $\pi(5\sqrt{5} - 1)/6$
35. $\pi^2/8$
37. $(\pi/4)\ln(2)$
39. $V(0, 0, R) = (4.71 \times 10^{22})G/R$
41. (a) $32/3$
    (b) 64π
43. $\sqrt{\pi/5}$
45. (a) It is equal to the average of the values of the function at the vertices.
    (b) Same as in (a).
47. Using d for density, the potential is given by $2\pi dG[(\rho_2)^2 - R^2/3 - 2(\rho_1)^3/3R]$.
49. Show that the double integral of $f_{xy} - f_{yx}$ over every rectangle is zero.
51. $u_x^2 + u_y^2 = 1$

Chapter 18 Answers

18.1 Line Integrals

1. 5
3. 0
5. $3/2$
7. -2π
9. π
11. 2π
13. $2\pi^2$
15. -1
17. 0
19. $2 + e$
21. 1
23. $2/3$
25. $-1/2$
27. $(\cos 3)/3 + 5/12$
29. 0
31. $(5\sqrt{5} - 1)/12$
33. $52\sqrt{14}$
35. Use the chain rule and make a change of variables in the integral.

18.2 Path Independence

1. 3
3. 8
5. $1/2$
7. π
9. 0
11. 256
13. $JMm(r_2^{-3} - r_1^{-3})/3$
15. No
17. Yes; $x^2y + \sin y + C$
19. Yes; $x^2 + x^2\cos y + C$
21. Calculate $f_x = -y/(x^2 + y^2)$ and $f_y = x/(x^2 + y^2)$.
23. $2\pi^2$
25. 0
27. (a) The integrals along the four sides are 0, 1, -1, and 0.
    (b) $f(x, y, z) = z^3x + x^2y + C$.
29. $(1/2)^3 - (\sqrt{2}/2)^3 + \sin(3\sqrt{2}\,\pi/4)$
31. The field is not conservative.
33. $x\exp(yz) + C$
35. If $\nabla f = \nabla g$ then $\nabla(f - g) = 0$. Show that $f - g$ is constant, using the second box on p. 896.
37. Pick a point P and draw an arc from P to each region. If the arc crosses the circles an even number of times, color the region red, otherwise color the region blue.

18.3 Exact Differentials

1. No
3. No
5. Not exact
7. Exact
9. f is the integral of $P\,dx + Q\,dy$ along the line segment from $(0, 0)$ to $(0, y)$ followed by the segment from $(0, y)$ to (x, y).
11. The two line integrals are equal since $P\,dx + Q\,dy$ is exact.
13. $xe^y + ye^x = 2$
15. $x^3 + x^2y + y^3/3 = 17/3$.
17. Not exact
19. $2x^3y + x^2y^2 = C$
21. (a) -1
    (b) $y = x$
23. $y = \pm x$
25. $\mu = x$
27. $x^2/2 + x/y = C$
29. $\ln \mu = \int [(N_x - M_y)/M]\,dy$
31. If f is an antiderivative of $P\,dx + Q\,dy$, then $f_x = \frac{\partial}{\partial x}\int P\,dx$; now integrate.

18.4 Green's Theorem

1. Each side gives $1/12$.
3. Each side gives $-\pi$.
5.

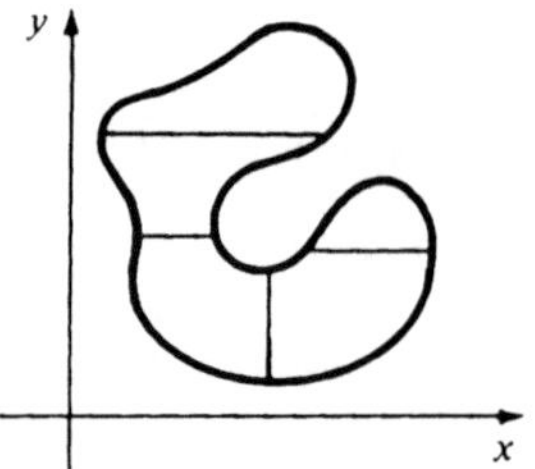

7.

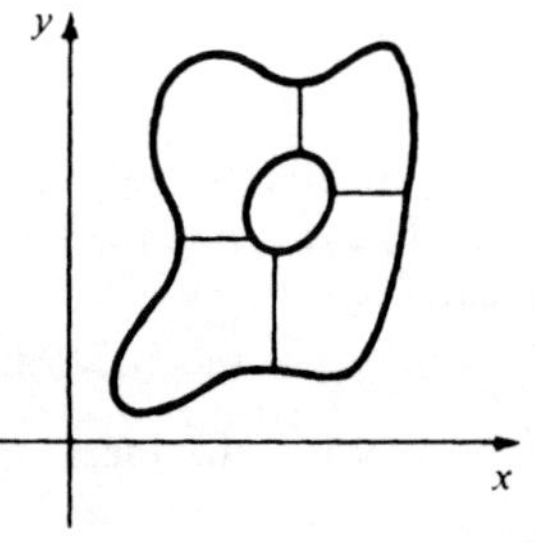

9. $-4\int_{-b}^{b}\int_{(a/b)\sqrt{b^2-y^2}}^{(a/b)\sqrt{b^2-y^2}} xy\,dx\,dy = 0$
11. $67/6$
13. -20
15. 0
17. (a) Recall that a zero dot product implies orthogonality.
(b) Use Green's Theorem.
19. In applying the hint, note that the terms involving $\nabla u \cdot \nabla v$ cancel.
21. 9
23. Simplify $A = \frac{1}{2}\int_C x\,dy - y\,dx$ to $\frac{3a^2}{8}\int_0^{2\pi}\sin^2 2\theta\,d\theta$; now use the double angle formula to integrate.
25. $5/12$
27. A horizontal line segment divides the region into three regions to which Green's theorem applies; see Example 2.
29. 0
31. $\partial\hat{u}/\partial y = Q$ by the [illegible]damental theorem of calculus; similarly for [illegible]$x = P$.
33. The device is run around the perimeter of the region and the mechanism evaluates the integral $\frac{1}{2}\int_{\partial D}(x\,dy - y\,dx)$.

18.5 Circulation and Stokes' Theorem

1. -2
3. $4xy/(x^2+y^2)^2$
5. 8
7. 0
9. $z^3\mathbf{i} + e^z\mathbf{j} + (y\sin xy)\mathbf{k}$.
11. $(2/(x^2+y^2+z^2)^2)(x^3\mathbf{i} + (x^2y - yz^2)\mathbf{j} - z^3\mathbf{k})$
13. Let $\mathbf{\Phi} = P\mathbf{i} + Q\mathbf{j} + R\mathbf{k}$, write out curl $(f\mathbf{\Phi})$, and use the product rule for derivatives.
15. $\nabla \times \nabla f = 0$ is a vector identity; the integral of an exact differential about a closed loop is zero.
17. Compute that $\nabla \times \mathbf{\Phi} = 0$ and use Stokes' theorem.
19. $-1/2$
21. Use Stokes' theorem and $\mathbf{\Phi}(\boldsymbol{\sigma}(s)) \cdot \boldsymbol{\sigma}'(s) = 0$.
23. $-17\pi\sqrt{2}$
25. $\int_C \mathbf{H}(\mathbf{r}) \cdot d\mathbf{r} = \int\int_S (\nabla \times \mathbf{H}) \cdot \mathbf{n}\,dA = \int\int_S \mathbf{J} \cdot \mathbf{n}\,dA$
27. The component of the curl of the velocity of the fluid along a vector $\mathbf{n}$ is the circulation around $\mathbf{n}$ per unit area (p. 921); this is maximized when the curl and $\mathbf{n}$ are aligned.
29. Partition the surface into the upper and lower hemispherical pieces; apply Stokes' theorem to each piece and add.

18.6 Flux and the Divergence Theorem

1. $3x^2 - x^2\cos(xy)$
3. $y\cos(xy) - x^2\sin(x^2y)$
5. -4
7. 0
9. (a) A, C
(b) B, D
11. $y\exp(xy) - x\exp(xy) + y\exp(yz)$
13. 3
15. $12\pi/5$
17. 1
19. Use Gauss' theorem and $\mathbf{V} \cdot \mathbf{n} = 0$.
21. Apply the divergence theorem to $f\mathbf{\Phi}$ using $\nabla \cdot (f\mathbf{\Phi}) = \nabla f \cdot \mathbf{\Phi} + f\nabla \cdot \mathbf{\Phi}$.
23. (i) Use the vector identity $\nabla \cdot (\nabla \times \mathbf{H}) = 0$.
(ii) Since charge is conserved, the rate at which charge is entering equals the rate at which charge is leaving; the total flux is therefore zero. By Gauss' theorem, the integral of $\nabla \cdot \mathbf{J}$ over any region is zero, so $\nabla \cdot \mathbf{J} = 0$.
25. (a) Calculate div $\nabla\phi$ on a region excluding a small ball near $\mathbf{q}$.
(b) Use (a).

Review Exercises for Chapter 18

1. $10/3 - 2\cos 2 + 2\sin 2 - 2\sin 1$
3. $2e$
5. -8
7. $1 - e$
9. $-\cos 5 + \cos 3 + (\ln 4)^2/4 + (44\ln 2)/3 + 83/18$
11. (a) $\sin(\ln(5/4)) - \sin(\ln(3/4)) - (\ln(5/4))^2 + (\ln(3/4))^2$
(b) 0
(c) 0
13. $43/54$
15. -1
17. $(e-1)/3$
19. Not conservative
21. Not conservative
23. Yes; $3z^3y + x^2y + C$
25. Exact; $xe^y\sin x + C$
27. Not exact
29. $y\sin x + x^2e^y + 2y = \pi^2/4$
31. $-x + x^2y + y = 1$
33. Not exact
35. Not exact
37. (a) $\int\int_D (1)\,dx\,dy$, the area of D
(b) $\int\int_D (-1)\,dy\,dx$, the negative of the area of D
(c) $\int\int_D (0)\,dy\,dx = 0$
39. curl $\mathbf{\Phi} = [y/(x+z)^2]\mathbf{i} - [y/(x+z)^2]\mathbf{k}$; div $\mathbf{\Phi} = 1/(x+z)$
41. curl $\mathbf{\Phi} = -4x\mathbf{i} - 2y\mathbf{j} + 2z\mathbf{k}$; div $\mathbf{\Phi} = 0$.

43. (a) $\nabla \times \mathbf{F} = 2x^3yz\mathbf{i} - 3x^2y^2z\mathbf{j} + 2\mathbf{k}$, $\nabla \cdot \mathbf{F} = x^3y^2$
(b) 2π
(c) $1/12$

45. $\int\int_S (\nabla \times \mathbf{F}) \cdot \mathbf{n}\, dA$, where C is the boundary of S.

47. 0

49. $-2\pi/3$

51. Use Gauss' theorem.

53. Use Gauss' theorem over a small region; divide by the volume of the region and use the mean value theorem for integrals.

55. -8π

57. (a) Let $u' = \int_0^x a(t,0,0)\,dt + \int_0^z c(x,0,t)\,dt + \int_0^y b(x,t,z)\,dt$, so $\partial u'/\partial y = Q$. Permute x, y, z to give u'' with $\partial u''/\partial z = R$, and u''' with $\partial u'''/\partial x = P$. Use Stokes' theorem to show that $u' = u'' = u'''$.
(b) $x^2yz - \cos x + C$

59. $\int\int_{\partial W} f(\nabla f) \cdot \mathbf{n}\, dA = \int\int\int_W (f\nabla^2 f + \nabla f \cdot \nabla f)\, dx\, dy\, dz$ gives $0 = \int\int\int_W (\nabla f \cdot \nabla f)\, dx\, dy\, dz$ $= \int\int\int_W \|\nabla f\|^2\, dx\, dy\, dz$, so $\nabla f = 0$ and thus f is constant.

61. $\int\int_S (\nabla \times \mathbf{\Phi}) \cdot \mathbf{n}\, dA = 0$ if S is the union of the two surfaces.

Index

Includes Volumes I, II and III

Note: Pages 1–336 are in Volume I; pages 337–644 are in Volume II; pages 645–934 are in Volume III.

Undergraduate Texts in Mathematics

(continued from page ii)

Halmos: Naive Set Theory.
Hämmerlin/Hoffmann: Numerical Mathematics.
Readings in Mathematics.
Harris/Hirst/Mossinghoff: Combinatorics and Graph Theory.
Hartshorne: Geometry: Euclid and Beyond.
Hijab: Introduction to Calculus and Classical Analysis.
Hilton/Holton/Pedersen: Mathematical Reflections: In a Room with Many Mirrors.
Iooss/Joseph: Elementary Stability and Bifurcation Theory. Second edition.
Isaac: The Pleasures of Probability.
Readings in Mathematics.
James: Topological and Uniform Spaces.
Jänich: Linear Algebra.
Jänich: Topology.
Jänich: Vector Analysis.
Kemeny/Snell: Finite Markov Chains.
Kinsey: Topology of Surfaces.
Klambauer: Aspects of Calculus.
Lang: A First Course in Calculus. Fifth edition.
Lang: Calculus of Several Variables. Third edition.
Lang: Introduction to Linear Algebra. Second edition.
Lang: Linear Algebra. Third edition.
Lang: Undergraduate Algebra. Second edition.
Lang: Undergraduate Analysis.
Lax/Burstein/Lax: Calculus with Applications and Computing. Volume 1.
LeCuyer: College Mathematics with APL.
Lidl/Pilz: Applied Abstract Algebra. Second edition.
Logan: Applied Partial Differential Equations.
Macki-Strauss: Introduction to Optimal Control Theory.
Malitz: Introduction to Mathematical Logic.
Marsden/Weinstein: Calculus I, II, III. Second edition.
Martin: The Foundations of Geometry and the Non-Euclidean Plane.
Martin: Geometric Constructions.
Martin: Transformation Geometry: An Introduction to Symmetry.
Millman/Parker: Geometry: A Metric Approach with Models. Second edition.
Moschovakis: Notes on Set Theory.
Owen: A First Course in the Mathematical Foundations of Thermodynamics.
Palka: An Introduction to Complex Function Theory.
Pedrick: A First Course in Analysis.
Peressini/Sullivan/Uhl: The Mathematics of Nonlinear Programming.
Prenowitz/Jantosciak: Join Geometries.
Priestley: Calculus: A Liberal Art. Second edition.
Protter/Morrey: A First Course in Real Analysis. Second edition.
Protter/Morrey: Intermediate Calculus. Second edition.
Roman: An Introduction to Coding and Information Theory.
Ross: Elementary Analysis: The Theory of Calculus.
Samuel: Projective Geometry.
Readings in Mathematics.
Scharlau/Opolka: From Fermat to Minkowski.
Schiff: The Laplace Transform: Theory and Applications.
Sethuraman: Rings, Fields, and Vector Spaces: An Approach to Geometric Constructability.
Sigler: Algebra.
Silverman/Tate: Rational Points on Elliptic Curves.
Simmonds: A Brief on Tensor Analysis. Second edition.

Singer: Geometry: Plane and Fancy.
Singer/Thorpe: Lecture Notes on Elementary Topology and Geometry.
Smith: Linear Algebra. Third edition.
Smith: Primer of Modern Analysis. Second edition.
Stanton/White: Constructive Combinatorics.
Stillwell: Elements of Algebra: Geometry, Numbers, Equations.
Stillwell: Mathematics and Its History.
Stillwell: Numbers and Geometry. *Readings in Mathematics.*
Strayer: Linear Programming and Its Applications.
Toth: Glimpses of Algebra and Geometry. *Readings in Mathematics.*
Troutman: Variational Calculus and Optimal Control. Second edition.
Valenza: Linear Algebra: An Introduction to Abstract Mathematics.
Whyburn/Duda: Dynamic Topology.
Wilson: Much Ado About Calculus.

29. $\int \operatorname{csch} x\, dx = \ln\left|\tanh \frac{x}{2}\right| = -\frac{1}{2} \ln \frac{\cosh x + 1}{\cosh x - 1}$

30. $\int \sinh^2 x\, dx = \frac{1}{4} \sinh 2x - \frac{1}{2} x$

31. $\int \cosh^2 x\, dx = \frac{1}{4} \sinh 2x + \frac{1}{2} x$

32. $\int \operatorname{sech}^2 x\, dx = \tanh x$

33. $\int \sinh^{-1} \frac{x}{a}\, dx = x \sinh^{-1} \frac{x}{a} - \sqrt{x^2 + a^2} \qquad (a > 0)$

34. $\int \cosh^{-1} \frac{x}{a}\, dx = \begin{cases} x \cosh^{-1} \frac{x}{a} - \sqrt{x^2 - a^2} & \left[\cosh^{-1}\left(\frac{x}{a}\right) > 0, a > 0\right] \\ x \cosh^{-1} \frac{x}{a} + \sqrt{x^2 - a^2} & \left[\cosh^{-1}\left(\frac{x}{a}\right) < 0, a > 0\right] \end{cases}$

35. $\int \tanh^{-1} \frac{x}{a}\, dx = x \tanh^{-1} \frac{x}{a} + \frac{a}{2} \ln|a^2 - x^2|$

36. $\int \frac{1}{\sqrt{a^2 + x^2}}\, dx = \ln(x + \sqrt{a^2 + x^2}) = \sinh^{-1} \frac{x}{a} \qquad (a > 0)$

37. $\int \frac{1}{a^2 + x^2}\, dx = \frac{1}{a} \tan^{-1} \frac{x}{a} \qquad (a > 0)$

38. $\int \sqrt{a^2 - x^2}\, dx = \frac{x}{2} \sqrt{a^2 - x^2} + \frac{a^2}{2} \sin^{-1} \frac{x}{a} \qquad (a > 0)$

39. $\int (a^2 - x^2)^{3/2}\, dx = \frac{x}{8} (5a^2 - 2x^2)\sqrt{a^2 - x^2} + \frac{3a^4}{8} \sin^{-1} \frac{x}{a} \qquad (a > 0)$

40. $\int \frac{1}{\sqrt{a^2 - x^2}}\, dx = \sin^{-1} \frac{x}{a} \qquad (a > 0)$

41. $\int \frac{1}{a^2 - x^2}\, dx = \frac{1}{2a} \ln\left|\frac{a + x}{a - x}\right|$

42. $\int \frac{1}{(a^2 - x^2)^{3/2}}\, dx = \frac{x}{a^2\sqrt{a^2 - x^2}}$

43. $\int \sqrt{x^2 \pm a^2}\, dx = \frac{x}{2} \sqrt{x^2 \pm a^2} \pm \frac{a^2}{2} \ln|x + \sqrt{x^2 \pm a^2}|$

44. $\int \frac{1}{\sqrt{x^2 - a^2}}\, dx = \ln|x + \sqrt{x^2 - a^2}| = \cosh^{-1} \frac{x}{a} \qquad (a > 0)$

45. $\int \frac{1}{x(a + bx)}\, dx = \frac{1}{a} \ln\left|\frac{x}{a + bx}\right|$

46. $\int x\sqrt{a + bx}\, dx = \frac{2(3bx - 2a)(a + bx)^{3/2}}{15b^2}$

47. $\int \frac{\sqrt{a + bx}}{x}\, dx = 2\sqrt{a + bx} + a \int \frac{1}{x\sqrt{a + bx}}\, dx$

48. $\int \frac{x}{\sqrt{a + bx}}\, dx = \frac{2(bx - 2a)\sqrt{a + bx}}{3b^2}$

49. $\int \frac{1}{x\sqrt{a + bx}}\, dx = \frac{1}{\sqrt{a}} \ln\left|\frac{\sqrt{a + bx} - \sqrt{a}}{\sqrt{a + bx} + \sqrt{a}}\right| \qquad (a > 0)$

$\qquad = \frac{2}{\sqrt{-a}} \tan^{-1}\sqrt{\frac{a + bx}{-a}} \qquad (a < 0)$

50. $\int \frac{\sqrt{a^2 - x^2}}{x}\, dx = \sqrt{a^2 - x^2} - a \ln\left|\frac{a + \sqrt{a^2 - x^2}}{x}\right|$

51. $\int x\sqrt{a^2 - x^2}\, dx = -\frac{1}{3} (a^2 - x^2)^{3/2}$

52. $\int x^2\sqrt{a^2 - x^2}\, dx = \frac{x}{8} (2x^2 - a^2)\sqrt{a^2 - x^2} + \frac{a^4}{8} \sin^{-1} \frac{x}{a} \qquad (a > 0)$

Continued on overleaf.

53. $\int \frac{1}{x\sqrt{a^2-x^2}}\,dx = -\frac{1}{a}\ln\left|\frac{a+\sqrt{a^2-x^2}}{x}\right|$

54. $\int \frac{x}{\sqrt{a^2-x^2}}\,dx = -\sqrt{a^2-x^2}$

55. $\int \frac{x^2}{\sqrt{a^2-x^2}}\,dx = -\frac{x}{2}\sqrt{a^2-x^2} + \frac{a^2}{2}\sin^{-1}\frac{x}{a} \qquad (a>0)$

56. $\int \frac{\sqrt{x^2+a^2}}{x}\,dx = \sqrt{x^2+a^2} - a\ln\left|\frac{a+\sqrt{x^2+a^2}}{x}\right|$

57. $\int \frac{\sqrt{x^2-a^2}}{x}\,dx = \sqrt{x^2-a^2} - a\cos^{-1}\frac{a}{|x|}$

$= \sqrt{x^2-a^2} - a\sec^{-1}\left(\frac{x}{a}\right) \qquad (a>0)$

58. $\int x\sqrt{x^2\pm a^2}\,dx = \frac{1}{3}(x^2\pm a^2)^{3/2}$

59. $\int \frac{1}{x\sqrt{x^2+a^2}}\,dx = \frac{1}{a}\ln\left|\frac{x}{a+\sqrt{x^2+a^2}}\right|$

60. $\int \frac{1}{x\sqrt{x^2-a^2}}\,dx = \frac{1}{a}\cos^{-1}\frac{a}{|x|} \qquad (a>0)$

61. $\int \frac{1}{x^2\sqrt{x^2\pm a^2}}\,dx = \mp\frac{\sqrt{x^2\pm a^2}}{a^2x}$

62. $\int \frac{x}{\sqrt{x^2\pm a^2}}\,dx = \sqrt{x^2\pm a^2}$

63. $\int \frac{1}{ax^2+bx+c}\,dx = \frac{1}{\sqrt{b^2-4ac}}\ln\left|\frac{2ax+b-\sqrt{b^2-4ac}}{2ax+b+\sqrt{b^2-4ac}}\right| \qquad (b^2>4ac)$

$= \frac{2}{\sqrt{4ac-b^2}}\tan^{-1}\frac{2ax+b}{\sqrt{4ac-b^2}} \qquad (b^2<4ac)$

64. $\int \frac{x}{ax^2+bx+c}\,dx = \frac{1}{2a}\ln|ax^2+bx+c| - \frac{b}{2a}\int\frac{1}{ax^2+bx+c}\,dx$

65. $\int \frac{1}{\sqrt{ax^2+bx+c}}\,dx = \frac{1}{\sqrt{a}}\ln|2ax+b+2\sqrt{a}\sqrt{ax^2+bx+c}| \qquad (a>0)$

$= \frac{1}{\sqrt{-a}}\sin^{-1}\frac{-2ax-b}{\sqrt{b^2-4ac}} \qquad (a<0)$

66. $\int \sqrt{ax^2+bx+c}\,dx = \frac{2ax+b}{4a}\sqrt{ax^2+bx+c} + \frac{4ac-b^2}{8a}\int\frac{1}{\sqrt{ax^2+b+c}}\,dx$

67. $\int \frac{x}{\sqrt{ax^2+bx+c}}\,dx = \frac{\sqrt{ax^2+bx+c}}{a} - \frac{b}{2a}\int\frac{1}{\sqrt{ax^2+bx+c}}\,dx$

68. $\int \frac{1}{x\sqrt{ax^2+bx+c}}\,dx = \frac{-1}{\sqrt{c}}\ln\left|\frac{2\sqrt{c}\sqrt{ax^2+bx+c}+bx+2c}{x}\right| \qquad (c>0)$

$= \frac{1}{\sqrt{-c}}\sin^{-1}\frac{bx+2c}{|x|\sqrt{b^2-4ac}} \qquad (c<0)$

69. $\int x^3\sqrt{x^2+a^2}\,dx = \left(\frac{1}{5}x^2 - \frac{2}{15}a^2\right)\sqrt{(a^2+x^2)^3}$

70. $\int \frac{\sqrt{x^2\pm a^2}}{x^4}\,dx = \frac{\mp\sqrt{(x^2\pm a^2)^3}}{3a^2x^3}$

71. $\int \sin ax\sin bx\,dx = \frac{\sin(a-b)x}{2(a-b)} - \frac{\sin(a+b)x}{2(a+b)} \qquad (a^2\neq b^2)$

Continued on inside back co

Lightning Source UK Ltd.
Milton Keynes UK
10 March 2010

151150UK00002B/6/A